SELECTED
SOLUTIONS MANUAL

Kathleen Thrush Shaginaw
Community College of Philadelphia
Particular Solutions, Inc.

CHEMISTRY
STRUCTURE AND PROPERTIES
Second Edition

NIVALDO J. TRO

Courseware Portfolio Management Director: Jeanne Zalesky
Executive Courseware Portfolio Manager: Terry Haugen
Courseware Director, Content Development: Jennifer Hart
Development Editor: Erin Mulligan
Courseware Analyst: Coleen Morrison
Portfolio Management Assistant: Lindsey Pruett
Portfolio Management Assistant: Shercian Kinosian
VP, Product Strategy & Development: Lauren Fogel
Content Producers: Lisa Pierce, Mae Lum
Managing Producer: Kristen Flathman
Director, Production & Digital Studio: Laura Tommasi
Editorial Content Producer: Jackie Jakob
Director, Production & Digital Studio: Katie Foley
Senior Mastering Media Producer: Jayne Sportelli
Rights and Permissions Manager: Ben Ferrini
Rights and Permissions Management: Cenveo Publisher Services
Photo Researcher: Eric Schrader
Production Management and Composition: codeMantra
Design Managers: Marilyn Perry, Maria Guglielmo Walsh
Cover and Interior Designer: Jeff Puda
Contributing Illustrators: Lachina
Printer/Binder: LSC Communications
Manufacturing Buyer: Maura Zaldivar-Garcia
Product Marketer: Elizabeth Bell
Executive Field Marketing Manager: Chris Barker

Cover Art: Quade Paul

7 2020

ISBN-10: 0-134-46067-7; ISBN-13: 978-0-134-46067-3

www.pearsonhighered.com

Contents

Student Guide to Using This Solutions Manual

The vision behind this solutions manual is to provide guidance that is useful for both the struggling student and the advanced student.

An important feature of this solutions manual is that answers for the review questions are given. This will help in the review of the major concepts in the chapter.

The format of the solutions follows the format in the textbook very closely. Each mathematical problem includes **Given, Find, Conceptual Plan, Solution,** and **Check** sections.

> **Given and Find:** Many students struggle with taking the written problem, parsing the information into categories, and determining the goal of the problem. It is also important to know which pieces of information in the problem are not necessary to solve the problem and if additional information needs to be gathered from sources such as tables in the textbook.

> **Conceptual Plan:** The conceptual plan shows a step-by-step method to solve the problem. In many cases, the given quantities need to be converted to a different unit. Under each of the arrows is the equation, constant, or conversion factor needed to complete this portion of the problem. In the "Problems by Topic" section of the end-of-chapter exercises, the odd-numbered and even-numbered problems are paired. This allows you to use a conceptual plan from an odd-numbered problem in this manual as a starting point to solve the following even-numbered problem. Students should keep in mind that the examples shown are one way to solve the problems. Other mathematically equivalent solutions may be possible.

12.45 **Given:** $m \, (CO_2) = 28.8$ g, $P = 742$ mmHg, and $T = 22\,°C$ **Find:** V

Conceptual Plan: $°C \rightarrow K$ and $mmHg \rightarrow atm$ and $g \rightarrow mol$ then $n, P, T \rightarrow V$

$$K = °C + 273.15 \qquad \frac{1 \text{ atm}}{760 \text{ mmHg}} \qquad \frac{1 \text{ mol}}{44.01 \text{ g}} \qquad PV = nRT$$

Solution: $T_1 = 22\,°C + 273.15 = 295$ K, $P = 742 \, \text{mmHg} \times \dfrac{1 \text{ atm}}{760 \, \text{mmHg}} = 0.976316$ atm

$n = 28.8 \, \text{g} \times \dfrac{1 \text{ mol}}{44.01 \, \text{g}} = 0.654397$ mol $PV = nRT$ Rearrange to solve for V.

$$V = \frac{nRT}{P} = \frac{0.654397 \, \text{mol} \times 0.08206 \, \dfrac{\text{L} \cdot \text{atm}}{\text{mol} \cdot \text{K}} \times 295 \text{ K}}{0.976316 \, \text{atm}} = 16.2 \text{ L}$$

Check: The units (L) are correct. The magnitude of the answer (16 L) makes sense because one mole of an ideal gas under standard conditions (273 K and 1 atm) occupies 22.4 L. Although these are not standard conditions, they are close enough for a ballpark check of the answer. Because this gas sample contains 0.65 mole, a volume of 16 L is reasonable.

> **Solution:** The Solution section walks you through solving the problem after the conceptual plan. Equations are rearranged to solve for the appropriate quantity. Intermediate results are shown with additional digits to minimize round-off error. The units are canceled in each appropriate step.

> **Check:** The Check section confirms that the units in the answer are correct. This section also challenges the student to think about whether the magnitude of the answer makes sense. Thinking about what is a reasonable answer can help uncover errors such as calculation errors.

E | Essentials: Units, Measurement, and Problem Solving

Review Questions

E.1   The SI base units include the meter (m) for length, the kilogram (kg) for mass, the second (s) for time, and the kelvin (K) for temperature.

E.3   Prefix multipliers are used with the standard units of measurement to change the value of the unit by powers of 10.
For example, the kilometer has the prefix "kilo," meaning 1000 or 10^3. Therefore:
1 kilometer = 1000 meters = 10^3 meters
Similarly, the millimeter has the prefix "milli," meaning 0.001 or 10^{-3}.
1 millimeter = 0.001 meters = 10^{-3} meters

E.5   The reliability of a measurement depends on the instrument used to make the measurement. The more precise the instrument, the more reliably we can differentiate between two measurements. We must consider the reliability of measurements when reporting and manipulating them.

E.7   Accuracy refers to how close the measured value is to the actual value. Precision refers to how close a series of measurements are to one another or how reproducible they are. A series of measurements can be precise (close to one another in value and reproducible) but not accurate (not close to the true value).

E.9   In multiplication or division, the result carries the same number of significant figures as the factor with the fewest significant figures.

E.11   The density (d) of a substance is the ratio of its mass (m) to its volume (V):
$$\text{Density} = \frac{\text{Mass}}{\text{Volume}} \text{ or } d = \frac{m}{V}$$
The density of a substance is an example of an intensive property, one that is independent of the amount of the substance. Mass is one of the properties used to calculate the density of a substance. Mass, in contrast to density, is an extensive property, one that depends on the amount of the substance.

E.13   Energy is the capacity to do work. Kinetic energy is associated with the motion of an object. Potential energy is associated with the position or composition of an object. Examples of kinetic energy are a moving billiard ball, moving gas molecules, and a raging river. Examples of potential energy are a billiard ball raised above the surface of a billiard table, a compressed spring, and bonds in molecules.

E.15   Chemical energy is potential energy. It is the energy that is contained in the bonds that hold the molecules together. This energy arises primarily from electrostatic forces between the electrically charged particles (protons and electrons) that compose atoms and molecules. Some of these arrangements—such as the one within the molecules that compose gasoline—have a much higher potential energy than others. When gasoline undergoes combustion, the arrangement of these particles changes, creating molecules with much lower potential energy and transferring a great deal of energy (mostly in the form of heat) to the surroundings. A raised weight has a certain amount of potential energy (dependent on how high the weight is raised) that can be converted to kinetic energy when the weight is released.

E.17   Using units as a guide to solving problems is often called dimensional analysis. Units should always be included in calculations; they are multiplied, divided, and canceled like any other algebraic quantity.

Problems by Topic

The Units in Measurement

E.19   (a)   To convert from °F to °C, first find the equation that relates these two quantities. $°C = \dfrac{°F - 32}{1.8}$ Now substitute °F

into the equation and compute the answer. Note: The number of digits reported in this answer

follows significant figure conventions, covered in Section 1.6. $°C = \dfrac{°F - 32}{1.8} = \dfrac{0.}{1.8} = 0.\,°C$

   (b)   To convert from K to °F, first find the equations that relate these two quantities.

$K = °C + 273.15$ and $°C = \dfrac{°F - 32}{1.8}$

Because these equations do not directly express K in terms of °F, you must combine the equations and then solve the equation for °F. Substituting for °C:

$K = \dfrac{°F - 32}{1.8} + 273.15$; rearrange $K - 273.15 = \dfrac{°F - 32}{1.8}$;

rearrange $1.8(K - 273.15) = (°F - 32)$; finally, $°F = 1.8(K - 273.15) + 32$. Now substitute K into the equation and compute the answer.

$°F = 1.8(77 - 273.15) + 32 = 1.8(-196) + 32 = -353 + 32 = -321\,°F$

Alternatively, you could first convert to K to °C, using the formula $K = 273.15 + °C$, and then convert °C to °F. So, $77\,K = 273.15 + °C$. Rearranging to solve for °C we get $°C = 77K - 273.15 = -196.15\,°C$. Then substitute

the temperature in °C in the formula $°C = \dfrac{°F - 32}{1.8}$ getting $-19\underline{6}.15\,°C = \dfrac{°F - 32}{1.8}$. Multiplying both sides

by 1.8, we get $-196.15\,°C \times 1.8 = °F - 32$ and so $-353.07\,°C = °F - 32$. Adding 32 to both sides of the equation, we get $°F = -32\underline{1}.07\,°F$ or $-321\,°F$, with the proper number of significant figures.

   (c)   To convert from °F to °C, first find the equation that relates these two quantities. $°C = \dfrac{°F - 32}{1.8}$

Now substitute °F into the equation and compute the answer.

$°C = \dfrac{-109\,°F - 32\,°F}{1.8} = \dfrac{-141}{1.8} = -78.3\,°C$

   (d)   To convert from °F to K, first find the equations that relate these two quantities.

$K = °C + 273.15$ and $°C = \dfrac{°F - 32}{1.8}$

Because these equations do not directly express K in terms of °F, you must combine the equations and then

solve the equation for K. Substituting for °C: $K = \dfrac{°F - 32}{1.8} + 273.15$.

Now substitute °F into the equation and compute the answer.

$K = \dfrac{(98.6 - 32)}{1.8} + 273.15 = \dfrac{66.6}{1.8} + 273.15 = 37.0 + 273.15 = 310.2\,K$

Alternatively, you could first convert to °F to °C, using the formula $°C = \dfrac{°F - 32}{1.8}$, and then convert

°C to K, using $K = 273.15 + °C$. So, $°C = \dfrac{98.6\,°F - 32}{1.8} = \dfrac{66.6\,°F}{1.8} = 37.0\,°C$. Then $K = 273.15 + 37.0\,°C = 310.2\,K$.

E.21   To convert from °F to °C, first find the equation that relates these two quantities: $°C = \dfrac{°F - 32}{1.8}$. Now substitute °F

into the equation and compute the answer. Note: The number of digits reported in this answer follows significant figure

conventions, covered in Section 1.6. $°C = \dfrac{-80.\,°F - 32\,°F}{1.8} = \dfrac{-112}{1.8} = -62.2\,°C$

Begin by finding the equation that relates the quantity that is given (°C) and the quantity you are trying to find (K). $K = °C + 273.15$. Because this equation gives the temperature in K directly, simply substitute the correct value for the temperature in °C and compute the answer. $K = -62.2\,°C + 273.15 = 210.9\,K$

E.23 Use Table E.2 to determine the appropriate prefix multiplier and substitute the meaning into the expressions.

(a) 10^{-9} is equivalent to "nano," so 1.2×10^{-9} m = 1.2 nanometers = 1.2 nm.

(b) 10^{-15} is equivalent to "femto," so 22×10^{-15} s = 22 femtoseconds = 22 fs.

(c) 10^{9} is equivalent to "giga," so 1.5×10^{9} g = 1.5 gigagrams = 1.5 Gg.

(d) 10^{6} is equivalent to "mega," so 3.5×10^{6} L = 3.5 megaliters = 3.5 ML.

E.25 Use Table E.2 to determine the appropriate prefix multiplier and substitute the meaning into the expressions.

(a) 10^{-9} is equivalent to "nano," so 4.5 ns = 4.5 nanoseconds = 4.5×10^{-9} s.

(b) 10^{-15} is equivalent to "femto," so 18 fs = 18 femtoseconds = 18×10^{-15} s = 1.8×10^{-14} s.
Remember that in scientific notation, the first number should be smaller than 10.

(c) 10^{-12} is equivalent to "pico," so 128 pm = 128×10^{-12} m = 1.28×10^{-10} m.
Remember that in scientific notation, the first number should be smaller than 10.

(d) 10^{-6} is equivalent to "micro," so 35 μm = 35 micrometers = 35×10^{-6} m = 3.5×10^{-5} m.
Remember that in scientific notation, the first number should be smaller than 10.

E.27 (b) **Given:** 515 km   **Find:** dm
Conceptual Plan: km → m → dm

$$\frac{1000\ m}{1\ km} \quad \frac{10\ dm}{1\ m}$$

Solution: $515 \text{ km} \times \dfrac{1000\ m}{1\ km} \times \dfrac{10\ dm}{1\ m} = 5.15 \times 10^6$ dm

Check: The units (dm) are correct. The magnitude of the answer (10^6) makes physical sense because a decimeter is a much smaller unit than a kilometer.

Given: 515 km   **Find:** cm
Conceptual Plan: km → m → cm

$$\frac{1000\ m}{1\ km} \quad \frac{100\ dm}{1\ m}$$

Solution: $515 \text{ km} \times \dfrac{1000\ m}{1\ km} \times \dfrac{100\ cm}{1\ m} = 5.15 \times 10^7$ cm

Check: The units (cm) are correct. The magnitude of the answer (10^7) makes physical sense because a centimeter is a much smaller unit than a kilometer or a decimeter.

(c) **Given:** 122.355 s   **Find:** ms
Conceptual Plan: s → ms

$$\frac{1000\ ms}{1\ s}$$

Solution: $122.355 \text{ s} \times \dfrac{1000\ ms}{1\ s} = 1.22355 \times 10^5$ ms

Check: The units (ms) are correct. The magnitude of the answer (10^5) makes physical sense because a millisecond is a much smaller unit than a second.

Given: 122.355 s   **Find:** ks
Conceptual Plan: s → ks

$$\frac{1\ ks}{1000\ s}$$

Solution: $122.355 \text{ s} \times \dfrac{1\ ks}{1000\ s} = 1.22355 \times 10^{-1}$ ks = 0.122355 ks

Check: The units (ks) are correct. The magnitude of the answer (10^{-1}) makes physical sense because a kilosecond is a much larger unit than a second.

(d) **Given:** 3.345 kJ   **Find:** J
Conceptual Plan: kJ → J

$$\frac{1000\ J}{1\ kJ}$$

Solution: $3.345 \text{ kJ} \times \dfrac{1000\ J}{1\ kJ} = 3.345 \times 10^3$ J

Check: The units (J) are correct. The magnitude of the answer (10^3) makes physical sense because a joule is a much smaller unit than a kilojoule.

Given: 3.345×10^3 J (from above)   **Find:** mJ

Conceptual Plan: J → mJ

$$\frac{1000\ \text{mJ}}{1\ \text{J}}$$

Solution: 3.345×10^3 J $\times \dfrac{1000\ \text{mJ}}{1\ \text{J}} = 3.345 \times 10^6$ mJ

Check: The units (mJ) are correct. The magnitude of the answer (10^6) makes physical sense because a milli-joule is a much smaller unit than a joule.

E.29   (a)   **Given:** 254,998 m   **Find:** km

Conceptual Plan: m → km

$$\frac{1\ \text{km}}{1000\ \text{m}}$$

Solution: 254,998 m $\times \dfrac{1\ \text{km}}{1000\ \text{m}} = 2.54998 \times 10^2$ km $= 254.998$ km

Check: The units (km) are correct. The magnitude of the answer (10^2) makes physical sense because a kilometer is a much larger unit than a meter.

(b)   **Given:** 254,998 m   **Find:** Mm

Conceptual Plan: m → Mm

$$\frac{1\ \text{Mm}}{10^6\ \text{m}}$$

Solution: 254,998 m $\times \dfrac{1\ \text{Mm}}{10^6\ \text{m}} = 2.54998 \times 10^{-1}$ Mm $= 0.254998$ Mm

Check: The units (Mm) are correct. The magnitude of the answer (10^{-1}) makes physical sense because a megameter is a much larger unit than a meter or a kilometer.

(c)   **Given:** 254,998 m   **Find:** mm

Conceptual Plan: m → mm

$$\frac{1000\ \text{mm}}{1\ \text{m}}$$

Solution: 254,998 m $\times \dfrac{1000\ \text{mm}}{1\ \text{m}} = 2.54998 \times 10^8$ mm

Check: The units (mm) are correct. The magnitude of the answer (10^8) makes physical sense because a millimeter is a much smaller unit than a meter.

(d)   **Given:** 254,998 m   **Find:** cm

Conceptual Plan: m → cm

$$\frac{100\ \text{cm}}{1\ \text{m}}$$

Solution: 254,998 m $\times \dfrac{100\ \text{cm}}{1\ \text{m}} = 2.54998 \times 10^7$ cm

Check: The units (cm) are correct. The magnitude of the answer (10^7) makes physical sense because a centimeter is a much smaller unit than a meter but larger than a millimeter.

E.31   **Given:** 1 square meter $\left(1\ \text{m}^2\right)$ **Find:** cm^2

Conceptual Plan: $1\ \text{m}^2 \rightarrow \text{cm}^2$

$$\frac{100\ \text{cm}}{1\ \text{m}}$$

Notice that for squared units, the conversion factors must be squared.

Solution: $1\ \text{m}^2 \times \dfrac{(100\ \text{cm})^2}{(1\ \text{m})^2} = 1 \times 10^4\ \text{cm}^2$

Check: The units of the answer are correct, and the magnitude makes sense. The unit centimeter is smaller than a meter, so the value in square centimeters should be larger than in square meters.

E.33   (a)   **Given:** 15.0 L   **Find:** mL

Conceptual Plan: L → mL

$$\frac{1000\ \text{mL}}{1\ \text{L}}$$

Solution: 15.0 L $\times \dfrac{1000\ \text{mL}}{1\ \text{L}} = 1.50 \times 10^4$ mL

Check: The units (mL) are correct. The magnitude of the answer (10^4) makes physical sense because a milliliter is much smaller than a liter, so the answer increases.

(b)   **Given:** 15.0 L   **Find:** cm^3

Conceptual Plan: L → cm^3

$$\frac{1000 \ cm^3}{1 \ L}$$

Solution: $15.0 \ L \times \dfrac{1000 \ cm^3}{1 \ L} = 1.50 \times 10^4 \ cm^3$

Check: The units (cm^3) are correct. The magnitude of the answer (10^4) makes physical sense because a cubic centimeter is much smaller than a liter, so the answer increases.

(c)   **Given:** 15.0 L   **Find:** gal

Conceptual Plan: L → gal

$$\frac{1 \ gal}{3.7854 \ L}$$

Solution: $15.0 \ L \times \dfrac{1 \ gal}{3.7854 \ L} = 3.96 \ gal$

Check: The units (gal) are correct. The magnitude of the answer (4) makes physical sense because a gallon is larger than a liter, so the answer decreases.

(d)   **Given:** 15.0 L   **Find:** qt

Conceptual Plan: L → qt

$$\frac{1.0567 \ qt}{1 \ L}$$

Solution: $15.0 \ L \times \dfrac{1.0567 \ qt}{1 \ L} = 15.9 \ qt$

Check: The units (qt) are correct. The magnitude of the answer (16) makes physical sense because a quart is slightly smaller than a liter, so the answer increases.

The Reliability of a Measurement and Significant Figures

E.35   (c)   19.1 mm. Since the ruler has markings every 1 mm, the measurement is to be reported to ± 0.1 mm.

E.37   To obtain the readings, look to see where the bottom of the meniscus lies. Estimate the distance between two markings on the device.

(a)   73.2 mL—the bottom of the meniscus appears to be sitting just above the 73 mL mark.

(b)   88.2 °C—the mercury is between the 88 °C mark and the 89 °C mark, but it is closer to the lower number.

(c)   645 mL—the meniscus appears to be just above the 640 mL mark.

E.39   Remember that

1.   interior zeros (zeros between two numbers) are significant.

2.   leading zeros (zeros to the left of the first nonzero number) are not significant. They only serve to locate the decimal point.

3.   trailing zeros (zeros at the end of a number) are categorized as follows:

•   Trailing zeros after a decimal point are always significant.

•   Trailing zeros before an implied decimal point are ambiguous and should be avoided by using scientific notation or by inserting a decimal point at the end of the number.

(a)   1,05̄0,501 km

(b)   0̸.0̸0̸2̄0̄ m

(c)   0̸.0̸0̸0̸0̸0̸0̸0̸0̸0̸0̸0̸0̸0̸0̸0̸0̸2 s

(d)   0̸.0̸0̸1̄0̄9̄0 cm

E.41   Remember the following rules from Section 1.7.

(a)   Three significant figures. The 3, 1, and 2 are significant (rule 1). The leading zeros only mark the decimal place and are, therefore, not significant (rule 3).

(b)   Ambiguous. The 3, 1, and 2 are significant (rule 1). The trailing zeros occur before an implied decimal point and are, therefore, ambiguous (rule 4). Without more information, we would assume three significant figures.

It is better to write this as 3.12×10^5 to indicate three significant figures or as 3.12000×10^5 to indicate six (rule 4).

(c)   Three significant figures. The 3, 1, and 2 are significant (rule 1).

(d)   Five significant figures. The 1s, 3, 2, and 7 are significant (rule 1).

(e)   Ambiguous. The 2 is significant (rule 1). The trailing zeros occur before an implied decimal point and are, therefore, ambiguous (rule 4). Without more information, we would assume one significant figure. It is better to write this as 2×10^3 to indicate one significant figure or as 2.000×10^3 to indicate four (rule 4).

E.43   (a)   This is not exact because π is an irrational number. The number 3.14 only shows three of the infinite number of significant figures that π has.

(b)   This is an exact conversion because it comes from a definition of the units and so has an unlimited number of significant figures.

(c)   This is a measured number, so it is not an exact number. There are two significant figures.

(d)   This is an exact conversion because it comes from a definition of the units and so has an unlimited number of significant figures.

E.45   (a)   156.9—The 8 is rounded up because the next digit is a 5.

(b)   156.8—The last two digits are dropped because 4 is less than 5.

(c)   156.8—The last two digits are dropped because 4 is less than 5.

(d)   156.9—The 8 is rounded up because the next digit is a 9, which is greater than 5.

Significant Figures in Calculations

E.47   (a) $9.15 \div 4.970 = 1.84$—Three significant figures are allowed to reflect the three significant figures in the least precisely known quantity (9.15).

(b) $1.54 \times 0.03060 \times 0.69 = 0.033$—Two significant figures are allowed to reflect the two significant figures in the least precisely known quantity (0.69). The intermediate answer (0.03251556) is rounded up because the first nonsignificant digit is a 5.

(c) $27.5 \times 1.82 \div 100.04 = 0.500$—Three significant figures are allowed to reflect the three significant figures in the least precisely known quantity (27.5 and 1.82). The intermediate answer (0.50029988) is truncated because the first nonsignificant digit is a 2, which is less than 5.

(d) $(2.290 \times 10^6) \div (6.7 \times 10^4) = 34$—Two significant figures are allowed to reflect the two significant figures in the least precisely known quantity (6.7×10^4). The intermediate answer (34.17910448) is truncated because the first nonsignificant digit is a 1, which is less than 5.

E.49   (a) $\begin{array}{r} 43.7 \\ -2.341 \\ \hline 41.359 \end{array} = 41.4$

Round the intermediate answer to one decimal place to reflect the quantity with the fewest decimal places (43.7). Round the last digit up because the first nonsignificant digit is 5.

(b) $\begin{array}{r} 17.6 \\ +2.838 \\ +2.3 \\ +110.77 \\ \hline 133.508 \end{array} = 133.5$

Round the intermediate answer to one decimal place to reflect the quantity with the fewest decimal places (2.3). Truncate nonsignificant digits because the first nonsignificant digit is 0.

(c) $\begin{array}{r} 19.6 \\ +58.33 \\ -4.974 \\ \hline 72.956 \end{array} = 73.0$

Round the intermediate answer to one decimal place to reflect the quantity with the fewest decimal places (19.6). Round the last digit up because the first nonsignificant digit is 5.

(d)
$$
\begin{array}{r}
5.99 \\
-5.572 \\
\hline
0.418 = 0.42
\end{array}
$$

Round the intermediate answer to two decimal places to reflect the quantity with the fewest decimal places (5.99). Round the last digit up because the first nonsignificant digit is 8.

E.51 Perform operations in parentheses first. Keep track of significant figures in each step by noting the last significant digit in an intermediate result.

(a) $(24.6681 \times 2.38) + 332.58 = 58.7\underline{1}0078$

$$
\begin{array}{r}
+332.58 \\
\hline
391.290078 = 391.3
\end{array}
$$

The first intermediate answer has one significant digit to the right of the decimal because it is allowed three significant figures [reflecting the quantity with the fewest significant figures (2.38)]. Underline the most significant digit in this answer. Round the next intermediate answer to one decimal place to reflect the quantity with the fewest decimal places (58.7). Round the last digit up because the first nonsignificant digit is 9.

(b) $\dfrac{(85.3 - 21.489)}{0.0059} = \dfrac{63.8\underline{1}1}{0.0059} = 1.08\underline{1}542 \times 10^4 = 1.1 \times 10^4$

The first intermediate answer has one significant digit to the right of the decimal to reflect the quantity with the fewest decimal places (85.3). Underline the most significant digit in this answer. Round the next intermediate answer to two significant figures to reflect the quantity with the fewest significant figures (0.0059). Round the last digit up because the first nonsignificant digit is 8.

(c) $(512 \div 986.7) + 5.44 = 0.5\underline{1}89014$

$$
\begin{array}{r}
+5.44 \\
\hline
5.9589014 = 5.96
\end{array}
$$

The first intermediate answer has three significant figures and three significant digits to the right of the decimal, reflecting the quantity with the fewest significant figures (512). Underline the most significant digit in this answer. Round the next intermediate answer to two decimal places to reflect the quantity with the fewest decimal places (5.44). Round the last digit up because the first nonsignificant digit is 8.

(d) $[(28.7 \times 10^5) \div 48.533] + 144.99 = 59\underline{1}35.02$

$$
\begin{array}{r}
+144.99 \\
\hline
59280.01 = 59300 = 5.93 \times 10^4
\end{array}
$$

The first intermediate answer has three significant figures, reflecting the quantity with the fewest significant figures (28.7×10^5). Underline the most significant digit in this answer. Because the number is so large, when the addition is performed, the most significant digit is the hundreds place. Round the next intermediate answer to the hundreds place and put in scientific notation to remove any ambiguity. Note that the last digit is rounded up because the first nonsignificant digit is 8.

Density

E.53 **Given:** $m = 2.49$ g; $V = 0.349$ cm^3 **Find:** d in g/cm^3 and compare to pure copper.
Conceptual Plan: $m, V \rightarrow d$

$$d = m/V$$

Compare to the published value: d (pure copper) $= 8.96$ g/cm^3. (This value is in Table E.4.)

Solution: $d = \dfrac{2.49 \text{ g}}{0.349 \text{ cm}^3} = 7.13 \dfrac{\text{g}}{\text{cm}^3}$

The density of the penny is much smaller than the density of pure copper (7.13 g/cm^3; 8.96 g/cm^3), so the penny is not pure copper.

Check: The units (g/cm^3) are correct. The magnitude of the answer seems correct. Many coins are layers of metals, so it is not surprising that the penny is not pure copper.

E.55 **Given:** $m = 4.10 \times 10^3$ g; $V = 3.25$ L   **Find:** d in g/cm^3
Conceptual Plan: $m, V \rightarrow d$ then L $\rightarrow$ cm^3

$$d = m/V \qquad \dfrac{1000 \text{ cm}^3}{1 \text{ L}}$$

Solution: $d = \dfrac{4.10 \times 10^3 \, g}{3.25 \, L} \times \dfrac{1 \, L}{1000 \, cm^3} = 1.26 \, \dfrac{g}{cm^3}$

Check: The units (g/cm^3) are correct. The magnitude of the answer seems correct.

E.57   (a)   **Given:** $d = 1.11 \, g/cm^3$; $V = 417 \, mL$ **Find:** m

           Conceptual Plan: $d, V \rightarrow m$ then $cm^3 \rightarrow mL$

$$d = m/V \qquad\qquad \dfrac{1 \, mL}{1 \, cm^3}$$

           Solution: $d = m/V$ Rearrange by multiplying both sides of the equation by V. $m = d \times V$

$$m = 1.11 \, \dfrac{g}{cm^3} \times \dfrac{1 \, cm^3}{1 \, mL} \times 417 \, mL = 4.63 \times 10^2 \, g$$

           Check: The units (g) are correct. The magnitude of the answer seems correct considering that the value of the density is about $1 \, g/cm^3$.

   (b)   **Given:** $d = 1.11 \, g/cm^3$; $m = 4.1 \, kg$ **Find:** V in L

           Conceptual Plan: $d, V \rightarrow m$ then $kg \rightarrow g$ and $cm^3 \rightarrow L$

$$d = m/V \qquad\qquad \dfrac{1000 \, g}{1 \, kg} \qquad\qquad \dfrac{1 \, L}{1000 \, cm^3}$$

           Solution: $d = m/V$ Rearrange by multiplying both sides of the equation by V and dividing both sides of the equation by d.

$$V = \dfrac{m}{d} = \dfrac{4.1 \, kg}{1.11 \, \dfrac{g}{cm^3}} \times \dfrac{1000 \, g}{1 \, kg} = 3.7 \times 10^3 \, cm^3 \times \dfrac{1 \, L}{1000 \, cm^3} = 3.7 \, L$$

           Check: The units (L) are correct. The magnitude of the answer seems correct considering that the value of the density is about $1 \, g/cm^3$.

E.59   **Given:** $V = 245 \, L$, $d = 0.821 \, g/mL$ **Find:** m

     Conceptual Plan: $g/mL \rightarrow g/L$ then $d, V \rightarrow m$

$$\dfrac{1000 \, mL}{1 \, L} \qquad\qquad d = m/V$$

     Solution: $d = m/V$ Rearrange by multiplying both sides of the equation by V. $m = d \times V$

$$m = 245 \, L \times \dfrac{1000 \, mL}{1 \, L} \times \left(0.821 \, \dfrac{g}{mL} \right) = 2.01 \times 10^5 \, g$$

     Check: The units (g) are correct. The magnitude of the answer seems correct considering the value of the density is less than $1 \, g/mL$ and the volume is very large.

Unit Conversions

E.61   (a)   **Given:** $27.8 \, L$ **Find:** cm^3

           Conceptual Plan: $L \rightarrow cm^3$

$$\dfrac{1000 \, cm^3}{1 \, L}$$

           Solution: $27.8 \, L \times \dfrac{1000 \, cm^3}{1 \, L} = 2.78 \times 10^4 \, cm^3$

           Check: The units (cm^3) are correct. The magnitude of the answer (10^4) makes physical sense because cm^3 is much smaller than a liter; so the answer should go up several orders of magnitude. Three significant figures are allowed because of the limitation of $27.8 \, L$ (three significant figures).

   (b)   **Given:** $1898 \, mg$ **Find:** kg

           Conceptual Plan: $mg \rightarrow g \rightarrow kg$

$$\dfrac{1 \, g}{1000 \, mg} \qquad \dfrac{1 \, kg}{1000 \, g}$$

           Solution: $1898 \, mg \times \dfrac{1 \, g}{1000 \, mg} \times \dfrac{1 \, kg}{1000 \, g} = 1.898 \times 10^{-3} \, kg$

Check: The units (kg) are correct. The magnitude of the answer (10^{-3}) makes physical sense because a kilogram is a much larger unit than a milligram. Four significant figures are allowed because 1898 mg has four significant figures.

(c)   **Given:** 198 km   **Find:** cm
Conceptual Plan: km → m → cm

$$\frac{1000\ m}{1\ m} \qquad \frac{100\ cm}{1\ m}$$

Solution: $198\ \text{km} \times \dfrac{1000\ m}{1\ \text{km}} \times \dfrac{100\ cm}{1\ m} = 1.98 \times 10^7\ cm$

Check: The units (cm) are correct. The magnitude of the answer (10^7) makes physical sense because a kilometer is a much larger unit than a centimeter. Three significant figures are allowed because 198 km has three significant figures.

E.63   (a)   **Given:** 154 cm   **Find:** in
Conceptual Plan: cm → in

$$\frac{1\ in}{2.54\ cm}$$

Solution: $154\ \text{cm} \times \dfrac{1\ in}{2.54\ \text{cm}} = 60.62992\ in = 60.6\ in$

Check: The units (in) are correct. The magnitude of the answer (60.6) makes physical sense because an inch is a larger unit than a centimeter. Three significant figures are allowed because 154 cm has three significant figures.

(b)   **Given:** 3.14 kg   **Find:** g
Conceptual Plan: kg → g

$$\frac{1000\ g}{1\ kg}$$

Solution: $3.14\ \text{kg} \times \dfrac{1000\ g}{1\ \text{kg}} = 3.14 \times 10^3\ g$

Check: The units (g) are correct. The magnitude of the answer (10^3) makes physical sense because a kilogram is a much larger unit than a gram. Three significant figures are allowed because 3.14 kg has three significant figures.

(c)   **Given:** 3.5 L   **Find:** qt
Conceptual Plan: L → qt

$$\frac{1.057\ qt}{1\ L}$$

Solution: $3.5\ \text{L} \times \dfrac{1.057\ qt}{1\ \text{L}} = 3.6995\ qt = 3.7\ qt$

Check: The units (qt) are correct. The magnitude of the answer (3.7) makes physical sense because a liter is a smaller unit than a quart. Two significant figures are allowed because 3.5 L has two significant figures. Round the last digit up because the first nonsignificant digit is a 9.

(d)   **Given:** 109 mm   **Find:** in
Conceptual Plan: mm → m → in

$$\frac{1\ m}{1000\ mm} \qquad \frac{39.37\ in}{1\ m}$$

Solution: $109\ \text{mm} \times \dfrac{1\ m}{1000\ \text{mm}} \times \dfrac{39.37\ in}{1\ m} = 4.29133\ in = 4.29\ in$

Check: The units (in) are correct. The magnitude of the answer (4) makes physical sense because a millimeter is a much smaller unit than an inch. Three significant figures are allowed because 109 mm has three significant figures.

E.65   **Given:** 10.0 km   **Find:** minutes   **Other:** running pace = 7.5 miles per hour
Conceptual Plan: km → mi → hr → min

$$\frac{0.6214\ mi}{1\ km} \qquad \frac{1\ hr}{7.5\ mi} \qquad \frac{60\ min}{1\ hr}$$

Solution: $10.0\ \text{km} \times \dfrac{0.6214\ mi}{1\ \text{km}} \times \dfrac{1\ hr}{7.5\ mi} \times \dfrac{60\ min}{1\ hr} = 49.712\ min = 50.\ min = 5.0 \times 10^1\ min$

Check: The units (min) are correct. The magnitude of the answer (50) makes physical sense because she is running almost 7.5 miles (which would take her 60 min = 1 hr). Two significant figures are allowed because of the limitation of 7.5 mi/hr (two significant figures). Round the last digit up because the first nonsignificant digit is a 7.

E.67   **Given:** 17 km/L   **Find:** miles per gallon

Conceptual Plan: $\dfrac{km}{L} \rightarrow \dfrac{mi}{L} \rightarrow \dfrac{mi}{gal}$

$$\frac{0.6214\ mi}{1\ km} \qquad \frac{3.785\ L}{1\ gallon}$$

Solution: $\dfrac{17\ \cancel{km}}{1\ \cancel{L}} \times \dfrac{0.6214\ mi}{1\ \cancel{km}} \times \dfrac{3.785\ \cancel{L}}{1\ gal} = \underline{3}9.98398\ \dfrac{mi}{gal} = 40.\ \dfrac{mi}{gal}$

Check: The units (mi/gal) are correct. The magnitude of the answer (40) makes physical sense because the dominating factor is that a liter is much smaller than a gallon; so the answer should go up. Two significant figures are allowed because of the limitation of 17 km/L (two significant figures). Round the last digit up because the first nonsignificant digit is a 9.

E.69   (a)   **Given:** 195 m^2 **Find:** km^2

Conceptual Plan: m$^2 \rightarrow$ km^2

$$\frac{(1\ km)^2}{(1000\ m)^2}$$

Notice that for squared units, the conversion factors must be squared.

Solution: $195\ \cancel{m^2} \times \dfrac{(1\ km)^2}{(1000\ \cancel{m})^2} = 1.95 \times 10^{-4}\ km^2$

Check: The units (km^2) are correct. The magnitude of the answer (10^{-4}) makes physical sense because a kilometer is a much larger unit than a meter.

(b)   **Given:** 195 m^2 **Find:** dm^2

Conceptual Plan: m$^2 \rightarrow$ dm^2

$$\frac{(10\ dm)^2}{(1\ m)^2}$$

Notice that for squared units, the conversion factors must be squared.

Solution: $195\ \cancel{m^2} \times \dfrac{(10\ dm)^2}{(1\ \cancel{m})^2} = 1.95 \times 10^{4}\ dm^2$

Check: The units (dm^2) are correct. The magnitude of the answer (10^4) makes physical sense because a decimeter is a much smaller unit than a meter.

(c)   **Given:** 195 m^2 **Find:** cm^2

Conceptual Plan: m$^2 \rightarrow$ cm^2

$$\frac{(100\ cm)^2}{(1\ m)^2}$$

Notice that for squared units, the conversion factors must be squared.

Solution: $195\ \cancel{m^2} \times \dfrac{(100\ cm)^2}{(1\ \cancel{m})^2} = 1.95 \times 10^{6}\ cm^2$

Check: The units (cm^2) are correct. The magnitude of the answer (10^6) makes physical sense because a centimeter is a much smaller unit than a meter.

E.71   **Given:** 435 acres   **Find:** square miles   **Other:** 1 acre = 43,560 ft^2; 1 mile = 5280 ft

Conceptual Plan: acres $\rightarrow$ ft$^2 \rightarrow$ mi^2

$$\frac{43560\ ft^2}{1\ acre} \qquad \frac{(1\ mi)^2}{(5280\ ft)^2}$$

Notice that for squared units, the conversion factors must be squared.

Solution: $435 \text{ acres} \times \dfrac{43560 \text{ ft}^2}{1 \text{ acres}} \times \dfrac{(1 \text{ mi})^2}{(5280 \text{ ft})^2} = 0.6796875 \text{ mi}^2 = 0.680 \text{ mi}^2$

Check: The units (mi^2) are correct. The magnitude of the answer (0.7) makes physical sense because an acre is much smaller than a square mile; so the answer should go down several orders of magnitude. Three significant figures are allowed because of the limitation of 435 acres (three significant figures). Round the last digit up because the first non-significant digit is a 7.

E.73   **Given:** 14 lb   **Find:** mL   **Other:** 80 mg/0.80 mL; 15 mg/kg body
      Conceptual Plan: lb → kg body → mg → mL

$$\dfrac{1 \text{ kg body}}{2.205 \text{ lb}} \qquad \dfrac{15 \text{ mg}}{1 \text{ kg body}} \qquad \dfrac{0.80 \text{ mL}}{80 \text{ mg}}$$

Solution: $14 \text{ lb} \times \dfrac{1 \text{ kg body}}{2.205 \text{ lb}} \times \dfrac{15 \text{ mg}}{1 \text{ kg body}} \times \dfrac{0.80 \text{ mL}}{80 \text{ mg}} = 0.9523809524 \text{ mL} = 0.95 \text{ mL}$

Check: The units (mL) are correct. The magnitude of the answer (1 mL) makes physical sense because it is a reasonable amount of liquid to give to a baby. Two significant figures are allowed because of the statement in the problem. Truncate after the last significant digit because the first nonsignificant digit is a 2.

E.75   (a)   **Given:** 534 kWh   **Find:** J
          Conceptual Plan: kWh → J

$$\dfrac{3.60 \times 10^6 \text{ J}}{1 \text{ kWh}}$$

          Solution: $534 \text{ kWh} \times \dfrac{3.60 \times 10^6 \text{ J}}{1 \text{ kWh}} = 1.92 \times 10^9 \text{ J}$

          Check: The units (J) are correct. The magnitude of the answer (10^9) makes physical sense because a kilowatt hour is much larger than a joule; so the answer increases.

   (b)   **Given:** 215 kJ   **Find:** Cal
          Conceptual Plan: kJ → J → Cal

$$\dfrac{1000 \text{ J}}{1 \text{ kJ}} \qquad \dfrac{1 \text{ Cal}}{4184 \text{ J}}$$

          Solution: $215 \text{ kJ} \times \dfrac{1000 \text{ J}}{1 \text{ kJ}} \times \dfrac{1 \text{ Cal}}{4184 \text{ J}} = 51.4 \text{ Cal}$

          Check: The units (Cal) are correct. The magnitude of the answer (51) makes physical sense because a Calorie is about $\frac{1}{4}$ of a kilojoule; so the answer decreases by a factor of about four.

   (c)   **Given:** 567 Cal   **Find:** J
          Conceptual Plan: Cal → J

$$\dfrac{4184 \text{ J}}{1 \text{ Cal}}$$

          Solution: $567 \text{ Cal} \times \dfrac{4184 \text{ J}}{1 \text{ Cal}} = 2.37 \times 10^6 \text{ J}$

          Check: The units (J) are correct. The magnitude of the answer (10^6) makes physical sense because a Calorie is much larger than a joule; so the answer increases.

   (d)   **Given:** 2.85×10^3 J   **Find:** cal
          Conceptual Plan: J → cal

$$\dfrac{1 \text{ cal}}{4.184 \text{ J}}$$

          Solution: $2.85 \times 10^3 \text{ J} \times \dfrac{1 \text{ cal}}{4.184 \text{ J}} = 681 \text{ cal}$

          Check: The units (cal) are correct. The magnitude of the answer (680) makes physical sense because a joule is about $\frac{1}{4}$ the size of a calorie; so the answer decreases by a factor of about four.

E.77   (a)   **Given:** 2387 Cal   **Find:** J
             Conceptual Plan: Cal → J

$$\frac{4184\,J}{1\,Cal}$$

Solution: $2387\,\cancel{Cal} \times \dfrac{4184\,J}{1\,\cancel{Cal}} = 9.987 \times 10^6\,J$

Check: The units (J) are correct. The magnitude of the answer (10^7) makes physical sense because a Calorie is much larger than a joule; so the answer increases.

        (b)   **Given:** 2387 Cal   **Find:** kJ
             Conceptual Plan: Cal → J → kWh

$$\frac{4184\,J}{1\,Cal} \quad \frac{1\,kJ}{1000\,J}$$

Solution: $2387\,\cancel{Cal} \times \dfrac{4184\,\cancel{J}}{1\,\cancel{Cal}} \times \dfrac{1\,kJ}{1000\,\cancel{J}} = 9.987 \times 10^3\,kJ$

Check: The units (kJ) are correct. The magnitude of the answer (10^4) makes physical sense because a Calorie is larger than a kilojoule; so the answer increases.

        (c)   **Given:** 2387 Cal   **Find:** kWh
             Conceptual Plan: Cal → J → kWh

$$\frac{4184\,J}{1\,Cal} \quad \frac{1\,kWh}{3.60 \times 10^6\,J}$$

Solution: $2387\,\cancel{Cal} \times \dfrac{4184\,\cancel{J}}{1\,\cancel{Cal}} \times \dfrac{1\,kWh}{3.60 \times 10^6\,\cancel{J}} = 2.774\,kWh$

Check: The units (kWh) are correct. The magnitude of the answer (3) makes physical sense because a Calorie is much smaller than a kilowatt hour; so the answer decreases.

E.79   **Given:** $145 electricity bill; electricity cost = $0.120/kWh   **Find:** J
       Conceptual Plan: Cost ($) → kWh → J

$$\frac{1\,kWh}{\$0.120} \quad \frac{3.60 \times 10^6\,J}{1\,kWh}$$

Solution: $\$\,145 \times \dfrac{1\,\cancel{kWh}}{\$\,0.120} \times \dfrac{3.60 \times 10^6\,J}{1\,\cancel{kWh}} = 4.35 \times 10^{10}\,J$

Check: The units (J) are correct. The magnitude of the answer (10^{10}) makes physical sense because a kilowatt hour is much larger than a joule; so the answer increases.

Cumulative Problems

E.81   (a)   **Given:** cylinder dimensions: length = 22 cm; radius = 3.8 cm; d(gold) = 19.3 g/cm³; d(sand) = 3.00 g/cm³   **Find:** m(gold) and m(sand)
             Conceptual Plan: l, r → V then d, V → m

$$V = l\pi r^2 \qquad d = m/V$$

Solution: V(gold) = V(sand) = $(22\,cm)(\pi)(3.8\,cm)^2 = 998.0212\,cm^3$ $d = m/V$
Rearrange by multiplying both sides of equation by V. → $m = d \times V$

$$m(\text{gold}) = \left(19.3\,\frac{g}{cm^3}\right) \times (998.0212\,\cancel{cm^3}) = 1.926181 \times 10^4\,g = 1.9 \times 10^4\,g$$

Check: The units (g) are correct. The magnitude of the answer seems correct considering that the value of the density is ~20 g/cm³. Two significant figures are allowed to reflect the significant figures in 22 cm and 3.8 cm. Truncate the nonsignificant digits because the first nonsignificant digit is a 2.

$$m(\text{sand}) = \left(3.00\,\frac{g}{cm^3}\right) \times (998.0212\,\cancel{cm^3}) = 2.99406 \times 10^3\,g = 3.0 \times 10^3\,g$$

Check: The units (g) are correct. The magnitude of the answer seems correct considering that the value of the density is 3 g/cm³. This number is much lower than the gold mass. Two significant figures are allowed to reflect the significant figures in 22 cm and 3.8 cm. Round the last digit up because the first nonsignificant digit is a 9.

(b)   Comparing the two values 1.9×10^4 g versus 3.0×10^3 g shows a difference in weight of almost a factor of 10. This difference should be enough to trip the alarm and alert the authorities to the presence of the thief.

E.83   **Given:** 3.5 lb of titanium   **Find:** volume in in^3 **Other:** density of titanium is 4.51 g/cm^3
     Conceptual Plan: lb → g then m, d → V then cm^3 → in^3

$$\frac{453.6 \text{ g}}{1 \text{ lb}} \qquad d = m/V \qquad \frac{(1 \text{ in})^3}{(2.54 \text{ cm})^3}$$

Solution: $3.5 \text{ lb} \times \dfrac{453.6 \text{ g}}{1 \text{ lb}} = 1.5\underline{8}76 \times 10^3 \text{ g}$

$d = m/V$ Rearrange by multiplying both sides of the equation by V and dividing both sides of the equation by d.

$$V = \frac{m}{d} = \frac{1.5\underline{8}76 \times 10^3 \text{ g}}{4.51 \dfrac{\text{g}}{\text{cm}^3}} = 3.\underline{5}20 \times 10^2 \text{ cm}^3 = 3.5 \times 10^2 \text{ cm}^3 \times \frac{(1 \text{ in})^3}{(2.54 \text{ cm})^3} = 21 \text{ in}^3$$

Check: The units (in^3) are correct. The magnitude of the answer seems correct considering the number of grams. Two significant figures are allowed to reflect the significant figures in 3.5 lb. Truncate the nonsignificant digits because the first nonsignificant digit is a 2.

E.85   **Given:** cylinder dimensions: length = 2.16 in; radius = 0.22 in; m = 41 g   **Find:** density (g/m^3)
     Conceptual Plan: in → cm then l, r → V then m, V → d

$$\frac{2.54 \text{ cm}}{1 \text{ in}} \qquad V = l\pi r^2 \qquad d = m/V$$

Solution: $2.16 \text{ in} \times \dfrac{2.54 \text{ cm}}{1 \text{ in}} = 5.4\underline{8}64 \text{ cm} = l \qquad\qquad 0.22 \text{ in} \times \dfrac{2.54 \text{ cm}}{1 \text{ in}} = 0.5\underline{5}88 \text{ cm} = r$

$V = l\pi r^2 = (5.4\underline{8}64 \text{ cm})(\pi)(0.5\underline{5}88 \text{ cm})^2 = 5.3\underline{8}20798 \text{ cm}^3$

$d = \dfrac{m}{V} = \dfrac{41 \text{ g}}{5.3\underline{8}20798 \text{ cm}^3} = 7.\underline{6}178729 \dfrac{\text{g}}{\text{cm}^3} = 7.6 \dfrac{\text{g}}{\text{cm}^3}$

Check: The units (g/cm^3) are correct. The magnitude of the answer seems correct considering that the value of the density of iron (a major component in steel) is 7.86 g/cm^3. Two significant figures are allowed to reflect the significant figures in 0.22 in and 41 g. Truncate the nonsignificant digits because the first nonsignificant digit is a 1.

E.87   **Given:** 185 cubic yards (yd^3) of H_2O **Find:** mass of the H_2O (pounds)
     Other: $d(H_2O) = 1.00 \text{ g}/cm^3$ at 4 °C
     Conceptual Plan: yd^3 → m^3 → cm^3 → g → lb

$$\frac{(1 \text{ m})^3}{(1.094 \text{ yd})^3} \qquad \frac{(100 \text{ cm})^3}{(1 \text{ m})^3} \qquad \frac{1.00 \text{ g}}{1.00 \text{ cm}^3} \qquad \frac{1 \text{ lb}}{453.59 \text{ g}}$$

Solution: $185 \text{ yd}^3 \times \dfrac{(1 \text{ m})^3}{(1.094 \text{ yd})^3} \times \dfrac{(100 \text{ cm})^3}{(1 \text{ m})^3} \times \dfrac{1.00 \text{ g}}{1.00 \text{ cm}^3} \times \dfrac{1 \text{ lb}}{453.59 \text{ g}} = 3.1\underline{1}4987377 \times 10^5 \text{ lb} = 3.11 \times 10^5 \text{ lb}$

Check: The units (lb) are correct. The magnitude of the answer (10^5) makes physical sense because a pool is not a small object. Three significant figures are allowed because the conversion factor with the least precision is the density $[1.00 \text{ g}/cm^3 - (3 \text{ significant figures})]$ and the initial size has three significant figures. Truncate after the last digit because the first nonsignificant digit is a 4.

E.89   **Given:** 15 liters of gasoline   **Find:** kilometers   **Other:** 52 mi/gal in the city
     Conceptual Plan: L → gal → mi → km

$$\frac{1 \text{ gal}}{3.785 \text{ L}} \qquad \frac{52 \text{ mi}}{1 \text{ gal}} \qquad \frac{1 \text{ km}}{0.6214 \text{ mi}}$$

Solution: $15 \text{ L} \times \dfrac{1 \text{ gal}}{3.785 \text{ L}} \times \dfrac{52 \text{ mi}}{1 \text{ gal}} \times \dfrac{1 \text{ km}}{0.6214 \text{ mi}} = 3.\underline{3}16327941 \times 10^2 \text{ km} = 3.3 \times 10^2 \text{ km}$

Check: The units (km) are correct. The magnitude of the answer (10^2) makes physical sense because the dominating conversion factor is the mileage, which increases the answer. Two significant figures are allowed because the conversion factor with the least precision is 52 mi/gal (two significant figures) and the initial volume (15 L) has two significant figures. Truncate the last digit because the first nonsignificant digit is a 1. It is best to put the answer in scientific notation so that it is clear how many significant figures are expressed.

E.91   **Given:** radius of nucleus of the hydrogen atom $= 1.0 \times 10^{-13}$ cm; radius of the hydrogen atom $= 52.9$ pm
Find: percent of volume occupied by nucleus (%)
Conceptual Plan: cm $\rightarrow$ m then pm $\rightarrow$ m then $r \rightarrow V$ then $V_{atom}, V_{nucleus} \rightarrow$ % $V_{nucleus}$

$$\frac{1\text{ m}}{100\text{ cm}} \qquad \frac{1\text{ m}}{10^{12}\text{ pm}} \qquad V = (4/3)\pi r^3 \qquad \% \ V_{nucleus} = \frac{V_{nucleus}}{V_{atom}} \times 100\%$$

Solution: 1.0×10^{-13} cm $\times \dfrac{1\text{ m}}{100\text{ cm}} = 1.0 \times 10^{-15}$ m and 52.9 pm $\times \dfrac{1\text{ m}}{10^{12}\text{ pm}} = 5.29 \times 10^{-11}$ m

$V = (4/3)\pi r^3$ Substitute into % V equation.

$\% \ V_{nucleus} = \dfrac{V_{nucleus}}{V_{atom}} \times 100\% \qquad \rightarrow \% \ V_{nucleus} = \dfrac{(4/3)\ \pi r^3_{nucleus}}{(4/3)\ \pi r^3_{atom}} \times 100\%$ Simplify equation.

$\% \ V_{nucleus} = \dfrac{r^3_{nucleus}}{r^3_{atom}} \times 100\%$ Substitute numbers and calculate result.

$\% \ V_{nucleus} = \dfrac{(1.0 \times 10^{-15}\text{ m})^3}{(5.29 \times 10^{-11}\text{ m})^3} \times 100\% = (1.890359168 \times 10^{-5})^3 \times 100\% = 6.7\underline{5}5118686 \times 10^{-13}$

$= 6.8 \times 10^{-13}$

Check: The units (none) are correct. The magnitude of the answer seems correct (10^{-13}) because a proton is so small. Two significant figures are allowed to reflect the significant figures in 1.0×10^{-13} cm. Round the last digits up because the first nonsignificant digit is a 5.

E.93   **Given:** radius of hydrogen $= 212$ pm; radius of Ping-Pong ball $= 4.0$ cm, 6.02×10^{23} atoms and balls in a row
Find: row length (km)
Conceptual Plan: atoms $\rightarrow$ pm $\rightarrow$ m $\rightarrow$ km and ball $\rightarrow$ cm $\rightarrow$ m $\rightarrow$ km

$$\frac{212\text{ pm}}{1\text{ atom}} \quad \frac{1\text{ m}}{10^{12}\text{ pm}} \quad \frac{1\text{ km}}{1000\text{ m}} \qquad \frac{4.0\text{ cm}}{1\text{ ball}} \quad \frac{100\text{ cm}}{1\text{ m}} \quad \frac{1\text{ km}}{1000\text{ m}}$$

Solution: 6.02×10^{23} atoms $\times \dfrac{212\text{ pm}}{1\text{ atom}} \times \dfrac{1\text{ m}}{10^{12}\text{ pm}} \times \dfrac{1\text{ km}}{1000\text{ m}} = 1.28 \times 10^{11}$ km

6.02×10^{23} balls $\times \dfrac{4.0\text{ cm}}{1\text{ ball}} \times \dfrac{1\text{ m}}{100\text{ cm}} \times \dfrac{1\text{ km}}{1000\text{ m}} = 2.4 \times 10^{19}$ km

Check: The units (km) are correct. The magnitude of the answers seems correct (10^{11} and 10^{19}). The answers are driven by the large number of atoms or balls. The Ping-Pong ball row is 10^8 times longer. Three significant figures are allowed to reflect the significant figures in 212 pm. Two significant figures are allowed to reflect the significant figures in 4.0 cm.

E.95   **Given:** 39.33 g sodium/100 g salt; 1.25 g salt/100 g snack mix; FDA maximum 2.40 g sodium/day
Find: g snack mix
Conceptual Plan: g sodium $\rightarrow$ g salt $\rightarrow$ g snack mix

$$\frac{100\text{ g salt}}{39.33\text{ g sodium}} \qquad \frac{100\text{ g snack mix}}{1.25\text{ g salt}}$$

Solution: $\dfrac{2.40\text{ g sodium}}{1\text{ day}} \times \dfrac{100\text{ g salt}}{39.33\text{ g sodium}} \times \dfrac{100\text{ g snack mix}}{1.25\text{ g salt}} = 488.1\underline{7}70$ g snack mix/day

$= 488$ g snack mix/day

Check: The units (g) are correct. The magnitude of the answer seems correct (500) because salt is less than half sodium and there is a little over a gram of salt per 100 grams of snack mix. Three significant figures are allowed to reflect the significant figures in the FDA maximum and in the amount of salt in the snack mix.

E.97   **Given:** 24.0 kg copper wire; wire is a cylinder of radius $= 1.63$ mm   **Find:** resistance (Ω)
Other: d(copper) $= 8.96$ g/cm^3; resistance $= 2.061\ \Omega/$km
Conceptual Plan: mm $\rightarrow$ m $\rightarrow$ cm and kg $\rightarrow$ g then

$$\frac{10^{-3}\text{ m}}{1\text{ mm}} \quad \frac{100\text{ cm}}{1\text{ m}} \quad \frac{1000\text{ g}}{1\text{ kg}}$$

$d, m \rightarrow V$ then $V, r \rightarrow l$(cm) $\rightarrow$ m $\rightarrow$ km $\rightarrow \Omega$

$$d = m/V \quad V = l\ \pi\ r^2 \quad \frac{1\text{ m}}{100\text{ cm}} \quad \frac{1\text{ km}}{1000\text{ m}} \quad \frac{2.061\ \Omega}{1\text{ km}}$$

Solution: 1.63 mm $\times \dfrac{10^{-3}\text{ m}}{1\text{ mm}} \times \dfrac{100\text{ cm}}{1\text{ m}} = 0.163$ cm

$24.0 \text{ kg} \times \dfrac{1000 \text{ g}}{1 \text{ kg}} = 2.40 \times 10^4 \text{ g}$; then $d = m/V$. Rearrange by multiplying both sides of equation by V to get

$m = d \times V$; then divide both sides by d. $V = \dfrac{m}{d} = \dfrac{2.40 \times 10^4 \text{ g}}{8.96 \text{ g}} = 2.6785714 \times 10^3 \text{ cm}^3$; then $V = l\pi r^2$
$\phantom{V = \dfrac{m}{d} = }\dfrac{}{1 \text{ cm}^3}$

Rearrange by dividing both sides of the equation by πr^2 to get

$l = \dfrac{V}{\pi r^2} = \dfrac{2.6785714 \times 10^3 \text{ cm}^3}{\pi (0.163 \text{ cm})^2} = 3.2090623 \times 10^4 \text{ cm}$

then

$3.2090623 \times 10^4 \text{ cm} \times \dfrac{1 \text{ m}}{100 \text{ cm}} \times \dfrac{1 \text{ km}}{1000 \text{ m}} \times \dfrac{2.061 \ \Omega}{1 \text{ km}} = 0.6613877 \ \Omega = 0.661 \ \Omega$

Check: The units (Ω) are correct. The magnitude of the answer seems correct because we expect a small resistance for a material that is commonly used for electrical wiring. Three significant figures are allowed to reflect the significant figures in 1.63 mm and 24.0 kg. Truncate the nonsignificant digits because the first nonsignificant digit is a 3.

E.99   **Given:** d(liquid nitrogen) $= 0.808$ g/mL; d(gaseous nitrogen) $= 1.15$ g/L; 175 L liquid nitrogen; 10.00 m $\times$ 10.00 m $\times$ 2.50 m room   **Find:** fraction of room displaced by nitrogen gas

Conceptual Plan: $L \rightarrow$ mL then $V_{\text{liquid}}, d_{\text{liquid}} \rightarrow m_{\text{liquid}}$ then set $m_{\text{liquid}} = m_{\text{gas}}$ then $m_{\text{gas}}, d_{\text{gas}} \rightarrow V_{\text{gas}}$ then

$\dfrac{1000 \text{ mL}}{1 \text{ L}}d = m/vd = m/v$

calculate the $V_{\text{room}} \rightarrow$ cm$^3 \rightarrow$ L then calculate the fraction displaced

$$V = l \times w \times h \qquad \dfrac{(100 \text{ cm})^3}{(1 \text{ m})^3} \qquad \dfrac{1 \text{ L}}{1000 \text{ cm}^3} \qquad \dfrac{V_{\text{gas}}}{V_{\text{room}}}$$

Solution: $175 \text{ L} \times \dfrac{1000 \text{ mL}}{1 \text{ L}} = 1.75 \times 10^5 \text{ mL}$. Solve for m by multiplying both sides of the equation by V.

$m = V \times d = 1.75 \times 10^5 \text{ mL} \times \dfrac{0.808 \text{ g}}{1 \text{ mL}} = 1.414 \times 10^5 \text{ g nitrogen liquid} = 1.414 \times 10^5 \text{ g nitrogen gas}$

$d = m/V$ Rearrange by multiplying both sides of the equation by V and dividing both sides of the equation by d.

$V = \dfrac{m}{d} = \dfrac{1.414 \times 10^5 \text{ g}}{1.15 \ \frac{\text{g}}{\text{L}}} = 1.229565 \times 10^5 \text{ L nitrogen gas}$

$V_{\text{room}} = l \times w \times h = 10.00 \text{ m} \times 10.00 \text{ m} \times 2.50 \text{ m} \times \dfrac{(100 \text{ cm})^3}{(1 \text{ m})^3} \times \dfrac{1 \text{ L}}{1000 \text{ cm}^3} = 2.50 \times 10^5 \text{ L}$

$\dfrac{V_{\text{gas}}}{V_{\text{room}}} = \dfrac{1.229565 \times 10^5 \text{ L}}{2.50 \times 10^5 \text{ L}} = 0.491826 = 0.492$

Check: The units (none) are correct. The magnitude of the answer seems correct (0.5) because there is a large volume of liquid and the density of the gas is about a factor of 1000 less than the density of the liquid. Three significant figures are allowed to reflect the significant figures in the densities and the volume of the liquid given.

E.101   **Given:** $r_{\text{nucleus}} = 2.7$ fm; $r_{\text{atom}} = 70$ pm (assume two significant figures)
Find: V_{nucleus}; V_{atom}; $\% \ V_{\text{nucleus}}$

Conceptual Plan:

$r_{\text{nucleus}}(\text{fm}) \rightarrow r_{\text{nucleus}}(\text{pm}) \rightarrow V_{\text{nucleus}}$ and then $r_{\text{atom}} \rightarrow V_{\text{atom}}$ and then $\% \ V$

$\dfrac{10^{-15} \text{ m}}{1 \text{ fm}} \ \dfrac{1 \text{ pm}}{10^{-12} \text{ m}} \qquad V = \dfrac{4}{3}\pi r^3 \qquad\qquad V = \dfrac{4}{3}\pi r^3 \qquad \dfrac{V_{\text{nucleus}}}{V_{\text{atom}}} \times 100\%$

Solution:

$2.7 \text{ fm} \times \dfrac{10^{-15} \text{ m}}{1 \text{ fm}} \times \dfrac{1 \text{ pm}}{10^{-12} \text{ m}} = 2.7 \times 10^{-3} \text{ pm} \qquad V_{\text{nucleus}} = \dfrac{4}{3}\pi (2.7 \times 10^{-3} \text{ pm})^3 = 8.2 \times 10^{-8} \text{ pm}^3$

$V_{\text{atom}} = \dfrac{4}{3}\pi (70 \text{ pm})^3 = 1.4 \times 10^6 \text{ pm}^3 \qquad\qquad \dfrac{8.2 \times 10^{-8} \text{ pm}^3}{1.4 \times 10^6 \text{ pm}^3} \times 100\% = 5.9 \times 10^{-12}\% \ V$

Check: The units of the answer ($\% V$) are correct. The magnitude of the answer is reasonable because the nucleus occupies only a very small percentage of the V of the atom.

Challenge Problems

E.103   **Given:** 15.0 ppm CO; 8-hour period   **Find:** milligrams of carbon monoxide
Other: 0.50 L of air per breath; 20 breaths per minute; carbon monoxide has a density of 1.2 g/L; 15.0 ppm CO means 15.0 L CO per 10^6 L air
Conceptual Plan: hr $\rightarrow$ min $\rightarrow$ breaths $\rightarrow$ L_{air} $\rightarrow$ L_{CO} $\rightarrow$ g_{CO} $\rightarrow$ mg_{CO}

$$\frac{60\ min}{1\ hr} \qquad \frac{20\ breath}{1\ min} \qquad \frac{0.50\ L_{air}}{1\ breath} \qquad \frac{15.0\ L_{CO}}{1\times10^6\ L_{air}} \qquad \frac{1.2\ g_{CO}}{1\ L_{CO}} \qquad \frac{1000\ mg_{CO}}{1\ g_{CO}}$$

Solution:

$$8\ hr \times \frac{60\ min}{1\ hr} \times \frac{20\ breath}{1\ min} \times \frac{0.50\ L_{air}}{1\ breath} \times \frac{15.0\ L_{CO}}{1\times10^6\ L_{air}} \times \frac{1.2\ g_{CO}}{1\ L_{CO}} \times \frac{1000\ mg_{CO}}{1\ g_{CO}} = \underline{8}6.4\ mg_{CO} = 9\times10^1\ mg_{CO}$$

Check: The units (mg) are correct. The magnitude of the answer (10^2) makes physical sense because there are more than 6 powers of 10 visible in these conversion factors in the numerator and one factor of 10^6 in the denominator. This means that most of the conversions cancel each other out, but there is still some left in the numerator. One significant figure is allowed because the conversion factor with the least precision is 20 breaths/minute (one significant figure); the starting time (8 hours) also has one significant figure. Round the last digit up because the first nonsignificant digit is a 6.

E.105   Because the person weighs 155 lb and has a density of 1.0 g/cm^3, the volume of the person can be calculated as

$$155\ lb \times \frac{453.59\ g}{1\ lb} \times \frac{1\ cm^3}{1.0\ g} = 7.\underline{0}30645\times10^4\ cm^3.$$

Approximating the volume of a person as a cylinder 4.0 feet tall, $V = l\pi r^2$. Rearranging the equation, solve for r.

$$r = \left(\frac{V}{l\pi}\right)^{\frac{1}{2}} = \sqrt{\frac{7.\underline{0}30645\times10^4\ cm^3}{4.0\ ft \times \dfrac{30.48\ cm}{1\ ft} \times \pi}} = 13.\underline{5}4831\ cm$$

The circumference is $2\pi r = 2\pi(13.\underline{5}4831\ cm) = 8\underline{5}.12655\ cm$. When the person gains 40.0 lb of fat, the volume

increase is $40.0\ lb \times \dfrac{453.59\ g}{1\ lb} \times \dfrac{1\ cm^3}{0.918\ g} = 1.9\underline{7}643\times10^4\ cm^3.$

Thus, the new volume is $7.\underline{0}30645\times10^4\ cm^3 + 1.9\underline{7}643\times10^4\ cm^3 = 9.0\underline{0}708\times10^4\ cm^3$. So the new

radius is $r = \left(\dfrac{V}{l\pi}\right)^{\frac{1}{2}} = \sqrt{\dfrac{9.0\underline{0}708\times10^4\ cm^3}{4.0\ ft \times \dfrac{30.48\ cm}{1\ ft} \times \pi}} = 15.\underline{3}3485\ cm$, and the new circumference is

$2\pi r = 2\pi(15.\underline{3}3485\ cm) = 9\underline{6}.35170\ cm.$

The percent increase in circumference $\dfrac{9\underline{6}.35170\ cm - 8\underline{5}.12655\ cm}{8\underline{5}.12655\ cm} \times 100\% = 13.\underline{1}864\% = 13\%.$

Conceptual Problems

E.107   This problem is similar to Problem E.32 except that the dimension is changed to 7 cm on each edge.
Given: 7 cm on each edge cube   **Find:** cm^3
Conceptual Plan: Read the information carefully. The cube is 7 cm on each side.
$l, w, h \rightarrow V$
     $V = l\,w\,h$
   in a cube $l = w = h$
Solution: $7\ cm \times 7\ cm \times 7\ cm = (7\ cm)^3 = 343\ cm^3$ or 343 cubes

E.109   Remember that density = mass/volume.

(a)   The darker-colored box has a heavier mass but a smaller volume, so it is denser than the lighter-colored box.

(b)   The lighter-colored box is heavier than the darker-colored box, and both boxes have the same volume; so the lighter-colored box is denser.

(c)   The larger box is the heavier box, so it cannot be determined with this information which box is denser.

Questions for Group Work

E.111   All of these values can be true because when we round 4.73297 km to three significant figures we get 4.73 km; and if we round this value to one significant figure we get 5 km. The number of digits reported indicates how accurately the distance was measured. The odometer on a car could be used for the first measurement. A trip meter on a car could be used for the second measurement. A long tape measure could be used for the last measurement.

E.113   To convert the height of a student from feet and inches into meters: first convert the feet to inches; then add this to the number of inches, and finally convert the inches to centimeters and to meters. For example, if the height of a student is 5'8.5": $5 \text{ ft} \times \dfrac{12 \text{ in}}{1 \text{ ft}} = 60 \text{ in (exactly)}$; then 60 in + 8.5 in = 68.5 in, and finally $68.5 \text{ in} \times \dfrac{2.54 \text{ cm}}{1 \text{ in}} \times \dfrac{1 \text{ m}}{100 \text{ cm}} = 1.7399 \text{ m} = \underline{1.74} \text{ m}$.

Add all of the heights together to get the sum for your group.

Data Interpretation and Analysis

E.114   (a)   The initial value (liquid water) is 1000. kg/m^3 and the final value (solid water – ice) is 917 kg/m^3 at 0°C.

$$\% \text{ change} = \frac{\text{final value} - \text{initial value}}{\text{initial value}} \times 100\% = \frac{917 \text{ kg/m}^3 - 1000. \text{ kg/m}^3}{1000. \text{ kg/m}^3} \times 100\%$$

$$= \frac{-83}{1,000.} \times 100\% = -8.3\%$$

(c)   The density of solid ice at −20°C is 919 kg/m^3, and the density of liquid water at 4°C is 999.97 kg/m^3.

The volume of the water from Antarctica will be $26.5 \times 10^6 \text{ km}^3 \times \dfrac{919 \text{ kg/m}^3}{999.97 \text{ kg/m}^3} = 2.435423 \times 10^7 \text{ km}^3 = \underline{2.44} \times 10^7 \text{ km}^3$.

1 Atoms

Review Questions

1.1   "The properties of the substances around us depend on the atoms, ions, or molecules that compose them" means that the specific types of atoms and molecules that compose something tell us a great deal about which properties to expect from a substance. A material composed of only sodium and chloride ions will have the properties of table salt. A material composed of molecules with one carbon atom and two oxygen atoms will have the properties of the gas carbon dioxide. If the atoms and molecules change, so do the properties that we expect the material to have.

1.3   Matter can be classified according to its state—solid, liquid, or gas—and according to its composition.

1.5   A pure substance is composed of only one type of atom or molecule. In contrast, a mixture is a substance composed of two or more different types of atoms or molecules that can be combined in variable proportions.

1.7   A homogeneous mixture has the same composition throughout, while a heterogeneous mixture has different compositions in different regions.

1.9   A hypothesis is a tentative interpretation or explanation of the observed phenomena. A law is a concise statement that summarizes observed behaviors and observations and predicts future observations. A theory attempts to explain why the observed behavior is happening.

1.11   John Dalton formulated the atomic theory of matter. Dalton explained the law of conservation of mass, as well as other laws and observations of the time, by proposing that matter was composed of small, indestructible particles called atoms. Because these particles were merely rearranged in chemical changes (and not created or destroyed), the total amount of mass would remain the same.

1.13   The first people to propose that matter was composed of small, indestructible particles were Leucippus and Democritus. These Greek philosophers theorized that matter was ultimately composed of small, indivisible particles called *atomos*. In the sixteenth century, modern science began to emerge. A greater emphasis on observation brought rapid advancement as the scientific method became the established way to learn about the physical world. By the early 1800s, certain observations led the English chemist John Dalton to offer convincing evidence that supported the early atomic ideas of Leucippus and Democritus. The theory that all matter is composed of atoms grew out of observations and laws. The three most important laws that led to the development and acceptance of the atomic theory were the law of conservation of mass, the law of definite proportions, and the law of multiple proportions. John Dalton explained the laws with his atomic theory.

1.15   The law of definite proportions states the following: All samples of a given compound, regardless of their source or how they were prepared, have the same proportions of their constituent elements. This means that elements composing a given compound always occur in fixed (or definite) proportions in all samples of the compound.

1.17   The main ideas of John Dalton's atomic theory are as follows: (1) Each element is composed of tiny, indestructible particles called atoms. (2) All atoms of a given element have the same mass and other properties that distinguish them from the atoms of other elements. (3) Atoms combine in simple whole-number ratios to form compounds. (4) Atoms of one element cannot change into atoms of another element. In a chemical reaction, atoms change the way they are bound together with other atoms to form a new substance. The law of conservation of mass is explained by the fourth idea. Because the atoms cannot change into another element, and just change how they are bound together, the total mass will remain constant. The law of constant composition is supported by ideas 2 and 3. Because the atoms of a given element always have the same mass and other distinguishing properties, and they combine in simple whole-number ratios, different samples of the same compound will have the same properties and the same composition. The law of multiple proportions is also supported by ideas 2 and 3 because the atoms can combine in simple whole-number ratios; the ratio of the mass of B in one compound to the mass of B in a second compound will also be a small whole number.

1.19   In Millikan's oil drop experiment, oil was sprayed into fine droplets using an atomizer. The droplets were allowed to fall under the influence of gravity through a small hole into the lower portion of the apparatus where they could be viewed. During their fall, the drops would acquire electrons that had been produced by the interaction of high-energy radiation with air. These charged drops interacted with two electrically charged plates within the apparatus. The negatively charged plate at the bottom of the apparatus repelled the negatively charged drops. By varying the voltage on the plates, the fall of the charged drops could be slowed, stopped, or even reversed. From the voltage required to halt the free fall of the drops and from the masses of the drops themselves, Millikan calculated the charge of each drop. He then reasoned that because each drop must contain an integral number of electrons, the charge of each drop must be a whole-number multiple of the electron's charge. The magnitude of the charge of the electron is of tremendous importance because it determines how strongly an atom holds its electrons.

1.21   Rutherford's gold foil experiment directed positively charged α particles at an ultrathin sheet of gold foil. These particles were to act as probes of the gold atoms' structures. If the gold atoms were indeed like plum pudding—with their mass and charge spread throughout the entire volume of the atom—these speeding probes should pass right through the gold foil with minimum deflection. A majority of the particles did pass directly through the foil, but some particles were deflected, and some even bounced back. Rutherford realized that to account for the deflections, the mass and positive charge of an atom must all be concentrated in a space much smaller than the size of the atom itself.

1.23   Matter appears solid because the variation in its density is on such a small scale that our eyes cannot see it.

1.25   The number of protons in the nucleus defines the identity of an element.

1.27   The names of the elements were often given to describe their properties. For example, argon originates from the Greek word *argos*, meaning "inactive." Other elements were named after figures from Greek or Roman mythology or astronomical bodies. Still others were named for the places where they were discovered or where their discoverer was born. More recently, elements have been named after scientists.

1.29   Isotopes can be symbolized as $^A_Z X$, where A is the mass number, Z is the atomic number, and X is the chemical symbol. A second notation is the chemical symbol (or chemical name) followed by a hyphen and the mass number of the isotope, such as X-A, where X is the chemical symbol or name and A is the mass number. The carbon isotope with a mass of 12 would have the symbol $^{12}_6 C$ or C-12 or carbon-12.

1.31   Atomic mass represents the average mass of the isotopes that compose the element. The average calculated atomic mass is weighted according to the natural abundance of each isotope.

$$\text{Atomic mass} = \sum_n (\text{fraction of isotope } n) \times (\text{mass of isotope } n)$$

1.33   A mole is an amount of material. It is defined as the amount of material containing 6.0221421×10^{23} particles (Avogadro's number). The numerical value of the mole is defined as being equal to the number of atoms in exactly 12 grams of pure carbon-12. It is useful for converting number of atoms to moles of atoms and moles of atoms to number of atoms.

Problems by Topic

The Classification of Matter

1.35   (a)   The single type of blue cubes indicates that this is a pure substance.
        (b)   The single type of red cylinders indicates that this is a pure substance.
        (c)   The random and uniform distribution of the blue cubes and red cylinders indicates that this is a homogeneous mixture.
        (d)   The region of only red cylinders above a region of only blue cubes indicates that this is a heterogeneous mixture.

1.37   (a)   Sweat is a homogeneous mixture of water, sodium chloride, and other components.
       (b)   Carbon dioxide is a pure substance that is a compound (two or more elements bonded together).
       (c)   Aluminum is a pure substance that is an element (element 13 in the periodic table).
       (d)   Vegetable soup is a heterogeneous mixture of broth, chunks of vegetables, and extracts from the vegetables.

1.39
Substance	Pure or Mixture	Type (element or compound)
aluminum	pure	element
apple juice	mixture	neither—mixture
hydrogen peroxide	pure	compound
chicken soup	mixture	neither—mixture

1.41   (a)   pure substance that is a compound (one type of molecule that contains two different elements)
       (b)   heterogeneous mixture (two different molecules that are segregated into regions)
       (c)   homogeneous mixture (two different molecules that are randomly mixed)
       (d)   pure substance that is an element (individual atoms of one type)

The Scientific Approach to Knowledge

1.43   (a)   This statement is a theory because it attempts to explain why. It is not possible to observe individual atoms.
       (b)   This statement is an observation.
       (c)   This statement is a law because it summarizes many observations and can explain future behavior.
       (d)   This statement is an observation.

1.45   (a)   If we divide the mass of the oxygen by the mass of the carbon, the result is always 4/3.
       (b)   If we divide the mass of the oxygen by the mass of the hydrogen, the result is always 16.
       (c)   These observations suggest that the masses of elements in molecules are ratios of whole numbers [4:3 and 16:1, respectively, for parts (a) and (b)].
       (d)   Atoms combine in small whole-number ratios and not as random weight ratios.

The Laws of Conservation of Mass, Definite Proportions, and Multiple Proportions

1.47   **Given:** 1.50 g hydrogen; 12.0 g oxygen   **Find:** grams of water vapor
       Conceptual Plan: total mass reactants = total mass products
       Solution: Mass of reactants = 1.50 g hydrogen + 12.0 g oxygen = 13.5 g
                   Mass of products = mass of reactants = 13.5 g water vapor

       Check: According to the law of conservation of mass, matter is neither created nor destroyed in a chemical reaction. Since water vapor is the only product, the masses of hydrogen and oxygen must combine to form the mass of water vapor.

1.49   **Given:** sample 1: 38.9 g carbon, 448 g chlorine; sample 2: 14.8 g carbon, 134 g chlorine
       Find: consistent with definite proportions
       Conceptual Plan: Determine mass ratio of samples 1 and 2 and compare.

$$\frac{\text{mass of chlorine}}{\text{mass of carbon}}$$

       Solution: sample 1: $\dfrac{448 \text{ g chlorine}}{38.9 \text{ g carbon}} = 11.5$ sample 2: $\dfrac{134 \text{ g chlorine}}{14.8 \text{ g carbon}} = 9.05$

       Results are not consistent with the law of definite proportions because the ratio of chlorine to carbon is not the same.

       Check: According to the law of definite proportions, the mass ratio of one element to another is the same for all samples of the compound.

1.51   **Given:** mass ratio sodium to fluorine = 1.21:1; sample = 28.8 g sodium   **Find:** g fluorine
       Conceptual Plan: g sodium → g fluorine

$$\frac{\text{mass of fluorine}}{\text{mass of sodium}}$$

       Solution: $28.8 \text{ g sodium} \times \dfrac{1 \text{ g fluorine}}{1.21 \text{ g sodium}} = 23.8 \text{ g fluorine}$

Check: The units of the answer (g fluorine) are correct. The magnitude of the answer is reasonable because it is less than the grams of sodium.

1.53   **Given:** 1 g osmium: sample 1 = 0.168 g oxygen; sample 2 = 0.3369 g oxygen
Find: consistent with multiple proportions
Conceptual Plan: Determine mass ratio of oxygen.
$$\frac{\text{mass of oxygen sample 2}}{\text{mass of oxygen sample 1}}$$

Solution: $\dfrac{0.3369 \text{ g oxygen}}{0.168 \text{ g oxygen}} = 2.00$ Ratio is a small whole number. Results are consistent with multiple proportions.

Check: According to the law of multiple proportions, when two elements form two different compounds, the masses of element B that combine with 1 g of element A can be expressed as a ratio of small whole numbers.

1.55   **Given:** sulfur dioxide = 3.49 g oxygen and 3.50 g sulfur; sulfur trioxide = 6.75 g oxygen and 4.50 g sulfur
Find: mass oxygen per g S for each compound and then determine the mass ratio of oxygen
$$\frac{\text{mass of oxygen in sulfur dioxide}}{\text{mass of sulfur in sulfur dioxide}} \quad \frac{\text{mass of oxygen in sulfur trioxide}}{\text{mass of sulfur in sulfur trioxide}} \quad \frac{\text{mass of oxygen in sulfur trioxide}}{\text{mass of oxygen in sulfur dioxide}}$$

Solution: sulfur dioxide $= \dfrac{3.49 \text{ g oxygen}}{3.50 \text{ g sulfur}} = \dfrac{0.997 \text{ g oxygen}}{1 \text{ g sulfur}}$ sulfur trioxide $= \dfrac{6.75 \text{ g oxygen}}{4.50 \text{ g sulfur}} = \dfrac{1.50 \text{ g oxygen}}{1 \text{ g sulfur}}$

$\dfrac{1.50 \text{ g oxygen in sulfur trioxide}}{0.997 \text{ g oxygen in sulfur dioxide}} = \dfrac{1.50}{1} = \dfrac{3}{2}$.   The ratio is converted from 1.50:1 to 3:2 because the law of multiple proportions states that the ratio is in small whole numbers. Ratio is in small whole numbers and is consistent with multiple proportions.

Check: According to the law of multiple proportions, when two elements form two different compounds, the masses of element B that combine with 1 g of element A can be expressed as a ratio of small whole numbers.

Atomic Theory, Nuclear Theory, and Subatomic Particles

1.57   (a)   Sulfur and oxygen atoms have the same mass. *Inconsistent* with Dalton's atomic theory because only atoms of the same element have the same mass.

(b)   All cobalt atoms are identical. *Consistent* with Dalton's atomic theory because all atoms of a given element have the same mass and other properties that distinguish them from atoms of other elements.

(c)   Potassium and chlorine atoms combine in a 1:1 ratio to form potassium chloride. *Consistent* with Dalton's atomic theory because atoms combine in simple whole-number ratios to form compounds.

(d)   Lead atoms can be converted into gold. *Inconsistent* with Dalton's atomic theory because atoms of one element cannot change into atoms of another element.

1.59   (a)   The volume of an atom is mostly empty space. *Consistent* with Rutherford's nuclear theory because most of the volume of the atom is empty space, throughout which tiny, negatively charged electrons are dispersed.

(b)   The nucleus of an atom is small compared to the size of the atom. *Consistent* with Rutherford's nuclear theory because most of the atom's mass and all of its positive charge are contained in a small core called the nucleus.

(c)   Neutral lithium atoms contain more neutrons than protons. *Inconsistent* with Rutherford's nuclear theory because it did not distinguish where the mass of the nucleus came from other than from the protons.

(d)   Neutral lithium atoms contain more protons than electrons. *Inconsistent* with Rutherford's nuclear theory because there are as many negatively charged particles outside the nucleus as there are positively charged particles in the nucleus.

1.61   **Given:** drop A = -6.9×10^{-19} C;   drop B = -9.2×10^{-19} C;   drop C = -11.5×10^{-19} C; drop D = -4.6×10^{-19} C
Find: the charge on a single electron
Conceptual Plan: Determine the ratio of charge for each set of drops.
$$\frac{\text{charge on drop 1}}{\text{charge on drop 2}}$$

Solution: $\dfrac{-6.9 \times 10^{-19} \text{ C drop A}}{-4.6 \times 10^{-19} \text{ C drop D}} = 1.5$ $\dfrac{-9.2 \times 10^{-19} \text{ C drop B}}{-4.6 \times 10^{-19} \text{ C drop D}} = 2$ $\dfrac{-11.5 \times 10^{-19} \text{ C drop C}}{-4.6 \times 10^{-19} \text{ C drop D}} = 2.5$

The ratios obtained are not whole numbers, but they can be converted to whole numbers by multiplying by 2.

Therefore, the charge on the electron has to be 1/2 the smallest value experimentally obtained. The charge on the electron $= -2.3 \times 10^{-19}$ C.

Check: The units of the answer (coulombs) are correct. The magnitude of the answer is reasonable because all of the values experimentally obtained are integer multiples of -2.3×10^{-19}.

1.63   (a)   True: Protons and electrons have equal and opposite charges.
       (b)   True: Protons and electrons have opposite charges, so they will attract each other.
       (c)   True: The mass of the electron is much less than the mass of the neutron.
       (d)   False: The mass of the proton and the mass of the neutron are about the same.

Isotopes and Ions

1.65   For each of the isotopes determine Z (the number of protons) from the periodic table and determine A (protons + neutrons). Then write the symbol in the form X-A.
       (a)   The silver isotope with 60 neutrons: $Z = 47$; $A = 47 + 60 = 107$;   Ag-107
       (b)   The silver isotope with 62 neutrons: $Z = 47$; $A = 47 + 62 = 109$;   Ag-109
       (c)   The uranium isotope with 146 neutrons: $Z = 92$; $A = 92 + 146 = 238$;   U-238
       (d)   The hydrogen isotope with 1 neutron: $Z = 1$; $A = 1 + 1 = 2$;   H-2

1.67   (a) $^{14}_{7}\text{N}$: $Z = 7$; $A = 14$; protons $= Z = 7$; neutrons $= A - Z = 14 - 7 = 7$
       (b) $^{23}_{11}\text{Na}$: $Z = 11$; $A = 23$; protons $= Z = 11$; neutrons $= A - Z = 23 - 11 = 12$
       (c) $^{222}_{86}\text{Rn}$: $Z = 86$; $A = 222$; protons $= Z = 86$; neutrons $= A - Z = 222 - 86 = 136$
       (d) $^{208}_{82}\text{Pb}$: $Z = 82$; $A = 208$; protons $= Z = 82$; neutrons $= A - Z = 208 - 82 = 126$

1.69   Carbon-14: $A = 14$, $Z = 6$; $^{14}_{6}\text{C}$ # protons $= Z = 6$ # neutrons $= A - Z = 14 - 6 = 8$

1.71   In a neutral atom, the number of protons $=$ the number of electrons $= Z$. For an ion, electrons are lost (cations) or gained (anions).
       (a) Ni^{2+}: $Z = 28 =$ protons; $Z - 2 = 26 =$ electrons
       (b) S^{2-}: $Z = 16 =$ protons; $Z + 2 = 18 =$ electrons
       (c) Br^{-}: $Z = 35 =$ protons; $Z + 1 = 36 =$ electrons
       (d) Cr^{3+}: $Z = 24 =$ protons; $Z - 3 = 21 =$ electrons

       Check: Ions with a positive charge (Ni^{2+} and Cr^{3+}) have more protons than electrons. Ions with a negative charge (S^{2-} and Br^{-}) have more electrons than protons.

Atomic Mass and Mass Spectrometry

1.73   **Given:** Ga-69; mass $= 68.92558$ amu; 60.108%: Ga-71; mass $= 70.92470$ amu; 39.892%   **Find:** sketch the mass spectrum
       Conceptual Plan: $\rightarrow$ show relative % abundance versus mass on the graph
       Solution:

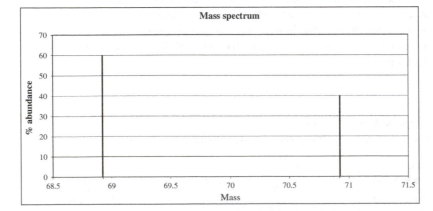

Check: The mass spectrum is reasonable because it has two mass lines corresponding to the two isotopes and the line at 68.92588 is about 1.5 times larger than the line at 70.92470.

1.75   Fluorine has an isotope F-19 with a very large abundance so that the mass of fluorine is very close to the mass of the isotope and the line in the mass spectrum reflects the abundance of F-19. Chlorine has two isotopes (Cl-35 and Cl-37), and the mass of 35.45 amu is the weighted average of these two isotopes; so there is no line at 35.45 amu.

1.77   **Given:** isotope 1 mass $= 120.903$ amu, 57.4%; isotope 2 mass $= 122.9042$ amu
Find: atomic mass of the element and identify the element
Conceptual Plan: % abundance isotope 2 → and then % abundance → fraction and then find atomic mass

$$100\% - \% \text{ abundance isotope 1} \qquad \frac{\% \text{ abundance}}{100} \qquad \text{Atomic mass} = \sum_n (\text{fraction of isotope } n) \times (\text{mass of isotope } n)$$

Solution: $100.0\% - 57.4\%$ isotope $1 = 42.6\%$ isotope 2

$$\text{Fraction isotope 1} = \frac{57.4}{100} = 0.574 \qquad \text{Fraction isotope 2} = \frac{42.6}{100} = 0.426$$

$$\text{Atomic mass} = \sum_n (\text{fraction of isotope } n) \times (\text{mass of isotope } n)$$

$$= 0.574(120.9038 \text{ amu}) + 0.426(122.9042 \text{ amu}) = 121.8 \text{ amu}$$

From the periodic table, Sb has a mass of 121.757 amu; so it is the closest mass, and the element is antimony.

Check: The units of the answer (amu) are correct. The magnitude of the answer is reasonable because it lies between 120.9038 and 122.9042 and is slightly less than halfway between the two values because the lower value has a slightly greater abundance.

1.79   **Given:** Br-81; mass $= 80.9163$ amu; 49.31%: atomic mass Br $= 79.904$ amu
Find: mass and abundance
Conceptual Plan: % abundance Br-79 → then % abundance → fraction → mass Br-79

$$100\% - \% \text{ Br-81} \qquad \frac{\% \text{ abundance}}{100} \qquad \text{Atomic mass} = \sum_n (\text{fraction of isotope } n) \times (\text{mass of isotope } n)$$

Solution: $100.00\% - 49.31\% = 50.69\%$

$$\text{Fraction Br-79} = \frac{50.69}{100} = 0.5069 \qquad \text{Fraction Br-81} = \frac{49.31}{100} = 0.4931$$

Let x be the mass of Br-79.

$$\text{Atomic mass} = \sum_n (\text{fraction of isotope } n) \times (\text{mass of isotope } n)$$

$$79.904 \text{ amu} = 0.5069(x \text{ amu}) + 0.4931(80.9163 \text{ amu})$$

$$x = 78.92 \text{ amu} = \text{mass Br-79}$$

Check: The units of the answer (amu) are correct. The magnitude of the answer is reasonable because it is less than the mass of the atom and the second isotope (Br-81) has a mass greater than the mass of the atom.

1.81   **Given:** graph of mass and % abundance of Eu   **Find:** atomic mass of Eu
Conceptual Plan: % abundance → relative fraction abundance → mass Eu

$$\frac{\% \text{ abundance of each isotope}}{\text{sum of all } \% \text{ abundance}} \qquad \text{Atomic mass} = \sum_n (\text{fraction of isotope } n) \times (\text{mass of isotope } n)$$

Solution: Fraction ^{151}Eu $= \dfrac{91.6}{191.6} = 0.4781$ Fraction ^{153}Eu $= \dfrac{100.0}{191.6} = 0.5219$

$$\text{Atomic mass} = \sum_n (\text{fraction of isotope } n) \times (\text{mass of isotope } n)$$

$$= 0.4781(151 \text{ amu}) + 0.5219(153 \text{ amu})$$

$$= 152.04 \text{ amu} = 152 \text{ amu}$$

Check: The units of the answer (amu) are correct. The magnitude of the answer is reasonable because the fraction of each isotope is about 0.5.

The Mole Concept

1.83   **Given:** 5.52 mol sulfur   **Find:** atoms of sulfur
Conceptual Plan: mol S → atoms S

$$\frac{6.022 \times 10^{23} \text{ atoms}}{1 \text{ mol}}$$

Solution: $5.52 \, \text{mol S} \times \dfrac{6.022 \times 10^{23} \text{ atoms S}}{1 \, \text{mol S}} = 3.32 \times 10^{24} \text{ atoms S}$

Check: The units of the answer (atoms S) are correct. The magnitude of the answer is reasonable because more than 1 mole of material is present.

1.85   (a)   **Given:** 11.8 g Ar   **Find:** mol Ar
               Conceptual Plan: g Ar → mol Ar

$$\dfrac{1 \text{ mol Ar}}{39.95 \text{ g Ar}}$$

Solution: $11.8 \, \text{g Ar} \times \dfrac{1 \text{ mol Ar}}{39.95 \text{ g Ar}} = 0.295 \text{ mol Ar}$

Check: The units of the answer (mol Ar) are correct. The magnitude of the answer is reasonable because less than the mass of 1 mol is present.

(b)   **Given:** 3.55 g Zn   **Find:** mol Zn
       Conceptual Plan: g Zn → mol Zn

$$\dfrac{1 \text{ mol Zn}}{65.41 \text{ g Zn}}$$

Solution: $3.55 \, \text{g Zn} \times \dfrac{1 \text{ mol Zn}}{65.41 \text{ g Zn}} = 0.0543 \text{ mol Zn}$

Check: The units of the answer (mol Zn) are correct. The magnitude of the answer is reasonable because less than the mass of 1 mol is present.

(c)   **Given:** 26.1 g Ta   **Find:** mol Ta
       Conceptual Plan: g Ta → mol Ta

$$\dfrac{1 \text{ mol Ta}}{180.95 \text{ g Ta}}$$

Solution: $26.1 \, \text{g Ta} \times \dfrac{1 \text{ mol Ta}}{180.95 \text{ g Ta}} = 0.144 \text{ mol Ta}$

Check: The units of the answer (mol Ta) are correct. The magnitude of the answer is reasonable because less than the mass of 1 mol is present.

(d)   **Given:** 0.211 g Li   **Find:** mol Li
       Conceptual Plan: g Li → mol Li

$$\dfrac{1 \text{ mol Li}}{6.941 \text{ g Li}}$$

Solution: $0.211 \, \text{g Li} \times \dfrac{1 \text{ mol Li}}{6.941 \text{ g Li}} = 0.0304 \text{ mol Li}$

Check: The units of the answer (mol Li) are correct. The magnitude of the answer is reasonable because less than the mass of 1 mol is present.

1.87   **Given:** 3.78 g silver   **Find:** atoms Ag
        Conceptual Plan: g Ag → mol Ag → atoms Ag

$$\dfrac{1 \text{ mol Ag}}{107.87 \text{ g Ag}} \qquad \dfrac{6.022 \times 10^{23} \text{ atoms}}{1 \text{ mol}}$$

Solution: $3.78 \, \text{g Ag} \times \dfrac{1 \text{ mol Ag}}{107.87 \text{ g Ag}} \times \dfrac{6.022 \times 10^{23} \text{ atoms Ag}}{1 \text{ mol Ag}} = 2.11 \times 10^{22} \text{ atoms Ag}$

Check: The units of the answer (atoms Ag) are correct. The magnitude of the answer is reasonable because less than the mass of 1 mol of Ag is present.

1.89   (a)   **Given:** 5.18 g P   **Find:** atoms P
               Conceptual Plan: g P → mol P → atoms P

$$\dfrac{1 \text{ mol P}}{30.97 \text{ g P}} \qquad \dfrac{6.022 \times 10^{23} \text{ atoms}}{1 \text{ mol}}$$

Solution: $5.18 \text{ g P} \times \dfrac{1 \text{ mol P}}{30.97 \text{ g P}} \times \dfrac{6.022 \times 10^{23} \text{ atoms P}}{1 \text{ mol P}} = 1.01 \times 10^{23} \text{ atoms P}$

Check: The units of the answer (atoms P) are correct. The magnitude of the answer is reasonable because less than the mass of 1 mol of P is present.

(b) **Given:** 2.26 g Hg   **Find:** atoms Hg
Conceptual Plan: g Hg → mol Hg → atoms Hg

$$\dfrac{1 \text{ mol Hg}}{200.59 \text{ g Hg}} \quad \dfrac{6.022 \times 10^{23} \text{ atoms}}{1 \text{ mol}}$$

Solution: $2.26 \text{ g Hg} \times \dfrac{1 \text{ mol Hg}}{200.59 \text{ g Hg}} \times \dfrac{6.022 \times 10^{23} \text{ atoms Hg}}{1 \text{ mol Hg}} = 6.78 \times 10^{21} \text{ atoms Hg}$

Check: The units of the answer (atoms Hg) are correct. The magnitude of the answer is reasonable because much less than the mass of 1 mol of Hg is present.

(c) **Given:** 1.87 g Bi   **Find:** atoms Bi
Conceptual Plan: g Bi → mol Bi → atoms Bi

$$\dfrac{1 \text{ mol Bi}}{208.98 \text{ g Bi}} \quad \dfrac{6.022 \times 10^{23} \text{ atoms}}{1 \text{ mol}}$$

Solution: $1.87 \text{ g Bi} \times \dfrac{1 \text{ mol Bi}}{208.98 \text{ g Bi}} \times \dfrac{6.022 \times 10^{23} \text{ atoms Bi}}{1 \text{ mol Bi}} = 5.39 \times 10^{21} \text{ atoms Bi}$

Check: The units of the answer (atoms Bi) are correct. The magnitude of the answer is reasonable because less than the mass of 1 mol of Bi is present.

(d) **Given:** 0.082 g Sr   **Find:** atoms Sr
Conceptual Plan: g Sr → mol Sr → atoms Sr

$$\dfrac{1 \text{ mol Sr}}{87.62 \text{ g Sr}} \quad \dfrac{6.022 \times 10^{23} \text{ atoms}}{1 \text{ mol}}$$

Solution: $0.082 \text{ g Sr} \times \dfrac{1 \text{ mol Sr}}{87.62 \text{ g Sr}} \times \dfrac{6.022 \times 10^{23} \text{ atoms Sr}}{1 \text{ mol Sr}} = 5.6 \times 10^{20} \text{ atoms Sr}$

Check: The units of the answer (atoms Sr) are correct. The magnitude of the answer is reasonable because less than the mass of 1 mol of Sr is present.

1.91 (a) **Given:** 1.1×10^{23} gold atoms   **Find:** grams Au
Conceptual Plan: atoms Au → mol Au → g Au

$$\dfrac{1 \text{ mol}}{6.022 \times 10^{23} \text{ atoms}} \quad \dfrac{196.97 \text{ g Au}}{1 \text{ mol Au}}$$

Solution: $1.1 \times 10^{23} \text{ atoms Au} \times \dfrac{1 \text{ mol Au}}{6.022 \times 10^{23} \text{ atoms Au}} \times \dfrac{196.97 \text{ g Au}}{1 \text{ mol Au}} = 36 \text{ g Au}$

Check: The units of the answer (g Au) are correct. The magnitude of the answer is reasonable because fewer than Avogadro's number of atoms are in the sample.

(b) **Given:** 2.82×10^{22} helium atoms   **Find:** grams He
Conceptual Plan: atoms He → mol He → g He

$$\dfrac{1 \text{ mol}}{6.022 \times 10^{23} \text{ atoms}} \quad \dfrac{4.002 \text{ g He}}{1 \text{ mol He}}$$

Solution: $2.82 \times 10^{22} \text{ atoms He} \times \dfrac{1 \text{ mol He}}{6.022 \times 10^{23} \text{ atoms He}} \times \dfrac{4.002 \text{ g He}}{1 \text{ mol He}} = 0.187 \text{ g He}$

Check: The units of the answer (g He) are correct. The magnitude of the answer is reasonable because fewer than Avogadro's number of atoms are in the sample.

(c)   **Given:** 1.8×10^{23} lead atoms   **Find:** grams Pb
        Conceptual Plan: atoms Pb $\rightarrow$ mol Pb $\rightarrow$ g Pb

$$\frac{1 \text{ mol}}{6.022 \times 10^{23} \text{ atoms}} \qquad \frac{207.2 \text{ g Pb}}{1 \text{ mol Pb}}$$

Solution: $1.8 \times 10^{23} \text{ atoms Pb} \times \dfrac{1 \text{ mol Pb}}{6.022 \times 10^{23} \text{ atoms Pb}} \times \dfrac{207.2 \text{ g Pb}}{1 \text{ mol Pb}} = 62 \text{ g Pb}$

Check: The units of the answer (g Pb) are correct. The magnitude of the answer is reasonable because fewer than Avogadro's number of atoms are in the sample.

(d)   **Given:** 7.9×10^{21} uranium atoms   **Find:** grams U
        Conceptual Plan: atoms U $\rightarrow$ mol U $\rightarrow$ g U

$$\frac{1 \text{ mol}}{6.022 \times 10^{23} \text{ atoms}} \qquad \frac{238.029 \text{ g U}}{1 \text{ mol U}}$$

Solution: $7.9 \times 10^{21} \text{ atoms U} \times \dfrac{1 \text{ mol U}}{6.022 \times 10^{23} \text{ atoms U}} \times \dfrac{238.029 \text{ g U}}{1 \text{ mol U}} = 3.1 \text{ g U}$

Check: The units of the answer (g U) are correct. The magnitude of the answer is reasonable because fewer than Avogadro's number of atoms are in the sample.

1.93   **Given:** 52 mg diamond (carbon)   **Find:** atoms C
         Conceptual Plan: mg C $\rightarrow$ g C $\rightarrow$ mol C $\rightarrow$ atoms C

$$\frac{1 \text{ g C}}{1000 \text{ mg C}} \quad \frac{1 \text{ mol C}}{12.011 \text{ g C}} \quad \frac{6.022 \times 10^{23} \text{ atoms}}{1 \text{ mol}}$$

Solution: $52 \text{ mg C} \times \dfrac{1 \text{ g C}}{1000 \text{ mg C}} \times \dfrac{1 \text{ mol C}}{12.011 \text{ g C}} \times \dfrac{6.022 \times 10^{23} \text{ atoms C}}{1 \text{ mol C}} = 7.2 \times 10^{21} \text{ atoms C}$

Check: The units of the answer (atoms C) are correct. The magnitude of the answer is reasonable because less than the mass of 1 mol of C is present.

1.95   **Given:** 1 atom platinum   **Find:** g Pt
         Conceptual Plan: atoms Pt $\rightarrow$ mol Pt $\rightarrow$ g Pt

$$\frac{1 \text{ mol}}{6.0221 \times 10^{23} \text{ atoms}} \qquad \frac{195.08 \text{ g Pt}}{1 \text{ mol Pt}}$$

Solution: $1 \text{ atom Pt} \times \dfrac{1 \text{ mol Pt}}{6.0221 \times 10^{23} \text{ atoms Pt}} \times \dfrac{195.08 \text{ g Pt}}{1 \text{ mol Pt}} = 3.2394 \times 10^{-22} \text{ g Pt}$

Check: The units of the answer (g Pt) are correct. The magnitude of the answer is reasonable because only one atom is in the sample.

Cumulative Problems

1.97   **Given:** 7.83 g HCN sample 1: 0.290 g H; 4.06 g N; 3.37 g HCN sample 2   **Find:** g C in sample 2
         Conceptual Plan: g HCN sample 1 $\rightarrow$ g C in HCN sample 1 $\rightarrow$ ratio g C to g HCN $\rightarrow$ g C in HCN sample 2

$$\text{g HCN} - \text{g H} - \text{g N} \qquad \frac{\text{g C}}{\text{g HCN}} \qquad \text{g HCN} \times \frac{\text{g C}}{\text{g HCN}}$$

Solution: $7.83 \text{ g HCN} - 0.290 \text{ g H} - 4.06 \text{ g N} = 3.48 \text{ g C}$

$$3.37 \text{ g HCN} \times \frac{3.48 \text{ g C}}{7.83 \text{ g HCN}} = 1.50 \text{ g C}$$

Check: The units of the answer (g C) are correct. The magnitude of the answer is reasonable because the sample size is about half the original sample size and the g C are about half the original g C.

1.99   **Given:** graph of mass and % abundance of Pb   **Find:** atomic mass of Pb
         Conceptual Plan: % abundance $\rightarrow$ relative fraction abundance $\rightarrow$ mass Eu

$$\frac{\text{\% abundance of each isotope}}{\text{sum of all \% abundance}} \qquad \text{Atomic mass} = \sum_{n} (\text{fraction of isotope } n) \times (\text{mass of isotope } n)$$

Solution: Fraction $^{204}\text{Pb} = \dfrac{2.7}{190.9} = 0.0141$ Fraction $^{206}\text{Pb} = \dfrac{46.0}{190.9} = 0.2410$

$$\text{Fraction } ^{207}\text{Pb} = \frac{42.2}{190.9} = 0.221\underline{1} \quad \text{Fraction } ^{208}\text{Pb} = \frac{100.0}{190.9} = 0.523\underline{8}$$

$$\text{Atomic mass} = \sum_n (\text{fraction of isotope } n) \times (\text{mass of isotope } n)$$

$$= 0.014\underline{1}(204 \text{ amu}) + 0.241\underline{0}(206 \text{ amu}) + 0.221\underline{1}(207 \text{ amu}) + 0.523\underline{8}(208 \text{ amu})$$

$$= 207.\underline{2}4 \text{ amu} = 207 \text{ amu}$$

Check: The units of the answer (amu) are correct. The magnitude of the answer is reasonable because the fraction of the heavier isotopes is greater than that of the lighter isotopes.

1.101 $\quad ^{236}_{90}\text{Th } A - Z = $ number of neutrons. $236 - 90 = 146$ neutrons. So any nucleus with 146 neutrons is an isotope of $^{236}_{90}\text{Th}$. Some would be $^{238}_{92}\text{U}$, $^{239}_{93}\text{Np}$, $^{241}_{95}\text{Am}$, $^{237}_{91}\text{Pa}$, $^{235}_{89}\text{Ac}$, and $^{244}_{98}\text{Cf}$.

1.103 $\quad$ **Given:** 1 penny $= 1.0$ mm $\quad$ **Find:** height in km of Avogadro's number of pennies

Conceptual Plan: height of 1 penny $\rightarrow$ height of Avogadro's number of pennies

$$6.022 \times 10^{23}$$

Solution: $\dfrac{1.0 \text{ mm}}{\text{penny}} \times 6.022 \times 10^{23} \text{ pennies} \times \dfrac{1 \text{ m}}{1000 \text{ mm}} \times \dfrac{1 \text{ km}}{1000 \text{ m}} = 6.0 \times 10^{17} \text{ km}$

Check: The units of the answer (km) are correct. The magnitude of the answer shows just how large Avogadro's number is.

1.105 $\quad$ **Given:** Cu sphere: $r = 0.935$ in; $d = 8.96$ g/cm^3 $\quad$ **Find:** number of Cu atoms

Conceptual Plan: r **(inch) $\rightarrow r$(cm) $\rightarrow V_{\text{sphere}} \rightarrow$ g Cu $\rightarrow$ mol Cu $\rightarrow$ atoms Cu**

$$\frac{2.54 \text{ cm}}{1 \text{ in}} \qquad V = \frac{4}{3}\pi r^3 \qquad \frac{8.96 \text{ g}}{\text{cm}^3} \qquad \frac{1 \text{ mol Cu}}{63.546 \text{ g}} \qquad \frac{6.022 \times 10^{23} \text{ atoms}}{1 \text{ mol}}$$

Solution: $\qquad 0.935 \text{ in} \times \dfrac{2.54 \text{ cm}}{1 \text{ in}} = 2.374\underline{9} \text{ cm}$

$$\frac{4}{3}\pi(2.374\underline{9} \text{ cm})^3 \times \frac{8.96 \text{ g}}{\text{cm}^3} \times \frac{1 \text{ mol Cu}}{63.546 \text{ g}} \times \frac{6.022 \times 10^{23} \text{ atoms Cu}}{1 \text{ mol Cu}} = 4.76 \times 10^{24} \text{ atoms Cu}$$

Check: The units of the answer (atoms Cu) are correct. The magnitude of the answer is reasonable because about 8 mol Cu is present.

1.107 $\quad$ **Given:** Alloy of Au and Pd $= 67.2$ g; 2.49×10^{23} atoms $\quad$ **Find:** % composition by mass

Conceptual Plan: atoms Au and Pd $\rightarrow$ mol Au and Pd $\rightarrow$ g Au and Pd $\rightarrow$ g Au

$$\frac{1 \text{ mol}}{6.022 \times 10^{23} \text{ atoms}} \qquad \frac{196.97 \text{g Au}}{1 \text{ mol Au}} \qquad \frac{106.42 \text{ g Pd}}{1 \text{ mol Pd}}$$

Solution: Let $x =$ atoms Au and $y =$ atoms Pd; develop expressions that will permit atoms to be related to moles and then to grams.

$$(x \text{ atoms Au})\left(\frac{1 \text{ mol Au}}{6.022 \times 10^{23} \text{ atoms Au}}\right) = \frac{x}{6.022 \times 10^{23}} \text{ mol Au}$$

$$(y \text{ atoms Pd})\left(\frac{1 \text{ mol Pd}}{6.022 \times 10^{23} \text{ atoms Pd}}\right) = \frac{y}{6.022 \times 10^{23}} \text{ mol Pd}$$

$$x + y = 2.49 \times 10^{23} \text{atoms}; \, y = 2.49 \times 10^{23} - x$$

$$\left(\frac{x}{6.022 \times 10^{23}} \text{ mol Au}\right)\left(\frac{196.97 \text{ g Au}}{1 \text{ mol Au}}\right) = \frac{196.97x}{6.022 \times 10^{23}} \text{ g Au}$$

$$\left(\frac{2.49 \times 10^{23} - x}{6.022 \times 10^{23}} \text{ mol Pd}\right)\left(\frac{106.42 \text{ g Pd}}{1 \text{ mol Pd}}\right) = \frac{106.42(2.49 \times 10^{23} - x)}{6.022 \times 10^{23}} \text{ g Pd}$$

$$\text{g Au} + \text{g Pd} = 67.2 \text{ g total}$$

$$\frac{196.97x}{6.022 \times 10^{23}} \text{g Au} + \frac{106.42(2.49 \times 10^{23} - x)}{6.022 \times 10^{23}} \text{g Pd} = 67.2 \text{ g}$$

$$X = 1.542\underline{6} \times 10^{23} \text{ atoms Au}$$

$$(1.54 \times 10^{23} \text{ atoms Au})\left(\frac{1 \text{ mol Au}}{6.022 \times 10^{23} \text{ atoms Au}}\right)\left(\frac{196.97 \text{ g Au}}{1 \text{ mol Au}}\right) = 50.\underline{3}7 \text{ g Au}$$

$$\left(\frac{50.\underline{3}7 \text{ g Au}}{67.2 \text{ g sample}}\right) \times 100 = 74.\underline{9}6\% \text{ Au} = 75.0\% \text{ Au}$$

$$\% \text{ Pd} = 100.0\% - 75.0\% \text{ Au} = 25.0\% \text{ Pd}$$

Check: Units of the answer (% composition) are correct.

1.109   **Given:** air contains 1.5 μg Pb/m^3; lung volume = 5.50 L   **Find:** atoms of Pb in lungs

Conceptual Plan: lung in L $\rightarrow$ mL $\rightarrow$ cm^3 $\rightarrow$ m^3 $\rightarrow$ μg Pb $\rightarrow$ g Pb $\rightarrow$ mol Pb $\rightarrow$ atoms Pb

$$\frac{1000 \text{ mL}}{L} \quad \frac{1 \text{ cm}^3}{\text{mL}} \quad \frac{1 \text{ m}^3}{(100 \text{ cm})^3} \quad \frac{1.5 \mu g \text{ Pb}}{\text{m}^3} \quad \frac{1 \text{ g Pb}}{10^6 \mu g \text{ Pb}} \quad \frac{1 \text{ mol Pb}}{207.2 \text{ g Pb}} \quad \frac{6.022 \times 10^{23} \text{ atoms Pb}}{1 \text{ mol Pb}}$$

Solution: $5.50 \text{ L} \times \left(\dfrac{1000 \text{ mL}}{L}\right) \times \left(\dfrac{1 \text{ cm}^3}{1 \text{ mL}}\right) \times \left(\dfrac{1 \text{ m}^3}{(100 \text{ cm})^3}\right) \times \left(\dfrac{1.5 \mu g \text{ Pb}}{\text{m}^3}\right) \times \left(\dfrac{1 \text{ g Pb}}{10^6 \mu g \text{ Pb}}\right)$

$$\times \left(\frac{1 \text{ mol Pb}}{207.2 \text{ g Pb}}\right) \times \left(\frac{6.022 \times 10^{23} \text{ atoms Pb}}{1 \text{ mol Pb}}\right) = 2.398 \times 10^{13} \text{ atoms Pb} = 2.4 \times 10^{13} \text{ atoms Pb}$$

Check: The units of the answer (atoms Pb) are correct. The magnitude of the answer is reasonable because about 400 nmol of Pb is present.

Challenge Problems

1.111   **Given:** Ag-107, 51.839%, Ag-109, $\dfrac{\text{mass Ag-109}}{\text{mass Ag-107}} = 1.0187$ **Find:** mass Ag-107

Conceptual Plan: % abundance Ag-107 $\rightarrow$ % abundance Ag-109 $\rightarrow$ fraction $\rightarrow$ mass Ag-107

$$100\% - (\% \text{ Ag-107}) \qquad \frac{\% \text{ abundance}}{100} \qquad \text{Atomic mass} = \sum_n (\text{fraction of isotope } n) \times (\text{mass of isotope } n)$$

Solution: $100.00\% - 51.839\% = 48.161\% \text{ Ag} - 109$

$$\text{Fraction Ag-107} = \frac{51.839}{100.00} = 0.51839 \qquad \text{Fraction Ag-109} = \frac{48.161}{100.00} = 0.48161$$

Let x be the mass of Ag-107, then mass Ag-109 = $1.0187x$

$$\text{Atomic mass} = \sum_n (\text{fraction of isotope } n) \times (\text{mass of isotope } n)$$

$$107.87 \text{ amu} = 0.51839(x \text{ amu}) + 0.48161(1.0187x \text{ amu})$$

$$x = 106.907 \text{ amu} = 106.91 \text{ amu mass Ag-107}$$

Check: The units of the answer (amu) are correct. The answer is reasonable because it is close to the atomic mass number of Ag-107.

1.113   **Given:** $\dfrac{\text{mass 2 O}}{\text{mass 1 N}} = \dfrac{2.29}{1.00}$; $\dfrac{\text{mass 3 F}}{\text{mass 1 N}} = \dfrac{4.07}{1.00}$ **Find:** $\dfrac{\text{mass O}}{\text{mass 2 F}}$

Conceptual Plan: Mass O/N and mass F/N $\rightarrow$ mass O/F $\rightarrow$ mass O/2F

$$\frac{\frac{\text{mass 2 O}}{\text{mass 1 N}}}{\frac{\text{mass 3 F}}{\text{mass 1 N}}} \qquad \frac{\text{mass 2 O}}{\text{mass 3 F}}$$

Solution: $\dfrac{\text{mass 2 O}}{\text{mass 1 N}} = \dfrac{2.29}{1.00}$; $\dfrac{\text{mass 3 F}}{\text{mass 1 N}} = \dfrac{4.07}{1.00}$ $\left(\dfrac{2.29 \text{ mass 2 O}}{4.07 \text{ mass 3 F}}\right)\left(\dfrac{1 \text{ O}}{2 \text{ O}}\right)\left(\dfrac{3 \text{ F}}{2 \text{ F}}\right) = \dfrac{0.422 \text{ mass O}}{\text{mass 2 F}}$

Check: Mass ratio of O to F is reasonable because the mass of O is slightly less than the mass of fluorine.

1.115   **Given:** 7.36 g Cu; 0.51 g Zn   **Find:** atomic mass of sample

Conceptual Plan: fraction Cu and Zn $\rightarrow$ atomic mass

$$\text{Atomic mass} = \sum_n (\text{fraction of atom } n) \times (\text{mass of atom } n)$$

Solution: 7.36 g Cu + 0.51 g Zn = 7.87 g sample

$$\left(\frac{7.36 \text{ g Cu}}{7.87 \text{ g sample}}\right)\left(\frac{63.55 \text{ g Cu}}{\text{mol Cu}}\right) + \left(\frac{0.51 \text{ g Zn}}{7.87 \text{ g sample}}\right)\left(\frac{65.41 \text{ g Zn}}{\text{mol Zn}}\right) = 63.67 \text{ g/mol}$$

Check: The units of the answer (g/mol) are correct. The magnitude of the answer is reasonable because it is between the mass of Cu (63.55 g/mol) and Zn (65.41 g/mol) and is closer to the mass of Cu.

1.117 **Given:** Mg = 24.312 amu, ^{24}Mg = 23.98504, 78.99%, ^{26}Mg = 25.98259 amu, $\dfrac{\text{abundance } ^{25}\text{Mg}}{\text{abundance } ^{26}\text{Mg}} = \dfrac{0.9083}{1}$

Find: mass ^{25}Mg

Conceptual Plan: abundance of ^{24}Mg and ratio ^{25}Mg/^{26}Mg → abundance ^{25}Mg and ^{26}Mg → mass ^{25}Mg

$$\text{Atomic mass} = \sum_n (\text{fraction of isotope } n) \times (\text{mass of isotope } n)$$

Solution: 100.00% − % abundance ^{24}Mg = % abundance ^{25}Mg and ^{26}Mg

100.00% − 78.99% = 21.01% ^{25}Mg and ^{26}Mg

$$\text{fraction } ^{25}\text{Mg and } ^{26}\text{Mg} = \frac{21.01}{100.0} = 0.2101$$

$$\frac{\text{abundance } ^{25}\text{Mg}}{\text{abundance } ^{26}\text{Mg}} = \frac{0.9083}{1}$$

Let x = fraction ^{26}Mg, $0.9083x$ = fraction ^{25}Mg

fraction ^{25}Mg and ^{26}Mg = x + $0.9083x$ = 0.2101

x = fraction ^{26}Mg = 0.1101, $0.9083x$ = fraction ^{25}Mg = 0.1000

$$\text{Atomic mass} = \sum_n (\text{fraction of isotope } n) \times (\text{mass of isotope } n)$$

$24.312 = (0.7899)(23.98504 \text{ amu}) + (0.1000)(\text{mass } ^{25}\text{Mg}) + (0.1101)(25.98259 \text{ amu})$

mass ^{25}Mg = 25.0$\underline{5}$6 amu = 25.06 amu

Check: The units of the answer (amu) are correct. The magnitude of the answer is reasonable because it is between the masses of ^{24}Mg and ^{26}Mg.

1.119 **Given:** $d = 1.4$ g/cm^3, $r = 7 \times 10^8$ m; 100 billion stars/galaxy; 10 billion galaxies/universe

Find: number of atoms in the universe

Conceptual Plan: $r_{(star)}$ (m) → $r_{(star)}$ (cm) → $V_{(star)}$ → g H/star → mol H/star → atoms H/star

$$\frac{100 \text{ cm}}{\text{m}} \qquad V = \frac{4}{3}\pi r^3 \qquad \frac{1.4 \text{ g H}}{\text{cm}^3} \qquad \frac{1 \text{ mol H}}{1.008 \text{ gH}} \qquad \frac{6.022 \times 10^{23} \text{ atoms}}{1 \text{ mol}}$$

atoms H/star → atoms H/galaxy → atoms H/universe

$$\frac{100 \times 10^9 \text{ stars}}{\text{galaxy}} \qquad \frac{10 \times 10^9 \text{ galaxies}}{\text{universe}}$$

Solution: $7 \times 10^8 \, \cancel{\text{m}} \times \dfrac{100 \text{ cm}}{\cancel{\text{m}}} = 7 \times 10^{10}$ cm

$$\frac{4}{3}\pi \frac{(7 \times 10^{10} \text{ cm})^3}{\text{star}} \times \frac{1.4 \text{ gH}}{\text{cm}^3} \times \frac{1 \text{ mol H}}{1.008 \text{ gH}} \times \frac{6.022 \times 10^{23} \text{ atoms H}}{1 \text{ mol H}} \times \frac{100 \times 10^9 \text{ stars}}{\text{galaxy}} \times \frac{10 \times 10^9 \text{ galaxies}}{\text{universe}}$$

$= 1 \times 10^{78}$ atoms/universe

Check: The units of the answer (atoms/universe) are correct.

Conceptual Problems

1.121 (c) is the best representation. When solid carbon dioxide (dry ice) sublimes, it changes phase from a solid to a gas. Phase changes are physical changes, so no molecular bonds are broken. This diagram shows molecules with one carbon atom and two oxygen atoms bonded together in every molecule. The other diagrams have no carbon dioxide molecules.

1.123 Remember that an observation is the information collected when studying phenomena. A law is a concise statement that summarizes observed behaviors and observations and predicts future observations. A theory attempts to explain why the observed behavior is happening.

    (a)   This statement is most like a law because it summarizes many observations and can explain future behavior—many places and many days.

    (b)   This statement is a theory because it attempts to explain why (gravitational forces).

(c) This statement is most like an observation because it is information collected to understand tidal behavior.

(d) This statement is most like a law because it summarizes many observations and can explain future behavior—many places and many days and months.

1.125   **Given:** (a) Cr: 55.0 g; atomic mass = 52 g/mol
(b) Ti: 45.0 g; atomic mass = 48 g/mol
(c) Zn: 60.0 g; atomic mass = 65 g/mol
Find: which has the greatest amount in moles and which has the greatest mass
Conceptual Plan: Without calculation, compare grams of material to g/mol for each.
Solution: Cr would have the greatest mole amount of the elements. It is the only one whose mass is greater than the molar mass. Zn would be the greatest mass amount because it is the largest mass value.

Questions for Group Work

1.127   Each boron atom has 5 protons. The mass of the boron atoms with 5 neutrons is ~10 amu, while the mass of the boron atoms with 6 neutrons is ~11 amu.

The average mass of boron in the drawing is

$$\text{Atomic mass} = \sum_n (\text{fraction of isotope } n) \times (\text{mass of isotope } n) =$$

$$0.2(10.01294 \text{ amu}) + 0.8(11.00931 \text{ amu}) = 10.810036 \text{ amu}$$

The average mass of boron is

$$\text{Atomic mass} = \sum_n (\text{fraction of isotope } n) \times (\text{mass of isotope } n) =$$

$$0.198(10.01294 \text{ amu}) + 0.802(11.00931 \text{ amu}) = 10.8 \text{ amu}$$

1.129   Electrons: $\dfrac{9.10938 \times 10^{-28} \text{ g}}{1 \text{ electron}} \times \dfrac{6.02214179 \times 10^{23} \text{ electrons}}{1 \text{ mole electrons}} = 5.48580 \times 10^{-4} \text{ g/mol}$

Protons: $\dfrac{1.67262 \times 10^{-24} \text{ g}}{1 \text{ proton}} \times \dfrac{6.02214179 \times 10^{23} \text{ electrons}}{1 \text{ mole electrons}} = 1.00728 \text{ g/mol}$

Neutrons: $\dfrac{1.67493 \times 10^{-24} \text{ g}}{1 \text{ neutron}} \times \dfrac{6.02214179 \times 10^{23} \text{ neutrons}}{1 \text{ mole neutrons}} = 1.00867 \text{ g/mol}$

Carbon-12: $\dfrac{1.992646 \times 10^{-23} \text{ g}}{1 \text{ carbon-12}} \times \dfrac{6.02214179 \times 10^{23} \text{ carbon-12}}{1 \text{ mole carbon-12}} = 12.00000 \text{ g/mol}$

$$\text{mass of components parts} = \left(6 \text{ mol protons} \times \frac{1.00728 \text{ g}}{1 \text{ mol protons}} \right) + \left(6 \text{ mol neutrons} \times \frac{1.00867 \text{ g}}{1 \text{ mol neutrons}} \right) +$$

$$\left(6 \text{ mol electrons} \times \frac{5.48580 \times 10^{-4} \text{ g}}{1 \text{ mol electrons}} \right) = 6.04368 \text{ g} + 6.05202 \text{ g} + 0.00329148 \text{ g} = 12.09899 \text{ g}$$

The total mass of the protons, neutrons, and electrons weighs 0.09899 g more than the mass of 1 mol of carbon-12 atoms.

Doughnut: $\dfrac{74 \text{ g}}{1 \text{ doughnut}} \times \dfrac{6.02214179 \times 10^{23} \text{ doughnut}}{1 \text{ mole doughnut}} = 4.5 \times 10^{25} \text{ g/mol}$

$74 \text{ g carbon} \times \dfrac{1 \text{ carbon atom}}{1.992646 \times 10^{-23} \text{ g carbon}} = 3.7 \times 10^{24} \text{ carbon atoms}$

Data Interpretation and Analysis

1.130   (a)   The lead concentration was 0.24 μg/m^3 in 2000 and 0.04 μg/m^3 in 2014.

       (c)   The lead concentration was 0.10 μg/m^3 in 2011, so

$$1 \text{ cm}^3 \times \left(\dfrac{1 \text{ m}}{100 \text{ cm}}\right)^3 \times \dfrac{0.10 \text{ } \mu\text{g Pb}}{1 \text{ m}^3} \times \dfrac{10^{-6} \text{ g Pb}}{1 \text{ } \mu\text{g Pb}} \times \dfrac{1 \text{ mol Pb}}{207.2 \text{ g Pb}} \times \dfrac{6.022 \times 10^{23} \text{ Pb atoms}}{1 \text{ mol Pb}}$$

$$= 2.9 \times 10^8 \text{ Pb atoms}$$

2 The Quantum-Mechanical Model of the Atom

Review Questions

2.1   The quantum-mechanical model of the atom is important because it explains how electrons exist in atoms and how those electrons determine the chemical and physical properties of elements.

2.3   The wavelength (λ) of the wave is the distance in space between adjacent crests and is measured in units of distance. The amplitude of the wave is the vertical height of a crest. The more closely spaced the waves (i.e., the shorter the wavelength), the more energy there is. The amplitudes of the electric and magnetic field waves in light determine the intensity or brightness of the light. The higher the amplitude, the more energy the wave has.

2.5   For visible light, wavelength determines the color. Red light has a wavelength of 750 nm, the longest wavelength of visible light, and blue has a wavelength of 500 nm.

2.7   (a)   Gamma rays (γ)—the wavelength range is 10^{-11} to 10^{-15} m. Gamma rays are produced by the sun, other stars, and certain unstable atomic nuclei on Earth. Human exposure to gamma rays is dangerous because the high energy of gamma rays can damage biological molecules.

   (b)   X-rays—the wavelength range is 10^{-8} to 10^{-11} m. X-rays are used in medicine. X-rays pass through many substances that block visible light and are, therefore, used to image bones and internal organs. X-rays are sufficiently energetic to damage biological molecules, so while several yearly exposures to X-rays are harmless, excessive exposure increases cancer risk.

   (c)   Ultraviolet radiation (UV)—the wavelength range is 0.4×10^{-6} to 10^{-8} m. Ultraviolet radiation is most familiar as the component of sunlight that produces a sunburn or suntan. While not as energetic as gamma rays or X-rays, ultraviolet light still carries enough energy to damage biological molecules. Excessive exposure to ultraviolet light increases the risk of skin cancer and cataracts and causes premature wrinkling of the skin.

   (d)   Visible light—the wavelength range is 0.75×10^{-6} to 0.4×10^{-6} m (750 nm to 400 nm). Visible light, as long as the intensity is not too high, does not carry enough energy to damage biological molecules. It does, however, cause certain molecules in our eyes to change their shape, sending a signal to the brain that results in vision.

   (e)   Infrared radiation (IR)—the wavelength range is 0.75×10^{-6} to 10^{-3} m. The heat you feel when you place your hand near a hot object is infrared radiation. All warm objects, including human bodies, emit infrared light. Although infrared light is invisible to our eyes, infrared sensors can detect it and are often used in night vision technology to "see" in the dark.

   (f)   Microwave radiation—the wavelength range is 10^{-3} to 10^{-1} m. Microwave radiation is used in radar and in microwave ovens. Microwave radiation is efficiently absorbed by water and can, therefore, heat substances that contain water.

   (g)   Radio waves—the wavelength range is 10^{-1} to 10^{5} m. Radio waves are used to transmit the signals responsible for AM and FM radio, cell telephones, television, and other forms of communication.

2.9   Diffraction occurs when a wave encounters an obstacle or a slit that is comparable in size to its wavelength. The wave bends around the slit. The diffraction of light through two slits separated by a distance comparable to the wavelength of the light results in an interference pattern. Each slit acts as a new wave source, and the two new waves interfere with each other. This results in a pattern of bright and dark lines.

Interference from Two Slits

2.11   Because of the results of the experiments with the photoelectric effect, Einstein proposed that light energy must come in packets. The amount of energy in a light packet depends on its frequency (wavelength). The emission of electrons depends on whether a single photon has sufficient energy to dislodge a single electron.

2.13   An emission spectrum occurs when an atom absorbs energy and re-emits that energy as light. The light emitted contains distinct wavelengths for each element. The emission spectrum of a particular element is always the same and can be used to identify the element. A white light spectrum is continuous, meaning that there are no sudden interruptions in the intensity of the light as a function of wavelengths. It consists of all wavelengths. Emission spectra are not continuous. They consist of bright lines at specific wavelengths, with complete darkness in between.

2.15   Electron diffraction occurs when an electron beam is aimed at two closely spaced slits and a series of detectors is arranged to detect the electrons after they pass through the slits. An interference pattern similar to that observed for light is recorded behind the slits. Electron diffraction is evidence of the wave nature of electrons.

2.17   Complementary properties exclude one another. The more you know about one, the less you know about the other. Which of two complementary properties you observe depends on the experiment you perform. In electron diffraction, when you try to observe which hole the electron goes through (particle nature), you lose the interference pattern (wave nature). When you try to observe the interference pattern, you cannot determine which hole the electron goes through.

2.19   A trajectory is a path that is determined by the particle's velocity (the speed and direction of travel), its position, and the forces acting on it. Both position and velocity are required to predict a trajectory.

2.21   Deterministic means that the present determines the future. That means that under the identical condition, identical results will occur.

2.23   A probability distribution map is a statistical map that shows where an electron is likely to be found under a given set of conditions.

2.25   An orbital is a probability distribution map showing where the electron is likely to be found.

2.27   The principal quantum number (n) is an integer that has possible values of 1, 2, 3, and so on. The principal quantum number determines the overall size and energy of an orbital.

2.29   The magnetic quantum number (m_l) is an integer ranging from $-l$ to $+l$. For example, if $l = 1$, $m_l = -1, 0, +1$. The magnetic quantum number specifies the orientation of the orbital.

2.31   The probability density is the probability per unit volume of finding the electron at a point in space. The radial distribution function represents the total probability of finding the electron within a thin spherical shell at a distance r from the nucleus. In contrast to probability density, which has a maximum at the nucleus for an s orbital, the radial distribution function has a value of zero at the nucleus. It increases to a maximum and then decreases again with increasing r.

2.33   The sublevels are s ($l = 0$), which can hold a maximum of 2 electrons; p ($l = 1$), which can hold a maximum of 6 electrons; d ($l = 2$), which can hold a maximum of 10 electrons; and f ($l = 3$), which can hold a maximum of 14 electrons.

Problems by Topic

Electromagnetic Radiation

2.35   **Given:** distance to sun $= 1.496 \times 10^8$ km   **Find:** time for light to travel from sun to Earth
        Conceptual Plan: distance km → distance m → time

$$\frac{1000 \text{ m}}{\text{km}} \qquad \text{time} = \frac{\text{distance}}{2.998 \times 10^8 \text{ m/s}}$$

Solution: $1.496 \times 10^8 \text{ km} \times \dfrac{1000 \text{ m}}{\text{km}} \times \dfrac{\text{s}}{2.998 \times 10^8 \text{ m}} = 499.0 \text{ s}$

Check: The units of the answer (s) are correct. The magnitude of the answer is reasonable because it corresponds to about 8 minutes.

2.37   (i) By increasing wavelength, the order is (d) ultraviolet < (c) infrared < (b) microwave < (a) radio waves.
        (ii) By increasing energy, the order is (a) radio waves < (b) microwaves < (c) infrared < (d) ultraviolet.

2.39   (a)   **Given:** $\lambda = 632.8$ nm   **Find:** frequency (ν)
               Conceptual Plan: nm → m → ν

$$\frac{1 \text{ m}}{10^9 \text{ nm}} \qquad \nu = \frac{c}{\lambda}$$

Solution:

$$632.8 \text{ nm} \times \frac{1 \text{ m}}{10^9 \text{ nm}} = 6.328 \times 10^{-7} \text{ m}; \ \nu = \frac{2.998 \times 10^8 \text{ m}}{\text{s}} \times \frac{1}{6.328 \times 10^{-7} \text{ m}} = 4.738 \times 10^{14} \text{ s}^{-1}$$

Check: The units of the answer (s^{-1}) are correct. The magnitude of the answer seems reasonable because wavelength and frequency are inversely proportional.

        (b)   **Given:** $\lambda = 503$ nm   **Find:** frequency (ν)
               Conceptual Plan: nm → m → ν

$$\frac{1 \text{ m}}{10^9 \text{ nm}} \qquad \nu = \frac{c}{\lambda}$$

Solution: $503 \text{ nm} \times \dfrac{1 \text{ m}}{10^9 \text{ nm}} = 5.03 \times 10^{-7} \text{ m}; \ \nu = \dfrac{3.00 \times 10^8 \text{ m}}{\text{s}} \times \dfrac{1}{5.03 \times 10^{-7} \text{ m}} = 5.96 \times 10^{14} \text{ s}^{-1}$

Check: The units of the answer (s^{-1}) are correct. The magnitude of the answer seems reasonable because wavelength and frequency are inversely proportional.

        (c)   **Given:** $\lambda = 0.052$ nm   **Find:** frequency (ν)
               Conceptual Plan: nm → m → ν

$$\frac{1 \text{ m}}{10^9 \text{ nm}} \qquad \nu = \frac{c}{\lambda}$$

Solution: $0.052 \text{ nm} \times \dfrac{1 \text{ m}}{10^9 \text{ nm}} = 5.2 \times 10^{-11} \text{ m}; \nu = \dfrac{3.00 \times 10^8 \text{ m}}{\text{s}} \times \dfrac{1}{5.2 \times 10^{-11} \text{ m}} = 5.8 \times 10^{18} \text{ s}^{-1}$

Check: The units of the answer (s^{-1}) are correct. The magnitude of the answer seems reasonable because wavelength and frequency are inversely proportional.

2.41   (a)   **Given:** frequency (ν) from Problem 2.39a $= 4.74 \times 10^{14} \text{ s}^{-1}$ **Find:** Energy
Conceptual Plan: $\nu \rightarrow E$

$\qquad E = h\nu \quad h = 6.626 \times 10^{-34} \text{ J} \cdot \text{s}$

Solution: $6.626 \times 10^{-34} \text{ J} \cdot \text{s} \times \dfrac{4.74 \times 10^{14}}{\text{s}} = 3.14 \times 10^{-19} \text{ J}$

Check: The units of the answer (J) are correct. The magnitude of the answer is reasonable because we are talking about the energy of one photon.

(b)   **Given:** frequency (ν) from Problem 2.39b $= 5.96 \times 10^{14} \text{ s}^{-1}$ **Find:** Energy
Conceptual Plan: $\nu \rightarrow E$

$\qquad E = h\nu \quad h = 6.626 \times 10^{-34} \text{ J} \cdot \text{s}$

Solution: $6.626 \times 10^{-34} \text{ J} \cdot \text{s} \times \dfrac{5.96 \times 10^{14}}{\text{s}} = 3.95 \times 10^{-19} \text{ J}$

Check: The units of the answer (J) are correct. The magnitude of the answer is reasonable because we are talking about the energy of one photon.

(c)   **Given:** frequency (ν) from Problem 2.39c $= 5.8 \times 10^{18} \text{ s}^{-1}$ **Find:** Energy
Conceptual Plan: $\nu \rightarrow E$

$\qquad E = h\nu \quad h = 6.626 \times 10^{-34} \text{ J} \cdot \text{s}$

Solution: $6.626 \times 10^{-34} \text{ J} \cdot \text{s} \times \dfrac{5.8 \times 10^{18}}{\text{s}} = 3.8 \times 10^{-15} \text{ J}$

Check: The units of the answer (J) are correct. The magnitude of the answer is reasonable because we are talking about the energy of one photon.

2.43   **Given:** $\lambda = 532 \text{ nm}$ and $E_{\text{pulse}} = 3.85 \text{ mJ}$ **Find:** number of photons
Conceptual Plan: $\text{nm} \rightarrow \text{m} \rightarrow E_{\text{photon}} \rightarrow$ **number of photons**

$\qquad \dfrac{1 \text{ m}}{10^9 \text{ nm}} \quad E = \dfrac{hc}{\lambda}; h = 6.626 \times 10^{-34} \text{ J} \cdot \text{s} \quad \dfrac{E_{\text{pulse}}}{E_{\text{photon}}}$

Solution:

$532 \text{ nm} \times \dfrac{1 \text{ m}}{10^9 \text{ nm}} = 5.32 \times 10^{-7} \text{ m}; \quad E = \dfrac{6.626 \times 10^{-34} \text{ J} \cdot \text{s} \times \dfrac{3.00 \times 10^8 \text{ m}}{\text{s}}}{5.32 \times 10^{-7} \text{ m}} = 3.7365 \times 10^{-19} \text{ J/photon}$

$3.85 \text{ mJ} \times \dfrac{1 \text{ J}}{1000 \text{ mJ}} \times \dfrac{1 \text{ photon}}{3.7365 \times 10^{-19} \text{ J}} = 1.03 \times 10^{16} \text{ photons}$

Check: The units of the answer (number of photons) are correct. The magnitude of the answer is reasonable for the amount of energy involved.

2.45   (a)   **Given:** $\lambda = 1500 \text{ nm}$ (assume 3 significant figures)   **Find:** E for 1 mol photons
Conceptual Plan: $\text{nm} \rightarrow \quad \text{m} \quad \rightarrow \quad E_{\text{photon}} \quad \rightarrow \quad E(\text{J})_{\text{mol}} \quad \rightarrow \quad E(\text{kJ})_{\text{mol}}$

$\qquad \dfrac{1 \text{ m}}{10^9 \text{ nm}} \quad E = \dfrac{hc}{\lambda}; h = 6.626 \times 10^{-34} \text{ J} \cdot \text{s} \quad \dfrac{6.022 \times 10^{23} \text{ photons}}{\text{mol}} \quad \dfrac{1 \text{ kJ}}{1000 \text{ J}}$

Solution: $1500 \text{ nm} \times \dfrac{1 \text{ m}}{10^9 \text{ nm}} = 1.50 \times 10^{-6} \text{ m};$

$E = \dfrac{6.626 \times 10^{-34} \text{ J} \cdot \text{s} \times \dfrac{3.00 \times 10^8 \text{ m}}{\text{s}}}{1.500 \times 10^{-6} \text{ m}} = 1.3252 \times 10^{-19} \text{ J/photon}$

$$\frac{1.3252 \times 10^{-19} \text{ J}}{\text{photon}} \times \frac{6.022 \times 10^{23} \text{ photons}}{\text{mol}} \times \frac{1 \text{ kJ}}{1000 \text{ J}} = 79.8 \text{ kJ/mol}$$

Check: The units of the answer (kJ/mol) are correct. The magnitude of the answer is reasonable for a wavelength in the infrared region.

(b)   **Given:** $\lambda = 500.$ nm   **Find:** E for 1 mol photons
        Conceptual Plan: nm $\rightarrow$ m $\rightarrow$ E_{photon} $\rightarrow$ $E(J)_{\text{mol}}$ $\rightarrow$ $E(kJ)_{\text{mol}}$

$$\frac{1 \text{ m}}{10^9 \text{ nm}} \qquad E = \frac{hc}{\lambda}; h = 6.626 \times 10^{-34} \text{ J} \cdot \text{s} \qquad \frac{6.022 \times 10^{23} \text{ photons}}{\text{mol}} \qquad \frac{1 \text{ kJ}}{1000 \text{ J}}$$

Solution:

$$500. \text{ nm} \times \frac{1 \text{ m}}{10^9 \text{ nm}} = 5.00 \times 10^{27} \text{ m}; \quad E = \frac{6.626 \times 10^{-34} \text{ J} \cdot \text{s} \times \dfrac{3.00 \times 10^8 \text{ m}}{\text{s}}}{5.00 \times 10^{27} \text{ m}} = 3.9756 \times 10^{-19} \text{ J/photon}$$

$$\frac{3.9756 \times 10^{-19} \text{ J}}{\text{photon}} \times \frac{6.022 \times 10^{23} \text{ photons}}{\text{mol}} \times \frac{1 \text{ kJ}}{1000 \text{ J}} = 239 \text{ kJ/mol}$$

Check: The units of the answer (kJ/mol) are correct. The magnitude of the answer is reasonable for a wavelength in the visible region.

(c)   **Given:** $\lambda = 150.$ nm   **Find:** E for 1 mol photons
        Conceptual Plan: nm $\rightarrow$ m $\rightarrow$ E_{photon} $\rightarrow$ $E(J)_{\text{mol}}$ $\rightarrow$ $E(kJ)_{\text{mol}}$

$$\frac{1 \text{ m}}{10^9 \text{ nm}} \qquad E = \frac{hc}{\lambda}; h = 6.626 \times 10^{-34} \text{ J} \cdot \text{s} \qquad \frac{6.022 \times 10^{23} \text{ photons}}{\text{mol}} \qquad \frac{1 \text{ kJ}}{1000 \text{ J}}$$

Solution:

$$150. \text{ nm} \times \frac{1 \text{ m}}{10^9 \text{ nm}} = 1.50 \times 10^{-7} \text{ m}; \quad E = \frac{6.626 \times 10^{-34} \text{ J} \cdot \text{s} \times \dfrac{3.00 \times 10^8 \text{ m}}{\text{s}}}{1.50 \times 10^{-7} \text{ m}} = 1.3252 \times 10^{-18} \text{ J/photon}$$

$$\frac{1.3252 \times 10^{-18} \text{ J}}{\text{photon}} \times \frac{6.022 \times 10^{23} \text{ photons}}{\text{mol}} \times \frac{1 \text{ kJ}}{1000 \text{ J}} = 798 \text{ kJ/mol}$$

Check: The units of the answer (kJ/mol) are correct. The magnitude of the answer is reasonable for a wavelength in the ultraviolet region. Note: The energy increases from the IR to the Vis to the UV as expected.

The Wave Nature of Matter and the Uncertainty Principle

2.47   The interference pattern would be a series of light and dark lines.

2.49   **Given:** $m = 9.109 \times 10^{-31}$ kg, $\lambda = 0.20$ nm   **Find:** v
         Conceptual Plan: $m, \lambda \rightarrow v$

$$v = \frac{h}{m\lambda}$$

Solution: $\dfrac{6.626 \times 10^{-34} \dfrac{\text{kg} \cdot \text{m}^2}{\text{s}^2} \cdot \text{s}}{(9.109 \times 10^{-31} \text{ kg})(0.20 \text{ nm})\left(\dfrac{1 \text{ m}}{10^9 \text{ nm}}\right)} = 3.6 \times 10^6 \text{ m/s}$

Check: The units of the answer (m/s) are correct. The magnitude of the answer is large, as would be expected for the speed of the electron.

2.51   **Given:** $m = 9.109 \times 10^{-31}$ kg; $v = 1.35 \times 10^5$ m/s   **Find:** λ
Conceptual Plan: $m, v \rightarrow \lambda$

$$\lambda = \frac{h}{mv}$$

Solution: $\dfrac{6.626 \times 10^{-34} \frac{\text{kg} \cdot \text{m}^2}{\text{s}^2} \cdot \text{s}}{(9.109 \times 10^{-31} \text{ kg})\left(\dfrac{1.35 \times 10^5 \text{ m}}{\text{s}}\right)} = 5.39 \times 10^{-9} \text{ m} = 5.39 \text{ nm}$$

Check: The units of the answer (m) are correct. The magnitude is reasonable because we are looking at an electron.

2.53   **Given:** $m = 143$ g; $v = 95$ mph   **Find:** λ
Conceptual Plan: $m, v \rightarrow \lambda$

$$\lambda = \frac{h}{mv}$$

Solution: $\dfrac{6.626 \times 10^{-34} \frac{1 \text{ kg} \cdot \text{m}^2}{\text{s}^2} \cdot \text{s}}{(143 \text{ g})\left(\dfrac{1 \text{ kg}}{1000 \text{ g}}\right)\left(\dfrac{95 \text{ mi}}{\text{hr}}\right)\left(\dfrac{1.609 \text{ km}}{\text{mi}}\right)\left(\dfrac{1000 \text{ m}}{1 \text{ km}}\right)\left(\dfrac{1 \text{ hr}}{3600 \text{ s}}\right)} = 1.1 \times 10^{-34} \text{ m}$$

The value of the wavelength $(1.1 \times 10^{-34}$ m) is so small that it will not have an effect on the trajectory of the baseball.

Check: The units of the answer (m) are correct. The magnitude of the answer is very small, as would be expected for the de Broglie wavelength of a baseball.

2.55   **Given:** $\Delta x = 552$ pm, $m = 9.109 \times 10^{-31}$ kg   **Find:** Δv
Conceptual Plan: $\Delta x, m \rightarrow \Delta v$

$$\Delta x \times m\Delta v \geq \frac{h}{4\pi}$$

Solution: $\dfrac{6.626 \times 10^{-34} \frac{\text{kg} \cdot \text{m}^2}{\text{s}^2} \cdot \text{s}}{4(3.141)(9.109 \times 10^{-31} \text{ kg})(552 \text{ pm})\left(\dfrac{1 \text{ m}}{10^{12} \text{ pm}}\right)} = 1.05 \times 10^5 \text{ m/s}$$

Check: The units of the answer (m/s) are correct. The magnitude is reasonable for the uncertainty in the speed of an electron.

Orbitals and Quantum Numbers

2.57   Because the size of the orbital is determined by the n quantum, with the size increasing with increasing n, an electron in a 2s orbital is closer, on average, to the nucleus than is an electron in a 3s orbital.

2.59   The value of l is an integer that lies between 0 and $n - 1$.
(a)   When $n = 1$, l can only be $l = 0$.
(b)   When $n = 2$, l can be $l = 0$ or $l = 1$.
(c)   When $n = 3$, l can be $l = 0$, $l = 1$, or $l = 2$.
(d)   When $n = 4$, l can be $l = 0$, $l = 1$, $l = 2$, or $l = 3$.

2.61   Set c cannot occur together as a set of quantum numbers to specify an orbital. l must lie between 0 and $n - 1$; so for $n = 3$, l can only be as high as 2.

2.63   The 2s orbital would be the same shape as the 1s orbital but would be larger in size, and the 3p orbitals would have the same shape as the 2p orbitals but would be larger in size. Also, the 2s and 3p orbitals would have more nodes.

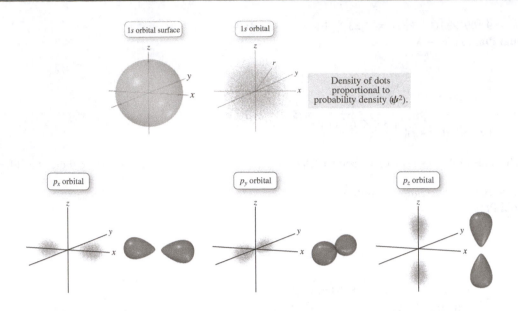

1s orbital surface

1s orbital

Density of dots proportional to probability density (ψ^2).

p_x orbital

p_y orbital

p_z orbital

Atomic Spectroscopy

2.65   When the atom emits the photon of energy that was needed to raise the electron to the $n = 2$ level, the photon has the same energy as the energy absorbed to move the electron to the excited state. Therefore, the electron has to be in $n = 1$ (the ground state) following the emission of the photon.

2.67   According to the quantum-mechanical model, the higher the n level, the higher the energy. So the transition from $3p \rightarrow 1s$ would be a greater energy difference than a transition from $2p \rightarrow 1s$. The lower energy transition would have the longer wavelength. Therefore, the $2p \rightarrow 1s$ transition would produce a longer wavelength.

2.69   (a)   **Given:** $n = 2 \rightarrow n = 1$ **Find:** λ
        Conceptual Plan: $n = 1, n = 2 \rightarrow \Delta E_{atom} \rightarrow \Delta E_{photon} \rightarrow \lambda$

$$\Delta E_{atom} = E_1 - E_2 \quad \Delta E_{atom} \rightarrow -\Delta E_{photon} \quad E = \frac{hc}{\lambda}$$

Solution:
$$\Delta E = E_1 - E_2$$

$$= -2.18 \times 10^{-18} \text{ J}\left(\frac{1}{1^2}\right) - \left[-2.18 \times 10^{-18} \text{ J}\left(\frac{1}{2^2}\right)\right] = -2.18 \times 10^{-18} \text{ J}\left[\left(\frac{1}{1^2}\right) - \left(\frac{1}{2^2}\right)\right] = -1.635 \times 10^{-18} \text{ J}$$

$$\Delta E_{photon} = -\Delta E_{atom} = 1.635 \times 10^{-18} \text{ J}; \lambda = \frac{hc}{E} = \frac{(6.626 \times 10^{-34} \text{ J} \cdot \text{s})(3.00 \times 10^8 \text{ m/s})}{1.635 \times 10^{-18} \text{ J}} = 1.22 \times 10^{-7} \text{ m}$$

This transition would produce a wavelength in the UV region.

Check: The units of the answer (m) are correct. The magnitude of the answer is reasonable because it is in the region of UV radiation.

        (b)   **Given:** $n = 3 \rightarrow n = 1$ **Find:** λ
        Conceptual Plan: $n = 1, n = 3 \rightarrow \Delta E_{atom} \rightarrow \Delta E_{photon} \rightarrow \lambda$

$$\Delta E_{atom} = E_1 - E_3 \quad \Delta E_{atom} \rightarrow -\Delta E_{photon} \quad E = \frac{hc}{\lambda}$$

Solution:
$$\Delta E = E_1 - E_3$$

$$= -2.18 \times 10^{-18} \text{ J}\left(\frac{1}{1^2}\right) - \left[-2.18 \times 10^{-18} \text{ J}\left(\frac{1}{3^2}\right)\right] = -2.18 \times 10^{-18} \text{ J}\left[\left(\frac{1}{1^2}\right) - \left(\frac{1}{3^2}\right)\right] = -1.938 \times 10^{-18} \text{ J}$$

$$\Delta E_{photon} = -\Delta E_{atom} = 1.938 \times 10^{-18} \text{ J}; \lambda = \frac{hc}{E} = \frac{(6.626 \times 10^{-34} \text{ J} \cdot \text{s})(3.00 \times 10^8 \text{ m/s})}{1.938 \times 10^{-18} \text{ J}} = 1.03 \times 10^{-7} \text{ m}$$

This transition would produce a wavelength in the UV region.

Check: The units of the answer (m) are correct. The magnitude of the answer is reasonable because it is in the region of UV radiation.

(c)   **Given:** $n = 4 \rightarrow n = 2$ **Find:** λ
     Conceptual Plan: $n = 2, n = 4 \rightarrow \Delta E_{atom} \rightarrow \Delta E_{photon} \rightarrow \lambda$

$$\Delta E_{atom} = E_2 - E_4 \quad \Delta E_{atom} \rightarrow -\Delta E_{photon} \quad E = \frac{hc}{\lambda}$$

Solution:
$$\Delta E = E_2 - E_4$$

$$= 2.18 \times 10^{-18}\,\text{J}\left(\frac{1}{2^2}\right) - \left[-2.18 \times 10^{-18}\,\text{J}\left(\frac{1}{4^2}\right)\right] = -2.18 \times 10^{-18}\,\text{J}\left[\left(\frac{1}{2^2}\right) - \left(\frac{1}{4^2}\right)\right] = -4.088 \times 10^{-19}\,\text{J}$$

$$\Delta E_{photon} = -\Delta E_{atom} = 4.088 \times 10^{-19}\,\text{J} \quad \lambda = \frac{hc}{E} = \frac{(6.626 \times 10^{-34}\,\text{J}\cdot\text{s})(3.00 \times 10^8\,\text{m/s})}{4.088 \times 10^{-19}\,\text{J}} = 4.86 \times 10^{-7}\,\text{m}$$

This transition would produce a wavelength in the visible region.

Check: The units of the answer (m) are correct. The magnitude of the answer is reasonable because it is in the region of visible light.

(d)   **Given:** $n = 5 \rightarrow n = 2$ **Find:** λ
     Conceptual Plan: $n = 2, n = 5 \rightarrow \Delta E_{atom} \rightarrow \Delta E_{photon} \rightarrow \lambda$

$$\Delta E_{atom} = E_2 - E_5 \quad \Delta E_{atom} \rightarrow -\Delta E_{photon} \quad E = \frac{hc}{\lambda}$$

Solution:
$$\Delta E = E_2 - E_5$$

$$= -2.18 \times 10^{-18}\,\text{J}\left(\frac{1}{2^2}\right) - \left[-2.18 \times 10^{-18}\,\text{J}\left(\frac{1}{5^2}\right)\right] = -2.18 \times 10^{-18}\,\text{J}\left[\left(\frac{1}{2^2}\right) - \left(\frac{1}{5^2}\right)\right] = -4.578 \times 10^{-19}\,\text{J}$$

$$\Delta E_{photon} = -\Delta E_{atom} = 4.578 \times 10^{-19}\,\text{J} \quad \lambda = \frac{hc}{E} = \frac{(6.626 \times 10^{-34}\,\text{J}\cdot\text{s})(3.00 \times 10^8\,\text{m/s})}{4.578 \times 10^{-19}\,\text{J}} = 4.34 \times 10^{-7}\,\text{m}$$

This transition would produce a wavelength in the visible region.

Check: The units of the answer (m) are correct. The magnitude of the answer is reasonable because it is in the region of visible light.

2.71   **Given:** $n(\text{initial}) = 7; \lambda = 397$ nm   **Find:** $n(\text{final})$
     Conceptual Plan: $\lambda \rightarrow \Delta E_{photon} \rightarrow \Delta E_{atom} \rightarrow n = x, n = 7$

$$E = \frac{hc}{\lambda} \quad \Delta E_{photon} \rightarrow -\Delta E_{atom} \quad \Delta E_{atom} = E_x - E_7$$

Solution: $E = \dfrac{hc}{\lambda} = \dfrac{(6.626 \times 10^{-34}\,\text{J}\cdot\text{s})(3.00 \times 10^8\,\text{m/s})}{(397\,\text{nm})\left(\dfrac{1\,\text{m}}{10^9\,\text{nm}}\right)} = 5.007 \times 10^{-19}\,\text{J}$

$$\Delta E_{atom} = -\Delta E_{photon} = -5.007 \times 10^{-19}\,\text{J}$$

$$\Delta E = E_x - E_7 = -5.007 \times 10^{-19} = -2.18 \times 10^{-18}\,\text{J}\left(\frac{1}{x^2}\right) - \left[-2.18 \times 10^{-18}\,\text{J}\left(\frac{1}{7^2}\right)\right] = -2.18 \times 10^{-18}\,\text{J}\left[\left(\frac{1}{x^2}\right) - \left(\frac{1}{7^2}\right)\right]$$

$$0.2297 = \left(\frac{1}{x^2}\right) - \left(\frac{1}{7^2}\right) \quad 0.25011 = \left(\frac{1}{x^2}\right) \quad x^2 = 3.998 \quad x = 2$$

Check: The answer is reasonable because it is an integer less than the initial value of 7.

Cumulative Problems

2.73   **Given:** 348 kJ/mol   **Find:** λ

Conceptual Plan: $\textbf{kJ/mol} \rightarrow \textbf{kJ/molec} \rightarrow \textbf{kJ/molec} \rightarrow \boldsymbol{\lambda}$

$$\frac{\text{mol C} - \text{C bonds}}{6.022 \times 10^{23}\,\text{C} - \text{C bonds}} \qquad \frac{1000\,\text{J}}{1\,\text{kJ}} \qquad E = \frac{hc}{\lambda}$$

Solution: $\dfrac{348\,\text{kJ}}{\text{mol C} - \text{C bonds}} \times \dfrac{\text{mol C} - \text{C bonds}}{6.022 \times 10^{23}\,\text{C} - \text{C bonds}} \times \dfrac{1000\,\text{J}}{1\,\text{kJ}} = 5.7\underline{7}9 \times 10^{-19}\,\text{J}$

$$\lambda = \frac{(6.626 \times 10^{-34}\,\text{J} \cdot \text{s})(3.00 \times 10^8\,\text{m/s})}{5.7\underline{7}9 \times 10^{-19}\,\text{J}} = 3.44 \times 10^{-7}\,\text{m} = 344\,\text{nm}$$

Check: The units of the answer (m or nm) are correct. The magnitude of the answer is reasonable because this wavelength is in the UV region.

2.75   **Given:** $E_{\text{pulse}} = 5.0$ watts; $d = 5.5$ mm; hole $= 1.2$ mm; $\lambda = 532$ nm   **Find:** photons/s

Conceptual Plan: fraction of beam through hole $\rightarrow$ fraction of power and then $E_{\text{photon}} \rightarrow$ number photons/s

$$\frac{\text{area hole}}{\text{area beam}} \qquad \text{fraction} \times \text{power} \qquad E = \frac{hc}{\lambda} \qquad \frac{\text{power/s}}{E/\text{photon}}$$

Solution: $A = \pi r^2$ $\dfrac{\pi (0.60\,\text{mm})^2}{\pi (2.75\,\text{mm})^2} = 0.0476$ $0.0476 \times 5.0\,\text{watts} \times \dfrac{1\,\text{J/s}}{1\,\text{watt}} = 0.2\underline{3}8\,\text{J/s}$

$$E_{\text{photon}} = \frac{(6.626 \times 10^{-34}\,\text{J} \cdot \text{s})(3.00 \times 10^8\,\text{m/s})}{(532\,\text{nm})\left(\dfrac{1\,\text{m}}{10^9\,\text{nm}}\right)} = 3.7\underline{3}6 \times 10^{-19}\,\text{J/photon}$$

$$\frac{0.2\underline{3}8\,\text{J/s}}{3.7\underline{3}6 \times 10^{-19}\,\text{J/photon}} = 6.4 \times 10^{17}\,\text{photons/s}$$

Check: The units of the answer (number of photons/s) are correct. The magnitude of the answer is reasonable.

2.77   **Given:** KE $= 506$ eV   **Find:** λ

Conceptual Plan: $\textbf{KE}_{\textbf{eV}} \quad \rightarrow \quad \textbf{KE}_{\textbf{J}} \quad \rightarrow \quad \boldsymbol{v} \quad \rightarrow \quad \boldsymbol{\lambda}$

$$\frac{1.602 \times 10^{-19}\,\text{J}}{1\,\text{eV}} \qquad \text{KE} = 1/2\,mv^2 \qquad \lambda = \frac{h}{mv}$$

Solution:

$$506\,\text{eV}\left(\frac{1.602 \times 10^{-19}\,\text{J}}{1\,\text{eV}}\right)\left(\frac{\text{kg} \cdot \text{m}^2}{\dfrac{\text{s}^2}{\text{J}}}\right) = \frac{1}{2}(9.11 \times 10^{-31}\,\text{kg})\,v^2$$

$$v^2 = \frac{506\,1\,\text{eV}\left(\dfrac{1.602 \times 10^{-19}\,\text{J}}{1\,\text{eV}}\right)\left(\dfrac{\text{kg} \cdot \text{m}^2}{\dfrac{\text{s}^2}{\text{J}}}\right)}{\dfrac{1}{2}(9.11 \times 10^{-31}\,\text{kg})} = 1.7796 \times 10^{14}\,\frac{\text{m}^2}{\text{s}^2}$$

$$v = 1.33 \times 10^7\,\text{m/s} \qquad \lambda = \frac{h}{mv} = \frac{6.626 \times 10^{-34}\,\dfrac{\text{kg} \cdot \text{m}^2}{\text{s}^2} \cdot \text{s}}{(9.11 \times 10^{-31}\,\text{kg})(1.33 \times 10^7\,\text{m/s})} = 5.47 \times 10^{-11}\,\text{m} = 0.0547\,\text{nm}$$

Check: The units of the answer (m or nm) are correct. The magnitude of the answer is reasonable because a de Broglie wavelength is usually a very small number.

2.79   **Given:** $n = 1 \rightarrow n = \infty$ **Find:** E; λ

Conceptual Plan: $n = \infty, n = 1 \quad \rightarrow \quad \Delta E_{\text{atom}} \quad \rightarrow \quad \Delta E_{\text{photon}} \quad \rightarrow \quad \lambda$

$$\Delta E_{\text{atom}} = E_\infty - E_1 \qquad \Delta E_{\text{atom}} = \Delta E_{\text{photon}} \qquad E = \frac{hc}{\lambda}$$

Solution: $\Delta E = E_\infty - E_1 = 0 - \left[-2.18 \times 10^{-18}\,\text{J}\left(\dfrac{1}{1^2}\right)\right] = +2.18 \times 10^{-18}\,\text{J}$

$$\Delta E_{photon} = -\Delta E_{atom} = +2.18 \times 10^{-18}\, J$$

$$\lambda = \frac{hc}{E} = \frac{(6.626 \times 10^{-34}\, J \cdot s)(3.00 \times 10^{8}\, m/s)}{2.18 \times 10^{-18}\, J} = 9.12 \times 10^{-8}\, m = 91.2\, nm$$

Check: The units of the answers (J for E and m or nm for the first part) are correct. The magnitude of the answer is reasonable because it would require more energy to completely remove the electron than just moving it to a higher n level. This results in a shorter wavelength.

2.81   (a)   **Given:** $n = 1$ **Find:** number of orbitals if $l = 0 \rightarrow n$

         Conceptual Plan: value $n \rightarrow$ values $l \rightarrow$ values $m_l \rightarrow$ number of orbitals

                               $l = 0 \rightarrow n$ $m_l = -1 \rightarrow +1$ total m_l

         Solution: $n =$ 1

                            $l =$ 0   1

                            $m_l =$ 0 $-1, 0, +1$

                            total 4 orbitals

         Check: The total orbitals will be equal to the (number of l sublevels)^2.

   (b)   **Given:** $n = 2$ **Find:** number of orbitals if $l = 0 \rightarrow n$

         Conceptual Plan: value $n \rightarrow$ values $l \rightarrow$ values $m_l \rightarrow$ number of orbitals

                               $l = 0 \rightarrow n$ $m_l = -1 \rightarrow +1$ total m_l

         Solution: $n =$ 2

                            $l =$ 0   1   2

                            $m_l =$ 0 $-1, 0, +1$ $-2, -1, 0, 1, 2$

                            total 9 orbitals

         Check: The total orbitals will be equal to the (number of l sublevels)^2.

   (c)   **Given:** $n = 3$ **Find:** number of orbitals if $l = 0 \rightarrow n$

         Conceptual Plan: value $n \rightarrow$ values $l \rightarrow$ values $m_l \rightarrow$ number of orbitals

                               $l = 0 \rightarrow n$ $m_l = -1 \rightarrow +1$ total m_l

         Solution:

                            $n =$ 3

                            $l =$ 0   1   2   3

                            $m_l =$ 0 $-1, 0, +1$ $-2, -1, 0, 1, 2$ $-3, -2, -1, 0, 1, 2, 3$

                            total 16 orbitals

         Check: The total orbitals will be equal to the (number of l sublevels)^2.

2.83   **Given:** $\lambda = 1875$ nm; 1282 nm; 1093 nm   **Find:** equivalent transitions

     Conceptual Plan: $\lambda \rightarrow E_{photon} \rightarrow E_{atom} \rightarrow n$

                 $E = \dfrac{hc}{\lambda}$ $E_{photon} = -E_{atom}$ $E = -2.18 \times 10^{-18}\, J \left(\dfrac{1}{n_f^2} - \dfrac{1}{n_i^2} \right)$

Solution: Because the wavelengths of the transitions are longer wavelengths than those obtained in the visual region, the electron must relax to a higher n level. Therefore, we can assume that the electron returns to the $n = 3$ level.

For $\lambda = 1875$ nm: $E = \dfrac{(6.626 \times 10^{-34}\, J \cdot s)(3.00 \times 10^{8}\, m/s)}{1875\, nm \left(\dfrac{1\, m}{10^{9}\, nm} \right)} = 1.060 \times 10^{-19}\, J$ $1.060 \times 10^{-19}\, J = -1.060 \times 10^{-19}\, J$

$$-1.060 \times 10^{-19}\, J = -2.18 \times 10^{-18}\, J \left(\frac{1}{3^2} - \frac{1}{n^2} \right); n = 4$$

For $\lambda = 1282$ nm: $E = \dfrac{(6.626 \times 10^{-34}\, J \cdot s)(3.00 \times 10^{8}\, m/s)}{1282\, 1\, nm \left(\dfrac{1\, m}{10^{9}\, nm} \right)} = 1.551 \times 10^{-19}\, J$ $1.551 \times 10^{-19}\, J = -1.551 \times 10^{-19}\, J$

$$-1.551 \times 10^{-19}\, J = -2.18 \times 10^{-18}\, J \left(\frac{1}{3^2} - \frac{1}{n^2} \right); n = 5$$

For $\lambda = 1093$ nm: $E = \dfrac{(6.626 \times 10^{-34}\,\text{J}\cdot\text{s})(3.00 \times 10^8\,\text{m/s})}{1093\,\text{nm}\left(\dfrac{1\,\text{m}}{10^9\,\text{nm}}\right)} = 1.819 \times 10^{-19}\,\text{J};\quad 1.819 \times 10^{-19}\,\text{J} = -1.819 \times 10^{-19}\,\text{J}$

$-1.819 \times 10^{-19}\,\text{J} = -2.18 \times 10^{-18}\,\text{J}\left(\dfrac{1}{3^2} - \dfrac{1}{n^2}\right); n = 6$

Check: The values obtained are all integers, which is correct. The values of n (4, 5, and 6) are reasonable. The values of n increase as the wavelength decreases because the two n levels involved are farther apart and more energy is released as the electron relaxes to the $n = 3$ level.

2.85   **Given:** $\Phi = 193$ kJ/mol   **Find:** threshold frequency (ν)
Conceptual Plan: Φ **kJ/mol** $\rightarrow$ Φ **kJ/atom** $\rightarrow$ Φ **J/atom** $\rightarrow$ ν

$$\frac{\text{mol}}{6.022 \times 10^{23}\,\text{atoms}} \qquad \frac{1000\,\text{J}}{1\,\text{kJ}} \qquad \Phi = h\nu$$

Solution: $\nu = \dfrac{\Phi}{h} = \dfrac{\left(\dfrac{193\,\text{kJ}}{\text{mol}}\right)\left(\dfrac{\text{mol}}{6.022 \times 10^{23}\,\text{atoms}}\right)\left(\dfrac{1000\,\text{J}}{1\,\text{kJ}}\right)}{6.626 \times 10^{-34}\,\text{J}\cdot\text{s}} = 4.84 \times 10^{14}\,\text{s}^{-1}$

Check: The units of the answer (s^{-1}) are correct. The magnitude of the answer puts the frequency in the infrared range and is a reasonable answer.

2.87   **Given:** $\nu_{low} = 30\,\text{s}^{-1}\ \nu_{hi} = 1.5 \times 10^4\,\text{s}^{-1}$; speed $= 344$ m/s   **Find:** $\lambda_{low} - \lambda_{hi}$
Conceptual Plan: $\nu_{low} \rightarrow \lambda_{low}$ **and** $\nu_{hi} = \lambda_{hi}$ **then** $\lambda_{low} - \lambda_{hi}$

$$\lambda\nu = \text{speed}$$

Solution: $\lambda = \dfrac{\text{speed}}{\nu};\quad \lambda_{low} = \dfrac{344\,\text{m/s}}{30\,\text{s}^{-1}} = 11\,\text{m};\quad \lambda_{hi} = \dfrac{344\,\text{m/s}}{1.5 \times 10^4\,\text{s}^{-1}} = 0.023\,\text{m};\quad 11\,\text{m} - 0.023\,\text{m} = 11\,\text{m}$

Check: The units of the answer (m) are correct. The magnitude is reasonable because the value is only determined by the low-frequency value because of significant figures.

2.89   **Given:** $\lambda = 792$ nm; $V = 100.0$ mL; $P = 55.7$ mtorr; $T = 25\,°\text{C}$ **Find:** E to dissociate 15.0%
Conceptual Plan: $\lambda \rightarrow E/$ **molecule and then** $P, V, T \rightarrow n \rightarrow$ **molecules**

$$E = \frac{hc}{\lambda} \qquad\qquad n = \frac{PV}{RT}\ \frac{6.022 \times 10^{23}\,\text{molecules}}{1\,\text{mol}}$$

Solution: $E = \dfrac{(6.626 \times 10^{-34}\,\text{J}\cdot\text{s})(3.00 \times 10^8\,\text{m/s})}{792\,\text{nm}\left(\dfrac{1\,\text{m}}{10^9\,1\,\text{nm}}\right)} = 2.51 \times 10^{-19}\,\text{J/molecule}$

$PV = nRT$. Rearrange to solve for n, so $n = PV/RT =$

$$\dfrac{(55.7\,\text{mtorr})\left(\dfrac{1\,\text{torr}}{1000\,\text{mtorr}}\right)\left(\dfrac{1\,\text{atm}}{760\,\text{torr}}\right)(100.0\,\text{mL})\left(\dfrac{1\,\text{L}}{1000\,\text{mL}}\right)}{\left(\dfrac{0.0821\,\text{L}\cdot\text{atm}}{\text{mol}\cdot\text{K}}\right)(298\,\text{K})} = 2.75503 \times 10^{-4}\,\text{mol}$$

$2.75503 \times 10^{-4}\,\text{mol} \times \left(\dfrac{6.022 \times 10^{23}\,\text{molecules}}{1\,\text{mol}}\right) = 1.80 \times 10^{17}\,\text{molecules}$

$(1.80 \times 10^{17}\,\text{molecules})(0.15\%) = 2.70 \times 10^{16}\,\text{molecules dissociated}$

$(2.51 \times 10^{-19}\,\text{J/molecule})(2.70 \times 10^{16}\,\text{molecules}) = 6.777 \times 10^{-3}\,\text{J} = 6.78 \times 10^{-3}\,\text{J}$

Check: The units of the answer (J) are correct. The magnitude is reasonable because it is for a part of a mole of molecules.

2.91   **Given:** 20.0 mW; 1.00 hr; 2.29×10^{20} photons   **Find:** λ
Conceptual Plan: **mW** $\rightarrow$ **W** $\rightarrow$ **J** $\rightarrow$ **J/photon** $\rightarrow$ λ

$$\frac{1\,\text{W}}{1000\,\text{mW}} \qquad E = \text{W} \times \text{s} \qquad \frac{E}{\text{number of photons}} \qquad \lambda = \frac{hc}{E}$$

Solution: $(20.0 \text{ mW})\left(\dfrac{1 \text{ W}}{1000 \text{ mW}}\right)\left(\dfrac{\frac{1 \text{ J}}{1 \text{ s}}}{1 \text{ W}}\right)\left(\dfrac{3600 \text{ s}}{2.29 \times 10^{20} \text{ photons}}\right) = 3.14 \times 10^{-19} \text{ J/photon}$

$$\dfrac{(6.626 \times 10^{-34} \text{ J} \cdot \text{s})(3.00 \times 10^{8} \text{ m/s})\left(\dfrac{10^{9} \text{ nm}}{1 \text{ m}}\right)}{3.14 \times 10^{-19} \text{ J}} = 632 \text{ nm}$$

Check: The units of the answer (nm) are correct. The magnitude is reasonable because it is in the red range.

Challenge Problems

2.93   (a)   **Given:** $n = 1; n = 2; n = 3; L = 155 \text{ pm}$ **Find:** E_1, E_2, E_3
Conceptual Plan: $n \rightarrow E$

$$E_n = \dfrac{n^2 h^2}{8 \, m \, L^2}$$

Solution:

$$E_1 = \dfrac{1^2(6.626 \times 10^{-34} \text{ J} \cdot \text{s})^2}{8(9.11 \times 10^{-31} \text{ kg})(155 \text{ pm})^2\left(\dfrac{1 \text{ m}}{10^{12} \text{ pm}}\right)^2} = \dfrac{1(6.626 \times 10^{-34})^2 \text{ J}^2\text{s}^2}{8(9.11 \times 10^{-31} \text{ kg})(155 \times 10^{-12})^2 \text{ m}^2}$$

$$= \dfrac{1(6.626 \times 10^{-34})^2\left(\dfrac{\text{kg} \cdot \text{m}^2}{\text{s}^2}\right)\text{J s}^2}{8(9.11 \times 10^{-31} \text{ kg})(155 \times 10^{-12})^2 \text{ m}^2} = 2.51 \times 10^{-18} \text{ J}$$

$$E_2 = \dfrac{2^2(6.626 \times 10^{-34} \text{ J} \cdot \text{s})^2}{8(9.11 \times 10^{-31} \text{ kg})(155 \text{ pm})^2\left(\dfrac{1 \text{ m}}{10^{12} \text{ pm}}\right)^2} = \dfrac{4(6.626 \times 10^{-34})^2 \text{ J}^2\text{ s}^2}{8(9.11 \times 10^{-31} \text{ kg})(155 \times 10^{-12})^2 \text{ m}^2}$$

$$= \dfrac{4(6.626 \times 10^{-34})^2\left(\dfrac{\text{kg} \cdot \text{m}^2}{\text{s}^2}\right)\text{J s}^2}{8(9.11 \times 10^{-31} \text{ kg})(155 \times 10^{-12})^2 \text{ m}^2} = 1.00 \times 10^{-17} \text{ J}$$

$$E_3 = \dfrac{3^2(6.626 \times 10^{-34} \text{ J} \cdot \text{s})^2}{8(9.11 \times 10^{-31} \text{ kg})(155 \text{ pm})^2\left(\dfrac{1 \text{ m}}{10^{12} \text{ pm}}\right)^2} = \dfrac{9(6.626 \times 10^{-34})^2 \text{ J}^2\text{ s}^2}{8(9.11 \times 10^{-31} \text{ kg})(155 \times 10^{-12})^2 \text{ m}^2}$$

$$= \dfrac{9(6.626 \times 10^{-34})^2\left(\dfrac{\text{kg} \cdot \text{m}^2}{\text{s}^2}\right)\text{J s}^2}{8(9.11 \times 10^{-31} \text{ kg})(155 \times 10^{-12})^2 \text{ m}^2} = 2.26 \times 10^{-17} \text{ J}$$

Check: The units of the answers (J) are correct. The answers seem reasonable because the energy is increasing with increasing n level.

(b)   **Given:** $n = 1 \rightarrow n = 2$ and $n = 2 \rightarrow n = 3$ **Find:** λ
Conceptual Plan: $n = 1, n = 2 \rightarrow \Delta E_{atom} \rightarrow \Delta E_{photon} \rightarrow \lambda$

$$\Delta E_{atom} = E_2 - E_1 \qquad \Delta E_{atom} \rightarrow -\Delta E_{photon} \qquad E = \dfrac{hc}{\lambda}$$

Solution: Use the energies calculated in part a.

$E_2 - E_1 = (1.00 \times 10^{-17} \text{ J} - 2.51 \times 10^{-18} \text{ J}) = 7.49 \times 10^{-18} \text{ J}$

$\lambda = \dfrac{(6.626 \times 10^{-34} \text{ J} \cdot \text{s})(3.00 \times 10^{8} \text{ m/s})}{7.49 \times 10^{-18} \text{ J}} = 2.65 \times 10^{-8} \text{ m} = 26.5 \text{ nm}$

$E_3 - E_2 = (2.26 \times 10^{-17} \text{ J} - 1.00 \times 10^{-17} \text{ J}) = 1.26 \times 10^{-17} \text{ J}$

$\lambda = \dfrac{(6.626 \times 10^{-34} \text{ J} \cdot \text{s})(3.00 \times 10^{8} \text{ m/s})}{1.26 \times 10^{-17} \text{ J}} = 1.58 \times 10^{-8} \text{ m} = 15.8 \text{ nm}$

These wavelengths would lie in the UV region.

Check: The units of the answers (m) are correct. The magnitude of the answers is reasonable based on the energies obtained for the levels.

2.95 For the $1s$ orbital in the Excel® spreadsheet, label the columns as follows: column A as r and column B as $\psi(1s)$. Make the values for r column A as follows: 0–200. In column B, write the equation for the wave function as follows: $= (\text{POWER}(1/3.1415,1/2))*(1/\text{POWER}(53,3/2))*(\text{EXP}(-A2/53))$. Go to make chart, choose xy scatter.

e.g., sample values	
r	$\psi\,(1s)$
0	7.000146224
1	7.000143491
2	7.000140809
3	7.000138177
4	7.000135594
5	7.000133060
6	7.000130573

For the $2s$ orbital in the same Excel® spreadsheet, label the columns as follows: column A as r and column C as $\psi\,(2s)$. Use the same values for r in column A: 0–200. In column C, write the equation for the wave function as follows: $= (\text{POWER}(1/((32)*(3.1415)),1/2))*(1/\text{POWER}(53,3/2))*(2-(A2/53))*(\text{EXP}(-A2/53))$. Go to make chart, choose xy scatter.

e.g., sample values	
r	$\psi\,(2s)$
0	7.0000516979
1	7.0000502530
2	7.0000488441
3	7.0000474702
4	7.0000461307
5	7.0000448247
6	7.0000435513

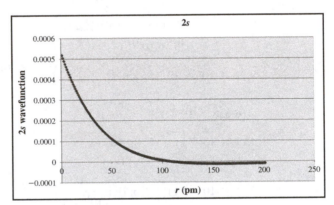

Note: The plot for the $2s$ orbital extends below the x-axis. The x-intercept represents the radial node of the orbital.

2.97 **Given:** threshold frequency $= 2.25 \times 10^{14}\ \text{s}^{-1}$; $\lambda = 5.00 \times 10^{-7}\ \text{m}$ **Find:** v of electron
Conceptual Plan: $v \rightarrow \Phi$ and then $\lambda \rightarrow E$ and then $\rightarrow$ KE $\rightarrow$ v

$$\Phi = h v \qquad E = \frac{hc}{\lambda} \qquad \text{KE} = E - \Phi \quad \text{KE} = 1/2\,mv^2$$

Solution:

$$\Phi = (6.626 \times 10^{-34}\ \text{J} \cdot \text{s})(2.25 \times 10^{14}\ \text{s}^{-1}) = 1.491 \times 10^{-19}\ \text{J}; \quad E = \frac{(6.626 \times 10^{-34}\ \text{J} \cdot \text{s})(3.00 \times 10^{8}\ \text{m/s})}{5.00 \times 10^{-7}\ \text{m}} = 3.976 \times 10^{-19}\ \text{J}$$

$$\text{KE} = 3.976 \times 10^{-19}\ \text{J} - 1.491 \times 10^{-19}\ \text{J} = 2.485 \times 10^{-19}\ \text{J}; \text{KE} = 1/2\,mv^2; \text{Substituting and rearranging gives}$$

$$v^2 = \frac{2.485 \times 10^{-19}\ \dfrac{\text{kg} \cdot \text{m}^2}{\text{s}^2}}{\dfrac{1}{2}(9.11 \times 10^{-31}\ \text{kg})} = 5.455 \times 10^{11}\ \frac{\text{m}^2}{\text{s}^2}$$

$$v = 7.39 \times 10^{5}\ \text{m/s}$$

Check: The units of the answer (m/s) are correct. The magnitude of the answer is reasonable for the speed of an electron.

2.99   **Given:** $t = 5.0 \text{ fs}$; $\lambda_{low} = 722 \text{ nm}$ **Find:** ΔE and λ_{high}

Conceptual Plan: $t \rightarrow \Delta E$ and then $\lambda_{low} \rightarrow E_{high} \rightarrow E_{low} \rightarrow \lambda_{high}$

$$\Delta t \times \Delta E \geq \frac{h}{4\pi} \qquad E = \frac{hc}{\lambda} \qquad E - \Delta E \qquad \lambda = \frac{hc}{E}$$

Solution: $\dfrac{6.626 \times 10^{-34} \text{ J} \cdot \text{s}}{4(3.141)(5.0 \text{ fs})\left(\dfrac{\text{s}}{10^{15} \text{ fs}}\right)} = 1.0\underline{5}5 \times 10^{-20} \text{ J}$

$$E = \frac{(6.626 \times 10^{-34} \text{ J} \cdot \text{s})(3.00 \times 10^8 \text{ m/s})}{722 \text{ nm}\left(\dfrac{1 \text{ m}}{10^9 \text{ nm}}\right)} = 2.7\underline{5} \times 10^{-19} \text{ J}$$

$2.75 \times 10^{-19} \text{ J} - 1.06 \times 10^{-20} \text{ J} = 2.6\underline{4} \times 10^{-19} \text{ J}$

$$\frac{(6.626 \times 10^{-34} \text{ J} \cdot \text{s})(3.00 \times 10^8 \text{ m/s})\left(\dfrac{10^9 \text{ nm}}{1 \text{ m}}\right)}{(2.64 \times 10^{-19} \text{ J})} = 751.8 \text{ nm} = 7.5 \times 10^2 \text{ nm}$$

Check: The units of the answer (nm) are correct. The magnitude of the answer is reasonable because it is a longer wavelength, but it is close to the original wavelength.

2.101   **Given:** $r = 1.8 \text{ m}$ **Find:** λ

Conceptual Plan: $r \rightarrow C$

$$C = 2\pi r$$

Solution: $(2)(3.141)(1.8\text{m}) = 11.\underline{3} \text{ m} =$ the circumference of the orbit. So the largest wavelength that would fit the orbit would be 11 m.

Check: The units of the answer (m) are correct. The magnitude of the wave is about the circumference of the orbit.

Conceptual Problems

2.103   In the Bohr model of the atom, the electron travels in a circular orbit around the nucleus. It is a two-dimensional model. The electron is constrained to move only from one orbit to another orbit. But the electron is treated as a particle that behaves according to the laws of classical physics.

The quantum-mechanical model of the atom is three dimensional. In this model, we treat the electron, an absolutely small particle, differently than we treat particles with classical physics. The electron is in an orbital, which gives us the probability of finding the electron within a volume of space.

Because the electron in the Bohr model is constrained to a circular orbit, it would theoretically be possible to know both the position and the velocity of the electron simultaneously. This contradicts the Heisenberg uncertainty principle, which states that position and velocity are complementary terms that cannot both be known with precision. So in the quantum-mechanical model we can only know the position or the velocity of an electron at any given time, never both.

2.105   (a)   Because the interference pattern is caused by single electrons interfering with themselves, the pattern remains the same even when the rate of the electrons passing through the slits is one electron per minute. It will simply take longer for the full pattern to develop.

(b)   When a light is placed behind the slits, it flashes to indicate which hole the electron passed through, but the interference pattern is now absent. With the laser on, the electrons hit positions directly behind each slit, as if they were ordinary particles.

(c)   Diffraction occurs when a wave encounters an obstacle of a slit that is comparable in size to its wavelength. The wave bends around the slit. The diffraction of light through two slits separated by a distance comparable to the wavelength of the light results in an interference pattern. Each slit acts as a new wave source, and the two new waves interfere with each other, which results in a pattern of bright and dark lines.

(d)   Because the mass of the bullets and their particle size are not absolutely small, the bullets will not produce an interference pattern when they pass through the slits. The de Broglie wavelength produced by the bullets will not be sufficiently large to interfere with the bullet trajectory, and no interference pattern will be observed.

Questions for Group Work

2.107   Light is electromagnetic radiation, a type of energy embodied in oscillating electric and magnetic fields. Light in a vacuum travels at 3.00×10^8 m/s.

Electromagnetic radiation can be characterized by wavelength, amplitude, and frequency. The wavelength (λ) of the wave is the distance in space between adjacent crests and is measured in units of distance. The amplitude of the wave is the vertical height of a crest. The more closely spaced the waves, that is, the shorter the wavelength, the more energy there is. The amplitude of the electric and magnetic field wave in light determines the intensity or brightness of the light. The higher the amplitude, the more photons are in the wave and so the more energy the wave has. The frequency, (ν), is the number of cycles (or wave crests) that pass through a stationary point in a given period of time. The units of frequency are cycles per second. The frequency is inversely proportional to the wavelength (λ). Frequency and wavelength are related by the equation: $\nu = \dfrac{c}{\lambda}$.

For visible light, wavelength determines the color. Red light has a wavelength of 750 nm, the longest wavelength of visible light, and blue has a wavelength of 500 nm. The presence of a variety of wavelengths in white light is responsible for the way we perceive colors in objects. When a substance absorbs some colors while reflecting others, it appears colored. Grass appears green because it reflects primarily the wavelength associated with green light and absorbs the others.

2.109   The magnetic quantum number (m_l) is an integer ranging from $-l$ to $+l$.
    (a)   When $l = 0$, $m_l = 0$.
    (b)   When $l = 1$, $m_l = -1, 0,$ or $+1$.
    (c)   When $l = 2$, $m_l = -2, -1, 0, +1$ or $+2$.
    (d)   When $l = 20$, $m_l = -20, -19, ..., 0, ...19,$ or 20, so there are 41 values for m_l.
   In the general case, the number of possible values for $m_l = 2l + 1$.

2.111   The $1s$ orbital does not have any nodes. The $2p$ orbitals have a node at the nucleus. Four of the $3d$ orbitals have a cloverleaf shape, with four lobes of electron density around the nucleus and two perpendicular nodal planes. The f orbitals are even more complex. The number of nodes (and nodal planes) depends on the specific orbital, but there will be more than for s, p, or d orbitals.

Data Interpretation and Analysis

2.112   (a) $335 \text{ nm} \times \dfrac{1 \text{ m}}{10^9 \text{ nm}} = 3.35 \times 10^{-7} \text{ m}; \quad E = \dfrac{6.626 \times 10^{-34} \text{ J} \cdot \text{s} \dfrac{3.00 \times 10^8 \text{ m}}{\text{s}}}{3.35 \times 10^{-7} \text{ m}}$

$= 5.9\underline{3}3731 \times 10^{-19} \text{ J / photon} = 5.93 \times 10^{-19} \text{ J / photon}$

    (c)   2-EHMC

    (e) $330. \text{ nm} \times \dfrac{1 \text{ m}}{10^9 \text{ nm}} = 3.30 \times 10^{-7} \text{ m}; \quad E = \dfrac{6.626 \times 10^{-34} \text{ J} \cdot \text{s} \dfrac{3.00 \times 10^8 \text{ m}}{\text{s}}}{3.30 \times 10^{-7} \text{ m}}$

$= 6.0\underline{2}3636 \times 10^{-19} \text{ J / photon};$

$1 \text{ h} \times \dfrac{60 \text{ min}}{1 \text{ h}} \times \dfrac{60 \text{ s}}{1 \text{ min}} \times \dfrac{3.066 \times 10^{21} \text{ photons}}{1 \text{ m}^2 \cdot \text{s}} \times \dfrac{6.0\underline{2}3636 \times 10^{-19} \text{ J}}{1 \text{ photon}} \times 0.42 \text{ m}^2 \times 0.50 = 1.4 \times 10^6 \text{ J}$

3 Periodic Properties of the Elements

Review Questions

3.1   A periodic property is predictable based on the element's position within the periodic table.

3.3   The first attempt to organize the elements according to similarities in their properties was made by the German chemist Johann Döbereiner. He grouped elements into triads: three elements with similar properties. The English chemist John Newlands attempted a more complex approach. He organized elements into octaves, analogous to musical notes. When arranged this way, the properties of every eighth element were similar.

3.5   Julius Lothar Meyer proposed an organization of the known elements based on some periodic properties. Henry Moseley listed elements according to the atomic number rather than the atomic mass. This resolved the problems in Mendeleev's table where an increase in atomic mass did not correlate with similar properties.

3.7   An electron configuration shows the particular orbitals that are occupied by electrons in an atom. Some examples are $H = 1s^1$, $He = 1s^2$, and $Li = 1s^2 2s^1$.

3.9   Shielding or screening occurs when one electron is blocked from the full effects of the nuclear charge, so that the electron experiences only a part of the nuclear charge. It is the inner (core) electrons that shield the outer electrons from the full nuclear charge.

3.11   The sublevels within a principal level split in multielectron atoms because of the penetration of the outer electrons into the region of the core electrons. The sublevels in hydrogen are not split because they are empty in the ground state.

3.13   The Pauli exclusion principle states the following: No two electrons in an atom can have the same four quantum numbers. Because two electrons occupying the same orbital have three identical quantum numbers (n, l, m_l), they must have different spin quantum numbers. The Pauli exclusion principle implies that each orbital can have a maximum of only two electrons, with opposing spins.

3.15   In order of increasing energy, the orbitals are $1s < 2s < 2p < 3s < 3p < 4s < 3d < 4p < 5s$. The $4s$ orbital fills before the $3d$ and the $5s$ fills before the $4d$. They are lower in energy because of greater penetration of the $4s$ and $5s$ orbitals.

3.17

Periodic table (d-block, s-block, p-block, f-block elements with ground-state outer electron configurations)

Legend: s-block elements, p-block elements, d-block elements, f-block elements

Group	1 (1A)	2 (2A)	3 (3B)	4 (4B)	5 (5B)	6 (6B)	7 (7B)	8 (8B)	9 (8B)	10 (8B)	11 (1B)	12 (2B)	13 (3A)	14 (4A)	15 (5A)	16 (6A)	17 (7A)	18 (8A)
1	1 H $1s^1$																	2 He $1s^2$
2	3 Li $2s^1$	4 Be $2s^2$											5 B $2s^2 2p^1$	6 C $2s^2 2p^2$	7 N $2s^2 2p^3$	8 O $2s^2 2p^4$	9 F $2s^2 2p^5$	10 Ne $2s^2 2p^6$
3	11 Na $3s^1$	12 Mg $3s^2$											13 Al $3s^2 3p^1$	14 Si $3s^2 3p^2$	15 P $3s^2 3p^3$	16 S $3s^2 3p^4$	17 Cl $3s^2 3p^5$	18 Ar $3s^2 3p^6$
4	19 K $4s^1$	20 Ca $4s^2$	21 Sc $4s^2 3d^1$	22 Ti $4s^2 3d^2$	23 V $4s^2 3d^3$	24 Cr $4s^1 3d^5$	25 Mn $4s^2 3d^5$	26 Fe $4s^2 3d^6$	27 Co $4s^2 3d^7$	28 Ni $4s^2 3d^8$	29 Cu $4s^1 3d^{10}$	30 Zn $4s^2 3d^{10}$	31 Ga $4s^2 4p^1$	32 Ge $4s^2 4p^2$	33 As $4s^2 4p^3$	34 Se $4s^2 4p^4$	35 Br $4s^2 4p^5$	36 Kr $4s^2 4p^6$
5	37 Rb $5s^1$	38 Sr $5s^2$	39 Y $5s^2 4d^1$	40 Zr $5s^2 4d^2$	41 Nb $5s^1 4d^4$	42 Mo $5s^1 4d^5$	43 Tc $5s^2 4d^5$	44 Ru $5s^1 4d^7$	45 Rh $5s^1 4d^8$	46 Pd $4d^{10}$	47 Ag $5s^1 4d^{10}$	48 Cd $5s^2 4d^{10}$	49 In $5s^2 5p^1$	50 Sn $5s^2 5p^2$	51 Sb $5s^2 5p^3$	52 Te $5s^2 5p^4$	53 I $5s^2 5p^5$	54 Xe $5s^2 5p^6$
6	55 Cs $6s^1$	56 Ba $6s^2$	57 La $6s^2 5d^1$	72 Hf $6s^2 5d^2$	73 Ta $6s^2 5d^3$	74 W $6s^2 5d^4$	75 Re $6s^2 5d^5$	76 Os $6s^2 5d^6$	77 Ir $6s^2 5d^7$	78 Pt $6s^1 5d^9$	79 Au $6s^1 5d^{10}$	80 Hg $6s^2 5d^{10}$	81 Tl $6s^2 6p^1$	82 Pb $6s^2 6p^2$	83 Bi $6s^2 6p^3$	84 Po $6s^2 6p^4$	85 At $6s^2 6p^5$	86 Rn $6s^2 6p^6$
7	87 Fr $7s^1$	88 Ra $7s^2$	89 Ac $7s^2 6d^1$	104 Rf $7s^2 6d^2$	105 Db $7s^2 6d^3$	106 Sg $7s^2 6d^4$	107 Bh	108 Hs	109 Mt	110 Ds	111 Rg	112	113	114	115	116		

Lanthanides	58 Ce $6s^2 4f^1 5d^1$	59 Pr $6s^2 4f^3$	60 Nd $6s^2 4f^4$	61 Pm $6s^2 4f^5$	62 Sm $6s^2 4f^6$	63 Eu $6s^2 4f^7$	64 Gd $6s^2 4f^7 5d^1$	65 Tb $6s^2 4f^9$	66 Dy $6s^2 4f^{10}$	67 Ho $6s^2 4f^{11}$	68 Er $6s^2 4f^{12}$	69 Tm $6s^2 4f^{13}$	70 Yb $6s^2 4f^{14}$
													71 Lu $6s^2 4f^{14} 6d^1$
Actinides	90 Th $7s^2 6d^2$	91 Pa $7s^2 5f^2 6d^1$	92 U $7s^2 5f^3 6d^1$	93 Np $7s^2 5f^4 6d^1$	94 Pu $7s^2 5f^6$	95 Am $7s^2 5f^7$	96 Cm $7s^2 5f^7 6d^1$	97 Bk $7s^2 5f^9$	98 Cf $7s^2 5f^{10}$	99 Es $7s^2 5f^{11}$	100 Fm $7s^2 5f^{12}$	101 Md $7s^2 5f^{13}$	102 No $7s^2 5f^{14}$
													103 Lr $7s^2 5f^{14} 6d^1$

3.19   The rows in the periodic table grow progressively longer because you are adding sublevels as the n level increases.

3.21   The row number of a main-group element is equal to the highest principal quantum number of that element. However, the principal quantum number of the d orbital being filled across each row in the transition series is equal to the row number minus one. For the inner transition elements, the principal quantum number of the f orbital being filled across each row is the row number minus two.

3.23   To use the periodic table to write the electron configuration, find the noble gas that precedes the element. The element has the inner electron configuration of that noble gas. Place the symbol for the noble gas in [ ]. Obtain the outer electron configuration by tracing the element across the period and assigning electrons in the appropriate orbitals.

3.25   (a)   The alkali metals (group 1A) have one valence electron and are among the most reactive metals because their outer electron configuration (ns^1) is one electron beyond a noble gas configuration. They react to lose the ns^1 electron, obtaining a noble gas configuration. This is why the group 1A metals tend to form 1+ cations.

(b)   The alkaline earth metals (group 2A) have two valence electrons and an outer electron configuration of ns^2, and also tend to be reactive metals. They lose their ns^2 electrons to form 2+ cations.

(c)   The halogens (group 7A) have seven valence electrons and an outer electron configuration of $ns^2 np^5$. They are among the most reactive nonmetals. They are only one electron short of a noble gas configuration and tend to react to gain that one electron, forming 1− anions.

(d)   The oxygen family (group 6A) has six valence electrons and an outer electron configuration of $ns^2 np^4$. They are two electrons short of a noble gas configuration and tend to react to gain those two electrons, forming 2− anions.

3.27   The effective nuclear charge (Z_{eff}) is the average or net charge from the nucleus experienced by the electrons in the outermost levels. Shielding is the blocking of nuclear charge from the outermost electrons. The shielding is primarily due to the inner (core) electrons, although there is some interaction and shielding from the electron repulsions of the outer electrons with each other.

3.29   When a halogen forms an ion, it gains one valence electron and forms an ion with a 1− charge. When an atom in the oxygen family forms an ion, it gains two valence electrons and forms an ion with a 2−. Nitrogen will gain three electrons, forming a 3− ion. Aluminum will lose three electrons, forming a 3+ ion.

3.31   (a)   The radii of transition elements stay roughly constant across each row instead of decreasing in size as in the main-group elements. The difference is that across a row of transition elements, the number of electrons in the outermost principal energy level is nearly constant. As another proton is added to the nucleus with each successive element, another electron is added as well, but the electron goes into an $n_{highest} - 1$ orbital. The number of outermost electrons stays constant, and they experience a roughly constant effective nuclear charge, keeping the radius approximately constant.

(b) As you go down the first two rows of a column within the transition metals, the elements follow the same general trend in atomic radii and the main-group elements; that is, the radii get larger because you are adding outermost electrons into higher n levels.

3.33 An important exception to simply subtracting the number of electrons occurs for transition metal cations. When writing the electron configuration of a transition metal cation, remove the electrons in the highest n-value orbitals first, even if this does not correspond to the reverse order of filling. Normally, even though the d orbital electrons add after the s orbital electrons, the s orbital electrons are lost first. This is because (1) the ns and $(n - 1)d$ orbitals are extremely close in energy and, depending on the exact configuration, can vary in relative energy ordering and (2) as the $(n - 1)d$ orbitals begin to fill in the first transition series, the increasing nuclear charge stabilizes the $(n - 1)d$ orbitals relative to the ns orbitals. This happens because the $(n - 1)d$ orbitals are not outermost orbitals and, therefore, are not effectively shielded from the increasing nuclear charge by the ns orbitals.

3.35 The ionization energy (IE) of an atom or ion is the energy required to remove an electron from the atom or ion in the gaseous state. The ionization energy is always positive because removing an electron takes energy. The energy required to remove the first electron is called the first ionization energy (IE_1). The energy required to remove the second electron is called the second ionization energy (IE_2). The second IE is always greater than the first IE.

3.37 Exceptions occur with elements Be, Mg, and Ca in group 2A having a higher first ionization energy than elements B, Al, and Ga in group 3A. This exception is caused by the change in going from the s block to the p block. The result is that the electrons in the s orbital shield the electron in the p orbital from nuclear charge, making it easier to remove.

Another exception occurs with N, P, and As in group 5A having a higher first ionization energy than O, S, and Se in group 6A. This exception is caused by the repulsion between electrons when they must occupy the same orbital. Group 5A has $3p$ electrons, while group 6A has $4p$ electrons. In the group 5A elements, the p orbitals are half-filled, which makes the configuration particularly stable. The fourth group 6A electron must pair with another electron, making it easier to remove.

3.39 The electron affinity (EA) of an atom or ion is the energy change associated with the gaining of an electron by the atom in the gaseous state. The electron affinity is usually—though not always—negative because an atom or ion usually releases energy when it gains an electron. The trends in electron affinity are not as regular as trends in other properties. For main-group elements, electron affinity generally becomes more negative as you move to the right across a row in the periodic table. There is no corresponding trend in electron affinity going down a column, with the exception of group IA which becomes more positive as you go down the column.

Problems by Topic

The Periodic Table

3.41			
	(a)	K	Potassium is a metal.
	(b)	Ba	Barium is a metal.
	(c)	I	Iodine is a nonmetal.
	(d)	O	Oxygen is a nonmetal.
	(e)	Sb	Antimony is a metalloid.

3.43			
	(a)	tellurium	Te is in group 6A and is a main-group element.
	(b)	potassium	K is in group 1A and is a main-group element.
	(c)	vanadium	V is in group 5B and is a transition element.
	(d)	manganese	Mn is in group 7B and is a transition element.

Electron Configurations

3.45

(a) Si   Silicon has 14 electrons. Distribute two of these into the $1s$ orbital, two into the $2s$ orbital, six into the $2p$ orbital, two into the $3s$ orbital, and two into the $3p$ orbital. $1s^2 2s^2 2p^6 3s^2 3p^2$

(b) O   Oxygen has 8 electrons. Distribute two of these into the $1s$ orbital, two into the $2s$ orbital, and four into the $2p$ orbital. $1s^2 2s^2 2p^4$

(c) K   Potassium has 19 electrons. Distribute two of these into the $1s$ orbital, two into the $2s$ orbital, six into the $2p$ orbital, two into the $3s$ orbital, six into the $3p$ orbital, and one into the $4s$ orbital. $1s^2 2s^2 2p^6 3s^2 3p^6 4s^1$

(d) Ne   Neon has ten electrons. Distribute two of these into the $1s$ orbital, two into the $2s$ orbital, and six into the $2p$ orbital. $1s^2 2s^2 2p^6$

3.47

(a) N   Nitrogen has seven electrons and has the electron configuration $1s^2 2s^2 2p^3$. Draw a box for each orbital, putting the lowest energy orbital $(1s)$ on the far left and proceeding to orbitals of higher energy to the right. Distribute the seven electrons into the boxes representing the orbitals, allowing a maximum of two electrons per orbital and remembering Hund's rule. You can see from the diagram that nitrogen has three unpaired electrons.

$$\underset{1s}{[\uparrow\downarrow]} \quad \underset{2s}{[\uparrow\downarrow]} \quad \underset{2p}{[\uparrow\,|\,\uparrow\,|\,\uparrow]}$$

(b) F   Fluorine has nine electrons and has the electron configuration $1s^2 2s^2 2p^5$. Draw a box for each orbital, putting the lowest energy orbital $(1s)$ on the far left and proceeding to orbitals of higher energy to the right. Distribute the nine electrons into the boxes representing the orbitals, allowing a maximum of two electrons per orbital and remembering Hund's rule. You can see from the diagram that fluorine has one unpaired electron.

$$\underset{1s}{[\uparrow\downarrow]} \quad \underset{2s}{[\uparrow\downarrow]} \quad \underset{2p}{[\uparrow\downarrow\,|\,\uparrow\downarrow\,|\,\uparrow]}$$

(c) Mg   Magnesium has 12 electrons and has the electron configuration $1s^2 2s^2 2p^6 3s^2$. Draw a box for each orbital, putting the lowest energy orbital $(1s)$ on the far left and proceeding to orbitals of higher energy to the right. Distribute the 12 electrons into the boxes representing the orbitals, allowing a maximum of two electrons per orbital and remembering Hund's rule. You can see from the diagram that magnesium has no unpaired electrons.

$$\underset{1s}{[\uparrow\downarrow]} \quad \underset{2s}{[\uparrow\downarrow]} \quad \underset{2p}{[\uparrow\downarrow\,|\,\uparrow\downarrow\,|\,\uparrow\downarrow]} \quad \underset{3s}{[\uparrow\downarrow]}$$

(d) Al   Aluminum has 13 electrons and has the electron configuration $1s^2 2s^2 2p^6 3s^2 3p^1$. Draw a box for each orbital, putting the lowest energy orbital $(1s)$ on the far left and proceeding to orbitals of higher energy to the right. Distribute the 13 electrons into the boxes representing the orbitals, allowing a maximum of two electrons per orbital and remembering Hund's rule. You can see from the diagram that aluminum has one unpaired electron.

$$\underset{1s}{[\uparrow\downarrow]} \quad \underset{2s}{[\uparrow\downarrow]} \quad \underset{2p}{[\uparrow\downarrow\,|\,\uparrow\downarrow\,|\,\uparrow\downarrow]} \quad \underset{3s}{[\uparrow\downarrow]} \quad \underset{3p}{[\uparrow\,|\,\,|\,\,]}$$

3.49

(a) P   The atomic number of P is 15. The noble gas that precedes P in the periodic table is neon, so the inner electron configuration is [Ne]. Obtain the outer electron configuration by tracing the elements between Ne and P and assigning electrons to the appropriate orbitals. Begin with [Ne]. Because P is in row 3, add two $3s$ electrons. Next, add three $3p$ electrons as you trace across the p block to P, which is in the third column of the p block.

         P $[Ne]3s^2 3p^3$

(b) Ge   The atomic number of Ge is 32. The noble gas that precedes Ge in the periodic table is argon, so the inner electron configuration is [Ar]. Obtain the outer electron configuration by tracing the elements between Ar and Ge and assigning electrons to the appropriate orbitals. Begin with [Ar]. Because Ge is in row 4, add two $4s$ electrons. Next, add ten $3d$ electrons as you trace across the d block. Finally, add two $4p$ electrons as you trace across the p block to Ge, which is in the second column of the p block.

         Ge $[Ar]4s^2 3d^{10} 4p^2$

    (c)   Zr   The atomic number of Zr is 40. The noble gas that precedes Zr in the periodic table is krypton, so the inner electron configuration is [Kr]. Obtain the outer electron configuration by tracing the elements between Kr and Zr and assigning electrons to the appropriate orbitals. Begin with [Kr]. Because Zr is in row 5, add two 5s electrons. Next, add two 4d electrons as you trace across the d block to Zr, which is in the second column.

$$\text{Zr} \quad [\text{Kr}]5s^2 4d^2$$

    (d)   I   The atomic number of I is 53. The noble gas that precedes I in the periodic table is krypton, so the inner electron configuration is [Kr]. Obtain the outer electron configuration by tracing the elements between Kr and I and assigning electrons to the appropriate orbitals. Begin with [Kr]. Because I is in row 5, add two 5s electrons. Next, add ten 4d electrons as you trace across the d block. Finally, add five 5p electrons as you trace across the p block to I, which is in the fifth column of the p block.

$$\text{I} \quad [\text{Kr}]5s^2 4d^{10} 5p^5$$

3.51   (a)   Li is in period 2 and the first column in the s block, so Li has one 2s electron.

    (b)   Cu is in period 4 and the ninth column in the d block $(n - 1)$, so Cu should have nine 3d electrons. However, it is one of our exceptions; so it has ten 3d electrons.

    (c)   Br is in period 4 and the fifth column of the p block, so Br has five 4p electrons.

    (d)   Zr is in period 5 and the second column of the d block $(n - 1)$, so Zr has two 4d electrons.

3.53   (a)   In period 4, an element with five valence electrons could be V or As.

    (b)   In period 4, an element with four 4p electrons would be in the fourth column of the p block and is Se.

    (c)   In period 4, an element with three 3d electrons would be in the third column of the d block $(n - 1)$ and is V.

    (d)   In period 4, an element with a complete outer shell would be in the sixth column of the p block and is Kr.

Valence Electrons and Simple Chemical Behavior from the Periodic Table

3.55   (a)   Ba is in column 2A, so it has two valence electrons.

    (b)   Cs is in column 1A, so it has one valence electron.

    (c)   Ni is in column 8 of the d block, so it has ten valence electrons (eight from the d block and two from the s block).

    (d)   S is in column 6A, so it has six valence electrons.

3.57   (a)   The outer electron configuration ns^2 would belong to a reactive metal in the alkaline earth family.

    (b)   The outer electron configuration $ns^2 np^6$ would belong to an unreactive nonmetal in the noble gas family.

    (c)   The outer electron configuration $ns^2 np^5$ would belong to a reactive nonmetal in the halogen family.

    (d)   The outer electron configuration $ns^2 np^2$ would belong to an element in the carbon family. If $n = 2$, the element is a nonmetal; if $n = 3$ or 4, the element is a metalloid; and if $n = 5$ or 6, the element is a metal.

3.59   (a)   sodium   Na is in group 1A, has 1 valence electron, and is an alkali metal.

    (b)   iodine   I is in group 7A, has 7 valence electrons, and is a halogen.

    (c)   calcium   Ca is in group 2A, has 2 valence electrons, and is an alkaline earth metal.

    (d)   barium   Ba is in group 2A, has 2 valence electrons, and is an alkaline earth metal.

    (e)   krypton   Kr is in group 8A, has 8 valence electrons, and is a noble gas.

3.61   (a)   N and Ni would not be most similar. Nitrogen is a nonmetal; nickel is a metal.

    (b)   Mo and Sn would not be most similar. Although both are metals, molybdenum is a transition metal and tin is a main-group metal.

    (c)   Na and Mg would not be most similar. Although both are main-group metals, sodium is in group 1A and magnesium is in group 2A.

    (d)   Cl and F would be most similar. Chlorine and fluorine are both in group 7A. Elements in the same group have similar chemical properties.

    (e)   Si and P would not be most similar. Silicon is a metalloid, and phosphorus is a nonmetal.

3.63   Main-group metal atoms will lose electrons to form a cation with the same number of electrons as the nearest previous noble gas. Nonmetal atoms will gain electrons to form an anion with the same number of electrons as the nearest noble gas.

(a) O^{2-} O is a nonmetal and has eight electrons. It will gain electrons to form an anion. The nearest noble gas is neon with ten electrons, so O will gain two electrons.

(b) K^+ K is a main-group metal and has 19 electrons. It will lose electrons to form a cation. The nearest noble gas is argon with 18 electrons, so K will lose 1 electron.

(c) Al^{3+} Al is a main-group metal and has 13 electrons. It will lose electrons to form a cation. The nearest noble gas is neon with 10 electrons, so Al will lose 3 electrons.

(d) Rb^+ Rb is a main-group metal and has 37 electrons. It will lose electrons to form a cation. The nearest noble gas is krypton with 36 electrons, so Rb will lose 1 electron.

Coulomb's Law and Effective Nuclear Charge

3.65   Coulomb's law states that the potential energy (E) of two charge particles depends on their charges (q_1 and q_2) and on their separation (r). $E = \dfrac{1}{4\pi\varepsilon_o} \dfrac{q_1 q_2}{r}$. The potential energy is positive for charges of the same sign and negative for charges of opposite signs. The magnitude of the potential energy depends inversely on the separation between the charged particles.

(a)   **Given:** $q_1 = 1-$; $q_2 = 2+$; $r = 150.\,\mathrm{pm}$ **Find:** E
Conceptual Plan: Magnitude of Potential Energy depends on the charge and the separation.

$$E = \dfrac{1}{4\pi\varepsilon_o} \dfrac{q_1 q_2}{r}$$

Solution: $E = \dfrac{1}{4\pi\varepsilon_o} \dfrac{(1-)(2+)}{150} = \dfrac{-0.013}{4\pi\varepsilon_o}$

(b)   **Given:** $q_1 = 1-$; $q_2 = 1+$; $r = 150.\,\mathrm{pm}$ **Find:** E
Conceptual Plan: Magnitude of Potential Energy depends on the charge and the separation.

$$E = \dfrac{1}{4\pi\varepsilon_o} \dfrac{q_1 q_2}{r}$$

Solution: $E = \dfrac{1}{4\pi\varepsilon_o} \dfrac{(1-)(1+)}{150} = \dfrac{-0.0067}{4\pi\varepsilon_o}$

(c)   **Given:** $q_1 = 1-$; $q_2 = 3+$; $r = 150.\,\mathrm{pm}$ **Find:** E
Conceptual Plan: Magnitude of Potential Energy depends on the charge and the separation.

$$E = \dfrac{1}{4\pi\varepsilon_o} \dfrac{q_1 q_2}{r}$$

Solution: $E = \dfrac{1}{4\pi\varepsilon_o} \dfrac{(1-)(3+)}{150.} = \dfrac{-0.030}{4\pi\varepsilon_o}$

(c) is most negative and will have the lowest potential energy.

3.67   The valence electrons in nitrogen would experience a greater effective nuclear charge. Be has four protons, and N has seven protons. Both atoms have two core electrons that predominately contribute to the shielding, while the valence electrons will contribute a slight shielding effect. So Be has an effective nuclear charge of slightly more than $2+$, and N has an effective nuclear charge of slightly more than $5+$.

3.69   (a) $K(19) = [Ar]4s^1$ $Z_{\mathrm{eff}} = Z - \text{core electrons} = 19 - 18 = 1+$
(b) $Ca(20) = [Ar]4s^2$ $Z_{\mathrm{eff}} = Z - \text{core electrons} = 20 - 18 = 2+$
(c) $O(8) = [He]2s^2 2p^4$ $Z_{\mathrm{eff}} = Z - \text{core electrons} = 8 - 2 = 6+$
(d) $C(6) = [He]2s^2 2p^2$ $Z_{\mathrm{eff}} = Z - \text{core electrons} = 6 - 2 = 4+$

Atomic Radius

3.71   (a)   Al or In   In atoms are larger than Al atoms because as you trace the path between Al and In on the periodic table, you move down a column. Atomic size increases as you move down a column because the outermost electrons occupy orbitals with a higher principal quantum number that are larger, resulting in a larger atom.

        (b)   Si or N   Si atoms are larger than N atoms because as you trace the path between N and Si on the periodic table, you move down a column (atomic size increases) and then to the left across a period (atomic size increases). These effects add together for an overall increase.

        (c)   P or Pb   Pb atoms are larger than P atoms because as you trace the path between P and Pb on the periodic table, you move down a column (atomic size increases) and then to the left across a period (atomic size increases). These effects add together for an overall increase.

        (d)   C or F   C atoms are larger than F atoms because as you trace the path between C and F on the periodic table, you move to the right within the same period. As you move to the right across a period, the effective nuclear charge experienced by the outermost electrons increases, which results in a smaller size.

3.73   Ca, Rb, S, Si, Ge, F   F is above and to the right of the other elements, so you start with F as the smallest atom. As you trace a path from F to S, you move to the left (size increases) and down (size increases). Next, you move left from S to Si (size increases), then down to Ge (size increases), then to the left to Ca (size increases), and then to the left and down to Rb (size increases). So in order of increasing atomic radii, F < S < Si < Ge < Ca < Rb.

Ionic Electron Configurations, Ionic Radii, Magnetic Properties, and Ionization Energy

3.75   (a) O^{2-} Begin by writing the electron configuration of the neutral atom.
                O $1s^2 2s^2 2p^4$
                Because this ion has a 2− charge, add two electrons to write the electron configuration of the ion.
                O^{2-} $1s^2 2s^2 2p^6$ This is isoelectronic with Ar.

        (b) Br^- Begin by writing the electron configuration of the neutral atom.
                Br $[Ar]4s^2 3d^{10} 4p^5$
                Because this ion has a 1− charge, add one electron to write the electron configuration of the ion.
                Br^- $[Ar]4s^2 3d^{10} 4p^6$ This is isoelectronic with Kr.

        (c) Sr^{2+} Begin by writing the electron configuration of the neutral atom.
                Sr $[Kr]5s^2$
                Because this ion has a 2+ charge, remove two electrons to write the electron configuration of the ion.
                Sr^{2+} $[Kr]$

        (d) Co^{3+} Begin by writing the electron configuration of the neutral atom.
                Co $[Ar]4s^2 3d^7$
                Because this ion has a 3+ charge, remove three electrons to write the electron configuration of the ion. Because it is a transition metal, remove the electrons from the $4s$ orbital before removing electrons from the $3d$ orbitals.
                Co^{3+} $[Ar]4s^0 3d^6$ or $[Ar]3d^6$

        (e) Cu^{2+} Begin by writing the electron configuration of the neutral atom. Remember, Cu is one of our exceptions.
                Cu $[Ar]4s^1 3d^{10}$
                Because this ion has a 2+ charge, remove two electrons to write the electron configuration of the ion. Because it is a transition metal, remove the electrons from the $4s$ orbital before removing electrons from the $3d$ orbitals.
                Cu^{2+} $[Ar]4s^0 3d^9$ or $[Ar]3d^9$

3.77   (a) V^{5+} Begin by writing the electron configuration of the neutral atom.

V $[Ar]4s^2 3d^3$

Because this ion has a 5+ charge, remove five electrons to write the electron configuration of the ion. Because it is a transition metal, remove the electrons from the 4s orbital before removing electrons from the 3d orbitals.

V^{5+} $[Ar]4s^0 3d^0 = [Ne]3s^2 3p^6$

V^{5+} is diamagnetic.

(b) Cr^{3+} Begin by writing the electron configuration of the neutral atom. Remember, Cr is one of our exceptions.

Cr $[Ar]4s^1 3d^5$

Because this ion has a 3+ charge, remove three electrons to write the electron configuration of the ion. Because it is a transition metal, remove the electrons from the 4s orbital before removing electrons from the 3d orbitals.

Cr^{3+} $[Ar]4s^0 3d^3$

[Ar] $\boxed{}$ $\boxed{\uparrow\,|\,\uparrow\,|\,\uparrow\,||}$
                   4s   3d

Cr^{3+} is paramagnetic.

(c) Ni^{2+} Begin by writing the electron configuration of the neutral atom.

Ni $[Ar]4s^2 3d^8$

Because this ion has a 2+ charge, remove two electrons to write the electron configuration of the ion. Because it is a transition metal, remove the electrons from the 4s orbital before removing electrons from the 3d orbitals.

Ni^{2+} $[Ar]4s^2 3d^8$

[Ar] $\boxed{}$ $\boxed{\uparrow\downarrow|\uparrow\downarrow|\uparrow\downarrow|\uparrow\,|\,\uparrow}$
                   4s   3d

Ni^{2+} is paramagnetic.

(d) Fe^{3+} Begin by writing the electron configuration of the neutral atom.

Fe $[Ar]4s^2 3d^6$

Because this ion has a 3+ charge, remove three electrons to write the electron configuration of the ion. Because it is a transition metal, remove the electrons from the 4s orbital before removing electrons from the 3d orbitals.

Fe^{3+} $[Ar]4s^0 3d^5$

[Ar] $\boxed{}$ $\boxed{\uparrow\,|\,\uparrow\,|\,\uparrow\,|\,\uparrow\,|\,\uparrow}$
                   4s   3d

Fe^{3+} is paramagnetic.

3.79   (a)   Li or Li^+ A Li atom is larger than Li^+ because cations are smaller than the atoms from which they are formed.

(b) I^- or Cs^+ An I^- ion is larger than a Cs^+ ion because, although they are isoelectronic, I^- has two fewer protons than Cs^+, resulting in a lesser pull on the electrons and, therefore, a larger radius.

(c)   Cr or Cr^{3+} A Cr atom is larger than Cr^{3+} because cations are smaller than the atoms from which they are formed.

(d)   O or O^{2-} An O^{2-} ion is larger than an O atom because anions are larger than the atoms from which they are formed.

3.81   Because all of the species are isoelectronic, the radius will depend on the number of protons in each species. The fewer the protons, the larger the radius.

F: $Z = 9$; Ne: $Z = 10$; O: $Z = 8$; Mg: $Z = 12$; Na: $Z = 11$

So the radii of the ions is: $O^{2-} > F^- > Ne > Na^+ > Mg^{2+}$

3.83   (a)   Br or Bi   Br has a higher ionization energy than Bi because as you trace the path between Br and Bi on the periodic table, you move down a column (ionization energy decreases) and then to the left across a period (ionization energy decreases). These effects sum together for an overall decrease.

(b)   Na or Rb   Na has a higher ionization energy than Rb because as you trace a path between Na and Rb on the periodic table, you move down a column. Ionization energy decreases as you go down a column because of the increasing size of orbitals with increasing n.

(c)   As or At   Based on periodic trends alone, it is impossible to tell which one has a higher ionization energy because as you trace the path between As and At, you move to the right across a period (ionization energy increases) and then down a column (ionization energy decreases). These effects tend to oppose each other, and it is not obvious which one will dominate.

(d)   P or Sn   P has a higher ionization energy than Sn because as you trace the path between P and Sn on the periodic table, you move down a column (ionization energy decreases) and then to the left across a period (ionization energy decreases). These effects sum together for an overall decrease.

3.85   Because ionization energy increases as you move to the right across a period and increases as you move up a column, the element with the smallest first ionization energy would be the element farthest to the left and lowest on the periodic table. So In has the smallest ionization energy. As you trace a path to the right and up on the periodic table, the next element you reach is Si; continuing up and to the right, you reach N; and continuing to the right, you reach F. So in the order of increasing first ionization energy, the elements are In $<$ Si $<$ N $<$ F.

3.87   The jump in ionization energy occurs when you change from removing a valence electron to removing a core electron. To determine where this jump occurs, you need to look at the electron configuration of the atom.

(a)   Be $1s^2 2s^2$ The first and second ionization energies involve removing $2s$ electrons, while the third ionization energy removes a core electron; so the jump will occur between the second and third ionization energies.

(b)   N $1s^2 2s^2 2p^3$ The first five ionization energies involve removing the $2p$ and $2s$ electrons, while the sixth ionization energy removes a core electron; so the jump will occur between the fifth and sixth ionization energies.

(c)   O $1s^2 2s^2 2p^4$ The first six ionization energies involve removing the $2p$ and $2s$ electrons, while the seventh ionization energy removes a core electron; so the jump will occur between the sixth and seventh ionization energies.

(d)   Li $1s^2 2s^1$ The first ionization energy involves removing a $2s$ electron, while the second ionization energy removes a core electron; so the jump will occur between the first and second ionization energies.

Electron Affinities and Metallic Character

3.89   (a)   Na or Rb   Na has a more negative electron affinity than Rb. In column 1A, electron affinity becomes less negative as you go down the column.

(b)   B or S   S has a more negative electron affinity than B. As you trace from B to S in the periodic table, you move to the right, which shows the value of the electron affinity becoming more negative. Also, as you move from period 2 to period 3, the value of the electron affinity becomes more negative. Both of these trends sum together for the value of the electron affinity to become more negative.

(c)   C or N   C has the more negative electron affinity. As you trace from C to N across the periodic table, you normally expect N to have the more negative electron affinity. However, N has a half-filled p sublevel, which lends it extra stability; therefore, it is harder to add an electron.

(d)   Li or F   F has the more negative electron affinity. As you trace from Li to F on the periodic table, you move to the right in the period. As you move to the right across a period, the value of the electron affinity generally becomes more negative.

3.91   (a)   Sr or Sb   Sr is more metallic than Sb because as we trace the path between Sr and Sb on the periodic table, we move to the right within the same period. Metallic character decreases as we move to the right.

(b)   As or Bi   Bi is more metallic because as we trace a path between As and Bi on the periodic table, we move down a column in the same family (metallic character increases).

(c)   Cl or O   Based on periodic trends alone, we cannot tell which is more metallic because as we trace the path between O and Cl, we move to the right across a period (metallic character decreases) and then down a column (metallic character increases). These effects tend to oppose each other, and it is not easy to tell which will predominate.

   (d)   S or As   As is more metallic than S because as we trace the path between S and As on the periodic table, we move down a column (metallic character increases) and then to the left across a period (metallic character increases). These effects add together for an overall increase.

3.93   The order of increasing metallic character is S $<$ Se $<$ Sb $<$ In $<$ Ba $<$ Fr.

Metallic character decreases as you move left to right across a period and decreases as you move up a column; therefore, the element with the least metallic character will be to the top right of the periodic table. So of these elements, S has the least metallic character. As you move down the column, the next element is Se. As you continue down and to the left, you reach Sb; continuing to the left, you reach In; moving down the column and to the left, you come to Ba; and down the column and to the left is Fr.

Cumulative Problems

3.95   Br: $1s^2 2s^2 2p^6 3s^2 3p^6 4s^2 3d^{10} 4p^5$
         Kr: $1s^2 2s^2 2p^6 3s^2 3p^6 4s^2 3d^{10} 4p^6$
         Krypton has a completely filled p sublevel, giving it chemical stability. Bromine needs one electron to achieve a completely filled p sublevel and thus has a highly negative electron affinity. Therefore, it easily takes on an electron and is reduced to the bromide ion, giving it the added stability of the filled p sublevel.

3.97   Write the electron configuration of vanadium.
         V: $[Ar]4s^2 3d^3$
         Because this ion has a 3+ charge, remove three electrons to write the electron configuration of the ion. Because it is a transition metal, remove the electrons from the $4s$ orbital before removing electrons from the $3d$ orbitals.
         V^{3+}: $[Ar]4s^0 3d^2$ or $[Ar]3d^2$
         Both vanadium and the V^{3+} ion have unpaired electrons and are paramagnetic.

3.99   Because K^+ has a 1+ charge, we would need a cation with a similar size and a 1+ charge. Looking at the ions in the same family, we see that Na^+ would be too small and Rb^+ would be too large. If we consider Ar^+ and Ca^+, we would have ions of similar size and charge. Between these two, Ca^+ would be easier to achieve because the first ionization energy of Ca is similar to that of K, while the first ionization energy of Ar is much larger. However, the second ionization energy of Ca is relatively low, making it easy to lose the second electron.

3.101   C has an outer shell electron configuration of $ns^2 np^2$; we would, therefore, expect Si and Ge, which are in the same family, to be most like carbon. Ionization energies for both Si and Ge are similar and tend to be slightly lower than that of C, but all are intermediate in the range of first ionization energies. The electron affinities of Si and Ge are close to that of C.

3.103   (a)   N: $[He]2s^2 2p^3$ Mg: $[Ne]3s^2$ O: $[He]2s^2 2p^4$
                 F: $[He]2s^2 2p^5$ Al: $[Ne]3s^2 3p^1$
         (b)   Mg $>$ Al $>$ N $>$ O $>$ F
         (c)   Al $<$ Mg $<$ O $<$ N $<$ F   (from the table)
         (d)   Mg and Al would have the largest radius because they are in period $n = 3$; Al is smaller than Mg because radius decreases as you move to the right across the period. F is smaller than O and O is smaller than N because as you move to the right across the period, radius decreases.

                 The first ionization energy of Al is smaller than the first ionization energy of Mg because Al loses the electron from the $3p$ orbital, which is shielded by the electrons in the $3s$ orbital. Mg loses the electron from the filled $3s$ orbital, which has added stability because it is a filled orbital. The first ionization energy of O is lower than the first ionization energy of N because N has a half-filled $2p$ orbital, which adds extra stability, thus making it harder to remove the electron. The fourth electron in the O $2p$ orbitals experiences added electron-electron repulsion because it must pair with another electron in the same $2p$ orbital, thus making it easier to remove.

3.105   As you move to the right across a row in the periodic table for the main-group elements, the effective nuclear charge (Z_{eff}) experienced by the electrons in the outermost principal energy level increases, resulting in a stronger attraction between the outermost electrons and the nucleus and, therefore, a smaller atomic radii.

Across the row of transition elements, the number of electrons in the outermost principal energy level (highest n value) is nearly constant. As another proton is added to the nucleus with each successive element, another electron is added, but that electron goes into an $n_{highest} - 1$ orbital (a core level). The number of outermost electrons stays constant, and they experience a roughly constant effective nuclear charge, keeping the radius approximately constant after the first couple of elements in the series.

3.107   All of the noble gases have a filled outer quantum level, very high first ionization energies, and positive values for the electron affinity; thus, the noble gases are particularly unreactive. The lighter noble gases will not form any compounds because the ionization energies of both He and Ne are over 2000 kJ/mol. Because ionization energy decreases as you move down a column, you find that the heavier noble gases (Ar, Kr, and Xe) do form some compounds. They have ionization energies that are close to the ionization energy of H and thus can be forced to lose an electron.

3.109   Group 6A: ns^2np^4 Group 7A: ns^2np^5
The electron affinity of the group 7A elements is more negative than that of the group 6A elements in the same period because group 7A requires only one electron to achieve the noble gas configuration ns^2np^6, while the group 6A elements require two electrons. Adding one electron to the group 6A element will not give them any added stability and leads to extra electron-electron repulsions; so the value of the electron affinity is less negative than that for group 7A.

3.111 $Br(35) = [Ar]4s^23d^{10}4p^5$ $I(53) = [Kr]5s^24d^{10}5p^5$
Br and I are both halogens with an outermost electron configuration of ns^2np^5; the next element with the same outermost electron configuration is 85, At.

3.113   (a) $10 - 2 = 8 \rightarrow O$; $12 - 2 = 10 \rightarrow Ne$; $58 - 5 = 53 \rightarrow I$; $11 - 2 = 9 \rightarrow F$; $7 - 2 = 5 \rightarrow B$; $44 - 5 = 39$
                   $\rightarrow Y$; $63 - 6 = 57 \rightarrow La$; $66 - 6 = 60 \rightarrow Nd$ One If by Land
          (b) $9 - 2 = 7 \rightarrow N$; $99 - 7 = 92 \rightarrow U$; $30 - 4 = 26 \rightarrow Fe$; $95 - 7 = 88 \rightarrow Ra$; $19 - 3 = 16 \rightarrow S$; $47 - 5$
                   $= 42 \rightarrow Mo$; $79 - 6 = 73 \rightarrow Ta$ (backwards) Atoms Are Fun

3.115   **Given:** $r = 100.00$ pm, $q_{proton} = 1.60218 \times 10^{-19}$ C, and $q_{electron} = -1.60218 \times 10^{-19}$ C
          Find: IE in kJ/mol and λ of ionization
          Conceptual Plan: $r, q_{proton}, q_{electron}, \rightarrow E_{atom} \rightarrow E_{mol}$ and then $E_{atom} \rightarrow \lambda$

$$E = \frac{1}{4\pi_0} \frac{q_p q_e}{r} \qquad \frac{1 \text{ kJ}}{(1000 \text{ J})} \qquad \frac{6.022 \times 10^{23} \text{ atoms}}{1 \text{ mol}} \qquad\qquad \lambda = \frac{hc}{E}$$

Solution:

$$E = \frac{1}{(4)(3.141)\left(8.85 \times 10^{-12} \frac{C^2}{J \cdot m}\right)} \times \frac{(1.60218 \times 10^{-19} \, C)(-1.60218 \times 10^{-19} \, C)}{(100.00 \text{ pm})\left(\frac{1 \text{ m}}{1 \times 10^{12} \text{ pm}}\right)} = -2.3089 \times 10^{-18} \text{ J/atom}$$

$$-2.3\underline{0}9 \times 10^{-18} \text{ J/atom} \times \frac{6.022 \times 10^{23} \text{ atoms}}{1 \text{ mol}} \times \frac{1 \text{ kJ}}{(1000 \text{ J})} = -1.39 \times 10^3 \text{ kJ/mol}$$

$$\text{IE} = 0 - (-1.39 \times 10^3 \text{ kJ/mol}) = 1.39 \times 10^3 \text{ kJ/mol}$$

$$\lambda = \frac{(6.626 \times 10^{-34} \text{ J} \cdot s)(3.00 \times 10^8 \text{ m} \cdot s)\left(\frac{1 \times 10^9 \text{ nm}}{1 \text{ m}}\right)}{(2.3\underline{0}9 \times 10^{-18} \text{ J})} = 86.1 \text{ nm}$$

Check: The units of the answer (kJ/mol) are correct. The magnitude of the answer is reasonable because the value is positive and energy must be added to the atom to remove the electron. The units of the wavelength (nm) are correct and the magnitude is reasonable based on the ionization energy.

Challenge Problems

3.117   (a)   Using Excel, make a table of radius, atomic number, and density. Using xy scatter, make a chart of radius versus density. With an exponential trendline, estimate the density of argon and xenon. Also make a chart of atomic number versus density. With a linear trendline, estimate the density of argon and xenon.

Element	Atomic Radius (pm)	Atomic Number	Density (g/mL)
He	32	2	0.18
Ne	70	10	0.90
Ar	98	18	–
Kr	112	36	3.75
Xe	130	54	–
Rn	–	86	9.73

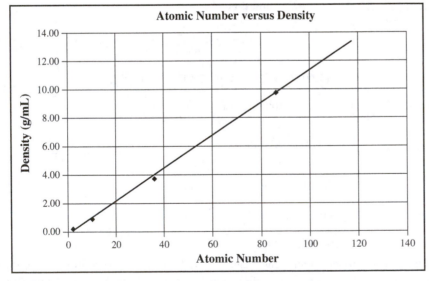

From the radius versus density chart, Ar has a density of ~2 g/L and Xe has a density of ~7.7 g/L. From the atomic number versus density chart, Ar has a density of ~1.8 g/L and Xe has a density of ~6 g/L.

(b)   Using the chart of atomic number versus density, we could predict element 118 to have a density of ~13 g/L.

(c)   **Given:** Ne: $M = 20.18$ g/mol; $r = 70.$ pm   **Find:** mass of neon; d neon

Conceptual Plan: $M \rightarrow m_{\text{atom}}$ and then $r \rightarrow \text{vol}_{\text{atom}}$ and then $\rightarrow d$

$$\frac{1 \text{ mol}}{6.022 \times 10^{23} \text{ atoms}} \qquad V = \frac{4}{3}\pi r^3 \qquad d = \frac{m}{V}$$

Solution: $\dfrac{20.18 \text{ g}}{\text{mol}} \times \dfrac{1 \text{ mol}}{6.022 \times 10^{23} \text{ atoms}} = 3.35 \times 10^{-23}$ g/atom

$$V = \frac{4}{3} \times 3.14 \times (70 \text{ pm})^3 \times \left(\frac{1 \text{ m}}{1 \times 10^{12} \text{ pm}}\right)^3 \times \frac{1 \text{ L}}{0.0010 \text{ m}^3} = 1.44 \times 10^{-27} \text{ L}$$

$$d = \frac{3.35 \times 10^{-23} \text{ g}}{1.44 \times 10^{-27} \text{ L}} = 2.\underline{33} \times 10^4 = 2.3 \times 10^4 \text{ g/L}$$

Check: The units of the answer (g/L) are correct. This density is significantly larger than the actual density of neon gas. This suggests that a liter of neon is composed primarily of empty space.

(d) **Given:** Ne: $M = 20.18$ g/mol, $d = 0.90$ g/L; Kr: $M = 83.30$ g/mol, $d = 3.75$ g/L; Ar: $M = 39.95$ g/mol
Find: d of argon in g/L
Conceptual Plan: $d \rightarrow$ mol/L $\rightarrow$ atoms/L for Kr and Ne and then atoms/L $\rightarrow$ mol/L $\rightarrow d$ for Ar

$$\text{mol} = \frac{\text{mass}}{\text{molar mass}} \qquad \frac{6.022 \times 10^{23} \text{ atoms}}{1 \text{ mol}} \qquad\qquad \frac{1 \text{ mol}}{6.022 \times 10^{23} \text{ atoms}} \qquad \frac{39.95 \text{ g}}{\text{mol}}$$

Solution: for Ne: $\dfrac{0.90 \text{ g}}{\text{L}} \times \dfrac{\text{mol}}{20.18 \text{ g}} \times \dfrac{6.022 \times 10^{23} \text{ atoms}}{1 \text{ mol}} = 2.7 \times 10^{22}$ atoms/L

for Kr: $\dfrac{3.75 \text{ g}}{\text{L}} \times \dfrac{\text{mol}}{83.80 \text{ g}} \times \dfrac{6.022 \times 10^{23} \text{ atoms}}{1 \text{ mol}} = 2.69 \times 10^{22}$ atoms/L

for Ar: $\dfrac{2.69 \times 10^{22} \text{ atoms}}{\text{L}} \times \dfrac{1 \text{ mol}}{6.022 \times 10^{23} \text{ atoms}} \times \dfrac{39.95 \text{ g}}{\text{mol}} = 1.78$ g/L

This value is similar to the value calculated in part a. The value of the density calculated from the radius was 2 g/L, and the value of the density calculated from the atomic number was 1.8 g/L.

Check: The units of the answer (g/L) are correct. The value of the answer agrees with the published value.

3.119 The density increases as you move to the right across the first transition series. For the first transition series, the mass increases as you move to the right across the periodic table. However, the radius of the transition series elements stays nearly constant as you move to the right across the periodic table; thus, the volume will remain nearly constant. Because density is mass/volume, the density of the elements increases.

3.121 The longest wavelength would be associated with the lowest energy state next to the ground state of carbon, which has two unpaired electrons.
Ground state of carbon:

1s	2s	2p
↑↓	↑↓	↑ ↑ _

Longest wavelength: One of the p electrons flipped in its orbital, which requires the least amount of energy.

1s	2s	2p
↑↓	↑↓	↑ ↓ _

The next wavelength would be associated with the pairing of the two p electrons in the same orbital because this requires energy and raises the energy.

1s	2s	2p
↑↓	↑↓	↑↓ _ _

The next wavelength would be associated with the energy needed to promote one of the s electrons to a p orbital.

1s	2s	2p
↑↓	↑	↑ ↑ ↑

3.123 The element that would fill the 8s and 8p orbitals would have atomic number 168. The element is in the noble gas family and would have the properties of noble gases. It would have the electron configuration of $[118]8s^2 5g^{18} 6f^{14} 7d^{10} 8p^6$. The outer shell electron (highest n level) configuration would be $8s^2 8p^6$. The element would be relatively inert, have a first ionization energy less than 1037 kJ/mol (the first ionization energy of Rn), and have a positive electron affinity. It would be difficult to form compounds with most elements, but it would be possible to form compounds with fluorine.

3.125   When you move down the column from Al to Ga, the size of the atom actually decreases because the $3d$ electrons in the Ga atom do not contribute much shielding, while there is a large increase in the nuclear charge. Therefore, the effective nuclear charge is greater for Ga than for Al; so the ionization energy does not decrease. As you go from In to Tl, the ionization energy actually increases because the $4f$ electrons do not contribute to the shielding of the outermost electrons and there is a large increase in the effective nuclear charge.

3.127   The second electron is added to an ion with a $1-$ charge, so a large repulsive force must be overcome to add the second electron. Thus, it will require energy to add the second electron, and the second electron affinity will have a positive value.

3.129   **Given:** Ra, $Z = 88$ **Find:** Z for next two alkaline earth metals
         Solution: The next element would lie in period 8, column 2A. The largest currently known element is 116 in period 7, column 6A. To reach period 8, column 2A, you need to add four protons and would have $Z = 120$.

         The alkaline earth metal following 120 would lie in period 9, column 2A. To reach this column, you need to add 18 g-block element protons, 10 d-block element protons, 14 f-block element protons, 6 p-block element protons, and 2 s-block element protons. This would give $Z = 170$.

Conceptual Problems

3.131   If six electrons rather than eight electrons led to a stable configuration, the electron configuration of the stable configuration would be ns^2np^4.
   (a)   A noble gas would have the electron configuration ns^2np^4. This could correspond to the O atom.
   (b)   A reactive nonmetal would have one less electron than the stable configuration. This would have the electron configuration ns^2np^3. This could correspond to the N atom.
   (c)   A reactive metal would have one more electron than the stable configuration. This would have the electron configuration of ns^1. This could correspond to the Li atom.

3.133   (a)   True: An electron in a $3s$ orbital is more shielded than an electron in a $2s$ orbital. This is true because there are more core electrons below a $3s$ orbital.
   (b)   True: An electron in a $3s$ orbital penetrates the region occupied by the core electrons more than electrons in a $3p$ orbital. Examine Figure 4.8, the radial distribution functions for the $3s$, $3p$, and $3d$ orbitals. You will see that the $3s$ electrons penetrate more deeply than the $3p$ electrons and more than the $3d$ electrons.
   (c)   False: An electron in an orbital that penetrates closer to the nucleus will experience *less* shielding than an electron in an orbital that does not penetrate as far.
   (d)   True: An electron in an orbital that penetrates close to the nucleus will tend to experience a higher effective nuclear charge than one that does not. Because the orbital penetrates closer to the nucleus, the electron will experience less shielding and therefore a higher effective nuclear charge.

3.135   The $4s$ electrons in calcium have relatively low ionization energies ($IE_1 = 590$ kJ/mol; $IE_2 = 1145$ kJ/mol) because they are valence electrons. The energetic cost for calcium to lose a third electron is extraordinarily high because the next electron to be lost is a core electron. Similarly, the electron affinity of fluorine to gain one electron (-328 kJ/mol) is highly exothermic because the added electron completes fluoride's valence shell. The gain of a second electron by the negatively charged fluoride anion would not be favorable. Therefore, we would expect calcium and fluoride to combine in a 1:2 ratio.

Questions for Group Work

3.137   The orbitals fill in order of increasing energy of the orbitals, which is $1s<2s<2p<3s<3p<4s<3d<4p<5s<4d<5p<6s$. Keep in mind that s subshells can contain two electrons, p subshells can contain six electrons, and d subshells can contain ten electrons.

3.139   (a)   The Br^- and Se^{2-} ions are about the same size because the ions are an isoelectronic pair.
   (b)   The Br^- ion is smaller than the Se^{2-} ion because the Br^- ion has a larger nuclear charge and fewer extra electrons than the Se^{2-} ion.

    (c)   The Fr^+ (194 pm) ion is the singly charged cation that is closest in size to the Br^- (195 pm) and Se^{2-} (198 pm) ions because cations are much smaller than their corresponding atoms.

Data Interpretation and Analysis

3.141   (a)   The ionization energy generally increases as you move from left to right across the periodic table. While the number of protons increases, moving left to right, the electrons enter the same shell, Coulomb's force law dictates that the distance between the protons in the nucleus and the outermost electrons will decrease, and the force of attraction will increase. The outer electrons are closer to the nucleus, requiring more energy to eject the outermost electrons.

    (c)   The electron affinities generally become more negative (i.e., more exothermic) because as an electron is added the electron configuration becomes or approaches the electron configuration of a noble gas. Ar has a positive electron affinity because the valence shell is full.

    (e)   The energy to remove an electron from Na is 496 kJ/mol, and the energy to add an electron to Cl is −349 kJ/mol. The net energy when combining these two processes is 496 kJ/mol − 349 kJ/mol = 147 kJ/mol, making this an endothermic process. This result is consistent with the results for the Born-Haber Cycle for NaCl, which is discussed in Figure 9.12.

4 Molecules and Compounds

Review Questions

4.1   The properties of compounds are generally very different from the properties of the elements that compose them. When two elements combine to form a compound, an entirely new substance results.

4.3   Ionic bonds, which occur between metals and nonmetals, involve the transfer of electrons from one atom to another. Covalent bonds, which occur between two or more nonmetals, involve the sharing of electrons between two atoms.

4.5   An empirical formula gives the relative number of atoms of each element in a compound.

       A molecular formula gives the actual number of atoms of each element in a molecule of a compound.

4.7   Bonds are formed when atoms attain a stable electron configuration. Because the stable configuration usually has eight electrons in the outermost shell, this is known as the octet rule.

4.9   To use Lewis theory to determine the formula of an ionic compound, determine the number of valence electrons lost by the metal to form an octet and the number of electrons gained by the nonmetal to form an octet. Consider the ionic compound formed between sodium and sulfur. The Lewis structures for sodium and sulfur are as follows:

$$Na\cdot \quad \cdot \ddot{\underset{\cdot\cdot}{S}}:$$

       Sodium must lose one electron to achieve an octet, while sulfur must gain two electrons to achieve an octet. So the compound requires two sodium atoms for each sulfur atom, giving the formula Na_2S.

4.11   The formation of the crystalline NaCl lattice from sodium cations and chloride anions is highly exothermic and more than compensates for the endothermicity of the electron transfer process. In other words, the formation of ionic compounds is not exothermic because sodium "wants" to lose electrons and chlorine "wants" to gain them; rather, it is exothermic because of the large amount of heat released when sodium and chlorine ions coalesce to form a crystal lattice.

4.13   Binary ionic compounds are named using the name of the cation (metal) and the base name of the anion (nonmetal) + the suffix -ide. Ionic compounds that contain a polyatomic anion are named using the name of the cation (metal) and the name of the polyatomic anion.

4.15   A pair of electrons that is shared between two atoms is called a bonding pair, while a pair of electrons that is associated with only one atom—and therefore, not involved in bonding—is called a lone pair.

4.17   Generally, combinations of atoms that can satisfy the octet rule on each atom are stable, while those combinations that do not satisfy the octet rule are not stable.

4.19   To name a binary molecular inorganic compound, list the name of the first element with a prefix to indicate the number of atoms in the compound if there is more than one, followed by the base name of the second element with a prefix to indicate the number of atoms in the compound if there is more than one, followed by the suffix -ide.

4.21 The formula mass is the average mass of the molecule (or formula unit) of a compound. The formula mass allows the conversion between the mass of molecules and the number of molecules present.

4.23 Mass percent composition is the mass of an element of the compound divided by the total mass of the compound times 100. Mass percent composition is used as a conversion factor between the mass of the element and the mass of the compound.

4.25 The experimental data showing the relative masses of the elements in a compound can be used to obtain an empirical formula.

4.27 In combustion analysis, the unknown compound undergoes combustion (burning) in the presence of pure oxygen. All of the carbon in the sample is converted to CO_2, and all of the hydrogen is converted to H_2O. Combustion analysis can be used to determine the empirical formula of a hydrocarbon.

Problems by Topic

Types of Compounds and Chemical Formulas

4.29 (a) CO_2 is a compound composed of a nonmetal and a nonmetal; therefore, it is a molecular compound.
   (b) $NiCl_2$ is a compound composed of a metal and a nonmetal; therefore, it is an ionic compound.
   (c) NaI is a compound composed of a metal and a nonmetal; therefore, it is an ionic compound.
   (d) PCl_3 is a compound composed of a nonmetal and a nonmetal; therefore, it is a molecular compound.

4.31 To determine the empirical formula from a molecular formula, divide the subscripts by the greatest common factor (the largest number that divides exactly into all of the subscripts).
   (a) For N_2O_4, the greatest common factor is 2. The empirical formula is therefore NO_2.
   (b) For C_5H_{12}, the only common factor is 1, so the empirical formula and the molecular formula are the same.
   (c) For C_4H_{10}, the greatest common factor is 2. The empirical formula is therefore C_2H_5.

4.33 The chemical formula gives you the kind of atom and the number of each atom in the compound.
   (a) $Mg_3(PO_4)_2$ contains: 3 magnesium atoms, 2 phosphorus atoms, and 8 oxygen atoms
   (b) $BaCl_2$ contains: 1 barium atom and 2 chlorine atoms
   (c) $Fe(NO_2)_2$ contains: 1 iron atom, 2 nitrogen atoms, and 4 oxygen atoms
   (d) $Ca(OH)_2$ contains: 1 calcium atom, 2 oxygen atoms, and 2 hydrogen atoms

4.35 (a) 1 blue = nitrogen, 3 white = hydrogen: NH_3
   (b) 2 black = carbon, 6 white = hydrogen: C_2H_6
   (c) 1 yellow = sulfur, 3 red = oxygen: SO_3

Valence Electrons and Lewis Dot Structures

4.37 N: $1s^2 2s^2 2p^3$ $\cdot \ddot{N} :$ The electrons included in the Lewis structure are $2s^2 2p^3$.

4.39 (a) Al: $1s^2 2s^2 2p^6 3s^2 3p^1$
$\cdot \dot{Al} \cdot$
   (b) Na^+: $1s^2 2s^2 2p^6$
Na^+
   (c) Cl: $1s^2 2s^2 2p^6 3s^2 3p^5$
$: \ddot{Cl} \cdot$
   (d) Cl^-: $1s^2 2s^2 2p^6 3s^2 3p^6$
$\left[: \ddot{\ddot{Cl}} : \right]^-$

Ionic Bonding and Lattice Energy

4.41   (a)   NaF: Draw the Lewis structures for Na and F based on their valence electrons.

Na: $3s^1$ F: $2s^22p^5$

Na· :F̈·

Sodium must lose one electron and be left with the octet from the previous shell, while fluorine needs to gain one electron to get an octet.

Na^+ $[:\ddot{F}:]^-$

(b)   CaO: Draw the Lewis structures for Ca and O based on their valence electrons. Ca: $4s^2$ O: $2s^22p^4$

Ca: :Ö·

Calcium must lose two electrons and be left with two $1s$ electrons from the previous shell, while oxygen needs to gain two electrons to get an octet.

Ca^{2+} $[:\ddot{O}:]^{2-}$

(c)   SrBr$_2$:   Draw the Lewis structures for Sr and Br based on their valence electrons. Sr: $5s^2$ Br: $4s^24p^5$

Sr: :B̈r·

Strontium must lose two electrons and be left with the octet from the previous shell, while bromine needs to gain one electron to get an octet.

Sr^{2+} $2[:\ddot{B}r:]^-$

(d)   K$_2$O: Draw the Lewis structures for K and O based on their valence electrons. K: $4s^1$ O: $2s^22p^4$

K· :Ö·

Potassium must lose one electron and be left with the octet from the previous shell, while oxygen needs to gain two electrons to get an octet.

$2K^+$ $[:\ddot{O}:]^{2-}$

4.43   (a)   Sr and Se: Draw the Lewis structures for Sr and Se based on their valence electrons.

Sr: $5s^2$ Se: $4s^24p^4$

Sr: :S̈e·

Strontium must lose two electrons and be left with the octet from the previous shell, while selenium needs to gain two electrons to get an octet.

Sr^{2+} $[:\ddot{S}e:]^{2-}$

Thus, we need one Sr^{2+} and one Sr^{2-}. Write the formula with subscripts (if necessary) to indicate the number of atoms.

SrSe

(b)   Ba and Cl: Draw the Lewis structures for Ba and Cl based on their valence electrons.

Ba: $6s^2$ Cl: $3s^23p^5$

Ba: :C̈l·

Barium must lose two electrons and be left with the octet from the previous shell, while chlorine needs to gain one electron to get an octet.

Ba^{2+} $2[:\ddot{C}l:]^-$

Thus, we need one Ba^{2+} and two Cl^-. Write the formula with subscripts (if necessary) to indicate the number of atoms.

BaCl$_2$

(c) Na and S: Draw the Lewis structures for Na and S based on their valence electrons.

Na: $3s^1$ S: $3s^23p^4$

Na· $:\!\ddot{\text{S}}\!\cdot$

Sodium must lose one electron and be left with the octet from the previous shell, while sulfur needs to gain two electrons to get an octet.

$$2\,\text{Na}^+ \quad \left[:\!\ddot{\text{S}}\!:\right]^{2-}$$

Thus, we need two Na^+ and one S^{2-}. Write the formula with subscripts (if necessary) to indicate the number of atoms.

Na_2S

(d) Al and O: Draw the Lewis structures for Al and O based on their valence electrons.

Al: $3s^23p^1$ O: $2s^22p^4$

$\ddot{\text{Al}}\!\cdot$ $:\!\ddot{\text{O}}\!\cdot$

Aluminum must lose three electrons and be left with the octet from the previous shell, while oxygen needs to gain two electrons to get an octet.

$$2\,\text{Al}^{3+} \quad 3\left[:\!\ddot{\text{O}}\!:\right]^{2-}$$

Thus, we need two Al^{3+} and three O^{2-} to lose and gain the same number of electrons. Write the formula with subscripts (if necessary) to indicate the number of atoms.

Al_2O_3

4.45 Cesium is slightly larger than barium, but oxygen is slightly larger than fluorine; so we cannot use size to explain the difference in the lattice energy. However, the charge on the cesium ion is $1+$ and the charge on the fluoride ion is $1-$, while the charge on the barium ion is $2+$ and the charge on the oxide ion is $2-$. The coulombic equation states that the magnitude of the potential also depends on the product of the charges. Because the product of the charges for CsF $= 1-$ and the product of the charges for BaO $= 4-$, the stabilization for BaO relative to CsF should be about four times greater, which is what we see in its more exothermic lattice energy.

Formulas and Names for Ionic Compounds

4.47 To write the formula for an ionic compound, do the following: (1) Write the symbol for the metal cation and its charge and the symbol for the nonmetal anion and its charge. (2) Adjust the subscript on each cation and anion to balance the overall charge. (3) Check that the sum of the charges of the cations equals the sum of the charges of the anions.

(a) calcium and oxygen: Ca^{2-} O^{2-} CaO   cations $2+$, anions $2-$
(b) zinc and sulfur: Zn^{2+} S^{2-} ZnS   cations $2+$, anions $2-$
(c) rubidium and bromine: Rb^+ Br^- RbBr   cation $+$, anions $-$
(d) aluminum and oxygen: Al^{3+} O^{2-} Al_2O_3 cation $2(3+) = 6+$, anions $3(2-) = 6-$

4.49 To write the formula for an ionic compound, do the following: (1) Write the symbol for the metal cation and its charge and the symbol for the polyatomic anion and its charge. (2) Adjust the subscript on each cation and anion to balance the overall charge. (3) Check that the sum of the charges of the cations equals the sum of the charges of the anions.

Cation $=$ calcium: Ca^{2+}

(a) hydroxide: OH^- Ca(OH)_2 cation $2+$, anion $2(1-) = 2-$
(b) chromate: CrO_4^{2-} CaCrO_4 cation $2+$, anion $2-$
(c) phosphate: PO_4^{3-} $\text{Ca}_3(\text{PO}_4)_2$ cation $3(2+) = 6+$, anion $2(3-) = 6-$
(d) cyanide: CN^- Ca(CN)_2 cation $2+$, anion $2(1-) = 2-$

4.51 To name a binary ionic compound, name the metal cation followed by the base name of the anion $+$ -*ide*.
(a) Mg_3N_2: The cation is magnesium; the anion is from nitrogen, which becomes nitride: magnesium nitride.
(b) KF: The cation is potassium; the anion is from fluorine, which becomes fluoride: potassium fluoride.
(c) Na_2O: The cation is sodium; the anion is from oxygen, which becomes oxide: sodium oxide.
(d) Li_2S: The cation is lithium; the anion is from sulfur, which becomes sulfide: lithium sulfide.

     (e)   CsF: The cation is cesium; the anion is from fluorine, which becomes fluoride: cesium fluoride.

     (f)   KI: The cation is potassium; the anion is from iodine, which becomes iodide: potassium iodide.

4.53   To name these compounds, you must first decide whether the metal cation is invariant or can have more than one charge. Then, name the metal cation followed by the base name of the anion + -ide.

     (a)   SnO: Sn can have more than one charge. The charge on Sn must be 2+ for the compound to be charge neutral: The cation is tin(II); the anion is from oxygen, which becomes oxide: tin(II) oxide.

     (b) Cr_2S_3:   Cr can have more than one charge. The charge on Cr must be 3+ for the compound to be charge neutral: The cation is chromium(III); the anion is from sulfur, which becomes sulfide: chromium(III) sulfide.

     (c)   RbI: Rb is invariant: The cation is rubidium; the anion is from iodine, which becomes iodide: rubidium iodide.

     (d) $BaBr_2$:   Ba is invariant: The cation is barium; the anion is from bromine, which becomes bromide: barium bromide.

4.55   To name these compounds, you must first decide whether the metal cation is invariant or can have more than one charge. Then, name the metal cation followed by the name of the polyatomic anion.

     (a) $CuNO_2$:   Cu can have more than one charge. The charge on Cu must be 1+ for the compound to be charge neutral: The cation is copper(I); the anion is nitrite: copper(I) nitrite.

     (b) $Mg(C_2H_3O_2)_2$:   Mg is invariant: The cation is magnesium; the anion is acetate: magnesium acetate.

     (c) $Ba(NO_3)_2$:   Ba is invariant: The cation is barium; the anion is nitrate: barium nitrate.

     (d) $Pb(C_2H_3O_2)_2$:   Pb can have more than one charge. The charge on Pb must be 2+ for the compound to be charge neutral: The cation is lead(II); the anion is acetate: lead(II) acetate.

4.57   To write the formula for an ionic compound, do the following: (1) Write the symbol for the metal cation and its charge and the symbol for the nonmetal anion or polyatomic anion and its charge. (2) Adjust the subscript on each cation and anion to balance the overall charge. (3) Check that the sum of the charges of the cations equals the sum of the charges of the anions.

     (a)   sodium hydrogen sulfite: Na^+ HSO^{3-} $NaHSO_3$ cation 1+, anion 1−

     (b)   lithium permanganate: Li^+ MnO_4^- $LiMnO_4$ cation 1+, anion 1−

     (c)   silver nitrate: Ag^+ NO_3^- $AgNO_3$ cation 1+, anion 1−

     (d)   potassium sulfate: K^+ SO_4^{2-} K_2SO_4 cation 2(1+) = 2+, anion 2−

     (e)   rubidium hydrogen sulfate: Rb^+ HSO_4^- $RbHSO_4$ cation 1+, anion 1−

     (f)   potassium hydrogen carbonate: K^+ HCO_3^- $KHCO_3$ cation 1+, anion 1−

4.59   Hydrates are named the same way as other ionic compounds with the addition of the term *prefix*hydrate, where the prefix is the number of water molecules associated with each formula unit.

     (a) $CoSO_4 \cdot 7H_2O$ cobalt(II) sulfate heptahydrate

     (b)   iridium(III) bromide tetrahydrate $IrBr_3 \cdot 4H_2O$

     (c) $Mg(BrO_3)_2 \cdot 6H_2O$ magnesium bromate hexahydrate

     (d)   potassium carbonate dihydrate $K_2CO_3 \cdot 2H_2O$

Simple Lewis Structures, Formulas, and Names for Molecular Compounds

4.61   (a)   Hydrogen: Write the Lewis structure of each atom based on the number of valence electrons.

               H· ·H

          When the two hydrogen atoms share their electrons, they each get a duet, which is a stable configuration for hydrogen.

               H—H

     (b)   The halogens: Write the Lewis structure of each atom based on the number of valence electrons.

               :Ẍ· ·Ẍ:

          If the two halogens pair together, each can achieve an octet, which is a stable configuration. So the halogens are predicted to exist as diatomic molecules.

               :Ẍ—Ẍ:

(c)   Oxygen: Write the Lewis structure of each atom based on the number of valence electrons.

$$:\ddot{O}\cdot \quad \cdot\ddot{O}:$$

To achieve a stable octet on each oxygen, the oxygen atoms will need to share two electron pairs. So oxygen is predicted to exist as a diatomic molecule with a double bond.

$$:\ddot{O}{=}\ddot{O}:$$

(d)   Nitrogen: Write the Lewis structure of each atom based on the number of valence electrons.

$$\cdot\ddot{N}\cdot \quad \cdot\ddot{N}\cdot$$

To achieve a stable octet on each nitrogen, the nitrogen atoms will need to share three electron pairs. So nitrogen is predicted to exist as a diatomic molecule with a triple bond.

$$\ddot{N}{\equiv}\ddot{N}$$

4.63   (a)   CO   The name of the compound is the name of the first element, *carbon*, followed by the base name of the second element, *ox*, prefixed by *mono-* to indicate 1 and given the suffix *-ide*. Because the prefix ends with "o" and the base name begins with "o," the first "o" is dropped: carbon monoxide.

          (b) NI_3 The name of the compound is the name of the first element, *nitrogen*, followed by the base name of the second element, *iod*, prefixed by *tri-* to indicate 3 and given the suffix *-ide*: nitrogen triiodide.

          (c) $SiCl_4$ The name of the compound is the name of the first element, *silicon*, followed by the base name of the second element, *chlor*, prefixed by *tetra-* to indicate 4 and given the suffix *-ide*: silicon tetrachloride.

          (d) N_4Se_4 The name of the compound is the name of the first element, *nitrogen*, prefixed by *tetra-* to indicate 4 followed by the base name of the second element, *selen*, prefixed by *tetra-* to indicate 4 and given the suffix *-ide*: tetranitrogen tetraselenide.

4.65   (a)   phosphorus trichloride: The name of the compound begins with the name of the first element, *phosphorous (P)*, followed by the base name of the second element, *chlor*, prefixed by *tri-* to indicate 3 chlorine (Cl) atoms, so the formula is PCl_3.

          (b)   chlorine monoxide: The name of the compound begins with the name of the first element, *chlorine (Cl)*, followed by the base name of the second element, *ox*, prefixed by *mono-* to indicate 1 oxygen (O) atom, so the formula is ClO.

          (c)   disulfur tetrafluoride: The name of the compound begins with the name of the first element, *sulfur (S)* prefixed by *di-* to indicate 2 S atoms, followed by the base name of the second element, *fluor*, prefixed by *tetra-* to indicate 4 fluorine (F) atoms, so the formula is S_2F_4.

          (d)   phosphorus pentafluoride: The name of the compound begins with the name of the first element, *phosphorous (P)*, followed by the base name of the second element, *fluor*, prefixed by *penta-* to indicate 5 fluorine (F) atoms, so the formula is PF_5.

Naming Compounds (When the Type Is Not Specified)

4.67   To use the flowchart, begin by determining the type of compound you are trying to name—ionic, molecular, or acid.

          (a) $SrCl_2$ is composed of a metal and a nonmetal, so it is ionic. Begin at the box labeled "IONIC" at the far left of the flowchart. The metal in the compound can only form one type of ion; therefore, take the left branch in the flowchart. Then name the compound according to the blocks at the end of the path in the flowchart. Write the name of the cation followed by the base name of the anion appended with the ending *-ide*. The full name is strontium chloride.

          (b) SnO_2 is composed of a metal and a nonmetal, so it is ionic. Begin at the box labeled "IONIC" at the far left of the flowchart. The metal in the compound forms more than one type of ion; therefore, take the right branch in the flowchart. Then name the compound according to the blocks at the end of the path in the flowchart. Write the name of the cation followed by the charge of the cation in Roman numerals followed by the base name of the anion appended with the ending *-ide*. The full name is tin(IV) oxide.

          (c) P_2S_5 is composed of two nonmetals and is a molecular compound. Begin at the box labeled "MOLECULAR" in the middle of the flowchart. Write the prefix to indicate the number of atoms of the first element, then the name of the first element followed by the prefix to indicate the number of atoms of the second element, then the base name of the second element followed by the ending *-ide*. The full name is diphorphorus pentasulfide.

4.69   To use the flowchart, begin by determining the type of compound you are trying to name—ionic, molecular, or acid.

(a) $KClO_3$ is composed of a metal and an oxyanion, so it is ionic. Begin at the box labeled "IONIC" at the far left of the flowchart. The metal in the compound can only form one type of ion; therefore, take the left branch in the flowchart. Then name the compound according to the blocks at the end of the path in the flowchart. Write the name of the cation followed by the base name of the anion appended with the ending -ate. ClO_3^- is the polyatomic ion chlorate, so the name of the compound is potassium chlorate.

(b) I_2O_5 is composed of two nonmetals and is a molecular compound. Begin at the box labeled "MOLECULAR" in the middle of the flowchart. Write the prefix to indicate the number of atoms of the first element, then the name of the first element followed by the prefix to indicate the number of atoms of the second element, then the base name of the second element followed by the ending -ide. The full name is diiodine pentoxide.

(c) $PbSO_4$ is composed of a metal and an oxyanion, so it is ionic. Begin at the box labeled "IONIC" at the far left of the flowchart. The metal in the compound forms more than one type of ion; therefore, take the right branch in the flowchart. Then name the compound according to the blocks at the end of the path in the flowchart. Write the name of the cation followed by the charge of the cation in Roman numerals followed by the base name of the anion appended with the ending -ate. SO_4^{2-} is the polyatomic ion sulfate, so the name of the compound is lead (II) sulfate.

Formula Mass and the Mole Concept for Compounds

4.71   To find the formula mass, sum the atomic masses of each atom in the chemical formula.

(a) NO_2 formula mass $= 1 \times (\text{atomic mass N}) + 2 \times (\text{atomic mass O})$
$$= 1 \times (14.01 \text{ amu}) + 2 \times (16.00 \text{ amu})$$
$$= 46.01 \text{ amu}$$

(b) C_4H_{10} formula mass $= 4 \times (\text{atomic mass C}) + 10 \times (\text{atomic mass H})$
$$= 4 \times (12.01 \text{ amu}) + 10 \times (1.008 \text{ amu})$$
$$= 58.12 \text{ amu}$$

(c) $C_6H_{12}O_6$ formula mass $= 6 \times (\text{atomic mass C}) + 12 \times (\text{atomic mass H}) + 6 \times (\text{atomic mass O})$
$$= 6 \times (12.01 \text{ amu}) + 12 \times (1.008 \text{ amu}) + 6 \times (16.00 \text{ amu})$$
$$= 180.16 \text{ amu}$$

(d) $Cr(NO_3)_3$ formula mass $= 1 \times (\text{atomic mass Cr}) + 3 \times (\text{atomic mass N}) + 9 \times (\text{atomic mass O})$
$$= 1 \times (52.00 \text{ amu}) + 3 \times (14.01 \text{ amu}) + 9 \times (16.00 \text{ amu})$$
$$= 238.03 \text{ amu}$$

Check: The units (amu) are correct. Each formula mass can be reported to two places to the right of the decimal point because each atom in the compound is reported to two or three places to the right of the decimal point.

4.73   (a)   **Given:** 72.5 g CCl_4 **Find:** number of moles
Conceptual Plan: g $CCl_4 \rightarrow$ mole CCl_4
$$\frac{1 \text{ mol}}{153.81 \text{ g } CCl_4}$$
Solution: $72.5 \text{ g } CCl_4 \times \dfrac{1 \text{ mol } CCl_4}{153.81 \text{ g } CCl_4} = 0.471\underline{4} \text{ mol } CCl_4 = 0.471 \text{ mol } CCl_4$

Check: Units of the answer (mole CCl_4) are correct. The magnitude is appropriate because it is less than 1 mole of NO_2.

(b)   **Given:** 12.4 g $C_{12}H_{22}O_{11}$ **Find:** number of moles
Conceptual Plan: g $C_{12}H_{22}O_{11} \rightarrow$ mole KNO_3
$$\frac{1 \text{ mol}}{342.296 \text{ g } C_{12}H_{22}O_{11}}$$
Solution: $12.4 \text{ g } C_{12}H_{22}O_{11} \times \dfrac{1 \text{ mol } C_{12}H_{22}O_{11}}{342.296 \text{ g } C_{12}H_{22}O_{11}} = 0.0362\underline{3} \text{ mol } C_{12}H_{22}O_{11} = 0.0362 \text{ mol } C_{12}H_{22}O_{11}$

Check: Units of the answer (mole $C_{12}H_{22}O_{11}$) are correct. The magnitude is appropriate because there is less than 1 mole of $C_{12}H_{22}O_{11}$.

(c)   **Given:** 25.2 kg C_2H_2 **Find:** number of moles
Conceptual Plan: kg $C_2H_2 \rightarrow$ g $C_2H_2 \rightarrow$ mole C_2H_2
$$\frac{1000 \text{ g } C_2H_2}{1 \text{ kg } C_2H_2} \quad \frac{1 \text{ mol}}{26.036 \text{ g } C_2H_2}$$

Solution: $25.2 \text{ kg } C_2H_2 \times \dfrac{1000 \text{ g } C_2H_2}{1 \text{ kg } C_2H_2} \times \dfrac{1 \text{ mol } C_2H_2}{26.036 \text{ g } C_2H_2} = 967.9 \text{ mol } C_2H_2 = 968 \text{ mol } C_2H_2$

Check: Units of the answer (mole C_2H_2) are correct. The magnitude is appropriate because more than a kilogram of C_2H_2 is present.

(d)   **Given:** 12.3 g dinitrogen monoxide   **Find:** number of moles

Conceptual Plan: Dinitrogen monoxide → formula and then g N_2O → mole N_2O

$$\dfrac{1 \text{ mol}}{44.02 \text{ g } N_2O}$$

Solution: Dinitrogen monoxide is N_2O.

$12.3 \text{ g } N_2O \times \dfrac{1 \text{ mol } N_2O}{44.02 \text{ g } N_2O} = 0.2794 \text{ mol } N_2O = 0.279 \text{ mol } N_2O$

Check: Units of the answer (mole N_2O) are correct. The magnitude is appropriate because less than a mole of N_2O is present.

4.75   (a)   **Given:** 25.5 g NO_2 **Find:** number of moles

Conceptual Plan: g NO_2 → mole NO_2

$$\dfrac{1 \text{ mol}}{46.01 \text{ g } NO_2}$$

Solution: $25.5 \text{ g } NO_2 \times \dfrac{1 \text{ mol } NO_2}{46.01 \text{ g } NO_2} = 0.554 \text{ mol } NO_2$

Check: The units of the answer (mole NO_2) are correct. The magnitude is appropriate because it is less than 1 mole of NO_2.

(b)   **Given:** 1.25 kg CO_2 **Find:** number of moles

Conceptual Plan: kg CO_2 → g CO_2 → mole CO_2

$$\dfrac{1000 \text{ g } CO_2}{1 \text{ kg } CO_2} \qquad \dfrac{1 \text{ mol}}{44.01 \text{ g } CO_2}$$

Solution: $1.25 \text{ kg } CO_2 \times \dfrac{1000 \text{ g } CO_2}{1 \text{ kg } CO_2} \times \dfrac{1 \text{ mol } CO_2}{44.01 \text{ g } CO_2} = 28.4 \text{ mol } CO_2$

Check: The units of the answer (mole CO_2) are correct. The magnitude is appropriate because over a kg of CO_2 is present.

(c)   **Given:** 38.2 g KNO_3 **Find:** number of moles

Conceptual Plan: g KNO_3 → mole KNO_3

$$\dfrac{1 \text{ mol}}{101.11 \text{ g } KNO_3}$$

Solution: $38.2 \text{ g } KNO_3 \times \dfrac{1 \text{ mol } KNO_3}{101.11 \text{ g } KNO_3} = 0.378 \text{ mol } KNO_3$

Check: The units of the answer (mole KNO_3) are correct. The magnitude is appropriate because there is less than 1 mole of KNO_3.

(d)   **Given:** 155.2 kg Na_2SO_4 **Find:** number of moles

Conceptual Plan: kg Na_2SO_4 → g Na_2SO_4 → mole Na_2SO_4

$$\dfrac{1000 \text{ g } Na_2SO_4}{1 \text{ kg } Na_2SO_4} \qquad \dfrac{1 \text{ mol}}{142.05 \text{ g } Na_2SO_4}$$

Solution: $155.2 \text{ kg } Na_2SO_4 \times \dfrac{1000 \text{ g } Na_2SO_4}{1 \text{ kg } Na_2SO_4} \times \dfrac{1 \text{ mol } Na_2SO_4}{142.05 \text{ g } Na_2SO_4} = 1092 \text{ mol } Na_2SO_4$

Check: The units of the answer (mole Na_2SO_4) are correct. The magnitude is appropriate because over 100 kg of Na_2SO_4 is present.

4.77   (a)   **Given:** 6.5 g H_2O **Find:** number of molecules
               Conceptual Plan: g H_2O → mole H_2O → number H_2O molecules

$$\frac{1 \text{ mol}}{18.02 \text{ g } H_2O} \qquad \frac{6.022 \times 10^{23} \ H_2O \text{ molecules}}{1 \text{ mol } H_2O}$$

Solution: $6.5 \text{ g } H_2O \times \dfrac{1 \text{ mol } H_2O}{18.02 \text{ g } H_2O} \times \dfrac{6.022 \times 10^{23} \ H_2O \text{ molecules}}{1 \text{ mol } H_2O} = 2.2 \times 10^{23} \ H_2O \text{ molecules}$

Check: The units of the answer (H_2O molecules) are correct. The magnitude is appropriate; it is smaller than Avogadro's number, as expected, because we have less than 1 mole of H_2O.

        (b)   **Given:** 389 g CBr_4 **Find:** number of molecules
               Conceptual Plan: g CBr_4 → mole CBr_4 → number CBr_4 molecules

$$\frac{1 \text{ mol}}{331.6 \text{ g } CBr_4} \qquad \frac{6.022 \times 10^{23} \ CBr_4 \text{ molecules}}{1 \text{ mol } CBr_4}$$

Solution: $389 \text{ g } CBr_4 \times \dfrac{1 \text{ mol } CBr_4}{331.6 \text{ g } CBr_4} \times \dfrac{6.022 \times 10^{23} \ CBr_4 \text{ molecules}}{1 \text{ mol } CBr_4} = 7.06 \times 10^{23} \ CBr_4 \text{ molecules}$

Check: The units of the answer (CBr_4 molecules) are correct. The magnitude is appropriate; it is larger than Avogadro's number, as expected, because we have more than 1 mole of CBr_4.

        (c)   **Given:** 22.1 g O_2 **Find:** number of molecules
               Conceptual Plan: g O_2 → mole O_2 → number O_2 molecules

$$\frac{1 \text{ mol}}{32.00 \text{ g } O_2} \qquad \frac{6.022 \times 10^{23} \ O_2 \text{ molecules}}{1 \text{ mol } O_2}$$

Solution: $22.1 \text{ g } O_2 \times \dfrac{1 \text{ mol } O_2}{32.00 \text{ g } O_2} \times \dfrac{6.022 \times 10^{23} \ O_2 \text{ molecules}}{1 \text{ mol } O_2} = 4.16 \times 10^{23} \ O_2 \text{ molecules}$

Check: The units of the answer (O_2 molecules) are correct. The magnitude is appropriate; it is smaller than Avogadro's number, as expected, because we have less than 1 mole of O_2.

        (d)   **Given:** 19.3 g C_8H_{10} **Find:** number of molecules
               Conceptual Plan: g C_8H_{10} → mole C_8H_{10} → number C_8H_{10} molecules

$$\frac{1 \text{ mol}}{106.16 \text{ g } C_8H_{10}} \qquad \frac{6.022 \times 10^{23} \ C_8H_{10} \text{ molecules}}{1 \text{ mol } C_8H_{10}}$$

Solution:

$$19.3 \text{ g } C_8H_{10} \times \frac{1 \text{ mol } C_8H_{10}}{106.16 \text{ g } C_8H_{10}} \times \frac{6.022 \times 10^{23} \ C_8H_{10} \text{ molecules}}{1 \text{ mol } C_8H_{10}} = 1.09 \times 10^{23} \ C_8H_{10} \text{ molecules}$$

Check: The units of the answer (C_8H_{10} molecules) are correct. The magnitude is appropriate; it is smaller than Avogadro's number, as expected, because we have less than 1 mole of C_8H_{10}.

4.79   (a)   **Given:** 5.94×10^{20} SO_3 molecules   **Find:** mass in g
               Conceptual Plan: Number SO_3 molecules → mole SO_3 → g SO_3

$$\frac{1 \text{ mol } SO_3}{6.022 \times 10^{23} \ SO_3 \text{ molecules}} \qquad \frac{80.07 \text{ g } SO_3}{1 \text{ mol } SO_3}$$

Solution: $5.94 \times 10^{20} \ SO_3 \text{ molecules} \times \dfrac{1 \text{ mol } SO_3}{6.022 \times 10^{23} \ SO_3 \text{ molecules}} \times \dfrac{80.07 \text{ g } SO_3}{1 \text{ mol } SO_3} = 0.0790 \text{ g } SO_3$

Check: The units of the answer (grams SO_3) are correct. The magnitude is appropriate; there is less than Avogadro's number of molecules, so we have less than 1 mole of SO_3.

        (b)   **Given:** 2.8×10^{22} H_2O molecules   **Find:** mass in g
               Conceptual Plan: Number H_2O molecules → mole H_2O → g H_2O

$$\frac{1 \text{ mol } H_2O}{6.022 \times 10^{23} \ H_2O \text{ molecules}} \qquad \frac{18.02 \text{ g } H_2O}{1 \text{ mol } H_2O}$$

Solution: $2.8 \times 10^{22} \ H_2O \text{ molecules} \times \dfrac{1 \text{ mol } H_2O}{6.022 \times 10^{23} \ H_2O \text{ molecules}} \times \dfrac{18.02 \text{ g } H_2O}{1 \text{ mol } H_2O} = 0.84 \text{ g } H_2O$

Check: The units of the answer (grams H_2O) are correct. The magnitude is appropriate; there is less than Avogadro's number of molecules, so we have less than 1 mole of H_2O.

(c)   **Given:** 1 $C_6H_{12}O_6$ molecule   **Find:** mass in g

Conceptual Plan: Number $C_6H_{12}O_6$ molecules → mole $C_6H_{12}O_6$ → g $C_6H_{12}O_6$

$$\frac{1\ \text{mol}\ C_6H_{12}O_6}{6.022 \times 10^{23}\ C_6H_{12}O_6\ \text{molecules}} \qquad \frac{180.16\ \text{g}\ C_6H_{12}O_6}{1\ \text{mol}\ C_6H_{12}O_6}$$

Solution:

$$1\ C_6H_{12}O_6\ \text{molecule} \times \frac{1\ \text{mol}\ C_6H_{12}O_6}{6.022 \times 10^{23}\ C_6H_{12}O_6\ \text{molecules}} \times \frac{180.16\ \text{g}\ C_6H_{12}O_6}{1\ \text{mol}\ C_6H_{12}O_6} = 2.992 \times 10^{-22}\ \text{g}\ C_6H_{12}O_6$$

Check: The units of the answer (grams $C_6H_{12}O_6$) are correct. The magnitude is appropriate; there is much less than Avogadro's number of molecules, so we have much less than 1 mole of $C_6H_{12}O_6$.

4.81   **Given:** 1.8×10^{17} $C_{12}H_{22}O_{11}$ molecules   **Find:** mass in mg

Conceptual Plan: Number $C_{12}H_{22}O_{11}$ molecules → mole $C_{12}H_{22}O_{11}$ → g $C_{12}H_{22}O_{11}$ → mg $C_{12}H_{22}O_{11}$

$$\frac{1\ \text{mol}\ C_{12}H_{22}O_{11}}{6.022 \times 10^{23}\ C_{12}H_{22}O_{11}\ \text{molecules}} \qquad \frac{342.3\ \text{g}\ C_{12}H_{22}O_{11}}{1\ \text{mol}\ C_{12}H_{22}O_{11}} \qquad \frac{1 \times 10^3\ \text{mg}\ C_{12}H_{22}O_{11}}{1\ \text{g}\ C_{12}H_{22}O_{11}}$$

Solution:

$$1.8 \times 10^{17}\ C_{12}H_{22}O_{11}\ \text{molecules} \times \frac{1\ \text{mol}\ C_{12}H_{22}O_{11}}{6.022 \times 10^{23}\ C_{12}H_{22}O_{11}\ \text{molecules}} \times \frac{342.3\ \text{g}\ C_{12}H_{22}O_{11}}{1\ \text{mol}\ C_{12}H_{22}O_{11}} \times \frac{1 \times 10^3\ \text{mg}\ C_{12}H_{22}O_{11}}{1\ \text{g}\ C_{12}H_{22}O_{11}}$$

$$= 0.10\ \text{mg}\ C_{12}H_{22}O_{11}$$

Check: The units of the answer (mg $C_{12}H_{22}O_{11}$) are correct. The magnitude is appropriate; there is much less than Avogadro's number of molecules, so we have much less than 1 mole of $C_{12}H_{22}O_{11}$.

Composition of Compounds

4.83   (a)   **Given:** CH_4 **Find:** mass percent C

Conceptual Plan: Mass % C $= \dfrac{1 \times \text{molar mass C}}{\text{molar mass } CH_4} \times 100\%$

Solution:

$$1 \times \text{molar mass C} = 1(12.01\ \text{g/mol}) = 12.01\ \text{g C}$$

$$\text{molar mass } CH_4 = 1(12.01\ \text{g/mol}) + 4(1.008\ \text{g/mol}) = 16.04\ \text{g/mol}$$

$$\text{mass \% C} = \frac{1 \times \text{molar mass C}}{\text{molar mass } CH_4} \times 100\%$$

$$= \frac{12.01\ \text{g/mol}}{16.04\ \text{g/mol}} \times 100\%$$

$$= 74.87\%$$

Check: The units of the answer (%) are correct. The magnitude is reasonable because it is between 0 and 100% and carbon is the heaviest element.

(b)   **Given:** C_2H_6 **Find:** mass percent C

Conceptual Plan: Mass % C $= \dfrac{2 \times \text{molar mass C}}{\text{molar mass } C_2H_6} \times 100\%$

Solution:

$$2 \times \text{molar mass C} = 2(12.01\ \text{g/mol}) = 24.02\ \text{g C}$$

$$\text{molar mass } C_2H_6 = 2(12.01\ \text{g/mol}) + 6(1.008\ \text{g/mol}) = 30.07\ \text{g/mol}$$

$$\text{mass \% C} = \frac{2 \times \text{molar mass C}}{\text{molar mass } C_2H_6} \times 100\%$$

$$= \frac{24.02\ \text{g/mol}}{30.07\ \text{g/mol}} \times 100\%$$

$$= 79.88\%$$

Check: The units of the answer (%) are correct. The magnitude is reasonable because it is between 0 and 100% and carbon is the heaviest element.

(c) **Given:** C_2H_2 **Find:** mass percent C

Conceptual Plan: Mass % C $= \dfrac{2 \times \text{molar mass C}}{\text{molar mass } C_2H_2} \times 100\%$

Solution:

$$2 \times \text{molar mass C} = 2(12.01 \text{ g/mol}) = 24.02 \text{ g C}$$

$$\text{molar mass } C_2H_2 = 2(12.01 \text{ g/mol}) + 2(1.008 \text{ g/mol}) = 26.04 \text{ g/mol}$$

$$\text{mass % C} = \dfrac{2 \times \text{molar mass C}}{\text{molar mass } C_2H_2} \times 100\%$$

$$= \dfrac{24.02 \text{ g/mol}}{26.04 \text{ g/mol}} \times 100\%$$

$$= 92.24\%$$

Check: The units of the answer (%) are correct. The magnitude is reasonable because it is between 0 and 100% and carbon is the heaviest element.

(d) **Given:** C_2H_5Cl **Find:** mass percent C

Conceptual Plan: Mass % C $= \dfrac{2 \times \text{molar mass C}}{\text{molar mass } C_2H_5Cl} \times 100\%$

Solution:

$$2 \times \text{molar mass C} = 2(12.01 \text{ g/mol}) = 24.02 \text{ g C}$$

$$\text{molar mass } C_2H_5Cl = 2(12.01 \text{ g/mol}) + 5(1.008 \text{ g/mol}) + 1(35.45 \text{ g/mol}) = 64.51 \text{ g/mol}$$

$$\text{mass % C} = \dfrac{2 \times \text{molar mass C}}{\text{molar mass } C_2H_5Cl} \times 100\%$$

$$= \dfrac{24.02 \text{ g/mol}}{64.51 \text{ g/mol}} \times 100\%$$

$$= 37.23\%$$

Check: The units of the answer (%) are correct. The magnitude is reasonable because it is between 0 and 100% and chlorine is heavier than carbon.

4.85   **Given:** NH_3 **Find:** mass percent N

Conceptual Plan: Mass % N $= \dfrac{1 \times \text{molar mass N}}{\text{molar mass } NH_3} \times 100\%$

Solution:

$$1 \times \text{molar mass N} = 1(14.01 \text{ g/mol}) = 14.01 \text{ g N}$$

$$\text{molar mass } NH_3 = 3(1.008 \text{ g/mol}) + (14.01 \text{ g/mol}) = 17.03 \text{ g/mol}$$

$$\text{mass % N} = \dfrac{1 \times \text{molar mass N}}{\text{molar mass } NH_3} \times 100\%$$

$$= \dfrac{14.01 \text{ g/mol}}{17.03 \text{ g/mol}} \times 100\%$$

$$= 82.27\%$$

Check: The units of the answer (%) are correct. The magnitude is reasonable because it is between 0 and 100% and nitrogen is the heaviest atom present.

Given: $CO(NH_2)_2$ **Find:** mass percent N

Conceptual Plan: Mass % N $= \dfrac{2 \times \text{molar mass N}}{\text{molar mass } CO(NH_2)_2} \times 100\%$

Solution:

$$2 \times \text{molar mass N} = 1(14.01 \text{ g/mol}) = 28.02 \text{ g N}$$

$$\text{molar mass } CO(NH_2)_2 = (12.01 \text{ g/mol}) + (16.00 \text{ g/mol}) + 2(14.01 \text{ g/mol}) + 4(1.008 \text{ g/mol}) = 60.06 \text{ g/mol}$$

$$\text{mass \% N} = \frac{2 \times \text{molar mass N}}{\text{molar mass CO(NH}_2)_2} \times 100\%$$

$$= \frac{28.02 \text{ g/mol}}{60.06 \text{ g/mol}} \times 100\%$$

$$= 46.65\%$$

Check: The units of the answer (%) are correct. The magnitude is reasonable because it is between 0 and 100% and there are two nitrogens and only one carbon and one oxygen per molecule.

Given: NH_4NO_3 **Find:** mass percent N

Conceptual Plan: Mass % N $= \dfrac{2 \times \text{molar mass N}}{\text{molar mass NH}_4NO_3} \times 100\%$

Solution:

$$2 \times \text{molar mass N} = 2(14.01 \text{ g/mol}) = 28.02 \text{ g N}$$

$$\text{molar mass NH}_4NO_3 = 2(14.01 \text{ g/mol}) + 4(1.008 \text{ g/mol}) + 3(16.00 \text{ g/mol}) = 80.05 \text{ g/mol}$$

$$\text{mass \% N} = \frac{2 \times \text{molar mass N}}{\text{molar mass NH}_4NO_3} \times 100\%$$

$$= \frac{28.02 \text{ g/mol}}{80.05 \text{ g/mol}} \times 100\%$$

$$= 35.00\%$$

Check: The units of the answer (%) are correct. The magnitude is reasonable because it is between 0 and 100%. The mass of nitrogen is less than the mass of oxygen, and there are two nitrogens and three oxygens per molecule.

Given: $(NH_4)_2SO_4$ **Find:** mass percent N

Conceptual Plan: Mass % N $= \dfrac{2 \times \text{molar mass N}}{\text{molar mass (NH}_4)_2SO_4} \times 100\%$

Solution:

$$2 \times \text{molar mass N} = 2(14.01 \text{ g/mol}) = 28.02 \text{ g N}$$

$$\text{molar mass (NH}_4)_2SO_4 = 2(14.01 \text{ g/mol}) + 8(1.008 \text{ g/mol}) + (32.07 \text{ g/mol}) + 4(16.00 \text{ g/mol}) = 132.15 \text{ g/mol}$$

$$\text{mass \% N} = \frac{2 \times \text{molar mass N}}{\text{molar mass (NH}_4)_2SO_4} \times 100\%$$

$$= \frac{28.02 \text{ g/mol}}{132.15 \text{ g/mol}} \times 100\%$$

$$= 21.20\%$$

Check: The units of the answer (%) are correct. The magnitude is reasonable because it is between 0 and 100% and the mass of nitrogen is less than the mass of oxygen and sulfur.

The fertilizer with the highest nitrogen content is NH_3 with a N content of 82.27%.

4.87 **Given:** 55.5 g CuF_2; 37.42 **Find:** g F in CuF_2

Conceptual Plan: g $CuF_2 \rightarrow$ g F

$$\frac{37.42 \text{ g F}}{100.0 \text{ g CuF}_2}$$

Solution: $55.5 \text{ g } CuF_2 \times \dfrac{37.42 \text{ g F}}{100.0 \text{ g } CuF_2} = 20.77 = 20.8 \text{ g F}$

Check: The units of the answer (g F) are correct. The magnitude is reasonable because it is less than the original mass.

4.89 **Given:** 150 μg I; 76.45% I in KI **Find:** μg KI

Conceptual Plan: μg I $\rightarrow$ g I $\rightarrow$ g KI $\rightarrow$ μg KI

$$\frac{1 \text{ g I}}{1 \times 10^6 \text{ } \mu\text{g I}} \quad \frac{100.0 \text{ g KI}}{76.45 \text{ g I}} \quad \frac{1 \times 10^6 \text{ } \mu\text{g KI}}{1 \text{ g KI}}$$

Solution: $150 \text{ } \mu g I \times \dfrac{1 \text{ g I}}{1 \times 10^6 \text{ } \mu g I} \times \dfrac{100.0 \text{ g KI}}{76.45 \text{ g I}} \times \dfrac{1 \times 10^6 \text{ } \mu g \text{ KI}}{1 \text{ g KI}} = 196.207 \text{ } \mu g \text{ KI} = 2.0 \times 10^2 \text{ } \mu g \text{ KI}$

Check: The units of the answer (μg KI) are correct. The magnitude is reasonable because it is greater than the original mass.

4.91　(a)　red = oxygen, white = hydrogen:　　　2H:O　　　H_2O

　　　(b)　black = carbon, white = hydrogen:　　　C:4H　　　CH_4

　　　(c)　black = carbon, white = hydrogen, red = oxygen:　　　2C:6H:O　　　CH_3CH_2OH or C_2H_6O

4.93　(a)　**Given:** 0.0885 mol C_4H_{10}　　**Find:** mol H atoms

　　　Conceptual Plan: Mol C_4H_{10} → mole H atom

$$\frac{10 \text{ mol H}}{1 \text{ mol } C_4H_{10}}$$

　　　Solution: 0.0885 mol C_4H_{10} × $\dfrac{10 \text{ mol H}}{1 \text{ mol } C_4H_{10}}$ = 0.885 mol H atoms

　　　Check: The units of the answer (mol H atoms) are correct. The magnitude is reasonable because it is greater than the original mol C_4H_{10}.

　　　(b)　**Given:** 1.3 mol CH_4　　**Find:** mol H atoms

　　　Conceptual Plan: Mol CH_4 → mole H atom

$$\frac{4 \text{ mol H}}{1 \text{ mol } CH_4}$$

　　　Solution: 1.3 mol CH_4 × $\dfrac{4 \text{ mol H}}{1 \text{ mol } CH_4}$ = 5.2 mol H atoms

　　　Check: The units of the answer (mol H atoms) are correct. The magnitude is reasonable because it is greater than the original mol CH_4.

　　　(c)　**Given:** 2.4 mol C_6H_{12}　　**Find:** mol H atoms

　　　Conceptual Plan: Mol C_6H_{12} → mole H atom

$$\frac{12 \text{ mol H}}{1 \text{ mol } C_6H_{12}}$$

　　　Solution: 2.4 mol C_6H_{12} × $\dfrac{12 \text{ mol H}}{1 \text{ mol } C_6H_{12}}$ = 29 mol H atoms

　　　Check: The units of the answer (mol H atoms) are correct. The magnitude is reasonable because it is greater than the original mol C_6H_{12}.

　　　(d)　**Given:** 1.87 mol C_8H_{18}　　**Find:** mol H atoms

　　　Conceptual Plan: Mol C_8H_{18} → mole H atom

$$\frac{18 \text{ mol H}}{1 \text{ mol } C_8H_{18}}$$

　　　Solution: 1.87 mol C_8H_{18} × $\dfrac{18 \text{ mol H}}{1 \text{ mol } C_8H_{18}}$ = 33.7 mol H atoms

　　　Check: The units of the answer (mol H atoms) are correct. The magnitude is reasonable because it is greater than the original mol C_8H_{18}.

4.95　(a)　**Given:** 8.5 g NaCl　　**Find:** g Na

　　　Conceptual Plan: g NaCl → mole NaCl → mol Na → g Na

$$\frac{1 \text{ mol NaCl}}{58.44 \text{ g NaCl}} \qquad \frac{1 \text{ mol Na}}{1 \text{ mol NaCl}} \qquad \frac{22.99 \text{ g Na}}{1 \text{ mol Na}}$$

　　　Solution: 8.5 g NaCl × $\dfrac{1 \text{ mol NaCl}}{58.44 \text{ g NaCl}}$ × $\dfrac{1 \text{ mol Na}}{1 \text{ mol NaCl}}$ × $\dfrac{22.99 \text{ g Na}}{1 \text{ mol Na}}$ = 3.3 g Na

　　　Check: The units of the answer (g Na) are correct. The magnitude is reasonable because it is less than the original g NaCl.

　　　(b)　**Given:** 8.5 g Na_3PO_4　　**Find:** g Na

　　　Conceptual Plan: g Na_3PO_4 → mole Na_3PO_4 → mol Na → g Na

$$\frac{1 \text{ mol } Na_3PO_4}{163.94 \text{ g } Na_3PO_4} \qquad \frac{3 \text{ mol Na}}{1 \text{ mol } Na_3PO_4} \qquad \frac{22.99 \text{ g Na}}{1 \text{ mol Na}}$$

　　　Solution: 8.5 g Na_3PO_4 × $\dfrac{1 \text{ mol } Na_3PO_4}{163.94 \text{ g } Na_3PO_4}$ × $\dfrac{3 \text{ mol Na}}{1 \text{ mol } Na_3PO_4}$ × $\dfrac{22.99 \text{ g Na}}{1 \text{ mol Na}}$ = 3.6 g Na

Check: The units of the answer (g Na) are correct. The magnitude is reasonable because it is less than the original g Na_3PO_4.

(c)   **Given:** 8.5 g $NaC_7H_5O_2$ **Find:** g Na
Conceptual Plan: g $NaC_7H_5O_2 \rightarrow$ mole $NaC_7H_5O_2 \rightarrow$ mol Na $\rightarrow$ g Na

$$\frac{1 \text{ mol } NaC_7H_5O_2}{144.10 \text{ g } NaC_7H_5O_2} \qquad \frac{1 \text{ mol Na}}{1 \text{ mol } NaC_7H_5O_2} \qquad \frac{22.99 \text{ g Na}}{1 \text{ mol Na}}$$

Solution: $8.5 \text{ g } NaC_7H_5O_2 \times \dfrac{1 \text{ mol } NaC_7H_5O_2}{144.10 \text{ g } NaC_7H_5O_2} \times \dfrac{1 \text{ mol Na}}{1 \text{ mol } NaC_7H_5O_2} \times \dfrac{22.99 \text{ g Na}}{1 \text{ mol Na}} = 1.4 \text{ g Na}$

Check: The units of the answer (g Na) are correct. The magnitude is reasonable because it is less than the original g $NaC_7H_5O_2$.

(d)   **Given:** 8.5 g $Na_2C_6H_6O_7$ **Find:** g Na
Conceptual Plan: g $Na_2C_6H_6O_7 \rightarrow$ mole $Na_2C_6H_6O_7 \rightarrow$ mol Na $\rightarrow$ g Na

$$\frac{1 \text{ mol } Na_2C_6H_6O_7}{236.1 \text{ g } Na_2C_6H_6O_7} \qquad \frac{2 \text{ mol Na}}{1 \text{ mol } Na_2C_6H_6O_7} \qquad \frac{22.99 \text{ g Na}}{1 \text{ mol Na}}$$

Solution: $8.5 \text{ g } Na_2C_6H_6O_7 \times \dfrac{1 \text{ mol } Na_2C_6H_6O_7}{236.1 \text{ g } Na_2C_6H_6O_7} \times \dfrac{2 \text{ mol Na}}{1 \text{ mol } Na_2C_6H_6O_7} \times \dfrac{22.99 \text{ g Na}}{1 \text{ mol Na}} = 1.7 \text{ g Na}$

Check: The units of the answer (g Na) are correct. The magnitude is reasonable because it is less than the original g $Na_2C_6H_6O_7$.

Chemical Formulas from Experimental Data

4.97   (a)   **Given:** 1.651 g Ag; 0.1224 g O   **Find:** empirical formula
Conceptual Plan:
Convert mass to mol of each element $\rightarrow$ write pseudoformula $\rightarrow$ write empirical formula

$$\frac{1 \text{ mol Ag}}{107.9 \text{ g Ag}} \qquad \frac{1 \text{ mol O}}{16.00 \text{ g O}} \qquad \text{divide by smallest number}$$

Solution: $1.651 \text{ g Ag} \times \dfrac{1 \text{ mol Ag}}{107.9 \text{ g Ag}} = 0.01530 \text{ mol Ag}$

$0.1224 \text{ g O} \times \dfrac{1 \text{ mol O}}{16.00 \text{ g O}} = 0.007650 \text{ mol O}$

$Ag_{0.01530} O_{0.007650}$

$Ag_{\frac{0.01530}{0.007650}} O_{\frac{0.007650}{0.007650}} \rightarrow Ag_2O$

The correct empirical formula is Ag_2O.

Check: The formula makes sense since silver forms a 1+ ion and oxygen typically forms a 2− ion.

(b)   **Given:** 0.672 g Co; 0.569 g As; 0.486 g O   **Find:** empirical formula
Conceptual Plan: Convert mass to mol of each element $\rightarrow$ write pseudoformula $\rightarrow$ write empirical formula

$$\frac{1 \text{ mol Co}}{58.93 \text{ g Co}} \qquad \frac{1 \text{ mol As}}{74.92 \text{ g As}} \qquad \frac{1 \text{ mol O}}{16.00 \text{ g O}} \qquad \text{divide by smallest number}$$

Solution: $0.672 \text{ g Co} \times \dfrac{1 \text{ mol Co}}{58.93 \text{ g Co}} = 0.0114 \text{ mol Co}$

$0.569 \text{ g As} \times \dfrac{1 \text{ mol As}}{74.92 \text{ g As}} = 0.00759 \text{ mol As}$

$0.486 \text{ g O} \times \dfrac{1 \text{ mol O}}{16.00 \text{ g O}} = 0.0304 \text{ mol O}$

$Co_{0.0114} As_{0.00759} O_{0.0304}$

$Co_{\frac{0.0114}{0.00759}} As_{\frac{0.00759}{0.00759}} O_{\frac{0.0304}{0.00759}} \rightarrow Co_{1.5}As_1O_4$

$Co_{1.5}As_1O_4 \times 2 \rightarrow Co_3As_2O_8$

The correct empirical formula is $Co_3As_2O_8$.

(c)   **Given:** 1.443 g Se; 5.841 g Br   **Find:** empirical formula

Conceptual Plan:

Convert mass to mol of each element → write pseudoformula → write empirical formula

$$\frac{1\ mol\ Se}{78.96\ g\ Se} \quad \frac{1\ mol\ Br}{79.90\ g\ Br} \qquad \text{divide by smallest number}$$

Solution: $1.443\ \cancel{g\ Se} \times \dfrac{1\ mol\ Se}{78.96\ \cancel{g\ Se}} = 0.01828\ mol\ Se$

$5.841\ \cancel{g\ Br} \times \dfrac{1\ mol\ Br}{79.90\ \cancel{g\ Br}} = 0.07310\ mol\ Br$

$Se_{0.01828}Br_{0.07310}$

$Se_{\frac{0.01828}{0.01828}}\ Br_{\frac{0.07310}{0.01828}} \rightarrow SeBr_4$

The correct empirical formula is $SeBr_4$.

4.99   (a)   **Given:** in a 100 g sample of nicotine: 74.03 g C; 8.70 g H; 17.27 g N   **Find:** empirical formula

Conceptual Plan:

Convert mass to mol of each element → write pseudoformula → write empirical formula

$$\frac{1\ mol\ C}{12.01\ g\ C} \quad \frac{1\ mol\ H}{1.008\ g\ H} \quad \frac{1\ mol\ N}{14.01\ g\ N} \qquad \text{divide by smallest number}$$

Solution: $74.03\ \cancel{g\ C} \times \dfrac{1\ mol\ C}{12.01\ \cancel{g\ C}} = 6.164\ mol\ C$

$8.70\ \cancel{g\ H} \times \dfrac{1\ mol\ H}{1.008\ \cancel{g\ H}} = 8.63\ mol\ H$

$17.27\ \cancel{g\ N} \times \dfrac{1\ mol\ N}{14.01\ \cancel{g\ N}} = 1.233\ mol\ N$

$C_{6.164}H_{8.63}N_{1.233}$

$C_{\frac{6.164}{1.233}}\ H_{\frac{8.63}{1.233}}\ N_{\frac{1.233}{1.233}} \rightarrow C_5H_7N$

The correct empirical formula is C_5H_7N.

(b)   **Given:** in a 100 g sample of caffeine: 49.48 g C; 5.19 g H; 28.85 g N; 16.48 g O   **Find:** empirical formula

Conceptual Plan:

Convert mass to mol of each element → write pseudoformula → write empirical formula

$$\frac{1\ mol\ C}{12.01\ g\ C} \quad \frac{1\ mol\ H}{1.008\ g\ H} \quad \frac{1\ mol\ N}{14.01\ g\ N} \quad \frac{1\ mol\ O}{16.00\ g\ O} \qquad \text{divide by smallest number}$$

Solution $49.48\ \cancel{g\ C} \times \dfrac{1\ mol\ C}{12.01\ \cancel{g\ C}} = 4.120\ mol\ C$

$5.19\ \cancel{g\ H} \times \dfrac{1\ mol\ H}{1.008\ \cancel{g\ H}} = 5.15\ mol\ H$

$28.85\ \cancel{g\ N} \times \dfrac{1\ mol\ N}{14.01\ \cancel{g\ N}} = 2.059\ mol\ N$

$16.48\ \cancel{g\ O} \times \dfrac{1\ mol\ O}{16.00\ \cancel{g\ O}} = 1.030\ mol\ O$

$C_{4.120}H_{5.15}N_{2.059}O_{1.030}$

$C_{\frac{4.120}{1.030}}\ H_{\frac{5.15}{1.030}}\ N_{\frac{2.059}{1.030}}\ O_{\frac{1.030}{1.030}} \rightarrow C_4H_5N_2O$

The correct empirical formula is $C_4H_5N_2O$.

4.101   **Given:** in a 100 g sample of ibuprofen 75.69 g C; 8.80 g H; 15.51 g O   **Find:** empirical formula

Conceptual Plan: Convert mass to mol of each element → write pseudoformula → write empirical formula

$$\frac{1\ mol\ C}{12.01\ g\ C} \quad \frac{1\ mol\ H}{1.008\ g\ H} \quad \frac{1\ mol\ O}{16.00\ g\ O} \qquad \text{divide by smallest number}$$

Solution: $75.69\ \cancel{g\ C} \times \dfrac{1\ mol\ C}{12.01\ \cancel{g\ C}} = 6.302\ mol\ C$

$$8.80 \text{ g H} \times \frac{1 \text{ mol H}}{1.008 \text{ g H}} = 8.73 \text{ mol H}$$

$$15.51 \text{ g O} \times \frac{1 \text{ mol O}}{16.00 \text{ g O}} = 0.9694 \text{ mol O}$$

$$C_{6.302}H_{8.73}O_{0.9694}$$

$$C_{\frac{6.302}{0.9694}} H_{\frac{8.73}{0.9694}} O_{\frac{0.9694}{0.9694}} \rightarrow C_{6.50}H_{9.01}O$$

$$C_{6.50}H_{9.01}O \times 2 = C_{13}H_{18}O_2$$

The correct empirical formula is $C_{13}H_{18}O_2$.

4.103   **Given:** 0.77 mg N; 6.61 mg N_xCl_y **Find:** empirical formula

Conceptual Plan: Find mg Cl → convert mg to g for each element → convert mass to mol of each element →

$$\text{mg N}_x\text{Cl}_y - \text{mg N} \qquad \frac{1 \text{ g}}{1000 \text{ mg}} \qquad \frac{1 \text{ mol N}}{14.01 \text{ g N}} \qquad \frac{1 \text{ mol Cl}}{35.45 \text{ g Cl}}$$

write pseudoformula → write empirical formula

divide by smallest number

Solution: 6.61 mg N_xCl_y − 0.77 mg N = 5.84 mg Cl

$$0.77 \text{ mg N} \times \frac{1 \text{ g N}}{1000 \text{ mg N}} \times \frac{1 \text{ mol N}}{14.01 \text{ g N}} = 5.5 \times 10^{-5} \text{ mol N}$$

$$5.84 \text{ mg Cl} \times \frac{1 \text{ g Cl}}{1000 \text{ mg Cl}} \times \frac{1 \text{ mol Cl}}{35.45 \text{ g Cl}} = 1.70 \times 10^{-4} \text{ mol Cl}$$

$$N_{5.5\times10^{-5}} Cl_{1.6\times10^{-4}}$$

$$N_{\frac{5.5\times10^{-5}}{5.5\times10^{-5}}} Cl_{\frac{1.6\times10^{-4}}{5.5\times10^{-5}}} \rightarrow NCl_3$$

The correct empirical formula is NCl_3.

4.105   (a)   **Given:** empirical formula = C_6H_7N; molar mass = 186.24 g/mol   **Find:** molecular formula

Conceptual Plan: Molecular formula = empirical formula $\times n$ $n = \dfrac{\text{molar mass}}{\text{empirical formula mass}}$

Solution: empirical formula mass = 6(12.01 g/mol) + 7(1.008 g/mol) + 1(14.01 g/mol) = 93.13 g/mol

$$n = \frac{\text{molar mass}}{\text{formula molar mass}} = \frac{186.24 \text{ g/mol}}{93.13 \text{ g/mol}} = 1.9998 = 2$$

molecular formula $= C_6H_7N \times 2$
                      $= C_{12}H_{14}N_2$

Check: The units (g/mol) are correct. Check the answer by calculating the molar mass of the formula as follows: 12(12.01 g/mol) + 14(1.008 g/mol) + 2(14.01 g/mol) = 186.24 g/mol. The calculated molar mass is in agreement with the given molar mass.

(b)   **Given:** empirical formula = C_2HCl; molar mass = 181.44 g/mol   **Find:** molecular formula

Conceptual Plan: Molecular formula = empirical formula $\times n$ $n = \dfrac{\text{molar mass}}{\text{empirical formula mass}}$

Solution: empirical formula mass = 2(12.01 g/mol) + 1(1.008 g/mol) + 1(35.45 g/mol) = 60.48 g/mol

$$n = \frac{\text{molar mass}}{\text{formula molar mass}} = \frac{181.44 \text{ g/mol}}{60.48 \text{ g/mol}} = 3$$

molecular formula $= C_2HCl \times 3$
                      $= C_6H_3Cl_3$

Check: The units (g/mol) are correct. Check the answer by calculating the molar mass of the formula as follows: 6(12.01 g/mol) + 3(1.008 g/mol) + 3(35.45 g/mol) = 181.44 g/mol. The calculated molar mass is in agreement with the given molar mass.

(c) **Given:** empirical formula $= C_5H_{10}NS_2$; molar mass $= 296.54$ g/mol   **Find:** molecular formula

Conceptual Plan: Molecular formula $=$ empirical formula $\times n$ $n = \dfrac{\text{molar mass}}{\text{empirical formula mass}}$

Solution: empirical formula mass $= 5(12.01 \text{ g/mol}) + 10(1.008 \text{ g/mol})$
$+ 1(14.01 \text{ g/mol}) + 2(32.07) = 148.28 \text{ g/mol}$

$n = \dfrac{\text{molar mass}}{\text{formula molar mass}} = \dfrac{296.54 \text{ g/mol}}{148.28 \text{ g/mol}} = 2$

molecular formula $= C_5H_{10}NS_2 \times 2$
$= C_{10}H_{20}N_2S_4$

Check: The units (g/mol) are correct. Check the answer by calculating the molar mass of the formula as follows: $10(12.01 \text{ g/mol}) + 20(1.008 \text{ g/mol}) + 2(14.01 \text{ g/mol}) + 1(32.07 \text{ g/mol}) = 296.54 \text{ g/mol}$. The calculated molar mass is in agreement with the given molar mass.

4.107   **Given:** 33.01 g CO_2; 13.51 g H_2O **Find:** empirical formula
Conceptual Plan: Mass $CO_2, H_2O \rightarrow$ mol $CO_2, H_2O \rightarrow$ mol C, mol H $\rightarrow$ pseudoformula $\rightarrow$ empirical formula

$\dfrac{1 \text{ mol } CO_2}{44.01 \text{ g } CO_2}$ $\dfrac{1 \text{ mol } H_2O}{18.02 \text{ g } H_2O}$ $\dfrac{1 \text{ mol C}}{1 \text{ mol } CO_2}$ $\dfrac{2 \text{ mol H}}{1 \text{ mol } H_2O}$ divide by smallest number

Solution:

$33.01 \text{ g } CO_2 \times \dfrac{1 \text{ mol } CO_2}{44.01 \text{ g } CO_2} = 0.7501 \text{ mol } CO_2$

$13.51 \text{ g } H_2O \times \dfrac{1 \text{ mol } H_2O}{18.02 \text{ g } H_2O} = 0.7497 \text{ mol } H_2O$

$0.7501 \text{ mol } CO_2 \times \dfrac{1 \text{ mol C}}{1 \text{ mol } CO_2} = 0.7501 \text{ mol C}$

$0.7497 \text{ mol } H_2O \times \dfrac{2 \text{ mol H}}{1 \text{ mol } H_2O} = 1.499 \text{ mol H}$

$C_{0.7501} H_{1.499}$

$C_{\frac{0.7501}{0.7501}} H_{\frac{1.499}{0.7501}} \rightarrow CH_2$

The correct empirical formula is CH_2.

4.109   **Given:** 4.30 g sample; 8.59 g CO_2; 3.52 g H_2O **Find:** empirical formula
Conceptual Plan: Mass $CO_2, H_2O \rightarrow$ mol $CO_2, H_2O \rightarrow$ mol C, mol H $\rightarrow$ mass C, mass H, mass O $\rightarrow$ mol O $\rightarrow$

$\dfrac{1 \text{ mol } CO_2}{44.01 \text{ g } CO_2}$ $\dfrac{1 \text{ mol } H_2O}{18.02 \text{ g } H_2O}$ $\dfrac{1 \text{ mol C}}{1 \text{ mol } CO_2}$ $\dfrac{2 \text{ mol H}}{1 \text{ mol } H_2O}$ $\dfrac{12.01 \text{ g C}}{1 \text{ mol C}}$ $\dfrac{1.008 \text{ g H}}{1 \text{ mol H}}$ g sample $-$ g C $-$ g H $\dfrac{1 \text{ mol O}}{16.00 \text{ g O}}$

pseudoformula $\rightarrow$ empirical formula

divide by smallest number

Solution:

$8.59 \text{ g } CO_2 \times \dfrac{1 \text{ mol } CO_2}{44.01 \text{ g } CO_2} = 0.195 \text{ mol } CO_2$

$3.52 \text{ g } H_2O \times \dfrac{1 \text{ mol } H_2O}{18.02 \text{ g } H_2O} = 0.195 \text{ mol } H_2O$

$0.195 \text{ mol } CO_2 \times \dfrac{1 \text{ mol C}}{1 \text{ mol } CO_2} = 0.195 \text{ mol C}$

$0.195 \text{ mol } H_2O \times \dfrac{2 \text{ mol H}}{1 \text{ mol } H_2O} = 0.390 \text{ mol H}$

$0.195 \text{ mol C} \times \dfrac{12.01 \text{ g C}}{1 \text{ mol C}} = 2.34 \text{ g C}$

$0.390 \text{ mol H} \times \dfrac{1.008 \text{ g H}}{1 \text{ mol H}} = 0.393 \text{ g H}$

$4.30 \text{ g} - 2.34 \text{ g} - 0.393 \text{ g} = 1.57 \text{ g O}$

$$1.57 \text{ g O} \times \frac{1 \text{ mol O}}{16.00 \text{ g O}} = 0.0981 \text{ mol O}$$

$$C_{0.195} H_{0.390} O_{0.0981}$$

$$C_{\frac{0.195}{0.0981}} H_{\frac{0.390}{0.0981}} O_{\frac{0.0981}{0.0981}} \rightarrow C_2H_4O$$

The correct empirical formula is C_2H_4O.

Organic Compounds

4.111   (a)   composed of metal cation and polyatomic anion—inorganic compound
        (b)   composed of carbon and hydrogen—organic compound
        (c)   composed of carbon, hydrogen, and oxygen—organic compound
        (d)   composed of metal cation and nonmetal anion—inorganic compound

4.113   (a)   This compound contains oxygen, in addition to carbon and hydrogen, so it is not a hydrocarbon.
        (b)   This compound contains only carbon and hydrogen, so it is a hydrocarbon.
        (c)   This compound contains oxygen, in addition to carbon and hydrogen, so it is not a hydrocarbon.
        (d)   This compound contains nitrogen, in addition to carbon and hydrogen, so it is not a hydrocarbon.

Cumulative Problems

4.115   **Given:** 145 mL C_2H_5OH; $d = 0.789 \text{ g/cm}^3$ **Find:** number of molecules
        Conceptual Plan: $\text{cm}^3 \rightarrow \text{mL} : \text{mL } C_2H_5OH \rightarrow \text{g } C_2H_5OH \rightarrow \text{mol } C_2H_5OH \rightarrow \text{molecules } C_2H_5OH$

$$\frac{1 \text{ cm}^3}{1 \text{ mL}} \qquad \frac{0.789 \text{ g } C_2H_5OH}{\text{cm}^3} \qquad \frac{1 \text{ mol } C_2H_5OH}{46.07 \text{ g } C_2H_5OH} \qquad \frac{6.022 \times 10^{23} \text{ molecules } C_2H_5OH}{1 \text{ mol } C_2H_5OH}$$

Solution:

$$145 \text{ mL } C_2H_5OH \times \frac{1 \text{ cm}^3}{1 \text{ mL}} \times \frac{0.789 \text{ g } C_2H_5OH}{1 \text{ cm}^3} \times \frac{1 \text{ mol } C_2H_5OH}{46.07 \text{ g } C_2H_5OH} \times \frac{6.022 \times 10^{23} \text{ molecules } C_2H_5OH}{1 \text{ mol } C_2H_5OH}$$

$$= 1.50 \times 10^{24} \text{ molecules } C_2H_5OH$$

Check: The units of the answer (molecules C_2H_5OH) are correct. The magnitude is reasonable because we had more than 2 moles of C_2H_5OH and we have more than two times Avogadro's number of molecules.

4.117   (a)   To write the formula for an ionic compound, do the following: (1) Write the symbol for the metal cation and its charge and the symbol for the nonmetal anion or polyatomic anion and its charge. (2) Adjust the subscript on each cation and anion to balance the overall charge. (3) Check that the sum of the charges of the cations equals the sum of the charges of the anions.
        potassium chromate: $K^+ \ CrO_4^{2-}$; K_2CrO_4 cation $2(1+) = 2+$; anion $2-$
        Given: K_2CrO_4 **Find:** mass percent of each element
        Conceptual Plan: % K, then % Cr, then % O

$$\text{mass \% K} = \frac{2 \times \text{molar mass K}}{\text{molar mass } K_2CrO_4} \times 100\% \quad \text{mass \% Cr} = \frac{1 \times \text{molar mass Cr}}{\text{molar mass } K_2CrO_4} \times 100\% \quad \text{mass \% O} = \frac{4 \times \text{molar mass O}}{\text{molar mass } K_2CrO_4} \times 100\%$$

molar mass K = 39.10 g/mol; molar mass Cr = 52.00 g/mol; molar mass O = 16.00 g/mol

Solution: molar mass K_2CrO_4 = 2(39.10 g/mol) + 1(52.00 g/mol) + 4(16.00 g/mol) = 194.20 g/mol
2 × molar mass K = 2(39.10 g/mol) = 78.20 g K   1 × molar mass Cr = 1(52.00 g/mol) = 52.00 g Cr

$$\text{mass \% K} = \frac{2 \times \text{molar mass K}}{\text{molar mass } K_2CrO_4} \times 100\% \qquad \text{mass \% Cr} = \frac{1 \times \text{molar mass Cr}}{\text{molar mass } K_2CrO_4} \times 100\%$$

$$= \frac{78.20 \text{ g/mol}}{194.20 \text{ g/mol}} \times 100\% \qquad\qquad = \frac{52.00 \text{ g/mol}}{194.20 \text{ g/mol}} \times 100\%$$

$$= 40.27\% \qquad\qquad\qquad = 26.78\%$$

$$4 \times \text{molar mass O} = 4(16.00 \text{ g/mol}) = 64.00 \text{ g O}$$

$$\text{mass \% O} = \frac{4 \times \text{molar mass O}}{\text{molar mass } K_2CrO_4} \times 100\%$$

$$= \frac{64.00 \text{ g/mol}}{194.20 \text{ g/mol}} \times 100\%$$

$$= 32.96\%$$

Check: The units of the answer (%) are correct. The magnitude is reasonable because each is between 0 and 100% and the total is 100%.

(b)   To write the formula for an ionic compound, do the following: (1) Write the symbol for the metal cation and its charge and the symbol for the nonmetal anion or polyatomic anion and its charge. (2) Adjust the subscript on each cation and anion to balance the overall charge. (3) Check that the sum of the charges of the cations equals the sum of the charges of the anions.

Lead(II)phosphate: Pb^{2+} PO_4^{3-}; $Pb_3(PO_4)_2$ cation $3(2+) = 6+$; anion $2(3-) = 6-$

Given: $Pb_3(PO_4)_2$ **Find:** mass percent of each element

Conceptual Plan: % Pb, then % P, then % O

$$\text{mass \% Pb} = \frac{3 \times \text{molar mass Pb}}{\text{molar mass } Pb_3(PO_4)_2} \times 100\% \quad \text{mass \% P} = \frac{2 \times \text{molar mass P}}{\text{molar mass } Pb_3(PO_4)_2} \times 100\% \quad \text{mass \% O} = \frac{8 \times \text{molar mass O}}{\text{molar mass } Pb_3(PO_4)_2} \times 100\%$$

Solution: molar mass $Pb_3(PO_4)_2 = 3(207.2 \text{ g/mol}) + 2(30.97 \text{ g/mol}) + 8(16.00 \text{ g/mol}) = 811.5 \text{ g/mol}$

$3 \times$ molar mass Pb $= 3(207.2 \text{ g/mol}) = 621.6 \text{ g Pb}$ $2 \times$ molar mass P $= 2(30.97 \text{ g/mol}) = 61.94 \text{ g P}$

$$\text{mass \% Pb} = \frac{3 \times \text{molar mass Pb}}{\text{molar mass } Pb_3(PO_4)_2} \times 100\% \qquad\qquad \text{mass \% P} = \frac{2 \times \text{molar mass P}}{\text{molar mass } Pb_3(PO_4)_2} \times 100\%$$

$$= \frac{621.6 \text{ g/mol}}{811.5 \text{ g/mol}} \times 100\% \qquad\qquad\qquad = \frac{61.94 \text{ g/mol}}{811.5 \text{ g/mol}} \times 100\%$$

$$= 76.60\% \qquad\qquad\qquad\qquad\qquad = 7.633\%$$

$$4 \times \text{molar mass O} = 8(16.00 \text{ g/mol}) = 128.0 \text{ g O}$$

$$\text{mass \% O} = \frac{8 \times \text{molar mass O}}{\text{molar mass } Pb_3(PO_4)_2} \times 100\%$$

$$= \frac{128.0 \text{ g/mol}}{811.5 \text{ g/mol}} \times 100\%$$

$$= 15.77\%$$

Check: The units of the answer (%) are correct. The magnitude is reasonable because each is between 0 and 100% and the total is 100%.

(c)   To write the formula for an ionic compound, do the following: (1) Write the symbol for the metal cation and its charge and the symbol for the nonmetal anion or polyatomic anion and its charge. (2) Adjust the subscript on each cation and anion to balance the overall charge. (3) Check that the sum of the charges of the cations equals the sum of the charges of the anions.

cobalt(II)bromide: Co^{2+} Br^-; $CoBr_2$ cation $2+ = 2+$; anion $2(1-) = 2-$

Given: $CoBr_2$ **Find:** mass percent of each element

Conceptual Plan: % Co, then % Br

$$\text{mass \% Co} = \frac{1 \times \text{molar mass Co}}{\text{molar mass } CoBr_2} \times 100\% \qquad \text{mass \% Br} = \frac{2 \times \text{molar mass Br}}{\text{molar mass } CoBr_2} \times 100\%$$

Solution: molar mass $CoBr_2 = (58.93 \text{ g/mol}) + 2(79.90 \text{ g/mol}) = 218.73 \text{ g/mol}$

$2 \times$ molar mass Co $= 1(58.93 \text{ g/mol}) = 58.93 \text{ g Co}$ $1 \times$ molar mass Br $= 2(79.90 \text{ g/mol}) = 159.80 \text{ g Br}$

$$\text{mass \% Co} = \frac{1 \times \text{molar mass Co}}{\text{molar mass } CoBr_2} \times 100\% \qquad\qquad \text{mass \% Br} = \frac{2 \times \text{molar mass Br}}{\text{molar mass } CoBr_2} \times 100\%$$

$$= \frac{58.93 \text{ g/mol}}{218.73 \text{ g/mol}} \times 100\% \qquad\qquad\qquad = \frac{159.80 \text{ g/mol}}{218.73 \text{ g/mol}} \times 100\%$$

$$= 26.94\% \qquad\qquad\qquad\qquad\qquad = 73.058\%$$

Check: The units of the answer (%) are correct. The magnitude is reasonable because each is between 0 and 100% and the total is 100%.

4.119 **Given:** 25 g CF_2Cl_2/mo **Find:** g Cl_2/yr
Conceptual Plan: g CF_2Cl_2/mo → g Cl_2/mo → g Cl_2/yr

$$\frac{70.90 \text{ g } Cl_2}{120.91 \text{ g } CF_2Cl_2} \qquad \frac{12 \text{ mo}}{1 \text{ yr}}$$

Solution: $\dfrac{25 \text{ g } CF_2Cl_2}{\text{mo}} \times \dfrac{70.90 \text{ g } Cl_2}{120.91 \text{ g } CF_2Cl_2} \times \dfrac{12 \text{ mo}}{1 \text{ yr}} = 1.8 \times 10^2 \text{ g } Cl_2/\text{yr}$

Check: The units of the answer (g Cl_2/yr) are correct. Magnitude is reasonable because it is less than the total CF_2Cl_2/yr.

4.121 **Given:** MCl_3; 65.57% Cl **Find:** identify M
Conceptual Plan: g Cl → mol Cl → mol M → atomic mass M

$$\frac{1 \text{ mol Cl}}{35.45 \text{ g Cl}} \qquad \frac{1 \text{ mol M}}{3 \text{ mol Cl}} \qquad \frac{g \text{ M}}{\text{mol M}}$$

Solution: in 100 g sample: 65.57 g Cl; 34.43 g M

$$65.57 \text{ g Cl} \times \frac{1 \text{ mol Cl}}{35.45 \text{ g Cl}} \times \frac{1 \text{ mol M}}{3 \text{ mol Cl}} = 0.6165 \text{ mol M} \qquad \frac{34.43 \text{ g M}}{0.6165 \text{ mol M}} = 55.85 \text{ g/mol M}$$

molar mass of 55.85 = Fe
The identity of M = Fe.

4.123 **Given:** in a 100 g sample of estradiol: 79.37 g C; 8.88 g H; 11.75 g O; molar mass = 272.37 g/mol **Find:** molecular formula
Conceptual Plan:
Convert mass to mol of each element → pseudoformula → empirical formula → molecular formula

$$\frac{1 \text{ mol C}}{12.01 \text{ g C}} \qquad \frac{1 \text{ mol H}}{1.008 \text{ g H}} \qquad \frac{1 \text{ mol O}}{16.00 \text{ g O}} \qquad \text{divide by smallest number} \qquad \text{empirical formula} \times n$$

Solution: $79.37 \text{ g C} \times \dfrac{1 \text{ mol C}}{12.01 \text{ g C}} = 6.609 \text{ mol C}$

$8.88 \text{ g H} \times \dfrac{1 \text{ mol H}}{1.008 \text{ g H}} = 8.81 \text{ mol H}$

$11.75 \text{ g O} \times \dfrac{1 \text{ mol O}}{16.00 \text{ g O}} = 0.7344 \text{ mol O}$

$C_{6.609} H_{8.81} O_{0.7344}$

$C_{\frac{6.609}{0.7344}} H_{\frac{8.81}{0.7344}} O_{\frac{0.7344}{0.7344}} \rightarrow C_9H_{12}O$

The correct empirical formula is $C_9H_{12}O$.

empirical formula mass = 9(12.01 g/mol) + 12(1.008 g/mol) + 1(16.00 g/mol) = 136.19 g/mol

$$n = \frac{\text{molar mass}}{\text{formula molar mass}} = \frac{272.37 \text{ g/mol}}{136.19 \text{ g/mol}} = 2$$

molecular formula $= C_9H_{12}O \times 2 = C_{18}H_{24}O_2$

4.125 **Given:** 13.42 g sample of equilin; 39.61 g CO_2; 9.01 g H_2O; molar mass = 268.34 g/mol
Find: molecular formula

Conceptual Plan:
Mass CO_2, H_2O → mol CO_2, H_2O → mol C, mol H → mass C, mass H, mass O → mol O →

$$\frac{1 \text{ mol } CO_2}{44.01 \text{ g } CO_2} \quad \frac{1 \text{ mol } H_2O}{18.02 \text{ g } H_2O} \quad \frac{1 \text{ mol C}}{1 \text{ mol } CO_2} \quad \frac{2 \text{ mol H}}{1 \text{ mol } H_2O} \quad \frac{12.01 \text{ g C}}{1 \text{ mol C}} \quad \frac{1.008 \text{ g H}}{1 \text{ mol H}} \quad \text{g sample} - \text{g C} - \text{g H} \quad \frac{1 \text{ mol O}}{16.00 \text{ g O}}$$

pseudoformula → empirical formula → molecular formula

$$\text{divide by smallest number} \qquad \text{empirical formula} \times n$$

$$39.61 \text{ g } CO_2 \times \frac{1 \text{ mol } CO_2}{44.01 \text{ g } CO_2} = 0.9000 \text{ mol } CO_2$$

$$9.01 \text{ g } H_2O \times \frac{1 \text{ mol } H_2O}{18.02 \text{ g } H_2O} = 0.500 \text{ mol } H_2O$$

$$0.9000 \text{ mol } CO_2 \times \frac{1 \text{ mol } C}{1 \text{ mol } CO_2} = 0.9000 \text{ mol } C$$

$$0.500 \text{ mol } H_2O \times \frac{2 \text{ mol } H}{1 \text{ mol } H_2O} = 1.00 \text{ mol } H$$

$$0.9000 \text{ mol } C \times \frac{12.01 \text{ g } C}{1 \text{ mol } C} = 10.81 \text{ g } C$$

$$1.00 \text{ mol } H \times \frac{1.008 \text{ g } H}{1 \text{ mol } H} = 1.01 \text{ g } H$$

$$13.42 \text{ g} - 10.81 \text{ g} - 1.01 \text{ g} = 1.60 \text{ g } O$$

$$1.60 \text{ g } O \times \frac{1 \text{ mol } O}{16.00 \text{ g } O} = 0.100 \text{ mol } O$$

$$C_{0.9000} H_{1.000} O_{0.100}$$

$$C_{\frac{0.9000}{0.100}} H_{\frac{1.00}{0.100}} O_{\frac{0.100}{0.100}} \rightarrow C_9H_{10}O$$

The correct empirical formula is $C_9H_{10}O$.

$$\text{empirical formula mass} = 9(12.01 \text{ g/mol}) + 10(1.008 \text{ g/mol}) + 1(16.00 \text{ g/mol}) = 134.2 \text{ g/mol}$$

$$n = \frac{\text{molar mass}}{\text{formula molar mass}} = \frac{268.34 \text{ g/mol}}{134.2 \text{ g/mol}} = 2$$

$$\text{molecular formula} = C_9H_{10}O \times 2 = C_{18}H_{20}O_2$$

4.127   **Given:** 4.93 g $MgSO_4 \cdot xH_2O$; 2.41 g $MgSO_4$ **Find:** value of x

Conceptual Plan: g $MgSO_4 \rightarrow$ mol $MgSO_4$ g $H_2O \rightarrow$ mol H_2O Determine mole ratio.

$$\frac{1 \text{ mol } MgSO_4}{120.38 \text{ g } MgSO_4} \qquad \frac{1 \text{ mol } H_2O}{18.02 \text{ g } H_2O} \qquad \frac{\text{mol } HO_2}{\text{mol } MgSO_4}$$

Solution:

$$2.41 \text{ g } MgSO_4 \times \frac{1 \text{ mol } MgSO_4}{120.38 \text{ g } MgSO_4} = 0.0200 \text{ mol } MgSO_4$$

Determine g H_2O: 4.93 g $MgSO_4 \cdot xH_2O$ − 2.41 g $MgSO_4$ = 2.52 g H_2O

$$2.52 \text{ g } H_2O \times \frac{1 \text{ mol } H_2O}{18.02 \text{ g } H_2O} = 0.140 \text{ mol } H_2O$$

$$\frac{0.140 \text{ mol } H_2O}{0.0200 \text{ mol } MgSO_4} = 7$$

$$x = 7$$

4.129   **Given:** molar mass = 177 g/mol; g C = 8(g H)   **Find:** molecular formula

Conceptual Plan: C_xH_yBrO

Solution: in 1 mol compound, let x = mol C and y = mol H, assume mol Br = 1, assume mol O = 1

$$177 \text{ g/mol} = x(12.01 \text{ g/mol}) + y(1.008 \text{ g/mol}) + 1(79.90 \text{ g/mol}) + 1(16.00 \text{ g/mol})$$

$$x(12.01 \text{ g/mol}) = 8[y(1.008 \text{ g/mol})]$$

$$177 \text{ g/mol} = 8y(1.008 \text{ g/mol}) + y(1.008 \text{ g/mol}) + 79.90 \text{ g/mol} + 16.00 \text{ g/mol}$$

$$81 = 9y(1.008)$$

$$y = 9 = \text{mol H}$$

$$x(12.01) = 8 \times 9(1.008)$$

$$x = 6 = \text{mol C}$$

$$\text{molecular formula} = C_6H_9BrO$$

Check: molar mass $= 6(12.01 \text{ g/mol}) + 9(1.008 \text{ g/mol}) + 1(79.90 \text{ g/mol}) + 1(16.00 \text{ g/mol}) = 177.0 \text{ g/mol}$

4.131 **Given:** 23.5 mg $C_{17}H_{22}ClNO_4$ **Find:** total number of atoms

Conceptual Plan: mg compound → g compound → mol compound → mol atoms → number of atoms

$$\frac{1 \text{ g}}{1000 \text{ mg}} \qquad \frac{1 \text{ mol compound}}{339.8 \text{ g}} \qquad \frac{45 \text{ mol atoms}}{1 \text{ mol compound}} \qquad \frac{6.022 \times 10^{23} \text{ atoms}}{1 \text{ mol atoms}}$$

Solution:

$$23.5 \text{ mg} \times \frac{1 \text{ g}}{1000 \text{ mg}} \times \frac{1 \text{ mol compound}}{339.8 \text{ g}} \times \frac{45 \text{ mol atoms}}{1 \text{ mol compound}} \times \frac{6.022 \times 10^{23} \text{ atoms}}{1 \text{ mol atoms}} = 1.87 \times 10^{21} \text{ atoms}$$

Check: The units of the answer (number of atoms) are correct. The magnitude of the answer is reasonable because the molecule is so complex.

4.133 **Given:** MCl_3; 2.395 g sample; 3.606×10^{-2} mol Cl **Find:** atomic mass M

Conceptual Plan: mol Cl → g Cl → g X

$$\frac{35.45 \text{ g Cl}}{1 \text{ mol Cl}} \qquad \text{g sample} - \text{g Cl} = \text{g M}$$

mol Cl → mol M → atomic mass M

$$\frac{1 \text{ mol M}}{3 \text{ mol Cl}} \qquad \frac{\text{g M}}{\text{mol M}}$$

Solution:

$$3.606 \times 10^{-2} \text{ mol Cl} \times \frac{35.45 \text{ g Cl}}{1 \text{ mol Cl}} = 1.278 \text{ g Cl}$$

$$2.395 \text{ g} - 1.278 \text{ g} = 1.117 \text{ g M}$$

$$3.606 \times 10^{-2} \text{ mol Cl} \times \frac{1 \text{ mol M}}{3 \text{ mol Cl}} = 1.202 \times 10^{-2} \text{ mol M}$$

$$\frac{1.117 \text{ g M}}{0.01202 \text{ mol M}} = 92.93 \text{ g/mol M}$$

$$\text{molar mass of M} = 92.93 \text{ g/mol}$$

4.135 **Given:** $Fe_xCr_yO_4$; 28.59% O **Find:** x and y

Conceptual Plan: % O → molar mass $Fe_xCr_yO_4$ → mass Fe + Cr

$$\frac{\text{mass O}}{\text{molar mass compound}} \times 100\% = \% \text{ O} \qquad \text{mass cpd} - \text{mass O} = \text{mass Fe} + \text{Cr}$$

Solution: $\dfrac{28.59 \text{ g O}}{100.0} = \dfrac{64.00 \text{ g O}}{\text{molar mass cpd}}$ molar mass $= 223.9 \text{ g/mol}$

Mass Fe + Cr = molar mass $- (4 \times \text{molar mass O}) = 223.9 - 64.00 = 159.9 \text{ g}$

Molar mass Fe = 55.85; molar mass Cr = 52.00

Because the mass of the two metals is close, the average mass can be used to determine the total moles of Fe and

Cr present in the compound. Average mass of Fe and Cr = 53.9. $\dfrac{159.9 \text{ g}}{53.9 \text{ g/mol}} = 2.97 = 3 \text{ mol metal}$

Let $x = \text{mol Fe}$ and $y = \text{mol Cr}$

x mol Fe $+ y$ mol Cr $= 3$ mol total

x mol Fe$(55.85 \text{ g Fe/mol}) + y$ mol Cr$(52.00 \text{ gCr/mol}) = 159.8$

y mol Cr $= 3 - x$ mol Fe

$x(55.85) + (3 - x)(52.00) = 159.8$

So $x = 1$ and $y = 2$.

Check: Formula $= FeCr_2O_4$ would have a molar mass of Fe $+ 2Cr + 4O = 55.85 + 2(52.00) + 4(16.00) = 223.85$, and the molar mass of the compound is 223.8.

4.137   **Given:** 0.0552% $NaNO_2$; 8.00 oz bag   **Find:** mass Na in bag

Conceptual Plan: oz bag → g bag → g $NaNO_2$ → g Na → mg Na

$$\frac{453.6\ g}{16.00\ oz} \quad \frac{0.0552\ g\ NaNO_2}{100.0\ g\ bag} \quad \frac{22.99\ g\ Na}{69.00\ g\ NaNO_2} \quad \frac{1000\ mg\ Na}{1\ mg\ Na}$$

Solution: $8.00\ \cancel{oz\ bag} \times \dfrac{453.6\ g\ \cancel{bag}}{16.00\ \cancel{oz\ bag}} \times \dfrac{0.0552\ g\ \cancel{NaNO_2}}{100.0\ g\ \cancel{bag}} \times \dfrac{22.99\ g\ \cancel{Na}}{69.00\ g\ \cancel{NaNO_2}} \times \dfrac{1000\ mg\ Na}{1\ \cancel{mg\ Na}} = 41.7\ mg\ Na$

Check: The units of the answer (mg Na) are correct. The magnitude of the answer is reasonable because only a small percentage of the total mass is Na.

Challenge Problems

4.139   **Given:** g NaCl + g NaBr = 2.00 g; g Na = 0.75 g   **Find:** g NaBr

Conceptual Plan:

let x = mol NaCl, y = mol NaBr, then x(molar mass NaCl) = g NaCl, y(molar mass NaBr) = g NaBr

Solution: $x(58.4) + y(102.9) = 2.00$

$\quad\quad x(23.0) + y(23.0) = 0.75 \quad\quad y = 0.0326 - x$

$\quad\quad 58.4x + 102.9(0.0326 - x) = 2.00$

$\quad\quad 58.4x + 3.354 - 102.9x = 2.00$

$\quad\quad 44.5x = 1.354$

$\quad\quad\quad\quad x = 0.03043$ mol NaCl

$\quad\quad\quad\quad y = 0.0326 - 0.03043 = 0.00217$ mol NaBr

$\quad\quad$ g NaBr = $(0.00217)(102.9g/mol) = 0.223$ g NaBr

Check: The units of the answer (g NaBr) are correct. The magnitude is reasonable because it is less than the total mass.

4.141   **Given:** sample of $CaCO_3$ and $(NH_4)_2CO_3$ is 61.9% CO_3^{2-} **Find:** % $CaCO_3$

Conceptual Plan: Let x = $CaCO_3$, y = $(NH_4)_2CO_3$, then x(molar mass $CaCO_3$) = g $CaCO_3$,

$\quad\quad y$[molar mass $(NH_4)_2CO_3$] = g $(NH_4)_2CO_3$

$\quad\quad$ then a 100.0 g sample contains $x(100.0)$ g $CaCO_3$; $y(96.1)$ g $(NH_4)_2CO_3$; 61.9 g CO_3^{2-}

Solution: $x(100.0) + y(96.1) = 100.0$

$\quad\quad\quad x(60.0) + y(60.0) = 61.9 \quad\quad y = 1.0\underline{3}167 - x$

$\quad\quad\quad 100.0x + 96.1(1.03 - x) = 100$

$\quad\quad\quad 100.0x + 99.0 - 96.1x = 100$

$\quad\quad\quad\quad 3.9x \quad = 1.0$

$\quad\quad\quad\quad\quad x = 0.26$ mol $CaCO_3$

$\quad\quad\quad\quad\quad y = 1.032 - 0.26 = 0.77$ mol $(NH_4)_2CO_3$

$\quad\quad$ g $CaCO_3$ = $(0.26\ mol)(100.0g/mol) = 26$ g $CaCO_3$ in a 100 g sample mass % $CaCO_3$ = 26%

Check: The units of the answer (mass % $CaCO_3$) are correct. The magnitude is reasonable because it is between 0 and 100%.

4.143   **Given:** 1.1 kg CF_2Cl_2/automobile; 25% leaked/year; 100×10^6 automobiles   **Find:** kg Cl/yr

Conceptual Plan: kg CF_2Cl_2/auto → kg CF_2Cl_2 leaked/yr → kg Cl/yr/auto → kg Cl

$$\frac{25\ kg\ CF_2Cl_2}{100\ kg\ CF_2Cl_2} \quad\quad \frac{70.9\ kg\ Cl}{120.91\ kg\ CF_2Cl_2} \quad\quad 100 \times 10^6\ auto$$

Solution: $\dfrac{1.1\ kg\ \cancel{CF_2Cl_2}}{\cancel{auto}} \times \dfrac{25\ kg\ \cancel{CF_2Cl_2}}{100\ kg\ \cancel{CF_2Cl_2}} \times \dfrac{70.9\ kg\ Cl}{120.91\ kg\ \cancel{CF_2Cl_2}} \times 100 \times 10^6\ \cancel{auto} = 1.6 \times 10^7$ kg Cl/yr

Check: The units of the answer (kg Cl) are correct. The magnitude is reasonable because it is less than the kilogram CF_2Cl_2 leaked per year.

4.145   **Given:** rock contains: 38.0% PbS; 25.0% $PbCO_3$; 17.4% $PbSO_4$ **Find:** kg rock needed for 5.0 metric ton Pb

Conceptual Plan: Determine kg Pb/100 kg rock then ton Pb → kg Pb → kg rock

$$\frac{1000\ kg}{1\ metric\ ton} \quad \frac{100\ kg\ rock}{64.2\ kg\ rock}$$

Solution: in 100 kg rock:

$$\left(38.0\ \cancel{kg\ PbS} \times \frac{207.2\ kg\ Pb}{239.3\ \cancel{kg\ PbS}}\right) + \left(25.0\ \cancel{kg\ PbCO_3} \times \frac{207.2\ kg\ Pb}{267.2\ \cancel{kg\ PbCO_3}}\right) + \left(17.4\ \cancel{kg\ PbSO_4} \times \frac{207.2\ kg\ Pb}{303.3\ \cancel{kg\ PbSO_4}}\right)$$

$$= 64.2 \text{ kg Pb}$$

$$5.0 \text{ metric ton Pb} \times \frac{1000 \text{ kg Pb}}{1 \text{ metric ton Pb}} \times \frac{100 \text{ kg rock}}{64.2 \text{ kg Pb}} = 7.8 \times 10^3 \text{ kg rock}$$

Check: The units of the answer (kg rock) are correct. Magnitude is reasonable because it is greater than the amount of Pb needed.

4.147   **Given:** molar mass $= 229 \text{ g/mol}$, 6 times mass C as H   **Find:** molecular formula
          Conceptual Plan: Let $x =$ **mass of C, then** $6x =$ **mass of C**
          Solution: in 1 mol of the compound: $g \text{ C} + g \text{ H} + g \text{ S} + g \text{ I} = 229 \text{ g}$

               Because the molar mass of I $= 127$, there cannot be more than 1 mol of I in the compound; so

               $x + 6x + g \text{ S} + 127 = 229.$

               $x + 6x + g \text{ S} = 102$

               If the compound contains 1 mol S, then $7x = 102 - 32 = 70$ and $x = 10 \text{ g H}$ and $6x = 60 \text{ g C}$.

               $10 \text{ g H} \times \dfrac{1 \text{ mol H}}{1.0 \text{ g H}} = 10 \text{ mol H}$

               $60 \text{ g C} \times \dfrac{1 \text{ mol C}}{12 \text{ g C}} = 5 \text{ mol C}$

               1 mol I and 1 mol S; so empirical formula is $C_5H_{10}SI$

          Check: Molar mass of $C_5H_{10}SI = 5(12) + 10(1.0) + 32 + 127 = 229 \text{ g/mol}$, which is the mass given.

4.149   **Given:** compound is 1/3 X by mass; atomic mass X is 1/3 atomic mass Y   **Find:** empirical formula
          Conceptual Plan: Mass X and Y $\rightarrow$ **mass ratio X: Y and then g X** $\rightarrow$ **mol X and g Y** $\rightarrow$ **mol Y and then**

$$\frac{g \text{ X}}{\text{atomic mass X}} \qquad \frac{g \text{ Y}}{\text{atomic mass Y}}$$

          mole ratio

          Solution: $\dfrac{\text{mass X}}{\text{mass Y}} = \dfrac{\frac{1}{3}}{\frac{2}{3}} = \dfrac{1}{2}.$ $\text{mol X} = \dfrac{1 \text{ g}}{\text{atomic mass X}}$ and $\text{mol Y} = \dfrac{2 \text{ g}}{\text{atomic mass Y}}$

               But atomic mass X $= 1/3$ (atomic mass Y)

               $\text{mol X} = \dfrac{1 \text{ g}}{1/3 \text{ (atomic mass Y)}}$ and $\text{mol Y} = \dfrac{2 \text{ g}}{\text{atomic mass Y}}$

               $\dfrac{\text{mol X}}{\text{mol Y}} = \dfrac{\frac{1 \text{ g}}{1/3 \text{ (atomic mass Y)}}}{\frac{2 \text{ g}}{\text{atomic mass Y}}} = 1/6$ empirical formula $= XY_6$

Conceptual Problems

4.151   The sphere in the molecular models represents the electron cloud of the atom. On this scale, the nucleus would be too small to see.

4.153   The statement is incorrect because a chemical formula is based on the ratio of atoms combined, not the ratio of grams combined. The statement should read as follows: The chemical formula for ammonia (NH_3) indicates that ammonia contains three hydrogen atoms to each nitrogen atom.

Questions for Group Work

4.155   In an ionic bond, an electron is transferred from the Na atom to the Cl atom, so a student would move from the Na atom to the Cl atom.
          For the covalent bond, a pair of electrons is shared between the O atom and each of the two H atoms.

4.157   Some examples of similarities are:
- The systematic names of both binary ionic compounds and binary molecular compounds contain two words.
- The systematic names of both binary ionic compounds and binary molecular compounds end in *-ide.*
- The systematic names of both binary ionic compounds and binary molecular compounds indicate how many atoms of each element are in the formula.
- The systematic names of both binary ionic compounds and binary molecular compounds name the element listed first first.

Some examples of differences are:
- Only the systematic names of binary molecular compounds contain Greek prefixes.
- There are no polyatomic group names used in generating the systematic name of a binary molecular compound.
- The systematic names of only ionic compounds can have a variety of endings (*-ide*, *-ate*, and *-ite*).
- Binary molecular compounds end in *-ide.*

Data Interpretation and Analysis

4.159   (a)   The CO_2 concentration was 286 ppmv in 1880 and 402 ppmv in the present day (2015). The increase is 402 ppmv $-$ 286 ppmv $=$ 116 ppmv.

(c) $5.1 \times 10^9 \, \text{km}^3 \, \text{air} \times \dfrac{(10^3 \, \text{m})^3}{(1 \, \text{km})^3} \times \dfrac{0.399 \, \text{cm}^3 \, CO_2}{1 \, \text{m}^3} \times \dfrac{(10^{-2} \, \text{m})^3}{(1 \, \text{cm})^3} \times \dfrac{(10^{-3} \, \text{km})^3}{(1 \, \text{m})^3} = 2.0 \times 10^3 \, \text{km}^3 \, CO_2$

        $= 2.0 \times 10^{12}$ metric tons CO_2

(e)   Based on the information presented in the figure, the average yearly increase in the concentration of CO_2 in the atmosphere is estimated at about 2 ppm CO_2 (looking at the fairly linear increase in CO_2 concentration from 1980 to 2010). Since the concentration of CO_2 in 2010 is about 390 ppm, in 2040 the concentration would be 450 ppm CO_2 [390 ppm $+$ (30 years $\times$ 2 ppm/year)].

5 Chemical Bonding I: Drawing Lewis Structures and Determining Molecular Shapes

Review Questions

5.1   Electronegativity is the ability of an atom to attract electrons to itself in a chemical bond. This results in a polar bond. Electronegativity generally increases across a period in the periodic table. And electronegativity generally decreases down a column (group) in the periodic table. The most electronegative element is fluorine.

5.3   Percent ionic character is defined as the ratio of a bond's actual dipole moment to the dipole moment it would have if the electron were completely transferred from one atom to the other, multiplied by 100.

       A bond in which an electron is completely transferred from one atom to another would have 100% ionic character. However, no bond is 100% ionic. Percent ionic character generally increases as the electronegativity difference increases. In general, bonds with greater than 50% ionic character are referred to as ionic bonds.

5.5   To calculate the dipole moment, we use $\mu = qr$:

       For 100 pm: $\mu = 1.6 \times 10^{-19} \, C \times 100 \, pm \times \dfrac{1 \, m}{10^{12} \, pm} \times \dfrac{1 \, D}{3.34 \times 10^{-30} \, C \cdot m} = 4.8 \, D$

       For 200 pm: $\mu = 1.6 \times 10^{-19} \, C \times 200 \, pm \times \dfrac{1 \, m}{10^{12} \, pm} \times \dfrac{1 \, D}{3.34 \times 10^{-30} \, C \cdot m} = 9.6 \, D$

5.7   The total number of electrons for a Lewis structure of a molecule is the sum of the valence electrons of each atom in the molecule.

       The total number of electrons for the Lewis structure of an ion is found by summing the number of valence electrons for each atom and then subtracting one electron for each positive charge or adding one electron for each negative charge.

5.9   In some cases, we can write resonance structures that are not equivalent. One possible resonance structure may be somewhat better than another. In such cases, the true structure may still be represented as an average of the resonance structures, but with the better resonance structure contributing more to the true structure. Multiple nonequivalent resonance structures may be weighted differently in their contributions to the true overall structure of a molecule.

5.11   The octet rule has some exceptions because not all atoms have eight electrons surrounding them. The three major categories are (1) odd octets—electron species, molecules, or ions with an odd number of electron (e.g., NO); (2) incomplete octets—molecules or ions with fewer than eight electrons around an atom (e.g., BF_3); and (3) expanded octets—molecules or ions with more than eight electrons around an atom (e.g., AsF_5).

5.13   The bond energy of a chemical bond is the energy required to break 1 mole of the bond in the gas phase. Because breaking bonds is endothermic and forming bonds is exothermic, we can calculate the overall enthalpy change as a sum of the enthalpy changes associated with breaking the required bonds in the reactants and forming the required bonds in the products.

5.15   The properties of molecules are directly related to their shape. The sensation of taste, immune response, the sense of smell, and many types of drug action all depend on shape-specific interactions between molecules and proteins.

5.17   The five basic electron geometries are as follows:
1.   Linear, which has two electron groups
2.   Trigonal planar, which has three electron groups
3.   Tetrahedral, which has four electron groups
4.   Trigonal bipyramidal, which has five electron groups
5.   Octahedral, which has six electron groups
An electron group is defined as a lone pair of electrons, a single bond, a multiple bond, or even a single electron.

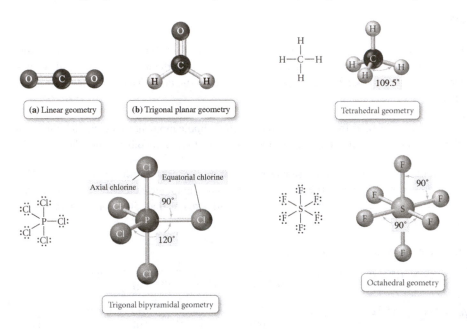

5.19   (a)   Four electron groups give a tetrahedral electron geometry, while three bonding groups and one lone pair give a trigonal pyramidal molecular geometry.
(b)   Four electron groups give a tetrahedral electron geometry, while two bonding groups and two lone pairs give a bent molecular geometry.
(c)   Five electron groups give a trigonal bipyramidal electron geometry, while four bonding groups and one lone pair give a seesaw molecular geometry.
(d)   Five electron groups give a trigonal bipyramidal electron geometry, while three bonding groups and two lone pairs give a T-shaped molecular geometry.
(e)   Five electron groups give a trigonal bipyramidal electron geometry, while two bonding groups and three lone pairs give a linear geometry.
(f)   Six electron groups give an octahedral electron geometry, while five bonding groups and one lone pair give a square pyramidal molecular geometry.
(g)   Six electron groups give an octahedral electron geometry, while four bonding groups and two lone pairs give a square planar molecular geometry.

5.21   To determine whether a molecule is polar, do the following:
1.   Draw the Lewis structure for the molecule and determine the molecular geometry.
2.   Determine whether the molecule contains polar bonds.
3.   Determine whether the polar bonds add together to form a net dipole moment.

Problems by Topic

Electronegativity and Bond Polarity

5.23   (a)   Br and Br: pure covalent   From Figure 5.3, we find that the electronegativity of Br is 2.5. Because both atoms are the same, the electronegativity difference (Δ EN) = 0, and using Table 5.1, we classify this bond as pure covalent.

   (b)   C and Cl: polar covalent   From Figure 5.3, we find that the electronegativity of C is 2.5 and Cl is 3.0. The electronegativity difference (Δ EN) is Δ EN = 3.0 − 2.5 = 0.5. Using Table 5.1, we classify this bond as polar covalent.

   (c)   C and S: pure covalent   From Figure 5.3, we find that the electronegativity of C is 2.5 and S is 2.5. The electronegativity difference (Δ EN) is Δ EN = 2.5 − 2.5 = 0. Using Table 5.1, we classify this bond as pure covalent.

   (d)   Sr and O: ionic   From Figure 5.3, we find that the electronegativity of Sr is 1.0 and O is 3.5. The electronegativity difference (Δ EN) is Δ EN = 3.5 − 1.0 = 2.5. Using Table 5.1, we classify this bond as ionic.

5.25   CO: Write the Lewis structure for each atom based on the number of valence electrons.

$\cdot \ddot{C} \cdot \quad : \ddot{O} \cdot$

The carbon will share three electron pairs with oxygen to achieve a stable octet.
The oxygen atom is more electronegative than the carbon atom; so the oxygen will have a partial negative charge, and the carbon will have a partial positive charge.

$: C \equiv O :$

To estimate the percent ionic character, determine the difference in electronegativity between carbon and oxygen. From Figure 5.3, we find that the electronegativity of C is 2.5 and O is 3.5. The electronegativity difference (Δ EN) is Δ EN = 3.5 − 2.5 = 1.0. From Figure 5.5, we can estimate a percent ionic character of 25%.

Writing Lewis Structures, Resonance, and Formal Charge

5.27   (a) PH_3:   Write the Lewis structure for each atom based on the number of valence electrons.

$\cdot \ddot{P} \cdot \quad \cdot H$

Phosphorus will share an electron pair with each hydrogen to achieve a stable octet.

$H - \ddot{P} - H$
$\quad\ \ |$
$\quad\ \ H$

Check: The phosphorous atom has an octet, and the structure has 1(5) + 3(1) = 8 electrons.

   (b) SCl_2:   Write the Lewis structure for each atom based on the number of valence electrons.

$: \ddot{S} \cdot \quad : \ddot{C}l \cdot$

The sulfur will share an electron pair with each chlorine to achieve a stable octet.

$: \ddot{S} - \ddot{C}l :$
$\ \ |$
$: \ddot{C}l :$

Check: All of the atoms have octets, and the structure has 1(6) + 2(7) = 20 electrons.

   (c)   HI: Write the Lewis structure for each atom based on the number of valence electrons.

$H \cdot \quad \cdot \ddot{I} :$

The iodine will share an electron pair with hydrogen to achieve a stable octet.

$H - \ddot{I} :$

Check: The iodine atom has an octet, and the structure has 1(1) + 1(7) = 8 electrons.

(d) CH_4: Write the Lewis structure for each atom based on the number of valence electrons.

$\cdot \overset{\displaystyle\cdot}{C} \cdot$ H$\cdot$

The carbon will share an electron pair with each hydrogen to achieve a stable octet.

$$\begin{array}{c} H \\ | \\ H-C-H \\ | \\ H \end{array}$$

Check: The carbon atom has an octet, and the structure has $1(4) + 4(1) = 8$ electrons.

5.29 (a) SF_2: Write the Lewis structure for each atom based on the number of valence electrons.

$: \overset{\displaystyle\cdot\cdot}{\underset{\displaystyle\cdot}{S}} \cdot$ $: \overset{\displaystyle\cdot\cdot}{\underset{\displaystyle\cdot\cdot}{F}} \cdot$

The sulfur will share an electron pair with each fluorine to achieve a stable octet.

$$\begin{array}{c} : \overset{\displaystyle\cdot\cdot}{S}-\overset{\displaystyle\cdot\cdot}{\underset{\displaystyle\cdot\cdot}{F}}: \\ | \\ : \overset{\displaystyle\cdot\cdot}{\underset{\displaystyle\cdot\cdot}{F}}: \end{array}$$

Check: All of the atoms have octets, and the structure has $1(6) + 2(7) = 20$ electrons.

(b) SiH_4: Write the Lewis structure for each atom based on the number of valence electrons.

H$\cdot$ $\cdot \overset{\displaystyle\cdot}{\underset{\displaystyle\cdot}{Si}} \cdot$

The silicon will share an electron pair with each hydrogen to achieve a stable octet.

$$\begin{array}{c} H \\ | \\ H-Si-H \\ | \\ H \end{array}$$

Check: The silicon atom has an octet, and the structure has $1(4) + 4(1) = 8$ electrons.

(c) HCOOH: Write the Lewis structure for each atom based on the number of valence electrons.

H$\cdot$ $\cdot \overset{\displaystyle\cdot}{C} \cdot$ $: \overset{\displaystyle\cdot\cdot}{O} \cdot$

The carbon will share an electron pair with hydrogen, an electron pair with the interior oxygen, and two electron pairs with the terminal oxygen to achieve a stable octet. The terminal oxygen will share two electron pairs with carbon to achieve a stable octet. The interior oxygen will share an electron pair with carbon and an electron pair with hydrogen to achieve a stable octet.

$$\begin{array}{c} :O: \\ \| \\ H-C-\overset{\displaystyle\cdot\cdot}{O}-H \end{array}$$

Check: The carbon and oxygen atoms have octets, and the structure has $1(4) + 2(1) + 2(6) = 18$ electrons.

(d) CH_3SH: Write the Lewis structure for each atom based on the number of valence electrons.

H$\cdot$ $\cdot \overset{\displaystyle\cdot}{C} \cdot$ $: \overset{\displaystyle\cdot\cdot}{S} \cdot$

The carbon will share an electron pair with each hydrogen and an electron pair with sulfur to achieve a stable octet. The sulfur will share an electron pair with carbon and an electron pair with hydrogen to achieve a stable octet.

$$\begin{array}{c} H \\ | \\ H-C-\overset{\displaystyle\cdot\cdot}{\underset{\displaystyle\cdot\cdot}{S}}-H \\ | \\ H \end{array}$$

Check: The carbon and sulfur atoms have octets, and the structure has $1(4) + 1(6) + 4(1) = 14$ electrons.

5.31   (a) CI_4:   Write the correct skeletal structure for the molecule.

$$\begin{array}{c} I \\ | \\ I-C-I \\ | \\ I \end{array}$$

Calculate the total number of electrons for the Lewis structure by summing the number of valence electrons of each atom in the molecule.

$$(\text{number of valence } e^- \text{ for C}) + 4(\text{number of valence } e^- \text{ for I}) = 4 + 4(7) = 32$$

Distribute the electrons among the atoms, giving octets to as many atoms as possible. Begin with the bonding electrons; then proceed to lone pairs on terminal atoms and finally to lone pairs on the central atom.

$$\begin{array}{c} :\ddot{I}: \\ | \\ :\ddot{I}-C-\ddot{I}: \\ | \\ :\ddot{I}: \end{array}$$

All 32 valence electrons are used.
If any atom lacks an octet, form double or triple bonds as necessary to give them octets.
All atoms have octets; the structure is complete.

(b) N_2O: Write the correct skeletal structure for the molecule.
N is less electronegative, so it is central.

$$N-N-O$$

Calculate the total number of electrons for the Lewis structure by summing the number of valence electrons of each atom in the molecule.

$$2(\text{number of valence } e^- \text{ for N}) + (\text{number of valence } e^- \text{ for O}) = 2(5) + 6 = 16$$

Distribute the electrons among the atoms, giving octets to as many atoms as possible. Begin with the bonding electrons; then proceed to lone pairs on terminal atoms and finally to lone pairs on the central atom.

$$:\ddot{N}-N-\ddot{O}:$$

All 16 valence electrons are used.
If any atom lacks an octet, form double or triple bonds as necessary.

$$:N\equiv N-\ddot{O}:$$

All atoms have octets; the structure is complete. The formal charges are 0, +1, and −1 for the N, N, and O, respectively.
An alternate structure with two double bonds is not preferred because the formal charges would be −1, +1, and 0, and the most negative formal charges should be on the O atom.

(c) SiH_4:   Write the correct skeletal structure for the molecule.
H is always terminal, so Si is the central atom.

$$\begin{array}{c} H \\ | \\ H-Si-H \\ | \\ H \end{array}$$

Calculate the total number of electrons for the Lewis structure by summing the number of valence electrons of each atom in the molecule.

$$(\text{number of valence } e^- \text{ for Si}) + 4(\text{number of valence } e^- \text{ for H}) = 4 + 4(1) = 8$$

Distribute the electrons among the atoms, giving octets (or duets for H) to as many atoms as possible. Begin with the bonding electrons; then proceed to lone pairs on terminal atoms and finally to lone pairs on the central atom.

$$
\begin{array}{c}
H \\
| \\
H-Si-H \\
| \\
H
\end{array}
$$

All eight valence electrons are used.

If any atom lacks an octet, form double or triple bonds as necessary to give them octets.

All atoms have octets; the structure is complete.

(d) Cl_2CO: Write the correct skeletal structure for the molecule.

C is least electronegative, so it is the central atom.

$$
\begin{array}{c}
O \\
| \\
Cl-C-Cl
\end{array}
$$

Calculate the total number of electrons for the Lewis structure by summing the number of valence electrons of each atom in the molecule.

(number of valence e⁻ for C) + 2(number of valence e⁻ for Cl) + (number of valence e⁻ for O) = 4 + 2(7) + 6 = 24

Distribute the electrons among the atoms, giving octets to as many atoms as possible. Begin with the bonding electrons; then proceed to lone pairs on terminal atoms and finally to lone pairs on the central atom.

$$
\begin{array}{c}
:\ddot{O}: \\
| \\
:\ddot{Cl}-C-\ddot{Cl}:
\end{array}
$$

All 24 valence electrons are used.

If any atom lacks an octet, form double or triple bonds as necessary.

$$
\begin{array}{c}
:O: \\
\| \\
:\ddot{Cl}-C-\ddot{Cl}:
\end{array}
$$

All atoms have octets; the structure is complete.

5.33   (a) N_2H_2:   Write the correct skeletal structure for the molecule.

H—N—N—H

Calculate the total number of electrons for the Lewis structure by summing the number of valence electrons of each atom in the molecule.

2(number of valence e⁻ for N) + 2(number of valence e⁻ for H) = 2(5) + 2(1) = 12

Distribute the electrons among the atoms, giving octets (or duets for H) to as many atoms as possible. Begin with the bonding electrons; then proceed to lone pairs on terminal atoms and finally to lone pairs on the central atom.

H—$\ddot{N}$—$\ddot{N}$—H

All 12 valence electrons are used.

If any atom lacks an octet, form double or triple bonds as necessary.

H—$\ddot{N}$=$\ddot{N}$—H

All atoms have octets (duets for H); the structure is complete.

(b) N_2H_4:   Write the correct skeletal structure for the molecule.

$$
\begin{array}{c}
H \diagdown \qquad \diagup H \\
\quad N-N \\
H \diagup \qquad \diagdown H
\end{array}
$$

Calculate the total number of electrons for the Lewis structure by summing the valence electrons of each atom in the molecule.

2(number of valence e⁻ for N) + 4(number of valence e⁻ for H) = 2(5) + 4(1) = 14

Distribute the electrons among the atoms, giving octets (or duets for H) to as many atoms as possible. Begin with the bonding electrons; then proceed to lone pairs on terminal atoms and finally to lone pairs on the central atom.

$$H_2N-NH_2$$

All 14 valence electrons are used.

If any atom lacks an octet, form double or triple bonds as necessary to give them octets.

All atoms have octets (duets for H); the structure is complete.

(c) C_2H_2:   Write the correct skeletal structure for the molecule.

$$H-C-C-H$$

Calculate the total number of electrons for the Lewis structure by summing the number of valence electrons of each atom in the molecule.

$$2(\text{number of valence e}^- \text{ for C}) + 2(\text{number of valence e}^- \text{ for H}) = 2(4) + 2(1) = 10$$

Distribute the electrons among the atoms, giving octets (or duets for H) to as many atoms as possible. Begin with the bonding electrons; then proceed to lone pairs on terminal atoms and finally to lone pairs on the central atom.

$$H-C-\overset{..}{\underset{..}{C}}-H$$

All 10 valence electrons are used.

If any atom lacks an octet, form double or triple bonds as necessary.

$$H-C\equiv C-H$$

All atoms have octets (duets for H); the structure is complete.

(d) C_2H_4:   Write the correct skeletal structure for the molecule.

$$\underset{H}{\overset{H}{>}}C-C\underset{H}{\overset{H}{<}}$$

Calculate the total number of electrons for the Lewis structure by summing the number of valence electrons of each atom in the molecule.

$$2(\text{number of valence e}^- \text{ for C}) + 4(\text{number of valence e}^- \text{ for H}) = 2(4) + 4(1) = 12$$

Distribute the electrons among the atoms, giving octets (or duets for H) to as many atoms as possible. Begin with the bonding electrons; then proceed to lone pairs on terminal atoms and finally to lone pairs on the central atom.

$$\underset{H}{\overset{H}{>}}\overset{..}{C}-C\underset{H}{\overset{H}{<}}$$

All 12 valence electrons are used.

If any atom lacks an octet, form double or triple bonds as necessary.

$$\underset{H}{\overset{H}{>}}C=C\underset{H}{\overset{H}{<}}$$

All atoms have octets (duets for H); the structure is complete.

5.35   (a) SeO_2:   Write the correct skeletal structure for the molecule.

Se is less electronegative, so it is central.

$$O-Se-O$$

Calculate the total number of electrons for the Lewis structure by summing the number of valence electrons of each atom in the molecule.

$$(\text{number of valence e}^- \text{ for Se}) + 2(\text{number of valence e}^- \text{ for O}) = 6 + 2(6) = 18$$

Distribute the electrons among the atoms, giving octets to as many atoms as possible. Begin with the bonding electrons; then proceed to lone pairs on terminal atoms and finally to lone pairs on the central atom.

$$:\overset{..}{\underset{..}{O}}-\overset{..}{Se}-\overset{..}{\underset{..}{O}}:$$

All 18 valence electrons are used.

If any atom lacks an octet, form double or triple bonds as necessary.

$$:\overset{..}{\underset{..}{O}}-\overset{..}{Se}=\overset{..}{\underset{..}{O}}:$$

All atoms have octets; the structure is complete. However, the double bond can form from either oxygen atom, so there are two resonance forms.

$$:\ddot{O}—\ddot{S}e=\ddot{O}: \longleftrightarrow :\ddot{O}=\ddot{S}e—\ddot{O}:$$

Calculate the formal charge on each atom by finding the number of valence electrons and subtracting the number of lone pair electrons and one-half the number of bonding electrons.

$$:\ddot{O}—\ddot{S}e=\ddot{O}: \longleftrightarrow :\ddot{O}=\ddot{S}e—\ddot{O}:$$

number of valence electrons	6	6	6	6	6	6
− number of lone pair electrons	6	2	4	4	2	6
− 1/2(number of bonding electrons)	1	3	2	2	3	1
Formal charge	−1	+1	0	0	+1	−1

Check: All of the atoms have octets, and the structure has 18 electrons. Atoms with fewer bonds than is expected (oxygen typically has 2 bonds) will have a negative formal charge, and atoms with more bonds than is expected (selenium typically has 2 bonds) will have a positive formal charge.

(b) CO_3^{2-}:   Write the correct skeletal structure for the ion.

$$\begin{array}{c} O \\ | \\ O—C—O \end{array}$$

Calculate the total number of electrons for the Lewis structure by summing the number of valence electrons of each atom in the ion and adding 2 for the 2− charge.

$$3(\text{number of valence e}^- \text{ for O}) + (\text{number of valence e}^- \text{ for C}) + 2 = 3(6) + 4 + 2 = 24$$

Distribute the electrons among the atoms, giving octets to as many atoms as possible. Begin with the bonding electrons; then proceed to lone pairs on terminal atoms and finally to lone pairs on the central atom.

$$\begin{array}{c} :\ddot{O}: \\ | \\ :\ddot{O}—C—\ddot{O}: \end{array}$$

All 24 valence electrons are used.

If any atom lacks an octet, form double or triple bonds as necessary.

$$\begin{array}{c} :\ddot{O}: \\ | \\ :\ddot{O}—C=\ddot{O}: \end{array}$$

Finally, write the Lewis structure in brackets with the charge of the ion in the upper right-hand corner.

$$\left[\begin{array}{c} :\ddot{O}: \\ | \\ :\ddot{O}—C=\ddot{O}: \end{array}\right]^{2-}$$

All atoms have octets; the structure is complete. However, the double bond can form from any oxygen atom, so there are three resonance forms.

$$\left[\begin{array}{c} :\ddot{O}: \\ | \\ :\ddot{O}—C=\ddot{O}: \end{array}\right]^{2-} \longleftrightarrow \left[\begin{array}{c} :\ddot{O}: \\ | \\ :\ddot{O}=C—\ddot{O}: \end{array}\right]^{2-} \longleftrightarrow \left[\begin{array}{c} :O: \\ \| \\ :\ddot{O}—C—\ddot{O}: \end{array}\right]^{2-}$$

Calculate the formal charge on each atom by finding the number of valence electrons and subtracting the number of lone pair electrons and one-half the number of bonding electrons.

$$\left[\begin{array}{c} :\ddot{O}: \\ | \\ :\ddot{O}—C=\ddot{O}: \end{array}\right]^{2-}$$

	O_{left}	O_{top}	O_{right}	C
number of valence electrons	6	6	6	4
− number of lone pair electrons	6	6	4	0
− 1/2(number of bonding electrons)	1	1	2	4
Formal charge	−1	−1	0	0

The sum of the formal charges is −2, which is the overall charge of the ion. The other resonance forms would have the same values for the single- and double-bonded oxygen atoms.

Check: All of the atoms have octets, and the structure has 24 electrons. Atoms with fewer bonds than is expected (oxygen typically has 2 bonds) will have a negative formal charge, and atoms with more bonds than is expected will have a positive formal charge.

(c) ClO^-:   Write the correct skeletal structure for the ion.

$$Cl—O$$

Calculate the total number of electrons for the Lewis structure by summing the number of valence electrons of each atom in the ion and adding 1 for the 1− charge.

$$(\text{number of valence } e^- \text{ for O}) + (\text{number of valence } e^- \text{ for Cl}) + 1 = 6 + 7 + 1 = 14$$

Distribute the electrons among the atoms, giving octets to as many atoms as possible. Begin with the bonding electrons; then proceed to lone pairs on terminal atoms and finally to lone pairs on the central atom.

$$:\!\ddot{C}l—\ddot{O}\!:$$

All 14 valence electrons are used.

If any atom lacks an octet, form double or triple bonds as necessary to give them octets.

Finally, write the Lewis structure in brackets with the charge of the ion in the upper right-hand corner.

$$\left[:\!\ddot{C}l—\ddot{O}\!:\right]^-$$

All atoms have octets; the structure is complete.

Calculate the formal charge on each atom by finding the number of valence electrons and subtracting the number of lone pair electrons and one-half the number of bonding electrons.

	Cl	O
number of valence electrons	7	6
− number of lone pair electrons	6	6
− 1/2(number of bonding electrons)	1	1
Formal charge	0	−1

The sum of the formal charges is −1, which is the overall charge of the ion.

Check: All of the atoms have octets, and the structure has 14 electrons. Atoms with fewer bonds than is expected (oxygen typically has 2 bonds) will have a negative formal charge.

(d) NO_2^-:   Write the correct skeletal structure for the ion.

$$O—N—O$$

Calculate the total number of electrons for the Lewis structure by summing the number of valence electrons of each atom in the ion and adding 1 for the 1− charge.

$$2(\text{number of valence } e^- \text{ for O}) + (\text{number of valence } e^- \text{ for N}) + 1 = 2(6) + 5 + 1 = 18$$

Distribute the electrons among the atoms, giving octets to as many atoms as possible. Begin with the bonding electrons; then proceed to lone pairs on terminal atoms and finally to lone pairs on the central atom.

$$:\!\ddot{O}—\ddot{N}—\ddot{O}\!:$$

All 18 valence electrons are used.

If any atom lacks an octet, form double or triple bonds as necessary.

$$:\ddot{O}=\ddot{N}-\ddot{O}:$$

Finally, write the Lewis structure in brackets with the charge of the ion in the upper right-hand corner.

$$\left[:\ddot{O}=\ddot{N}-\ddot{O}:\right]^{-}$$

All atoms have octets; the structure is complete. However, the double bond can form from either oxygen atom, so there are two resonance forms.

$$\left[:\ddot{O}=\ddot{N}-\ddot{O}:\right]^{-} \longleftrightarrow \left[:\ddot{O}-\ddot{N}=\ddot{O}:\right]^{-}$$

Calculate the formal charge on each atom by finding the number of valence electrons and subtracting the number of lone pair electrons and one-half the number of bonding electrons. Using the left-side structure:

	O	N	O
number of valence electrons	6	5	6
− number of lone pair electrons	4	2	6
− 1/2(number of bonding electrons)	2	3	1
Formal charge	0	0	−1

The sum of the formal charges is −1, which is the overall charge of the ion.

Check: All of the atoms have octets, and the structure has 18 electrons. Atoms with fewer bonds than is expected (oxygen typically has 2 bonds) will have a negative formal charge.

5.37

$$\begin{array}{cc} H & H \\ | & | \\ H-C=\ddot{S} & H-S=\ddot{C} \\ I & II \end{array}$$

Calculate the formal charge on each atom in structure I by finding the number of valence electrons and subtracting the number of lone pair electrons and one-half the number of bonding electrons.

	H_{left}	H_{top}	C	S
number of valence electrons	1	1	4	6
− number of lone pair electrons	0	0	0	4
− 1/2(number of bonding electrons)	1	1	4	2
Formal charge	0	0	0	0

The sum of the formal charges is 0, which is the overall charge of the molecule.

Calculate the formal charge on each atom in structure II by finding the number of valence electrons and subtracting the number of lone pair electrons and one-half the number of bonding electrons.

	H_{left}	H_{top}	S	C
number of valence electrons	1	1	6	4
− number of lone pair electrons	0	0	0	4
− 1/2(number of bonding electrons)	1	1	4	2
Formal charge	0	0	+2	−2

The sum of the formal charges is 0, which is the overall charge of the molecule.

Structure I is the better Lewis structure because it has the least amount of formal charge on each atom.

5.39 $:O\equiv C-\ddot{O}:$

This resonance structure does not provide a significant contribution to the resonance hybrid as it has a +1 formal charge on a very electronegative oxygen.

	O_{left}	O_{right}	C
number of valence electrons	6	6	4
− number of lone pair electrons	2	6	0
− 1/2(number of bonding electrons)	3	1	4
Formal charge	+1	−1	0

Check: Atoms with fewer bonds than is expected (oxygen typically has 2 bonds) will have a negative formal charge, and atoms with more bonds than is expected will have a positive formal charge.

5.41 CH_3COO^-:   Write the correct skeletal structure for the ion.

$$\begin{array}{ccc} H & & O \\ | & & | \\ H-C- & \!\!\!\!C & \\ | & & | \\ H & & O \end{array}$$

Calculate the total number of electrons for the Lewis structure by summing the valence electrons of each atom in the ion and adding 1 for the 1− charge.

3(number of valence e^- for H) + 2(number of valence e^- for C) + 2(number of valence e^- for O) + 1 = 3(1) + 2(4) + 2(6) + 1 = 24

Distribute the electrons among the atoms, giving octets (or duets for H) to as many atoms as possible. Begin with the bonding electrons; then proceed to lone pairs on terminal atoms and finally to lone pairs of the central atom.

All 24 valence electrons are used.

If any atom lacks an octet, form double or triple bonds as necessary to give them octets.

Finally, write the Lewis structure in brackets with the charge of the ion in the upper right-hand corner.

All atoms have octets (duets for H); the structure is complete. However, the double bond can form from any oxygen atom, so there are two resonance forms.

Calculate the formal charge on each atom by finding the number of valence electrons and subtracting the number of lone pair electrons and one-half the number of bonding electrons.

	C_{left}	C_{right}	O_{top}	O_{bottom}	C_{left}	C_{right}	O_{top}	O_{bottom}
number of valence electrons	4	4	6	6	4	4	6	6
− number of lone pair electrons	0	0	6	4	0	0	4	6
− 1/2(number of bonding electrons)	4	4	1	2	4	4	2	1
Formal charge	0	0	−1	0	0	0	0	−1

The sum of the formal charges is −1, which is the overall charge of the ion.

5.43

$$CH_3—N—\ddot{\underset{..}{O}}:$$

with CH_3 groups above and below the N.

	N	O
number of valence electrons	5	6
− number of lone pair electrons	0	6
− 1/2(number of bonding electrons)	4	1
Formal charge	+1	−1

Check: Atoms with fewer bonds than is expected (oxygen typically has 2 bonds) will have a negative formal charge, and atoms with more bonds than is expected (nitrogen typically has 3 bonds) will have a positive formal charge.

Odd-Electron Species, Incomplete Octets, and Expanded Octets

5.45   (a) BCl_3:   Write the correct skeletal structure for the molecule.

B is less electronegative, so it is central.

$$Cl—B—Cl$$

with Cl above the B.

Calculate the total number of electrons for the Lewis structure by summing the number of valence electrons of each atom in the molecule.

(number of valence e⁻ for B) + 3(number of valence e⁻ for Cl) = 3 + 3(7) = 24

Distribute the electrons among the atoms, giving octets to as many atoms as possible. Begin with the bonding electrons; then proceed to lone pairs on terminal atoms and finally to lone pairs on the central atom.

$$:\ddot{\underset{..}{Cl}}—B—\ddot{\underset{..}{Cl}}:$$

with $:\ddot{\underset{..}{Cl}}:$ above the B.

All 24 valence electrons are used.

B has an incomplete octet. If we complete the octet, there is a formal charge of −1 on the B, which is less electronegative than Cl.

(b) NO_2:   Write the correct skeletal structure for the molecule.

N is less electronegative, so it is central.

$$O—N—O$$

Calculate the total number of electrons for the Lewis structure by summing the number of valence electrons of each atom in the molecule.

(number of valence e⁻ for N) + 2(number of valence e⁻ for O) = 5 + 2(6) = 17

Distribute the electrons among the atoms, giving octets to as many atoms as possible. Begin with the bonding electrons; then proceed to lone pairs on terminal atoms and finally to lone pairs on the central atom.

$$\ddot{\underset{..}{O}}=\dot{N}—\ddot{\underset{..}{O}}: \longleftrightarrow :\ddot{\underset{..}{O}}—\dot{N}=\ddot{O}$$

All 17 valence electrons are used.

N has an incomplete octet. It has seven electrons because we have an odd number of valence electrons.

(c) BH_3:   Write the correct skeletal structure for the molecule.

B is less electronegative, so it is central.

$$H—B—H$$

with H above the B.

Calculate the total number of electrons for the Lewis structure by summing the number of valence electrons of each atom in the molecule.

(number of valence e⁻ for B) + 3(number of valence e⁻ for H) = 3 + 3(1) = 6

Distribute the electrons among the atoms, giving octets (or duets for H) to as many atoms as possible. Begin with the bonding electrons; then proceed to lone pairs on terminal atoms and finally to lone pairs on the central atom.

$$\begin{array}{c} H \\ | \\ H\!-\!B\!-\!H \end{array}$$

All 6 valence electrons are used.

B has an incomplete octet. H cannot double-bond, so it is not possible to complete the octet on B with a double bond.

5.47   (a) PO_4^{3-}:   Write the correct skeletal structure for the ion.

$$\begin{array}{c} O \\ | \\ O\!-\!P\!-\!O \\ | \\ O \end{array}$$

Calculate the total number of electrons for the Lewis structure by summing the number of valence electrons of each atom in the ion and adding 3 for the 3− charge.

$$4(\text{number of valence e}^- \text{ for O}) + (\text{number of valence e}^- \text{ for P}) + 3 = 4(6) + 5 + 3 = 32$$

Distribute the electrons among the atoms, giving octets to as many atoms as possible. Begin with the bonding electrons; then proceed to lone pairs on terminal atoms and finally to lone pairs on the central atom.

$$\begin{array}{c} :\ddot{O}: \\ | \\ :\ddot{O}\!-\!P\!-\!\ddot{O}: \\ | \\ :\ddot{O}: \end{array}$$

All 32 valence electrons are used.

Finally, write the Lewis structure in brackets with the charge of the ion in the upper right-hand corner.

$$\left[\begin{array}{c} :\ddot{O}: \\ | \\ :\ddot{O}\!-\!P\!-\!\ddot{O}: \\ | \\ :\ddot{O}: \end{array}\right]^{3-}$$

All atoms have octets (duets for H); the structure is complete.

Calculate the formal charge on each atom by finding the number of valence electrons and subtracting the number of lone pair electrons and one-half the number of bonding electrons.

$$\left[\begin{array}{c} :\ddot{O}: \\ | \\ :\ddot{O}\!-\!P\!-\!\ddot{O}: \\ | \\ :\ddot{O}: \end{array}\right]^{3-}$$

	O_{left}	O_{top}	O_{right}	O_{bottom}	P
number of valence electrons	6	6	6	6	5
− number of lone pair electrons	6	6	6	6	0
− 1/2(number of bonding electrons)	1	1	1	1	4
Formal charge	−1	−1	−1	−1	+1

The sum of the formal charges is −3, which is the overall charge of the ion. However, we can write a resonance structure with a double bond to an oxygen because P can expand its octet. This leads to lower formal charges on P and O.

$$\left[\begin{array}{c} :\ddot{O}: \\ | \\ \ddot{O}=P-\ddot{O}: \\ | \\ :\ddot{O}: \end{array}\right]^{3-} \longleftrightarrow \left[\begin{array}{c} :O: \\ || \\ :\ddot{O}-P-\ddot{O}: \\ | \\ :\ddot{O}: \end{array}\right]^{3-} \longleftrightarrow \left[\begin{array}{c} :\ddot{O}: \\ | \\ :\ddot{O}-P=\ddot{O} \\ | \\ :\ddot{O}: \end{array}\right]^{3-} \longleftrightarrow \left[\begin{array}{c} :\ddot{O}: \\ | \\ :\ddot{O}-P-\ddot{O}: \\ || \\ :O: \end{array}\right]^{3-}$$

Using the leftmost structure, calculate the formal charge on each atom by finding the number of valence electrons and subtracting the number of lone pair electrons and one-half the number of bonding electrons.

	O_{left}	O_{top}	O_{right}	O_{bottom}	P
number of valence electrons	6	6	6	6	5
− number of lone pair electrons	4	6	6	6	0
− 1/2(number of bonding electrons)	2	1	1	1	5
Formal charge	0	−1	−1	−1	0

The sum of the formal charges is −3, which is the overall charge of the ion. These resonance forms would all have the lower formal charges associated with the double-bonded O and P.

(b)   CN⁻:   Write the correct skeletal structure for the ion.

$$C-N$$

Calculate the total number of electrons for the Lewis structure by summing the number of valence electrons of each atom in the ion and adding 1 for the 1− charge.

(number of valence e⁻ for C) + (number of valence e⁻ for N) + 1 = 4 + 5 + 1 = 10

Distribute the electrons among the atoms, giving octets to as many atoms as possible. Begin with the bonding electrons; then proceed to lone pairs on terminal atoms and finally to lone pairs on the central atom.

$$:C-\ddot{N}:$$

All 10 valence electrons are used.

If any atom lacks an octet, form double or triple bonds as necessary.

$$:C\equiv N:$$

Finally, write the Lewis structure in brackets with the charge of the ion in the upper right-hand corner.

$$\left[:C\equiv N:\right]^{-}$$

All atoms have octets; the structure is complete.

Calculate the formal charge on each atom by finding the number of valence electrons and subtracting the number of lone pair electrons and one-half the number of bonding electrons.

$$\left[:C\equiv N:\right]^{-}$$

	C	N
number of valence electrons	4	5
− number of lone pair electrons	2	2
− 1/2(number of bonding electrons)	3	3
Formal charge	−1	0

The sum of the formal charges is −1, which is the overall charge of the ion.

(c) SO_3^{2-}:   Write the correct skeletal structure for the ion.

$$
\begin{array}{c}
O \\
| \\
O-S-O
\end{array}
$$

Calculate the total number of electrons for the Lewis structure by summing the valence electrons of each atom in the ion and adding 2 for the 2− charge.

$$3(\text{number of valence e}^- \text{ for O}) + (\text{number of valence e}^- \text{ for S}) + 2 = 3(6) + 6 + 2 = 26$$

Distribute the electrons among the atoms, giving octets to as many atoms as possible. Begin with the bonding electrons; then proceed to lone pairs on terminal atoms and finally to lone pairs on the central atom.

$$
\begin{array}{c}
:\ddot{O}: \\
| \\
:\ddot{O}-S-\ddot{O}:
\end{array}
$$

All 26 valence electrons are used.

Finally, write the Lewis structure in brackets with the charge of the ion in the upper right-hand corner.

$$
\left[\begin{array}{c}
:\ddot{O}: \\
| \\
:\ddot{O}-\ddot{S}-\ddot{O}:
\end{array}\right]^{2-}
$$

Calculate the formal charge on each atom by finding the number of valence electrons and subtracting the number of lone pair electrons and one-half the number of bonding electrons.

$$
\left[\begin{array}{c}
:\ddot{O}: \\
| \\
:\ddot{O}-S-\ddot{O}:
\end{array}\right]^{2-}
$$

	O_{left}	O_{top}	O_{right}	S
number of valence electrons	6	6	6	6
− number of lone pair electrons	6	6	6	2
− 1/2(number of bonding electrons)	1	1	1	3
Formal charge	−1	−1	−1	+1

The sum of the formal charges is −2, which is the overall charge of the ion. However, we can write a resonance structure with a double bond to an oxygen because S can expand its octet. This leads to a lower formal charge.

$$
\left[\begin{array}{c}
:\ddot{O}: \\
\| \\
\ddot{O}=S-\ddot{O}:
\end{array}\right]^{2-}
\longleftrightarrow
\left[\begin{array}{c}
:O: \\
\| \\
:\ddot{O}-S-\ddot{O}:
\end{array}\right]^{2-}
\longleftrightarrow
\left[\begin{array}{c}
:\ddot{O}: \\
| \\
:\ddot{O}-S=\ddot{O}
\end{array}\right]^{2-}
$$

Using the leftmost resonance form, calculate the formal charge on each atom by finding the number of valence electrons and subtracting the number of lone pair electrons and one-half the number of bonding electrons.

	O_{left}	O_{top}	O_{right}	S
number of valence electrons	6	6	6	6
− number of lone pair electrons	4	6	6	2
− 1/2(number of bonding electrons)	2	1	1	4
Formal charge	0	−1	−1	0

The sum of the formal charges is −2, which is the overall charge of the ion. These resonance forms would all have the lower formal charge on the double-bonded O and S.

(d) ClO_2^-:   Write the correct skeletal structure for the ion.

$$O-Cl-O$$

Calculate the total number of electrons for the Lewis structure by summing the number of valence electrons of each atom in the ion and adding 1 for the 1− charge.

$$2(\text{number of valence } e^- \text{ for O}) + (\text{number of valence } e^- \text{ for Cl}) + 1 = 2(6) + 7 + 1 = 20$$

Distribute the electrons among the atoms, giving octets (or duets for H) to as many atoms as possible. Begin with the bonding electrons; then proceed to lone pairs on terminal atoms and finally to lone pairs on the central atom.

$$:\ddot{\text{O}}—\ddot{\text{C}}\text{l}—\ddot{\text{O}}:$$

All 20 valence electrons are used.

Finally, write the Lewis structure in brackets with the charge of the ion in the upper right-hand corner.

$$\left[:\ddot{\text{O}}—\ddot{\text{C}}\text{l}—\ddot{\text{O}}:\right]^-$$

All atoms have octets; the structure is complete.

Calculate the formal charge on each atom by finding the number of valence electrons and subtracting the number of lone pair electrons and one-half the number of bonding electrons.

$$\left[:\ddot{\text{O}}—\ddot{\text{C}}\text{l}—\ddot{\text{O}}:\right]^-$$

	O_{left}	O_{right}	Cl
number of valence electrons	6	6	7
− number of lone pair electrons	6	6	4
− 1/2(number of bonding electrons)	1	1	2
Formal charge	−1	−1	+1

The sum of the formal charges is −1, which is the overall charge of the ion. However, we can write a resonance structure with a double bond to an oxygen because Cl can expand its octet. This leads to a lower formal charge.

$$\left[\ddot{\text{O}}=\ddot{\text{C}}\text{l}—\ddot{\text{O}}:\right]^- \longleftrightarrow \left[:\ddot{\text{O}}—\ddot{\text{C}}\text{l}=\ddot{\text{O}}\right]^-$$

Using the leftmost resonance form, calculate the formal charge on each atom by finding the number of valence electrons and subtracting the number of lone pair electrons and one-half the number of bonding electrons.

	O_{left}	O_{right}	Cl
number of valence electrons	6	6	7
− number of lone pair electrons	4	6	4
− 1/2(number of bonding electrons)	2	1	3
Formal charge	0	−1	0

The sum of the formal charges is −1, which is the overall charge of the ion. These resonance forms would all have the lower formal charge on the double-bonded O and Cl.

5.49   (a) PF_5:   Write the correct skeletal structure for the molecule.

Calculate the total number of electrons for the Lewis structure by summing the number of valence electrons of each atom in the molecule.

$$(\text{number of valence } e^- \text{ for P}) + 5(\text{number of valence } e^- \text{ for F}) = 5 + 5(7) = 40$$

Distribute the electrons among the atoms, giving octets to as many atoms as possible. Begin with the bonding electrons; then proceed to lone pairs on terminal atoms and finally to lone pairs on the central atom. Arrange additional electrons around the central atom, giving it an expanded octet of up to 12 electrons.

$$:\overset{\displaystyle ..}{\underset{\displaystyle ..}{F}}:$$

$$:\overset{..}{\underset{..}{F}}-P\overset{\displaystyle \overset{..}{\underset{..}{F}}:}{\underset{\displaystyle :\overset{..}{\underset{..}{F}}:}{}}$$

(b) I_3^-:   Write the correct skeletal structure for the ion.

$$I-I-I$$

Calculate the total number of electrons for the Lewis structure by summing the number of valence electrons of each atom in the ion and adding 1 for the 1− charge.

$$3(\text{number of valence } e^- \text{ for I}) + 1 = 3(7) + 1 = 22$$

Distribute the electrons among the atoms, giving octets to as many atoms as possible. Begin with the bonding electrons; then proceed to lone pairs on terminal atoms and finally to lone pairs on the central atom. Arrange additional electrons around the central atom, giving it an expanded octet of up to 12 electrons.

$$:\overset{..}{\underset{..}{I}}-\overset{..}{\underset{..}{I}}-\overset{..}{\underset{..}{I}}:$$

Finally, write the Lewis structure in brackets with the charge of the ion in the upper right-hand corner.

$$\left[:\overset{..}{\underset{..}{I}}-\overset{..}{\underset{..}{I}}-\overset{..}{\underset{..}{I}}:\right]^-$$

(c) SF_4:   Write the correct skeletal structure for the molecule.

$$\underset{\displaystyle F}{\overset{\displaystyle F}{F-S-F}}$$

Calculate the total number of electrons for the Lewis structure by summing the number of valence electrons of each atom in the molecule.

$$(\text{number of valence } e^- \text{ for S}) + 4(\text{number of valence } e^- \text{ for F}) = 6 + 4(7) = 34$$

Distribute the electrons among the atoms, giving octets (or duets for H) to as many atoms as possible. Begin with the bonding electrons; then proceed to lone pairs on terminal atoms and finally to lone pairs on the central atom. Arrange additional electrons around the central atom, giving it an expanded octet of up to 12 electrons.

$$\underset{\displaystyle :\overset{..}{\underset{..}{F}}:}{\overset{\displaystyle :\overset{..}{F}:}{:\overset{..}{\underset{..}{F}}-\overset{..}{S}-\overset{..}{\underset{..}{F}}:}}$$

(d) GeF_4:   Write the correct skeletal structure for the molecule.

$$\underset{\displaystyle F}{\overset{\displaystyle F}{F-Ge-F}}$$

Calculate the total number of electrons for the Lewis structure by summing the number of valence electrons of each atom in the molecule.

$$(\text{number of valence } e^- \text{ for Ge}) + 4(\text{number of valence } e^- \text{ for F}) = 4 + 4(7) = 32$$

Distribute the electrons among the atoms, giving octets to as many atoms as possible. Begin with the bonding electrons; then proceed to lone pairs on terminal atoms and finally to lone pairs on the central atom.

$$\underset{\displaystyle :\overset{..}{\underset{..}{F}}:}{\overset{\displaystyle :\overset{..}{F}:}{:\overset{..}{\underset{..}{F}}-Ge-\overset{..}{\underset{..}{F}}:}}$$

Bond Energies and Bond Lengths

5.51   Bond strength: $H_3CCH_3 < H_2CCH_2 < HCCH$
Bond length: $H_3CCH_3 > H_2CCH_2 > HCCH$
Write the Lewis structures for the three compounds. Compare the C—C bonds. Triple bonds are stronger than double bonds; double bonds are stronger than single bonds. Also, single bonds are longer than double bonds, which are longer than triple bonds.

$HCCH$ (10 e$^-$) H_2CCH_2 (12 e$^-$) H_3CCH_3 (14 e$^-$)

$H—C{\equiv}C—H$

VSEPR Theory and Molecular Geometry

5.53   Four electron groups: A trigonal pyramidal molecular geometry has three bonding groups and one lone pair of electrons, so there are four electron groups on atom A.

5.55   (a)   4 total electron groups, 4 bonding groups, 0 lone pairs
A tetrahedral molecular geometry has four bonding groups and no lone pairs. So there are four total electron groups, four bonding groups, and no lone pairs.

(b)   5 total electron groups, 3 bonding groups, 2 lone pairs
A T-shaped molecular geometry has three bonding groups and two lone pairs. So there are five total electron groups, three bonding groups, and two lone pairs.

(c)   6 total electron groups, 5 bonding groups, 1 lone pair
A square pyramidal molecular geometry has five bonding groups and one lone pair. So there are six total electron groups, five bonding groups, and one lone pair.

5.57   (a) PF_3:   Electron geometry—tetrahedral; molecular geometry—trigonal pyramidal; bond angle = 109.5°
Because of the lone pair, the bond angle will be less than 109.5°.
Draw a Lewis structure for the molecule:
PF_3 has 26 valence electrons.

Determine the total number of electron groups around the central atom:
There are four electron groups on P.
Determine the number of bonding groups and the number of lone pairs around the central atom:
There are three bonding groups and one lone pair.
Use Table 5.5 to determine the electron geometry, molecular geometry, and bond angles:
Four electron groups give a tetrahedral electron geometry; three bonding groups and one lone pair give a trigonal pyramidal molecular geometry; the idealized bond angles for tetrahedral geometry are 109.5°. The lone pair will make the bond angle less than idealized.

(b) SBr_2:   Electron geometry—tetrahedral; molecular geometry—bent; bond angle = 109.5°
Because of the lone pairs, the bond angle will be less than 109.5°.
Draw a Lewis structure for the molecule:
SBr_2 has 20 valence electrons.

Determine the total number of electron groups around the central atom:
There are four electron groups on S.
Determine the number of bonding groups and the number of lone pairs around the central atom:

There are two bonding groups and two lone pairs.

Use Table 5.5 to determine the electron geometry, molecular geometry, and bond angles:

Four electron groups give a tetrahedral electron geometry; two bonding groups and two lone pairs give a bent molecular geometry; the idealized bond angles for tetrahedral geometry are 109.5°. The lone pairs will make the bond angle less than idealized.

(c) $CHCl_3$:   Electron geometry—tetrahedral; molecular geometry—tetrahedral; bond angle = 109.5°

Because there are no lone pairs, the bond angle will be 109.5°.

Draw a Lewis structure for the molecule:

$CHCl_3$ has 26 valence electrons.

$$:\ddot{C}l-\underset{\underset{\displaystyle :\ddot{C}l:}{|}}{\overset{\overset{\displaystyle H}{|}}{C}}-\ddot{C}l:$$

Determine the total number of electron groups around the central atom:

There are four electron groups on C.

Determine the number of bonding groups and the number of lone pairs around the central atom:

There are four bonding groups and no lone pairs.

Use Table 5.5 to determine the electron geometry, molecular geometry, and bond angles:

Four electron groups give a tetrahedral electron geometry; four bonding groups and no lone pairs give a tetrahedral molecular geometry; the idealized bond angles for tetrahedral geometry are 109.5°. However, because the attached atoms have different electronegativities, the bond angles are less than idealized.

(d) CS_2:   Electron geometry—linear; molecular geometry—linear; bond angle = 180°

Because there are no lone pairs, the bond angle will be 180°.

Draw a Lewis structure for the molecule:

CS_2 has 16 valence electrons.

$$\ddot{S}=C=\ddot{S}$$

Determine the total number of electron groups around the central atom:

There are two electron groups on C.

Determine the number of bonding groups and the number of lone pairs around the central atom:

There are two bonding groups and no lone pairs.

Use Table 5.5 to determine the electron geometry, molecular geometry, and bond angles:

Two electron groups give a linear geometry; two bonding groups and no lone pairs give a linear molecular geometry; the idealized bond angle is 180°. The molecule will not deviate from this.

5.59 H_2O will have the smaller bond angle because lone pair–lone pair repulsions are greater than lone pair–bonding pair repulsions.

Draw the Lewis structures for both structures:

H_3O^+ has 8 valence electrons. H_2O has 8 valence electrons.

$$\left[\underset{\underset{\displaystyle \ddot{}}{H-\overset{\overset{\displaystyle H}{|}}{O}-H}}{}\right]^+ \qquad\qquad H-\ddot{O}-H$$

There are three bonding groups and   There are two bonding groups and
one lone pair.   two lone pairs.

Both have four electron groups, but the two lone pairs in H_2O will cause the bond angle to be smaller because of the lone pair–lone pair repulsions.

5.61   (a) SF_4 Draw a Lewis structure for the molecule:

SF_4 has 34 valence electrons.

$$:\ddot{F}-\underset{\underset{\displaystyle :\ddot{F}:}{|}}{\overset{\overset{\displaystyle :\ddot{F}:}{|}}{S}}-\ddot{F}:$$

Determine the total number of electron groups around the central atom:

There are five electron groups on S.

Determine the number of bonding groups and the number of lone pairs around the central atom:

There are four bonding groups and one lone pair.

Use Table 5.5 to determine the electron geometry and molecular geometry:

The electron geometry is trigonal bipyramidal, so the molecular geometry is seesaw.

Sketch the molecule:

```
    F
    |
    |   ,,,,F
    S
    |   `␙F
    |
    F
```

(b) ClF_3 Draw a Lewis structure for the molecule:

ClF_3 has 28 valence electrons.

```
        ..
       :F:
        |
  ..    |    ..
 :F — Cl — F:
  ..    ..   ..
```

Determine the total number of electron groups around the central atom:

There are five electron groups on Cl.

Determine the number of bonding groups and the number of lone pairs around the central atom:

There are three bonding groups and two lone pairs.

Use Table 5.5 to determine the electron geometry and molecular geometry:

The electron geometry is trigonal bipyramidal, so the molecular geometry is T-shaped.

Sketch the molecule:

```
      F
      |
      |
 F — Cl
      |
      |
      F
```

(c) IF_2^- Draw a Lewis structure for the ion:

IF_2^- has 22 valence electrons.

$$\left[:\overset{..}{\underset{..}{F}} - \overset{..}{\underset{..}{I}} - \overset{..}{\underset{..}{F}}: \right]^-$$

Determine the total number of electron groups around the central atom:

There are five electron groups on I.

Determine the number of bonding groups and the number of lone pairs around the central atom:

There are two bonding groups and three lone pairs.

Use Table 5.5 to determine the electron geometry and molecular geometry:

The electron geometry is trigonal bipyramidal, so the molecular geometry is linear.

Sketch the ion:

$$\left[F - I - F \right]^-$$

(d) IBr_4^- Draw a Lewis structure for the ion:

IBr_4^- has 36 valence electrons.

$$\left[\begin{array}{c} \overset{..}{:Br:} \\ | \\ :\overset{..}{\underset{..}{Br}} - \overset{|}{\underset{|}{I}} - \overset{..}{\underset{..}{Br}}: \\ | \\ :\overset{..}{\underset{..}{Br}}: \end{array} \right]^-$$

Determine the total number of electron groups around the central atom:

There are six electron groups on I.

Determine the number of bonding groups and the number of lone pairs around the central atom:

There are four bonding groups and two lone pairs.

Use Table 5.5 to determine the electron geometry and molecular geometry:

The electron geometry is octahedral, so the molecular geometry is square planar.

Sketch the ion:

$$\left[\begin{array}{c} Br \diagdown \quad \diagup Br \\ I \\ Br \diagup \quad \diagdown Br \end{array} \right]^{-}$$

5.63   (a) C_2H_2 Draw the Lewis structure:

$$H-C\equiv C-H$$

Atom	Number of Electron Groups	Number of Lone Pairs	Molecular Geometry
Left C	2	0	Linear
Right C	2	0	Linear

Sketch the molecule:

$$H-C\equiv C-H$$

(b) C_2H_4 Draw the Lewis structure:

$$\begin{array}{c} H \diagdown \qquad \diagup H \\ C=C \\ H \diagup \qquad \diagdown H \end{array}$$

Atom	Number of Electron Groups	Number of Lone Pairs	Molecular Geometry
Left C	3	0	Trigonal planar
Right C	3	0	Trigonal planar

Sketch the molecule:

$$\begin{array}{c} H \diagdown \qquad \diagup H \\ C=C \\ H \diagup \qquad \diagdown H \end{array}$$

(c) C_2H_6 Draw the Lewis structure:

$$\begin{array}{c} \quad H \quad H \\ \quad | \quad | \\ H-C-C-H \\ \quad | \quad | \\ \quad H \quad H \end{array}$$

Atom	Number of Electron Groups	Number of Lone Pairs	Molecular Geometry
Left C	4	0	Tetrahedral
Right C	4	0	Tetrahedral

Sketch the molecule:

5.65   (a)   Four pairs of electrons give a tetrahedral electron geometry. The lone pair would cause lone pair–bonded pair repulsions and would have a trigonal pyramidal molecular geometry.

(b)   Five pairs of electrons give a trigonal bipyramidal electron geometry. The lone pair occupies an equatorial position to minimize lone pair–bonded pair repulsions, and the molecule would have a seesaw molecular geometry.

(c)   Six pairs of electrons give an octahedral electron geometry. The two lone pairs would occupy opposite positions to minimize lone pair–lone pair repulsions. The molecular geometry would be square planar.

5.67   (a) CH_3OH Draw the Lewis structure and determine the geometry about each interior atom:

$$
\begin{array}{c}
:\overset{..}{O}-H \\
| \\
H-C-H \\
| \\
H
\end{array}
$$

Atom	Number of Electron Groups	Number of Lone Pairs	Molecular Geometry
C	4	0	Tetrahedral
O	4	2	Bent

Sketch the molecule:

(b) CH_3OCH_3 Draw the Lewis structure and determine the geometry about each interior atom:

$$
\begin{array}{ccc}
H & & H \\
| & & | \\
H-C-&\overset{..}{\underset{..}{O}}-&C-H \\
| & & | \\
H & & H
\end{array}
$$

Atom	Number of Electron Groups	Number of Lone Pairs	Molecular Geometry
C	4	0	Tetrahedral
O	4	2	Bent
C	4	0	Tetrahedral

Sketch the molecule:

(c) H_2O_2 Draw the Lewis structure and determine the geometry about each interior atom:

$$
H-\overset{..}{\underset{..}{O}}-\overset{..}{\underset{..}{O}}-H
$$

Atom	Number of Electron Groups	Number of Lone Pairs	Molecular Geometry
O	4	2	Bent
O	4	2	Bent

Sketch the molecule:

Molecular Shape and Polarity

5.69   Draw the Lewis structure for CO_2 and CCl_4; determine the molecular geometry and the polarity.

$$
\overset{..}{\underset{..}{O}}=C=\overset{..}{\underset{..}{O}}
$$

$$
\begin{array}{c}
:\overset{..}{\underset{..}{Cl}}: \\
| \\
:\overset{..}{\underset{..}{Cl}}-C-\overset{..}{\underset{..}{Cl}}: \\
| \\
:\overset{..}{\underset{..}{Cl}}:
\end{array}
$$

Number of electron groups on C	2	4
Number of lone pairs	0	0
Molecular geometry	linear	tetrahedral

Even though each molecule contains polar bonds, the sum of the bond dipoles gives a net dipole of zero for each molecule.

The linear molecular geometry of CO_2 will have bond vectors that are equal and opposite. ⟵———⟶

The tetrahedral molecular geometry of CCl_4 will have bond vectors that are equal and have a net dipole of zero.

5.71   (a) PF_3—polar

Use the Lewis structure and molecular geometry determined in Problem 5.57, which is trigonal pyramidal.

Determine whether the molecule contains polar bonds:

The electronegativity of P = 2.1 and F = 4.0. Therefore, the bonds are polar.

Determine whether the polar bonds add together to form a net dipole:

Because the molecule is trigonal pyramidal, the three dipole moments sum to a nonzero net dipole moment. The molecule is polar. Refer to Table 5.6 to see how dipole moments add to determine polarity.

       (b) SBr_2—nonpolar

Use the Lewis structure and molecular geometry determined in Problem 5.57, which is bent.

Determine whether the molecule contains polar bonds:

The electronegativity of S = 2.5 and Br = 2.8. Therefore, the bonds are nonpolar.

Even though the molecule is bent, because the bonds are nonpolar, the molecule is nonpolar.

       (c) $CHCl_3$—polar

Use the Lewis structure and molecular geometry determined in Problem 5.57, which is tetrahedral.

Determine whether the molecule contains polar bonds:

The electronegativity of C = 2.5, H = 2.1, and Cl = 3.0. Therefore, the bonds are polar.

Determine whether the polar bonds add together to form a net dipole:

Because the bonds have different dipole moments due to the different atoms involved, the four dipole moments sum to a nonzero net dipole moment. The molecule is polar. Refer to Table 5.6 to see how dipole moments add to determine polarity.

       (d) CS_2—nonpolar

Use the Lewis structure and molecular geometry determined in Problem 5.57, which is linear.

Determine whether the molecule contains polar bonds:

The electronegativity of C = 2.5 and S = 2.5. Therefore, the bonds are nonpolar. Also, the molecule is linear, which would result in a zero net dipole even if the bonds were polar.

The molecule is nonpolar. Refer to Table 5.6 to see how dipole moments add to determine polarity.

5.73   (a) ClO_3^-—polar

Draw the Lewis structure and determine the molecular geometry:

$$\left[\; :\!\overset{\displaystyle ..}{\underset{\displaystyle ..}{O}}\!-\!\overset{\displaystyle :\!\ddot{O}\!:}{\underset{\displaystyle |}{Cl}}\!-\!\ddot{O}\!: \;\right]^{-}$$

Four electron pairs with one lone pair give a trigonal pyramidal molecular geometry.

Determine whether the molecule contains polar bonds:

The electronegativities of Cl = 3.0 and O = 3.5. Therefore, the bonds are polar.

Determine whether the polar bonds add together to form a net dipole:

Because the molecular geometry is trigonal pyramidal, the three dipole moments sum to a nonzero net dipole moment. The molecule is polar. Refer to Table 5.6 to see how dipole moments add to determine polarity.

(b) SCl_2—polar

Draw the Lewis structure and determine the molecular geometry:

$$:\overset{..}{\underset{}{S}}—\overset{..}{\underset{..}{Cl}}:$$
$$\overset{|}{:\underset{..}{Cl}:}$$

Four electron pairs with two lone pairs give a bent molecular geometry.

Determine whether the molecule contains polar bonds:

The electronegativity of S = 2.5 and Cl = 3.0. Therefore, the bonds are polar.

Determine whether the polar bonds add together to form a net dipole:

Because the molecular geometry is bent, the two dipole moments sum to a nonzero net dipole moment. The molecule is polar. Refer to Table 5.6 to see how dipole moments add to determine polarity.

(c) SCl_4—polar

Draw the Lewis structure and determine the molecular geometry:

$$:\overset{..}{\underset{}{Cl}}:$$
$$|$$
$$:\overset{..}{\underset{..}{Cl}}—\overset{..}{\underset{}{S}}—\overset{..}{\underset{..}{Cl}}:$$
$$|$$
$$:\overset{}{\underset{..}{Cl}}:$$

Five electron pairs with one lone pair give a seesaw molecular geometry.

Determine whether the molecule contains polar bonds:

The electronegativity of S = 2.5 and Cl = 3.0. Therefore, the bonds are polar.

Determine whether the polar bonds add together to form a net dipole:

Because the molecular geometry is seesaw, the four equal dipole moments sum to a nonzero net dipole moment. The molecule is polar.

The seesaw molecular geometry will not have offsetting bond vectors.

(d) $BrCl_5$—nonpolar

Draw the Lewis structure and determine the molecular geometry.

$$:\overset{..}{\underset{}{Cl}}:$$
$$:\overset{..}{\underset{..}{Cl}}\diagdown\overset{|}{}\diagup\overset{..}{\underset{..}{Cl}}:$$
$$\overset{}{\underset{}{Br}}$$
$$:\overset{..}{\underset{..}{Cl}}\diagup\overset{..}{}\diagdown\overset{..}{\underset{..}{Cl}}:$$

Six electron pairs with one lone pair give square pyramidal molecular geometry.

Determine whether the molecule contains polar bonds:

The electronegativity of Br = 2.8 and Cl = 3.0. The difference is only 0.2; therefore, the bonds are nonpolar. Even though the molecular geometry is square pyramidal, the five bonds are nonpolar; so there is no net dipole. The molecule is nonpolar.

Cumulative Problems

5.75   (a) $BaCO_3$: Ba^{2+}

$$\left[\overset{:\overset{..}{\underset{}{O}}:}{\underset{}{}} \right]^{2-}$$
$$\left[\overset{..}{\underset{..}{O}}=C—\overset{..}{\underset{..}{O}}: \right]^{2-}$$

Determine the cation and anion.

    Ba^{2+} CO_3^{2-}

Write the Lewis structure for the barium cation based on the valence electrons.

    Ba $5s^2$ Ba^{2+} $5s^0$

    Ba: Ba^{2+}

Ba must lose two electrons and be left with the octet from the previous shell.

Write the Lewis structure for the covalent anion.

Write the correct skeletal structure for the ion.

$$O \overset{\displaystyle |}{\underset{\displaystyle |}{O-C-O}}$$

Calculate the total number of electrons for the Lewis structure by summing the number of valence electrons of each atom in the ion and adding two for the 2− charge.

(number of valence e⁻ for C) + 3(number of valence e⁻ for O) = 4 + 3(6) + 2 = 24

Distribute the electrons among the atoms, giving octets to as many atoms as possible. Begin with the bonding electrons; then proceed to lone pairs on terminal atoms and finally to lone pairs on the central atom.

$$:\ddot{O}: \\ | \\ :\ddot{O}-C-\ddot{O}:$$

If any atom lacks an octet, form double or triple bonds as necessary.

$$:\ddot{O}: \\ | \\ \ddot{O}=C-\ddot{O}:$$

Finally, write the Lewis structure in brackets with the charge of the ion in the upper right-hand corner.

$$Ba^{2+} \left[\begin{array}{c} :\ddot{O}: \\ | \\ \ddot{O}=C-\ddot{O}: \end{array} \right]^{2-}$$

The double bond can be between the C and any of the oxygen atoms, so there are resonance structures.

$$Ba^{2+} \left[\begin{array}{c} :\ddot{O}: \\ | \\ \ddot{O}=C-\ddot{O}: \end{array} \right]^{2-} \longleftrightarrow Ba^{2+} \left[\begin{array}{c} :O: \\ \| \\ :\ddot{O}-C-\ddot{O}: \end{array} \right]^{2-} \longleftrightarrow Ba^{2+} \left[\begin{array}{c} :\ddot{O}: \\ | \\ :\ddot{O}-C=\ddot{O} \end{array} \right]^{2-}$$

(b)   Ca(OH)₂: Ca^{2+}

$$2\left[:\ddot{O}-H \right]^{-}$$

Determine the cation and anion.

Ca^{2+} OH^-

Write the Lewis structure for the calcium cation based on the valence electrons.

Ca $4s^2$ Ca^{2+} $4s^0$
Ca: Ca^{2+}

Ca must lose two electrons and be left with the octet from the previous shell.
Write the Lewis structure for the covalent anion.
Write the correct skeletal structure for the ion.

O—H

Calculate the total number of electrons for the Lewis structure by summing the valence electrons of each atom in the ion and adding 1 for the 1− charge.

(number of valence e⁻ for H) + (number of valence e⁻ for O) + 1 = 1 + 6 + 1 = 8

Distribute the electrons among the atoms, giving octets (or duets for H) to as many atoms as possible. Begin with the bonding electrons; then proceed to lone pairs on terminal atoms and finally to lone pairs on the central atom.

$$:\ddot{O}-H$$

Finally, write the Lewis structure in brackets with the charge of the ion in the upper right-hand corner.

$$Ca^{2+}\ 2\left[:\ddot{O}-H \right]^{-}$$

(c)   KNO₃: K^+

$$\left[\begin{array}{c} :\ddot{O}: \\ | \\ \ddot{O}=N-\ddot{O}: \end{array} \right]^{-}$$

Determine the cation and anion.

$$K^+ \qquad NO_3^-$$

Write the Lewis structure for the potassium cation based on the valence electrons.

$$K \quad 4s^1 \qquad K^+ \quad 4s^0$$
$$K \cdot \qquad\qquad K^+$$

K must lose one electron and be left with the octet from the previous shell.

Write the Lewis structure for the covalent anion.

Write the correct skeletal structure for the ion.

$$\begin{array}{c} O \\ | \\ O-N-O \end{array}$$

Calculate the total number of electrons for the Lewis structure by summing the valence electrons of each atom in the ion and adding 1 for the 1− charge.

(number of valence e⁻ for N) + (number of valence e⁻ for O) = 5 + 3(6) + 1 = 24

Distribute the electrons among the atoms, giving octets to as many atoms as possible. Begin with the bonding electrons; then proceed to lone pairs on terminal atoms and finally to lone pairs on the central atom.

$$\begin{array}{c} :\ddot{O}: \\ | \\ :\ddot{O}-N-\ddot{O}: \end{array}$$

If any atom lacks an octet, form double or triple bonds as necessary.

$$\begin{array}{c} :\ddot{O}: \\ | \\ \ddot{O}=N-\ddot{O}: \end{array}$$

Finally, write the Lewis structure in brackets with the charge of the ion in the upper right-hand corner.

$$K^+ \left[\begin{array}{c} :\ddot{O}: \\ | \\ \ddot{O}=N-\ddot{O}: \end{array} \right]^-$$

The double bond can be between the N and any of the oxygen atoms, so there are resonance structures.

$$K^+ \left[\begin{array}{c} :\ddot{O}: \\ | \\ \ddot{O}=N-\ddot{O}: \end{array} \right]^- \longleftrightarrow K^+ \left[\begin{array}{c} :O: \\ \| \\ :\ddot{O}-N-\ddot{O}: \end{array} \right]^- \longleftrightarrow K^+ \left[\begin{array}{c} :\ddot{O}: \\ | \\ :\ddot{O}-N=\ddot{O} \end{array} \right]^-$$

(d) LiIO: Li⁺

$$\left[:\ddot{I}-\ddot{O}: \right]^-$$

Determine the cation and anion.

$$Li^+ \qquad IO^-$$

Write the Lewis structure for the lithium cation based on the valence electrons.

$$Li \quad 2s^1 \qquad Li^+ \quad 2s^0$$
$$Li \cdot \qquad\qquad Li^+$$

Li must lose one electron and be left with the octet from the previous shell.

Write the Lewis structure for the covalent anion.

Write the correct skeletal structure for the ion.

$$I-O$$

Calculate the total number of electrons for the Lewis structure by summing the number of valence electrons of each atom in the ion and adding 1 for the 1− charge.

(number of valence e⁻ for I) + (number of valence e⁻ for O) = 7 + 6 + 1 = 14

Distribute the electrons among the atoms, giving octets to as many atoms as possible. Begin with the bonding electrons; then proceed to lone pairs on terminal atoms and finally to lone pairs on the central atom.

$$:\ddot{I}-\ddot{O}:$$

Finally, write the Lewis structure in brackets with the charge of the ion in the upper right-hand corner.

$$Li^+ \left[:\ddot{I}-\ddot{O}: \right]^-$$

5.77   (a) C_4H_8:   Write the correct skeletal structure for the molecule.

$$
\begin{array}{c}
HH\\
||\\
H-C-C-H\\
||\\
H-C-C-H\\
||\\
HH
\end{array}
$$

Calculate the total number of electrons for the Lewis structure by summing the number of valence electrons of each atom in the molecule.

$$4(\text{number of valence } e^- \text{ for C}) + 8(\text{number of valence } e^- \text{ for H}) = 4(4) + 8(1) = 24$$

Distribute the electrons among the atoms, giving octets (or duets for H) to as many atoms as possible.

$$
\begin{array}{c}
HH\\
||\\
H-C-C-H\\
||\\
H-C-C-H\\
||\\
HH
\end{array}
$$

All atoms have octets or duets for H.

(b) C_4H_4:   Write the correct skeletal structure for the molecule.

$$
\begin{array}{c}
H-C-C-H\\
||\\
H-C-C-H
\end{array}
$$

Calculate the total number of electrons for the Lewis structure by summing the number of valence electrons of each atom in the molecule.

$$4(\text{number of valence } e^- \text{ for C}) + 4(\text{number of valence } e^- \text{ for H}) = 4(4) + 4(1) = 20$$

Distribute the electrons among the atoms, giving octets (or duets for H) to as many atoms as possible.

$$
\begin{array}{c}
H-\overset{\cdot\cdot}{C}-\overset{\cdot\cdot}{C}-H\\
||\\
H-C-C-H
\end{array}
$$

Complete octets by forming double bonds on alternating carbons; draw resonance structures.

$$
\begin{array}{cc}
H-C-C-H & \qquad H-C=C-H\\
\|\| & \qquad\qquad ||\\
H-C-C-H & \qquad H-C=C-H
\end{array}
$$

(c) C_6H_{12}:   Write the correct skeletal structure for the molecule.

$$
\begin{array}{c}
\quad H\ \ H\\
\quad \backslash\ /\\
H\backslash\ \ \ \ C\ \ \ \ /H\\
\ \ C\ \ \ \ C\\
H/\ \ |\ \ \ \ |\ \ \backslash H\\
H-C\ \ \ \ \ C-H\\
/\ \ \ \ C\ \ \ \ \backslash\\
H/\ \ /\ \backslash\ \ \backslash H\\
\quad H\ \ H
\end{array}
$$

Calculate the total number of electrons for the Lewis structure by summing the valence electrons of each atom in the molecule.

$$6(\text{number of valence } e^- \text{ for C}) + 12(\text{number of valence } e^- \text{ for H}) = 6(4) + 12(1) = 36$$

Distribute the electrons among the atoms, giving octets (or duets for H) to as many atoms as possible. Begin with the bonding.

$$
\begin{array}{c}
\quad H\ \ H\\
\quad \backslash\ /\\
H\backslash\ \ \ \ C\ \ \ \ /H\\
\ \ C\ \ \ \ C\\
H/\ \ |\ \ \ \ |\ \ \backslash H\\
H-C\ \ \ \ \ C-H\\
/\ \ \ \ C\ \ \ \ \backslash\\
H/\ \ /\ \backslash\ \ \backslash H\\
\quad H\ \ H
\end{array}
$$

All 36 electrons are used, and all atoms have octets or duets for H.

(d) C_6H_6:   Write the correct skeletal structure for the molecule.

Calculate the total number of electrons for the Lewis structure by summing the number of valence electrons of each atom in the molecule.

$$6(\text{number of valence e}^- \text{ for C}) + 6(\text{number of valence e}^- \text{ for H}) = 6(4) + 6(1) = 30$$

Distribute the electrons among the atoms, giving octets (or duets for H) to as many atoms as possible.

Complete octets by forming double bonds on alternating carbons; draw resonance structures.

5.79   **Given:** 26.01% C; 4.38% H; 69.52% O; molar mass = 46.02 g/mol
Find: molecular formula and Lewis structure
Conceptual Plan: Convert mass to mol of each element → pseudoformula → empirical formula

$$\frac{1 \text{ mol C}}{12.01 \text{ g C}} \qquad \frac{1 \text{ mol H}}{1.008 \text{ g H}} \qquad \frac{1 \text{ mol O}}{16.00 \text{ g O}} \qquad\qquad \text{divide by smallest number}$$

→ molecular formula → Lewis structure

empirical formula $\times$ n

Solution: $26.01 \text{ g C} \times \dfrac{1 \text{ mol C}}{12.01 \text{ g C}} = 2.166 \text{ mol C}$

$4.38 \text{ g H} \times \dfrac{1 \text{ mol H}}{1.008 \text{ g H}} = 4.345 \text{ mol H}$

$69.52 \text{ g O} \times \dfrac{1 \text{ mol O}}{16.00 \text{ g O}} = 4.345 \text{ mol O}$

$C_{2.166}H_{4.345}O_{4.345}$

$\dfrac{C_{2.166}H_{4.345}O_{4.345}}{2.166 \quad 2.166 \quad 2.166} \rightarrow CH_2O_2$

The correct empirical formula is CH_2O_2.

empirical formula mass = $(12.01 \text{ g/mol}) + 2(1.008 \text{ g/mol}) + 2(16.00 \text{ g/mol}) = 46.03 \text{ g/mol}$

$n = \dfrac{\text{molar mass}}{\text{formula molar mass}} = \dfrac{46.02 \text{ g/mol}}{46.03 \text{ g/mol}} = 1$

molecular formula $= CH_2O_2 \times 1$
          $= CH_2O_2$

Write the correct skeletal structure for the molecule.

$$
\begin{array}{c}
\text{O} \\
| \\
\text{H}-\text{C}-\text{O}-\text{H}
\end{array}
$$

Calculate the total number of electrons for the Lewis structure by summing the number of valence electrons of each atom in the molecule.

(number of valence e⁻ for C) + 2(number of valence e⁻ for O) + 2(number of valence e⁻ for H)

$= 4 + 2(6) + 2(1) = 18$

Distribute the electrons among the atoms, giving octets (or duets for H) to as many atoms as possible. Begin with the bonding electrons; then proceed to lone pairs on terminal atoms and finally to lone pairs on the central atoms.

$$
\begin{array}{c}
:\ddot{\text{O}}: \\
| \\
\text{H}-\text{C}-\ddot{\text{O}}-\text{H} \\
\phantom{\text{H}-\text{C}-}\ddot{}
\end{array}
$$

Complete the octet on C by forming a double bond.

$$
\begin{array}{c}
:\text{O}: \\
|| \\
\text{H}-\text{C}-\ddot{\text{O}}-\text{H} \\
\phantom{\text{H}-\text{C}-}\ddot{}
\end{array}
$$

5.81 HNO_3 Write the correct skeletal structure for the molecule.

$$
\begin{array}{c}
\text{O} \\
| \\
\text{H}-\text{O}-\text{N}-\text{O}
\end{array}
$$

Calculate the total number of electrons for the Lewis structure by summing the number of valence electrons of each atom in the molecule.

3(number of valence e⁻ for O) + (number of valence e⁻ for N) + (number of valence e⁻ for H) = 3(6) + 5 + 1 = 24

Distribute the electrons among the atoms, giving octets (or duets for H) to as many atoms as possible. Begin with the bonding electrons; then proceed to lone pairs on terminal atoms and finally to lone pairs on the central atom.

$$
\begin{array}{c}
:\ddot{\text{O}}: \\
| \\
\text{H}-\ddot{\text{O}}-\text{N}-\ddot{\text{O}}: \\
\phantom{\text{H}-}\ddot{}
\end{array}
$$

All 24 valence electrons are used.

If any atoms lack an octet, form double or triple bonds as necessary. The double bond can be formed to any of the three oxygen atoms, so there are three resonance forms.

$$
\begin{array}{ccc}
:\text{O}: & :\ddot{\text{O}}: & :\ddot{\text{O}}: \\
|| & | & | \\
\text{H}-\ddot{\text{O}}-\text{N}-\ddot{\text{O}}: \longleftrightarrow & \text{H}-\ddot{\text{O}}-\text{N}=\ddot{\text{O}} \longleftrightarrow & \text{H}-\ddot{\text{O}}=\text{N}-\ddot{\text{O}}: \\
\text{I} & \text{II} & \text{III}
\end{array}
$$

All atoms have octets (duets for H); the structure is complete.

To determine which resonance hybrid(s) is most important, calculate the formal charge on each atom in each structure by finding the number of valence electrons and subtracting the number of lone pair electrons and one-half the number of bonding electrons.

	Structure I					Structure II				
	O_{left}	O_{top}	O_{right}	N	H	O_{left}	O_{top}	O_{right}	N	H
number of valence electrons	6	6	6	5	1	6	6	6	5	1
− number of lone pair electrons	4	4	6	0	0	4	6	4	0	0
− 1/2(number of bonding electrons)	2	2	1	4	1	2	1	2	4	1
Formal charge	0	0	−1	+1	0	0	−1	0	+1	0

Structure III

	O_{left}	O_{top}	O_{right}	N	H
number of valence electrons	6	6	6	5	1
− number of lone pair electrons	2	6	6	0	0
− 1/2(number of bonding electrons)	3	1	1	4	1
Formal charge	+1	−1	−1	+1	0

The sum of the formal charges is 0 for each structure, which is the overall charge of the molecule. However, in structures I and II, the individual formal charges are lower. These two forms would contribute equally to the structure of HNO_3. Structure III would be less important because the individual formal charges are higher. In addition, oxygen is more electronegative than N, and in this structure N has the positive formal charge. So structure III will be unacceptable.

5.83 CNO^- Write the skeletal structure.

   C—N—O

Calculate the total number of electrons for the Lewis structure by summing the number of valence electrons of each atom in the molecule.

   (valence e^- from C) + (valence e^- from N) + (valence e^- from O) + 1 (from the negative charge) = 4 + 5 + 6 + 1 = 16

   :C̈—N—Ö:

Distribute the electrons to complete octets if possible.

   $\left[\ddot{\text{C}}=\text{N}=\ddot{\text{O}} \right]^- \longleftrightarrow \left[:\text{C}\equiv\text{N}-\ddot{\text{O}}: \right]^- \longleftrightarrow \left[:\ddot{\text{C}}-\text{N}\equiv\text{O}: \right]^-$

        I   II   III

Determine the formal charge on each atom for each structure.

Structure I

	C	N	O
number of valence electrons	4	5	6
− number of lone pair electrons	4	0	4
− 1/2(number of bonding electrons)	2	4	2
Formal charge	−2	+1	0

Structure II

	C	N	O
number of valence electrons	4	5	6
− number of lone pair electrons	2	0	6
− 1/2(number of bonding electrons)	3	4	1
Formal charge	−1	+1	−1

Structure III

	C	N	O
number of valence electrons	4	5	6
− number of lone pair electrons	6	0	2
− 1/2(number of bonding electrons)	1	4	3
Formal charge	−3	+1	+1

Structures I, II, and III all follow the octet rule but have varying degrees of negative formal charge on carbon, which is the least electronegative atom. Also, the amount of formal charge is very high in all three resonance forms. Although structure II is the best of the resonance forms, it has a −1 charge on the least electronegative atom, C, and a +1 charge on the more electronegative atom, N. Therefore, none of these resonance forms contributes strongly to the stability of the fulminate ion and the ion is not very stable.

5.85   (a) C_2H_4:   Write the correct skeletal structure for the molecule.

                H—C—C—H
                   | |
                   H   H

                Calculate the total number of electrons for the Lewis structure by summing the valence electrons of each atom in the molecule.

                2(number of valence e^- for C) + 4(number of valence e^- for H) = 2(4) + 4(1) = 12

Distribute the electrons among the atoms, giving octets (or duets for H) to as many atoms as possible. Begin with the bonding electrons; then proceed to lone pairs on terminal atoms and finally to lone pairs on the central atom.

$$H—\overset{..}{C}—C—H$$
$$\;\;\;\;|\;\;\;\;|$$
$$\;\;\;\;H\;\;\;H$$

All 12 valence electrons are used.

If any atom lacks an octet, form double or triple bonds to give them octets.

$$H—C{=}C—H$$
$$\;\;\;\;|\;\;\;\;|$$
$$\;\;\;\;H\;\;\;H$$

All atoms have octets (duets for hydrogen); the structure is complete.

(b) CH_3NH_2:   Write the correct skeletal structure for the molecule.

$$H$$
$$|$$
$$H—C—N—H$$
$$\;\;\;\;|\;\;\;\;|$$
$$\;\;\;\;H\;\;\;H$$

Calculate the total number of electrons for the Lewis structure by summing the valence electrons of each atom in the molecule.

(number of valence e^- for C) + 5(number of valence e^- for H) + (number of valence e^- for N) = 4 + 5(1) + 5 = 14

Distribute the electrons among the atoms, giving octets (or duets for H) to as many atoms as possible. Begin with the bonding electrons; then proceed to lone pairs on terminal atoms and finally to lone pairs on the central atom.

$$H$$
$$|$$
$$H—C—\overset{..}{N}—H$$
$$\;\;\;\;|\;\;\;\;|$$
$$\;\;\;\;H\;\;\;H$$

All 14 valence electrons are used.

All atoms have octets (duets for hydrogen); the structure is complete.

(c)   HCHO: Write the correct skeletal structure for the molecule.

$$O$$
$$|$$
$$H—C—H$$

Calculate the total number of electrons for the Lewis structure by summing the valence electrons of each atom in the molecule.

(number of valence e^- for C) + 2(number of valence e^- for H) + (number of valence e^- for O) = 4 + 2(1) + 6 = 12

Distribute the electrons among the atoms, giving octets (or duets for H) to as many atoms as possible. Begin with the bonding electrons; then proceed to lone pairs on terminal atoms and finally to lone pairs on the central atom.

$$:\overset{..}{O}:$$
$$|$$
$$H—C—H$$

All 12 valence electrons are used.

If any atom lacks an octet, form double or triple bonds to give them octets.

$$:O:$$
$$\|$$
$$H—C—H$$

All atoms have octets (duets for hydrogen); the structure is complete.

(d) CH$_3$CH$_2$OH: Write the correct skeletal structure for the molecule.

$$
\begin{array}{c}
\ \ \ \ \text{H} \ \ \ \text{H} \\
\ \ \ \ | \ \ \ \ | \\
\text{H}-\text{C}-\text{C}-\text{O}-\text{H} \\
\ \ \ \ | \ \ \ \ | \\
\ \ \ \ \text{H} \ \ \ \text{H}
\end{array}
$$

Calculate the total number of electrons for the Lewis structure by summing the valence electrons of each atom in the molecule.

2(number of valence e$^-$ for C) + 6(number of valence e$^-$ for H) + (number of valence e$^-$ for O) = 2(4) + 6(1) + 6 = 20

Distribute the electrons among the atoms, giving octets (or duets for H) to as many atoms as possible. Begin with the bonding electrons; then proceed to lone pairs on terminal atoms and finally to lone pairs on the central atom.

$$
\begin{array}{c}
\ \ \ \ \text{H} \ \ \ \text{H} \\
\ \ \ \ | \ \ \ \ | \\
\text{H}-\text{C}-\text{C}-\ddot{\text{O}}-\text{H} \\
\ \ \ \ | \ \ \ \ | \\
\ \ \ \ \text{H} \ \ \ \text{H}
\end{array}
$$

All 20 valence electrons are used.

All atoms have octets (or duets for H); the structure is complete.

(e) HCOOH: Write the correct skeletal structure for the molecule.

$$
\begin{array}{c}
\ \ \ \ \ \ \text{O} \\
\ \ \ \ \ \ | \\
\text{H}-\text{C}-\text{O}-\text{H}
\end{array}
$$

Calculate the total number of electrons for the Lewis structure by summing the valence electrons of each atom in the molecule.

(number of valence e$^-$ for C) + 2(number of valence e$^-$ for H) + 2(number of valence e$^-$ for O) = 4 + 2(1) + 2(6) = 18

Distribute the electrons among the atoms, giving octets (or duets for H) to as many atoms as possible. Begin with the bonding electrons; then proceed to lone pairs on terminal atoms and finally to lone pairs on the central atom.

$$
\begin{array}{c}
\ \ \ \ :\ddot{\text{O}}: \\
\ \ \ \ \ | \\
\text{H}-\text{C}-\ddot{\text{O}}-\text{H}
\end{array}
$$

All 18 valence electrons are used.

If any atom lacks an octet, form double or triple bonds to give them octets.

$$
\begin{array}{c}
\ \ \ \ :\text{O}: \\
\ \ \ \ \ \| \\
\text{H}-\text{C}-\ddot{\text{O}}-\text{H}
\end{array}
$$

All atoms have octets (or duets for H); the structure is complete.

5.87 HCSNH$_2$: Write the correct skeletal structure for the molecule.

$$
\begin{array}{c}
\ \ \ \ \ \ \text{S} \\
\ \ \ \ \ \ | \\
\text{H}-\text{C}-\text{N}\overset{\text{H}}{\underset{\text{H}}{<}}
\end{array}
$$

Calculate the total number of electrons for the Lewis structure by summing the number of valence electrons of each atom in the molecule.

(number of valence e$^-$ for N) + (number of valence e$^-$ for S) + (number of valence e$^-$ for C) + 3(number of valence e$^-$ for H) = 5 + 6 + 4 + 3(1) = 18

Distribute the electrons among the atoms, giving octets (or duets for H) to as many atoms as possible. Begin with the bonding electrons; then proceed to lone pairs on terminal atoms and finally to lone pairs on the central atoms.

$$
\begin{array}{c}
\ \ \ \ :\ddot{\text{S}}: \\
\ \ \ \ \ | \\
\text{H}-\text{C}-\ddot{\text{N}}\overset{\text{H}}{\underset{\text{H}}{<}}
\end{array}
$$

Complete the octet on C by forming a double bond.

$$:S:$$

$$\text{nonpolar} \parallel \quad \text{polar}$$
$$H—C—\ddot{N}<^{H}_{H}$$
$$\text{nonpolar} \quad \text{polar} \quad \text{polar}$$

5.89   (a) O_2^-:   Write the correct skeletal structure for the radical.

O—O

Calculate the total number of electrons for the Lewis structure by summing the number of valence electrons of each atom in the radical and adding 1 for the 1− charge.

2(number of valence e^- for O) + 1 = 2(6) + 1 = 13

Distribute the electrons among the atoms, giving octets to as many atoms as possible. Begin with the bonding electrons; then proceed to lone pairs on terminal atoms and finally to lone pairs on the central atom.

$$\left[\cdot\ddot{O}—\ddot{O}: \right]^-$$

All 13 valence electrons are used.

O has an incomplete octet. It has seven electrons because we have an odd number of valence electrons.

(b) O^-:   Write the Lewis structure based on the valence electrons $2s^2 2p^5$.

$$\left[\cdot\ddot{O}: \right]^-$$

(c)   OH: Write the correct skeletal structure for the molecule.

H—O

Calculate the total number of electrons for the Lewis structure by summing the number of valence electrons of each atom in the molecule.

(number of valence e^- for O) + (number of valence e^- for H) = 6 + 1 = 7

Distribute the electrons among the atoms, giving octets (or duets for H) to as many atoms as possible. Begin with the bonding electrons; then proceed to lone pairs on terminal atoms and finally to lone pairs on the central atom.

$$H—\ddot{O}\cdot$$

All 7 valence electrons are used.

O has an incomplete octet. It has seven electrons because we have an odd number of valence electrons.

(d) CH_3OO: Write the correct skeletal structure for the radical.
C is the least electronegative atom, so it is central.

$$H$$
$$|$$
$$H—C—O—O$$
$$|$$
$$H$$

Calculate the total number of electrons for the Lewis structure by summing the number of valence electrons of each atom in the molecule.

3(number of valence e^- for H) + (number of valence e^- for C) + 2(number of valence e^- for O) = 3(1) + 4 + 2(6) = 19

Distribute the electrons among the atoms, giving octets (or duets for H) to as many atoms as possible. Begin with the bonding electrons; then proceed to lone pairs on terminal atoms and finally to lone pairs on the central atoms.

$$H$$
$$|$$
$$H—C—\ddot{O}—\ddot{O}\cdot$$
$$|$$
$$H$$

All 19 valence electrons are used.

O has an incomplete octet. It has seven electrons because we have an odd number of valence electrons.

5.91 **Given:** 7.743% H **Find:** Lewis structure

Conceptual Plan: %H $\rightarrow$ %C $\rightarrow$ mass C, H $\rightarrow$ mol C, H $\rightarrow$ pseudoformula $\rightarrow$ empirical formula

$$100\% - \%H \quad \text{Assume 100 g sample} \quad \frac{1 \text{ mol C}}{12.01 \text{ g C}} \quad \frac{1 \text{ mol H}}{1.008 \text{ g H}} \qquad \text{divide by smallest number}$$

Solution: %C $= 100\% - 7.743\% = 92.568\%$ C

In a 100.00 g sample: 7.743 g H, 92.568 g C

$$7.743 \text{ g H} \times \frac{1 \text{ mol H}}{1.008 \text{ g H}} = 7.682 \text{ mol H}$$

$$92.568 \text{ g C} \times \frac{1 \text{ mol C}}{12.01 \text{ g C}} = 7.708 \text{ mol C}$$

$C_{7.708}H_{7.682}$

$\dfrac{C_{7.708}H_{7.682}}{7.682 \quad 7.682} \rightarrow CH$

The smallest molecular formula would be C_2H_2.

Write the correct skeletal structure for the molecule.

H—C—C—H

Calculate the total number of electrons for the Lewis structure by summing the valence electrons of each atom in the molecule.

$$2(\text{number of valence e}^- \text{ for C}) + 2(\text{number of valence e}^- \text{ for H}) = 2(4) + 2(1) = 10$$

Distribute the electrons among the atoms, giving octets (or duets for H) to as many atoms as possible. Begin with the bonding electrons; then proceed to lone pairs on terminal atoms and finally to lone pairs on the central atom.

H—C̈—C̈—H

Complete the octet on C by forming a triple bond.

H—C≡C—H

5.93 **(a)** serine

H—N—C₁—C₂—Ö₁—H (with H, H, :O: above, H—C₃—H below, and :Ö₂—H below that)

The Cs in positions 1 and 3 have four electron groups around the atom. Four electron groups give a tetrahedral electron geometry; tetrahedral electron geometry has sp^3 hybridization. Four bonding groups and zero lone pairs give a tetrahedral molecular geometry.

C in position 2 has three electron groups around the atom. Three electron groups give a trigonal planar geometry; trigonal planar geometry has sp^2 hybridization. Three bonding groups and zero lone pairs give a trigonal planar molecular geometry.

N has four electron groups around the atom. Four electron groups give a tetrahedral electron geometry; tetrahedral electron geometry has sp^3 hybridization. Three bonding groups and one lone pair give a trigonal pyramidal molecular geometry.

The Os in positions 1 and 2 each have four electron groups around the atom. Four electron groups give a tetrahedral electron geometry; tetrahedral electron geometry has sp^3 hybridization. Two bonding groups and two lone pairs give a bent molecular geometry.

(b)   asparagine

The Cs in positions 1 and 3 each have four electron groups around the atom. Four electron groups give a tetrahedral electron geometry; tetrahedral electron geometry has sp^3 hybridization. Four bonding groups and zero lone pairs give a tetrahedral molecular geometry.

The Cs in positions 2 and 4 each have three electron groups around the atom. Three electron groups give a trigonal planar geometry; trigonal planar geometry has sp^2 hybridization. Three bonding groups and zero lone groups give a trigonal planar molecular geometry.

The Ns in positions 1 and 2 each have four electron groups around the atom. Four electron groups give a tetrahedral electron geometry; tetrahedral electron geometry has sp^3 hybridization. Three bonding groups and one lone pair give a trigonal pyramidal molecular geometry.

O has four electron groups around the atom. Four electron groups give a tetrahedral electron geometry; tetrahedral electron geometry has sp^3 hybridization. Two bonding groups and two lone pairs give a bent molecular geometry.

(c)   cysteine

The Cs in positions 1 and 3 each have four electron groups around the atom. Four electron groups give a tetrahedral electron geometry; tetrahedral electron geometry has sp^3 hybridization. Four bonding groups and zero lone pairs give a tetrahedral molecular geometry.

C in position 2 has three electron groups around the atom. Three electron groups give a trigonal planar geometry; trigonal planar geometry has sp^2 hybridization. Three bonding groups and zero lone pairs give a trigonal planar molecular geometry.

N has four electron groups around the atom. Four electron groups give a tetrahedral electron geometry; tetrahedral electron geometry has sp^3 hybridization. Three bonding groups and one lone pair give a trigonal pyramidal molecular geometry.

O and S have four electron groups around the atom. Four electron groups give a tetrahedral electron geometry; tetrahedral electron geometry has sp^3 hybridization. Two bonding groups and two lone pairs give bent molecular geometry.

5.95 (a) Water-soluble: The four C—OH bonds, the C=O bond, and the C—O bonds in the ring make the molecule polar. Because of the large electronegativity difference between the C and O, each of the bonds will have a dipole moment. The sum of the dipole moments does *not* give a net zero dipole moment, so the molecule is polar. Because it is polar, it will be water-soluble.

(b) Fat-soluble: There is only one C—O bond in the molecule. The dipole moment from this bond is not enough to make the molecule polar because of all of the nonpolar components of the molecule. The C—H bonds in the structure lead to a net dipole of zero for most of the sites in the molecule. Because the molecule is nonpolar, it is fat-soluble.

(c) Water-soluble: The carboxylic acid function (COOH group) along with the N atom in the ring make the molecule polar. Because of the electronegativity difference between the C and O and the C and N atoms, the bonds will have a dipole moment and the net dipole moment of the molecule is *not* zero; so the molecule is polar. Because the molecule is polar, it is water-soluble.

(d) Fat-soluble: The two O atoms in the structure contribute a very small amount to the net dipole moment of this molecule. The majority of the molecule is nonpolar because there is no net dipole moment around the interior C atoms. Because the molecule is nonpolar, it is fat-soluble.

Challenge Problems

5.97 N_3^-: Write the correct skeletal structure for the ion.

$$N—N—N$$

Calculate the total number of electrons for the Lewis structure by summing the valence electrons of each atom in the ion and adding 1 for the 1− charge.

$$3(\text{number of valence } e^- \text{ for N}) + 1 = 3(5) + 1 = 16$$

Distribute the electrons among the atoms, giving octets to as many atoms as possible. Begin with the bonding electrons; then proceed to lone pairs on terminal atoms and finally to lone pairs on the central atom.

$$:\ddot{N}—N—\ddot{N}:$$

Complete octets with double or triple bonds.

$$\ddot{N}=N=\ddot{N}$$

Finally, write the ion in brackets with the charge in the upper right-hand corner.

$$\left[\ddot{N}=N=\ddot{N}\right]^-$$

Write the resonance forms.

$$\left[\ddot{N}=N=\ddot{N}\right]^- \longleftrightarrow \left[:N\equiv N—\ddot{N}:\right]^- \longleftrightarrow \left[:\ddot{N}—N\equiv N:\right]^-$$

5.99 **Given:** $\mu = 1.08$ D HCl, 20% ionic and $\mu = 1.82$ D HF, 45% ionic   **Find:** r

Conceptual Plan: $\mu \rightarrow \mu_{calc} \rightarrow r$

$$\% \text{ ionic character} = \frac{\mu}{\mu_{calc}} \times 100\%$$

Solution: For HCl $\mu_{calc} = \dfrac{1.08}{0.20} = 5.4$ D

$$\frac{5.4\ \cancel{D} \times \dfrac{3.34 \times 10^{-30}\ \cancel{C} \cdot \cancel{m}}{\cancel{D}} \times \dfrac{10^{12}\ \text{pm}}{\cancel{m}}}{1.6 \times 10^{-19}\ \cancel{C}} = 113\ \text{pm}$$

For HF $\mu_{calc} = \dfrac{1.82}{0.45} = 4.04$ D

$$\frac{4.04\ \cancel{D} \times \dfrac{3.34 \times 10^{-30}\ \cancel{C} \cdot \cancel{m}}{\cancel{D}} \times \dfrac{10^{12}\ \text{pm}}{\cancel{m}}}{1.6 \times 10^{-19}\ \cancel{C}} = 84\ \text{pm}$$

From Table 5.4, the bond length of HCl = 127 pm, and HF = 92 pm. Both of these values are slightly higher than the calculated values.

5.101 **Given:** C_8H_8 all C's identical, all H's identical   **Find:** Lewis structure

For the eight C atoms to be equivalent, they must all be in the same electronic environment; for the eight H atoms to be equivalent, they must all be in the same electronic environment as well. That is, they must all see the same number and kinds of bonds. One way to achieve this is with a cubic arrangement of the C atoms and then a H atom attached to each C.

5.103 CH_3N_3 Predict geometry and sketch the molecule.

	I		II			III		
Formal charge 0	+1	−1		−1	+1	0	+1	+1	−2

There are three resonance forms for the methyl azide molecule. Structures I and II would be most likely because they have the smaller formal charges. So we will look at structures I and II.

In structure I: C has four electron groups around the atom. Four electron groups give a tetrahedral electron geometry. Four bonding groups and zero lone pairs give a tetrahedral molecular geometry. N_1 has three electron groups around the atom. Three electron groups give a trigonal planar electron geometry. Two bonding groups and one lone pair give a bent molecular geometry. N_2 has two electron groups around the atom. Two electron groups give a linear electron geometry. Two bonding groups and zero lone pairs give a linear molecular geometry.

In structure II: C has four electron groups around the atom. Four electron groups give a tetrahedral electron geometry. Four bonding groups and zero lone pairs give a tetrahedral molecular geometry. N_1 has four electron groups around the atom. Four electron groups give a tetrahedral electron geometry. Two bonding groups and two lone pairs give a bent molecular geometry. N_2 has two electron groups around the atom. Two electron groups give a linear electron geometry. Two bonding groups and zero lone pairs give a linear molecular geometry.

If structure I is the resonance, the C—N—N bond angle will be closer to 120° and the N—N—N bond angle will be 180°. The two N—N bond lengths would be expected to be the same. If structure II is the resonance, the C—N—N bond angle will be closer to 109° and the N—N—N bond angle will be 180°. The N_1—N_2 bond length would be expected to be longer than the N_2—N_3 bond length.

Conceptual Problems

5.105   Statement a is the best statement.
Statement b neglects the lowering of potential energy that arises from the interaction of the lone pair electrons with the bonding electrons.
Statement c neglects the interaction of the electrons altogether. The molecular geometries are determined by the number and types of electron groups around the central atom.

Questions for Group Work

5.107   Each group member should have a coin for each of the valence electrons in the atom that they represent. The nonmetal atoms will be the most reluctant to give up a coin. The metal atoms will be less reluctant to give up a coin. A bond between a metal and a nonmetal would be an ionic bond, while a bond between two nonmetals will be a covalent bond, and a bond between two metals will be a metallic bond. Some examples could be:

Atom 1	Atom 2	Type of Bonding
Na	Br	Ionic bonding
Rb	O	Ionic bonding
Ca	Cl	Ionic bonding
Cl	F	Covalent bonding
C	O	Covalent bonding
Mg	S	Covalent bonding
Mg	Mg	Metallic bonding

Atom 1	Atom 2	Type of Bonding
Al	F	Ionic bonding
Li	N	Ionic bonding
Sr	Cl	Covalent bonding
N	O	Covalent bonding
P	I	Covalent bonding
Na	K	Metallic bonding
Pt	Au	Metallic bonding

5.109   Draw the Lewis structures for Al and O based on their valence electrons. Al: $3s^2 3p^1$ O: $2s^2 2p^4$.

·Al·   :Ö·

Aluminum must lose three electrons and be left with the octet from the previous shell, while oxygen needs to gain two electrons to get an octet.

$2\ Al^{3+}$ $3\left[:\ddot{O}:\right]^{2-}$

Thus, we need two Al^{3+} and three O^{2-} in order to lose and gain the same number of electrons. Write the formula with subscripts (if necessary) to indicate the number of atoms.

Al_2O_3

5.111 N_2H_4:   Write the correct skeletal structure for the molecule.

H, H / N—N / H, H

Calculate the total number of electrons for the Lewis structure by summing the valence electrons of each atom in the molecule

2(number of valence e^- for N) + 4(number of valence e^- for H) = 2(5) + 4(1) = 14

Distribute the electrons among the atoms, giving octets (or duets for H) to as many atoms as possible. Begin with the bonding electrons, and then proceed to lone pairs on terminal atoms and finally to lone pairs of the central atom.

H, ... H / N — N / H, H

All 14 valence electrons are used.
If any atom lacks an octet, form double or triple bonds as necessary to give them octets. All atoms have octets (duets for hydrogen); the structure is complete.

CCl_4:   Write the Lewis symbol for each atom based on the valence electron.

·C·   :Cl·

The carbon will share an electron pair with each chlorine in order to achieve a stable octet.

:Cl:
|
:Cl— C —Cl:
|
:Cl:

CO_3^{2-}:   Write the correct skeletal structure for the ion.

O
|
O — C — O

Calculate the total number of electrons for the Lewis structure by summing the valence electrons of each atom in the ion and adding 2 for the 2 −charge.

$$3(\text{number of valence } e^- \text{ for O}) + (\text{number of valence } e^- \text{ for C}) + 2 = 3(6) + 4 + 2 = 24$$

Distribute the electrons among the atoms, giving octets (or duets for H) to as many atoms as possible. Begin with the bonding electrons, and then proceed to lone pairs on terminal atoms and finally to lone pairs of the central atom.

All 24 valence electrons are used.

If any atom lacks an octet, form double or triple bonds as necessary to give them octets.

Lastly, write the Lewis structure in brackets with the charge of the ion in the upper right-hand corner.

All atoms have octets (duets for H); the structure is complete. However, the double bond can form from any oxygen atom, so there are three resonance forms.

Calculate the formal charge on each atom by finding the number of valence electrons and subtracting the number of lone pair electrons and one-half the number of bonding electrons.

	O_{left}	O_{top}	O_{right}	C
number of valence electrons	6	6	6	4
− number of lone pair electrons	6	6	4	0
− 1/2(number of bonding electrons)	1	1	2	4
Formal charge	−1	−1	0	0

The sum of the formal charges is −2, which is the overall charge of the ion. The other resonance forms would be the same.

NH_4^+:   Write the correct skeletal structure for the ion.

Calculate the total number of electrons for the Lewis structure by summing the valence electrons of each atom in the ion and subtracting 1 for the 1+ charge.

4(number of valence e⁻ for H) + (number of valence e⁻ for N) −1 = 4(1) + 5 − 1 = 8

Distribute the electrons among the atoms, giving octets (or duets for H) to as many atoms as possible. Begin with the bonding electrons, and then proceed to lone pairs on terminal atoms and finally to lone pairs of the central atom.

All eight valence electrons are used.

If any atom lacks an octet, form double or triple bonds as necessary to give them octets.

Lastly, write the Lewis structure in brackets with the charge of the ion in the upper right-hand corner.

$$\left[\begin{array}{c} H \\ | \\ H - N - H \\ | \\ H \end{array} \right]^{+}$$

All atoms have octets (duets for H); the structure is complete.

Calculate the formal charge on each atom by finding the number of valence electrons and subtracting the number of lone pair electrons and one-half the number of bonding electrons.

	H_{left}	H_{top}	H_{right}	H_{bottom}	N
number of valence electrons	1	1	1	1	5
− number of lone pair electrons	0	0	0	0	0
− 1/2(number of bonding electrons)	1	1	1	1	4
Formal charge	0	0	0	0	+1

The sum of the formal charges is +1, which is the overall charge of the ion.

5.113   Linear molecules can be derived from two different electron geometries:
- Two electron groups, with both groups being bonding groups; CO_2 is a common example of this geometry.
- Five electron groups, with two bonding groups and three lone pairs; XeF_2 is a common example of this geometry.

Data Interpretation and Analysis

5.115   (a)   For each write the Lewis structure:

NO_2 has 17 valence electrons. NO_2^+ has 16 valence electrons.

$$:\ddot{O} - \dot{N} = \ddot{O}$$ $$\left[\ddot{O} = N = \ddot{O} \right]^{+}$$

NO_2^- has 18 valence electrons.

$$\left[:\ddot{O} - \ddot{N} = \ddot{O} \right]^{-}$$

Note that the Lewis structures for NO_2 and NO_2^- have resonance structures, shifting the position of the N=O double bond.

(c)   For each write the Lewis structure:

PH_3 has 8 valence electrons. PF_3 has 26 valence electrons.

$$H - \ddot{P} - H$$
$$|$$
$$H$$

$$:\ddot{F} - \ddot{P} - \ddot{F}:$$
$$|$$
$$:\ddot{F}:$$

PCl$_3$ has 26 valence electrons.   PI$_3$ has 26 valence electrons.

:$\ddot{\text{C}}$l—$\overset{..}{\text{P}}$—$\ddot{\text{C}}$l: :$\ddot{\text{I}}$—$\overset{..}{\text{P}}$—$\ddot{\text{I}}$:

:$\ddot{\text{C}}$l: :$\ddot{\text{I}}$:

All of these Lewis structures have the same bonding pattern, with three bonding pairs and one lone pair on the central P atom. All of these molecules have a tetrahedral electron geometry and trigonal pyramidal molecular geometry.

(e)   As you move from H to F and move down the column from F to Cl and to I, the atomic radius of the atoms increases. As a result, the larger atoms cannot be accommodated with the smaller bond angle. The attached atoms themselves would begin to overlap their orbitals. So as the size of the attached atom increases, the bond angle becomes larger, approaching the hybridized 109.5° angle.

6 Chemical Bonding II: Valence Bond Theory and Molecular Orbital Theory

Review Questions

6.1 While the Lewis model is a simple model that can describe chemical bonding, it cannot predict all of the properties of a molecule or ion. More sophisticated models, such as valence bond theory and molecular orbital theory, are needed to adequately describe bonding and magnetism in certain species.

6.3 According to valence bond theory, the shape of the molecule is determined by the geometry of the overlapping orbitals.

6.5 Hybridization is a mathematical procedure in which the standard atomic orbitals are combined to form new atomic orbitals called hybrid orbitals. Hybrid orbitals are still localized on individual atoms, but they have different shapes and energies from those of standard atomic orbitals. They are necessary in valence bond theory because they correspond more closely to the actual distribution of electrons in chemically bonded atoms.

6.7 The number of standard atomic orbitals added together always equals the number of hybrid orbitals formed. The total number of orbitals is conserved.

6.9 The double bond in Lewis theory is simply two pairs of electrons that are shared between the same two atoms. However, in valence bond theory, we see that the double bond is made up of two different kinds of bonds. The double bond in valence bond theory consists of one σ bond and one π bond. Valence bond theory shows us that rotation about a double bond is severely restricted. Because of the side-by-side overlap of the p orbitals, the π bond must essentially break for rotation to occur. The σ bond consists of end-to-end overlap. Because the overlap is linear, rotation is not restricted.

6.11 In molecular orbital theory, atoms will bond when the electrons in the atoms can lower their energy by occupying the molecular orbitals of the resultant molecule.

6.13 A bonding molecular orbital is lower in energy than the atomic orbitals from which it is formed. There is an increased electron density in the internuclear region.

6.15 The electrons in orbitals behave like waves. The bonding molecular orbital arises from the constructive interference between the atomic orbitals and is lower in energy than the atomic orbitals. The antibonding molecular orbital arises from the destructive interference between the atomic orbitals and is higher in energy than the atomic orbitals.

6.17 Molecular orbitals can be approximated by a linear combination of atomic orbitals (AOs). The total number of MOs formed from a particular set of AOs will always equal the number of AOs used.

6.19

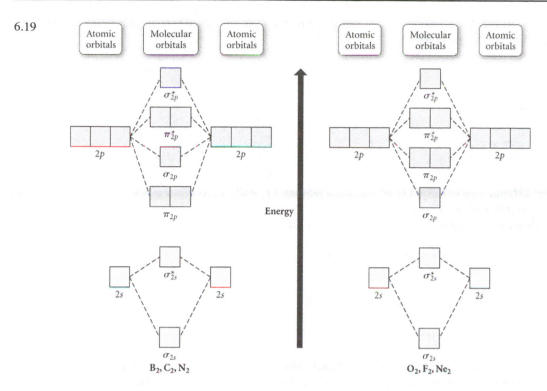

6.21　A paramagnetic species has unpaired electrons in one or more molecular orbitals. A paramagnetic species is attracted to a magnetic field. The magnetic property is a direct result of the unpaired electron(s). The spin and angular momentum of the electrons generate tiny magnetic fields. A diamagnetic species has all of its electrons paired. The magnetic fields caused by the electron spin, and orbital angular momentum tend to cancel each other. A diamagnetic species is not attracted to a magnetic field and is, in fact, slightly repelled.

6.23　Nonbonding orbitals are atomic orbitals not involved in a bond that remain localized on the atom.

Problems by Topic

Valence Bond Theory

6.25　(a)　Be $2s^2$　No bonds can form. Beryllium contains no unpaired electrons, so no bonds can form without hybridization.

(b)　P $3s^2 3p^3$　Three bonds can form. Phosphorus contains three unpaired electrons, so three bonds can form without hybridization.

(c)　F $2s^2 2p^5$　One bond can form. Fluorine contains one unpaired electron, so one bond can form without hybridization.

6.27　PH_3

The unhybridized bond angles should be 90°. So without hybridization, there is good agreement between valence bond theory and the actual bond angle of 93.3°.

6.29   C $2s^2 2p^2$

C

$\uparrow\downarrow$		$\uparrow$	$\uparrow$	
2s			2p	

$\uparrow$	$\uparrow$	$\uparrow$	$\uparrow$

sp^3

6.31 sp^2 Only sp^2 hybridization of this set of orbitals has a remaining p orbital to form a π bond.
sp^3 hybridization utilizes all $3p$ orbitals.
$sp^3 d^2$ hybridization utilizes all $3p$ orbitals and $2d$ orbitals.

6.33   (a) CCl_4 Write the Lewis structure for the molecule:

$$:\ddot{C}l:$$
$$|$$
$$:\ddot{C}l-C-\ddot{C}l:$$
$$|$$
$$:\ddot{C}l:$$

Use valence shell electron pair repulsion (VSEPR) to predict the electron geometry:
Four electron groups around the central atom give a tetrahedral electron geometry.
Select the correct hybridization for the central atom based on the electron geometry:
Tetrahedral electron geometry has sp^3 hybridization.
Sketch the molecule and label the bonds:

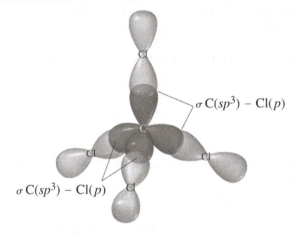

$\sigma\ C(sp^3) - Cl(p)$

$\sigma\ C(sp^3) - Cl(p)$

(b) NH_3 Write the Lewis structure for the molecule:

$$H$$
$$|$$
$$H-\underset{..}{N}-H$$

Use VSEPR to predict the electron geometry:
Four electron groups around the central atom give a tetrahedral electron geometry.
Select the correct hybridization for the central atom based on the electron geometry:
Tetrahedral electron geometry has sp^3 hybridization.
Sketch the molecule and label the bonds:

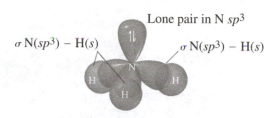

Lone pair in N sp^3

σ N(sp^3) – H(s) σ N(sp^3) – H(s)

(c)   OF$_2$ Write the Lewis structure for the molecule:

$$:\ddot{F}-\ddot{O}-\ddot{F}:$$

Use VSEPR to predict the electron geometry:
Four electron groups around the central atom give a tetrahedral electron geometry.
Select the correct hybridization for the central atom based on the electron geometry:
Tetrahedral electron geometry has sp^3 hybridization.
Sketch the molecule and label the bonds:

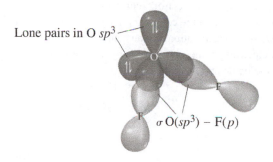

Lone pairs in O sp^3

σ O(sp^3) – F(p)

(d)   CO$_2$ Write the Lewis structure for the molecule:

$$\ddot{O}=C=\ddot{O}$$

Use VSEPR to predict the electron geometry:
Two electron groups around the central atom give a linear electron geometry.
Select the correct hybridization for the central atom based on the electron geometry:
Linear electron geometry has sp hybridization.
Sketch the molecule and label the bonds:

π C(p_y) – O(p_y) π C(p_z) – O(p_z)

σ C(sp) – O(p)

6.35   (a)   COCl$_2$ Write the Lewis structure for the molecule:

$$:\!O\!:$$
$$\|$$
$$:\ddot{C}l-C-\ddot{C}l:$$

Use VSEPR to predict the electron geometry:
Three electron groups around the central atom give a trigonal planar electron geometry.
Select the correct hybridization for the central atom based on the electron geometry:
Trigonal planar electron geometry has sp^2 hybridization.
Sketch the molecule and label the bonds:

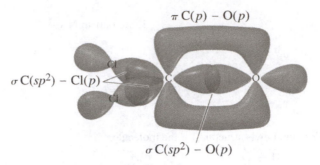

π C(p) – O(p)

σ C(sp^2) – Cl(p)

σ C(sp^2) – O(p)

(b) BrF_5 Write the Lewis structure for the molecule:

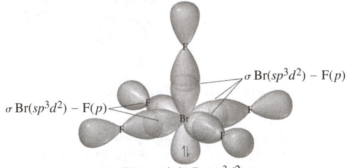

Use VSEPR to predict the electron geometry:
Six electron pairs around the central atoms give an octahedral electron geometry.
Select the correct hybridization for the central atom based on the electron geometry:
Octahedral electron geometry has sp^3d^2 hybridization.
Sketch the molecule and label the bonds:

σ Br(sp^3d^2) – F(p)

σ Br(sp^3d^2) – F(p)

Lone pair in Br(sp^3d^2)

(c) XeF_2 Write the Lewis structure for the molecule:

$:\ddot{F}\!-\!\ddot{X}\!e\!-\!\ddot{F}:$

Use VSEPR to predict the electron geometry:
Five electron groups around the central atom give a trigonal bipyramidal geometry.
Select the correct hybridization for the central atom based on the electron geometry:
Trigonal bipyramidal geometry has sp^3d hybridization.
Sketch the molecule and label the bonds:

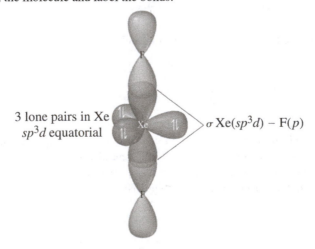

3 lone pairs in Xe
sp^3d equatorial

σ Xe(sp^3d) – F(p)

(d) I_3^- Write the Lewis structure for the molecule:

$$\left[:\overset{..}{\underset{..}{I}} - \overset{..}{\underset{..}{I}} - \overset{..}{\underset{..}{I}}: \right]^-$$

Use VSEPR to predict the electron geometry:
Five electron groups around the central atom give a trigonal bipyramidal geometry.
Select the correct hybridization for the central atom based on the electron geometry:
Trigonal bipyramidal geometry has sp^3d hybridization.
Sketch the molecule and label the bonds:

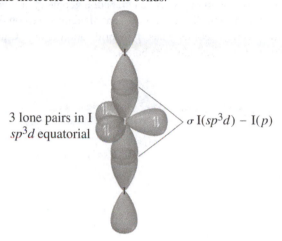

3 lone pairs in I
sp^3d equatorial

$\sigma\, I(sp^3d) - I(p)$

6.37   (a) N_2H_2 Write the Lewis structure for the molecule:

$$H - \overset{..}{N} = \overset{..}{N} - H$$

Use VSEPR to predict the electron geometry:
Three electron groups around each interior atom give a trigonal planar electron geometry.
Select the correct hybridization for the central atoms based on the electron geometry:
Trigonal planar electron geometry has sp^2 hybridization.
Sketch the molecule and label the bonds:

$\pi\, N(p) - N(p)$

Lone pair in $N(sp^2)$ Lone pair in $N(sp^2)$

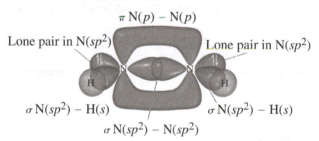

$\sigma\, N(sp^2) - H(s)$ $\sigma\, N(sp^2) - H(s)$

$\sigma\, N(sp^2) - N(sp^2)$

(b) N_2H_4 Write the Lewis structure for the molecule:

$$\overset{H}{\underset{H}{>}} \overset{..}{N} - \overset{..}{N} \overset{H}{\underset{H}{<}}$$

Use VSEPR to predict the electron geometry:
Four electron groups around each interior atom give tetrahedral electron geometry.
Select the correct hybridization for the central atoms based on the electron geometry:
Tetrahedral electron geometry has sp^3 hybridization.
Sketch the molecule and label the bonds:

2 lone pairs in
$N(sp^3)$

$\sigma\, N(sp^3) - H(s)$

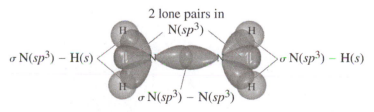

$\sigma\, N(sp^3) - H(s)$

$\sigma\, N(sp^3) - N(sp^3)$

(c) CH₃NH₂ Write the Lewis structure for the molecule:

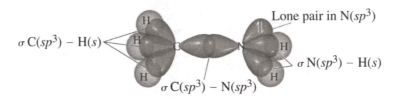

Use VSEPR to predict the electron geometry:
Four electron groups around the C give a tetrahedral electron geometry around the C atom, and four electron groups around the N give a tetrahedral geometry around the N atom.
Select the correct hybridization for the central atoms based on the electron geometry:
Tetrahedral electron geometry has sp^3 hybridization of both C and N.
Sketch the molecule and label the bonds:

6.39

The Cs in positions 1 and 2 have four electron groups around the atom, which is tetrahedral electron geometry. Tetrahedral electron geometry is sp^3 hybridization.
C in position 3 has three electron groups around the atom, which is trigonal planar electron geometry. Trigonal planar electron geometry is sp^2 hybridization.
O (bonded to C and H) has four electron groups around the atom, which is tetrahedral electron geometry. Tetrahedral electron geometry is sp^3 hybridization.
N has four electron groups around the atom, which is tetrahedral electron geometry. Tetrahedral electron geometry is sp^3 hybridization.

Molecular Orbital Theory

6.41 $1s + 1s$ constructive interference results in a bonding orbital:

6.43 Be_2^+ has seven electrons.

Be_2^- has nine electrons.

AO = Atomic Orbital; MO = Molecular Orbital

Bond order $= \dfrac{4-3}{2} = \dfrac{1}{2}$; stable

Bond order $= \dfrac{5-4}{2} = \dfrac{1}{2}$; stable

Both of these ions are able to exist in the gas phase because the bond order is positive.

6.45 The bonding and antibonding molecular orbitals from the combination of p_x and p_x atomic orbitals lie along the internuclear axis.

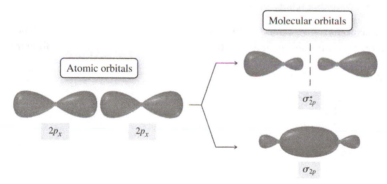

Molecular orbitals

Atomic orbitals

$2p_x$ $2p_x$

σ^*_{2p}

σ_{2p}

6.47 (a) 4 valence electrons   (b) 6 valence electrons   (c) 8 valence electrons   (d) 9 valence electrons

	(a)	(b)	(c)	(d)
σ^*_{2p}	___	___	___	___
π^*_{2p}	___ ___	___ ___	___ ___	___ ___
σ_{2p}	___	___	___	1
π_{2p}	___ ___	1 1	⇅ ⇅	⇅ ⇅
σ^*_{2s}	⇅	⇅	⇅	⇅
σ_{2s}	⇅	⇅	⇅	⇅

Bond order $= \dfrac{2-2}{2} = 0$

diamagnetic

Bond order $= \dfrac{4-2}{2} = 1$

paramagnetic

Bond order $= \dfrac{6-2}{2} = 2$

diamagnetic

Bond order $= \dfrac{7-2}{2} = 2.5$

paramagnetic

6.49   (a)   Write an energy-level diagram for the molecular orbitals in H_2^{2-}. The ion has four valence electrons. Assign the electrons to the molecular orbitals beginning with the lowest energy orbitals and following Hund's rule.

σ^*_{1s} ⇅

σ_{1s} ⇅   Bond order $= \dfrac{2-2}{2} = 0$. With a bond order of 0, the ion will not exist.

(b)   Write an energy-level diagram for the molecular orbitals in Ne_2. The molecule has 16 valence electrons. Assign the electrons to the molecular orbitals beginning with the lowest energy orbitals and following Hund's rule.

σ^*_{2p} ⇅

π^*_{2p} ⇅   ⇅

π_{2p} ⇅   ⇅

σ_{2p} ⇅

σ^*_{2s} ⇅

σ_{2s} ⇅

Bond order $= \dfrac{8-8}{2} = 0$. With a bond order of 0, the molecule will not exist.

(c)   Write an energy-level diagram for the molecular orbitals in He_2^{2+}. The ion has two valence electrons. Assign the electrons to the molecular orbitals beginning with the lowest energy orbitals and following Hund's rule.

σ^*_{1s} _____

σ_{1s} ⇅

Bond order $= \dfrac{2-0}{2} = 1$. With a bond order of 1, the ion will exist.

(d)   Write an energy-level diagram for the molecular orbitals in F_2^{2-}. The molecule has 16 valence electrons. Assign the electrons to the molecular orbitals beginning with the lowest energy orbitals and following Hund's rule.

σ^*_{2p} ⇅

π^*_{2p} ⇅   ⇅

π_{2p} ⇅   ⇅

σ_{2p} ⇅

σ^*_{2s} ⇅

σ_{2s} ⇅

Bond order $= \dfrac{8-8}{2} = 0$. With a bond order of 0, the ion will not exist.

6.51 C_2^- has the highest bond order, the highest bond energy, and the shortest bond.

Write an energy-level diagram for the molecular orbitals in each of the C_2 species.

Assign the electrons to the molecular orbitals beginning with the lowest energy orbitals and following Hund's rule for each of the species.

C$_2$ (8 valence electrons)   C$_2^+$ (7 valence electrons)   C$_2^-$ (9 valence electrons)

C$_2$ C$_2^+$ C$_2^-$

σ^*_{2p} _____   _____   _____   _____   _____   _____

π^*_{2p} _____   _____   _____   _____   _____   _____

σ_{2p} _____   _____ $\underline{\;1\;}$

π_{2p} $\underline{\uparrow\downarrow}$ $\underline{\uparrow\downarrow}$ $\underline{\uparrow\downarrow}$ $\underline{\;\uparrow\;}$ $\underline{\uparrow\downarrow}$ $\underline{\uparrow\downarrow}$

σ^*_{2s} $\underline{\uparrow\downarrow}$ $\underline{\uparrow\downarrow}$ $\underline{\uparrow\downarrow}$

σ_{2s} $\underline{\uparrow\downarrow}$ $\underline{\uparrow\downarrow}$ $\underline{\uparrow\downarrow}$

Bond order $= \dfrac{6-2}{2} = 2$ Bond order $= \dfrac{5-2}{2} = 1.5$ Bond order $= \dfrac{7-2}{2} = 2.5$

C$_2^-$ has the highest bond order at 2.5. Bond order is directly related to bond energy, so C$_2^-$ has the largest bond energy. Bond order is inversely related to bond length, so C$_2^-$ has the shortest bond length.

6.53   Write an energy-level diagram for the molecular orbitals in CO using O$_2$ energy ordering.
Assign the electrons to the molecular orbitals beginning with the lowest energy orbitals and following Hund's rule. CO has 10 valence electrons.

σ^*_{2p} _____

π^*_{2p} _____   _____

π_{2p} $\underline{\uparrow\downarrow}$ $\underline{\uparrow\downarrow}$

σ_{2p} $\underline{\uparrow\downarrow}$

σ^*_{2s} $\underline{\uparrow\downarrow}$

σ_{2s} $\underline{\uparrow\downarrow}$

Bond order $= \dfrac{8-2}{2} = 3$

The electron density is toward the O atom because it is more electronegative.

Cumulative Problems

6.55   (a)   COF$_2$ Write the Lewis structure for the molecule:

$$:\!\overset{\displaystyle :O:}{\underset{}{\overset{\|}{:\ddot{F}\!-\!C\!-\!\ddot{F}:}}}$$

Use VSEPR to predict the electron geometry:
Three electron groups around the central atom give a trigonal planar electron geometry. Three bonding pairs of electrons give a trigonal planar molecular geometry.
Determine whether the molecule contains polar bonds:
The electronegativity of C = 2.5, O = 3.5, and F = 4.0. Therefore, the bonds are polar.
Determine whether the polar bonds add together to form a net dipole:

Even though a trigonal planar molecular geometry normally is nonpolar, because the bonds have different dipole moments, the sum of the dipole moments is not zero. The molecule is polar. See Table 10.2 to see how dipole moments add to determine polarity.

Select the correct hybridization for the central atom based on the electron geometry:

Trigonal planar geometry has sp^2 hybridization.

Sketch the molecule and label the bonds:

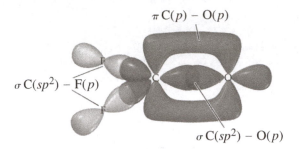

$\pi \, C(p) - O(p)$

$\sigma \, C(sp^2) - F(p)$

$\sigma \, C(sp^2) - O(p)$

(b) S_2Cl_2 Write the Lewis structure for the molecule:

$$:\!\ddot{\underset{..}{Cl}}\!-\!\ddot{\underset{..}{S}}\!-\!\ddot{\underset{..}{S}}\!-\!\ddot{\underset{..}{Cl}}\!:$$

Use VSEPR to predict the electron geometry:

Four electron groups around a central atom give a tetrahedral electron geometry. Two bonding pairs and two lone pairs of electrons give a bent molecular geometry.

Determine whether the molecule contains polar bonds:

The electronegativity of S = 2.5 and Cl = 3.0. Therefore, the bonds are polar.

Determine whether the polar bonds add together to form a net dipole:

In a bent molecular geometry, the sum of the dipole moments is not zero. The molecule is polar. See Table 10.2 to see how dipole moments add to determine polarity.

Select the correct hybridization for the central atom based on the electron geometry:

Tetrahedral geometry has sp^3 hybridization.

Sketch the molecule and label the bonds:

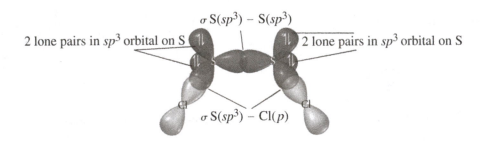

$\sigma \, S(sp^3) - S(sp^3)$

2 lone pairs in sp^3 orbital on S

2 lone pairs in sp^3 orbital on S

$\sigma \, S(sp^3) - Cl(p)$

(c) SF_4 Write the Lewis structure for the molecule:

$$\ddot{\underset{}{:\!\ddot{F}\!:}}$$
$$:\!\ddot{\underset{..}{F}}\!-\!S\!-\!\ddot{\underset{..}{F}}\!:$$
$$:\!\ddot{F}\!:$$

Use VSEPR to predict the electron geometry:

Five electron groups around the central atom give a trigonal bipyramidal electron geometry.

Four bonding pairs and one lone pair of electrons give a seesaw molecular geometry.

Determine whether the molecule contains polar bonds:

The electronegativity of S = 2.5 and F = 4.0. Therefore, the bonds are polar.

Determine whether the polar bonds add together to form a net dipole:

In a seesaw molecular geometry, the sum of the dipole moments is not zero. The molecule is polar. See Table 10.2 to see how dipole moments add to determine polarity.

Select the correct hybridization for the central atom based on the electron geometry:
Trigonal bipyramidal electron geometry has sp^3d hybridization.
Sketch the molecule and label the bonds:

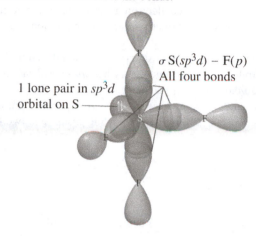

σ S(sp^3d) − F(p)
All four bonds

1 lone pair in sp^3d
orbital on S

6.57   (a)   serine

$$\begin{array}{c}
\text{H} \quad \text{H} \quad :\!\text{O}: \\
| \quad\quad | \quad\quad \| \\
\text{H}-\text{N}-\underset{1}{\text{C}}-\underset{2}{\text{C}}-\underset{\cdot\cdot\,1}{\ddot{\text{O}}}-\text{H} \\
\;\;\;\; \ddot{}\;\; | \\
\;\;\;\;\;\;\; \text{H}-\underset{3}{\text{C}}-\text{H} \\
\;\;\;\;\;\;\;\;\;\; | \\
\;\;\;\;\;\;\;\;\; :\!\underset{\cdot\cdot\,2}{\text{O}}-\text{H}
\end{array}$$

The Cs in positions 1 and 3 have four electron groups around the atom. Four electron groups give a tetrahedral electron geometry; tetrahedral electron geometry has sp^3 hybridization. Four bonding groups and zero lone pairs give a tetrahedral molecular geometry.

C in position 2 has three electron groups around the atom. Three electron groups give a trigonal planar geometry; trigonal planar geometry has sp^2 hybridization. Three bonding groups and zero lone pairs give a trigonal planar molecular geometry.

N has four electron groups around the atom. Four electron groups give a tetrahedral electron geometry; tetrahedral electron geometry has sp^3 hybridization. Three bonding groups and one lone pair give a trigonal pyramidal molecular geometry.

The Os in positions 1 and 2 each have four electron groups around the atom. Four electron groups give a tetrahedral electron geometry; tetrahedral electron geometry has sp^3 hybridization. Two bonding groups and two lone pairs give a bent molecular geometry.

$$\begin{array}{c}
\text{H} \quad\quad \text{OH} \\
\text{H}_2\text{N} \diagup\!\!\!\!\diagdown \diagup\!\!\!\!\!\diagdown \overset{\displaystyle O}{} \\
\text{OH}
\end{array}$$

(b)   asparagine

$$\begin{array}{c}
\text{H} \quad \text{H} \quad :\!\text{O}: \\
| \quad\quad | \quad\quad \| \\
\text{H}-\underset{\cdot\cdot\,1}{\text{N}}-\underset{1}{\text{C}}-\underset{2}{\text{C}}-\ddot{\text{O}}-\text{H} \\
\;\;\;\; | \quad\quad | \\
\;\;\;\;\;\;\; \text{H}-\underset{3}{\text{C}}-\text{H} \\
\;\;\;\;\;\;\;\;\;\; | \\
\;\;\;\;\;\;\;\;\;\; \underset{4}{\text{C}}=\ddot{\text{O}} \\
\;\;\;\;\;\;\;\;\;\; | \\
\;\;\;\;\;\;\; :\!\underset{2}{\text{N}}-\text{H} \\
\;\;\;\;\;\;\;\;\;\; | \\
\;\;\;\;\;\;\;\;\;\; \text{H}
\end{array}$$

The Cs in positions 1 and 3 each have four electron groups around the atom. Four electron groups give a tetrahedral electron geometry; tetrahedral electron geometry has sp^3 hybridization. Four bonding groups and zero lone pairs give a tetrahedral molecular geometry.

The Cs in positions 2 and 4 each have three electron groups around the atom. Three electron groups give a trigonal planar geometry; trigonal planar geometry has sp^2 hybridization. Three bonding groups and zero lone groups give a trigonal planar molecular geometry.

The Ns in positions 1 and 2 each have four electron groups around the atom. Four electron groups give a tetrahedral electron geometry; tetrahedral electron geometry has sp^3 hybridization. Three bonding groups and one lone pair give a trigonal pyramidal molecular geometry.

O has four electron groups around the atom. Four electron groups give a tetrahedral electron geometry; tetrahedral electron geometry has sp^3 hybridization. Two bonding groups and two lone pairs give a bent molecular geometry.

(c) cysteine

The Cs in positions 1 and 3 each have four electron groups around the atom. Four electron groups give a tetrahedral electron geometry; tetrahedral electron geometry has sp^3 hybridization. Four bonding groups and zero lone pairs give a tetrahedral molecular geometry.

C in position 2 has three electron groups around the atom. Three electron groups give a trigonal planar geometry; trigonal planar geometry has sp^2 hybridization. Three bonding groups and zero lone pairs give a trigonal planar molecular geometry.

N has four electron groups around the atom. Four electron groups give a tetrahedral electron geometry; tetrahedral electron geometry has sp^3 hybridization. Three bonding groups and one lone pair give a trigonal pyramidal molecular geometry.

O and S have four electron groups around the atom. Four electron groups give a tetrahedral electron geometry; tetrahedral electron geometry has sp^3 hybridization. Two bonding groups and two lone pairs give bent molecular geometry.

6.59 4π bonds, 25σ bonds; the lone pair on the Os and N in position 2 occupies sp^2 orbitals; the lone pairs on N in position 1, N in position 3, and N in position 4 occupy sp^3 orbitals.

caffeine

6.61   ClF has 14 valence electrons. Assign the electrons to the lowest energy MOs first and then follow Hund's rule. The MOs are formed from the $2s$ and $2p$ orbitals on F and the $3s$ and $3p$ orbitals on Cl.

$$\text{Bond order} = \frac{8-6}{2} = 1$$

6.63   BrF (14 valence electrons)

:B̈r—F̈: no central atom, no hybridization, no electron structure.

BrF$_2^-$ (22 valence electrons)

$\left[:\ddot{F}—\ddot{B}r—\ddot{F}:\right]^-$ There are five electron groups on the central atom, so the electron geometry is trigonal bipyramidal. The two bonding groups and three lone pairs give a linear molecular geometry. An electron geometry of trigonal bipyramidal has sp^3d hybridization.

BrF$_3$ (28 valence electrons)

:F̈:
|
:F̈—B̈r—F̈:   There are five electron groups on the central atom, so the electron geometry is trigonal bipyramidal. The three bonding groups and two lone pairs give a T-shaped molecular geometry. An electron geometry of trigonal bipyramidal has sp^3d hybridization.

BrF$_4^-$ (36 valence electrons)

$$\left[\begin{array}{c}:\ddot{F}: \\ | \\ :\ddot{F}—\ddot{B}r—\ddot{F}: \\ | \\ :\ddot{F}:\end{array}\right]^-$$

There are six electron groups on the central atom, so the electron geometry is octahedral. The four bonding groups and two lone pairs give a square planar molecular geometry. An electron geometry of octahedral has sp^3d^2 hybridization.

BrF$_5$ (42 valence electrons)

There are six electron groups on the central atom, so the electron geometry is octahedral. The five bonding groups and one lone pair give a square pyramidal molecular geometry. An electron geometry of octahedral has sp^3d^2 hybridization.

6.65   (a)   N$_2$O$_5$ Draw the Lewis structure: (40 valence electrons)

Each N has a trigonal planar electron geometry, so there are 3 sp^2 hybrid orbitals on each N. The central O has tetrahedral electron geometry, so there are 4 sp^3 hybrid orbitals. There are a total of 10 hybrid orbitals.

(b) C_2H_5NO Draw the Lewis structure: (24 valence electrons)

$$H_2C_A=\ddot{N}-C_B-\ddot{O}-H$$

with H on C_A, and H, H on C_B, as drawn

C_A has a trigonal planar electron geometry, so there are 3 sp^2 hybrid orbitals.
N has a trigonal planar electron geometry, so there are 3 sp^2 hybrid orbitals.
C_B has a tetrahedral electron geometry, so there are 4 sp^3 hybrid orbitals.
O has a tetrahedral electron geometry, so there are 4 sp^3 hybrid orbitals.
There are a total of 14 hybrid orbitals.

(c)   BrCN   Draw the Lewis structure: (16 valence electrons)

$$:\ddot{B}r-C\equiv N:$$

The C has linear electron geometry, so there are 2 sp hybrid orbitals.

Challenge Problems

6.67   According to valence bond theory, CH_4, NH_3, and H_2O are all sp^3 hybridized. This hybridization results in a tetrahedral electron group configuration with a 109.5° bond angle. NH_3 and H_2O deviate from this idealized bond angle because their lone electron pairs exist in their own sp^3 orbitals. The presence of lone pairs lowers the tendency for the central atom's orbitals to hybridize. As a result, as lone pairs are added, the bond angle moves from the 109.5° hybrid angle toward the 90° unhybridized angle.

6.69   Using the MO diagram for NH_3, assign the eight valence electrons to the molecular orbitals. Start with the lowest energy orbital first and follow Hund's rule.

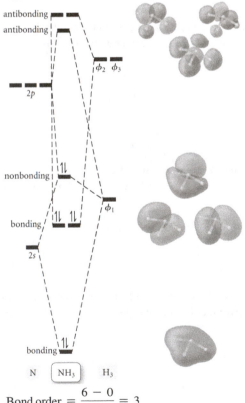

$$Bond\ order = \frac{6-0}{2} = 3$$

With a bond order of 3, the molecule is stable.

6.71   To accommodate the five σ bonds, you need five equal energy hybrid orbitals. So you need to combine five atomic orbitals. The valence electrons on C are in the $2s$ and $2p$ orbitals. You can, therefore, combine the $2s$ and $2p$ orbitals, but this would only give you four hybrid orbitals. The next lowest energy orbital available on C is the $3s$, so this would be the next atomic orbital added in. This gives a hybridization of s^2p^3. VSEPR theory would predict that the geometry is trigonal bipyramidal to accommodate the five bonds, as five identical substituents.

Conceptual Problems

6.73   In Lewis theory, a covalent bond comes from the sharing of electrons.

>   A single bond shares two electrons (one pair).
>   A double bond shares four electrons (two pairs).
>   A triple bond shares six electrons (three pairs).

In valence bond theory, a covalent bond forms when orbitals overlap. The orbitals can be unhybridized or hybridized orbitals.

>   A single bond forms when a σ bond is formed from the overlap of an s orbital with an s orbital, an s orbital with a p orbital, or a p orbital and a p orbital overlapping end to end. A σ can also form from the overlap of a hybridized orbital on the central atom with an s orbital or with a p orbital overlapping end to end.
>   A double bond is a combination of a σ bond and a π bond. The π bond forms from the sideways overlap of a p orbital on each of the atoms involved in the bond. The p orbitals must have the same orientation.
>   A triple bond is a combination of a σ bond and 2π bonds. The π bonds form from the sideways overlap of a p orbital on each of the atoms involved in the bond. The p orbitals must have the same orientation, so each π bond is formed from a different set of p orbitals.

In molecular orbital theory, molecular orbitals form. These are combinations of the atomic orbitals of the atoms involved in the bond. The bonds form when the valence electrons occupy more bonding molecular orbitals than anti-bonding molecular orbitals. This is calculated by the bond order.

>   A single bond has a bond order of 1.
>   A double bond has a bond order of 2.
>   A triple bond has a bond order of 3.

All three models show the formation of bonds between two atoms. All three models also show the formation of the same number of bonds between the atoms involved. Lewis theory tells us only about the number of bonds formed and, combined with VSEPR theory, allows us to predict the shape of the molecule. It does not, however, tell us anything about how the bonds are formed. Valence bond theory addresses the formation of the different types of bonds, sigma and pi. In valence bond theory, the bonds form from the overlap of atomic orbitals on the individual atoms involved in the bonds, and the atoms are localized between the two atoms involved in the bond. Molecular orbital theory approaches the formation of bonds by looking at the entire molecule. The electrons are not restricted to any two individual atoms, but are treated as belonging to the whole molecule. The electrons reside in molecular orbitals that are part of the entire molecule rather than being restricted to individual atoms. Each model gives us information about the molecule. The amount and type of information we need determine the model we choose to use.

6.75   Metals have a tendency to lose electrons, which means that they have relatively low ionization energies. When metal atoms bond together to form a solid, each metal atom donates one or more electrons to an *electron sea*. The metal cations are held together by their attraction to the sea of electrons. Metals are excellent conductors of heat, because of these highly mobile electrons, which help to disperse thermal energy throughout the metal.

Questions for Group Work

6.77 The number of standard atomic orbitals added together always equals the number of hybrid orbitals formed. The total number of orbitals is conserved.

Each sketch indicates the number of hybrid orbitals formed.

A linear electron geometry corresponds to sp hybridization, with bonding to two electron groups.
A trigonal planar electron geometry corresponds to sp^2 hybridization, with bonding to three electron groups.
A tetrahedral electron geometry corresponds to sp^3 hybridization, with bonding to four electron groups.
A trigonal bipyramidal electron geometry corresponds to sp^3d hybridization, with bonding to five electron groups.
An octahedral electron geometry corresponds to sp^3d^2 hybridization, with bonding to six electron groups.

Data Interpretation and Analysis

6.79 (a) $\dfrac{2.18 \times 10^{-8}\,\text{J}}{1\,\text{photon}} \times \dfrac{1\,\text{kJ}}{10^3\,\text{J}} \times \dfrac{6.022 \times 10^{23}\,\text{photons}}{1\,\text{mol photons}} = 1{,}3\underline{1}3.796\,\text{kJ/mol} = 1{,}310\,\text{kJ/mol}$

(c) $2 \times (1{,}5\underline{1}1.522\,\text{kJ/mol} - 1{,}3\underline{1}3.796\,\text{kJ/mol}) = 2 \times (1\underline{9}7.7\,\text{kJ/mol})$
$= 3\underline{9}5.4\,\text{kJ/mol} = 4.0 \times 10^2\,\text{kJ/mol}$

7

Chemical Reactions and Chemical Quantities

Review Questions

7.1   The greenhouse effect is the heating of the Earth's atmosphere that occurs when greenhouse gases trap some of the sunlight that penetrates the Earth's atmosphere. Carbon dioxide is the most significant greenhouse gas.

7.3   A physical property is one that a substance displays without changing its composition, whereas a chemical property is one that a substance displays only by changing its composition via a chemical change.

7.5   In chemistry, we represent chemical reactions with chemical equations where the substances on the left-hand side of a chemical equation are the reactants and the substances on the right-hand side are the products. Chemical equations are balanced when the number of each type of atom on the left side of the equation is equal to the number on the right side.

7.7   Reaction stoichiometry is the numerical relationship between chemical amounts in a balanced chemical equation. The coefficients in a chemical reaction specify the relative amounts in moles of each of the substances involved in the reaction.

7.9   The theoretical yield is the amount of product that can be made in a chemical reaction based on the amount of limiting reactant. The percent yield is calculated as $\dfrac{\text{actual yield}}{\text{theoretical yield}} \times 100\%$.

7.11 Most U.S. energy comes from the combustion of fossil fuels, which include petroleum, natural gas, and coal.

7.13 (a)   The reactions of the alkali metals with halogens result in the formation of metal halides.
              $2\,\text{M}(s) + \text{X}_2 \rightarrow 2\,\text{MX}(s)$
       (b)   Alkali metals react with water to form the dissolved alkali metal ion, the hydroxide ion, and hydrogen gas.
              $2\,\text{M}(s) + 2\,\text{H}_2\text{O}(l) \rightarrow 2\,\text{M}^+(aq) + 2\,\text{OH}^-(aq) + \text{H}_2(g)$

Problems by Topic

Chemical and Physical Changes

7.15 (a)   chemical change (new compounds are formed as methane and oxygen react to form carbon dioxide and water)
       (b)   physical change (vaporization is a phase change and does not involve the making or breaking of chemical bonds)
       (c)   chemical change (new compounds are formed as propane and oxygen react to form carbon dioxide and water)
       (d)   chemical change (new compounds are formed as the metal in the frame is converted to oxides)

7.17   (a)   physical change (vaporization is a phase change and does not involve the making or breaking of chemical bonds)

    (b)   chemical change (new compounds are formed)

    (c)   physical change (vaporization is a phase change and does not involve the making or breaking of chemical bonds)

7.19   (a)   physical property (color can be observed without making or breaking chemical bonds)

    (b)   chemical property (must observe by making or breaking chemical bonds)

    (c)   physical property (the phase can be observed without making or breaking chemical bonds)

    (d)   physical property (density can be observed without making or breaking chemical bonds)

    (e)   physical property (mixing does not involve making or breaking chemical bonds, so this can be observed without making or breaking chemical bonds)

7.21   (a)   chemical property (burning involves breaking and making bonds, so bonds must be broken and made to observe this property)

    (b)   physical property (shininess is a physical property and so can be observed without making or breaking chemical bonds)

    (c)   physical property (odor can be observed without making or breaking chemical bonds)

    (d)   chemical property (burning involves breaking and making bonds, so bonds must be broken and made to observe this property)

Writing and Balancing Chemical Equations

7.23   **Conceptual Plan: Write a skeletal reaction → balance atoms in more complex compounds → balance elements that occur as free elements → clear fractions**

    Solution: Skeletal reaction: $SO_2(g) + O_2(g) + H_2O(l) \rightarrow H_2SO_4(aq)$

                 Balance O: $SO_2(g) + 1/2\,O_2(g) + H_2O(l) \rightarrow H_2SO_4(aq)$

                 Clear fraction: $2\,SO_2(g) + O_2(g) + 2\,H_2O(l) \rightarrow 2\,H_2SO_4(aq)$

    Check:

left side	right side
2 S atoms	2 S atoms
8 O atoms	8 O atoms
4 H atoms	4 H atoms

7.25   **Conceptual Plan: Write a skeletal reaction → balance atoms in more complex compounds → balance elements that occur as free elements → clear fractions**

    Solution: Skeletal reaction: $Na(s) + H_2O(l) \rightarrow H_2(g) + NaOH(aq)$

                 Balance H: $Na(s) + H_2O(l) \rightarrow 1/2\,H_2(g) + NaOH(aq)$

                 Clear fraction: $2\,Na(s) + 2\,H_2O(l) \rightarrow H_2(g) + 2\,NaOH(aq)$

    Check:

left side	right side
2 Na atoms	2 Na atoms
4 H atoms	4 H atoms
2 O atoms	2 O atoms

7.27   **Conceptual Plan: Write a skeletal reaction → balance atoms in more complex compounds → balance elements that occur as free elements → clear fractions**

    Solution: Skeletal reaction: $C_{12}H_{22}O_{11}(aq) + H_2O(l) \rightarrow C_2H_5OH(aq) + CO_2(g)$

                 Balance H: $C_{12}H_{22}O_{11}(aq) + H_2O(l) \rightarrow 4\,C_2H_5OH(aq) + CO_2(g)$

                 Balance C: $C_{12}H_{22}O_{11}(aq) + H_2O(l) \rightarrow 4\,C_2H_5OH(aq) + 4\,CO_2(g)$

    Check:

left side	right side
12 C atoms	12 C atoms
24 H atoms	24 H atoms
12 O atoms	12 O atoms

7.29   (a)   **Conceptual Plan: Write a skeletal reaction → balance atoms in more complex compounds → balance elements that occur as free elements → clear fractions**

          Solution: Skeletal reaction: $PbS(s) + HBr(aq) \rightarrow PbBr_2(s) + H_2S(g)$

                     Balance Br: $PbS(s) + 2\,HBr(aq) \rightarrow PbBr_2(s) + H_2S(g)$

Check:

	left side	right side
	1 Pb atom	1 Pb atom
	1 S atom	1 S atom
	2 H atoms	2 H atoms
	2 Br atoms	2 Br atoms

(b) **Conceptual Plan: Write a skeletal reaction → balance atoms in more complex compounds → balance elements that occur as free elements → clear fractions**

Solution: Skeletal reaction: $CO(g) + H_2(g) \rightarrow CH_4(g) + H_2O(l)$

Balance H: $CO(g) + 3\,H_2(g) \rightarrow CH_4(g) + H_2O(l)$

Check:

left side	right side
1 C atom	1 C atom
1 O atom	1 O atom
6 H atoms	6 H atoms

(c) **Conceptual Plan: Write a skeletal reaction → balance atoms in more complex compounds → balance elements that occur as free elements → clear fractions**

Solution: Skeletal reaction: $HCl(aq) + MnO_2(s) \rightarrow MnCl_2(aq) + H_2O(l) + Cl_2(g)$

Balance Cl: $4\,HCl(aq) + MnO_2(s) \rightarrow MnCl_2(aq) + H_2O(l) + Cl_2(g)$

Balance O: $4\,HCl(aq) + MnO_2(s) \rightarrow MnCl_2(aq) + 2\,H_2O(l) + Cl_2(g)$

Check:

left side	right side
4 H atoms	4 H atoms
4 Cl atoms	4 Cl atoms
1 Mn atom	1 Mn atom
2 O atoms	2 O atoms

(d) **Conceptual Plan: Write a skeletal reaction → balance atoms in more complex compounds → balance elements that occur as free elements → clear fractions**

Solution: Skeletal reaction: $C_5H_{12}(l) + O_2(g) \rightarrow CO_2(g) + H_2O(l)$

Balance C: $C_5H_{12}(l) + O_2(g) \rightarrow 5\,CO_2(g) + H_2O(l)$

Balance H: $C_5H_{12}(l) + O_2(g) \rightarrow 5\,CO_2(g) + 6\,H_2O(l)$

Balance O: $C_5H_{12}(l) + 8\,O_2(g) \rightarrow 5\,CO_2(g) + 6\,H_2O(l)$

Check:

left side	right side
5 C atoms	5 C atoms
12 H atoms	12 H atoms
16 O atoms	16 O atoms

7.31 **Conceptual Plan: Write a skeletal reaction → balance atoms in more complex compounds → balance elements that occur as free elements → clear fractions**

Solution: Skeletal reaction: $Na_2CO_3(aq) + CuCl_2(aq) \rightarrow CuCO_3(s) + NaCl(aq)$

Balance Na: $Na_2CO_3(aq) + CuCl_2(aq) \rightarrow CuCO_3(s) + 2\,NaCl(aq)$

Check:

left side	right side
2 Na atoms	2 Na atoms
1 C atom	1 C atom
3 O atoms	3 O atoms
1 Cu atom	1 Cu atom
2 Cl atoms	2 Cl atoms

7.33 (a) **Conceptual Plan: Balance atoms in more complex compounds → balance elements that occur as free elements → clear fractions**

Solution: Skeletal reaction: $CO_2(g) + CaSiO_3(s) + H_2O(l) \rightarrow SiO_2(s) + Ca(HCO_3)_2(aq)$

Balance C: $2\,CO_2(g) + CaSiO_3(s) + H_2O(l) \rightarrow SiO_2(s) + Ca(HCO_3)_2(aq)$

Check:

left side	right side
2 C atoms	2 C atoms
8 O atoms	8 O atoms
1 Ca atom	1 Ca atom
1 Si atom	1 Si atom
2 H atoms	2 H atoms

(b)   **Conceptual Plan: Balance atoms in more complex compounds → balance elements that occur as free elements → clear fractions**

     Solution: Skeletal reaction: $Co(NO_3)_3(aq) + (NH_4)_2S(aq) \rightarrow Co_2S_3(s) + NH_4NO_3(aq)$

                         Balance S: $Co(NO_3)_3(aq) + 3\,(NH_4)_2S(aq) \rightarrow Co_2S_3(s) + NH_4NO_3(aq)$

                         Balance Co: $2\,Co(NO_3)_3(aq) + 3\,(NH_4)_2S(aq) \rightarrow Co_2S_3(s) + NH_4NO_3(aq)$

                         Balance N: $2\,Co(NO_3)_3(aq) + 3\,(NH_4)_2S(aq) \rightarrow Co_2S_3(s) + 6\,NH_4NO_3(aq)$

     Check:

	left side	right side
	2 Co atoms	2 Co atoms
	12 N atoms	12 N atoms
	18 O atoms	18 O atoms
	24 H atoms	24 H atoms
	3 S atoms	3 S atoms

(c)   **Conceptual Plan: Balance atoms in more complex compounds → balance elements that occur as free elements → clear fractions**

     Solution: Skeletal reaction: $Cu_2O(s) + C(s) \rightarrow Cu(s) + CO(g)$

                         Balance Cu: $Cu_2O(s) + C(s) \rightarrow 2\,Cu(s) + CO(g)$

     Check:

	left side	right side
	2 Cu atoms	2 Cu atoms
	1 O atom	1 O atom
	1 C atom	1 C atom

(d)   **Conceptual Plan: Balance atoms in more complex compounds → balance elements that occur as free elements → clear fractions**

     Solution: Skeletal reaction: $H_2(g) + Cl_2(g) \rightarrow HCl(g)$

                         Balance Cl: $H_2(g) + Cl_2(g) \rightarrow 2\,HCl(g)$

     Check:

	left side	right side
	2 H atoms	2 H atoms
	2 Cl atoms	2 Cl atoms

Reaction Stoichiometry

7.35   **Given:** 7.2 moles C_6H_{14} **Find:** balanced reaction, moles O_2 required

       Conceptual Plan: Balance the equation then mol C_6H_{14} → mol O_2

       $2\,C_6H_{14}(g) + 19\,O_2(g) \rightarrow 12\,CO_2(g) + 14\,H_2O(g)$ $\dfrac{19 \text{ mol } O_2}{2 \text{ mol } C_6H_{14}}$

       Solution: $7.2 \text{ mol } C_6H_{14} \times \dfrac{19 \text{ mol } O_2}{2 \text{ mol } C_6H_{14}} = 68.4 \text{ mol } O_2 = 68 \text{ mol } O_2$

       Check: The units of the answer $(\text{mol } O_2)$ are correct. The magnitude is reasonable because much more O_2 is needed than C_6H_{14}.

7.37   (a)   **Given:** 2.5 mol N_2O_5 **Find:** mol NO_2

           Conceptual Plan: mol N_2O_5 → mol NO_2

           $\dfrac{4 \text{ mol } NO_2}{2 \text{ mol } N_2O_5}$

           Solution: $2.5 \text{ mol } N_2O_5 \times \dfrac{4 \text{ mol } NO_2}{2 \text{ mol } N_2O_5} = 5.0 \text{ mol } NO_2$

           Check: The units of the answer $(\text{mol } NO_2)$ are correct. The magnitude is reasonable because it is greater than mol N_2O_5.

   (b)   **Given:** 6.8 mol N_2O_5 **Find:** mol NO_2

           Conceptual Plan: mol N_2O_5 → mol NO_2

           $\dfrac{4 \text{ mol } NO_2}{2 \text{ mol } N_2O_5}$

           Solution: $6.8 \text{ mol } N_2O_5 \times \dfrac{4 \text{ mol } NO_2}{2 \text{ mol } N_2O_5} = 13.6 \text{ mol } NO_2 = 14 \text{ mol } NO_2$

           Check: The units of the answer $(\text{mol } NO_2)$ are correct. The magnitude is reasonable because it is greater than mol N_2O_5.

(c)   **Given:** 15.2 g N_2O_5 **Find:** mol NO_2

Conceptual Plan: g N_2O_5 → mol N_2O_5 → mol NO_2

$$\frac{1 \text{ mol } N_2O_5}{108.02 \text{ g } N_2O_5} \qquad \frac{4 \text{ mol } NO_2}{2 \text{ mol } N_2O_5}$$

Solution: $15.2 \text{ g } \cancel{N_2O_5} \times \dfrac{1 \text{ mol } \cancel{N_2O_5}}{108.02 \text{ g } \cancel{N_2O_5}} \times \dfrac{4 \text{ mol } NO_2}{2 \text{ mol } \cancel{N_2O_5}} = 0.28\underline{1}4 \text{ mol } NO_2 = 0.281 \text{ mol } NO_2$

Check: The units of the answer $(\text{mol } NO_2)$ are correct. The magnitude is reasonable because 10 g is about 0.09 mol N_2O_5 and the answer is greater than mol N_2O_5.

(d)   **Given:** 2.87 kg N_2O_5 **Find:** mol NO_2

Conceptual Plan: kg N_2O_5 → g N_2O_5 → mol N_2O_5 → mol NO_2

$$\frac{1000 \text{ g } N_2O_5}{1 \text{ kg } N_2O_5} \qquad \frac{1 \text{ mol } N_2O_5}{108.02 \text{ g } N_2O_5} \qquad \frac{4 \text{ mol } NO_2}{2 \text{ mol } N_2O_5}$$

Solution:

$$2.87 \text{ kg } \cancel{N_2O_5} \times \frac{1000 \text{ g } \cancel{N_2O_5}}{1 \text{ kg } \cancel{N_2O_5}} \times \frac{1 \text{ mol } \cancel{N_2O_5}}{108.02 \text{ g } \cancel{N_2O_5}} \times \frac{4 \text{ mol } NO_2}{2 \text{ mol } \cancel{N_2O_5}} = 53.\underline{1}4 \text{ mol } NO_2 = 53.1 \text{ mol } NO_2$$

Check: The units of the answer $(\text{mol } NO_2)$ are correct. The magnitude is reasonable because 1.55 kg is about 14 mol N_2O_5 and the answer is greater than mol N_2O_5.

7.39   **Given:** 3 mol SiO_2 **Find:** mol C; mol SiC; mol CO

Conceptual Plan: mol SiO_2 → mol C → mol SiC → mol CO

$$\frac{3 \text{ mol } C}{1 \text{ mol } SiO_2} \qquad \frac{1 \text{ mol } SiC}{1 \text{ mol } SiO_2} \qquad \frac{2 \text{ mol } CO}{1 \text{ mol } SiO_2}$$

Solution: $3 \text{ mol } \cancel{SiO_2} \times \dfrac{3 \text{ mol } C}{1 \text{ mol } \cancel{SiO_2}} = 9 \text{ mol } C \qquad 3 \text{ mol } \cancel{SiO_2} \times \dfrac{1 \text{ mol } SiC}{1 \text{ mol } \cancel{SiO_2}} = 3 \text{ mol } SiC$

$$3 \text{ mol } \cancel{SiO_2} \times \frac{2 \text{ mol } CO}{1 \text{ mol } \cancel{SiO_2}} = 6 \text{ mol } CO$$

Given: 6 mol C   **Find:** mol SiO_2; mol SiC; mol CO

Conceptual Plan: mol C → mol SiO_2 → mol SiC → mol CO

$$\frac{1 \text{ mol } SiO_2}{3 \text{ mol } C} \qquad \frac{1 \text{ mol } SiC}{3 \text{ mol } C} \qquad \frac{2 \text{ mol } CO}{3 \text{ mol } C}$$

Solution: $6 \text{ mol } \cancel{C} \times \dfrac{1 \text{ mol } SiO_2}{3 \text{ mol } \cancel{C}} = 2 \text{ mol } SiO_2 \qquad 6 \text{ mol } \cancel{C} \times \dfrac{1 \text{ mol } SiC}{3 \text{ mol } \cancel{C}} = 2 \text{ mol } SiC$

$$6 \text{ mol } \cancel{C} \times \frac{2 \text{ mol } CO}{3 \text{ mol } \cancel{C}} = 4 \text{ mol } CO$$

Given: 10 mol CO   **Find:** mol SiO_2; mol C; mol SiC

Conceptual Plan: mol CO → mol SiO_2 → mol C → mol SiC

$$\frac{1 \text{ mol } SiO_2}{2 \text{ mol } CO} \qquad \frac{3 \text{ mol } C}{2 \text{ mol } CO} \qquad \frac{1 \text{ mol } SiC}{2 \text{ mol } CO}$$

Solution: $10 \text{ mol } \cancel{CO} \times \dfrac{1 \text{ mol } SiO_2}{2 \text{ mol } \cancel{CO}} = 5 \text{ mol } SiO_2 \qquad 10 \text{ mol } \cancel{CO} \times \dfrac{3 \text{ mol } C}{2 \text{ mol } \cancel{CO}} = \underline{1}5 \text{ mol } C = 20 \text{ mol } C$

$$10 \text{ mol } \cancel{CO} \times \frac{1 \text{ mol } SiC}{2 \text{ mol } \cancel{CO}} = 5 \text{ mol } SiC$$

Given: 2.8 mol SiO_2 **Find:** mol C; mol SiC; mol CO

Conceptual Plan: mol SiO_2 → mol C → mol SiC → mol CO

$$\frac{3 \text{ mol } C}{1 \text{ mol } SiO_2} \qquad \frac{1 \text{ mol } SiC}{1 \text{ mol } SiO_2} \qquad \frac{2 \text{ mol } CO}{1 \text{ mol } SiO_2}$$

Solution: $2.8 \text{ mol } \cancel{SiO_2} \times \dfrac{3 \text{ mol } C}{1 \text{ mol } \cancel{SiO_2}} = 8.4 \text{ mol } C \qquad 2.8 \text{ mol } \cancel{SiO_2} \times \dfrac{1 \text{ mol } SiC}{1 \text{ mol } \cancel{SiO_2}} = 2.8 \text{ mol } SiC$

$$2.8 \text{ mol } \cancel{SiO_2} \times \frac{2 \text{ mol } CO}{1 \text{ mol } \cancel{SiO_2}} = 5.6 \text{ mol } CO$$

Given: 1.55 mol C   **Find:** mol SiO_2; mol SiC; mol CO
Conceptual Plan: mol C → mol SiO_2 → mol SiC → mol CO

$$\frac{1 \text{ mol } SiO_2}{3 \text{ mol C}} \qquad \frac{1 \text{ mol SiC}}{3 \text{ mol C}} \qquad \frac{2 \text{ mol CO}}{3 \text{ mol C}}$$

Solution: $1.55 \text{ mol C} \times \dfrac{1 \text{ mol } SiO_2}{3 \text{ mol C}} = 0.517 \text{ mol } SiO_2 \qquad 1.55 \text{ mol C} \times \dfrac{1 \text{ mol SiC}}{3 \text{ mol C}} = 0.517 \text{ mol SiC}$

$$1.55 \text{ mol C} \times \frac{2 \text{ mol CO}}{3 \text{ mol C}} = 1.03 \text{ mol CO}$$

Mol SiO_2	Mol C	Mol SiC	Mol CO
3	9	3	6
2	**6**	2	4
5	20	5	**10**
2.8	8.4	2.8	5.6
0.517	**1.55**	0.517	1.03

7.41   **Given:** 3.2 g Fe   **Find:** g HBr; gH_2
Conceptual Plan: g Fe → mol Fe → mol HBr → g HBr

$$\frac{1 \text{ mol Fe}}{55.8 \text{ g Fe}} \qquad \frac{2 \text{ mol HBr}}{1 \text{ mol Fe}} \qquad \frac{80.9 \text{ g HBr}}{1 \text{ mol HBr}}$$

g Fe → mol Fe → mol H_2 → g H_2

$$\frac{1 \text{ mol Fe}}{55.8 \text{ g Fe}} \qquad \frac{1 \text{ mol } H_2}{1 \text{ mol Fe}} \qquad \frac{2.02 \text{ g } H_2}{1 \text{ mol } H_2}$$

Solution: $3.2 \text{ g Fe} \times \dfrac{1 \text{ mol Fe}}{55.8 \text{ g Fe}} \times \dfrac{2 \text{ mol HBr}}{1 \text{ mol Fe}} \times \dfrac{80.9 \text{ g HBr}}{1 \text{ mol HBr}} = 9.28 \text{ g HBr} = 9.3 \text{ g HBr}$

$$3.2 \text{ g Fe} \times \frac{1 \text{ mol Fe}}{55.8 \text{ g Fe}} \times \frac{1 \text{ mol } H_2}{1 \text{ mol Fe}} \times \frac{2.02 \text{ g } H_2}{1 \text{ mol } H_2} = 0.116 \text{ g } H_2 = 0.12 \text{ g } H_2$$

Check: The units of the answers $(\text{g HBr, g } H_2)$ are correct. The magnitude of the answers is reasonable because molar mass HBr is greater than Fe and molar mass H_2 is much less than Fe.

7.43   (a)   **Given:** 3.67 g Ba   **Find:** g $BaCl_2$
Conceptual Plan: g Ba → mol Ba → mol $BaCl_2$ → g $BaCl_2$

$$\frac{1 \text{ mol Ba}}{137.33 \text{ g Ba}} \qquad \frac{1 \text{ mol } BaCl_2}{1 \text{ mol Ba}} \qquad \frac{208.23 \text{ g } BaCl_2}{1 \text{ mol } BaCl_2}$$

Solution: $3.67 \text{ g Ba} \times \dfrac{1 \text{ mol Ba}}{137.33 \text{ g Ba}} \times \dfrac{1 \text{ mol } BaCl_2}{1 \text{ mol Ba}} \times \dfrac{208.23 \text{ g } BaCl_2}{1 \text{ mol } BaCl_2} = 5.5647 \text{ g } BaCl_2 = 5.56 \text{ g } BaCl_2$

Check: The units of the answer $(\text{g } BaCl_2)$ are correct. The magnitude of the answer is reasonable because it is larger than grams Ba.

(b)   **Given:** 3.67 g CaO   **Find:** g $CaCO_3$
Conceptual Plan: g CaO → mol CaO → mol $CaCO_3$ → g $CaCO_3$

$$\frac{1 \text{ mol CaO}}{56.08 \text{ g CaO}} \qquad \frac{1 \text{ mol } CaCO_3}{1 \text{ mol CaO}} \qquad \frac{100.09 \text{ g } CaCO_3}{1 \text{ mol } CaCO_3}$$

Solution:

$$3.67 \text{ g CaO} \times \frac{1 \text{ mol CaO}}{56.08 \text{ g CaO}} \times \frac{1 \text{ mol } CaCO_3}{1 \text{ mol CaO}} \times \frac{100.09 \text{ g } CaCO_3}{1 \text{ mol } CaCO_3} = 6.550 \text{ g } CaCO_3 = 6.55 \text{ g } CaCO_3$$

Check: Units of answer $(\text{g } CaCO_3)$ are correct. The magnitude of the answer is reasonable because it is larger than grams CaO.

(c)   **Given:** 3.67 g Mg   **Find:** g MgO
Conceptual Plan: g Mg → mol Mg → mol MgO → g MgO

$$\frac{1 \text{ mol Mg}}{24.30 \text{ g Mg}} \qquad \frac{1 \text{ mol MgO}}{1 \text{ mol Mg}} \qquad \frac{40.30 \text{ g MgO}}{1 \text{ mol MgO}}$$

Solution: $3.67 \text{ g Mg} \times \dfrac{1 \text{ mol Mg}}{24.30 \text{ g Mg}} \times \dfrac{1 \text{ mol MgO}}{1 \text{ mol Mg}} \times \dfrac{40.30 \text{ g MgO}}{1 \text{ mol MgO}} = 6.086 \text{ g MgO} = 6.09 \text{ g MgO}$

Check: The units of the answer (g MgO) are correct. The magnitude of the answer is reasonable because it is larger than grams Mg.

(d)　**Given:** 3.67 g Al　**Find:** g Al_2O_3
Conceptual Plan: g Al → mol Al → mol Al_2O_3 → g Al_2O_3

$$\dfrac{1 \text{ mol Al}}{26.98 \text{ g Al}} \quad \dfrac{2 \text{ mol Al}_2O_3}{4 \text{ mol Al}} \quad \dfrac{101.96 \text{ g Al}_2O_3}{1 \text{ mol Al}_2O_3}$$

Solution: $3.67 \text{ g Al} \times \dfrac{1 \text{ mol Al}}{26.98 \text{ g Al}} \times \dfrac{2 \text{ mol Al}_2O_3}{4 \text{ mol Al}} \times \dfrac{101.96 \text{ g Al}_2O_3}{1 \text{ mol Al}_2O_3} = 6.934 \text{ g Al}_2O_3 = 6.93 \text{ g Al}_2O_3$

Check: The units of the answer (g Al_2O_3) are correct. The magnitude of the answer is reasonable because it is larger than grams Al.

Limiting Reactant, Theoretical Yield, and Percent Yield

7.45　(a)　**Given:** 2 mol Na; 2 mol Br_2　**Find:** limiting reactant
Conceptual Plan: mol Na → mol NaBr

$$\dfrac{2 \text{ mol NaBr}}{2 \text{ mol Na}} \quad \rightarrow \text{ smaller mol amount determines limiting reactant}$$

mol Br_2 → mol NaBr

$$\dfrac{2 \text{ mol NaBr}}{1 \text{ mol Br}_2}$$

Solution: $2 \text{ mol Na} \times \dfrac{2 \text{ mol NaBr}}{2 \text{ mol Na}} = 2 \text{ mol NaBr}$

$2 \text{ mol Br}_2 \times \dfrac{2 \text{ mol NaBr}}{1 \text{ mol Br}_2} = 4 \text{ mol NaBr}$

Na is the limiting reactant.

Check: The answer is reasonable because Na produced the smallest amount of product.

(b)　**Given:** 1.8 mol Na; 1.4 mol Br_2　**Find:** limiting reactant
Conceptual Plan: mol Na → mol NaBr

$$\dfrac{2 \text{ mol NaBr}}{2 \text{ mol Na}} \quad \rightarrow \text{ smaller mol amount determines limiting reactant}$$

mol Br_2 → mol NaBr

$$\dfrac{2 \text{ mol NaBr}}{1 \text{ mol Br}_2}$$

Solution: $1.8 \text{ mol Na} \times \dfrac{2 \text{ mol NaBr}}{2 \text{ mol Na}} = 1.8 \text{ mol NaBr}$

$1.4 \text{ mol Br}_2 \times \dfrac{2 \text{ mol NaBr}}{1 \text{ mol Br}_2} = 2.8 \text{ mol NaBr}$
Na is the limiting reactant.

Check: The answer is reasonable because Na produced the smallest amount of product.

(c)　**Given:** 2.5 mol Na; 1 mol Br_2　**Find:** limiting reactant
Conceptual Plan: mol Na → mol NaBr

$$\dfrac{2 \text{ mol NaBr}}{2 \text{ mol Na}} \quad \rightarrow \text{ smaller mol amount determines limiting reactant}$$

mol Br_2 → mol NaBr

$$\dfrac{2 \text{ mol NaBr}}{1 \text{ mol Br}_2}$$

Solution: $2.5 \; \text{mol Na} \times \dfrac{2 \; \text{mol NaBr}}{2 \; \text{mol Na}} = 2.5 \; \text{mol NaBr}$

$1 \; \text{mol Br}_2 \times \dfrac{2 \; \text{mol NaBr}}{1 \; \text{mol Br}_2} = 2 \; \text{mol NaBr}$

Br_2 is the limiting reactant.

Check: The answer is reasonable because Br_2 produced the smallest amount of product.

(d) **Given:** 12.6 mol Na; 6.9 mol Br_2 **Find:** limiting reactant

Conceptual Plan: mol Na → mol NaBr

$\dfrac{2 \; \text{mol NaBr}}{2 \; \text{mol Na}}$ **→ smaller mol amount determines limiting reactant**

mol Br$_2$ → mol NaBr

$\dfrac{2 \; \text{mol NaBr}}{1 \; \text{mol Br}_2}$

Solution: $12.6 \; \text{mol Na} \times \dfrac{2 \; \text{mol NaBr}}{2 \; \text{mol Na}} = 12.6 \; \text{mol NaBr}$

$6.9 \; \text{mol Br}_2 \times \dfrac{2 \; \text{mol NaBr}}{1 \; \text{mol Br}_2} = 13.8 \; \text{mol NaBr} = 14 \; \text{mol NaBr}$

Na is the limiting reactant.

Check: The answer is reasonable because Na produced the smallest amount of product.

7.47 The greatest number of Cl_2 molecules will be formed from reaction mixture (b) and would be 3 molecules Cl_2.

(a) **Given:** 7 molecules HCl; 1 molecule O_2 **Find:** theoretical yield Cl_2

Conceptual Plan: molecules HCl → molecules Cl$_2$

$\dfrac{2 \; \text{molecules Cl}_2}{4 \; \text{molecules HCl}}$ **→ smaller molecule amount determines limiting reactant**

molecules O$_2$ → molecules Cl$_2$

$\dfrac{2 \; \text{molecules Cl}_2}{1 \; \text{molecule O}_2}$

Solution: $7 \; \text{molecules HCl} \times \dfrac{2 \; \text{molecules Cl}_2}{4 \; \text{molecules HCl}} = 3 \; \text{molecules Cl}_2$

$1 \; \text{molecule O}_2 \times \dfrac{2 \; \text{molecules Cl}_2}{1 \; \text{molecule O}_2} = 2 \; \text{molecules Cl}_2$

theoretical yield $= 2 \; \text{molecules Cl}_2$

(b) **Given:** 6 molecules HCl; 3 molecules O_2 **Find:** theoretical yield Cl_2

Conceptual Plan: molecules HCl → molecules Cl$_2$

$\dfrac{2 \; \text{molecules Cl}_2}{4 \; \text{molecules HCl}}$ **→ smaller molecule amount determines limiting reactant**

molecules O$_2$ → molecules Cl$_2$

$\dfrac{2 \; \text{molecules Cl}_2}{1 \; \text{molecule O}_2}$

Solution: $6 \; \text{molecules HCl} \times \dfrac{2 \; \text{molecules Cl}_2}{4 \; \text{molecules HCl}} = 3 \; \text{molecules Cl}_2$

$3 \; \text{molecules O}_2 \times \dfrac{2 \; \text{molecules Cl}_2}{1 \; \text{molecule O}_2} = 6 \; \text{molecules Cl}_2$

theoretical yield $= 3 \; \text{molecules}$

(c) **Given:** 4 molecules HCl; 5 molecules O_2 **Find:** theoretical yield Cl_2

Conceptual Plan: molecules HCl → molecules Cl$_2$

$\dfrac{2 \; \text{molecules Cl}_2}{4 \; \text{molecules HCl}}$ **→ smaller molecule amount determines limiting reactant**

molecules O$_2$ → molecules Cl$_2$

$\dfrac{2 \; \text{molecules Cl}_2}{1 \; \text{molecule O}_2}$

Solution: $4 \text{ molecules HCl} \times \dfrac{2 \text{ molecules Cl}_2}{4 \text{ molecules HCl}} = 2 \text{ molecules Cl}_2$

$5 \text{ molecules O}_2 \times \dfrac{2 \text{ molecules Cl}_2}{1 \text{ molecule O}_2} = 10 \text{ molecules Cl}_2$

theoretical yield $= 2$ molecules Cl_2

Check: The units of the answer (molecules Cl_2) are correct. The answer is reasonable based on the limiting reactant in each mixture.

7.49   (a)   **Given:** 4 mol Ti; 4 mol Cl_2 **Find:** theoretical yield $TiCl_4$
Conceptual Plan: mol Ti → mol TiCl$_4$

$\dfrac{1 \text{ mol TiCl}_4}{1 \text{ mol Ti}}$ → **smaller mol amount determines limiting reactant**

mol Cl$_2$ → mol TiCl$_4$

$\dfrac{1 \text{ mol TiCl}_4}{2 \text{ mol Cl}_2}$

Solution: $4 \text{ mol Ti} \times \dfrac{1 \text{ mol TiCl}_4}{1 \text{ mol Ti}} = 4 \text{ mol TiCl}_4$

$4 \text{ mol Cl}_2 \times \dfrac{1 \text{ mol TiCl}_4}{2 \text{ mol Cl}_2} = 2 \text{ mol TiCl}_4$

theoretical yield $= 2$ mol $TiCl_4$

Check: The units of the answer (mol $TiCl_4$) are correct. The answer is reasonable because Cl_2 produced the smallest amount of product and is the limiting reactant.

(b)   **Given:** 7 mol Ti; 17 mol Cl_2 **Find:** theoretical yield $TiCl_4$
Conceptual Plan: mol Ti → mol TiCl$_4$

$\dfrac{1 \text{ mol TiCl}_4}{1 \text{ mol Ti}}$ → **smaller mol amount determines limiting reactant**

mol Cl$_2$ → mol TiCl$_4$

$\dfrac{1 \text{ mol TiCl}_4}{2 \text{ mol Cl}_2}$

Solution: $7 \text{ mol Ti} \times \dfrac{1 \text{ mol TiCl}_4}{1 \text{ mol Ti}} = 7 \text{ mol TiCl}_4$

$17 \text{ mol Cl}_2 \times \dfrac{1 \text{ mol TiCl}_4}{2 \text{ mol Cl}_2} = 8.5 \text{ mol TiCl}_4$

theoretical yield $= 7$ mol $TiCl_4$

Check: The units of the answer (mol $TiCl_4$) are correct. The answer is reasonable because Ti produced the smallest amount of product and is the limiting reactant.

(c)   **Given:** 12.4 mol Ti; 18.8 mol Cl_2 **Find:** theoretical yield $TiCl_4$
Conceptual Plan: mol Ti → mol TiCl$_4$

$\dfrac{1 \text{ mol TiCl}_4}{1 \text{ mol Ti}}$ → **smaller mol amount determines limiting reactant**

mol Cl$_2$ → mol TiCl$_4$

$\dfrac{1 \text{ mol TiCl}_4}{2 \text{ mol Cl}_2}$

Solution: $12.4 \text{ mol Ti} \times \dfrac{1 \text{ mol TiCl}_4}{1 \text{ mol Ti}} = 12.4 \text{ mol TiCl}_4$

$18.8 \text{ mol Cl}_2 \times \dfrac{1 \text{ mol TiCl}_4}{2 \text{ mol Cl}_2} = 9.40 \text{ mol TiCl}_4$

theoretical yield $= 9.40$ mol $TiCl_4$

Check: The units of the answer (mol $TiCl_4$) are correct. The answer is reasonable because Cl_2 produced the smallest amount of product and is the limiting reactant.

7.51   **Given:** 4.2 mol ZnS; 6.8 mol O_2 **Find:** mole amount of excess reactant left
         Conceptual Plan: mol ZnS → mol ZnO

$$\frac{2 \text{ mol ZnO}}{2 \text{ mol ZnS}} \quad \rightarrow \text{ smaller mol amount determines limiting reactant}$$

mol O_2 → mol ZnO

$$\frac{2 \text{ mol ZnO}}{3 \text{ mol } O_2}$$

mol limiting reactant → mol excess reactant required → mol excess reactant left

$$\frac{3 \text{ mol } O_2}{2 \text{ mol ZnS}}$$

Solution: $4.2 \text{ mol ZnS} \times \dfrac{2 \text{ mol ZnO}}{2 \text{ mol ZnS}} = 4.2 \text{ mol ZnO}$

$6.8 \text{ mol } O_2 \times \dfrac{2 \text{ mol ZnO}}{3 \text{ mol } O_2} = 4.5 \text{ mol ZnO}$

ZnS is the limiting reactant; therefore, O_2 is the excess reactant.

$$4.2 \text{ mol ZnS} \times \dfrac{3 \text{ mol } O_2}{2 \text{ mol ZnS}} = 6.3 \text{ mol } O_2 \text{ required}$$

$$6.8 \text{ mol } O_2 - 6.3 \text{ mol } O_2 = 0.5 \text{ mol } O_2 \text{ left}$$

Check: The units of the answer (mol O_2) are correct. The magnitude is reasonable because it is less than the original amount of O_2.

7.53   (a)   **Given:** 2.0 g Al; 2.0 g Cl_2 **Find:** theoretical yield in g $AlCl_3$
                 Conceptual Plan: g Al → mol Al → mol $AlCl_3$

$$\frac{1 \text{ mol Al}}{26.98 \text{ g Al}} \quad \frac{2 \text{ mol AlCl}_3}{2 \text{ mol Al}} \quad \rightarrow \text{ smaller mol amount determines limiting reactant}$$

g Cl_2 → mol Cl_2 → mol $AlCl_3$

$$\frac{1 \text{ mol Cl}_2}{70.90 \text{ g Cl}_2} \quad \frac{2 \text{ mol AlCl}_3}{3 \text{ mol Cl}_2}$$

then mol $AlCl_3$ → g $AlCl_3$

$$\frac{133.3 \text{ g AlCl}_3}{1 \text{ mol AlCl}_3}$$

Solution: $2.0 \text{ g Al} \times \dfrac{1 \text{ mol Al}}{26.98 \text{ g Al}} \times \dfrac{2 \text{ mol AlCl}_3}{2 \text{ mol Al}} = 0.074 \text{ mol AlCl}_3$

$2.0 \text{ g Cl}_2 \times \dfrac{1 \text{ mol Cl}_2}{70.90 \text{ g Cl}_2} \times \dfrac{2 \text{ mol AlCl}_3}{3 \text{ mol Cl}_2} = 0.0188 \text{ mol AlCl}_3$

$0.0188 \text{ mol AlCl}_3 \times \dfrac{133.3 \text{ g AlCl}_3}{1 \text{ mol AlCl}_3} = 2.5 \text{ g AlCl}_3$

Check: The units of the answer (g $AlCl_3$) are correct. The answer is reasonable because Cl_2 produced the smallest amount of product and is the limiting reactant.

         (b)   **Given:** 7.5 g Al; 24.8 g Cl_2 **Find:** theoretical yield in g $AlCl_3$
                 Conceptual Plan: g Al → mol Al → mol $AlCl_3$

$$\frac{1 \text{ mol Al}}{26.98 \text{ g Al}} \quad \frac{2 \text{ mol AlCl}_3}{2 \text{ mol Al}} \quad \rightarrow \text{ smaller mol amount determines limiting reactant}$$

g Cl_2 → mol Cl_2 → mol $AlCl_3$

$$\frac{1 \text{ mol Cl}_2}{70.90 \text{ g Cl}_2} \quad \frac{2 \text{ mol AlCl}_3}{3 \text{ mol Cl}_2}$$

then mol $AlCl_3$ → g $AlCl_3$

$$\frac{133.3 \text{ g AlCl}_3}{1 \text{ mol AlCl}_3}$$

Solution: $7.5 \text{ g Al} \times \dfrac{1 \text{ mol Al}}{26.98 \text{ g Al}} \times \dfrac{2 \text{ mol AlCl}_3}{2 \text{ mol Al}} = 0.2780 \text{ mol AlCl}_3$

$24.8 \text{ g Cl}_2 \times \dfrac{1 \text{ mol Cl}_2}{70.90 \text{ g Cl}_2} \times \dfrac{2 \text{ mol AlCl}_3}{3 \text{ mol Cl}_2} = 0.2332 \text{ mol AlCl}_3$

$$0.23\underline{3}2 \text{ mol AlCl}_3 \times \frac{133.3 \text{ g AlCl}_3}{1 \text{ mol AlCl}_3} = 31.1 \text{ g AlCl}_3$$

Check: The units of the answer (g AlCl$_3$) are correct. The answer is reasonable because Cl$_2$ produced the smallest amount of product and is the limiting reactant.

(c)   **Given:** 0.235 g Al; 1.15 g Cl$_2$ **Find:** theoretical yield in g AlCl$_3$

Conceptual Plan: g A → mol Al → mol AlCl$_3$

$$\frac{1 \text{ mol Al}}{26.98 \text{ g Al}} \quad \frac{2 \text{ mol AlCl}_3}{2 \text{ mol Al}} \qquad \rightarrow \textbf{ smaller mol amount determines limiting reactant}$$

g Cl$_2$ → mol Cl$_2$ → mol AlCl$_3$

$$\frac{1 \text{ mol Cl}_2}{70.90 \text{ g Cl}_2} \quad \frac{2 \text{ mol AlCl}_3}{3 \text{ mol Cl}_2}$$

then mol AlCl$_3$ → g AlCl$_3$

$$\frac{133.3 \text{ g AlCl}_3}{1 \text{ mol AlCl}_3}$$

Solution: $0.235 \text{ g Al} \times \dfrac{1 \text{ mol Al}}{26.98 \text{ g Al}} \times \dfrac{2 \text{ mol AlCl}_3}{2 \text{ mol Al}} = 0.0087\underline{1}0 \text{ mol AlCl}_3$

$$1.15 \text{ g Cl}_2 \times \frac{1 \text{ mol Cl}_2}{70.90 \text{ g Cl}_2} \times \frac{2 \text{ mol AlCl}_3}{3 \text{ mol Cl}_2} = 0.01081 \text{ mol AlCl}_3$$

$$0.0087\underline{1}0 \text{ mol AlCl}_3 \times \frac{133.3 \text{ g AlCl}_3}{1 \text{ mol AlCl}_3} = 1.16 \text{ g AlCl}_3$$

Check: The units of the answer (g AlCl$_3$) are correct. The answer is reasonable because Al produced the smallest amount of product and is the limiting reactant.

7.55   **Given:** 22.55 g Fe$_2$O$_3$; 14.78 g CO   **Find:** mole amount of excess reactant left

Conceptual Plan: g Fe$_2$O$_3$ → mol Fe$_2$O$_3$ → mol Fe

$$\frac{1 \text{ mol Fe}_2\text{O}_3}{159.7 \text{ g Fe}_2\text{O}_3} \quad \frac{2 \text{ mol Fe}}{1 \text{ mol Fe}_2\text{O}_3} \qquad \rightarrow \textbf{ smaller mol amount determines limiting reactant}$$

g CO → mol CO → mol Fe

$$\frac{1 \text{ mol CO}}{28.01 \text{ g CO}} \quad \frac{2 \text{ mol Fe}}{3 \text{ mol CO}}$$

then mol limiting reactant → mol excess reactant required → mol excess reactant left → g excess reactant left

$$\frac{3 \text{ mol CO}}{1 \text{ mol Fe}_2\text{O}_3} \qquad\qquad \frac{1 \text{ mol Fe}_2\text{O}_3}{159.7 \text{ g Fe}_2\text{O}_3} \qquad\qquad \text{or } \frac{28.01 \text{ g CO}}{1 \text{ mol CO}}$$

Solution: $22.55 \text{ g Fe}_2\text{O}_3 \times \dfrac{1 \text{ mol Fe}_2\text{O}_3}{159.7 \text{ g Fe}_2\text{O}_3} \times \dfrac{2 \text{ mol Fe}}{1 \text{ mol Fe}_2\text{O}_3} = 0.2824 \text{ mol Fe}$

$$14.78 \text{ g CO} \times \frac{1 \text{ mol CO}}{28.01 \text{ g CO}} \times \frac{2 \text{ mol Fe}}{3 \text{ mol CO}} = 0.3518 \text{ mol Fe}$$

Fe$_2$O$_3$ is the limiting reactant; therefore, CO is the excess reactant.

$$22.55 \text{ g Fe}_2\text{O}_3 \times \frac{1 \text{ mol Fe}_2\text{O}_3}{159.7 \text{ g Fe}_2\text{O}_3} \times \frac{3 \text{ mol CO}}{1 \text{ mol Fe}_2\text{O}_3} \times \frac{28.01 \text{ g CO}}{1 \text{ mol CO}} = 11.8\underline{6}5 \text{ g CO required}$$

$$14.78 \text{ g CO} - 11.87 \text{ g CO} = 2.91 \text{ g CO left}$$

Check: The units of the answer (g CO) are correct. The magnitude is reasonable because it is less than the original amount of CO.

7.57   **Given:** 28.5 g KCl; 25.7 g Pb^{2+}; 29.4 g PbCl$_2$ **Find:** limiting reactant; theoretical yield PbCl$_2$; % yield

Conceptual Plan: g KCl → mol KCl → mol PbCl$_2$

$$\frac{1 \text{ mol KCl}}{74.55 \text{ g KCl}} \quad \frac{1 \text{ mol PbCl}_2}{2 \text{ mol KCl}} \qquad \rightarrow \textbf{ smaller mol amount determines limiting reactant}$$

g Pb^{2+} → mol Pb^{2+} → mol PbCl$_2$

$$\frac{1 \text{ mol Pb}^{2+}}{207.2 \text{ g Pb}^{2+}} \quad \frac{1 \text{ mol PbCl}_2}{1 \text{ mol Pb}^{2+}}$$

then mol PbCl$_2$ → g PbCl$_2$ then determine % yield

$$\frac{278.1 \text{ g PbCl}_2}{1 \text{ mol PbCl}_2} \qquad \frac{\text{actual yield g PbCl}_2}{\text{theoretical yield g PbCl}_2} \times 100\%$$

Solution: $28.5 \text{ g } \cancel{\text{KCl}} \times \dfrac{1 \text{ mol } \cancel{\text{KCl}}}{74.55 \text{ g } \cancel{\text{KCl}}} \times \dfrac{1 \text{ mol PbCl}_2}{2 \text{ mol } \cancel{\text{KCl}}} = 0.191\underline{1} \text{ mol PbCl}_2$

$25.7 \text{ g } \cancel{\text{Pb}^{2+}} \times \dfrac{1 \text{ mol } \cancel{\text{Pb}^{2+}}}{207.2 \text{ g } \cancel{\text{Pb}^{2+}}} \times \dfrac{1 \text{ mol PbCl}_2}{1 \text{ mol } \cancel{\text{Pb}^{2+}}} = 0.124\underline{0} \text{ mol PbCl}_2 \quad \text{Pb}^{2+}$ is the limiting reactant.

$0.124\underline{0} \text{ mol } \cancel{\text{PbCl}_2} \times \dfrac{278.1 \text{ g PbCl}_2}{1 \text{ mol } \cancel{\text{PbCl}_2}} = 34.\underline{5} \text{ g PbCl}_2$

$\dfrac{29.4 \text{ g } \cancel{\text{PbCl}_2}}{34.\underline{5} \text{ g } \cancel{\text{PbCl}_2}} \times 100\% = 85.2\%$

Check: The theoretical yield has the correct units (g PbCl_2) and has a reasonable magnitude compared to the mass of Pb^{2+}, the limiting reactant. The % yield is reasonable, under 100%.

7.59 **Given:** 136.4 kg NH_3; 211.4 kg CO_2; 168.4 kg $\text{CH}_4\text{N}_2\text{O}$ **Find:** limiting reactant; theoretical yield $\text{CH}_4\text{N}_2\text{O}$; % yield

Conceptual Plan: kg $\text{NH}_3 \rightarrow$ g $\text{NH}_3 \rightarrow$ mol $\text{NH}_3 \rightarrow$ mol $\text{CH}_4\text{N}_2\text{O}$

$$\frac{1000 \text{ g}}{1 \text{ kg}} \qquad \frac{1 \text{ mol NH}_3}{17.03 \text{ g NH}_3} \qquad \frac{1 \text{ mol CH}_4\text{N}_2\text{O}}{2 \text{ mol NH}_3} \qquad \rightarrow \textbf{ smaller amount determines limiting reactant}$$

kg $\text{CO}_2 \rightarrow$ g $\text{CO}_2 \rightarrow$ mol $\text{CO}_2 \rightarrow$ mol $\text{CH}_4\text{N}_2\text{O}$

$$\frac{1000 \text{ g}}{1 \text{ kg}} \qquad \frac{1 \text{ mol CO}_2}{44.01 \text{ g CO}_2} \qquad \frac{1 \text{ mol CH}_4\text{N}_2\text{O}}{1 \text{ mol CO}_2}$$

then mol $\text{CH}_4\text{N}_2\text{O} \rightarrow$ g $\text{CH}_4\text{N}_2\text{O} \rightarrow$ kg $\text{CH}_4\text{N}_2\text{O}$ then determine % yield

$$\frac{60.06 \text{ g CH}_4\text{N}_2\text{O}}{1 \text{ mol CH}_4\text{N}_2\text{O}} \qquad \frac{1 \text{ kg}}{1000 \text{ g}} \qquad \frac{\text{actual yield kg CH}_4\text{N}_2\text{O}}{\text{theoretical yield kg CH}_4\text{N}_2\text{O}} \times 100\%$$

Solution: $136.4 \text{ kg } \cancel{\text{NH}_3} \times \dfrac{1000 \text{ g}}{1 \text{ kg}} \times \dfrac{1 \text{ mol } \cancel{\text{NH}_3}}{17.03 \text{ g } \cancel{\text{NH}_3}} \times \dfrac{1 \text{ mol CH}_4\text{N}_2\text{O}}{2 \text{ mol } \cancel{\text{NH}_3}} = 4004.\underline{7} \text{ mol CH}_4\text{N}_2\text{O}$

$211.4 \text{ kg } \cancel{\text{CO}_2} \times \dfrac{1000 \text{ g}}{1 \text{ kg}} \times \dfrac{1 \text{ mol } \cancel{\text{CO}_2}}{44.01 \text{ g } \cancel{\text{CO}_2}} \times \dfrac{1 \text{ mol CH}_4\text{N}_2\text{O}}{1 \text{ mol } \cancel{\text{CO}_2}} = 4803.\underline{5} \text{ mol CH}_4\text{N}_2\text{O}$

NH_3 is the limiting reactant.

$4004.\underline{7} \text{ mol } \cancel{\text{CH}_4\text{N}_2\text{O}} \times \dfrac{60.06 \text{ g } \cancel{\text{CH}_4\text{N}_2\text{O}}}{1 \text{ mol } \cancel{\text{CH}_4\text{N}_2\text{O}}} \times \dfrac{1 \text{ kg}}{1000 \text{ g}} = 240.5\underline{2} \text{ kg CH}_4\text{N}_2\text{O}$

$\dfrac{168.4 \text{ kg } \cancel{\text{CH}_4\text{N}_2\text{O}}}{240.5\underline{2} \text{ kg } \cancel{\text{CH}_4\text{N}_2\text{O}}} \times 100\% = 70.01\%$

Check: The theoretical yield has the correct units $(\text{kg CH}_4\text{N}_2\text{O})$ and has a reasonable magnitude compared to the mass of NH_3, the limiting reactant. The % yield is reasonable, under 100%.

Combustion, Alkali Metal, and Halogen Reactions

7.61 (a) Skeletal reaction: $\text{S}(s) + \text{O}_2(g) \rightarrow \text{SO}_2(g)$

Balanced reaction: $\text{S}(s) + \text{O}_2(g) \rightarrow \text{SO}_2(g)$

Check:

left side	right side
1 S atom	1 S atom
2 O atoms	2 O atoms

(b) Skeletal reaction: $\text{C}_3\text{H}_6(g) + \text{O}_2(g) \rightarrow \text{CO}_2(g) + \text{H}_2\text{O}(g)$

Balance C: $\text{C}_3\text{H}_6(g) + \text{O}_2(g) \rightarrow 3 \text{ CO}_2(g) + \text{H}_2\text{O}(g)$

Balance H: $\text{C}_3\text{H}_6(g) + \text{O}_2(g) \rightarrow 3 \text{ CO}_2(g) + 3 \text{ H}_2\text{O}(g)$

Balance O: $\text{C}_3\text{H}_6(g) + 9/2 \text{ O}_2(g) \rightarrow 3 \text{ CO}_2(g) + 3 \text{ H}_2\text{O}(g)$

Clear fraction: $2 \text{ C}_3\text{H}_6(g) + 9 \text{ O}_2(g) \rightarrow 6 \text{ CO}_2(g) + 6 \text{ H}_2\text{O}(g)$

Check:

left side	right side
6 C atoms	6 C atoms
12 H atoms	12 H atoms
18 O atoms	18 O atoms

(c)   Skeletal reaction: $Ca(s) + O_2(g) \rightarrow CaO(s)$

               Balance O: $Ca(s) + O_2(g) \rightarrow 2\,CaO(s)$

               Balance Ca: $2\,Ca(s) + O_2(g) \rightarrow 2\,CaO(s)$

     Check:

	left side	right side
	2 Ca atoms	2 Ca atoms
	2 O atoms	2 O atoms

(d)   Skeletal reaction: $C_5H_{12}S(l) + O_2(g) \rightarrow CO_2(g) + H_2O(g) + SO_2(g)$

               Balance C: $C_5H_{12}S(l) + O_2(g) \rightarrow 5\,CO_2(g) + H_2O(g) + SO_2(g)$

               Balance H: $C_5H_{12}S(l) + O_2(g) \rightarrow 5\,CO_2(g) + 6\,H_2O(g) + SO_2(g)$

               Balance S: $C_5H_{12}S(l) + O_2(g) \rightarrow 5\,CO_2(g) + 6\,H_2O(g) + SO_2(g)$

               Balance O: $C_5H_{12}S(l) + 9\,O_2(g) \rightarrow 5\,CO_2(g) + 6\,H_2O(g) + SO_2(g)$

     Check:

	left side	right side
	5 C atoms	5 C atoms
	12 H atoms	12 H atoms
	18 O atoms	18 O atoms
	1 S atom	1 S atom

7.63   Alkaline earth metals react with halogens to form metal halides. Write the formulas for the reactants and the metal halide product.

$$Sr(s) + I_2(g) \rightarrow SrI_2(s)$$

7.65   Alkali metals react with water to form the dissolved metal ion, the hydroxide ion, and hydrogen gas. Write the skeletal equation including each of these and then balance it.

$$Li(s) + H_2O(l) \rightarrow Li^+(aq) + OH^-(aq) + H_2(g)$$
$$2\,Li(s) + 2\,H_2O(l) \rightarrow 2\,Li^+(aq) + 2\,OH^-(aq) + H_2(g)$$

7.67   The halogens react with hydrogen to form hydrogen halides. Write the skeletal reaction with each of the halogens and hydrogens as the reactants and the hydrogen halide compound as the product and balance the equation.

$$H_2(g) + Br_2(g) \rightarrow HBr(g)$$
$$H_2(g) + Br_2(g) \rightarrow 2\,HBr(g)$$

Cumulative Problems

7.69   **Given:** $1.0\text{ kg }C_8H_{18}$ **Find:** $kg\ CO_2$

     Conceptual Plan: $kg\ C_8H_{18} \rightarrow g\ C_8H_{18} \rightarrow mol\ C_8H_{18} \rightarrow mol\ CO_2 \rightarrow g\ CO_2 \rightarrow kg\ CO_2$

$$\frac{1000\text{ g}}{1\text{ kg}} \quad \frac{1\text{ mol }C_8H_{18}}{114.22\text{ g }C_8H_{18}} \quad \frac{16\text{ mol }CO_2}{2\text{ mol }C_8H_{18}} \quad \frac{44.01\text{ g }CO_2}{1\text{ mol }CO_2} \quad \frac{1\text{ kg}}{1000\text{ g}}$$

     Solution: $2\,C_8H_{18}(g) + 25\,O_2(g) \rightarrow 16\,CO_2(g) + 18\,H_2O(g)$

$$1.0\text{ kg }C_8H_{18} \times \frac{1000\text{ g}}{1\text{ kg}} \times \frac{1\text{ mol }C_8H_{18}}{114.22\text{ g }C_8H_{18}} \times \frac{16\text{ mol }CO_2}{2\text{ mol }C_8H_{18}} \times \frac{44.01\text{ g }CO_2}{1\text{ mol }CO_2}\ \frac{1\text{ kg}}{1000\text{ g}} = 3.1\text{ kg }CO_2$$

     Check: The units of the answer $(kg\ CO_2)$ are correct. The magnitude of the answer is reasonable because the mole ratio of CO_2 to C_8H_{18} is 8:1.

7.71   **Given:** $3.00\text{ mL }C_4H_6O_3$, $d = 1.08\text{ g/mL}$; $1.25\text{ g }C_7H_6O_3$; $1.22\text{ g }C_9H_8O_4$ **Find:** limiting reactant; theoretical yield $C_9H_8O_4$; % yield $C_9H_8O_4$

     Conceptual Plan: $mL\ C_4H_6O_3 \rightarrow g\ C_4H_6O_3 \rightarrow mol\ C_4H_6O_3 \rightarrow mol\ C_9H_8O_4$

$$\frac{1.08\text{ g }C_4H_6O_3}{1.00\text{ mL }C_4H_6O_3} \quad \frac{1\text{ mol }C_4H_6O_3}{102.09\text{ g }C_4H_6O_3} \quad \frac{1\text{ mol }C_9H_8O_4}{1\text{ mol }C_4H_6O_3}$$

         $g\ C_7H_6O_3 \rightarrow mol\ C_7H_6O_3 \rightarrow mol\ C_9H_8O_4$ → **smaller amount determines limiting reactant**

$$\frac{1\text{ mol }C_7H_6O_3}{138.12\text{ g }C_7H_6O_3} \quad \frac{1\text{ mol }C_9H_8O_4}{1\text{ mol }C_7H_6O_3}$$

     then $mol\ C_9H_8O_4 \rightarrow g\ C_9H_8O_4$ **then determine % yield**

$$\frac{180.1 \text{ g C}_9\text{H}_8\text{O}_4}{1 \text{ mol C}_9\text{H}_8\text{O}_4} \qquad \frac{\text{actual yield g C}_9\text{H}_8\text{O}_4}{\text{theoretical yield g C}_9\text{H}_8\text{O}_4} \times 100\%$$

Solution:

$$3.00 \text{ mL } C_4H_6O_3 \times \frac{1.08 \text{ g } C_4H_6O_3}{1 \text{ mL } C_4H_6O_3} \times \frac{1 \text{ mol } C_4H_6O_3}{102.09 \text{ g } C_4H_6O_3} \times \frac{1 \text{ mol C}_9\text{H}_8\text{O}_4}{1 \text{ mol } C_4H_6O_3} = 0.031\underline{7}4 \text{ mol C}_9\text{H}_8\text{O}_4$$

$$1.25 \text{ g } C_7H_6O_3 \times \frac{1 \text{ mol } C_7H_6O_3}{138.12 \text{ g } C_7H_6O_3} \times \frac{1 \text{ mol C}_9\text{H}_8\text{O}_4}{1 \text{ mol } C_7H_6O_3} = 0.009050 \text{ mol C}_9\text{H}_8\text{O}_4$$

Salicylic acid is the limiting reactant.

$$0.009050 \text{ mol } C_9H_8O_4 \times \frac{180.1 \text{ g C}_9\text{H}_8\text{O}_4}{1 \text{ mol } C_9H_8O_4} = 1.6\underline{3}0 \text{ g C}_9\text{H}_8\text{O}_4$$

$$\frac{1.22 \text{ g } C_9H_8O_4}{1.6\underline{3}0 \text{ g } C_9H_8O_4} \times 100\% = 74.8\%$$

Check: The theoretical yield has the correct units $(\text{g C}_9\text{H}_8\text{O}_4)$ and has a reasonable magnitude compared to the mass of $C_7H_6O_3$, the limiting reactant. The % yield is reasonable, under 100%.

7.73   **Given:** (a) 11 molecules H_2, 2 molecules O_2; (b) 8 molecules H_2, 4 molecules O_2; (c) 4 molecules H_2, 5 molecules O_2; (d) 3 molecules H_2, 6 molecules O_2 **Find:** loudest explosion based on equation
Conceptual Plan: Loudest explosion will occur in the balloon with the mol ratio closest to the balanced equation that contains the most H_2.
Solution: $2 H_2(g) + O_2(g) \rightarrow H_2O(l)$
Balloon (a) has enough O_2 to react with 4 molecules H_2; balloon (b) has enough O_2 to react with 8 molecules H_2; balloon (c) has enough O_2 to react with 10 molecules H_2; balloon (d) has enough O_2 for 3 molecules of H_2 to react. Balloon (b) also has the proper stoichiometric ratio of 2 H_2:1 O_2 unlike the other three. Therefore, balloon (b) will have the loudest explosion because it has the most H_2 that will react.

Check: The answer seems correct because it has the most H_2 with enough O_2 in the balloon to react completely.

7.75   **Given:** 1.00 g NH_3 **Find:** g PH_3
Conceptual Plan: Determine reaction sequence, then g $NH_3 \rightarrow$ mol $NH_3 \rightarrow$ mol $PH_3 \rightarrow$ g PH_3

$$\frac{1 \text{ mol NH}_3}{17.04 \text{ g NH}_3} \qquad \frac{34.00 \text{ g PH}_3}{1 \text{ mol PH}_3}$$

Solution: Balance the reaction sequence:

$$6 NH_3 + 7 \tfrac{1}{2} O_2 \rightarrow 6 NO + 9 H_2O$$
$$6 NO + P_4 \rightarrow P_4O_6 + 3 N_2$$
$$P_4O_6 + 6 H_2O \rightarrow 4 H_3PO_3$$
$$4 H_3PO_3 \rightarrow PH_3 + 3 H_3PO_4$$
$$6 NH_3 + 7 \tfrac{1}{2} O_2 + P_4 \rightarrow 3 H_2O + 3 N_2 + PH_3 + 3 H_3PO_4$$

Therefore, 6 mol NH_3 produces 1 mol PH_3.

$$1.00 \text{ g } NH_3 \times \frac{1 \text{ mol } NH_3}{17.04 \text{ g } NH_3} \times \frac{1 \text{ mol } PH_3}{6 \text{ mol } NH_3} \times \frac{34.00 \text{ g PH}_3}{1 \text{ mol } PH_3} = 0.333 \text{ g PH}_3$$

Check: The answer is g PH_3, which is correct. The magnitude is reasonable. Even though the molar mass of PH_3 is greater than NH_3, 6 mol of NH_3 are required to produce 1 mol PH_3.

7.77   **Given:** 10.0 kg mixture; 30.35% hexane; 15.85% heptane; 53.80% octane   **Find:** total mass CO_2
Conceptual Plan: kg hexane $\rightarrow$ kmol hexane $\rightarrow$ kmol $CO_2 \rightarrow$ kg CO_2

$$\frac{1 \text{ kmol } C_6H_{14}}{86.20 \text{ kg } C_6H_{14}} \qquad \frac{12 \text{ kmol CO}_2}{2 \text{ kmol } C_6H_{14}} \qquad \frac{44.01 \text{ kg CO}_2}{1 \text{ kmol CO}_2}$$

kg heptane $\rightarrow$ kmol heptane $\rightarrow$ kmol $CO_2 \rightarrow$ kg CO_2

$$\frac{1 \text{ kmol } C_7H_{16}}{100.23 \text{ kg } C_7H_{16}} \qquad \frac{7 \text{ kmol CO}_2}{1 \text{ kmol } C_7H_{16}} \qquad \frac{44.01 \text{ kg CO}_2}{1 \text{ kmol CO}_2}$$

kg octane $\rightarrow$ kmol octane $\rightarrow$ kmol $CO_2 \rightarrow$ kg CO_2

$$\frac{1 \text{ kmol } C_8H_{18}}{114.26 \text{ kg } C_8H_{18}} \qquad \frac{16 \text{ kmol CO}_2}{2 \text{ kmol } C_8H_{18}} \qquad \frac{44.01 \text{ kg CO}_2}{1 \text{ kmol CO}_2}$$

Solution: Balanced Reactions:

$$2\,C_6H_{14}(l) + 19\,O_2(g) \rightarrow 12\,CO_2(g) + 14\,H_2O(l)$$
$$C_7H_{16}(l) + 11\,O_2(g) \rightarrow 7\,CO_2(g) + 8\,H_2O(l)$$
$$2\,C_8H_{18}(l) + 25\,O_2(g) \rightarrow 16\,CO_2(g) + 18\,H_2O(l)$$

$$10.0\ \text{kg mix} \times \frac{30.35\ \text{kg C}_6\text{H}_{14}}{100.0\ \text{kg mix}} \times \frac{1\ \text{kmol C}_6\text{H}_{14}}{86.20\ \text{kg C}_6\text{H}_{14}} \times \frac{12\ \text{kmol CO}_2}{2\ \text{kmol C}_6\text{H}_{14}} \times \frac{44.01\ \text{kg CO}_2}{1\ \text{kmol CO}_2} = 9.2\underline{9}7\ \text{kg CO}_2$$

$$10.0\ \text{kg mix} \times \frac{15.85\ \text{kg C}_7\text{H}_{16}}{100.0\ \text{kg mix}} \times \frac{1\ \text{kmol C}_7\text{H}_{16}}{100.23\ \text{kg C}_7\text{H}_{16}} \times \frac{7\ \text{kmol CO}_2}{1\ \text{kmol C}_7\text{H}_{16}} \times \frac{44.01\ \text{kg CO}_2}{1\ \text{kmol CO}_2} = 4.872\ \text{kg CO}_2$$

$$10.0\ \text{kg mix} \times \frac{53.80\ \text{kg C}_8\text{H}_{18}}{100.0\ \text{kg mix}} \times \frac{1\ \text{kmol C}_8\text{H}_{18}}{114.26\ \text{kg C}_8\text{H}_{18}} \times \frac{16\ \text{kmol CO}_2}{2\ \text{kmol C}_8\text{H}_{18}} \times \frac{44.01\ \text{kg CO}_2}{1\ \text{kmol CO}_2} = 16.578\ \text{kg CO}_2$$

Total $CO_2 = 9.30\ \text{kg} + 4.87\ \text{kg} + 16.6\ \text{kg} = 30.8\ \text{kg CO}_2$

Check: The units of the answer (kg CO_2) are correct. The magnitude of the answer is reasonable because a large amount of CO_2 is produced per mole of hydrocarbon.

Challenge Problems

7.79   **Given:** g $C_3H_8 + C_2H_2 = 2.0$ g; mol $CO_2 = 1.5$ mol H_2O **Find:** original g C_2H_2

Conceptual Plan: mol $C_3H_8 \rightarrow$ mol CO_2 and mol H_2O and mol $C_2H_2 \rightarrow$ mol CO_2 and mol H_2O

Solution: Let $a = $ mol C_3H_8 and $b = $ mol C_2H_2

$$C_3H_8 + 5\,O_2 \rightarrow 3\,CO_2 + 4\,H_2O \qquad\qquad C_2H_2 + 3/2\,O_2 \rightarrow 2\,CO_2 + H_2O$$
$$\ a\qquad\ 3a\qquad 4a \qquad\qquad\qquad b\qquad\qquad\quad 2b\quad\ b$$

Total mol $CO_2 = 3a + 2b$ and total mol $H_2O = 4a + b$

mol $CO_2 = 1.5(\text{mol }H_2O)$

So $3a + 2b = 1.5(4a + b)$

$$\text{And}\ \left(a\ \text{mol C}_3\text{H}_8 \times \frac{44.11\ C_3H_8}{1\ \text{mol C}_3\text{H}_8}\right) + \left(b\ \text{mol C}_2\text{H}_2 \times \frac{26.01\ C_2H_2}{\text{mol C}_2\text{H}_2}\right) = 2.0\ \text{g}$$

Solve the simultaneous equations: $a = 9.98 \times 10^{-3}$ mol C_3H_8 and $b = 0.059\underline{9}$ mol C_2H_2.

Substitute for b and solve for grams C_2H_2.

$$0.06\underline{0}\ \text{mol C}_2\text{H}_2 \times \frac{26.01\ C_2H_2}{\text{mol C}_2\text{H}_2} = 1.\underline{5}6\ \text{g C}_2\text{H}_2 = 1.6\ \text{g C}_2\text{H}_2$$

Check: The units of the answer $(\text{g C}_2\text{H}_2)$ are correct. The magnitude is reasonable because it is less than the total mass.

7.81   **Given:** 45 μg Pb/dL blood; $V = 5.0$ L; 1 mol succimer$(C_4H_6O_4S_2) = 1$ mol Pb   **Find:** mass $C_4H_6O_4S_2$ in mg

Conceptual Plan: volume blood L $\rightarrow$ volume blood dL $\rightarrow$ μg Pb $\rightarrow$ g Pb $\rightarrow$ mol Pb $\rightarrow$

$$\frac{10\ \text{dL}}{1\ \text{L}} \qquad\qquad \frac{45\ \mu\text{g}}{\text{dL}} \quad \frac{1\ \text{g}}{10^6\ \mu\text{g}} \quad \frac{1\ \text{mol Pb}}{207.2\ \text{g Pb}} \quad \frac{1\ \text{mol succimer}}{1\ \text{mol Pb}}$$

mol succimer $\rightarrow$ g succimer $\rightarrow$ mg succimer

$$\frac{182.23\ \text{g succimer}}{1\ \text{mol succimer}} \quad \frac{1000\ \text{mg succimer}}{1\ \text{g succimer}}$$

Solution:

$$5.0\ \text{L blood} \times \frac{10\ \text{dL}}{1\ \text{L}} \times \frac{45\ \mu\text{g}}{\text{dL}} \times \frac{1\ \text{g}}{10^6\ \mu\text{g}} \times \frac{1\ \text{mol Pb}}{207.2\ \text{g}} \times \frac{1\ \text{mol succimer}}{1\ \text{mol Pb}} \times \frac{182.23\ \text{g succimer}}{1\ \text{mol succimer}} \times \frac{1000\ \text{mg}}{1\ \text{g}}$$

$$= 2.0\ \text{mg succimer}$$

Check: The units of the answer (mg succimer) are correct. The magnitude is reasonable for the volume of blood and the concentration.

7.83   **Given:** 250. g sample; 67.2 mol % Al   **Find:** theoretical yield in g of Mn

Conceptual Plan: mol Al $\rightarrow$ g Al and mol MnO$_2$ $\rightarrow$ g MnO$_2$, then mass %Al

$$\frac{26.98\ \text{g Al}}{1\ \text{mol Al}} \qquad\qquad \frac{86.94\ \text{g MnO}_2}{1\ \text{mol MnO}_2} \qquad\qquad \frac{\text{g Al}}{\text{total g}} \times 100\%$$

then sample → g Al → mol Al → mol Mn

$$\frac{38.86 \text{ g Al}}{100 \text{ g sample}} \quad \frac{1 \text{ mol Al}}{26.98 \text{ g Al}} \quad \frac{3 \text{ mol Mn}}{4 \text{ mol Al}} \rightarrow \text{smaller mol amount determines limiting reactant}$$

sample → g MnO_2 → mol MnO_2 → mol Mn

$$\frac{61.14 \text{ g MnO}_2}{100 \text{ g sample}} \quad \frac{1 \text{ mol MnO}_2}{86.94 \text{ g MnO}_2} \quad \frac{1 \text{ mol Mn}}{\text{mol MnO}_2}$$

then mol Mn → g Mn

$$\frac{54.94 \text{ g Mn}}{1 \text{ mol Mn}}$$

Solution: $4 \text{ Al}(s) + 3 \text{ MnO}_2(s) \rightarrow 3 \text{ Mn} + 2 \text{ Al}_2\text{O}_3(s)$

Assume 1 mole: $0.672 \text{ mol Al} \times \dfrac{26.98 \text{ g Al}}{1 \text{ mol Al}} = 18.13 \text{ g Al}$

$0.328 \text{ mol MnO}_2 \times \dfrac{86.94 \text{ g MnO}_2}{1 \text{ mol MnO}_2} = 28.52 \text{ g MnO}_2$

$\dfrac{18.13 \text{ g Al}}{(18.13 \text{ g Al} + 28.52 \text{ g MnO}_2)} \times 100\% = 38.86\% \text{ Al}$ So 61.14% MnO_2

$250 \text{ g sample} \times \dfrac{38.86 \text{ g Al}}{100 \text{ g sample}} \times \dfrac{1 \text{ mol Al}}{26.98 \text{ g Al}} \times \dfrac{3 \text{ mol Mn}}{4 \text{ mol Al}} = 2.701 \text{ mol Mn}$

$250 \text{ g sample} \times \dfrac{61.14 \text{ g MnO}_2}{100 \text{ g sample}} \times \dfrac{1 \text{ mol MnO}_2}{86.94 \text{ g MnO}_2} \times \dfrac{1 \text{ mol Mn}}{1 \text{ mol MnO}_2} = 1.758 \text{ mol Mn}$

$1.758 \text{ mol Mn} \times \dfrac{54.94 \text{ g Mn}}{1 \text{ mol Mn}} = 96.6 \text{ g Mn}$

Check: The units of the answer (g Mn) are correct. The magnitude of the answer is reasonable based on the amount of the limiting reactant, MnO_2.

Conceptual Problems

7.85   The correct answer is (d). The molar masses of K and O_2 are comparable. Because the stoichiometry has a ratio of 4 mol K to 1 mol O_2, K will be the limiting reactant when mass of K is less than four times the mass of O_2.

7.87   **Given:** 6 molecules N_2H_4; 4 molecules N_2O_4; (a) contains 9 molecules N_2, 12 molecules H_2O, and 1 molecule N_2H_4; solution (b) contains 12 molecules N_2, 16 molecules H_2O, and 2 molecules N_2H_4; solution (c) contains 9 molecules N_2 and 12 molecules H_2O **Find:** theoretical yield N_2, H_2O

Conceptual Plan: molecules N_2H_4 → molecules N_2

$$\frac{3 \text{ molecules N}_2}{2 \text{ molecules N}_2\text{H}_4} \rightarrow \text{smaller molecules amount determines limiting reactant}$$

molecules N_2O_4 → molecules N_2

$$\frac{3 \text{ molecules N}_2}{1 \text{ molecule N}_2\text{O}_4}$$

molecules N_2H_4 → molecules H_2O

$$\frac{4 \text{ molecules H}_2\text{O}}{2 \text{ molecules N}_2\text{H}_4}$$

molecules N_2H_4 → molecules N_2O_4

$$\frac{1 \text{ molecule N}_2\text{O}_4}{2 \text{ molecules N}_2\text{H}_4}$$

Solution: $6 \text{ molecules N}_2\text{H}_4 \times \dfrac{3 \text{ molecules N}_2}{2 \text{ molecules N}_2\text{H}_4} = 9 \text{ molecules N}_2$

$6 \text{ molecules N}_2\text{O}_4 \times \dfrac{3 \text{ molecules N}_2}{1 \text{ molecule N}_2\text{O}_4} = 18 \text{ molecules N}_2$

Limiting reactant = N_2H_4 because it produced the least molecules of N_2.

$$6 \text{ molecules } N_2H_4 \times \frac{4 \text{ molecules } H_2O}{2 \text{ molecules } N_2H_4} = 12 \text{ molecules } H_2O$$

$$6 \text{ molecules } N_2H_4 \times \frac{1 \text{ molecule } N_2O_4}{2 \text{ molecules } N_2H_4} = 3 \text{ molecules } N_2O_4 \text{ used}$$

Reaction mixture should contain 9 molecules N_2, 12 molecules H_2O, and 1 molecule N_2O_4; this is best represented by (a).

Questions for Group Work

7.89 (a) $CH_4(g) + 2 O_2(g) \rightarrow 2 H_2O(g) + CO_2(g)$

(b) Calculate the number of moles of methane and oxygen:

$$16.05 \text{ g } CH_4 \times \frac{1 \text{ mol } CH_4}{16.04 \text{ g } CH_4} = 1.001 \text{ mol } CH_4 \text{ and } 96.00 \text{ g } O_2 \times \frac{1 \text{ mol } O_2}{32.00 \text{ g } O_2} = 3.000 \text{ mol } O_2$$

● = carbon

○ = hydrogen

◐ = oxygen

(c) Calculate the number of moles of water and CO_2 that can be made if all of the methane reacts:

$$1.001 \text{ mol } CH_4 \times \frac{2 \text{ mol } H_2O}{1 \text{ mol } CH_4} = 2.002 \text{ mol } H_2O \text{ and } 1.001 \text{ mol } CH_4 \times \frac{1 \text{ mol } CO_2}{1 \text{ mol } CH_4} = 1.001 \text{ mol } CO_2$$

Calculate the number of moles of water and CO_2 that can be made if all of the oxygen reacts:

$$3.000 \text{ mol } O_2 \times \frac{2 \text{ mol } H_2O}{2 \text{ mol } O_2} = 3.000 \text{ mol } H_2O \text{ and } 3.000 \text{ mol } O_2 \times \frac{1 \text{ mol } CO_2}{2 \text{ mol } O_2} = 1.500 \text{ mol } CO_2$$

The lower set of quantities is the amount that can be made: 2.002 mol water and 1.001 mol CO_2.

(d) Since the amount of methane determines the amount of the products, there will be oxygen left over.

$$2.002 \text{ mol } H_2O \times \frac{2 \text{ mol } O_2}{2 \text{ mol } H_2O} = 2.002 \text{ mol } O_2 \text{ reacted.}$$

The amount of oxygen left over = 3.000 mol O_2 − 2.002 mol O_2 = 0.998 mol O_2 left over.

(e) The limiting reagent is methane; the excess reagent is oxygen; the theoretical yields are 2.002 mol water and 1.001 mol CO_2.

Data Interpretation and Analysis

7.90 (a) Since the mass of C increases as the mass of A increases and the mass of B is constant in Experiments 1, 2, and 3, A is the limiting reactant.

(c) Convert the mass of A and the mass of B for each experiment to moles of A and B:

$$2.51 \text{ g } A \times \frac{1 \text{ mol } A}{50.0 \text{ g } A} = 0.0502 \text{ mol } A \text{ and } 7.54 \text{ g } B \times \frac{1 \text{ mol } B}{75.0 \text{ g } B} = \underline{0.1005} \text{ mol } B$$

Experiment #	Mass A (g)	Mass B (g)	Mol A (mol)	Mol B (mol)
1	2.51	7.54	0.0502	0.101
2	5.03	7.51	0.101	0.100
3	7.55	7.52	0.151	0.100
4	12.53	7.49	0.251	0.100
5	15.04	7.47	0.301	0.100
6	19.98	7.51	0.400	0.100
7	20.04	9.95	0.401	0.133
8	20.02	12.55	0.400	0.167

Looking at the results, we see that A is the limiting reactant when it is half the number of moles of B, so the relative coefficients for A and B are 2 and 1, respectively.

(e)   Convert the mass of C to moles of C for each experiment:

$$3.76 \text{ g C} \times \frac{1 \text{ mol C}}{88.0 \text{ g C}} = 0.0427 \text{ mol C}$$

Experiment #	Mass A (g)	Mass B (g)	Mass C Obtained (g)	Limiting Reactant	Mol A (mol)	Mol B (mol)	Mol C (mol)	Mol C/ Mol A	Mol C/ Mol B
1	2.51	7.54	3.76	A	0.0502	0.101	0.0427	0.851	0.425
2	5.03	7.51	7.43	A	0.101	0.100	0.0844	0.839	0.843
3	7.55	7.52	11.13	A	0.151	0.100	0.1265	0.838	1.26
4	12.53	7.49	14.84	B	0.251	0.100	0.1686	0.673	1.69
5	15.04	7.47	14.94	B	0.301	0.100	0.1698	0.564	1.70
6	19.98	7.51	15.17	B	0.400	0.100	0.1724	0.431	1.72
7	20.04	9.95	19.31	B	0.401	0.133	0.2194	0.547	1.65
8	20.02	12.55	24.69	B	0.400	0.167	0.2806	0.701	1.68

It may help to calculate the ratios of the moles of C to moles of reactant to look for patterns. Looking at the results, we see that moles of C produced are just a little less than the number of moles of A in the experiment, so the coefficient for C is 2. The balanced reaction is $2A + B \rightarrow 2C$.

8 Introduction to Solutions and Aqueous Reactions

Review Questions

8.1 An aqueous solution is a solution in which water acts as the solvent. The solvent is the majority component of the mixture, and the solute is the minority component of the mixture.

8.3 Substances that completely dissociate into ions when they dissolve in water are called strong electrolytes and conduct electricity easily. Substances that do not completely dissociate in water are called weak electrolytes and conduct electricity only weakly. Compounds that do not dissociate into ions when dissolved in water are called nonelectrolytes and do not conduct electricity.

8.5 A compound is soluble if it dissolves in water. A compound is insoluble if it does not dissolve in water.

8.7 Cations that usually form soluble compounds are Li^+, Na^+, K^+, and NH_4^+. The anions that usually form soluble compounds are NO_3^- and $C_2H_3O_2^-$, which have no exceptions; Cl^-, Br^-, and I^- except when these ions pair with Ag^+, Hg_2^{2+}, or Pb^{2+}, which result in insoluble compounds; and SO_4^{2-} except with Sr^{2+}, Ba^{2+}, Pb^{2+}, Ag^+, or Ca^{2+}, which form insoluble compounds. The anions that usually form insoluble compounds are OH^- and S^{2-} except with Li^+, Na^+, K^+, and NH_4^+, which form soluble compounds; and when S^{2-} pairs with Ca^{2+}, Sr^{2+}, or Ba^{2+}, which form soluble compounds; CO_3^{2-} and PO_4^{3-} are insoluble except when paired with Li^+, Na^+, K^+, and NH_4^+.

8.9 The key to predicting precipitation reactions is to understand that only insoluble compounds form precipitates. In a precipitation reaction, two solutions containing soluble compounds combine and an insoluble compound precipitates.

8.11 An Arrhenius base is a substance that produces OH^- ions in aqueous solutions.

8.13 When an acid and a base are mixed, the $H^+(aq)$ from the acid combines with the $OH^-(aq)$ from the base to form $H_2O(l)$. An example is $HCl(aq) + NaOH(aq) \rightarrow H_2O(l) + NaCl(aq)$.

8.15 Aqueous reactions that form a gas upon mixing two solutions are called gas-evolution reactions. An example is $H_2SO_4(aq) + Li_2S(aq) \rightarrow H_2S(g) + Li_2SO_4(aq)$.

8.17 Oxidation–reduction reactions, or redox reactions, are reactions in which electrons are transferred from one reactant to the other. An example is $4\,Fe(s) + 3\,O_2(g) \rightarrow 2\,Fe_2O_3(s)$.

8.19 When a substance is oxidized, it loses electrons and there is an increase in oxidation state. When a substance is reduced, it gains electrons and there is a reduction in oxidation state.

Problems by Topic

Solution Concentration and Solution Stoichiometry

8.21 (a) **Given:** 3.25 mol LiCl; 2.78 L solution **Find:** molarity LiCl
Conceptual Plan: mol LiCl, L solution → molarity

$$\text{molarity (M)} = \frac{\text{amount of solute (in moles)}}{\text{volume of solution (in L)}}$$

163

Copyright © 2018 Pearson Education, Inc.

Solution: $\dfrac{3.25 \text{ mol LiCl}}{2.78 \text{ L solution}} = 1.16\underline{9}\text{ M} = 1.17\text{ M}$

Check: The units of the answer (M) are correct. The magnitude of the answer is reasonable. Concentrations are usually between 0 M and 18 M.

(b)   **Given:** 28.33 g $C_6H_{12}O_6$; 1.28 L solution   **Find:** molarity $C_6H_{12}O_6$
Conceptual Plan: g $C_6H_{12}O_6$ → mol $C_6H_{12}O_6$, L solution → molarity

$$\dfrac{1 \text{ mol } C_6H_{12}O_6}{180.16 \text{ g } C_6H_{12}O_6} \qquad \text{molarity (M)} = \dfrac{\text{amount of solute (in moles)}}{\text{volume of solution (in L)}}$$

Solution: $28.33 \text{ g } C_6H_{12}O_6 \times \dfrac{1 \text{ mol } C_6H_{12}O_6}{180.16 \text{ g } C_6H_{12}O_6} = 0.1572\underline{4} \text{ mol } C_6H_{12}O_6$

$\dfrac{0.1572\underline{4} \text{ mol } C_6H_{12}O_6}{1.28 \text{ L solution}} = 0.122\underline{8}\text{ M} = 0.123\text{ M}$

Check: The units of the answer (M) are correct. The magnitude of the answer is reasonable. Concentrations are usually between 0 M and 18 M.

(c)   **Given:** 32.4 mg NaCl; 122.4 mL solution   **Find:** molarity NaCl
Conceptual Plan: mg NaCl → g NaCl → mol NaCl, and mL solution → L solution then molarity

$$\dfrac{1 \text{ g NaCl}}{1000 \text{ mg NaCl}} \quad \dfrac{1 \text{ mol NaCl}}{58.45 \text{ g NaCl}} \quad \dfrac{1 \text{ L solution}}{1000 \text{ mL solution}} \quad \text{molarity (M)} = \dfrac{\text{amount of solute (in moles)}}{\text{volume of solution (in L)}}$$

Solution: $32.4 \text{ mg NaCl} \times \dfrac{1 \text{ g}}{1000 \text{ mg}} \times \dfrac{1 \text{ mol NaCl}}{58.45 \text{ g NaCl}} = 5.54\underline{3} \times 10^{-4} \text{ mol NaCl}$

$122.4 \text{ mL solution} \times \dfrac{1 \text{ L}}{1000 \text{ mL}} = 0.1224 \text{ L}$

$\dfrac{5.54\underline{3} \times 10^{-4} \text{ mol NaCl}}{0.1224 \text{ L}} = 0.00452\underline{8} \text{ M NaCl} = 0.00453 \text{ M NaCl}$

Check: The units of the answer (M) are correct. The magnitude of the answer is reasonable. Concentrations are usually between 0 M and 18 M.

8.23   (a)   **Given:** 0.150 M KNO_3 **Find:** M NO_3^-
Conceptual Plan: M KNO_3 → M NO_3^-

$$\dfrac{1 \text{ M } NO_3^-}{1 \text{ M } KNO_3}$$

Solution: $0.150 \text{ M } KNO_3 \times \dfrac{1 \text{ M } NO_3^-}{1 \text{ M } KNO_3} = 0.150 \text{ M } NO_3^-$

Check: The units of the answer, M, are correct. The answer is reasonable because it is greater than or equal to the concentration of KNO_3.

(b)   **Given:** 0.150 M $Ca(NO_3)_2$ **Find:** M NO_3^-
Conceptual Plan: M $Ca(NO_3)_2$ → M NO_3^-

$$\dfrac{2 \text{ M } NO_3^-}{1 \text{ M } Ca(NO_3)_2}$$

Solution: $0.150 \text{ M } Ca(NO_3)_2 \times \dfrac{2 \text{ M } NO_3^-}{1 \text{ M } Ca(NO_3)_2} = 0.300 \text{ M } NO_3^-$

Check: The units of the answer, M, are correct. The answer is reasonable because it is greater than or equal to the concentration of $Ca(NO_3)_2$.

(c)   **Given:** 0.150 M $Al(NO_3)_3$ **Find:** M NO_3^-
Conceptual Plan: M $Al(NO_3)_3$ → M NO_3^-

$$\dfrac{3 \text{ M } NO_3^-}{1 \text{ M } Al(NO_3)_3}$$

Solution: $0.150 \text{ M } Al(NO_3)_3 \times \dfrac{3 \text{ M } NO_3^-}{1 \text{ M } Al(NO_3)_3} = 0.450 \text{ M } NO_3^-$

Check: The units of the answer, M, are correct. The answer is reasonable because it is greater than or equal to the concentration of $Al(NO_3)_3$.

8.25   (a)   **Given:** 0.556 L; 2.3 M KCl   **Find:** mol KCl

         Conceptual Plan: volume solution × M = mol

$$\text{volume solution (L)} \times M = \text{mol}$$

         Solution: $0.556 \; \text{L solution} \times \dfrac{2.3 \; \text{mol KCl}}{\text{L solution}} = 1.3 \; \text{mol KCl}$

         Check: The units of the answer (mol KCl) are correct. The magnitude is reasonable because it is less than 1 L solution.

     (b)   **Given:** 1.8 L; 0.85 M KCl   **Find:** mol KCl

         Conceptual Plan: volume solution × M = mol

$$\text{volume solution (L)} \times M = \text{mol}$$

         Solution: $1.8 \; \text{L solution} \times \dfrac{0.85 \; \text{mol KCl}}{\text{L solution}} = 1.5 \; \text{mol KCl}$

         Check: The units of the answer (mol KCl) are correct. The magnitude is reasonable because it is less than 2 L solution.

     (c)   **Given:** 114 mL; 1.85 M KCl   **Find:** mol KCl

         Conceptual Plan: mL solution → L solution, then volume solution × M = mol

$$\dfrac{1 \; \text{L}}{1000 \; \text{mL}} \qquad\qquad \text{volume solution (L)} \times M = \text{mol}$$

         Solution: $114 \; \text{mL solution} \times \dfrac{1 \; \text{L}}{1000 \; \text{mL}} \times \dfrac{1.85 \; \text{mol KCl}}{\text{L solution}} = 0.211 \; \text{mol KCl}$

         Check: The units of the answer (mol KCl) are correct. The magnitude is reasonable because it is less than 1 L solution.

8.27   **Given:** 400.0 mL; 1.1 M $NaNO_3$ **Find:** g $NaNO_3$

     Conceptual Plan: mL solution → L solution, then volume solution × M = mol $NaNO_3$

$$\dfrac{1 \; \text{L solution}}{1000 \; \text{mL solution}} \qquad\qquad \text{volume solution (L)} \times M = \text{mol}$$

         then mol $NaNO_3$ → g $NaNO_3$

$$\dfrac{85.00 \; \text{g NaNO}_3}{1 \; \text{mol NaNO}_3}$$

     Solution: $400.0 \; \text{mL solution} \times \dfrac{1 \; \text{L}}{1000 \; \text{mL}} \times \dfrac{1.1 \; \text{mol NaNO}_3}{1 \; \text{L solution}} \times \dfrac{85.00 \; \text{g NaNO}_3}{1 \; \text{mol NaNO}_3} = 37.40 \; \text{g NaNO}_3 = 37 \; \text{g NaNO}_3$

     Check: The units of the answer (g $NaNO_3$) are correct. The magnitude is reasonable for the concentration and volume of solution.

8.29   **Given:** $V_1 = 123 \; \text{mL}$; $M_1 = 1.1 \; \text{M}$; $V_2 = 500.0 \; \text{mL}$ **Find:** M_2

     Conceptual Plan: mL → L then $V_1, M_1, V_2 → M_2$

$$\dfrac{1 \; \text{L}}{1000 \; \text{mL}} \qquad\qquad V_1 M_1 = V_2 M_2$$

     Solution: $123 \; \text{mL} \times \dfrac{1 \; \text{L}}{1000 \; \text{mL}} = 0.123 \; \text{L} \qquad 500.0 \; \text{mL} \times \dfrac{1 \; \text{L}}{1000 \; \text{mL}} = 0.5000 \; \text{L}$

$$M_2 = \dfrac{V_1 M_1}{V_2} = \dfrac{(0.123 \; \text{L})(1.1 \; \text{M})}{(0.5000 \; \text{L})} = 0.27 \; \text{M}$$

     Check: The units of the answer (M) are correct. The magnitude of the answer is reasonable because it is less than the original concentration.

8.31   **Given:** $V_1 = 50.0 \; \text{mL}$; $M_1 = 12 \; \text{M}$; $M_2 = 0.100 \; \text{M}$ **Find:** V_2

     Conceptual Plan: mL → L then $V_1, M_1, M_2 → V_2$

$$\dfrac{1 \; \text{L}}{1000 \; \text{mL}} \qquad\qquad V_1 M_1 = V_2 M_2$$

Solution: $50.0 \text{ mL} \times \dfrac{1 \text{ L}}{1000 \text{ mL}} = 0.050 \text{ L}$

$$V_2 = \frac{V_1 M_1}{M_2} = \frac{(0.0500 \text{ L})(12 \text{ M})}{(0.100 \text{ M})} = 6.0 \text{ L}$$

Check: The units of the answer (L) are correct. The magnitude of the answer is reasonable because the new concentration is much less than the original; the volume must be larger.

8.33   **Given:** 95.4 mL; 0.102 M $CuCl_2$; 0.175 M Na_3PO_4 **Find:** volume Na_3PO_4
Conceptual Plan: mL $CuCl_2$ → L $CuCl_2$ → mol $CuCl_2$ → mol Na_3PO_4 → L Na_3PO_4 → mL Na_3PO_4

$$\frac{1 \text{ L}}{1000 \text{ mL}} \qquad \frac{0.102 \text{ mol } CuCl_2}{1 \text{ L}} \qquad \frac{2 \text{ mol } Na_3PO_4}{3 \text{ mol } CuCl_2} \qquad \frac{1 \text{ L}}{0.175 \text{ mol } Na_3PO_4} \qquad \frac{1000 \text{ mL}}{1 \text{ L}}$$

Solution: $95.4 \text{ mL } CuCl_2 \times \dfrac{1 \text{ L}}{1000 \text{ mL}} \times \dfrac{0.102 \text{ mol } CuCl_2}{1 \text{ L}} \times \dfrac{2 \text{ mol } Na_3PO_4}{3 \text{ mol } CuCl_2} \times \dfrac{1 \text{ L}}{0.175 \text{ mol } Na_3PO_4} \times \dfrac{1000 \text{ mL}}{1 \text{ L}}$

$= 37.1 \text{ mL } Na_3PO_4$

Check: The units of the answer ($\text{mL } Na_3PO_4$) are correct. The magnitude of the answer is reasonable because the concentration of Na_3PO_4 is greater.

8.35   **Given:** 25.0 g H_2; 6.0 M H_2SO_4 **Find:** volume H_2SO_4
Conceptual Plan: g H_2 → mol H_2 → mol H_2SO_4 → L H_2SO_4

$$\frac{1 \text{ mol } H_2}{2.016 \text{ g } H_2} \qquad \frac{3 \text{ mol } H_2SO_4}{3 \text{ mol } H_2} \qquad \frac{1 \text{ L}}{6.0 \text{ mol } H_2SO_4}$$

Solution: $25.0 \text{ g } H_2 \times \dfrac{1 \text{ mol } H_2}{2.016 \text{ g } H_2} \times \dfrac{3 \text{ mol } H_2SO_4}{3 \text{ mol } H_2} \times \dfrac{1 \text{ L}}{6.0 \text{ mol } H_2SO_4} = 2.1 \text{ L } H_2SO_4$

Check: The units of the answer ($\text{L } H_2SO_4$) are correct. The magnitude is reasonable because there are approximately 12 mol H_2 and the mole ratio is 1:1.

8.37   **Given:** 25.0 mL, 1.20 M KCl; 15.0 mL, 0.900 M $Ba(NO_3)_2$; 2.45 g $BaCl_2$
Find: limiting reactant, theoretical yield $BaCl_2$, % yield
Conceptual Plan: volume KCl solution × M → mol KCl → mol $BaCl_2$

$$\text{volume solution (L)} \times \text{M} = \text{mol} \qquad \frac{1 \text{ mol } BaCl_2}{2 \text{ mol KCl}} \qquad \rightarrow \quad \textbf{smaller mol amount determines limiting reactant}$$

volume $Ba(NO_3)_2$ solution × M → mol $Ba(NO_3)_2$ → mol $BaCl_2$

$$\text{volume solution (L)} \times \text{M} = \text{mol} \qquad \frac{1 \text{ mol } BaCl_2}{1 \text{ mol } Ba(NO_3)_2}$$

then mol $BaCl_2$ → g $BaCl_2$ then determine % yield

$$\frac{208.23 \text{ g } BaCl_2}{1 \text{ mol } BaCl_2} \qquad \frac{\text{actual yield g } BaCl_2}{\text{theoretical yield g } BaCl_2} \times 100\%$$

Solution: $25.0 \text{ mL solution} \times \dfrac{1 \text{ L solution}}{1000 \text{ mL solution}} \times \dfrac{1.20 \text{ mol KCl}}{1 \text{ L solution}} \times \dfrac{1 \text{ mol } BaCl_2}{2 \text{ mol KCl}} = 0.015\underline{0} \text{ mol } BaCl_2$

$15.0 \text{ mL solution} \times \dfrac{1 \text{ L solution}}{1000 \text{ mL solution}} \times \dfrac{0.900 \text{ mol } Ba(NO_3)_2}{1 \text{ L solution}} \times \dfrac{1 \text{ mol } BaCl_2}{1 \text{ mol } Ba(NO_3)_2} = 0.013\underline{5} \text{ mol } BaCl_2$

$0.013\underline{5} \text{ mol } BaCl_2 \times \dfrac{208.23 \text{ g } BaCl_2}{1 \text{ mol } BaCl_2} = 2.8\underline{1}1 \text{ g } BaCl_2$

$\dfrac{2.45 \text{ g } BaCl_2}{2.811 \text{ g } BaCl_2} \times 100\% = 87.2\%$

Check: The units of the answer (%) are correct. The magnitude is reasonable since the value is between 0 and 100%. It is much closer to 100% because the theoretical yield is not much more than the actual yield.

Types of Aqueous Solutions and Solubility

8.39   (a)   CsCl is an ionic compound. An aqueous solution is an electrolyte solution, so it conducts electricity.

(b) CH_3OH is a molecular compound that does not dissociate. An aqueous solution is a nonelectrolyte solution, so it does not conduct electricity.

(c) $Ca(NO_2)_2$ is an ionic compound. An aqueous solution is an electrolyte solution, so it conducts electricity.

(d) $C_6H_{12}O_6$ is a molecular compound that does not dissociate. An aqueous solution is a nonelectrolyte solution, so it does not conduct electricity.

8.41   (a) $AgNO_3$ is soluble. Compounds containing NO_3^- are always soluble with no exceptions. The ions in the solution are $Ag^+(aq)$ and $NO_3^-(aq)$.

(b) $Pb(C_2H_3O_2)_2$ is soluble. Compounds containing $C_2H_3O_2^-$ are always soluble with no exceptions. The ions in the solution are $Pb^{2+}(aq)$ and $C_2H_3O_2^-(aq)$.

(c) KNO_3 is soluble. Compounds containing K^+ or NO_3^- are always soluble with no exceptions. The ions in solution are $K^+(aq)$ and $NO_3^-(aq)$.

(d) $(NH_4)_2S$ is soluble. Compounds containing NH_4^+ are always soluble with no exceptions. The ions in solution are $NH_4^+(aq)$ and $S^{2-}(aq)$.

Precipitation Reactions

8.43   (a) $LiI(aq) + BaS(aq) \rightarrow$ Possible products: Li_2S and BaI_2. Li_2S is soluble. Compounds containing S^{2-} are normally insoluble, but Li^+ is an exception. BaI_2 is soluble. Compounds containing I^- are normally soluble, and Ba^{2+} is not an exception. $LiI(aq) + BaS(aq) \rightarrow$ No Reaction

(b) $KCl(aq) + CaS(aq) \rightarrow$ Possible products: K_2S and $CaCl_2$. K_2S is soluble. Compounds containing S^{2-} are normally insoluble, but K^+ is an exception. $CaCl_2$ is soluble. Compounds containing Cl^- are normally soluble, and Ca^{2+} is not an exception. $KCl(aq) + CaS(aq) \rightarrow$ No Reaction

(c) $CrBr_2(aq) + Na_2CO_3(aq) \rightarrow$ Possible products: $CrCO_3$ and $NaBr$. $CrCO_3$ is insoluble. Compounds containing CO_3^{2-} are normally insoluble, and Cr^{2+} is not an exception. $NaBr$ is soluble. Compounds containing Br^- are normally soluble, and Na^+ is not an exception.
$CrBr_2(aq) + Na_2CO_3(aq) \rightarrow CrCO_3(s) + 2\,NaBr(aq)$

(d) $NaOH(aq) + FeCl_3(aq) \rightarrow$ Possible products: $NaCl$ and $Fe(OH)_3$. $NaCl$ is soluble. Compounds containing Na^+ are normally soluble—no exceptions. $Fe(OH)_3$ is insoluble. Compounds containing OH^- are normally insoluble, and Fe^{3+} is not an exception.
$3\,NaOH(aq) + FeCl_3(aq) \rightarrow 3\,NaCl(aq) + Fe(OH)_3(s)$

8.45   (a) $K_2CO_3(aq) + Pb(NO_3)_2(aq) \rightarrow$ Possible products: KNO_3 and $PbCO_3$. KNO_3 is soluble. Compounds containing K^+ are always soluble—no exceptions. $PbCO_3$ is insoluble. Compounds containing CO_3^{2-} are normally insoluble, and Pb^{2+} is not an exception.
$K_2CO_3(aq) + Pb(NO_3)_2(aq) \rightarrow 2\,KNO_3(aq) + PbCO_3(s)$

(b) $Li_2SO_4(aq) + Pb(C_2H_3O_2)_2(aq) \rightarrow$ Possible products: $LiC_2H_3O_2$ and $PbSO_4$. $LiC_2H_3O_2$ is soluble. Compounds containing Li^+ are always soluble—no exceptions. $PbSO_4$ is insoluble. Compounds containing SO_4^{2-} are normally soluble, but Pb^{2+} is an exception.
$Li_2SO_4(aq) + Pb(C_2H_3O_2)_2(aq) \rightarrow 2\,LiC_2H_3O_2(aq) + PbSO_4(s)$

(c) $Cu(NO_3)_2(aq) + MgS(s) \rightarrow$ Possible products: CuS and $Mg(NO_3)_2$. CuS is insoluble. Compounds containing S^{2-} are normally insoluble, and Cu^{2+} is not an exception. $Mg(NO_3)_2$ is soluble. Compounds containing NO_3^- are always soluble—no exceptions.
$Cu(NO_3)_2(aq) + MgS(s) \rightarrow CuS(s) + Mg(NO_3)_2(aq)$

(d) $Sr(NO_3)_2(aq) + KI(aq) \rightarrow$ Possible products: SrI_2 and KNO_3. SrI_2 is soluble. Compounds containing I^- are normally soluble, and Sr^{2+} is not an exception. KNO_3 is soluble. Compounds containing K^+ are always soluble—no exceptions. $Sr(NO_3)_2(aq) + KI(aq) \rightarrow$ No Reaction

Ionic and Net Ionic Equations

8.47   (a) $H^+(aq) + \cancel{Cl^-}(aq) + \cancel{Li^+}(aq) + OH^-(aq) \rightarrow H_2O(l) + \cancel{Li^+}(aq) + \cancel{Cl^-}(aq)$
$H^+(aq) + OH^-(aq) \rightarrow H_2O(l)$

(b) $\cancel{Ca^{2+}}(aq) + S^{2-}(aq) + Cu^{2+}(aq) + 2\,\cancel{Cl^-}(aq) \rightarrow CuS(s) + \cancel{Ca^{2+}}(aq) + 2\,\cancel{Cl^-}(aq)$
$S^{2-}(aq) + Cu^{2+}(aq) \rightarrow CuS(s)$

(c) $\cancel{Na^+}(aq) + OH^-(aq) + H^+(aq) + \cancel{NO_3^-}(aq) \rightarrow H_2O(l) + \cancel{Na^+}(aq) + \cancel{NO_3^-}(aq)$
$H^+(aq) + OH^-(aq) \rightarrow H_2O(l)$

(d) $6\,\overline{\text{Na}^+}(aq) + 2\,\text{PO}_4^{3-}(aq) + 3\,\text{Ni}^{2+}(aq) + 6\,\overline{\text{Cl}^-}(aq) \rightarrow \text{Ni}_3(\text{PO}_4)_2(s) + 6\,\overline{\text{Na}^+}(aq) + 6\,\overline{\text{Cl}^-}(aq)$

   $3\,\text{Ni}^{2+}(aq) + 2\,\text{PO}_4^{3-}(aq) \rightarrow \text{Ni}_3(\text{PO}_4)_2(s)$

8.49 $\text{Hg}_2^{2+}(aq) + 2\,\overline{\text{NO}_3^-}(aq) + 2\,\overline{\text{Na}^+}(aq) + 2\,\text{Cl}^-(aq) \rightarrow \text{Hg}_2\text{Cl}_2(s) + 2\,\overline{\text{Na}^+}(aq) + 2\,\overline{\text{NO}_3^-}(aq)$

   $\text{Hg}_2^{2+}(aq) + 2\,\text{Cl}^-(aq) \rightarrow \text{Hg}_2\text{Cl}_2(s)$

Naming Acids

8.51 (a) HI: The base name of I is *iod*, so the name is hydroiodic acid.

   (b) HNO$_3$: The oxyanion is *nitrate*, which ends in *-ate*; therefore, the name of the acid is nitric acid.

   (c) H$_2$CO$_3$: The oxyanion is *carbonate*, which ends in *-ate*; therefore, the name of the acid is carbonic acid.

8.53 (a) hydrofluoric acid: HF

   (b) hydrobromic acid: HBr

   (c) sulfurous acid: H$_2$SO$_3$

Acid–Base Reactions

8.55 Skeletal reaction: $\text{HBr}(aq) + \text{KOH}(aq) \rightarrow \text{H}_2\text{O}(l) + \text{KBr}(aq)$

   acid   base   water   salt

   Net ionic equation: $\text{H}^+(aq) + \text{OH}^-(aq) \rightarrow \text{H}_2\text{O}(l)$

8.57 (a) Skeletal reaction: $\text{H}_2\text{SO}_4(aq) + \text{Ca(OH)}_2(aq) \rightarrow \text{H}_2\text{O}(l) + \text{CaSO}_4(s)$

   acid   base   water   salt

   Balanced reaction: $\text{H}_2\text{SO}_4(aq) + \text{Ca(OH)}_2(aq) \rightarrow 2\,\text{H}_2\text{O}(l) + \text{CaSO}_4(s)$

   (b) Skeletal reaction: $\text{HClO}_4(aq) + \text{KOH}(aq) \rightarrow \text{H}_2\text{O}(l) + \text{KClO}_4(aq)$

   acid   base   water   salt

   Balanced reaction: $\text{HClO}_4(aq) + \text{KOH}(aq) \rightarrow \text{H}_2\text{O}(l) + \text{KClO}_4(aq)$

   (c) Skeletal reaction: $\text{H}_2\text{SO}_4(aq) + \text{NaOH}(aq) \rightarrow \text{H}_2\text{O}(l) + \text{Na}_2\text{SO}_4(aq)$

   acid   base   water   salt

   Balanced reaction: $\text{H}_2\text{SO}_4(aq) + 2\,\text{NaOH}(aq) \rightarrow 2\,\text{H}_2\text{O}(l) + \text{Na}_2\text{SO}_4(aq)$

8.59 (a) Skeletal reaction: $\text{HBr}(aq) + \text{NaOH}(aq) \rightarrow \text{H}_2\text{O}(l) + \text{NaBr}(aq)$

   acid   base   water   salt

   Balanced reaction: $\text{HBr}(aq) + \text{NaOH}(aq) \rightarrow \text{H}_2\text{O}(l) + \text{NaBr}(aq)$

   Complete ionic equation: $\text{H}^+(aq) + \text{Br}^-(aq) + \text{Na}^+(aq) + \text{OH}^-(aq) \rightarrow \text{H}_2\text{O}(l) + \text{Na}^+(aq) + \text{Br}^-(aq)$

   Net ionic equation: $\text{H}^+(aq) + \text{OH}^-(aq) \rightarrow \text{H}_2\text{O}(l)$

   (b) Skeletal reaction: $\text{HF}(aq) + \text{NaOH}(aq) \rightarrow \text{H}_2\text{O}(l) + \text{NaF}(aq)$

   acid   base   water   salt

   Balanced reaction: $\text{HF}(aq) + \text{NaOH}(aq) \rightarrow \text{H}_2\text{O}(l) + \text{NaF}(aq)$

   Complete ionic equation: $\text{H}^+(aq) + \text{F}^-(aq) + \text{Na}^+(aq) + \text{OH}^-(aq) \rightarrow \text{H}_2\text{O}(l) + \text{Na}^+(aq) + \text{F}^-(aq)$

   Net ionic equation: $\text{H}^+(aq) + \text{OH}^-(aq) \rightarrow \text{H}_2\text{O}(l)$

   (c) Skeletal reaction: $\text{HC}_2\text{H}_3\text{O}_2(aq) + \text{RbOH}(aq) \rightarrow \text{H}_2\text{O}(l) + \text{RbC}_2\text{H}_3\text{O}_2(aq)$

   acid   base   water   salt

   Balanced reaction: $\text{HC}_2\text{H}_3\text{O}_2(aq) + \text{RbOH}(aq) \rightarrow \text{H}_2\text{O}(l) + \text{RbC}_2\text{H}_3\text{O}_2(aq)$

   Complete ionic equation: $\text{HC}_2\text{H}_3\text{O}_2(aq) + \text{Rb}^+(aq) + \text{OH}^-(aq) \rightarrow \text{H}_2\text{O}(l) + \text{Rb}^+(aq) + \text{C}_2\text{H}_3\text{O}_2^-(aq)$

   Net ionic equation: $\text{HC}_2\text{H}_3\text{O}_2(aq) + \text{OH}^-(aq) \rightarrow \text{H}_2\text{O}(l) + \text{C}_2\text{H}_3\text{O}_2^-(aq)$

8.61 **Given:** 22.62 mL, 0.2000 M NaOH solution; 25.00 mL HClO$_4$ solution   **Find:** M HClO$_4$ solution

   Conceptual Plan: mL NaOH → L NaOH → mol NaOH → mol HClO$_4$

   $\dfrac{1\ \text{L}}{1000\ \text{mL}}$ $\dfrac{0.2000\ \text{mol NaOH}}{1\ \text{L NaOH}}$ $\dfrac{1\ \text{mol HClO}_4}{1\ \text{mol NaOH}}$

   mol HClO$_4$, volume HClO$_4$ solution → M

   $\text{M} = \dfrac{1\ \text{mol HClO}_4}{\text{L HClO}_4\ \text{solution}}$

Solution: $22.62 \text{ mL NaOH} \times \dfrac{1 \text{ L}}{1000 \text{ mL}} \times \dfrac{0.2000 \text{ mol NaOH}}{1 \text{ L NaOH}} \times \dfrac{1 \text{ mol HClO}_4}{1 \text{ mol NaOH}} = 0.004524 \text{ mol HClO}_4$

$\dfrac{0.004524 \text{ mol HClO}_4}{25.00 \text{ mL HClO}_4} \times \dfrac{1000 \text{ mL}}{1 \text{ L}} = 0.18096 \text{ M HClO}_4 = 0.1810 \text{ M HClO}_4$

Check: The units of the answer ($M \text{ HClO}_4$) are correct. The magnitude of the answer is reasonable because it is less than the M of NaOH.

Gas-Evolution Reactions

8.63 (a) Skeletal reaction: $\text{HBr}(aq) + \text{NiS}(s) \rightarrow \text{NiBr}_2(aq) + \underset{gas}{\text{H}_2\text{S}(g)}$

Balanced reaction: $2\,\text{HBr}(aq) + \text{NiS}(s) \rightarrow \text{NiBr}_2(aq) + \text{H}_2\text{S}(g)$

(b) Skeletal reaction:

$\text{NH}_4\text{I}(aq) + \text{NaOH}(aq) \rightarrow \underset{\text{decomposes}}{\text{NH}_4\text{OH}(aq)} + \text{NaI}(aq) \rightarrow \text{H}_2\text{O}(l) + \underset{\text{gas}}{\text{NH}_3(g)} + \text{NaI}(aq)$

Balanced reaction: $\text{NH}_4\text{I}(aq) + \text{NaOH}(aq) \rightarrow \text{H}_2\text{O}(l) + \text{NH}_3(g) + \text{NaI}(aq)$

(c) Skeletal reaction: $\text{HBr}(aq) + \text{Na}_2\text{S}(aq) \rightarrow \text{NaBr}(aq) + \underset{gas}{\text{H}_2\text{S}(g)}$

Balanced reaction: $2\,\text{HBr}(aq) + \text{Na}_2\text{S}(aq) \rightarrow 2\,\text{NaBr}(aq) + \text{H}_2\text{S}(g)$

8.65 Skeletal reaction:

$\text{HClO}_4(aq) + \text{Li}_2\text{CO}_3(aq) \rightarrow \underset{\text{decomposes}}{\text{H}_2\text{CO}_3(aq)} + \text{LiClO}_4(aq) \rightarrow \text{H}_2\text{O}(l) + \underset{\text{gas}}{\text{CO}_2(g)} + \text{LiClO}_4(aq)$

Balanced reaction: $2\,\text{HClO}_4(aq) + \text{Li}_2\text{CO}_3(aq) \rightarrow \text{H}_2\text{O}(l) + \text{CO}_2(g) + 2\,\text{LiClO}_4(aq)$

Oxidation and Reduction

8.67 (a) Ag. The oxidation state of $\text{Ag} = 0$. The oxidation state of an atom in a free element is 0.

(b) Ag^+. The oxidation state of $\text{Ag}^+ = +1$. The oxidation state of a monatomic ion is equal to its charge.

(c) CaF_2. The oxidation state of $\text{Ca} = +2$, and the oxidation state of $\text{F} = -1$. The oxidation state of a group 2A metal always has an oxidation state of $+2$, and the oxidation of F is -1 because the sum of the oxidation states in a neutral formula unit $= 0$.

(d) H_2S. The oxidation state of $\text{H} = +1$, and the oxidation state of $\text{S} = -2$. The oxidation state of H when listed first is $+1$, and the oxidation state of S is -2 because S is in group 6A and the sum of the oxidation states in a neutral molecular unit $= 0$.

(e) CO_3^{2-}. The oxidation state of $\text{C} = +4$, and the oxidation state of $\text{O} = -2$. The oxidation state of O is normally -2, and the oxidation state of C is deduced from the formula because the sum of the oxidation states must equal the charge on the ion. (C ox state) $+3$(O ox state) $= -2$; (C ox state) $+ 3(-2) = -2$, so C ox state $= +4$.

(f) CrO_4^{2-}. The oxidation state of $\text{Cr} = +6$, and the oxidation state of $\text{O} = -2$. The oxidation state of O is normally -2, and the oxidation state of Cr is deduced from the formula because the sum of the oxidation states must equal the charge on the ion. (Cr ox state) $+ 4$(O ox state) $= -2$; (Cr ox state) $+ 4(-2) = -2$, so Cr ox state $= +6$.

8.69 (a) CrO. The oxidation state of $\text{Cr} = +2$, and the oxidation state of $\text{O} = -2$. The oxidation state of O is normally -2, and the oxidation state of Cr is deduced from the formula because the sum of the oxidation states must $= 0$. (Cr ox state) $+$ (O ox state) $= 0$; (Cr ox state) $+ (-2) = 0$, so Cr $= +2$.

(b) CrO_3. The oxidation state of $\text{Cr} = +6$, and the oxidation state of $\text{O} = -2$. The oxidation state of O is normally -2, and the oxidation state of Cr is deduced from the formula because the sum of the oxidation states must $= 0$. (Cr ox state) $+ 3$(O ox state) $= 0$; (Cr ox state) $+ 3\,(-2) = 0$, so Cr $= +6$.

(c) Cr_2O_3. The oxidation state of $\text{Cr} = +3$, and the oxidation state of $\text{O} = -2$. The oxidation state of O is normally -2, and the oxidation state of Cr is deduced from the formula because the sum of the oxidation states must $= 0$. 2(Cr ox state) $+ 3$(O ox state) $= 0$; 2(Cr ox state) $+ 3(-2) = 0$, so Cr $= +3$.

8.71 (a)
$$4 \, Li(s) + O_2(g) \rightarrow 2 \, Li_2O(s)$$
Oxidation states; 0   0   +1 −2

This is a redox reaction because Li increases in oxidation number (oxidation) and O decreases in number (reduction). O_2 is the oxidizing agent, and Li is the reducing agent.

(b)
$$Mg(s) + Fe^{2+}(aq) \rightarrow Mg^{2+}(aq) + Fe(s)$$
Oxidation states; 0   +2   +2   0

This is a redox reaction because Mg increases in oxidation number (oxidation) and Fe decreases in number (reduction). Fe^{2+} is the oxidizing agent, and Mg is the reducing agent.

(c)
$$Pb(NO_3)_2(aq) + Na_2SO_4(aq) \rightarrow PbSO_4(s) + 2 \, NaNO_3(aq)$$
Oxidation states; +2 +5 −2   +1 +6 −2   +2 +6 −2   +1 +5 −2

This is not a redox reaction because none of the atoms undergoes a change in oxidation number.

(d)
$$HBr(aq) + KOH(aq) \rightarrow H_2O(l) + KBr(aq)$$
Oxidation states; +1 −1   +1 −2 +1   +1 −2   +1 −1

This is not a redox reaction because none of the atoms undergoes a change in oxidation number.

8.73 **Given:** overall reactions **Find:** spontaneity in forward direction

Conceptual Plan: Separate the overall reaction into two half-cell reactions and add electrons as needed to balance reactions. Look up half-reactions in Table 8.4 (Activity Series of Metals). If the reaction that is written as an oxidation is above the reverse of the reaction written as a reduction, the reaction is spontaneous.

Solution:

(a) $Ni(s) + Zn^{2+}(aq) \rightarrow Ni^{2+}(aq) + Zn(s)$ separates to $Ni(s) \rightarrow Ni^{2+}(aq)$ and $Zn^{2+}(aq) \rightarrow Zn(s)$. Add electrons. $Ni(s) \rightarrow Ni^{2+}(aq) + 2 \, e^-$ and $Zn^{2+}(aq) + 2 \, e^- \rightarrow Zn(s)$. Look up the half-reactions in Table 8.4. Since the oxidation of Zn is above the oxidation of Ni, the reaction is not spontaneous.

(b) $Ni(s) + Pb^{2+}(aq) \rightarrow Ni^{2+}(aq) + Pb(s)$ separates to $Ni(s) \rightarrow Ni^{2+}(aq)$ and $Pb^{2+}(aq) \rightarrow Pb(s)$. Add electrons. $Ni(s) \rightarrow Ni^{2+}(aq) + 2 \, e^-$ and $Pb^{2+}(aq) + 2 \, e^- \rightarrow Pb(s)$. Look up the half-reactions in Table 8.4. Since the oxidation of Ni is above the oxidation of Pb, the reaction is spontaneous.

(c) $Al(s) + 3 \, Ag^+(aq) \rightarrow Al^{3+}(aq) + 3 \, Ag(s)$ separates to $Al(s) \rightarrow Al^{3+}(aq)$ and $3 \, Ag^+(aq) \rightarrow 3 \, Ag(s)$. Add electrons. $Al(s) \rightarrow Al^{3+}(aq) + 3 \, e^-$ and $3 \, Ag^+(aq) + 3 \, e^- \rightarrow 3 \, Ag(s)$. Simplify the Ag reaction to $Ag^+(aq) + e^- \rightarrow Ag(s)$. Look up the half-reactions in Table 8.4. Since the oxidation of Al is above the oxidation of Ag, the reaction is spontaneous.

(d) $Pb(s) + Mn^{2+}(aq) \rightarrow Pb^{2+}(aq) + Mn(s)$ separates to $Pb(s) \rightarrow Pb^{2+}(aq)$ and $Mn^{2+}(aq) \rightarrow Mn(s)$. Add electrons. $Pb(s) \rightarrow Pb^{2+}(aq) + 2 \, e^-$ and $Mn^{2+}(aq) + 2 \, e^- \rightarrow Mn(s)$. Look up the half-reactions in Table 8.4. Since the oxidation of Mn is above the oxidation of Pb, the reaction is not spontaneous.

8.75 For a metal to be able to reduce an ion, it must be below it in Table 8.4. So we need a metal that is above Ni^{2+}. Lithium, potassium, calcium, sodium, magnesium, aluminum, manganese, zinc, chromium, and iron meet this criterion.

8.77 For a metal to be able to reduce an ion, it must be below it in Table 8.4. So we need a metal that is below Na^+ but above Al^{3+}. Manganese is the only one in the table that meets those criteria.

Cumulative Problems

8.79 **Given:** Assume 100.0 g solution, 20.0 g $C_2H_6O_2$; density of solution = 1.03 g/mL **Find:** M of solution

Conceptual Plan: g $C_2H_6O_2$ → mol $C_2H_6O_2$ and g solution → mL solution → L solution

$$\frac{1 \, mol \, C_2H_6O_2}{62.07 \, g \, C_2H_6O_2} \qquad \frac{1.00 \, mL}{1.03 \, g} \qquad \frac{1 \, L}{1000 \, mL}$$

then M $C_2H_6O_2$

$$M = \frac{mol \, C_2H_6O_2}{L \, solution}$$

Solution:

$$20.0 \, \cancel{g \, C_2H_6O_2} \times \frac{1 \, mol \, C_2H_6O_2}{62.07 \, \cancel{g \, C_2H_6O_2}} = 0.3222 \, mol \, C_2H_6O_2$$

$$100.0 \, \cancel{g \, solution} \times \frac{1.00 \, \cancel{mL \, solution}}{1.03 \, \cancel{g \, solution}} \times \frac{1 \, L}{1000 \, \cancel{mL}} = 0.09709 \, L \, H_2O$$

$$M = \frac{0.3222 \text{ mol } C_2H_6O_2}{0.09709 \text{ L}} = 3.32 \text{ M } C_2H_6O_2$$

Check: The units of the answer (M) are correct. The magnitude of the answer is reasonable because the concentration of solutions is usually between 0 M and 18 M.

8.81 **Given:** 2.5 g $NaHCO_3$ **Find:** g HCl
Conceptual Plan: g $NaHCO_3$ → mol $NaHCO_3$ → mol HCl → g HCl

$$\frac{1 \text{ mol } NaHCO_3}{84.02 \text{ g } NaHCO_3} \qquad \frac{1 \text{ mol HCl}}{1 \text{ mol } NaHCO_3} \qquad \frac{36.46 \text{ g HCl}}{1 \text{ mol HCl}}$$

Solution: $HCl(aq) + NaHCO_3(aq) \rightarrow H_2O(l) + CO_2(g) + NaCl(aq)$

$$2.5 \text{ g } \cancel{NaHCO_3} \times \frac{1 \text{ mol } \cancel{NaHCO_3}}{84.02 \text{ g } \cancel{NaHCO_3}} \times \frac{1 \text{ mol } \cancel{HCl}}{1 \text{ mol } \cancel{NaHCO_3}} \times \frac{36.46 \text{ g HCl}}{1 \text{ mol } \cancel{HCl}} = 1.1 \text{ g HCl}$$

Check: The units of the answer (g HCl) are correct. The magnitude of the answer is reasonable because the molar mass of HCl is less than the molar mass of $NaHCO_3$.

8.83 **Given:** Beaker containing 4 ions H^+, 4 ions Cl^- **Find:** which NaOH beaker will just neutralize HCl beaker
Conceptual Plan: molecules H^+ → molecules OH^- then compare to four beakers

$$\frac{1 \text{ ion } OH^-}{1 \text{ ion } H^+}$$

Solution: $HCl(aq) + NaOH(aq) \rightarrow NaCl(aq) + H_2O(l)$
Net Ionic: $H^+(aq) + OH^-(aq) \rightarrow H_2O(l)$

$$4 \text{ ions } \cancel{H^+} \times \frac{1 \text{ ion } OH^-}{1 \text{ ion } \cancel{H^+}} = 4 \text{ ions } OH^-$$

Beaker (a) contains 2 ions OH^-; beaker (b) contains 4 ions OH^-; beaker (c) contains 5 ions OH^-; beaker (d) contains 8 ions OH^-. Beaker (b) will completely neutralize the HCl beaker with no excess.

Check: The answer is correct because it will completely neutralize the H^+ with no excess OH^-.

8.85 (a) Skeletal reaction: $\quad HCl(aq) + Hg_2(NO_3)_2(aq) \rightarrow Hg_2Cl_2(s) + HNO_3(aq)$
Balance Cl: $\quad 2\,HCl(aq) + Hg_2(NO_3)_2(aq) \rightarrow Hg_2Cl_2(s) + 2\,HNO_3(aq)$
(b) Skeletal reaction: $\quad KHSO_3(aq) + HNO_3(aq) \rightarrow H_2O(l) + SO_2(g) + KNO_3(aq)$
Balanced reaction: $\quad KHSO_3(aq) + HNO_3(aq) \rightarrow H_2O(l) + SO_2(g) + KNO_3(aq)$
(c) Skeletal reaction: $\quad NH_4Cl(aq) + Pb(NO_3)_2(aq) \rightarrow PbCl_2(s) + NH_4NO_3(aq)$
Balance Cl: $\quad 2\,NH_4Cl(aq) + Pb(NO_3)_2(aq) \rightarrow PbCl_2(s) + NH_4NO_3(aq)$
Balance N: $\quad 2\,NH_4Cl(aq) + Pb(NO_3)_2(aq) \rightarrow PbCl_2(s) + 2\,NH_4NO_3(aq)$
(d) Skeletal reaction: $\quad NH_4Cl(aq) + Ca(OH)_2(aq) \rightarrow NH_3(g) + H_2O(l) + CaCl_2(aq)$
Balance Cl: $\quad 2\,NH_4Cl(aq) + Ca(OH)_2(aq) \rightarrow NH_3(g) + H_2O(l) + CaCl_2(aq)$
Balance N: $\quad 2\,NH_4Cl(aq) + Ca(OH)_2(aq) \rightarrow 2\,NH_3(g) + H_2O(l) + CaCl_2(aq)$
Balance H: $\quad 2\,NH_4Cl(aq) + Ca(OH)_2(aq) \rightarrow 2\,NH_3(g) + 2\,H_2O(l) + CaCl_2(aq)$

8.87 **Given:** 1.5 L solution; 0.050 M $CaCl_2$; 0.085 M $Mg(NO_3)_2$ **Find:** g Na_3PO_4
Conceptual Plan: $V, M\, CaCl_2$ → mol $CaCl_2$ and $V, M\, Mg(NO_3)_2$ → mol $Mg(NO_3)_2$

$$V \times M = \text{mol} \qquad\qquad V \times M = \text{mol}$$

then (mol $CaCl_2$ + mol $Mg(NO_3)_2$) → Na_3PO_4 → g Na_3PO_4

$$\frac{2 \text{ mol } Na_3PO_4}{3 \text{ mol } (CaCl_2 + Mg(NO_3)_2)} \qquad \frac{163.97 \text{ g } Na_3PO_4}{1 \text{ mol } Na_3PO_4}$$

Solution: $3\,CaCl_2(aq) + 2\,Na_3PO_4(aq) \rightarrow Ca_3(PO_4)_2(s) + 6\,NaCl(aq)$
$3\,Mg(NO_3)_2(aq) + 2\,Na_3PO_4(aq) \rightarrow Mg_3(PO_4)_2(s) + 6\,NaCl(aq)$
$1.5 \text{ L} \times 0.050 \text{ M } CaCl_2 = 0.075 \text{ mol } CaCl_2$
$1.5 \text{ L} \times 0.085 \text{ M } Mg(NO_3)_2 = 0.1275 \text{ mol } Mg(NO_3)_2$

$$0.2025 \text{ mol } \cancel{CaCl_2 \text{ and } Mg(NO_3)_2} \times \frac{2 \text{ mol } \cancel{Na_3PO_4}}{3 \text{ mol } \cancel{CaCl_2 \text{ and } Mg(NO_3)_2}} \times \frac{163.97 \text{ g } Na_3PO_4}{1 \text{ mol } \cancel{Na_3PO_4}} = 22 \text{ g mol } Na_3PO_4$$

Check: The units of the answer (g Na_3PO_4) are correct. The magnitude of the answer is reasonable because it is needed to remove both the Ca and Mg ions.

8.89 **Given:** 1.0 L; 0.10 M OH^- **Find:** g Ba

Conceptual Plan: $V, M \rightarrow$ mol $OH^- \rightarrow$ mol $Ba(OH)_2 \rightarrow$ mol BaO $\rightarrow$ mol Ba $\rightarrow$ g Ba

$$V \times M = mol \quad \frac{1\ mol\ Ba(OH)_2}{2\ mol\ OH^-} \quad \frac{1\ mol\ BaO}{1\ mol\ Ba(OH)_2} \quad \frac{1\ mol\ Ba}{1\ mol\ BaO} \quad \frac{137.3\ g\ Ba}{1\ mol\ Ba}$$

Solution: $Ba(s) + H_2O(l) \rightarrow BaO(s); BaO(s) + H_2O(l) \rightarrow Ba(OH)_2(aq)$

$$1.0\ L \times \frac{0.10\ mol\ OH^-}{1\ L} \times \frac{1\ mol\ Ba(OH)_2}{2\ mol\ OH^-} \times \frac{1\ mol\ BaO}{1\ mol\ Ba(OH)_2} \times \frac{1\ mol\ Ba}{1\ mol\ BaO} \times \frac{137.3\ g\ Ba}{1\ mol\ Ba} = 6.9\ g\ Ba$$

Check: The units of the answer (g Ba) are correct. The magnitude is reasonable because the molar mass of Ba is large and there are 2 moles hydroxide per mole Ba.

8.91 **Given:** 0.110 M HCl; 1.52 g $Al(OH)_3$ **Find:** volume HCl needed to neutralize

Conceptual Plan: g $Al(OH)_3 \rightarrow$ mol $Al(OH)_3 \rightarrow$ mol HCl $\rightarrow$ vol HCl

$$\frac{1\ mol\ Al(OH)_3}{78.00\ g\ Al(OH)_3} \quad \frac{3\ mol\ HCl}{1\ mol\ Al(OH)_3} \quad \frac{mol\ HCl}{M\ HCl}$$

Solution: $3\ HCl(aq) + Al(OH)_3(aq) \rightarrow 3\ H_2O(l) + AlCl_3(aq)$

$$1.52\ g\ Al(OH)_3 \times \frac{1\ mol\ Al(OH)_3}{78.00\ g\ Al(OH)_3} \times \frac{3\ mol\ HCl}{1\ mol\ Al(OH)_3} \times \frac{1\ L\ HCl}{0.110\ mol\ HCl} = 0.531\ L\ HCl$$

Check: The units of the answer (L HCl) are correct. The magnitude (0.5 L) of the answer is reasonable because we have much less than one mole of $Al(OH)_3$ and the mole ratio of HCl to $Al(OH)_3$ is 3:1.

8.93 **Given:** 24.5 g Au; 24.5 g BrF_3; 24.5 g KF **Find:** g $KAuF_4$

Conceptual Plan: g Au $\rightarrow$ mol Au $\rightarrow$ mol $KAuF_4$

$$\frac{1\ mol\ Au}{196.97\ g\ Au} \quad \frac{2\ mol\ KAuF_4}{2\ mol\ Au}$$

g $BrF_3 \rightarrow$ mol $BrF_3 \rightarrow$ mol $KAuF_4$ $\rightarrow$ **smaller mol amount determines limiting reactant**

$$\frac{1\ mol\ BrF_3}{136.90\ g\ BrF_3} \quad \frac{2\ mol\ KAuF_4}{2\ mol\ BrF_3}$$

g KF $\rightarrow$ mol KF $\rightarrow$ mol $KAuF_4$

$$\frac{1\ mol\ KF}{58.10\ g\ KF} \quad \frac{2\ mol\ KAuF_4}{2\ mol\ KF}$$

then mol $KAuF_4 \rightarrow$ g $KAuF_4$

$$\frac{312.07\ g\ KAuF_4}{1\ mol\ KAuF_4}$$

$$2\ Au(s) + 2\ BrF_3(l) + 2\ KF(s) \rightarrow Br_2(l) + 2\ KAuF_4(s)$$

Oxidation states;   0   +3 −1   +1 −1   0   +1 +3 −1

This is a redox reaction because Au increases in oxidation number (oxidation) and Br decreases in number (reduction). BrF_3 is the oxidizing agent, and Au is the reducing agent.

Solution:

$$24.5\ g\ Au \times \frac{1\ mol\ Au}{196.97\ g\ Au} \times \frac{2\ mol\ KAuF_4}{2\ mol\ Au} = 0.1244\ mol\ KAuF_4$$

$$24.5\ g\ BrF_3 \times \frac{1\ mol\ BrF_3}{136.90\ g\ BrF_3} \times \frac{2\ mol\ KAuF_4}{2\ mol\ BrF_3} = 0.1790\ mol\ KAuF$$

$$24.5\ g\ KF \times \frac{1\ mol\ KF}{58.10\ g\ KF} \times \frac{2\ mol\ KAuF_4}{2\ mol\ KF} = 0.4217\ mol\ KAuF_4$$

$$0.1244\ mol\ KAuF_4 \times \frac{312.07\ g\ KAuF_4}{1\ mol\ KAuF_4} = 38.8\ g\ KAuF_4$$

Check: The units of the answer (g $KAuF_4$) are correct. The magnitude of the answer is reasonable compared to the mass of the limiting reactant Au.

Challenge Problems

8.95   **Given:** 0.100 L, 1.22 M NaI; total mass = 28.1 g   **Find:** g AgI

     Conceptual Plan: vol, M → mol NaI → mol I⁻; total mol I⁻ → mol AgI and HgI₂

     Solution: $0.100 \text{ L soln} \times \dfrac{1.22 \text{ mol NaI}}{1 \text{ L soln}} \times \dfrac{1 \text{ mol I}^-}{1 \text{ mol NaI}} = 0.122 \text{ mol I}^-$

         Let x = mol AgI   and y = mol HgI₂

         $x + 2y = 0.122 \text{ mol I}^-$ so $y = 0.061 - 0.5x$

         $\left(x \text{ mol AgI} \times \dfrac{234.77 \text{ g AgI}}{1 \text{ mol AgI}} \right) + \left(y \text{ mol HgI}_2 \times \dfrac{454.39 \text{ g HgI}_2}{1 \text{ mol HgI}_2} \right) = 28.1 \text{ g}$

         Solve the simultaneous equations and x = 0.0504 mol AgI.

         $0.0504 \text{ mol AgI} \times \dfrac{234.77 \text{ g AgI}}{1 \text{ mol AgI}} = 11.8 \text{ g AgI}$

     Check: The units of the answer (g AgI) are correct. The magnitude is reasonable because it is less than the total mass.

8.97   **Given:** 3.5×10^{-3} M Ca²⁺; 1.1×10^{-3} M Mg²⁺; 19.5 gal H₂O; 0.65 kg detergent/load   **Find:** % by mass Na₂CO₃

     Conceptual Plan: gal H₂O → L H₂O then V, M → mol Ca²⁺ and V, M → mol Mg²⁺

         $\dfrac{3.785 \text{ L}}{1 \text{ gal}}$ vol × M = mol   vol × M = mol

     then total moles ions → mol CO₃²⁻ → mol Na₂CO₃ → g Na₂CO₃ → kg Na₂CO₃ → % Na₂CO₃

         $\dfrac{1 \text{ mol CO}_3^{2-}}{1 \text{ mol ion}}$ $\dfrac{1 \text{ mol Na}_2\text{CO}_3}{1 \text{ mol CO}_3^{2-}}$ $\dfrac{106.01 \text{ g Na}_2\text{CO}_3}{1 \text{ mol Na}_2\text{CO}_3}$ $\dfrac{1 \text{ kg}}{1000 \text{ g}}$ $\dfrac{\text{kg Na}_2\text{CO}_3}{\text{kg detergent}} \times 100\%$

     Solution: $19.5 \text{ gal} \times \dfrac{3.785 \text{ L}}{1 \text{ gal}} \times \dfrac{3.5 \times 10^{-3} \text{ mol Ca}^{2+}}{1 \text{ L}} = 0.258 \text{ mol Ca}^{2+}$

         $19.5 \text{ gal} \times \dfrac{3.785 \text{ L}}{1 \text{ gal}} \times \dfrac{1.1 \times 10^{-3} \text{ mol Mg}^{2+}}{1 \text{ L}} = 0.08119 \text{ mol Mg}^{2+}$

         $0.3392 \text{ mol ions} \times \dfrac{1 \text{ mol CO}_3^{2-}}{1 \text{ mol ions}} \times \dfrac{1 \text{ mol Na}_2\text{CO}_3}{1 \text{ mol CO}_3^{2-}} \times \dfrac{106.01 \text{ g Na}_2\text{CO}_3}{1 \text{ mol Na}_2\text{CO}_3} \times \dfrac{1 \text{ kg Na}_2\text{CO}_3}{1000 \text{ g Na}_2\text{CO}_3} = 0.03596 \text{ kg Na}_2\text{CO}_3$

         $\dfrac{0.03596 \text{ kg Na}_2\text{CO}_3}{0.65 \text{ kg detergent}} \times 100\% = 5.5\% \text{ Na}_2\text{CO}_3$

     Check: The units of the answer (% Na₂CO₃) are correct. The magnitude of the answer is reasonable. The percent is less than 100%.

8.99   **Given:** Solution may contain Hg₂²⁺, Ba²⁺, and Fe²⁺   **Find:** which ions are present

     Conceptual Plan: Test the solution sequentially with KCl, K₂SO₄, and K₂CO₃ and see if precipitates form.

     Solution: Original solution + KCl yields a precipitate and solution 2. The precipitate is Hg₂Cl₂, so Hg₂²⁺ is present. Chlorides are normally soluble, but Hg₂²⁺ is an exception.

     Solution 2 with K₂SO₄ yields no precipitate, so Ba²⁺ is not present. Sulfates are normally soluble, but Ba²⁺ is an exception. Solution 2 with K₂CO₃ yields a precipitate. The precipitate is FeCO₃, so Fe²⁺ is present. All carbonates are insoluble.

     Net Ionic Equations:

         $\text{Hg}_2^{2+}(aq) + 2\text{Cl}^-(aq) \rightarrow \text{Hg}_2\text{Cl}_2(s)$

         $\text{Fe}^{2+}(aq) + \text{CO}_3^{2-}(aq) \rightarrow \text{FeCO}_3(s)$

     Check: The answer is reasonable because two different precipitates formed and all of the Hg₂²⁺ was removed before the carbonate was added.

Conceptual Problems

8.101   **Given:** A₂X soluble; BY₂ soluble; AY insoluble; BX soluble; 2 molecules A₂X mixed with 2 molecules BY₂

     Find: a molecular representation of the mixture; an equation for the reaction

Solution:

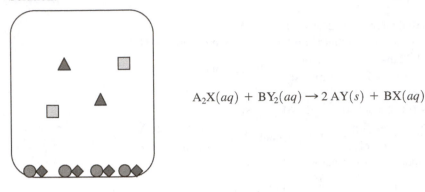

$$A_2X(aq) + BY_2(aq) \rightarrow 2\,AY(s) + BX(aq)$$

8.103   The charge of an atomic ion (O^{2-}) is associated with the individual atom (O) and is equal to its oxidation state. For a molecule, the oxidation states of the individual atoms must sum to 0 since molecules have no charge, but they are assigned to the individual atoms themselves. The oxidation states of atoms are not real physical properties, but an imposed electron bookkeeping scheme.

Questions for Group Work

8.105   Examples of soluble compounds are: NaCl, KBr, AgNO₃, **NH₄OH**, LiF, **Na₂S**.
Examples of insoluble compounds are: **AgBr**, Fe(OH)₃, ZnS, **BaSO₄**, MgCO₃, **PbCl₂**.
Exceptions to one of the rules are listed in **bold**.

8.107   When a substance is oxidized it loses electrons, and there is an increase in oxidation state. When a substance is reduced it gains electrons, and there is a reduction in oxidation state. In this example, $Zn(s)$ is being oxidized to $Zn^{2+}(aq)$, while $Fe^{2+}(aq)$ is being reduced to $Fe(s)$. Electrons would move from the student representing the $Zn(s)$ to the student representing the $Fe^{2+}(aq)$. A substance that causes the oxidation of another substance is called an oxidizing agent – $Fe^{2+}(aq)$. A substance that causes the reduction of another substance is called a reducing agent – $Zn(s)$.

Data Interpretation and Analysis

8.108   (a)   The average of the first draw samples is 10.3 ppb Pb. The average of the 45-second flush samples is 3.81 ppb Pb. The average of the 2-minute flush samples is 1.69 ppb Pb.

      (c)   Yes, if the water provider used the first draw samples to monitor the water quality, they would have been required to notify the customers and take steps to control the corrosiveness of the water. Three of the first draw samples (20%) tested above 15 ppb Pb (#7 at 40.63 ppb Pb, #12 at 24.37 ppb Pb, and #15 at 29.59 ppb Pb).

          No, if the water provider used the 2-minute flush samples to monitor the water quality, they would not have been required to notify the customers or take steps to control the corrosiveness of the water. None of the 2-minute flush samples tested above 15 ppb Pb (the highest value was sample #2 at 10.77 ppb Pb).

          The typical resident probably does not flush for as long as 2 minutes. A flush time of 0–30 seconds is more likely.

9 Thermochemistry

Review Questions

9.1   Thermochemistry is the study of the relationship between chemistry and energy. It is important because energy and its uses are critical to society. It is important to understand how much energy is required or released in a process.

9.3   Kinetic energy is associated with the motion of an object. Potential energy is associated with the position or composition of an object. Examples of kinetic energy are a moving billiard ball, movement of gas molecules, and a raging river. Examples of potential energy are a billiard ball raised above the surface of a billiard table, a compressed spring, and bonds in molecules.

9.5   According to the first law, a device that would continually produce energy with no energy input, sometimes known as a perpetual motion machine, cannot exist because the best we can do with energy is to break even.

9.7   The internal energy (E) of a system is the sum of the kinetic and potential energies of all of the particles that compose the system. Internal energy is a state function.

9.9   If the reactants have a lower internal energy than the products, ΔE_{sys} is positive and energy flows into the system from the surroundings.

9.11   The internal energy (E) of a system is the sum of the kinetic and potential energies of all of the particles that compose the system. The change in the internal energy of the system (ΔE) must be the sum of the heat transferred (q) and the work done (w): $\Delta E = q + w$.

9.13   The heat capacity of a system is usually defined as the quantity of heat required to change its temperature by 1 °C. Heat capacity (C) is a measure of the system's ability to hold thermal energy without undergoing a large change in temperature. The difference between heat capacity (C) and specific heat capacity (C_s) is that the specific heat capacity is the amount of heat required to raise the temperature of *1 gram* of the substance by 1 °C.

9.15   When two objects of different temperatures come in direct contact, heat flows from the higher temperature object to the lower temperature object. The amount of heat lost by the warmer object is equal to the amount of heat gained by the cooler object. The warmer object's temperature will drop and the cooler object's temperature will rise until they reach the same temperature. The magnitude of these temperature changes depends on the mass and heat capacities of the two objects.

9.17   In calorimetry, the thermal energy exchanged between the reaction (defined as the system) and the surroundings is measured by observing the change in temperature of the surroundings. A bomb calorimeter is used to measure the ΔE_{rxn} for combustion reactions. The calorimeter includes a tight-fitting, sealed container that forces the reaction to occur at constant volume. A coffee-cup calorimeter is used to measure ΔH_{rxn} for many aqueous reactions. The calorimeter consists of two Styrofoam® coffee cups, one inserted into the other, to provide insulation from the laboratory environment. Because the reaction happens under conditions of constant pressure (open to the atmosphere), $q_{rxn} = q_p = \Delta H_{rxn}$.

9.19   An endothermic reaction has a positive ΔH and absorbs heat from the surroundings. An endothermic reaction feels cold to the touch. An exothermic reaction has a negative ΔH and gives off heat to the surroundings. An exothermic reaction feels warm to the touch.

9.21   The internal energy of a chemical system is the sum of its kinetic energy and its potential energy. It is this potential energy that absorbs the energy in an endothermic chemical reaction. In an endothermic reaction, as some bonds break and others form, the protons and electrons go from an arrangement of lower potential energy to one of higher potential energy, absorbing thermal energy in the process. This absorption of thermal energy reduces the kinetic energy of the system. This is detected as a drop in temperature.

9.23   (a) If a reaction is multiplied by a factor, ΔH is multiplied by the same factor.
          (b) If a reaction is reversed, the sign of ΔH is reversed.
          The relationships hold because H is a state function. Twice as much energy is contained in twice the quantity of reactants or products. If the reaction is reversed, the final and initial states have been switched and the direction of heat flow is reversed.

9.25   The standard state is defined as follows: for a gas, the pure gas at a pressure of exactly 1 atm; for a liquid or solid, the pure substance in its most stable form at a pressure of 1 atm and the temperature of interest (often taken to be 25 °C); and for a substance in solution, a concentration of exactly 1 M. The standard enthalpy change $(\Delta H°)$ is the change in enthalpy for a process when all reactants and products are in their standard states. The superscript degree sign indicates standard states.

9.27   A reaction is exothermic when weak bonds break and strong bonds form. A reaction is endothermic when strong bonds break and weak bonds form.

9.29   To calculate $\Delta H°_{rxn}$, subtract the heats of formation of the reactants multiplied by their stoichiometric coefficients from the heats of formation of the products multiplied by their stoichiometric coefficients. In the form of an equation:

$$\Delta H°_{rxn} = \sum n_p \Delta H°_f(\text{products}) - \sum n_r \Delta H°_f(\text{reactants})$$

Problem by Topic

Internal Energy, Heat, and Work

9.31   (d) $\Delta E_{sys} = -\Delta E_{surr}$ If energy change of the system is negative, energy is being transferred from the system to the surroundings, decreasing the energy of the system and increasing the energy of the surroundings. The amount of energy lost by the system must go somewhere; so the amount gained by the surroundings is equal and opposite to that lost by the system.

9.33   (a) The energy exchange is primarily heat because the skin (part of the surroundings) is cooled. There is a small expansion (work) because water is being converted from a liquid to a gas. The sign of ΔE_{sys} is positive because the surroundings cool.
          (b) The energy exchange is primarily work. The sign of ΔE_{sys} is negative because the system is expanding (doing work on the surroundings).
          (c) The energy exchange is primarily heat. The sign of ΔE_{sys} is positive because the system is being heated by the flame.

9.35   **Given:** 622 kJ heat released; 105 kJ work done on surroundings   **Find:** ΔE_{sys}
          Conceptual Plan: Interpret language to determine the sign of the two terms then $q, w \rightarrow \Delta E_{sys}$
$$\Delta E = q + w$$

          Solution: Because heat is released from the system to the surroundings, $q = -622$ kJ; because the system is doing work on the surroundings, $w = -105$ kJ. $\Delta E = q + w = -622$ kJ $- 105$ kJ $= -727$ kJ $= -7.27 \times 10^2$ kJ.

          Check: The units (kJ) are correct. The magnitude of the answer (-730) makes physical sense because both terms are negative.

9.37   **Given:** 655 J heat absorbed; 344 J work done on surroundings   **Find:** ΔE_{sys}
          Conceptual Plan: Interpret language to determine the sign of the two terms then $q, w \rightarrow \Delta E_{sys}$
$$\Delta E = q + w$$

Solution: Because heat is absorbed by the system, $q = +655$ J; because the system is doing work on the surroundings, $w = -344$ J. $\Delta E = q + w = 655\,\text{J} - 344\,\text{J} = 311\,\text{J}$.

Check: The units (J) are correct. The magnitude of the answer $(+300)$ makes physical sense because the heat term dominates over the work term.

Heat, Heat Capacity, and Work

9.39   Cooler A had more ice after 3 hours because most of the ice in cooler B was melted to cool the soft drinks that started at room temperature. In cooler A, the drinks were already cold; so the ice only needed to maintain this cool temperature.

9.41   **Given:** 1.50 L water, $T_i = 25.0\,°\text{C}$, $T_f = 100.0\,°\text{C}$, $d = 1.0$ g/mL   **Find:** q
Conceptual Plan: $\text{L} \rightarrow \text{mL} \rightarrow \text{g}$ and pull C_s from Table 9.2 and $T_i, T_f \rightarrow \Delta T$ then $m, C_s, \Delta T \rightarrow q$

$$\frac{1000\ \text{mL}}{1\ \text{L}} \quad \frac{1.0\ \text{g}}{1.0\ \text{mL}} \qquad 4.18\,\frac{\text{J}}{\text{g}\cdot°\text{C}} \qquad\qquad m\Delta T = T_f - T_i \qquad\qquad q = mC_s\Delta T$$

Solution: $1.50\ \cancel{\text{L}} \times \dfrac{1000\ \cancel{\text{mL}}}{1\ \cancel{\text{L}}} \times \dfrac{1.0\ \text{g}}{1.0\ \cancel{\text{mL}}} = 1500$ g and $\Delta T = T_f - T_i = 100.0\,°\text{C} - 25.0\,°\text{C} = 75.0\,°\text{C}$

then $q = mC_s\Delta T = 1500\ \cancel{\text{g}} \times 4.18\dfrac{\text{J}}{\cancel{\text{g}}\cdot\cancel{°\text{C}}} \times 75.0\,\cancel{°\text{C}} = 4.7 \times 10^5$ J

Check: The units (J) are correct. The magnitude of the answer (10^5) makes physical sense because there are a large mass, a significant temperature change, and a high specific heat capacity material.

9.43   (a)   **Given:** 25 g gold; $T_i = 27.0\,°\text{C}$; $q = 2.35$ kJ   **Find:** T_f
Conceptual Plan: $\text{kJ} \rightarrow \text{J}$ and pull C_s from Table 9.2 then $m, C_s, q \rightarrow \Delta T$ then $T_i, \Delta T \rightarrow T_f$

$$\frac{1000\ \text{J}}{1\ \text{kJ}} \qquad\qquad 0.128\,\frac{\text{J}}{\text{g}\cdot°\text{C}} \qquad\qquad q = mC_s\Delta T \qquad\qquad \Delta T = T_f - T_i$$

Solution: $2.35\ \cancel{\text{kJ}} \times \dfrac{1000\ \text{J}}{1\ \cancel{\text{kJ}}} = 2350$ J then $q = mC_s\Delta T$.   Rearrange to solve for ΔT.

$\Delta T = \dfrac{q}{mC_s} = \dfrac{2350\ \text{J}}{25\ \cancel{\text{g}} \times 0.128\dfrac{\text{J}}{\cancel{\text{g}}\cdot°\text{C}}} = 734.375\,°\text{C}$ finally $\Delta T = T_f - T_i$.   Rearrange to solve for T_f.

$T_f = \Delta T + T_i = 734.375\,°\text{C} + 27.0\,°\text{C} = 760\,°\text{C}$

Check: The units $(°\text{C})$ are correct. The magnitude of the answer (760) makes physical sense because such a large amount of heat is absorbed and there are a small mass and specific heat capacity. The temperature change should be very large.

(b)   **Given:** 25 g silver; $T_i = 27.0\,°\text{C}$; $q = 2.35$ kJ   **Find:** T_f
Conceptual Plan: $\text{kJ} \rightarrow \text{J}$ and pull C_s from Table 9.2 then $m, C_s, q \rightarrow \Delta T$ then $T_i, \Delta T \rightarrow T_f$

$$\frac{1000\ \text{J}}{1\ \text{kJ}} \qquad\qquad 0.235\,\frac{\text{J}}{\text{g}\cdot°\text{C}} \qquad\qquad q = mC_s\Delta T \qquad \Delta T = T_f - T_i$$

Solution: $2.35\ \cancel{\text{kJ}} \times \dfrac{1000\ \text{J}}{1\ \cancel{\text{kJ}}} = 2350$ J then $q = mC_s\Delta T$.   Rearrange to solve for ΔT.

$\Delta T = \dfrac{q}{mC_s} = \dfrac{2350\ \text{J}}{25\ \cancel{\text{g}} \times 0.235\dfrac{\text{J}}{\cancel{\text{g}}\cdot°\text{C}}} = 400\,°\text{C}$ finally $\Delta T = T_f - T_i$.   Rearrange to solve for T_f.

$T_f = \Delta T + T_i = 400\,°\text{C} + 27.0\,°\text{C} = 430\,°\text{C}$

Check: The units $(°\text{C})$ are correct. The magnitude of the answer (430) makes physical sense because such a large amount of heat is absorbed and there are a small mass and specific heat capacity. The temperature change should be very large. The temperature change should be less than that of the gold because the specific heat capacity is greater.

(c)   **Given:** 25 g aluminum; $T_i = 27.0\,°\text{C}$; $q = 2.35$ kJ   **Find:** T_f
Conceptual Plan: $\text{kJ} \rightarrow \text{J}$ and pull C_s from Table 9.2 then $m, C_s, q \rightarrow \Delta T$ then $T_i, \Delta T \rightarrow T_f$

$$\frac{1000\ \text{J}}{1\ \text{kJ}} \qquad\qquad 0.903\,\frac{\text{J}}{\text{g}\cdot°\text{C}} \qquad\qquad q = mC_s\Delta T \qquad \Delta T = T_f - T_i$$

Solution: $2.35 \text{ kJ} \times \dfrac{1000 \text{ J}}{1 \text{ kJ}} = 2350 \text{ J}$ then $q = mC_s\Delta T$.   Rearrange to solve for ΔT.

$$\Delta T = \frac{q}{mC_s} = \frac{2350 \text{ J}}{25 \text{ g} \times 0.903\dfrac{\text{J}}{\text{g} \cdot {}^\circ\text{C}}} = 104.10 \,{}^\circ\text{C} \text{ finally } \Delta T = T_f - T_i. \text{ Rearrange to solve for } T_f.$$

$T_f = \Delta T + T_i = 104.10\,{}^\circ\text{C} + 27.0\,{}^\circ\text{C} = 130\,{}^\circ\text{C}$

Check: The units (°C) are correct. The magnitude of the answer (130) makes physical sense because such a large amount of heat is absorbed and there is such a small mass. The temperature change should be less than that of the silver because the specific heat capacity is greater.

(d)   **Given:** 25 g water; $T_i = 27.0\,{}^\circ\text{C}$; $q = 2.35 \text{ kJ}$ **Find:** T_f
Conceptual Plan: kJ $\rightarrow$ J and pull C_s from Table 9.2 then $m, C_s, q \rightarrow \Delta T$ then $T_i, \Delta T \rightarrow T_f$

$$\frac{1000 \text{ J}}{1 \text{ kJ}} \qquad\qquad 4.18\frac{\text{J}}{\text{g} \cdot {}^\circ\text{C}} \qquad\qquad q = mC_s\Delta T \qquad\qquad \Delta T = T_f - T_i$$

Solution: $2.35 \text{ kJ} \times \dfrac{1000 \text{ J}}{1 \text{ kJ}} = 2350 \text{ J}$ then $q = mC_s\Delta T$.   Rearrange to solve for ΔT.

$$\Delta T = \frac{q}{mC_s} = \frac{2350 \text{ J}}{25 \text{ g} \times 4.18\dfrac{\text{J}}{\text{g} \cdot {}^\circ\text{C}}} = 22.488 \,{}^\circ\text{C} \text{ finally } \Delta T = T_f - T_i. \text{ Rearrange to solve for } T_f.$$

$T_f = \Delta T + T_i = 22.488\,{}^\circ\text{C} + 27.0\,{}^\circ\text{C} = 49\,{}^\circ\text{C}$

Check: The units (°C) are correct. The magnitude of the answer (49) makes physical sense because such a large amount of heat is absorbed and there is such a small mass. The temperature change should be less than that of the aluminum because the specific heat capacity is greater.

9.45   **Given:** $V_i = 0.0 \text{ L}$; $V_f = 2.5 \text{ L}$; $P = 1.1 \text{ atm}$ **Find:** w (J)
Conceptual Plan: $V_i, V_f \rightarrow \Delta V$ then $P, \Delta V \rightarrow w$ (L atm) $\rightarrow w$ (J)

$$\Delta V = V_f - V_i \qquad\qquad w = -P\Delta V \qquad\qquad \frac{101.3 \text{ J}}{1 \text{ L} \cdot \text{atm}}$$

Solution: $\Delta V = V_f - V_i = 2.5 \text{ L} - 0.0 \text{ L} = 2.5 \text{ L}$ then

$$w = -P\Delta V = -1.1 \text{ atm} \times 2.5 \text{ L} \times \frac{101.3 \text{ J}}{1 \text{ L} \cdot \text{atm}} = -280 \text{ J or} -2.8 \times 10^2 \text{ J}$$

Check: The units (J) are correct. The magnitude of the answer (-280) makes physical sense because this is an expansion (negative work) and we have atmospheric pressure and a small volume of expansion.

9.47   **Given:** $q = 565 \text{ J}$ absorbed; $V_i = 0.10 \text{ L}$; $V_f = 0.85 \text{ L}$; $P = 1.0 \text{ atm}$ **Find:** ΔE_{sys}
Conceptual Plan: $V_i, V_f \rightarrow \Delta V$ and interpret language to determine the sign of the heat

$$\Delta V = V_f - V_i \qquad\qquad\qquad q = +565 \text{ J}$$

then $P, \Delta V \rightarrow w$ (L atm) $\rightarrow w$ (J) finally $q, w \rightarrow \Delta E_{sys}$

$$w = -P\Delta V \qquad \frac{101.3 \text{ J}}{1 \text{ L} \cdot \text{atm}} \qquad\qquad \Delta E = q + w$$

Solution: $\Delta V = V_f - V_i = 0.85 \text{ L} - 0.10 \text{ L} = 0.75 \text{ L}$ then

$$w = -P\Delta V = -1.0 \text{ atm} \times 0.75 \text{ L} \times \frac{101.3 \text{ J}}{1 \text{ L} \cdot \text{atm}} = -75.975 \text{ J} \qquad \Delta E = q + w = +565 \text{ J} - 75.975 \text{ J} = 489 \text{ J}$$

Check: The units (J) are correct. The magnitude of the answer (500) makes physical sense because the heat absorbed dominated the small expansion work (negative work).

Enthalpy and Thermochemical Stoichiometry

9.49   **Given:** 1 mol fuel, 3452 kJ heat produced; 11 kJ work done on surroundings   **Find:** $\Delta E_{sys}, \Delta H$
Conceptual Plan: Interpret language to determine the sign of the two terms then $q \rightarrow \Delta H$ and $q, w \rightarrow \Delta E_{sys}$

$$\Delta H = q_p \qquad\qquad\qquad \Delta E = q + w$$

Solution: Because heat is produced by the system to the surroundings, $q = -3452$ kJ; because the system is doing work on the surroundings, $w = -11$ kJ. $\Delta H = q_p = -3452$ kJ and
$\Delta E = q + w = -3452$ kJ $- 11$ kJ $= -3463$ kJ

Check: The units (kJ) are correct. The magnitude of the answer (-3500) makes physical sense because both terms are negative. We expect significant amounts of energy from fuels.

9.51   (a)   Combustion is an exothermic process; ΔH is negative.
        (b)   Evaporation requires an input of energy, so it is endothermic; ΔH is positive.
        (c)   Condensation is the reverse of evaporation, so it is exothermic; ΔH is negative.

9.53   **Given:** 177 mL acetone (C_3H_6O), $\Delta H^\circ_{rxn} = -1790$ kJ; $d = 0.788$ g/mL   **Find:** q
        Conceptual Plan: mL acetone $\rightarrow$ g acetone $\rightarrow$ mol acetone $\rightarrow$ q

$$\frac{0.788 \text{ g}}{1 \text{ mL}} \qquad \frac{1 \text{ mol}}{58.08 \text{ g}} \qquad \frac{-1790 \text{ kJ}}{1 \text{ mol}}$$

Solution: $177 \text{ mL} \times \dfrac{0.788 \text{ g}}{1 \text{ mL}} \times \dfrac{1 \text{ mol}}{58.08 \text{ g}} \times \dfrac{-1790 \text{ kJ}}{1 \text{ mol}} = -4.30 \times 10^3$ kJ or 4.30×10^3 kJ released

Check: The units (kJ) are correct. The magnitude of the answer (-10^3) makes physical sense because the enthalpy change is negative and we have more than a mole of acetone. We expect more than 1790 kJ to be released.

9.55   **Given:** 5.56 kg nitromethane (CH_3NO_2); $\Delta H^\circ_{rxn} = -1418$ kJ/2 mol nitromethane   **Find:** q
        Conceptual Plan: kg nitromethane $\rightarrow$ g nitromethane $\rightarrow$ mol nitromethane $\rightarrow$ q

$$\frac{1000 \text{ g}}{1 \text{ kg}} \qquad \frac{1 \text{ mol}}{61.04 \text{ g}} \qquad \frac{-1418 \text{ kJ}}{2 \text{ mol}}$$

Solution: $5.56 \text{ kg} \times \dfrac{1000 \text{ g}}{1 \text{ kg}} \times \dfrac{1 \text{ mol}}{61.04 \text{ g}} \times \dfrac{-1418 \text{ kJ}}{2 \text{ mol}} = -6.46 \times 10^4$ kJ or 6.46×10^4 kJ released

Check: The units (kJ) are correct. The magnitude of the answer (-10^4) makes physical sense because the enthalpy change is negative and we have more than a mole of acetone. We expect more than 1418 kJ to be released.

9.57   **Given:** pork roast, $\Delta H^\circ_{rxn} = -2217$ kJ/3 mol CO_2; q needed $= 1.6 \times 10^3$ kJ, 10% efficiency   **Find:** $m(CO_2)$
        Conceptual Plan: q used $\rightarrow$ q generated $\rightarrow$ mol CO_2 $\rightarrow$ g CO_2

$$\frac{100 \text{ kJ generated}}{10 \text{ kJ used}} \qquad \frac{3 \text{ mol}}{2217 \text{ kJ}} \qquad \frac{44.01 \text{ g}}{1 \text{ mol}}$$

Solution: $1.6 \times 10^3 \text{ kJ} \times \dfrac{100 \text{ kJ generated}}{10 \text{ kJ used}} \times \dfrac{3 \text{ mol } CO_2}{2217 \text{ kJ}} \times \dfrac{44.01 \text{ g } CO_2}{1 \text{ mol } CO_2} = 950 \text{ g } CO_2$

Check: The units (g) are correct. The magnitude of the answer ($\sim$1000) makes physical sense because the process is not very efficient and a great deal of energy is needed.

Thermal Energy Transfer

9.59   **Given:** silver block; $T_{Ag, i} = 58.5\,°C$; 100.0 g water; $T_{H_2O, i} = 24.8\,°C$; $T_f = 26.2\,°C$ **Find:** mass of silver block
        Conceptual Plan: Pull C_s values from table then H_2O : $m, C_s, T_i, T_f \rightarrow q$ and Ag : $C_s, T_i, T_f \rightarrow m$

$$\text{Ag: } 0.235\,\frac{J}{g \cdot °C} \qquad H_2O\text{: } 4.18\,\frac{J}{g \cdot °C} \qquad q = mC_s(T_f - T_i) \text{ then set } q_{Ag} = -q_{H_2O}$$

Solution: $q = mC_s(T_f - T_i)$ substitute in values and set $q_{Ag} = -q_{H_2O}$.

$$q_{Ag} = m_{Ag}C_{s, Ag}(T_f - T_{Ag, i}) = m_{Ag} \times 0.235\,\frac{J}{g \cdot °C} \times (26.2\,°C - 58.5\,°C) =$$

$$-q_{H_2O} = -m_{H_2O}C_{s, H_2O}(T_f - T_{H_2O, i}) = -100.0\,g \times 4.18\,\frac{J}{g \cdot °C} \times (26.2\,°C - 24.8\,°C)$$

Rearrange to solve for m_{Ag}.

$$m_{Ag} \times \left(-7.5\underline{9}05\,\frac{J}{g}\right) = -585.2\,J \rightarrow m_{Ag} = \frac{-585.2\,J}{-7.5\underline{9}05\,\dfrac{J}{g}} = 77.0\underline{9}64 \text{ g Ag} = 77.1 \text{ g Ag}$$

Check: The units (g) are correct. The magnitude of the answer (77) makes physical sense because the specific heat capacity of water is much greater than the specific heat capacity of silver.

9.61   **Given:** 31.1 g gold; $T_{Au,\,i} = 69.3\,°C$; 64.2 g water; $T_{H_2O,\,i} = 27.8\,°C$ **Find:** T_f
Conceptual Plan: Pull C_s values from table then $m, C_s, T_i \rightarrow T_f$

$$Au: 0.128\,\frac{J}{g\cdot°C} \quad H_2O: 4.18\,\frac{J}{g\cdot°C} \qquad q = mC_s(T_f - T_i) \text{ then set } q_{Au} = -q_{H_2O}$$

Solution: $q = mC_s(T_f - T_i)$ substitute in values and set $q_{Au} = -q_{H_2O}$.

$$q_{Au} = m_{Au}C_{s,\,Au}(T_f - T_{Au,\,i}) = 31.1\,\cancel{g} \times 0.128\,\frac{J}{\cancel{g}\cdot°C} \times (T_f - 69.3\,°C) =$$

$$-q_{H_2O} = -m_{H_2O}C_{s,\,H_2O}(T_f - T_{H_2O,\,i}) = -64.2\,\cancel{g} \times 4.18\,\frac{J}{\cancel{g}\cdot°C} \times (T_f - 27.8\,°C)$$

Rearrange to solve for T_f.

$$3.9808\,\frac{J}{°C} \times (T_f - 69.3\,°C) = -268.356\,\frac{J}{°C} \times (T_f - 27.8\,°C) \rightarrow$$

$$3.9808\,\frac{J}{°C}\,T_f - 275.8694\,J = -268.356\,\frac{J}{°C}\,T_f + 7460.2968\,J \rightarrow$$

$$268.356\,\frac{J}{°C}\,T_f + 3.9808\,\frac{J}{°C}\,T_f = 275.8694\,J + 7460.2968\,J \rightarrow 272.3368\,\frac{J}{°C}\,T_f = 7736.1662\,J \rightarrow$$

$$T_f = \frac{7736.1662\,J}{272.3368\,\frac{J}{°C}} = 28.4\,°C$$

Check: The units (°C) are correct. The magnitude of the answer (28) makes physical sense because the heat transfer is dominated by the water (larger mass and larger specific heat capacity). The final temperature should be closer to the initial temperature of water than of gold.

9.63   **Given:** 6.15 g substance A; $T_{A,\,i} = 20.5\,°C$; 25.2 g substance B; $T_{B,\,i} = 52.7\,°C$; $C_s = 1.17\,J/g\cdot°C$; $T_f = 46.7\,°C$
Find: specific heat capacity of substance A
Conceptual Plan: A: $m, T_i, T_f \rightarrow q$ B: $m, C_s, T_i, T_f \rightarrow q$ and solve for C_s

$$q = mC_s(T_f - T_i) \qquad \text{then set } q_A = -q_B$$

Solution: $q = mC_s(T_f - T_i)$ substitute in values and set $q_A = -q_B$.
$$q_A = m_A C_{s,\,A}(T_f - T_{A,\,i}) = 6.15\,g \times C_{s,\,A} \times (46.7\,°C - 20.5\,°C) =$$

$$-q_B = -m_B C_{s,\,B}(T_f - T_{B,\,i}) = -25.2\,\cancel{g} \times 1.17\,\frac{J}{\cancel{g}\cdot\cancel{°C}} \times (46.7\,\cancel{°C} - 52.7\,\cancel{°C})$$

Rearrange to solve for C_A.

$$C_{s,\,A} \times (161.13\,g\cdot°C) = 176.904\,J \rightarrow C_{s,\,A} = \frac{176.904\,J}{161.13\,g\cdot°C} = 1.097896\,\frac{J}{g\cdot°C} = 1.10\,\frac{J}{g\cdot°C}$$

Check: The units (J/g·°C) are correct. The magnitude of the answer (1) makes physical sense because the mass of substance B is greater than the mass of substance A by a factor of ~4.1 and the temperature change for substance A is greater than the temperature change of substance B by a factor of ~4.4; so the specific heat capacity of substance A will be slightly smaller.

Calorimetry

9.65 $\Delta H_{rxn} = q_p$ and $\Delta E_{rxn} = q_V = \Delta H - P\Delta V$. Because combustion always involves expansions, expansions do work and, therefore, have a negative value. Combustions are always exothermic and, therefore, have a negative value. This means that ΔE_{rxn} is more negative than ΔH_{rxn}; so A (−25.9 kJ) is the constant volume process, and B (−23.3 kJ) is the constant pressure process.

9.67   **Given:** 0.514 g biphenyl($C_{12}H_{10}$); bomb calorimeter; $T_i = 25.8\,°C$; $T_f = 29.4\,°C$; $C_{cal} = 5.86\,kJ/°C$ **Find:** ΔE_{rxn}
Conceptual Plan: $T_i, T_f \rightarrow \Delta T$ then $\Delta T, C_{cal} \rightarrow q_{cal} \rightarrow q_{rxn}$ then g $C_{12}H_{10} \rightarrow$ mol $C_{12}H_{10}$

$$\Delta T = T_f - T_i \qquad q_{cal} = C_{cal}\Delta T \qquad q_{cal} = -q_{rxn} \qquad \frac{1\,mol}{154.20\,g}$$

then q_{rxn}, mol $C_{12}H_{10} \rightarrow \Delta E_{rxn}$

$$\Delta E_{rxn} = \frac{q_v}{\text{mol } C_{12}H_{10}}$$

Solution: $\Delta T = T_f - T_i = 29.4\,°C - 25.8\,°C = 3.6\,°C$ then $q_{cal} = C_{cal}\Delta T = 5.86\,\frac{kJ}{°C} \times 3.6\,°C = 21.096\,kJ$

then $q_{cal} = -q_{rxn} = -21.096\,kJ$ and $0.514\,g\,C_{12}H_{10} \times \dfrac{1\,\text{mol }C_{12}H_{10}}{154.20\,g\,C_{12}H_{10}} = 0.00333333\,\text{mol }C_{12}H_{10}$ then

$$\Delta E_{rxn} = \frac{q_v}{\text{mol }C_{12}H_{10}} = \frac{-21.096\,kJ}{0.00333333\,\text{mol }C_{12}H_{10}} = -6.3 \times 10^3\,kJ/mol$$

Check: The units (kJ/mol) are correct. The magnitude of the answer (-6000) makes physical sense because a large amount of heat is generated from a very small amount of biphenyl.

9.69   **Given:** 0.103 g zinc; coffee-cup calorimeter; $T_i = 22.5\,°C$; $T_f = 23.7\,°C$; 50.0 mL solution; $d(\text{solution}) = 1.0\,g/mL$; $C_{s,\,soln} = 4.18\,kJ/g \cdot °C$ **Find:** ΔH_{rxn}

Conceptual Plan: $T_i, T_f \rightarrow \Delta T$ and mL soln $\rightarrow$ g soln then $\Delta T, C_{s,\,soln} \rightarrow q_{cal} \rightarrow q_{rxn}$ then

$$\Delta T = T_f - T_i \qquad \frac{1.0\,g}{1.0\,mL} \qquad q_{soln} = mC_{s,\,soln}\,\Delta T \quad q_{soln} = -q_{rxn}$$

g Zn $\rightarrow$ mol Zn then q_{rxn}, mol Zn $\rightarrow \Delta H_{rxn}$

$$\frac{1\,mol}{65.37\,g} \qquad \Delta H_{rxn} = \frac{q_p}{\text{mol Zn}}$$

Solution: $\Delta T = T_f - T_i = 23.7\,°C - 22.5\,°C = 1.2\,°C$ and $50.0\,mL \times \dfrac{1.0\,g}{1.0\,mL} = 50.0\,g$

then $q_{soln} = mC_{s,\,soln}\,\Delta T = 50.0\,g \times 4.18\,\dfrac{J}{g \cdot °C} \times 1.2\,°C = 250.8\,J$ then

$q_{soln} = -q_{rxn} = -250.8\,J$ and $0.103\,g\,Zn \times \dfrac{1\,\text{mol Zn}}{65.37\,g\,Zn} = 0.00157565\,\text{mol Zn}$ then

$$\Delta H_{rxn} = \frac{q_p}{\text{mol Zn}} = \frac{-250.8\,J}{0.00157565\,\text{mol Zn}} = -1.6 \times 10^5\,J/mol = -1.6 \times 10^2\,kJ/mol$$

Check: The units (kJ/mol) are correct. The magnitude of the answer (-160) makes physical sense because a large amount of heat is generated from a very small amount of zinc.

Quantitative Relationships Involving ΔH and Hess's Law

9.71   (a)   Because A + B $\rightarrow$ 2C has ΔH_1, 2C $\rightarrow$ A + B will have a $\Delta H_2 = -\Delta H_1$. When the reaction direction is reversed, it changes from exothermic to endothermic (or vice versa); so the sign of ΔH changes.

(b)   Because A + $\frac{1}{2}$B $\rightarrow$ C has ΔH_1, 2A + B $\rightarrow$ 2C will have a $\Delta H_2 = 2\,\Delta H_1$. When the reaction amount doubles, the amount of heat (or ΔH) doubles.

(c)   Because A $\rightarrow$ B + 2C has ΔH_1, $\frac{1}{2}$A $\rightarrow$ $\frac{1}{2}$B + C will have a $\Delta H_1' = \frac{1}{2}\Delta H_1$. When the reaction amount is cut in half, the amount of heat (or ΔH) is cut in half. Then $\frac{1}{2}$B + C $\rightarrow$ $\frac{1}{2}$A will have a $\Delta H_2 = -H_1' = -\frac{1}{2}\Delta H_1$. When the reaction direction is reversed, it changes from exothermic to endothermic (or vice versa); so the sign of ΔH changes.

9.73   Because the first reaction has Fe_2O_3 as a product and the reaction of interest has it as a reactant, we need to reverse the first reaction. When the reaction direction is reversed, ΔH changes.

$Fe_2O_3(s) \rightarrow 2\,Fe(s) + 3/2\,O_2(g)$ $\qquad\qquad\qquad \Delta H = +824.2\,kJ$

Because the second reaction has 1 mole CO as a reactant and the reaction of interest has 3 moles of CO as a reactant, we need to multiply the second reaction and the ΔH by 3.

$3[CO(g) + 1/2\,O_2(g) \rightarrow CO_2(g)]$ $\qquad\qquad \Delta H = 3(-282.7\,kJ) = -848.1\,kJ$

Hess's law states that the ΔH of the net reaction is the sum of the ΔH of the steps.

The rewritten reactions are as follows:

$Fe_2O_3(s) \rightarrow 2\,Fe(s) + 3/2\,O_2(g)$ $\qquad\qquad\qquad \Delta H = +824.2\,kJ$

$3\,CO(g) + 3/2\,O_2(g) \rightarrow 3\,CO_2(g)$ $\qquad\qquad\quad \Delta H = -848.1\,kJ$

$\overline{Fe_2O_3(s) + 3\,CO(g) \rightarrow 2\,Fe(s) + 3\,CO_2(g)} \qquad \Delta H_{rxn} = -23.9\,kJ$

9.75   Because the first reaction has C_5H_{12} as a reactant and the reaction of interest has it as a product, we need to reverse the first reaction. When the reaction direction is reversed, ΔH changes.

$$5\,CO_2(g) + 6\,H_2O(g) \rightarrow C_5H_{12}(l) + 8\,O_2(g) \qquad \Delta H = +3244.8\,kJ$$

Because the second reaction has 1 mole C as a reactant and the reaction of interest has 5 moles of C as a reactant, we need to multiply the second reaction and the ΔH by 5.

$$5[\,C(s) + O_2(g) \rightarrow CO_2(g)\,] \qquad\qquad \Delta H = 5(-393.5\,kJ) = -1967.5\,kJ$$

Because the third reaction has 2 moles H_2 as a reactant and the reaction of interest has 6 moles of H_2 as a reactant, we need to multiply the third reaction and the ΔH by 3.

$$3[\,2\,H_2(g) + O_2(g) \rightarrow 2\,H_2O(g)\,] \qquad \Delta H = 3(-483.5\,kJ) = -1450.5\,kJ$$

Hess's law states that the ΔH of the net reaction is the sum of the ΔH of the steps. The rewritten reactions are as follows:

$$\begin{aligned}
5\,CO_2(g) + 6\,H_2O(g) &\rightarrow C_5H_{12}(l) + 8\,O_2(g) && \Delta H = +3244.8\,kJ \\
5\,C(s) + 5\,O_2(g) &\rightarrow 5\,CO_2(g) && \Delta H = -1967.5\,kJ \\
6\,H_2(g) + 3\,O_2(g) &\rightarrow 6\,H_2O(g) && \Delta H = -1450.5\,kJ \\
\hline
5\,C(s) + 6\,H_2(g) &\rightarrow C_5H_{12}(l) && \Delta H_{rxn} = -173.2\,kJ
\end{aligned}$$

Using Bond Energies to Calculate ΔH_{rxn}

9.77   Rewrite the reaction using the Lewis structures of the molecules involved.

Determine which bonds are broken in the reaction and sum the bond energies of the following:

$$\begin{aligned}
\Sigma(\Delta H\text{'s bonds broken}) \\
= 4\,mol(C{-}H) + 1\,mol(C{=}C) + 1\,mol(H{-}H) \\
= 4\,mol(414\,kJ/mol) + 1\,mol(611\,kJ/mol) + 1\,mol(436\,kJ/mol) \\
= 2703\,kJ/mol
\end{aligned}$$

Determine which bonds are formed in the reaction and sum the negatives of the bond energies of the following:

$$\begin{aligned}
\Sigma(-\Delta H\text{'s bonds formed}) \\
= -6\,mol(C{-}H) - 1\,mol(C{-}C) \\
= -6\,mol(414\,kJ/mol) - 1\,mol(347\,kJ/mol) \\
= -2831\,kJ/mol
\end{aligned}$$

Find ΔH_{rxn} by summing the results of the two steps.

$$\begin{aligned}
\Delta H_{rxn} &= \Sigma(\Delta H\text{'s bonds broken}) + \Sigma(\Delta H{-}\Delta H\text{'s bonds formed}) \\
&= 2703\,kJ/mol - 2831\,kJ/mol \\
&= -128\,kJ/mol
\end{aligned}$$

9.79   Hydrogen, a potential future fuel, can be produced from carbon (from coal) and steam by this reaction:

$$C(s) + 2\,H_2O(g) \rightarrow 2\,H_2(g) + CO_2(g)$$

Use average bond energies to calculate ΔH_{rxn} for the reaction, and then use standard enthalpies of formation to calculate ΔH_{rxn}. Why are the two values different, and which value is more accurate?

Rewrite the reaction using the Lewis structures of the molecules involved.

Determine which bonds are broken in the reaction and sum the bond energies of the following:

$$\begin{aligned}
\Sigma(\Delta H\text{'s bonds broken}) \\
= 4(O{-}H) \\
= 4(464\,kJ/mol) \\
= 1856\,kJ/mol
\end{aligned}$$

Determine which bonds are formed in the reaction and sum the negatives of the bond energies of the following:

$$\begin{aligned}
\Sigma(-\Delta H\text{'s bonds formed}) \\
= -2(C{=}O) - 2(H{-}H)
\end{aligned}$$

$$= -2(799 \text{ kJ/mol}) - 2(436 \text{ kJ/mol})$$
$$= -2470 \text{ kJ/mol}$$

Find ΔH_{rxn} by summing the results of the two steps.

$$\Delta H_{rxn} = \Sigma(\Delta H\text{'s bonds broken}) + \Sigma(-\Delta H\text{'s bonds formed})$$
$$= 1856 \text{ kJ/mol} - 2470 \text{ kJ/mol}$$
$$= -614 \text{ kJ/mol}$$

$$C(s) + 2 \, H_2O(g) \rightarrow 2 \, H_2(g) + CO_2(g)$$
$$\Delta H_{rxn}^0 = \Sigma n_P \Delta H_f^0 (products) - \Sigma n_R \Delta H_f^0 (reactants)$$
$$= [2(\Delta H_f^0(H_2(g)) + 1(\Delta H_f^0(CO_2(g))] - [1(\Delta H_f^0(C(s)) + 2(\Delta H_f^0(H_2O(g))]$$
$$= [2(0.0 \text{kJ}) + 1(-393.5 \text{ kJ})] - [1(0.0 \text{ kJ}) + 2(-241.8 \text{ kJ})]$$
$$= [-393.5 \text{ kJ}] - [-483.6 \text{ kJ}]$$
$$= +90.1 \text{ kJ from Hess's law, which is significantly different than the value calculated using bond energies.}$$

The difference is a result of not considering the bonding between the carbon atoms in the coal. Taking these bonds into consideration would increase the enthalpy. Thus, the enthalpy calculated using the standard enthalpies of formation is more accurate.

Enthalpies of Formation and ΔH

9.81   (a) $\frac{1}{2} N_2(g) + \frac{3}{2} H_2(g) \rightarrow NH_3(g)$ $\Delta H_f^\circ = -45.9 \text{ kJ/mol}$

       (b) $C(s) + O_2(g) \rightarrow CO_2(g)$ $\Delta H_f^\circ = -393.5 \text{ kJ/mol}$

       (c) $2 \, Fe(s) + \frac{3}{2} O_2(g) \rightarrow Fe_2O_3(s)$ $\Delta H_f^\circ = -824.2 \text{ kJ/mol}$

       (d) $C(s) + 2 \, H_2(g) \rightarrow CH_4(g)$ $\Delta H_f^\circ = -74.6 \text{ kJ/mol}$

9.83   **Given:** $N_2H_4(l) + N_2O_4(g) \rightarrow 2 \, N_2O(g) + 2 \, H_2O(g)$ **Find:** ΔH_{rxn}°
      Conceptual Plan: $\Delta H_{rxn}^\circ = \Sigma n_p \Delta H_f^\circ(products) - \Sigma n_r \Delta H_f^\circ(reactants)$
      Solution:

Reactant/Product	ΔH_f° (kJ/mol from Appendix IIB)
$N_2H_4(l)$	50.6
$N_2O_4(g)$	9.16
$N_2O(g)$	81.6
$H_2O(g)$	−241.8

Be sure to pull data for the correct formula and phase.
$$\Delta H_{rxn}^\circ = \Sigma n_p \Delta H_f^\circ(products) - \Sigma n_r \Delta H_f^\circ(reactants)$$
$$= [2(\Delta H_f^\circ(N_2O(g))) + 2(\Delta H_f^\circ(H_2O(g)))] - [1(\Delta H_f^\circ(N_2H_4(l))) + 1(\Delta H_f^\circ(N_2O_4(g)))]$$
$$= [2(81.6 \text{ kJ}) + 2(-241.8 \text{ kJ})] - [1(50.6 \text{ kJ}) + 1(9.16 \text{ kJ})]$$
$$= [-320.4 \text{ kJ}] - [59.8 \text{ kJ}]$$
$$= -380.2 \text{ kJ}$$

Check: The units (kJ) are correct. The answer is negative, which means that the reaction is exothermic. The answer is dominated by the negative heat of formation of water.

9.85   (a)   **Given:** $C_2H_4(g) + H_2(g) \rightarrow C_2H_6(g)$ **Find:** ΔH_{rxn}°
          Conceptual Plan: $\Delta H_{rxn}^\circ = \Sigma n_p \Delta H_f^\circ(products) - \Sigma n_r \Delta H_f^\circ(reactants)$
          Solution:

Reactant/Product	ΔH_f° (kJ/mol from Appendix IIB)
$C_2H_4(g)$	52.4
$H_2(g)$	0.0
$C_2H_6(g)$	−84.68

Be sure to pull data for the correct formula and phase.

$$\Delta H^\circ_{rxn} = \sum n_p \Delta H^\circ_f(products) - \sum n_r \Delta H^\circ_f(reactants)$$
$$= [1(\Delta H^\circ_f(C_2H_6(g)))] - [1(\Delta H^\circ_f(C_2H_4(g))) + 1(\Delta H^\circ_f(H_2(g)))]$$
$$= [1(-84.68\ kJ)] - [1(52.4\ kJ) + 1(0.0\ kJ)]$$
$$= [-84.68\ kJ] - [52.4\ kJ]$$
$$= -137.1\ kJ$$

Check: The units (kJ) are correct. The answer is negative, which means that the reaction is exothermic. Both hydrocarbon terms are negative, so the final answer is negative.

(b)   **Given:** $CO(g) + H_2O(g) \rightarrow H_2(g) + CO_2(g)$ **Find:** ΔH°_{rxn}
Conceptual Plan: $\Delta H^\circ_{rxn} = \sum n_p \Delta H^\circ_f(products) - \sum n_r \Delta H^\circ_f(reactants)$
Solution:

Reactant/Product	ΔH°_f (kJ/mol from Appendix IIB)
$CO(g)$	-110.5
$H_2O(g)$	-241.8
$H_2(g)$	0.0
$CO_2(g)$	-393.5

Be sure to pull data for the correct formula and phase.

$$\Delta H^\circ_{rxn} = \sum n_p \Delta H^\circ_f(products) - \sum n_r \Delta H^\circ_f(reactants)$$
$$= [1(\Delta H^\circ_f(H_2(g))) + 1(\Delta H^\circ_f(CO_2(g)))] - [1(\Delta H^\circ_f(CO(g))) + 1(\Delta H^\circ_f(H_2O(g)))]$$
$$= [1(0.0\ kJ) + 1(-393.5\ kJ)] - [1(-110.5\ kJ) + 1(-241.8\ kJ)]$$
$$= [-393.5\ kJ] - [-352.3\ kJ]$$
$$= -41.2\ kJ$$

Check: The units (kJ) are correct. The answer is negative, which means that the reaction is exothermic.

(c)   **Given:** $3\ NO_2(g) + H_2O(l) \rightarrow 2\ HNO_3(aq) + NO(g)$ **Find:** ΔH°_{rxn}
Conceptual Plan: $\Delta H^\circ_{rxn} = \sum n_p \Delta H^\circ_f(products) - \sum n_r \Delta H^\circ_f(reactants)$
Solution:

Reactant/Product	ΔH°_f (kJ/mol from Appendix IIB)
$NO_2(g)$	33.2
$H_2O(l)$	-285.8
$HNO_3(aq)$	-207
$NO(g)$	91.3

Be sure to pull data for the correct formula and phase.

$$\Delta H^\circ_{rxn} = \sum n_p \Delta H^\circ_f(products) - \sum n_r \Delta H^\circ_f(reactants)$$
$$= [2(\Delta H^\circ_f(HNO_3(aq))) + 1(\Delta H^\circ_f(NO(g)))] - [3(\Delta H^\circ_f(NO_2(g))) + 1(\Delta H^\circ_f(H_2O(l)))]$$
$$= [2(-207\ kJ) + 1(91.3\ kJ)] - [3(33.2\ kJ) + 1(-285.8\ kJ)]$$
$$= [-322.7\ kJ] - [-186.2\ kJ]$$
$$= -137\ kJ$$

Check: The units (kJ) are correct. The answer is negative, which means that the reaction is exothermic.

(d)   **Given:** $Cr_2O_3(s) + 3\ CO(g) \rightarrow 2\ Cr(s) + 3\ CO_2(g)$ **Find:** ΔH°_{rxn}
Conceptual Plan: $\Delta H^\circ_{rxn} = \sum n_p \Delta H^\circ_f(products) - \sum n_r \Delta H^\circ_f(reactants)$
Solution:

Reactant/Product	ΔH°_f (kJ/mol from Appendix IIB)
$Cr_2O_3(s)$	-1139.7
$CO(g)$	-110.5
$Cr(s)$	0.0
$CO_2(g)$	-393.5

Be sure to pull data for the correct formula and phase.

$$\Delta H^\circ_{rxn} = \sum n_p \Delta H^\circ_f(products) - \sum n_r \Delta H^\circ_f(reactants)$$
$$= [2(\Delta H^\circ_f(Cr(s))) + 3(\Delta H^\circ_f(CO_2(g)))] - [1(\Delta H^\circ_f(Cr_2O_3(s))) + 3(\Delta H^\circ_f(CO(g)))]$$
$$= [2(0.0\,kJ) + 3(-393.5\,kJ)] - [1(-1139.7\,kJ) + 3(-110.5\,kJ)]$$
$$= [-1180.5\,kJ] - [-1471.2\,kJ]$$
$$= 290.7\,kJ$$

Check: The units (kJ) are correct. The answer is positive, which means that the reaction is endothermic.

9.87   **Given:** form glucose ($C_6H_{12}O_6$) and oxygen from sunlight, carbon dioxide, and water   **Find:** ΔH°_{rxn}
Conceptual Plan: Write balanced reaction then $\Delta H^\circ_{rxn} = \sum n_p \Delta H^\circ_f(products) - \sum n_r \Delta H^\circ_f(reactants)$
Solution: $6\,CO_2(g) + 6\,H_2O(l) \rightarrow C_6H_{12}O_6(s) + 6\,O_2(g)$

Reactant/Product	ΔH°_f (kJ/mol from Appendix IIB)
$CO_2(g)$	−393.5
$H_2O(l)$	−285.8
$C_6H_{12}O_6(s, glucose)$	−1273.3
$O_2(g)$	0.0

Be sure to pull data for the correct formula and phase.
$$\Delta H^\circ_{rxn} = \sum n_p \Delta H^\circ_f(products) - \sum n_r \Delta H^\circ_f(reactants)$$
$$= [1(\Delta H^\circ_f(C_6H_{12}O_6(s))) + 6(\Delta H^\circ_f(O_2(g)))] - [6(\Delta H^\circ_f(CO_2(g))) + 6(\Delta H^\circ_f(H_2O(l)))]$$
$$= [1(-1273.3\,kJ) + 6(0.0\,kJ)] - [6(-393.5\,kJ) + 6(-285.8\,kJ)]$$
$$= [-1273.3\,kJ] - [-4075.8\,kJ]$$
$$= +2802.5\,kJ$$

Check: The units (kJ) are correct. The answer is positive, which means that the reaction is endothermic. The reaction requires the input of light energy, so we expect that this will be an endothermic reaction.

9.89   **Given:** $2\,CH_3NO_2(l) + 3/2\,O_2(g) \rightarrow 2\,CO_2(g) + 3\,H_2O(l) + N_2(g)$ and $\Delta H^\circ_{rxn} = -1418.4\,kJ$
Find: $\Delta H^\circ_f(CH_3NO_2(l))$
Conceptual Plan: Fill known values into $\Delta H^\circ_{rxn} = \sum n_p \Delta H^\circ_f(products) - \sum n_r \Delta H^\circ_f(reactants)$ **and rearrange to solve for** $\Delta H^\circ_f(CH_3NO_2(l))$.
Solution:

Reactant/Product	ΔH°_f (kJ/mol from Appendix IIB)
$O_2(g)$	0.0
$CO_2(g)$	−393.5
$H_2O(l)$	−285.8
$N_2(g)$	0.0

Be sure to pull data for the correct formula and phase.
$$\Delta H^\circ_{rxn} = \sum n_p \Delta H^\circ_f(products) - \sum n_r \Delta H^\circ_f(reactants)$$
$$= [2(\Delta H^\circ_f(CO_2(g))) + 3(\Delta H^\circ_f(H_2O(l))) + 1(\Delta H^\circ_f(N_2(g)))] - [2(\Delta H^\circ_f(CH_3NO_2(l))) + 3/2(\Delta H^\circ_f(O_2(g)))]$$
$$(-1418.4\,kJ) = [2(-393.5\,kJ) + 3(-285.8\,kJ) + 1(0.0\,kJ)] - [2(\Delta H^\circ_f(CH_3NO_2(l)) + 3/2(0.0\,kJ)]$$
$$-1418.4\,kJ = [-1644.4\,kJ] - [2(\Delta H^\circ_f(CH_3NO_2(l)))]$$
$$\Delta H^\circ_f(CH_3NO_2(l)) = -113.0\,kJ/mol$$

Check: The units (kJ/mol) are correct. The answer is negative (but not as negative as water and carbon dioxide), which is consistent with an exothermic combustion reaction.

Lattice Energies

9.91   As the size of the alkaline metal ions increases down the column, so does the distance between the metal cation and the oxide anion. Therefore, the magnitude of the lattice energy of the oxides decreases, making the formation of the oxides less exothermic and the compounds less stable. Because the ions cannot get as close to each other, they do not release as much energy.

9.93   Cesium is slightly larger than barium, but oxygen is slightly larger than fluorine; so we cannot use size to explain the difference in the lattice energy. However, the charge on the cesium ion is 1+ and the charge on the fluoride ion is 1−, while the charge on the barium ion is 2+ and the charge on the oxide ion is 2−. The coulombic equation states that the magnitude of the potential also depends on the product of the charges. Because the product of the charges for

$CsF = 1-$ and the product of the charges for $BaO = 4-$, the stabilization for BaO relative to CsF should be about four times greater, which is what we see in its more exothermic lattice energy.

9.95 **Given:** $\Delta H_f^\circ(KCl) = -436.5 \text{ kJ/mol}$; $IE_1(K) = 419 \text{ kJ/mol}$; $\Delta H_{sub}(K) = 89.0 \text{ kJ/mol}$; $Cl_2(g)$ bond energy $= 243 \text{ kJ/mol}$; $EA(Cl) = -349 \text{ kJ/mol}$ **Find:** lattice energy

Conceptual Plan:

$$K(s) + 1/2\,Cl_2(g) \rightarrow K(g) + 1/2\,Cl_2(g) \rightarrow K^+(g) + 1/2\,Cl_2(g) \rightarrow K^+(g) + Cl(g) \rightarrow K^+(g) + Cl^-(g) \rightarrow KCl(s)$$

$$\underset{\Delta H_{sub}}{\qquad} \underset{IE_1}{\qquad} \underset{\text{bond energy}}{\qquad} \underset{EA}{\qquad} \underset{\text{lattice energy}}{\qquad}$$

$$\underset{\Delta H_f^\circ}{\qquad}$$

Solution: $\Delta H_f^\circ = \Delta H_{sub} + IE_1 + 1/2 \text{ bond energy} + EA + \text{lattice energy}$

$$-436.5 \frac{\text{kJ}}{\text{mol}} = +89.0 \frac{\text{kJ}}{\text{mol}} + 419 \frac{\text{kJ}}{\text{mol}} + \frac{1}{2}(243) \frac{\text{kJ}}{\text{mol}} + (-349) \frac{\text{kJ}}{\text{mol}} + \text{lattice energy}$$

$$\text{lattice energy} = -717 \text{ kJ/mol}$$

Cumulative Problems

9.97 **Given:**

billiard ball$_A$ = system: $m_A = 0.17 \text{ kg}$, $v_{A1} = 4.5 \text{ m/s}$ slows to $v_{A2} = 3.8 \text{ m/s}$ and $v_{A3} = 0$; ball$_B$: $m_B = 0.17 \text{ kg}$, $v_{B1} = 0$ and $v_{B2} = 3.8 \text{ m/s}$, and $KE = \frac{1}{2}mv^2$ **Find:** w, q, ΔE_{sys}

Conceptual Plan: $m, v \rightarrow KE$ then $KE_{A3}, KE_{A1} \rightarrow \Delta E_{sys}$ and $KE_{A2}, KE_{A1} \rightarrow q$ and $KE_{B2}, KE_{B1} \rightarrow w_B$

$$KE = \frac{1}{2}mv^2 \qquad \Delta E_{sys} = KE_{A3} - KE_{A1} \qquad q = KE_{A2} - KE_{A1} \qquad w_B = KE_{B2} - KE_{B1}$$

$\Delta E_{sys}, q \rightarrow w_A$ **verify that** $w_A = -w_B$ **so that no heat is transferred to ball$_B$**

$$\Delta E = q + w$$

Solution: $KE = \frac{1}{2}mv^2$ because m is in kg and v is in m/s; KE will be in $\text{kg} \cdot \text{m}^2/\text{s}^2$, which is the definition of a joule.

$$KE_{A1} = \frac{1}{2}(0.17 \text{ kg})\left(4.5 \frac{\text{m}}{\text{s}}\right)^2 = 1.\underline{7}213 \frac{\text{kg} \cdot \text{m}^2}{\text{s}^2} = 1.\underline{7}213 \text{ J}$$

$$KE_{A2} = \frac{1}{2}(0.17 \text{ kg})\left(3.8 \frac{\text{m}}{\text{s}}\right)^2 = 1.\underline{2}274 \frac{\text{kg} \cdot \text{m}^2}{\text{s}^2} = 1.\underline{2}274 \text{ J}$$

$$KE_{A3} = \frac{1}{2}(0.17 \text{ kg})\left(0 \frac{\text{m}}{\text{s}}\right)^2 = 0 \frac{\text{kg} \cdot \text{m}^2}{\text{s}^2} = 0 \text{ J}$$

$$KE_{B1} = \frac{1}{2}(0.17 \text{ kg})\left(0 \frac{\text{m}}{\text{s}}\right)^2 = 0 \frac{\text{kg} \cdot \text{m}^2}{\text{s}^2} = 0 \text{ J}$$

$$KE_{B2} = \frac{1}{2}(0.17 \text{ kg})\left(3.8 \frac{\text{m}}{\text{s}}\right)^2 = 1.\underline{2}274 \frac{\text{kg} \cdot \text{m}^2}{\text{s}^2} = 1.\underline{2}274 \text{ J}$$

$\Delta E_{sys} = KE_{A3} - KE_{A1} = 0 \text{ J} - 1.\underline{7}213 \text{ J} = -1.\underline{7}213 \text{ J} = -1.7 \text{ J}$

$q = KE_{A2} - KE_{A1} = 1.\underline{2}274 \text{ J} - 1.\underline{7}213 \text{ J} = -0.\underline{4}939 \text{ J} = -0.5 \text{ J}$

$w_B = KE_{B2} - KE_{B1} = 1.\underline{2}274 \text{ J} - 0 \text{ J} = 1.\underline{2}274 \text{ J}$

$w = \Delta E - q = -1.\underline{7}213 \text{ J} - (-0.\underline{4}939 \text{ J}) = -1.\underline{2}274 \text{ J} = -1.2 \text{ J}$

Because $w_A = -w_B$, no heat is transferred to ball$_B$.

Check: The units (J) are correct. Because the ball is initially moving and is stopped at the end, it has lost energy (negative ΔE_{sys}). As the ball slows due to friction, it is releasing heat (negative q). The kinetic energy is transferred to a second ball, so it does work (w negative).

9.99 **Given:** $H_2O(l) \rightarrow H_2O(g)$ $\Delta H_{rxn}^\circ = +44.01 \text{ kJ/mol}$; $\Delta T_{body} = -0.50 \,^\circ\text{C}$, $m_{body} = 95 \text{ kg}$, $C_{body} = 4.0 \text{ J/g}\,^\circ\text{C}$
Find: m_{H_2O}
Conceptual Plan: $\text{kg} \rightarrow \text{g}$ then $m_{body}, \Delta T, C_{s,\,body} \rightarrow q_{body} \rightarrow q_{rxn}(\text{J}) \rightarrow q_{rxn}(\text{kJ}) \rightarrow \text{mol } H_2O \rightarrow \text{g } H_2O$

$$\frac{1000 \text{ g}}{1 \text{ kg}} \qquad q_{body} = m_{body}C_{s,\,body}\Delta T_{body} \qquad q_{rxn} = -q_{body} \qquad \frac{1 \text{ kJ}}{1000 \text{ J}} \qquad \frac{1 \text{ mol}}{44.01 \text{ kJ}} \qquad \frac{18.02 \text{ g}}{1 \text{ mol}}$$

Solution: $95 \text{ kg} \times \dfrac{1000 \text{ g}}{1 \text{ kg}} = 95000 \text{ g}$ then

$$q_{\text{body}} = m_{\text{body}} C_{\text{s, body}} \Delta T_{\text{body}} = 95000 \text{ g} \times 4.0 \dfrac{\text{J}}{\text{g} \cdot {}^\circ\text{C}} \times (-0.50 \, {}^\circ\text{C}) = -190000 \text{ J} \text{ then}$$

$$q_{\text{rxn}} = -q_{\text{body}} = 190000 \text{ J} \times \dfrac{1 \text{ kJ}}{1000 \text{ J}} \times \dfrac{1 \text{ mol}}{44.01 \text{ kJ}} \times \dfrac{18.02 \text{ g}}{1 \text{ mol}} = 78 \text{ g } H_2O$$

Check: The units (g) are correct. The magnitude of the answer (78) makes physical sense because a person can sweat this much on a hot day.

9.101   **Given:** $H_2O(s) \rightarrow H_2O(l)$ $\Delta H_f^\circ(H_2O(s)) = -291.8 \text{ kJ/mol}$; 355 mL beverage $T_{\text{Bev, i}} = 25.0 \, {}^\circ\text{C}$, $T_{\text{Bev, f}} = 0.0 \, {}^\circ\text{C}$, $C_{\text{s, Bev}} = 4.184 \text{ J/g} \, {}^\circ\text{C}$, $d_{\text{Bev}} = 1.0 \text{ g/mL}$ **Find:** $\Delta H_{\text{rxn}}^\circ$ (ice melting) and m_{ice}
Conceptual Plan: $\Delta H_{\text{rxn}}^\circ = \sum n_p \Delta H_f^\circ (products) - \sum n_r \Delta H_f^\circ (reactants)$ **mL** $\rightarrow$ **g and** $T_i, T_f \rightarrow \Delta T$ **then**

$$\dfrac{1.0 \text{ g}}{1.0 \text{ mL}} \qquad\qquad \Delta T = T_f - T_i$$

$$m_{H_2O}, \Delta T_{H_2O}, C_{\text{s, } H_2O} \rightarrow q_{H_2O} \rightarrow q_{\text{rxn}}(\text{J}) \rightarrow q_{\text{rxn}} (\text{kJ}) \rightarrow \text{mol ice} \rightarrow \text{g ice}$$

$$q_{\text{Bev}} = m_{\text{Bev}} C_{\text{s, Bev}} \Delta T_{\text{Bev}} \qquad q_{\text{rxn}} = -q_{\text{Bev}} \qquad \dfrac{1 \text{ kJ}}{1000 \text{ J}} \quad \dfrac{1 \text{ mol}}{\Delta H_{\text{rxn}}^\circ} \quad \dfrac{18.02 \text{ g}}{1 \text{ mol}}$$

Solution:

Reactant/Product	ΔH_f° (kJ/mol from Appendix IIB)
$H_2O(s)$	-291.8
$H_2O(l)$	-285.8

Be sure to pull data for the correct formula and phase.
$$\begin{aligned}\Delta H_{\text{rxn}}^\circ &= \sum n_p \Delta H_f^\circ (products) - \sum n_r \Delta H_f^\circ (reactants) \\ &= [1(\Delta H_f^\circ(H_2O(l)))] - [1(\Delta H_f^\circ(H_2O(s)))] \\ &= [1(-285.8 \text{ kJ})] - [1(-291.8 \text{ kJ})] \\ &= +6.0 \text{ kJ}\end{aligned}$$

$$355 \text{ mL} \times \dfrac{1.0 \text{ g}}{1.0 \text{ mL}} = 355 \text{ g and } \Delta T = T_f - T_i = 0.0 \, {}^\circ\text{C} - 25.0 \, {}^\circ\text{C} = -25.0 \, {}^\circ\text{C} \text{ then}$$

$$q_{\text{Bev}} = m_{\text{Bev}} C_{\text{s, Bev}} \Delta T_{\text{Bev}} = 355 \text{ g} \times 4.184 \dfrac{\text{J}}{\text{g} \cdot {}^\circ\text{C}} \times (-25.0 \, {}^\circ\text{C}) = -37133 \text{ J} \text{ then}$$

$$q_{\text{rxn}} = -q_{\text{Bev}} = 37133 \text{ J} \times \dfrac{1 \text{ kJ}}{1000 \text{ J}} \times \dfrac{1 \text{ mol}}{6.0 \text{ kJ}} \times \dfrac{18.02 \text{ g}}{1 \text{ mol}} = 110 \text{ g ice}$$

Check: The units (J and g) are correct. The answer is positive, which means that the reaction is endothermic. We expect an endothermic reaction because we know that heat must be added to melt ice. The magnitude of the answer (110 g) makes physical sense because it is much smaller than the weight of the beverage and it would fit in a glass with the beverage.

9.103   **Given:** 25.5 g aluminum; $T_{\text{Al, i}} = 65.4 \, {}^\circ\text{C}$; 55.2 g water; $T_{H_2O, i} = 22.2 \, {}^\circ\text{C}$ **Find:** T_f
Conceptual Plan: Pull C_s values from table then $m, C_s, T_i \rightarrow T_f$

$$\text{Al: } 0.903 \dfrac{\text{J}}{\text{g} \cdot {}^\circ\text{C}} \quad H_2O: 4.18 \dfrac{\text{J}}{\text{g} \cdot {}^\circ\text{C}} \qquad q = mC_s(T_f - T) \text{ then set } q_{\text{Al}} = -q_{H_2O}$$

Solution: $q = mC_s(T_f - T_i)$ substitute in values and set $q_{\text{Al}} = -q_{H_2O}$.

$$q_{\text{Al}} = m_{\text{Al}} C_{\text{s, Al}} (T_f - T_{\text{Al, i}}) = 25.5 \text{ g} \times 0.903 \dfrac{\text{J}}{\text{g} \cdot {}^\circ\text{C}} \times (T_f - 65.4 \, {}^\circ\text{C}) =$$

$$-q_{H_2O} = -m_{H_2O} C_{\text{s, } H_2O} (T_f - T_{H_2O, i}) = -55.2 \text{ g} \times 4.18 \dfrac{\text{J}}{\text{g} \cdot {}^\circ\text{C}} \times (T_f - 22.2 \, {}^\circ\text{C})$$

Rearrange to solve for T_f.

$$23.0265 \dfrac{\text{J}}{{}^\circ\text{C}} \times (T_f - 65.4 \, {}^\circ\text{C}) = -230.736 \dfrac{\text{J}}{{}^\circ\text{C}} \times (T_f - 22.2 \, {}^\circ\text{C}) \rightarrow$$

$$23.0265 \dfrac{\text{J}}{{}^\circ\text{C}} T_f - 1505.93 \text{ J} = -230.736 \dfrac{\text{J}}{{}^\circ\text{C}} T_f + 5122.34 \text{ J} \rightarrow$$

$$+5122.34 \text{ J} + 1505.93 \text{ J} = +230.736 \frac{\text{J}}{^\circ\text{C}} T_f + 23.0265 \frac{\text{J}}{^\circ\text{C}} T_f \rightarrow$$

$$6628.27 \text{ J} = 253.7625 \frac{\text{J}}{^\circ\text{C}} T_f \rightarrow$$

$$T_f = \frac{6628.27 \text{ J}}{253.7625 \frac{\text{J}}{^\circ\text{C}}} = 26.1 \,^\circ\text{C}$$

Check: The units (°C) are correct. The magnitude of the answer (26) makes physical sense because the heat transfer is dominated by the water (larger mass and larger specific heat capacity). The final temperature should be closer to the initial temperature of water than of aluminum.

9.105   **Given:** palmitic acid ($C_{16}H_{32}O_2$) combustion $\Delta H_f^\circ(C_{16}H_{32}O_2\,(s)) = -208$ kJ/mol; sucrose ($C_{12}H_{22}O_{11}$) combustion $\Delta H_f^\circ(C_{12}H_{22}O_{11}(s)) = -2226.1$ kJ/mol   **Find:** ΔH_{rxn}° in kJ/mol and Cal/g
Conceptual Plan: Write balanced reaction then $\Delta H_{rxn}^\circ = \sum n_p \Delta H_f^\circ(products) - \sum n_r \Delta H_f^\circ(reactants)$ **then**
$\text{kJ/mol} \rightarrow \text{J/mol} \rightarrow \text{Cal/mol} \rightarrow \text{Cal/g}$

$$\frac{1000 \text{ J}}{1 \text{ kJ}} \qquad \frac{1 \text{ Cal}}{4184 \text{ J}} \qquad \text{PA:} \frac{1 \text{ mol}}{256.42 \text{ g}} \quad \text{S:} \frac{1 \text{ mol}}{342.30 \text{ g}}$$

Solution: Combustion is the combination with oxygen to form carbon dioxide and water (*l*):
$$C_{16}H_{32}O_2(s) + 23\,O_2(g) \rightarrow 16\,CO_2(g) + 16\,H_2O(l)$$

Reactant/Product	ΔH_f° (kJ/mol from Appendix IIB)
$C_{16}H_{32}O_2(s)$	-208
$O_2(g)$	0.0
$CO_2(g)$	-393.5
$H_2O(l)$	-285.8

Be sure to pull data for the correct formula and phase.

$$\Delta H_{rxn}^\circ = \sum n_p \Delta H_f^\circ(products) - \sum n_r \Delta H_f^\circ(reactants)$$
$$= [16(\Delta H_f^\circ(CO_2(g))) + 16(\Delta H_f^\circ(H_2O(l)))] - [1(\Delta H_f^\circ(C_{16}H_{32}O_2(s))) + 23(\Delta H_f^\circ(O_2(g)))]$$
$$= [16(-393.5 \text{ kJ}) + 16(-285.8 \text{ kJ})] - [1(-208 \text{ kJ}) + 23(0.0 \text{ kJ})]$$
$$= [-10868.8 \text{ kJ}] - [-208 \text{ kJ}]$$
$$= -10660.8 \text{ kJ/mol} = -10661 \text{ kJ/mol}$$

$$-10660.8 \,\frac{\text{kJ}}{\text{mol}} \times \frac{1000 \text{ J}}{1 \text{ kJ}} \times \frac{1 \text{ Cal}}{4184 \text{ J}} \times \frac{1 \text{ mol}}{256.42 \text{ g}} = -9.9378 \text{ Cal/g}$$

$$C_{12}H_{22}O_{11}(s) + 12\,O_2(g) \rightarrow 12\,CO_2(g) + 11\,H_2O(l)$$

Reactant/Product	ΔH_f° (kJ/mol from Appendix IIB)
$C_{12}H_{22}O_{11}(s)$	-2226.1
$O_2(g)$	0.0
$CO_2(g)$	-393.5
$H_2O(l)$	-285.8

Be sure to pull data for the correct formula and phase.

$$\Delta H_{rxn}^\circ = \sum n_p \Delta H_f^\circ(products) - \sum n_r \Delta H_f^\circ(reactants)$$
$$= [12(\Delta H_f^\circ(CO_2(g))) + 11(\Delta H_f^\circ(H_2O(l)))] - [1(\Delta H_f^\circ(C_{12}H_{22}O_{11}(s))) + 12(\Delta H_f^\circ(O_2(g)))]$$
$$= [12(-393.5 \text{ kJ}) + 11(-285.8 \text{ kJ})] - [1(-2226.1 \text{ kJ}) + 12(0.0 \text{ kJ})]$$
$$= [-7865.8 \text{ kJ}] - [-2226.1 \text{ kJ}]$$
$$= -5639.7 \text{ kJ/mol}$$

$$-5639.7 \,\frac{\text{kJ}}{\text{mol}} \times \frac{1000 \text{ J}}{1 \text{ kJ}} \times \frac{1 \text{ Cal}}{4184 \text{ J}} \times \frac{1 \text{ mol}}{342.30 \text{ g}} = -3.9378 \text{ Cal/g}$$

Check: The units (kJ/mol and Cal/g) are correct. The magnitudes of the answers are consistent with the food labels we see every day. Palmitic acid gives more Cal/g than does sucrose.

9.107   **Given:** 16 g peanut butter; bomb calorimeter; $T_i = 22.2\,^\circ\text{C}$; $T_f = 25.4\,^\circ\text{C}$; $C_{cal} = 120.0$ kJ/°C   **Find:** calories in peanut butter

Conceptual Plan: $T_i, T_f \rightarrow \Delta T$ then $\Delta T, C_{cal} \rightarrow q_{cal} \rightarrow q_{rxn}$ (kJ) $\rightarrow q_{rxn}$ (kJ) $\rightarrow q_{rxn}$ (Cal)

$$\Delta T = T_f - T_i \qquad q_{cal} = C_{cal}\Delta T \qquad q_{rxn} = -q_{cal} \qquad \frac{1000\text{ J}}{1\text{ kJ}} \qquad \frac{1\text{ Cal}}{4184\text{ J}}$$

then q_{rxn} (Cal) $\rightarrow$ Cal/g

$$\div 16\text{ g peanut butter}$$

Solution: $\Delta T = T_f - T_i = 25.4\,^\circ\text{C} - 22.2\,^\circ\text{C} = 3.2\,^\circ\text{C}$ then $q_{cal} = C_{cal}\Delta T = 120.0\,\dfrac{\text{kJ}}{^\circ\text{C}} \times 3.2\,^\circ\text{C} = 3\underline{8}4\text{ kJ}$

then $q_{rxn} = -q_{cal} = -3\underline{8}4\text{ kJ} \times \dfrac{1000\text{ J}}{1\text{ kJ}} \times \dfrac{1\text{ Cal}}{4184\text{ J}} = 91.778\text{ Cal}$ then $\dfrac{91.778\text{ Cal}}{16\text{ g}} = 5.7\text{ Cal/g}$

Check: The units (Cal/g) are correct. The magnitude of the answer (6) makes physical sense because there is a significant percentage of fat and sugar in peanut butter. The answer is in line with the answers in Problem 9.103.

9.109 **Given:** $V_1 = 20.0$ L at $P_1 = 3.0$ atm; $P_2 = 1.5$ atm; let it expand at constant T **Find:** $w, q, \Delta E_{sys}$
Conceptual Plan: $V_1, P_1, P_2 \rightarrow V_2$ then $V_1, V_2 \rightarrow \Delta V$ then $P, \Delta V \rightarrow w$ (L atm) $\rightarrow w$ (J)

$$P_1V_1 = P_2V_2 \qquad \Delta V = V_2 - V_1 \qquad w = -P\Delta V \qquad \frac{101.3\text{ J}}{1\text{ L atm}}$$

for an ideal gas $\Delta E_{sys} \alpha\, T$; so because this is a constant temperature process, $\Delta E_{sys} = 0$ finally $\Delta E_{sys}, w \rightarrow q$

$$\Delta E = q + w$$

Solution: $P_1V_1 = P_2V_2$. Rearrange to solve for V_2. $V_2 = V_1\dfrac{P_1}{P_2} = (20.0\text{ L}) \times \dfrac{3.0\text{ atm}}{1.5\text{ atm}} = 40.\text{ L}$ and

$\Delta V = V_2 - V_1 = 40.\text{ L} - 20.0\text{ L} = 20.\text{ L}$ then

$w = -P\Delta V = -1.5\text{ atm} \times 20.\text{ L} \times \dfrac{101.3\text{ J}}{1\text{ L}\cdot\text{atm}} = -30\underline{3}9\text{ J} = -3.0 \times 10^3\text{ J} \qquad \Delta E = q + w$

Rearrange to solve for q. $q = \Delta E_{sys} - w = +0\text{ J} - (-30\underline{3}9\text{ J}) = 3.0 \times 10^3\text{ J}$

Check: The units (J) are correct. Because there is no temperature change, we expect no energy change ($\Delta E_{sys} = 0$). The piston expands as it does work (negative work), so heat is absorbed (positive q).

9.111 The oxidation of S(g) to SO_3 can be written as follows:
$S(g) + 3/2\,O_2(g) \rightarrow SO_3(g)$ $\qquad \Delta H = -204$ kJ
The oxidation of $SO_2(g)$ to SO_3 can be written as follows:
$SO_2(g) + 1/2\,O_2(g) \rightarrow SO_3(g)$ $\qquad \Delta H = +89.5$ kJ
The enthalpy of formation reaction for $SO_2(g)$ under these conditions can be written as follows:
$S(g) + O_2(g) \rightarrow SO_2(g)$ $\qquad \Delta H = $??
Because the second reaction has 1 mole SO_2 as a reactant and the reaction of interest has 1 mole of SO_2 as a product, we need to reverse the second reaction. When the reaction direction is reversed, ΔH changes sign.
$SO_3(g) \rightarrow SO_2(g) + 1/2\,O_2(g)$ $\qquad \Delta H = -89.5$ kJ
Hess's law states that the ΔH of the net reaction is the sum of the ΔH of the steps.
The rewritten reactions are as follows:
$S(g) + 3/2\,O_2(g) \rightarrow SO_3(g)$ $\qquad \Delta H = -204$ kJ
$SO_3(g) \rightarrow SO_2(g) + 1/2\,O_2(g)$ $\qquad \Delta H = -89.5$ kJ
$S(g) + O_2(g) \rightarrow SO_2(g)$ $\qquad \Delta H = -294\text{ kJ} = \Delta H_f$
Note that this is not under standard conditions because S is not a solid.

9.113 **Given:** 1.55 cm copper cube and 1.62 cm aluminum cube, $T_{\text{Metals, i}} = 55.0\,^\circ\text{C}$, 100.0 mL water, $T_{\text{H}_2\text{O, i}} = 22.2\,^\circ\text{C}$
Other: density (water) $= 0.998$ g/mL **Find:** T_f
Conceptual Plan: Pull d values from Table 9.2 then edge length $\rightarrow V \rightarrow m$ then

$$\text{Cu: 8.96 g/mL Al: 2.70 g/mL} \qquad V = l^3\, d = m/V$$

pull C_s values from Table 9.2 then $\quad m, C_s, T_i \rightarrow T_f$

$$\text{Cu: 0.385}\,\frac{\text{J}}{\text{g}\cdot^\circ\text{C}}\,\text{Al: 0.903}\,\frac{\text{J}}{\text{g}\cdot^\circ\text{C}}\,\text{H}_2\text{O: 4.18}\,\frac{\text{J}}{\text{g}\cdot^\circ\text{C}} \qquad q = mC_s(T_f - T_i)\ \text{then set}\ q_{\text{Cu}} + q_{\text{Al}} = -q_{\text{H}_2\text{O}}$$

Solution: $V_{\text{Cu}} = l^3 = (1.55\text{ cm})^3 = 3.7\underline{2}3875\text{ cm}^3 = 3.723875\text{ mL}$ and

$V_{\text{Al}} = l^3 = (1.62\text{ cm})^3 = 4.2\underline{5}1528\text{ cm}^3 = 4.251528\text{ mL}$ then $d = m/V$. Rearrange to solve for m. $m = dV$

$m_{\text{Cu}} = 8.96\,\dfrac{\text{g}}{\text{mL}} \times 3.7\underline{2}3875\text{ mL} = 33.36592\text{ g Cu}$

$$m_{Al} = 2.70\,\frac{g}{mL} \times 4.2\underline{5}1528\,mL = 11.\underline{4}791256\,g\,Al \text{ and } m_{H_2O} = 0.998\,\frac{g}{mL} \times 100.0\,mL = 99.8\,g\,H_2O$$

$q = mC_s(T_f - T_i)$ substitute in values and set $q_{Cu} + q_{Al} = -q_{H_2O}$

$q_{Cu} + q_{Al} = m_{Cu}C_{s,\,Cu}(T_f - T_{Cu,\,i}) + m_{Al}C_{s,\,Al}(T_f - T_{Al,\,i}) =$

$$33.3\underline{6}592\,g \times 0.385\,\frac{J}{g\cdot°C} \times (T_f - 55.0\,°C) + 11.\underline{4}791256\,g \times 0.903\,\frac{J}{g\cdot°C} \times (T_f - 55.0\,°C) =$$

$$-q_{H_2O} = -m_{H_2O}C_{s,\,H_2O}(T_f - T_{H_2O,\,i}) = -99.8\,g \times 4.18\,\frac{J}{g\cdot°C} \times (T_f - 22.2\,°C)$$

Rearrange to solve for T_f.

$$12.8\underline{4}588\,\frac{J}{°C} \times (T_f - 55.0\,°C) + 10.\underline{3}6565\,\frac{J}{°C} \times (T_f - 55.0\,°C) = -41\underline{7}.164\,\frac{J}{°C} \times (T_f - 22.2\,°C) \rightarrow$$

$$12.8\underline{4}588\,\frac{J}{°C}T_f - 70\underline{6}.5234\,J + 10.\underline{3}6565\,\frac{J}{°C}T_f - 57\underline{0}.1108\,J = -41\underline{7}.164\,\frac{J}{°C}T_f + 92\underline{6}1.041\,J \rightarrow$$

$$12.8\underline{4}588\,\frac{J}{°C}T_f + 10.\underline{3}6565\,\frac{J}{°C}T_f + 41\underline{7}.164\,\frac{J}{°C}T_f = +70\underline{6}.5234\,J + 57\underline{0}.1108\,J + 92\underline{6}1.041\,J \rightarrow$$

$$44\underline{0}.3755\,\frac{J}{°C}T_f = 105\underline{3}7.675\,J \rightarrow T_f = \frac{105\underline{3}7.675\,J}{44\underline{0}.3755\,\frac{J}{°C}} = 23.\underline{9}2884\,°C = 23.9\,°C$$

Check: The units ($°C$) are correct. The magnitude of the answer (24) makes physical sense because the heat transfer is dominated by the water (larger mass and larger specific heat capacity). The final temperature should be closer to the initial temperature of water than of copper and aluminum.

9.115   To determine the values of the lattice energy, it is necessary to look them up online. The lattice energy of Al_2O_3 is $-15{,}916\,kJ/mol$; the value for Fe_2O_3 is $-14{,}774\,kJ/mol$. The thermite reaction is exothermic due to the energy released when the Al_2O_3 lattice forms. The lattice energy of Al_2O_3 is more negative than the lattice energy of Fe_2O_3.

9.117   Rewrite the reaction using the Lewis structures of the molecules involved.

$$H—H(g) + 1/2\,\ddot{O}{=}\ddot{O}(g) \longrightarrow H—\ddot{O}—H$$

Determine which bonds are broken in the reaction and sum the bond energies of the following:

$\Sigma(\Delta H\text{'s bonds broken})$

$= 1\,mol(H—H) + 1/2\,mol(O{=}O)$

$= 1\,mol(436\,kJ/mol) + 1/2\,mol(498)$

$= 685\,kJ/mol$

Determine which bonds are formed in the reaction and sum the negatives of the bond energies of the following:

$\Sigma(-\Delta H\text{'s bonds formed})$

$= -2\,mol(O—H)$

$= -2\,mol(464\,kJ/mol)$

$= -928\,kJ/mol$

Find ΔH_{rxn} by summing the results of the two steps.

$\Delta H_{rxn} = \Sigma(\Delta H\text{'s bonds broken}) + \Sigma(-\Delta H\text{'s bonds formed})$

$= 685\,kJ/mol - 928\,kJ/mol$

$= -243\,kJ/mol$

$CH_4(g) + 2\,O_2(g) \rightarrow CO_2(g) + 2\,H_2O(g)$

Rewrite the reaction using the Lewis structures of the molecules involved.

$$
\begin{array}{c}
H \\
| \\
H—C—H + 2\,\ddot{O}{=}\ddot{O} \longrightarrow \ddot{O}{=}C{=}\ddot{O} + 2\,H—\ddot{O}—H \\
| \\
H
\end{array}
$$

Determine which bonds are broken in the reaction and sum the bond energies of the following:

$\Sigma(\Delta H\text{'s bonds broken})$

$$= 4\,\text{mol}(\text{C}\!-\!\text{H}) + 2\,\text{mol}(\text{O}\!=\!\text{O})$$
$$= 4\,\text{mol}(414\,\text{kJ/mol}) + 2\,\text{mol}(498)$$
$$= 2652\,\text{kJ/mol}$$

Determine which bonds are formed in the reaction and sum the negatives of the bond energies of the following:

$$\Sigma(-\Delta H\text{'s bonds formed})$$
$$= -2\,\text{mol}(\text{C}\!=\!\text{O}) - 4\,\text{mol}(\text{O}\!-\!\text{H})$$
$$= -2\,\text{mol}(799\,\text{kJ/mol}) - 4\,\text{mol}(464\,\text{kJ/mol})$$
$$= -3454\,\text{kJ/mol}$$

Find ΔH_{rxn} by summing the results of the two steps.

$$\Delta H_{\text{rxn}} = \Sigma(\Delta H\text{'s bonds broken}) + \Sigma(-\Delta H\text{'s bonds formed})$$
$$= 2652\,\text{kJ/mol} - 3454\,\text{kJ/mol}$$
$$= -802\,\text{kJ/mol}$$

Compare the following:

	kJ/mol	kJ/g
H_2	-243	-121 (using a molar mass of 2.016)
CH_4	-802	-50.0 (using a molar mass of 16.04)

$$\frac{-243\,\text{kJ}}{1\,\text{mol}\,H_2} \times \frac{1\,\text{mol}\,H_2}{2.016\,\text{g}\,H_2} = \frac{-121\,\text{kJ}}{\text{g}\,H_2} \qquad \frac{-802\,\text{kJ}}{1\,\text{mol}\,CH_4} \times \frac{1\,\text{mol}\,CH_4}{16.04\,\text{g}\,CH_4} = \frac{-50.0\,\text{kJ}}{\text{g}\,CH_4}$$

So methane yields more energy per mole, but hydrogen yields more energy per gram.

9.119 **Given:** heat atomization $CH_4 = 1660\,\text{kJ/mol}$, $CH_2Cl_2 = 1495\,\text{kJ/mol}$ **Find:** bond energy C—Cl Write the reaction using the Lewis structure.

$$\text{H}-\!\!\!\overset{\displaystyle \text{H}}{\underset{\displaystyle \text{H}}{\text{C}}}\!\!\!-\text{H}(g) \longrightarrow \text{C}(g) + 4\,\text{H}(g)$$

Determine the number and kinds of bonds broken and then ΔH atomization $= \Sigma$ bonds broken.

ΔH atomization $= \Sigma\,4\,(\text{C}\!-\!\text{H})$ bonds broken

$$\frac{1660\,\text{kJ}}{1\,\text{mol}\,CH_4} \times \frac{1\,\text{mol}\,CH_4}{4\,\text{C}\!-\!\text{H bonds}} = 415\,\text{kJ/C}\!-\!\text{H bond}$$

Write the reaction using the Lewis structure.

$$:\!\overset{..}{\underset{..}{\text{Cl}}}\!-\!\!\!\overset{\displaystyle \text{H}}{\underset{\displaystyle :\!\overset{..}{\underset{..}{\text{Cl}}}\!:}{\text{C}}}\!\!\!-\text{H}(g) \longrightarrow \text{C}(g) + 2\,\text{H}(g) + 2\,\text{Cl}(g)$$

Determine the number and kinds of bonds broken and ΔH atomization $= \Sigma$ bonds broken.

ΔH atomization $= \Sigma\,2\,(\text{C}\!-\!\text{H})$ bonds broken $+ 2(\text{C}\!-\!\text{Cl})$ bonds broken

$1495\,\text{kJ/mol} = 2(415\,\text{kJ/mol}\,) + 2(x) \qquad x = 333\,\text{kJ/mol}$ for the C—Cl bond energy

Check: The bond energy found (333 kJ/mol) is very close to the table value of 339 kJ/mol.

Challenge Problems

9.121 **Given:** 655 kWh/yr; coal is 3.2% S; remainder is C; S emitted as $SO_2(g)$ and gets converted to H_2SO_4 when reacting with water **Find:** $m(H_2SO_4)/\text{yr}$

Conceptual Plan: Write balanced reaction then $\Delta H_{\text{rxn}}^{\circ} = \Sigma n_p \Delta H_f^{\circ}(products) - \Sigma n_r \Delta H_f^{\circ}(reactants)$

(because the form of sulfur is not given, assume that all heat is from combustion of only carbon) then

$$\text{kWh} \rightarrow \text{J} \rightarrow \text{kJ} \rightarrow \text{mol}\,(\text{C}) \rightarrow \text{g}\,(\text{C}) \rightarrow \text{g}\,(\text{S}) \rightarrow \text{mol}\,(H_2SO_4) \rightarrow \text{mol}\,(H_2SO_4) \rightarrow \text{g}\,(H_2SO_4)$$

| $\dfrac{3.60 \times 10^6\,\text{J}}{1\,\text{kWh}}$ | $\dfrac{1\,\text{kJ}}{1000\,\text{J}}$ | $\dfrac{\text{mol C}}{\Delta H_f^{\circ}(CO_2(g))}$ | $\dfrac{12.01\,\text{g}}{1\,\text{mol}}$ | $\dfrac{3.2\,\text{g S}}{(100.0-3.2)\,\text{g C}}$ | $\dfrac{1\,\text{mol}}{32.07\,\text{g}}$ | $\dfrac{1\,\text{mol}\,H_2SO_4}{1\,\text{mol S}}$ | $\dfrac{98.09\,\text{g}}{1\,\text{mol}}$ |

Solution: $C(s) + O_2(g) \rightarrow CO_2(g)$. This reaction is the heat of formation of $CO_2(g)$, so

$$\Delta H_{\text{rxn}}^{\circ} = \Delta H_f^{\circ}(CO_2(g)) = -393.5\,\text{kJ/mol then}$$

$$655 \text{ kWh} \times \frac{3.60 \times 10^6 \text{ J}}{1 \text{ kWh}} \times \frac{1 \text{ kJ}}{1000 \text{ J}} \times \frac{1 \text{ mol C}}{393.5 \text{ kJ}} \times \frac{12.01 \text{ g C}}{1 \text{ mol C}} \times \frac{3.2 \text{ g S}}{(100.0 - 3.2) \text{ g C}} \times \frac{1 \text{ mol S}}{32.07 \text{ g S}}$$

$$\times \frac{1 \text{ mol H}_2\text{SO}_4}{1 \text{ mol S}} \times \frac{98.09 \text{ g H}_2\text{SO}_4}{1 \text{ mol H}_2\text{SO}_4} = 7.3 \times 10^3 \text{ g H}_2\text{SO}_4$$

Check: The units (g) are correct. The magnitude (7300) is reasonable, considering this calculation is for only one home.

9.123   **Given:** methane combustion, 100% efficiency; $\Delta T = 10.0\,°C$; house $= 30.0 \text{ m} \times 30.0 \text{ m} \times 3.0 \text{ m}$; C_s (air) $=$ 30 J/K·mol; 1.00 mol air $= 22.4$ L   **Find:** m (CH_4)
Conceptual Plan: $l, w, h \to V(\text{m}^3) \to V(\text{cm}^3) \to V(\text{L}) \to$ mol (air) then $m, C_s, \Delta T \to q_{air}$ (J)

$$V = lwh \qquad \frac{(100 \text{ cm})^3}{(1 \text{ m})^3} \qquad \frac{1 \text{ L}}{1000 \text{ cm}^3} \qquad \frac{1 \text{ mol air}}{22.4 \text{ L}} \qquad\qquad q = mC_s\Delta T$$

then q_{air} (J) $\to q_{rxn}$ (J) $\to q$ (kJ) then write balanced reaction for methane combustion

$$q_{rxn} = -q_{air} \qquad \frac{1 \text{ kJ}}{1000 \text{ J}}$$

then $\Delta H^\circ_{rxn} = \sum n_p \Delta H^\circ_f(\text{products}) - \sum n_r \Delta H^\circ_f(\text{reactants})$, **and then** q (kJ) $\to$ mol (CH_4) $\to$ g (CH_4)

$$\Delta H^\circ_{rxn} \qquad\qquad \frac{16.04 \text{ g}}{1 \text{ mol}}$$

Solution: $V = lwh = 30.0 \text{ m} \times 30.0 \text{ m} \times 3.0 \text{ m} = 2700 \text{ m}^3$, then

$$2700 \text{ m}^3 \times \frac{(100 \text{ cm})^3}{(1 \text{ m}^3)} \times \frac{1 \text{ L}}{1000 \text{ cm}^3} \times \frac{1 \text{ mol air}}{22.4 \text{ L}} = 1.20536 \times 10^5 \text{ mol air, and then}$$

$$q = mC_s\Delta T = 1.20536 \times 10^5 \text{ mol} \times 30 \frac{\text{J}}{\text{mol} \cdot °C} \times 10.0 \,°C = 3.6161 \times 10^7 \text{ J} \times \frac{1 \text{ kJ}}{1000 \text{ J}} = 3.6161 \times 10^4 \text{ J needed}$$

$$CH_4(g) + 2 O_2(g) \to CO_2(g) + 2 H_2O(g)$$

Reactant/Product	ΔH°_f (kJ/mol from Appendix IIB)
$CH_4(g)$	−74.6
$O_2(g)$	0.0
$CO_2(g)$	−393.5
$H_2O(g)$	−241.8

Be sure to pull data for the correct formula and phase.

$$\Delta H^\circ_{rxn} = \sum n_p \Delta H^\circ_f(\text{products}) - \sum n_r \Delta H^\circ_f(\text{reactants})$$
$$= [1(\Delta H^\circ_f(CO_2(g))) + 2(\Delta H^\circ_f(H_2O(g)))] - [1(\Delta H^\circ_f(CH_4(g))) + 2(\Delta H^\circ_f(O_2(g)))]$$
$$= [(-393.5 \text{ kJ}) + 2(-241.8 \text{ kJ})] - [1(-74.6 \text{ kJ}) + 2(0.0 \text{ kJ})]$$
$$= [-877.1 \text{ kJ}] - [-74.6 \text{ kJ}]$$
$$= -802.5 \text{ kJ}$$

$$q_{rxn} = -q_{air} = -3.6161 \times 10^4 \text{ kJ} \times \frac{1 \text{ mol CH}_4}{-802.5 \text{ kJ}} \times \frac{16.04 \text{ g CH}_4}{1 \text{ mol CH}_4} = 722.8 \text{ g CH}_4 = 700 \text{ g CH}_4$$

Check: The units (g) are correct. The magnitude (700) is not surprising because the volume of a house is large.

9.125   **Given:** $m(\text{ice}) = 9.0$ g; coffee: $T_1 = 90.0\,°C$, $m = 120.0$ g, $C_s = C_{s, H_2O}$, $\Delta H^\circ_{fus} = 6.0$ kJ/mol   **Find:** T_f of coffee
Conceptual Plan: $q_{ice} = -q_{coffee}$ so g (ice) $\to$ mol (ice) $\to q_{fus}$ (kJ) $\to q_{fus}$ (J) $\to q_{coffee}$ (J) then

$$\frac{1 \text{ mol}}{18.01 \text{ g}} \qquad \frac{6.0 \text{ kJ}}{1 \text{ mol}} \qquad \frac{1000 \text{ J}}{1 \text{ kJ}} \qquad q_{coffee} = -q_{ice}$$

$q, m, C_s \to \Delta T$ then $T_i, \Delta T \to T_2$; **now we have slightly cooled coffee in contact with 0.0 °C water**

$$q = mC_s\Delta T \qquad\qquad \Delta T = T_2 - T_i$$

so $q_{ice} = -q_{coffee}$ with $m, C_s, T_i \to T_f$

$$q = mC_s(T_f - T_i) \text{ then set } q_{H_2O} = -q_{coffee}$$

Solution: $9.0 \text{ g} \times \frac{1 \text{ mol}}{18.01 \text{ g}} \times \frac{6.0 \text{ kJ}}{1 \text{ mol}} \times \frac{1000 \text{ J}}{1 \text{ kJ}} = 2.9983 \times 10^3 \text{ J}$, $q_{coffee} = -q_{ice} = -2.9983 \times 10^3 \text{ J}$

$q = mC_s\Delta T.$ Rearrange to solve for ΔT. $\Delta T = \dfrac{q}{mC_s} = \dfrac{-2.9983 \times 10^3 \text{ J}}{120.0 \text{ g} \times 4.18 \dfrac{\text{J}}{\text{g} \cdot {}^\circ\text{C}}} = -5.9775 \,{}^\circ\text{C}$ then

$\Delta T = T_2 - T_i.$ Rearrange to solve for T_2. $T_2 = \Delta T + T_i = -5.9775\,{}^\circ\text{C} + 90.0\,{}^\circ\text{C} = 84.0225\,{}^\circ\text{C}$

$q = mC_s(T_f - T_i)$ substitute in values and set $q_{H_2O} = -q_{coffee}$

$q_{H_2O} = m_{H_2O}C_{s,\,H_2O}(T_f - T_{H_2O,i}) = 9.0 \text{ g} \times 4.18 \dfrac{\text{J}}{\text{g} \cdot {}^\circ\text{C}} \times (T_f - 0.0\,{}^\circ\text{C}) =$

$-q_{coffee} = -m_{coffee}C_{s,\,coffee}(T_f - T_{coffee,\,i}) = -120.0 \text{ g} \times 4.18 \dfrac{\text{J}}{\text{g} \cdot {}^\circ\text{C}} \times (T_f - 84.0225\,{}^\circ\text{C})$

Rearrange to solve for T_f.

$9.0 \text{ g } T_f = -120.0 \text{ g } (T_f - 84.0225\,{}^\circ\text{C}) \rightarrow 9.0 \text{ g } T_f = -120.0 \text{ g } T_f + 10082.7 \text{ g} \rightarrow$

$+10082.7 \text{ g} = +129.0 \dfrac{\text{g}}{{}^\circ\text{C}} T_f \rightarrow T_f = \dfrac{+10082.7 \text{ g}}{+129.0 \dfrac{\text{g}}{{}^\circ\text{C}}} = 78.2\,{}^\circ\text{C}$

Check: The units (°C) are correct. The temperature is closer to the original coffee temperature because the mass of coffee is so much larger than the ice mass.

9.127 $q = \Delta H = 454 \text{ g} \times \dfrac{1 \text{ mol}}{18.02 \text{ g}} \times \dfrac{40.7 \text{ kJ}}{1 \text{ mol}} = 1025.405 \text{ kJ} = 1030 \text{ kJ}$ and $w = -P\Delta V$. Assume that $P = 1$ atm

(exactly) and $\Delta V = V_G - V_L$, where $V_L = 454 \text{ g} \times \dfrac{1 \text{ mL}}{0.9998 \text{ g}} \times \dfrac{1 \text{ L}}{1000 \text{ mL}} = 0.4540908 \text{ L}$ and $PV = nRT$.

Rearrange to solve for V_G. $V_G = \dfrac{nRT}{P} = \dfrac{454 \text{ g} \times \dfrac{1 \text{ mol}}{18.02 \text{ g}} \times 0.08206 \dfrac{\text{L} \cdot \text{atm}}{\text{mol} \cdot \text{K}} \times 373 \text{ K}}{1 \text{ atm}} = 771.1545 \text{ L}$

$\Delta V = V_G - V_L = 771.1545 \text{ L} - 0.4540908 \text{ L} = 770.7004 \text{ L}$ and so

$w = -P\Delta V = -1.0 \text{ atm} \times 770.7004 \text{ L} \times \dfrac{101.3 \text{ J}}{1 \text{ L} \cdot \text{atm}} = -78071.9515 \text{ J} = -7.81 \times 10^4 \text{ J} = -78.1 \text{ kJ}$

Finally, $\Delta E = q + w = 1025.405 \text{ kJ} - 78.0719515 \text{ kJ} = 947.333 \text{ kJ} = 950 \text{ kJ}$.

9.129

Step 1:

Bonds broken: $2 \text{ mol}(S{=}O) + 1 \text{ mol}(H{-}O) = 2 \text{ mol}(523 \text{ kJ/mol}) + 1 \text{ mol}(464 \text{ kJ/mol}) = 1510 \text{ kJ/mol}$

Bonds formed: $-2 \text{ mol}(S{-}O) - 1 \text{ mol}(S{=}O) - 1 \text{ mol}(O{-}H) =$

$-2 \text{ mol}(265 \text{ kJ/mol}) - 1 \text{ mol}(523 \text{ kJ/mol}) - 1 \text{ mol}(464 \text{ kJ/mol}) = -1517 \text{ kJ/mol}$

$\Delta H_{step} = -7 \text{ kJ/mol}$

Step 2:

Bonds broken: $2\,mol(S{-}O) + 1\,mol(S{=}O) + 1\,mol(O{-}H) + 1\,mol(O{=}O) =$
$2\,mol(265\,kJ/mol) + 1\,mol(523\,kJ/mol) + 1\,mol(464\,kJ/mol) + 1\,mol(498\,kJ/mol) =$
$2015\,kJ/mol$

Bonds formed: $-2\,mol(S{-}O) - 1\,mol(S{=}O) - 1\,mol(O{-}H) - 1\,mol(O{=}O) =$
$-2\,mol(265\,kJ/mol) - 1\,mol(523\,kJ/mol) - 1\,mol(464\,kJ/mol) - 1\,mol(142\,kJ/mol) =$
$-1659\,kJ/mol$

$\Delta H_{step} = +356\,kJ/mol$

Step 3:

Bonds broken: $2\,mol(S{-}O) + 1\,mol(S{=}O) + 2\,mol(O{-}H) =$
$2\,mol(265\,kJ/mol) + 1\,mol(523\,kJ/mol) + 2\,mol(464\,kJ/mol) = 1981\,kJ/mol$

Bonds formed: $-2\,mol(S{-}O) - 2\,mol(S{=}O) - 2\,mol(O{-}H) =$
$-2\,mol(265\,kJ/mol) + -2\,mol(523\,kJ/mol) + -2\,mol(464\,kJ/mol) = -2504\,kJ/mol$

$\Delta H_{step} = -523\,kJ/mol$

Hess's law states that ΔH for the reaction is the sum of ΔH of the steps:
$\Delta H_{rxn} = (-7\,kJ/mol) + (+356\,kJ/mol) + (-523\,kJ/mol) = -174\,kJ/mol$

9.131   **Given:** ΔH_f° $CaBr_2 = -675\,kJ/mol$; $IE_1(Ca) = 590\,kJ/mol$; $IE_2(Ca) = 1145\,kJ/mol$; $\Delta H_{sub}(Ca) = 178\,kJ/mol$;
$Br_2(g)$ bond energy $= 193\,kJ/mol$; $\Delta H_{vap}(Br_2(l)) = 31\,kJ/mol$; $EA(Br) = -325\,kJ/mol$ **Find:** lattice energy
Conceptual Plan:
$$Ca(s) + Br_2(l) \rightarrow Ca(g) + Br_2(l) \rightarrow Ca^+(g) + Br_2(l) \rightarrow Ca^{2+}(g) + Br_2(l) \rightarrow Ca^{2+}(g) + Br_2(g) \rightarrow Ca^{2+}(g) + 2Br(g)$$

$\quad\quad\quad\quad \Delta H_{sub} \quad\quad\quad\quad IE_1 \quad\quad\quad\quad IE_2 \quad\quad\quad\quad \Delta H_{vap} \quad\quad\quad$ bond energy

$$\rightarrow Ca^{2+}(g) + 2Br^-(g) \rightarrow CaBr_2(s)$$

$\quad\quad EA \quad\quad\quad\quad$ lattice energy

ΔH_f°

Solution: $\Delta H_f^\circ = \Delta H_{sub} + IE_1 + IE_2 + \Delta H_{vap} +$ **bond energy** $+$ **2 EA** $+$ **lattice energy**

$-675\,\dfrac{kJ}{mol} = +178\,\dfrac{kJ}{mol} + 590\,\dfrac{kJ}{mol} + 1145\,\dfrac{kJ}{mol} + 31\,\dfrac{kJ}{mol} + 193\,\dfrac{kJ}{mol} + 2(-325)\,\dfrac{kJ}{mol} +$ lattice energy

lattice energy $= -2162\,kJ/mol$

Conceptual Problems

9.133   (d)   Only one answer is possible: $\Delta E_{sys} = -\Delta E_{surr}$.

9.135   Refrigerator A contains only air, which will cool quickly but will not stabilize the temperature. Refrigerator B is filled with containers of water, which require a great deal of energy to cool on day 1 but will remain stable at a cold temperature on day 2.

9.137   Because $q = mC_s\Delta T$, $m_A = 2\,m_B$, $C_{s,\,B} = 4\,C_{s,\,A}$, $q_A = -q_B$, we can substitute into the equation to get $q_A = m_A C_{s,\,A}\Delta T_A =$
$-q_B = -m_B C_{s,\,B}\Delta T_B \rightarrow (2\,m_B)C_{s,\,A}\Delta T_A = -m_B(4\,C_{s,\,A})\Delta T_B$

$\rightarrow \Delta T_A = \dfrac{-\cancel{m_B}(4\,\cancel{C_{s,\,A}})}{(2\,\cancel{m_B})\cancel{C_{s,\,A}}}\Delta T_B = \dfrac{-4}{2}\Delta T_B = -2\Delta T_B$, or the temperature change for substance A is twice the magnitude of the temperature change for substance B.

9.139   The internal energy of a chemical system is the sum of its kinetic energy and its potential energy. This potential energy is the energy source in an exothermic chemical reaction. Under normal circumstances, chemical potential energy (or simply chemical energy) arises primarily from the electrostatic forces between the protons and electrons that compose the atoms and molecules within the system. In an exothermic reaction, some bonds break and new ones form, and the protons and electrons go from an arrangement of higher potential energy to one of lower potential energy. As they rearrange, their potential energy is converted into kinetic energy. Heat is emitted in the reaction, so it feels hot to the touch.

9.141   (a) is true: Strong bonds break, and weak bonds form. In an endothermic reaction, the energy required to break the bonds is greater than the energy given off when the bonds are formed ($\Delta H > 0$); therefore, in an endothermic reaction, the bonds that are breaking are stronger than the bonds that are forming.

Questions for Group Work

9.143 Examples of heat transfer problems can be seen in Problems 6.33–6.38. All of the problems use the equations $q = m\,C_s\,\Delta T$ where $\Delta T = T_f - T_i$. The difference can include different masses of the two object/substances, different initial and final temperatures, and different identities of the two object/substances (which changes the heat capacities).

9.145 (a) $C_3H_8(g) + 5\,O_2(g) \rightarrow 3\,CO_2(g) + 4\,H_2O(g)$

(b) $10.4 \times 10^3 \; g\,\cancel{C_3H_8} \times \dfrac{1\;mol\,\cancel{C_3H_8}}{44.09\;g\,\cancel{C_3H_8}} \times \dfrac{-2044\;kJ}{1\;mol\,\cancel{C_3H_8}} = -4.82 \times 10^5\;kJ$

(c) $q = m\,C_s\,\Delta T$ so $m = \dfrac{q}{C_s\,\Delta T} = \dfrac{-4.82 \times 10^5\;kJ \times \dfrac{1000\;J}{1\;kJ}}{4.184\,\dfrac{J}{g \times \degree C} \times (100.\,\degree C - 25\,\degree C)} = 1.5 \times 10^6\;g\;water$

9.147 The decomposition of hydrogen peroxide is $2\,H_2O_2(l) \rightarrow 2\,H_2O(l) + O_2(g)$.

(a) The heat of formation of hydrogen peroxide is $H_2(g) + O_2(g) \rightarrow H_2O_2(l)$ $\Delta H = -187.8$ kJ.

(b) The heat of formation of water is $H_2(g) + 1/2\,O_2(g) \rightarrow H_2O(l)$ $\Delta H = -285.8$ kJ.

(c) The heat of formation of oxygen gas is defined as zero, since it is an element in its most stable state at 25 °C.

(d) See above.

(e) $2\,H_2O_2(l) \rightarrow 2\,H_2O(l) + O_2(g)$.

(f) The heat of the reaction in part (e) is
$$\Delta H^\circ_{rxn} = \sum n_P \Delta H^\circ_f(products) - \sum n_R \Delta H^\circ_f(reactants)$$
$$= [2(\Delta H^\circ_f(H_2O(l))) + 1(\Delta H^\circ_f(O_2(g))] - [2(\Delta H^\circ_f(H_2O_2(l))]$$
$$= [2(-285.8\;kJ) + 1(0.0\;kJ)] - [2(-187.8\;kJ)]$$
$$= [-571.6\;kJ] - [-375.6\;kJ]$$
$$= -196.0\;kJ$$

(g)

Data Interpretation and Analysis

9.148 (a) (i) $CH_3CH_2CH_3(g) + 5\,O_2(g) \longrightarrow 3\,CO_2(g) + 4\,H_2O(l)$

(ii) $CH_3CH_2CH_3(g) + 5\,O_2(g) \longrightarrow 3\,CO_2(g) + 4\,H_2O(g)$

(c) Assuming most of the water stays in the gas phase, the heat released would be the LHV.

Reactant/Product	ΔH°_f (kJ/mol from Appendix IIB)
$C_2H_4(g)$	52.4
$H_2(g)$	0.0
$C_2H_6(g)$	−84.68

Reactant/Product	ΔH°_f (kJ/mol from Appendix IIB)
$CO(g)$	−110.5
$H_2O(g)$	−241.8
$H_2(g)$	0.0
$CO_2(g)$	−393.5

10 Gases

Review Questions

10.1 The basic postulates of kinetic molecular theory are as follows: (1) The size of a particle is negligibly small, (2) the average kinetic energy of a particle is proportional to the temperature in kelvins, and (3) the collision of one particle with another (or with the walls) is completely elastic. Pressure is defined as force divided by area. According to kinetic molecular theory, a gas is a collection of particles in constant motion. The motion results in collisions between the particles and the surfaces around them. As each particle collides with a surface, it exerts a force upon that surface. The result of many particles in a gas sample exerting forces on the surfaces around them is constant pressure.

10.3 The pressure drops as altitude increases because the gravitational forces are decreased at higher altitudes compared to sea level.

10.5 $101,325 \, \text{Pa} = 760 \, \text{torr} = 760 \, \text{mmHg} = 29.92 \, \text{in Hg} = 14.7 \, \text{psi} = 1 \, \text{atm}$

10.7 Boyle's law states that the volume of the gas varies inversely to the pressure on the gas, while temperature and number of moles are kept constant ($P_1 V_1 = P_2 V_2$). Charles's law states that the volume of a gas is directly proportional to the temperature of the gas, while pressure and number of moles are kept constant ($V_1/T_1 = V_2/T_2$). All temperatures must be in kelvins when used in gas law calculations. Avogadro's law states that the volume of a gas is directly proportional to the number of moles of the gas, while pressure and temperature are kept constant ($V_1/n_1 = V_2/n_2$).

10.9 For every 10 m of depth a diver descends in water, she experiences an additional 1 atm of pressure due to the weight of the water above her. The pressure regulator used in scuba diving delivers air at a pressure that matches the external pressure; otherwise, the diver could not inhale the air. For example, when a diver is at a depth of 20 m below the surface, the regulator delivers air at a pressure of 3 atm to match the 3 atm of pressure around the diver (1 atm due to normal atmospheric pressure and 2 additional atmospheres due to the weight of the water at 20 m). Suppose a diver inhaled a lungful of air at a pressure of 3 atm and swam quickly to the surface (where the pressure drops to 1 atm) while holding his breath. What would happen to the volume of air in the diver's lungs? Because the pressure decreases by a factor of three, the volume of the air in the diver's lungs would increase by a factor of three, severely damaging his lungs and possibly killing him.

10.11 Charles's law explains why a hot-air balloon can take flight. The gas that fills a hot-air balloon is warmed with a burner, increasing its volume and lowering its density and causing it to float in the colder, denser surrounding air.

10.13 We know that $V \propto 1/P$ (Boyle's law), $V \propto T$ (Charles's law), $V \propto n$ (Avogadro's law).

Combining these three expressions, we get $V \propto nT/P$. Replace the proportional sign with an equal sign by incorporating R (the ideal gas constant): $V = RnT/P$. Rearranging, we get $PV = nRT$.

10.15 Because $d = \dfrac{P \, \mathcal{M}}{R \, T}$, the density will decrease as temperature increases. It will increase as pressure increases or as the molar mass of the gas increases.

10.17   Boyle's law states that for a constant number of particles at constant temperature, the volume of a gas is inversely proportional to its pressure. If you decrease the volume of a gas, you force the gas particles to occupy a smaller space. It follows from kinetic molecular theory that as long as the temperature remains the same, the result is a greater number of collisions with the surrounding surfaces and, therefore, a greater pressure.

Charles's law states that for a constant number of particles at constant pressure, the volume of a gas is proportional to its temperature. According to kinetic molecular theory, when you increase the temperature of a gas, the average speed, and thus the average kinetic energy, of the particles increases. Because this greater kinetic energy results in more frequent collisions and more force per collision, the pressure of the gas would increase if its volume were held constant (Gay-Lussac's law). The only way for the pressure to remain constant is for the volume to increase. The greater volume spreads the collisions out over a greater area, so that the pressure (defined as force per unit area) is unchanged.

Avogadro's law states that at constant temperature and pressure, the volume of a gas is proportional to the number of particles. According to kinetic molecular theory, when you increase the number of particles in a gas sample, the number of collisions with the surrounding surfaces increases. Because the greater number of collisions results in a greater overall force on surrounding surfaces, the only way for the pressure to remain constant is for the volume to increase so that the number of particles per unit volume (and thus the number of collisions) remains constant.

Dalton's law states that the total pressure of a gas mixture is the sum of the partial pressures of its components. In other words, according to Dalton's law, the components in a gas mixture act identically to, and independently of, one another. According to kinetic molecular theory, the particles have negligible size and they do not interact. Consequently, the only property that would distinguish one type of particle from another is its mass. However, even particles of different masses have the same average kinetic energy at a given temperature; so they exert the same force upon a collision with a surface. Consequently, adding components to a gas mixture—even different *kinds* of gases—has the same effect as simply adding more particles. The partial pressures of all of the components sum to the overall pressure.

10.19   No, when collecting a gas over water, it will contain some water molecules. The vapor pressure of water can be found in Table 10.3. Therefore, $P_{Gas} = P_{Total} - P_{H_2O}$.

10.21   Gaseous particles travel at tremendous speeds along very haphazard paths. To a perfume molecule, the path from the perfume bottle in the bathroom to your nose 2 m away is much like the path through a busy shopping mall during a clearance sale. The molecule travels only a short distance before it collides with another molecule, changes direction, only to collide again, and so on. The average distance a molecule travels between collisions is called its mean free path.

10.23   In Chapter 7, we learned how the coefficients in chemical equations can be used as conversion factors between number of moles of reactants and number of moles of products in a chemical reaction, and how the molar mass can be used to convert the number of moles to the mass. At STP, each mole of gas occupies 22.414 L. The mass of the product will be as follows:

$$\text{Volume of limiting reagent (L)} \times \frac{1 \text{ mol limiting reagent}}{22.414 \text{ L}} \times \frac{c \text{ mol product}}{a \text{ mol limiting reagent}} \times \frac{g \text{ product}}{1 \text{ mol product}}$$

for the reaction:
a A + b B → c C + d D where A is the limiting reagent and C is the product of interest.

Problems by Topic

Converting between Pressure Units

10.25   (a)   **Given:** 24.9 in Hg   **Find:** atm
Conceptual Plan: in Hg → atm

$$\frac{1 \text{ atm}}{29.92 \text{ in Hg}}$$

Solution: $24.9 \text{ in Hg} \times \dfrac{1 \text{ atm}}{29.92 \text{ in Hg}} = 0.832 \text{ atm}$

Check: The units (atm) are correct. The magnitude of the answer (< 1) makes physical sense because we started with less than 29.92 in Hg.

(b)   **Given:** 24.9 in Hg   **Find:** mmHg
         Conceptual Plan: Use answer from part (a), then convert atm → mmHg

$$\frac{760 \text{ mmHg}}{1 \text{ atm}}$$

Solution: $0.832 \text{ atm} \times \dfrac{760 \text{ mmHg}}{1 \text{ atm}} = 632 \text{ mmHg}$

Check: The units (mmHg) are correct. The magnitude of the answer (< 760) makes physical sense because we started with less than 1 atm.

(c)   **Given:** 24.9 in Hg   **Find:** psi
         Conceptual Plan: Use answer from part (a), then convert atm → psi

$$\frac{14.7 \text{ psi}}{1 \text{ atm}}$$

Solution: $0.832 \text{ atm} \times \dfrac{14.7 \text{ psi}}{1 \text{ atm}} = 12.2 \text{ psi}$

Check: The units (psi) are correct. The magnitude of the answer (< 14.7) makes physical sense because we started with less than 1 atm.

(d)   **Given:** 24.9 in Hg   **Find:** Pa
         Conceptual Plan: Use answer from part (a), then convert atm → Pa

$$\frac{101,325 \text{ Pa}}{1 \text{ atm}}$$

Solution: $0.832 \text{ atm} \times \dfrac{101,325 \text{ Pa}}{1 \text{ atm}} = 8.43 \times 10^4 \text{ Pa}$

Check: The units (Pa) are correct. The magnitude of the answer $(< 101,325)$ makes physical sense because we started with less than 1 atm.

10.27   (a)   **Given:** 31.85 in Hg   **Find:** mmHg
                  Conceptual Plan: in Hg → mmHg

$$\frac{25.4 \text{ mmHg}}{1 \text{ in Hg}}$$

Solution: $31.85 \text{ in Hg} \times \dfrac{25.4 \text{ mmHg}}{1 \text{ in Hg}} = 809.0 \text{ mmHg}$

Check: The units (mmHg) are correct. The magnitude of the answer (809) makes physical sense because inches are larger than millimeters.

(b)   **Given:** 31.85 in Hg   **Find:** atm
         Conceptual Plan: Use answer from part (a), then convert mmHg → atm

$$\frac{1 \text{ atm}}{760 \text{ mmHg}}$$

Solution: $809.0 \text{ mmHg} \times \dfrac{1 \text{ atm}}{760 \text{ mmHg}} = 1.064 \text{ atm}$

Check: The units (atm) are correct. The magnitude of the answer (> 1) makes physical sense because we started with more than 760 mmHg.

(c)   **Given:** 31.85 in Hg   **Find:** torr
         Conceptual Plan: Use answer from part (a), then convert mmHg → torr

$$\frac{1 \text{ torr}}{1 \text{ mmHg}}$$

Solution: $809.0 \text{ mmHg} \times \dfrac{1 \text{ torr}}{1 \text{ mmHg}} = 809.0 \text{ torr}$

Check: The units (torr) are correct. The magnitude of the answer (809) makes physical sense because both units are of the same size.

(d)   **Given:** 31.85 in Hg   **Find:** kPa
        Conceptual Plan: Use answer from part (b), then convert atm → Pa → kPa

$$\frac{101{,}325 \text{ Pa}}{1 \text{ atm}} \quad \frac{1 \text{ kPa}}{1000 \text{ Pa}}$$

Solution: $1.064 \text{ atm} \times \dfrac{101{,}325 \text{ Pa}}{1 \text{ atm}} \times \dfrac{1 \text{ kPa}}{1000 \text{ Pa}} = 107.8 \text{ kPa}$

Check: The units (kPa) are correct. The magnitude of the answer (108) makes physical sense because we started with more than 1 atm and there are ~101 kPa in an atm.

10.29   (a)   **Given:** $P_{bar} = 762.4 \text{ mmHg}$ and figure   **Find:** P_{gas}
        Conceptual Plan: Measure height difference, then convert cm Hg → mmHg → mmHg

$$\frac{10 \text{ mmHg}}{1 \text{ cm Hg}} \quad P_{gas} = h + P_{bar}$$

Solution:

$$h = 7.0 \text{ cm Hg} \times \frac{10 \text{ mmHg}}{1 \text{ cm Hg}} = 70.\text{ mmHg} \quad P_{gas} = 70.\text{ mmHg} + 762.4 \text{ mmHg} = 832 \text{ mmHg}$$

Check: The units (mmHg) are correct. The magnitude of the answer (832) makes physical sense because the mercury column is higher on the right, indicating that the pressure is above barometric pressure. No significant figures to the right of the decimal point can be reported because the mercury height is known only to the ones place. One digit beyond the markings can be estimated (and is included above), and it would be considered significant.

(b)   **Given:** $P_{bar} = 762.4 \text{ mmHg}$ and figure   **Find:** P_{gas}
        Conceptual Plan: Measure height difference, then convert cm Hg → mmHg → mmHg

$$\frac{10 \text{ mmHg}}{1 \text{ cm Hg}} \quad P_{gas} = h + P_{bar}$$

Solution:

$$h = -4.4 \text{ cm Hg} \times \frac{10 \text{ mmHg}}{1 \text{ cm Hg}} = -44 \text{ mmHg} \quad P_{gas} = -44 \text{ mmHg} + 762.4 \text{ mmHg} = 718 \text{ mmHg}$$

Check: The units (mmHg) are correct. The magnitude of the answer (718) makes physical sense because the mercury column is higher on the left, indicating that the pressure is below barometric pressure. No significant figures to the right of the decimal point can be reported because the mercury height is known only to the ones place. One digit beyond the markings can be estimated (and is included above), and it would be considered significant.

Simple Gas Laws

10.31   **Given:** $V_1 = 5.6 \text{ L}$, $P_1 = 735 \text{ mmHg}$, and $V_2 = 9.4 \text{ L}$ **Find:** P_2
        Conceptual Plan: $V_1, P_1, V_2 \rightarrow P_2$

$$P_1 V_1 = P_2 V_2$$

Solution:

$P_1 V_1 = P_2 V_2$ Rearrange to solve for P_2.

$$P_2 = P_1 \frac{V_1}{V_2} = 735 \text{ mmHg} \times \frac{5.6 \text{ L}}{9.4 \text{ L}} = 437.872 \text{ mmHg} = 4.4 \times 10^2 \text{ mmHg}$$

Check: The units (mmHg) are correct. The magnitude of the answer (440) makes physical sense because Boyle's law indicates that as the volume increases, the pressure decreases.

10.33   **Given:** $V_1 = 48.3 \text{ mL}$, $T_1 = 22 \,^\circ\text{C}$, and $T_2 = 87 \,^\circ\text{C}$ **Find:** V_2
        Conceptual Plan: °C → K then $V_1, T_1, T_2 \rightarrow V_2$

$$K = \,^\circ\text{C} + 273.15 \quad \frac{V_1}{T_1} = \frac{V_2}{T_2}$$

Solution: $T_1 = 22 \,^\circ\text{C} + 273.15 = 295 \text{ K}$ and $T_2 = 87 \,^\circ\text{C} + 273.15 = 360.\text{ K}$

$$\frac{V_1}{T_1} = \frac{V_2}{T_2} \quad \text{Rearrange to solve for } V_2. \; V_2 = V_1 \frac{T_2}{T_1} = 48.3 \text{ mL} \times \frac{360 \text{ K}}{295 \text{ K}} = 58.9 \text{ mL}$$

Check: The units (mL) are correct. The magnitude of the answer (59) makes physical sense because Charles's law indicates that as the volume increases, the temperature increases.

10.35 **Given:** $V_1 = 2.46$ L, $n_1 = 0.158$ mol, and $\Delta n = 0.113$ mol **Find:** V_2

Conceptual Plan: $n_1 \rightarrow n_2$ then $V_1, n_1, n_2 \rightarrow V_2$

$$n_1 + \Delta n = n_2 \qquad \frac{V_1}{n_1} = \frac{V_2}{n_2}$$

Solution: $n_2 = 0.158$ mol $+ 0.113$ mol $= 0.271$ mol

$\frac{V_1}{n_1} = \frac{V_2}{n_2}$ Rearrange to solve for V_2. $V_2 = V_1 \frac{n_2}{n_1} = 2.46$ L $\times \frac{0.271 \text{ mol}}{0.158 \text{ mol}} = 4.21937$ L $= 4.22$ L

Check: The units (L) are correct. The magnitude of the answer (4) makes physical sense because Avogadro's law indicates that as the number of moles increases, the volume increases.

Ideal Gas Law

10.37 **Given:** $n = 0.118$ mol, $P = 0.97$ atm, and $T = 305$ K **Find:** V

Conceptual Plan: $n, P, T \rightarrow V$

$$PV = nRT$$

Solution: $PV = nRT$ Rearrange to solve for V. $V = \frac{nRT}{P} = \frac{0.118 \text{ mol} \times 0.08206 \frac{\text{L} \cdot \text{atm}}{\text{mol} \cdot \text{K}} \times 305 \text{ K}}{0.97 \text{ atm}} = 3.0$ L

The volume would be the same for argon gas because the ideal gas law does not care about the mass of the gas, only the number of moles of gas.

Check: The units (L) are correct. The magnitude of the answer (3) makes sense because, as you will see in the next section, 1 mole of an ideal gas under standard conditions (273 K and 1 atm) occupies 22.4 L. Although these are not standard conditions, they are close enough for a ballpark check of the answer. Because this gas sample contains 0.118 mole, a volume of 3 L is reasonable.

10.39 **Given:** $V = 10.0$ L, $n = 0.448$ mol, and $T = 315$ K **Find:** P

Conceptual Plan: $n, V, T \rightarrow P$

$$PV = nRT$$

Solution:

$PV = nRT$ Rearrange to solve for P. $P = \frac{nRT}{V} = \frac{0.448 \text{ mol} \times 0.08206 \frac{\text{L} \cdot \text{atm}}{\text{mol} \cdot \text{K}} \times 315 \text{ K}}{10.0 \text{ L}} = 1.16$ atm

Check: The units (atm) are correct. The magnitude of the answer (~ 1) makes sense because, as you will see in the next section, 1 mole of an ideal gas under standard conditions (273 K and 1 atm) occupies 22.4 L. Although these are not standard conditions, they are close enough for a ballpark check of the answer. Because this gas sample contains 0.448 mole in a volume of 10 L, a pressure of 1 atm is reasonable.

10.41 **Given:** $V = 28.5$ L, $P = 1.8$ atm, and $T = 298$ K **Find:** n

Conceptual Plan: $V, P, T \rightarrow n$

$$PV = nRT$$

Solution: $PV = nRT$ Rearrange to solve for n. $n = \frac{PV}{RT} = \frac{1.8 \text{ atm} \times 28.5 \text{ L}}{0.08206 \frac{\text{L} \cdot \text{atm}}{\text{mol} \cdot \text{K}} \times 298 \text{ K}} = 2.1$ mol

Check: The units (mol) are correct. The magnitude of the answer (2) makes sense because, as you will see in the next section, 1 mole of an ideal gas under standard conditions (273 K and 1 atm) occupies 22.4 L. Although these are not standard conditions, they are close enough for a ballpark check of the answer. Because this gas sample has a volume of 28.5 L and a pressure of 1.8 atm, ~ 2 mol is reasonable.

10.43 **Given:** $P_1 = 36.0$ psi (gauge P), $V_1 = 11.8$ L, $T_1 = 12.0\,°C$, $V_2 = 12.2$ L, and $T_2 = 65.0\,°C$
Find: P_2 and compare to $P_{\text{max}} = 38.0$ psi (gauge P)

Conceptual Plan: $°C \rightarrow K$ and gauge $P \rightarrow psi \rightarrow atm$ then $P_1, V_1, T_1, V_2, T_2 \rightarrow P_2$

$$K = °C + 273.15 \qquad psi = gauge\ P + 14.7 \frac{1\ atm}{14.7\ psi} \qquad \frac{P_1 V_1}{T_1} = \frac{P_2 V_2}{T_2}$$

Solution: $T_1 = 12.0\ °C + 273.15 = 285.2\ K$ and $T_2 = 65.0\ °C + 273.15 = 338.2\ K$

$$P_1 = 36.0\ psi\ (gauge\ P) + 14.7 = 50.7\ psi \times \frac{1\ atm}{14.7\ psi} = 3.44898\ atm$$

$$P_{max} = 38.0\ psi\ (gauge\ P) + 14.7 = 52.7\ psi \times \frac{1\ atm}{14.7\ psi} = 3.59\ atm$$

$$\frac{P_1 V_1}{T_1} = \frac{P_2 V_2}{T_2} \quad \text{Rearrange to solve for } P_2.\ P_2 = P_1 \frac{V_1}{V_2}\frac{T_2}{T_1} = 3.44898\ atm \times \frac{11.8\ L}{12.2\ L} \times \frac{338.2\ K}{285.2\ K} = 3.96\ atm$$

This exceeds the maximum tire rating of 3.59 atm or 38.0 psi (gauge P).

Check: The units (atm) are correct. The magnitude of the answer (3.96) makes physical sense because the relative increase in T is greater than the relative increase in V; so P should increase.

10.45   **Given:** $m_{CO_2} = 28.8\ g$, $P = 742\ mmHg$, and $T = 22\ °C$ **Find:** V

Conceptual Plan: $°C \rightarrow K$ and $mmHg \rightarrow atm$ and $g \rightarrow mol$ then $n, P, T \rightarrow V$

$$K = °C + 273.15 \qquad \frac{1\ atm}{760\ mmHg} \qquad \frac{1\ mol}{44.01\ g} \qquad PV = nRT$$

Solution: $T_1 = 22\ °C + 273.15 = 295\ K$, $P = 742\ mmHg \times \frac{1\ atm}{760\ mmHg} = 0.976316\ atm$

$$n = 28.8\ g \times \frac{1\ mol}{44.01\ g} = 0.654397\ mol\quad PV = nRT \quad \text{Rearrange to solve for } V.$$

$$V = \frac{nRT}{P} = \frac{0.654397\ mol \times 0.08206\ \frac{L \cdot atm}{mol \cdot K} \times 295\ K}{0.976316\ atm} = 16.2\ L$$

Check: The units (L) are correct. The magnitude of the answer (16) makes sense because 1 mole of an ideal gas under standard conditions (273 K and 1 atm) occupies 22.4 L. Although these are not standard conditions, they are close enough for a ballpark check of the answer. Because this gas sample contains 0.65 mole, a volume of 16 L is reasonable.

10.47   **Given:** 26.0 g argon, $V = 55.0\ mL$, and $T = 295\ K$ **Find:** P
Conceptual Plan: $m \rightarrow n$ and $mL \rightarrow L$ then $n, V, T \rightarrow P$

$$\frac{1\ mol}{39.95\ g} \qquad \frac{1\ L}{1000\ mL} \qquad PV = nRT$$

Solution: $26.0\ g\ Ar \times \frac{1\ mol\ Ar}{39.95\ g\ Ar} = 0.6508135\ mol\ Ar\quad 55.0\ mL \times \frac{1\ L}{1000\ mL} = 0.0550\ L$

$PV = nRT$ Rearrange to solve for P.

$$P = \frac{nRT}{V} = \frac{0.6508135\ mol \times 0.08206\ \frac{L \cdot atm}{mol \cdot K} \times 295\ K}{0.0550\ L} = 286.44906\ atm = 286\ atm$$

Check: The units (atm) are correct. The magnitude of the answer (300) makes sense because, as you will see in the next section, 1 mole of an ideal gas under standard conditions (273 K and 1 atm) occupies 22.4 L. Although these are not standard conditions, they can be used for a ballpark check of the answer. Because the volume is $\sim 1/400^{th}$ the molar volume and we have $\sim 2/3$ of a mole, the resulting pressure should be $(400)(2/3) = 270$ atm.

Given: $V_1 = 55.0\ mL$, $P_1 = 286\ atm$, and $P_2 = 1.20\ atm$ **Find:** V_2 (number of 750 mL bottles)
Conceptual Plan: $V_1, P_1, P_2 \rightarrow V_2$

$$P_1 V_1 = P_2 V_2$$

Solution: $P_1 V_1 = P_2 V_2$ Rearrange to solve for V_2.

$$V_2 = V_1 \frac{P_1}{P_2} = 55.0\ mL \times \frac{286.44906\ atm}{1.20\ atm} = 1.31289 \times 10^4\ mL \times \frac{1\ bottle}{750.0\ mL} = 17.5\ bottles$$

Check: The units (bottles) are correct. The magnitude of the answer (18) makes physical sense because Boyle's law indicates that as the volume decreases, the pressure increases. The pressure is decreasing by a factor of ~ 250, so the volume should increase by this factor.

10.49   **Given:** sample a $=$ 5 gas particles, sample b $=$ 10 gas particles, and sample c $=$ 8 gas particles, with all temperatures and volumes the same   **Find:** sample with largest P
Conceptual Plan: $n, V, T \rightarrow P$

$$PV = nRT$$

Solution: $PV = nRT$ Because V and T are constant, $P \propto n$. The sample with the largest number of gas particles will have the highest P. $P_b > P_c > P_a$

10.51   **Given:** $P_1 = 755$ mmHg, $T_1 = 25\,°C$, and $T_2 = 1155\,°C$ **Find:** P_2
Conceptual Plan: $°C \rightarrow K$ and mmHg $\rightarrow$ atm then $P_1, T_1, T_2 \rightarrow P_2$

$$K = °C + 273.15 \qquad \frac{1\ atm}{760\ mmHg} \qquad \frac{P_1}{T_1} = \frac{P_2}{T_2}$$

Solution: $T_1 = 25\,°C + 273.15 = 298$ K and $T_2 = 1155\,°C + 273.15 = 1428$ K

$$P = 755\ \text{mmHg} \times \frac{1\ atm}{760\ \text{mmHg}} = 0.993421\ atm \quad \frac{P_1}{T_1} = \frac{P_2}{T_2} \quad \text{Rearrange to solve for } P_2.$$

$$P_2 = P_1\frac{T_2}{T_1} = 0.993421\ atm \times \frac{1428\ K}{298\ K} = 4.76\ atm$$

Check: The units (atm) are correct. The magnitude of the answer (5) makes physical sense because there is a significant increase in T, which will increase P significantly.

Molar Volume, Density, and Molar Mass of a Gas

10.53   **Given:** STP and $m_{Ne} = 33.6$ g   **Find:** V
Conceptual Plan: g $\rightarrow$ mol $\rightarrow$ V

$$\frac{1\ mol}{20.18\ g} \qquad \frac{22.414\ L}{1\ mol}$$

Solution: $33.6\ g \times \dfrac{1\ mol}{20.18\ g} \times \dfrac{22.414\ L}{1\ mol} = 37.3\ L$

Check: The units (L) are correct. The magnitude of the answer (37) makes sense because 1 mole of an ideal gas under standard conditions (273 K and 1 atm) occupies 22.4 L and we have about 1.7 mol.

10.55   **Given:** H_2, $P = 1655$ psi, and $T = 20.0\,°C$ **Find:** d
Conceptual Plan: $°C \rightarrow K$ and psi $\rightarrow$ atm then $P, T, \mathcal{M} \rightarrow d$

$$K = °C + 273.15 \qquad \frac{1\ atm}{14.70\ psi} \qquad d = \frac{P\mathcal{M}}{RT}$$

Solution: $T = 20.0\,°C + 273.15 = 293.2$ K $P = 1655\ \text{psi} \times \dfrac{1\ atm}{14.70\ \text{psi}} = 112.585\ atm$

$$d = \frac{P\mathcal{M}}{RT} = \frac{112.585\ atm \times 2.016\ \dfrac{g}{mol}}{0.08206\dfrac{L \cdot atm}{K \cdot mol} \times 293.2\ K} = 9.434\ \frac{g}{L}$$

Check: The units (g/L) are correct. The magnitude of the answer (9) makes physical sense because this is a high pressure; so the gas density will be on the high side.

10.57   **Given:** $V = 248$ mL, $m = 0.433$ g, $P = 745$ mmHg, and $T = 28\,°C$ **Find:** $\mathcal{M}$
Conceptual Plan: $°C \rightarrow K$ mmHg $\rightarrow$ atm mL $\rightarrow$ L then $V, m \rightarrow d$ then $d, P, T \rightarrow \mathcal{M}$

$$K = °C + 273.15 \qquad \frac{1\ atm}{760\ mmHg} \qquad \frac{1\ L}{1000\ mL} \qquad d = \frac{m}{V} \qquad d = \frac{P\mathcal{M}}{RT}$$

Solution: $T = 28\,°C + 273.15 = 301\,K$ $P = 745\,\text{mmHg} \times \dfrac{1\,\text{atm}}{760\,\text{mmHg}} = 0.980263\,\text{atm}$

$V = 248\,\text{mL} \times \dfrac{1\,L}{1000\,\text{mL}} = 0.248\,L$ $d = \dfrac{m}{V} = \dfrac{0.433\,g}{0.248\,L} = 1.74597\,g/L$ $d = \dfrac{P\mathcal{M}}{RT}$ Rearrange to solve for $\mathcal{M}$.

$\mathcal{M} = \dfrac{dRT}{P} = \dfrac{1.74597\,\frac{g}{L} \times 0.08206\,\frac{L \cdot \text{atm}}{K \cdot \text{mol}} \times 301\,K}{0.980263\,\text{atm}} = 44.0\,g/\text{mol}$

Check: The units (g/mol) are correct. The magnitude of the answer (44) makes physical sense because this is a reasonable number for a molar mass of a gas.

10.59 **Given:** $m = 38.8\,\text{mg}$, $V = 224\,\text{mL}$, $T = 55\,°C$, and $P = 886\,\text{torr}$ **Find:** $\mathcal{M}$
Conceptual Plan: $\text{mg} \to \text{g mL} \to L\ °C \to K\ \text{torr} \to \text{atm then } V, m \to d \text{ then } d, P, T \to \mathcal{M}$

$\dfrac{1\,g}{1000\,\text{mg}}$ $\dfrac{1\,L}{1000\,\text{mL}}$ $K = °C + 273.15$ $\dfrac{1\,\text{atm}}{760\,\text{torr}}$ $d = \dfrac{m}{V}$ $d = \dfrac{P\mathcal{M}}{RT}$

Solution:

$m = 38.8\,\text{mg} \times \dfrac{1\,g}{1000\,\text{mg}} = 0.0388\,g$ $V = 224\,\text{mL} \times \dfrac{1\,L}{1000\,\text{mL}} = 0.224\,L$ $T = 55\,°C + 273.15 = 328\,K$

$P = 886\,\text{torr} \times \dfrac{1\,\text{atm}}{760\,\text{torr}} = 1.165789\,\text{atm}$ $d = \dfrac{m}{V} = \dfrac{0.0388\,g}{0.224\,L} = 0.173214\,g/L$ $d = \dfrac{P\mathcal{M}}{RT}$

Rearrange to solve for $\mathcal{M}$. $\mathcal{M} = \dfrac{dRT}{P} = \dfrac{0.173214\,\frac{g}{L} \times 0.08206\,\frac{L \cdot \text{atm}}{K \cdot \text{mol}} \times 328\,K}{1.165789\,\text{atm}} = 4.00\,g/\text{mol}$

Check: The units (g/mol) are correct. The magnitude of the answer (4) makes physical sense because this is a reasonable number for a molar mass of a gas, especially because the density is on the low side.

Partial Pressure

10.61 **Given:** $P_{N_2} = 215\,\text{torr}$, $P_{O_2} = 102\,\text{torr}$, $P_{He} = 117\,\text{torr}$, $V = 1.35\,L$, and $T = 25.0\,°C$ **Find:** $P_{\text{Total}}, m_{N_2}, m_{O_2}, m_{He}$
Conceptual Plan: $°C \to K$ and $\text{torr} \to \text{atm}$ and $P, V, T \to n$ then $n \to g$

$K = °C + 273.15$ $\dfrac{1\,\text{atm}}{760\,\text{torr}}$ $PV = nRT$ $\mathcal{M}$

and $P_{N_2}, P_{O_2}, P_{He} \to P_{\text{Total}}$

$P_{\text{Total}} = P_{N_2} + P_{O_2} + P_{He}$

Solution: $T_1 = 25.0\,°C + 273.15 = 298.2\,K$, $PV = nRT$ Rearrange to solve for n.

$n = \dfrac{PV}{RT}$; $P_{N_2} = 215\,\text{torr} \times \dfrac{1\,\text{atm}}{760\,\text{torr}} = 0.2828947\,\text{atm}$ $n_{N_2} = \dfrac{0.2828947\,\text{atm} \times 1.35\,L}{0.08206\,\frac{L \cdot \text{atm}}{\text{mol} \cdot K} \times 298.2\,K} = 0.01560700\,\text{mol}$

$0.01560700\,\text{mol} \times \dfrac{28.02\,g}{1\,\text{mol}} = 0.437\,g\ N_2$

$P_{O_2} = 102\,\text{torr} \times \dfrac{1\,\text{atm}}{760\,\text{torr}} = 0.1342105\,\text{atm}$ $n_{O_2} = \dfrac{0.1342105\,\text{atm} \times 1.35\,L}{0.08206\,\frac{L \cdot \text{atm}}{\text{mol} \cdot K} \times 298.2\,K} = 0.007404252\,\text{mol}$

$0.007404252\,\text{mol} \times \dfrac{32.00\,g}{1\,\text{mol}} = 0.237\,g\ O_2$

$P_{He} = 117\,\text{torr} \times \dfrac{1\,\text{atm}}{760\,\text{torr}} = 0.1539474\,\text{atm}$ $n_{He} = \dfrac{0.1539474\,\text{atm} \times 1.35\,L}{0.08206\,\frac{L \cdot \text{atm}}{\text{mol} \cdot K} \times 298.2\,K} = 0.008493113\,\text{mol}$

$0.008493113\,\text{mol} \times \dfrac{4.003\,g}{1\,\text{mol}} = 0.0340\,g\ He$ and

$P_{\text{Total}} = P_{N_2} + P_{O_2} + P_{He} = 0.283\,\text{atm} + 0.134\,\text{atm} + 0.154\,\text{atm} = 0.571\,\text{atm}$ or
$P_{\text{Total}} = P_{N_2} + P_{O_2} + P_{He} = 215\,\text{torr} + 102\,\text{torr} + 117\,\text{torr} = 434\,\text{torr}$

Check: The units (g and atm) are correct. The magnitude of the answers (<1 and < 1) makes sense because gases are not very dense and these pressures are <1 atm. Because all of the pressures are small, the total is <1 atm.

10.63 **Given:** $m_{CO_2} = 1.20$ g, $V = 755$ mL, $P_{N_2} = 725$ mmHg, and $T = 25.0\,°C$ **Find:** P_{Total}

Conceptual Plan: mL $\rightarrow$ L and °C $\rightarrow$ K and $m \rightarrow n$ and $n, P, T \rightarrow V$ then atm $\rightarrow$ mmHg

$$\frac{1\,L}{1000\,mL} \qquad K = °C + 273.15 \qquad \frac{1\,mol}{44.01\,g} \qquad PV = nRT \qquad \frac{760\,mmHg}{1\,atm}$$

finally $P_{CO_2}, P_{N_2} \rightarrow P_{Total}$

$$P_{Total} = P_{CO_2} + P_{N_2}$$

Solution: $V = 755\ \cancel{mL} \times \dfrac{1\,L}{1000\ \cancel{mL}} = 0.755$ L $\qquad T = 25.0\,°C + 273.15 = 298.2$ K

$n = 1.20\ \cancel{g} \times \dfrac{1\,mol}{44.01\ \cancel{g}} = 0.027\underline{2}665$ mol, $PV = nRT$ Rearrange to solve for P.

$$P = \frac{nRT}{V} = \frac{0.027\underline{2}665\ \cancel{mol} \times 0.08206\,\dfrac{\cancel{L} \cdot atm}{\cancel{mol} \cdot \cancel{K}} \times 298.2\ \cancel{K}}{0.755\ \cancel{L}} = 0.88\underline{3}734\ atm$$

$$P_{CO_2} = 0.88\underline{3}734\ \cancel{atm} \times \frac{760\,mmHg}{1\ \cancel{atm}} = 672\ mmHg$$

$$P_{Total} = P_{CO_2} + P_{N_2} = 672\,mmHg + 725\,mmHg = 1397\,mmHg \text{ or } 1397\ \cancel{torr} \times \frac{1\,atm}{760\ \cancel{torr}} = 1.84\ atm$$

Check: The units (mmHg) are correct. The magnitude of the answer (1400) makes sense because it must be greater than 725 mmHg.

10.65 **Given:** $m_{N_2} = 1.25$ g, $m_{O_2} = 0.85$ g, $V = 1.55$ L, and $T = 18\,°C$ **Find:** $\chi_{N_2}, \chi_{O_2}, P_{N_2}, P_{O_2}$

Conceptual Plan: g $\rightarrow$ mol °C $\rightarrow$ K, then $n_{N_2}, n_{O_2} \rightarrow \chi_{N_2}$ and $n_{N_2}, n_{O_2} \rightarrow \chi_{O_2}$

$$\mathcal{M} \qquad K = °C + 273.15 \qquad \chi_{N_2} = \frac{n_{N_2}}{n_{N_2} + n_{O_2}} \qquad \chi_{O_2} = \frac{n_{O_2}}{n_{N_2} + n_{O_2}}$$

then $n, V, T \rightarrow P$

$$PV = nRT$$

Solution: $n_{N_2} = 1.25\ \cancel{g} \times \dfrac{1\,mol}{28.02\ \cancel{g}} = 0.044\underline{6}110$ mol, $n_{O_2} = 0.85\ \cancel{g} \times \dfrac{1\,mol}{32.00\ \cancel{g}} = 0.02\underline{6}563$ mol,

$$T = 18\,°C + 273.15 = 291\ K, \ \chi_{N_2} = \frac{n_{N_2}}{n_{N_2} + n_{O_2}} = \frac{0.044\underline{6}110\ \cancel{mol}}{0.044\underline{6}110\ \cancel{mol} + 0.02\underline{6}563\ \cancel{mol}} = 0.62\underline{6}792 = 0.627,$$

$$\chi_{O_2} = \frac{n_{O_2}}{n_{N_2} + n_{O_2}} = \frac{0.02\underline{6}563\ \cancel{mol}}{0.044\underline{6}110\ \cancel{mol} + 0.02\underline{6}563\ \cancel{mol}} = 0.37\underline{3}212 \quad \text{We can also calculate this as}$$

$$\chi_{O_2} = 1 - \chi_{N_2} = 1 - 0.62\underline{6}792 = 0.37\underline{3}208 = 0.373\ PV = nRT \quad \text{Rearrange to solve for } P. \ P = \frac{nRT}{V}$$

$$P_{N_2} = \frac{0.044\underline{6}110\ \cancel{mol} \times 0.08206\dfrac{\cancel{L} \cdot atm}{\cancel{mol} \cdot \cancel{K}} \times 291\ \cancel{K}}{1.55\ \cancel{L}} = 0.687\ atm$$

$$P_{O_2} = \frac{0.02\underline{6}563\ \cancel{mol} \times 0.08206\dfrac{\cancel{L} \cdot atm}{\cancel{mol} \cdot \cancel{K}} \times 291\ \cancel{K}}{1.55\ \cancel{L}} = 0.409\ atm$$

Check: The units (none and atm) are correct. The magnitude of the answers (<1 and <1) makes sense because the mole fractions should total 1, and because the weight of N_2 is greater than O_2, its mole fraction is larger. The number of moles is << 1, so we expect the pressures to be < 1 atm, given the V (1.55 L).

10.67   **Given:** $T = 30.0\,°C$, $P_{Total} = 732\text{ mmHg}$, and $V = 722\text{ mL}$ **Find:** P_{H_2} and m_{H_2}
        Conceptual Plan: $T \rightarrow P_{H_2O}$ then $P_{Total}, P_{H_2O} \rightarrow P_{H_2}$ then $\text{mmHg} \rightarrow \text{atm}$ and $\text{mL} \rightarrow \text{L}$

                Table 10.3 $P_{Total} = P_{H_2O} + P_{H_2}$ $\dfrac{1\text{ atm}}{760\text{ mmHg}}$ $\dfrac{1\text{ L}}{1000\text{ mL}}$

        and $°C \rightarrow K$ $P, V, T \rightarrow n$ **then** $n \rightarrow m$

             $K = °C + 273.15$ $PV = nRT$ $\dfrac{2.016\text{ g}}{1\text{ mol}}$

        Solution: Table 10.3 states that at $30\,°C$, $P_{H_2O} = 31.86\text{ mmHg}$ $P_{Total} = P_{H_2O} + P_{H_2}$ Rearrange to solve for
        P_{H_2}. $P_{H_2} = P_{Total} - P_{H_2O} = 732\text{ mmHg} - 31.86\text{ mmHg} = 700.\text{ mmHg}$

$$P_{H_2} = 700.\text{ mmHg} \times \frac{1\text{ atm}}{760\text{ mmHg}} = 0.921052\text{ atm} \quad V = 722\text{ mL} \times \frac{1\text{ L}}{1000\text{ mL}} = 0.722\text{ L}$$

$$T = 30.0\,°C + 273.15 = 303.2\text{ K}, \quad PV = nRT \quad \text{Rearrange to solve for } n. \; n = \frac{PV}{RT}$$

$$n_{H_2} = \frac{0.921052\text{ atm} \times 0.722\text{ L}}{0.08206\,\dfrac{\text{L} \cdot \text{atm}}{\text{mol} \cdot \text{K}} \times 303.2\text{ K}} = 0.0267276\text{ mol} \quad \text{then } 0.0267276\text{ mol} \times \frac{2.016\text{ g}}{1\text{ mol}} = 0.0539\text{ g H}_2$$

        Check: The units (g) are correct. The magnitude of the answer $(\ll 1)$ makes sense because gases are not very
        dense, hydrogen is light, the volume is small, and the pressure is ~ 1 atm.

10.69   **Given:** $T = 25\,°C$, $P_{Total} = 748\text{ mmHg}$, and $V = 0.951\text{ L}$ **Find:** P_{H_2} and m_{H_2}
        Conceptual Plan: $T \rightarrow P_{H_2O}$ then $P_{Total}, P_{H_2O} \rightarrow P_{H_2}$ then $\text{mmHg} \rightarrow \text{atm}$ and $\text{mL} \rightarrow \text{L}$

                Table 10.3 $P_{Total} = P_{H_2O} + P_{H_2}$ $\dfrac{1\text{ atm}}{760\text{ mmHg}}$ $\dfrac{1\text{ L}}{1000\text{ mL}}$

        and $°C \rightarrow K$ $P, V, T \rightarrow n$ **then** $n \rightarrow m$

             $K = °C + 273.15$ $PV = nRT$ $\dfrac{2.016\text{ g}}{1\text{ mol}}$

        Solution: Table 10.3 states that at $25\,°C$, $P_{H_2O} = 23.78\text{ mmHg}$ $P_{Total} = P_{H_2O} + P_{H_2}$
        Rearrange to solve for P_{H_2}. $P_{H_2} = P_{Total} - P_{H_2O} = 748\text{ mmHg} - 23.78\text{ mmHg} = 724\text{ mmHg}$

$$P_{H_2} = 724\text{ mmHg} \times \frac{1\text{ atm}}{760\text{ mmHg}} = 0.952632\text{ atm} \quad T = 25\,°C + 273.15 = 298\text{ K} \quad PV = nRT$$

$$\text{Rearrange to solve for } n. \; n_{H_2} = \frac{PV}{RT} = \frac{0.952632\text{ atm} \times 0.951\text{ L}}{0.08206\,\dfrac{\text{L} \cdot \text{atm}}{\text{mol} \cdot \text{K}} \times 298\text{ K}} = 0.0370474\text{ mol}$$

$$0.0370474\text{ mol} \times \frac{2.016\text{ g}}{1\text{ mol}} = 0.0747\text{ g H}_2$$

        Check: The units (g) are correct. The magnitude of the answer $(\ll 1)$ makes sense because gases are not very
        dense, hydrogen is light, the volume is small, and the pressure is ~ 1 atm.

Kinetic Molecular Theory

10.71   (a)   Yes, because the average kinetic energy of a particle is proportional to the temperature in kelvins and the two
              gases are at the same temperature, they have the same average kinetic energy.
       (b)   No, because the helium atoms are lighter, they must move faster to have the same kinetic energy as argon
              atoms.
       (c)   No, because the Ar atoms are moving slower to compensate for their larger mass, they will exert the same
              pressure on the walls of the container.
       (d)   Because He is lighter, it has the faster rate of effusion.

10.73 **Given:** F_2, Cl_2, Br_2, and $T = 298$ K   **Find:** u_{rms} and KE_{avg} for each gas and relative rates of effusion

Conceptual Plan: $\mathcal{M}, T \rightarrow u_{rms} \rightarrow KE_{avg}$

$$u_{rms} = \sqrt{\frac{3RT}{\mathcal{M}}} \qquad KE_{avg} = \frac{1}{2}N_A m u_{rms}^2 = \frac{3}{2}RT$$

Solution:

$$F_2: \mathcal{M} = \frac{38.00 \text{ g}}{1 \text{ mol}} \times \frac{1 \text{ kg}}{1000 \text{ g}} = 0.03800 \text{ kg/mol}, u_{rms} = \sqrt{\frac{3RT}{\mathcal{M}}} = \sqrt{\frac{3 \times 8.314\dfrac{\text{J}}{\text{K} \cdot \text{mol}} \times 298 \text{ K}}{0.03800 \dfrac{\text{kg}}{\text{mol}}}} = 442 \text{ m/s}$$

$$Cl_2: \mathcal{M} = \frac{70.90 \text{ g}}{1 \text{ mol}} \times \frac{1 \text{ kg}}{1000 \text{ g}} = 0.07090 \text{ kg/mol}, u_{rms} = \sqrt{\frac{3RT}{\mathcal{M}}} = \sqrt{\frac{3 \times 8.314\dfrac{\text{J}}{\text{K} \cdot \text{mol}} \times 298 \text{ K}}{0.07090 \dfrac{\text{kg}}{\text{mol}}}} = 324 \text{ m/s}$$

$$Br_2: \mathcal{M} = \frac{159.80 \text{ g}}{1 \text{ mol}} \times \frac{1 \text{ kg}}{1000 \text{ g}} = 0.15980 \text{ kg/mol}, u_{rms} = \sqrt{\frac{3RT}{\mathcal{M}}} = \sqrt{\frac{3 \times 8.314\dfrac{\text{J}}{\text{K} \cdot \text{mol}} \times 298 \text{ K}}{0.15980 \dfrac{\text{kg}}{\text{mol}}}} = 216 \text{ m/s}$$

All molecules have the same average kinetic energy:

$$KE_{avg} = \frac{3}{2}RT = \frac{3}{2} \times 8.314 \frac{\text{J}}{\text{K} \cdot \text{mol}} \times 298 \text{ K} = 3.72 \times 10^3 \text{ J/mol}$$

Because rate of effusion is proportional to $\sqrt{\dfrac{1}{\mathcal{M}}}$, F_2 will have the fastest rate and Br_2 will have the slowest rate.

Check: The units (m/s) are correct. The magnitude of the answers $(200-450)$ makes sense because it is consistent with what was shown in the text and the heavier the molecule, the slower the molecule.

10.75 **Given:** $^{238}UF_6$ and $^{235}UF_6$ U-235 $= 235.054$ amu, U-238 $= 238.051$ amu

Find: ratio of effusion rates $^{238}UF_6/^{235}UF_6$

Conceptual Plan: $\mathcal{M}_{238_{UF_6}}, \mathcal{M}_{235_{UF_6}} \rightarrow Rate_{238_{UF_6}}/Rate_{235_{UF_6}}$

$$\frac{Rate_{238_{UF_6}}}{Rate_{235_{UF_6}}} = \sqrt{\frac{\mathcal{M}_{235_{UF_6}}}{\mathcal{M}_{238_{UF_6}}}}$$

Solution: $^{238}UF_6: \mathcal{M} = \dfrac{352.05 \text{ g}}{1 \text{ mol}} \times \dfrac{1 \text{ kg}}{1000 \text{ g}} = 0.35205 \text{ kg/mol}$

$^{235}UF_6: \mathcal{M} = \dfrac{349.05 \text{ g}}{1 \text{ mol}} \times \dfrac{1 \text{ kg}}{1000 \text{ g}} = 0.34905 \text{ kg/mol}$

$$\frac{Rate_{238_{UF_6}}}{Rate_{235_{UF_6}}} = \sqrt{\frac{\mathcal{M}_{235_{UF_6}}}{\mathcal{M}_{238_{UF_6}}}} = \sqrt{\frac{0.34905 \text{ kg/mol}}{0.35205 \text{ kg/mol}}} = 0.99573$$

Check: The units (none) are correct. The magnitude of the answer (< 1) makes sense because the heavier molecule has the lower effusion rate because it moves more slowly.

10.77 **Given:** Ne and unknown gas; Ne effusion in 76 s and unknown in 155 s   **Find:** identify unknown gas

Conceptual Plan: $\mathcal{M}_{Ne}, Rate_{Ne}, Rate_{Unk} \rightarrow \mathcal{M}_{Kr}$

$$\frac{Rate_{Ne}}{Rate_{Unk}} = \sqrt{\frac{\mathcal{M}_{Unk}}{\mathcal{M}_{Ne}}}$$

Solution: Ne: $\mathcal{M} = \dfrac{20.18 \text{ g}}{1 \text{ mol}} \times \dfrac{1 \text{ kg}}{1000 \text{ g}} = 0.02018 \text{ kg/mol}$, $\dfrac{Rate_{Ne}}{Rate_{Unk}} = \sqrt{\dfrac{\mathcal{M}_{unk}}{\mathcal{M}_{Ne}}}$

Rearrange to solve for $\mathcal{M}_{Unk}$. $\mathcal{M}_{Unk} = \mathcal{M}_{Ne}\left(\dfrac{Rate_{Ne}}{Rate_{Unk}}\right)^2$ Because Rate α 1/(effusion time),

$$\mathcal{M}_{\text{Unk}} = \mathcal{M}_{\text{Ne}} \left(\frac{Time_{\text{Unk}}}{Time_{\text{Ne}}}\right)^2 = 0.02018 \frac{\text{kg}}{\text{mol}} \times \left(\frac{155 \text{ s}}{76 \text{ s}}\right)^2 = 0.084 \frac{\text{kg}}{\text{mol}} \times \frac{1000 \text{ g}}{1 \text{ kg}} = 84 \text{ g/mol or Kr}$$

Check: The units (g/mol) are correct. The magnitude of the answer ($>$ Ne) makes sense because Ne effused faster; so it must be lighter.

10.79 Gas A has the higher molar mass because it has slower average velocity. Gas B will have the higher effusion rate because it has the higher average velocity.

Reaction Stoichiometry Involving Gases

10.81 **Given:** $m_C = 15.7$ g, $P = 1.0$ atm, and $T = 355$ K $\quad$ **Find:** V

Conceptual Plan: $m_C \rightarrow n_C \rightarrow n_{H_2}$ then $n_{H_2}, P, T \rightarrow V$

$$\frac{1 \text{ mol}}{12.01 \text{ g C}} \qquad \frac{1 \text{ mol H}_2}{1 \text{ mol C}} \qquad\qquad PV = nRT$$

Solution: $15.7 \text{ g C} \times \dfrac{1 \text{ mol C}}{12.01 \text{ g C}} \times \dfrac{1 \text{ mol H}_2}{1 \text{ mol C}} = 1.3\underline{0}724 \text{ mol H}_2 \quad PV = nRT \quad$ Rearrange to solve for V.

$$V = \frac{nRT}{P} = \frac{1.3\underline{0}724 \text{ mol} \times 0.08206 \dfrac{\text{L} \cdot \text{atm}}{\text{mol} \cdot \text{K}} \times 355 \text{ K}}{1.0 \text{ atm}} = 38 \text{ L}$$

Check: The units (L) are correct. The magnitude of the answer (38) makes sense because we have more than 1 mole of gas; so we expect more than 22 L.

10.83 **Given:** $P = 748$ mmHg, $T = 86\,°C$, and $m_{CH_3OH} = 25.8$ g $\quad$ **Find:** V_{H_2} and V_{CO}

Conceptual Plan: $m_{CH_3OH} \rightarrow n_{CH_3OH} \rightarrow n_{H_2}$ and mmHg $\rightarrow$ atm and $°C \rightarrow$ K

$$\frac{1 \text{ mol CH}_3\text{OH}}{32.04 \text{ g CH}_3\text{OH}} \qquad \frac{2 \text{ mol H}_2}{1 \text{ mol CH}_3\text{OH}} \qquad \frac{1 \text{ atm}}{760 \text{ mmHg}} \qquad K = °C + 273.15$$

then $n_{H_2}, P, T \rightarrow V$ and $n_{H_2} \rightarrow n_{CO}$ then $n_{CO}, P, T \rightarrow V$

$$PV = nRT \qquad \frac{1 \text{ mol CO}}{2 \text{ mol H}_2} \qquad\qquad PV = nRT$$

Solution: $25.8 \text{ g CH}_3\text{OH} \times \dfrac{1 \text{ mol CH}_3\text{OH}}{32.04 \text{ g CH}_3\text{OH}} \times \dfrac{2 \text{ mol H}_2}{1 \text{ mol CH}_3\text{OH}} = 1.6\underline{1}049 \text{ mol H}_2$

$P_{H_2} = 748 \text{ mmHg} \times \dfrac{1 \text{ atm}}{760 \text{ mmHg}} = 0.98\underline{4}211 \text{ atm}, \; T = 86\,°C + 273.15 = 359 \text{ K}, \; PV = nRT$

Rearrange to solve for V. $V = \dfrac{nRT}{P}, \; V_{H_2} = \dfrac{1.6\underline{1}049 \text{ mol} \times 0.08206 \dfrac{\text{L} \cdot \text{atm}}{\text{mol} \cdot \text{K}} \times 359 \text{ K}}{0.98\underline{4}211 \text{ atm}} = 48.2 \text{ L H}_2$

$1.6\underline{1}049 \text{ mol H}_2 \times \dfrac{1 \text{ mol CO}}{2 \text{ mol H}_2} = 0.80\underline{5}25 \text{ mol CO}, \; V_{CO} = \dfrac{0.80\underline{5}25 \text{ mol} \times 0.08206 \dfrac{\text{L} \cdot \text{atm}}{\text{mol} \cdot \text{K}} \times 359 \text{ K}}{0.98\underline{4}211 \text{ atm}} = 24.1 \text{ L CO}$

Check: The units (L) are correct. The magnitude of the answers (48 and 24) makes sense because we have more than 1 mole of hydrogen gas and half that of CO; so we expect significantly more than 22 L for hydrogen and half the volume of hydrogen for CO.

10.85 **Given:** $V = 11.8$ L at STP $\quad$ **Find:** m_{NaN_3}

Conceptual Plan: $V_{N_2} \rightarrow n_{N_2} \rightarrow n_{NaN_3} \rightarrow m_{NaN_3}$

$$\frac{1 \text{ mol N}_2}{22.414 \text{ L N}_2} \qquad \frac{2 \text{ mol NaN}_3}{3 \text{ mol N}_2} \qquad \frac{65.02 \text{ g NaN}_3}{1 \text{ mol NaN}_3}$$

Solution: $11.8 \text{ L N}_2 \times \dfrac{1 \text{ mol N}_2}{22.414 \text{ L N}_2} \times \dfrac{2 \text{ mol NaN}_3}{3 \text{ mol N}_2} \times \dfrac{65.02 \text{ g NaN}_3}{1 \text{ mol NaN}_3} = 22.8 \text{ g NaN}_3$

Check: The units (g) are correct. The magnitude of the answer (23 g) makes sense because we have about half a mole of nitrogen gas, which translates to even fewer moles of NaN_3; so we expect significantly less than 65 g.

10.87 **Given:** $V_{CH_4} = 25.5$ L, $P_{CH_4} = 732$ torr, and $T = 25\,°C$; mixed with $V_{H_2O} = 22.8$ L, $P_{H_2O} = 702$ torr, and $T = 125\,°C$; forms $P_{H_2} = 26.2$ L at STP **Find:** % yield

Conceptual Plan: $CH_4 : torr \rightarrow atm$ and $°C \rightarrow K$ and $P, V, T \rightarrow n_{CH_4} \rightarrow n_{H_2}$

$$\frac{1\ atm}{760\ torr} \qquad K = °C + 273.15 \qquad PV = nRT \qquad \frac{3\ mol\ H_2}{1\ mol\ CH_4}$$

$H_2O : torr \rightarrow atm$ and $°C \rightarrow K$ and $P, V, T \rightarrow n_{H_2O} \rightarrow n_{H_2}$

$$\frac{1\ atm}{760\ torr} \qquad K = °C + 273.15 \qquad PV = nRT \qquad \frac{3\ mol\ H_2}{1\ mol\ H_2O}$$

Select smaller n_{H_2} as theoretical yield,

then $V_{H_2} \rightarrow n_{H_2}$ (actual yield) then actual yield, theoretical yield $\rightarrow$ % yield

$$\frac{1\ mol\ H_2}{22.414\ L\ H_2} \qquad\qquad \text{\% yield} = \frac{\text{actual yield}}{\text{theoretical yield}} \times 100\%$$

Solution: CH_4: $P_{CH_4} = 732\ torr \times \dfrac{1\ atm}{760\ torr} = 0.96\underline{3}158\ atm$, $T = 25\,°C + 273.15 = 298\ K$, $PV = nRT$

Rearrange to solve for n. $n = \dfrac{PV}{RT}$ $\qquad n_{CH_4} = \dfrac{0.96\underline{3}158\ atm \times 25.5\ L}{0.08206\ \dfrac{L \cdot atm}{mol \cdot K} \times 298\ K} = 1.00\underline{4}36\ mol\ CH_4$

$1.00\underline{4}36\ mol\ CH_4 \times \dfrac{3\ mol\ H_2}{1\ mol\ CH_4} = 3.01\underline{3}08\ mol\ H_2$

H_2O: $P_{H_2O} = 702\ torr \times \dfrac{1\ atm}{760\ torr} = 0.92\underline{3}684\ atm$, $T = 125\,°C + 273.15 = 398\ K$, $n = \dfrac{PV}{RT}$

$n_{H_2O} = \dfrac{0.92\underline{3}684\ atm \times 22.8\ L}{0.08206\ \dfrac{L \cdot atm}{mol \cdot K} \times 398\ K} = 0.64\underline{4}828\ mol\ H_2O$ $\quad 0.64\underline{4}828\ mol\ H_2O \times \dfrac{3\ mol\ H_2}{1\ mol\ H_2O} = 1.93\underline{4}48\ mol\ H_2$

Water is the limiting reagent because the moles of hydrogen generated are lower.

theoretical yield $= 1.93\underline{4}48\ mol\ H_2$

$26.2\ L\ H_2 \times \dfrac{1\ mol\ H_2}{22.414\ L\ H_2} = 1.1\underline{6}891\ mol\ H_2 =$ actual yield

% yield $= \dfrac{\text{actual yield}}{\text{theoretical yield}} \times 100\% = \dfrac{1.1\underline{6}891\ mol\ H_2}{1.93\underline{4}48\ mol\ H_2} \times 100\% = 60.4\%$

Check: The units (%) are correct. The magnitude of the answer (60) makes sense because it is between 0 and 100%.

10.89 **Given:** $V = 2.00$ L, $P_{Cl_2} = 337$ mmHg, $P_{F_2} = 729$ mmHg, $T = 298$ K **Find:** limiting reactant and m_{ClF_3}

Conceptual Plan: Determine limiting reactant by comparing the pressures of each reactant in the same way we would compare moles then

$$\frac{3\ mmHg\ F_2}{1\ mmHg\ Cl_2}$$

mmHg limiting reactant $\rightarrow$ mmHg ClF₃ and mmHg $\rightarrow$ atm then

$$\frac{2\ mmHg\ ClF_3}{1\ mmHg\ Cl_2} \ \text{or} \ \frac{2\ mmHg\ ClF_3}{3\ mmHg\ F_2} \qquad \frac{1\ atm}{760\ mmHg}$$

$P, V, T \rightarrow n \rightarrow m$

$$PV = nRT \quad \frac{92.45\ g\ ClF_3}{1\ mol\ ClF_3}$$

Solution: To determine the limiting reactant, calculate the pressure of fluorine needed to react all of the chlorine and compare to the pressure of fluorine available.

$337\ mmHg\ Cl_2 \times \dfrac{3\ mmHg\ F_2}{1\ mmHg\ Cl_2} = 101\underline{1}\ mmHg\ F_2$ needed.

Because only 729 mmHg F_2 is available, F_2 is the limiting reactant; then

$P_{ClF_3} = 729\ mmHg\ F_2 \times \dfrac{2\ mmHg\ ClF_3}{3\ mmHg\ F_2} \times \dfrac{1\ atm\ ClF_3}{760\ mmHg\ ClF_3} = 0.63\underline{9}4737\ atm\ ClF_3$ $PV = nRT$ Rearrange to

$$\text{solve for } n. \ n = \frac{PV}{RT} \quad n_{ClF_3} = \frac{0.6394737 \text{ atm } ClF_3 \times 2.00 \, L}{0.08206 \dfrac{L \cdot atm}{mol \cdot K} \times 298 \text{ K}} = 0.052300387 \text{ mol } ClF_3$$

$$\text{Finally, } 0.052300387 \text{ mol } ClF_3 \times \frac{92.45 \text{ g } ClF_3}{1 \text{ mol } ClF_3} = 4.8351708 \text{ g } ClF_3 = 4.84 \text{ g } ClF_3$$

Check: The units (g) are correct. The magnitude of the answer (5) makes sense because we have much less than a mole of fluorine; so we expect a mass much less than the molar mass of ClF_3.

Real Gases

10.91   The postulate that the volume of the gas particles is small compared to the space between them breaks down at high pressure. At high pressures, the number of molecules per unit volume increases; so the volume of the gas particles becomes more significant. Because the spacing between the particles is reduced, the molecules themselves occupy a significant portion of the volume.

10.93   **Given:** Ne, $n = 1.000 \text{ mol}$, $P = 500.0 \text{ atm}$, and $T = 355.0 \text{ K}$ **Find:** V(ideal) and V(van der Waals)

Conceptual Plan: $n, P, T \rightarrow V$ **and** $n, P, T \rightarrow V$

$$PV = nRT \qquad \left(P + \frac{an^2}{V^2}\right)(V - nb) = nRT$$

Solution: $PV = nRT$ Rearrange to solve for V.

$$V = \frac{nRT}{P} = \frac{1.000 \text{ mol} \times 0.08206 \dfrac{L \cdot atm}{mol \cdot K} \times 355.0 \text{ K}}{500.0 \text{ atm}} = 0.05826 \text{ L}$$

$$\left(P + \frac{an^2}{V^2}\right)(V - nb) = nRT \quad \text{Rearrange to solve for } V = \frac{nRT}{\left(P + \dfrac{an^2}{V^2}\right)} + nb$$

Using $a = 0.211 \, L^2 \, atm/mol^2$ and $b = 0.0171 \, L/mol$ from Table 10.4 and the V from the ideal gas law calculation above, solve for V by successive approximations.

$$V = \frac{1.000 \text{ mol} \times 0.08206 \dfrac{L \cdot atm}{mol \cdot K} \times 355.0 \text{ K}}{500.0 \text{ atm} + \dfrac{0.211 \dfrac{L^2 \cdot atm}{mol^2} \times (1.000 \text{ mol})^2}{(0.05826 \, L)^2}} + \left(1.000 \text{ mol} \times 0.0171 \dfrac{L}{mol}\right) = 0.068920 \text{ L}$$

Plug in this new value.

$$V = \frac{1.000 \text{ mol} \times 0.08206 \dfrac{L \cdot atm}{mol \cdot K} \times 355.0 \text{ K}}{500.0 \text{ atm} + \dfrac{0.211 \dfrac{L^2 \cdot atm}{mol^2} \times (1.000 \text{ mol})^2}{(0.068920 \, L)^2}} + \left(1.000 \text{ mol} \times 0.0171 \dfrac{L}{mol}\right) = 0.070609 \text{ L}$$

Plug in this new value.

$$V = \frac{1.000 \text{ mol} \times 0.08206 \dfrac{L \cdot atm}{mol \cdot K} \times 355.0 \text{ K}}{500.0 \text{ atm} + \dfrac{0.211 \dfrac{L^2 \cdot atm}{mol^2} \times (1.000 \text{ mol})^2}{(0.070609 \, L)^2}} + \left(1.000 \text{ mol} \times 0.0171 \dfrac{L}{mol}\right) = 0.070816 \text{ L}$$

Plug in this new value.

$$V = \frac{1.000 \text{ mol} \times 0.08206 \frac{\text{L} \cdot \text{atm}}{\text{mol} \cdot \text{K}} \times 355.0 \text{ K}}{500.0 \text{ atm} + \dfrac{0.211 \frac{\text{L}^2 \cdot \text{atm}}{\text{mol}^2} \times (1.000 \text{ mol})^2}{(0.070816 \text{ L})^2}} + \left(1.000 \text{ mol} \times 0.0171 \frac{\text{L}}{\text{mol}} \right) = 0.070840 \text{ L} = 0.0708 \text{ L}$$

The two values are different because we are at very high pressures. The pressure is corrected from 500.0 atm to 542.1 atm, and the final volume correction is 0.0171 L.

Check: The units (L) are correct. The magnitude of the answer (0.07) makes sense because we are at such a high pressure and have 1 mole of gas.

Cumulative Problems

10.95   **Given:** $m_{penny} = 2.482$ g, $T = 25\,°C$, $V = 0.899$ L, and $P_{Total} = 791$ mmHg   **Find:** % Zn in penny

         Conceptual Plan: $T \rightarrow P_{H_2O}$ then $P_{Total}, P_{H_2O} \rightarrow P_{H_2}$ then mmHg $\rightarrow$ atm and $°C \rightarrow$ K

                 Table 10.3 $P_{Total} = P_{H_2O} + P_{H_2}$ $\dfrac{1 \text{ atm}}{760 \text{ mmHg}}$ $K = °C + 273.15$

         and $P, V, T \rightarrow n_{H_2} \rightarrow n_{Zn} \rightarrow g_{Zn} \rightarrow$ **% Zn**

             $PV = nRT$ $\dfrac{1 \text{ mol Zn}}{1 \text{ mol H}_2}$ $\dfrac{65.39 \text{ g Zn}}{1 \text{ mol Zn}}$ $\% \text{ Zn} = \dfrac{g_{Zn}}{g_{penny}} \times 100\%$

         Solution: Table 10.3 states that $P_{H_2O} = 23.78$ mmHg at 25 °C. $P_{Total} = P_{H_2O} + P_{H_2}$ Rearrange to solve for P_{H_2}.

$$P_{H_2} = P_{Total} - P_{H_2O} = 791 \text{ mmHg} - 23.78 \text{ mmHg} = 767 \text{ mmHg} \quad P_{H_2} = 767 \text{ mmHg} \times \frac{1 \text{ atm}}{760 \text{ mmHg}} = 1.0092 \text{ atm}$$

         then $T = 25\,°C + 273.15 = 298$ K, $PV = nRT$

         Rearrange to solve for n. $n_{H_2} = \dfrac{PV}{RT} = \dfrac{1.0092 \text{ atm} \times 0.899 \text{ L}}{0.08206 \frac{\text{L} \cdot \text{atm}}{\text{mol} \cdot \text{K}} \times 298 \text{ K}} = 0.0371013$ mol

$$0.0371013 \text{ mol H}_2 \times \frac{1 \text{ mol Zn}}{1 \text{ mol H}_2} \times \frac{65.39 \text{ g Zn}}{1 \text{ mol Zn}} = 2.42605 \text{ g Zn}$$

$$\% \text{ Zn} = \frac{g_{Zn}}{g_{penny}} \times 100\% = \frac{2.42605 \text{ g}}{2.482 \text{ g}} \times 100\% = 97.7\% \text{ Zn}$$

         Check: The units (% Zn) are correct. The magnitude of the answer (98) makes sense because it should be between 0 and 100%. We expect about 1/22 moles of gas because our conditions are close to STP and we have ~1 L of gas.

10.97   **Given:** $V = 255$ mL, $m_{flask} = 143.187$ g, $m_{flask+gas} = 143.289$ g, $P = 267$ torr, and $T = 25\,°C$ **Find:** $\mathcal{M}$

         Conceptual Plan: $°C \rightarrow$ K torr $\rightarrow$ atm mL $\rightarrow$ L $m_{flask}, m_{flask+gas} \rightarrow m_{gas}$

             $K = °C + 273.15$ $\dfrac{1 \text{ atm}}{760 \text{ torr}}$ $\dfrac{1 \text{ L}}{1000 \text{ mL}}$ $m_{gas} = m_{flask+gas} - m_{flask}$

         then $V, m \rightarrow d$ **then** $d, P, T, \rightarrow \mathcal{M}$

             $d = \dfrac{m}{V}$ $d = \dfrac{P\mathcal{M}}{RT}$

         Solution: $T = 25\,°C + 273.15 = 298$ K, $P = 267 \text{ torr} \times \dfrac{1 \text{ atm}}{760 \text{ torr}} = 0.351316$ atm

$$V = 255 \text{ mL} \times \frac{1 \text{ L}}{1000 \text{ mL}} = 0.255 \text{ L}$$

$$m_{gas} = m_{flask+gas} - m_{flask} = 143.289 \text{ g} - 143.187 \text{ g} = 0.102 \text{ g}$$

$$d = \frac{m}{V} = \frac{0.102 \text{ g}}{0.255 \text{ L}} = 0.400 \text{ g/L}, d = \frac{P\mathcal{M}}{RT} \quad \text{Rearrange to solve for } \mathcal{M}.$$

$$\mathcal{M} = \frac{dRT}{P} = \frac{0.400 \frac{\text{g}}{\text{L}} \times 0.08206 \frac{\text{L} \cdot \text{atm}}{\text{K} \cdot \text{mol}} \times 298 \text{ K}}{0.351316 \text{ atm}} = 27.8 \text{ g/mol}$$

An alternative method to solve this problem would be to calculate the moles of gas using the ideal gas law (0.00366 mole) and then taking the mass (0.102 g) and dividing it by the number of moles.

Check: The units (g/mol) are correct. The magnitude of the answer (28) makes physical sense because this is a reasonable number for a molar mass of a gas.

10.99 **Given:** $V = 158$ mL, $m_{gas} = 0.275$ g, $P = 556$ mmHg, $T = 25\,°C$, gas = 82.66% C and 17.34% H
Find: molecular formula
Conceptual Plan: $°C \rightarrow K$ mmHg $\rightarrow$ atm mL $\rightarrow$ L then $V, m \rightarrow d$

$$K = °C + 273.15 \qquad \frac{1\,atm}{760\,mmHg} \qquad \frac{1\,L}{1000\,mL} \qquad d = \frac{m}{V}$$

then $d, P, T, \rightarrow \mathcal{M}$ **then** % **C,** % **H,** $\mathcal{M} \rightarrow$ **formula**

$$d = \frac{P\mathcal{M}}{RT} \qquad \#C = \frac{\mathcal{M}\,0.8266\,g\,C}{12.01\frac{g\,C}{mol\,C}} \qquad \#H = \frac{\mathcal{M}\,0.1734\,g\,H}{1.008\frac{g\,H}{mol\,H}}$$

Solution: $T = 25\,°C + 273.15 = 298$ K, $P = 556\,\cancel{mmHg} \times \dfrac{1\,atm}{760\,\cancel{mmHg}} = 0.731579$ atm

$$V = 158\,\cancel{mL} \times \frac{1\,L}{1000\,\cancel{mL}} = 0.158\,L,\ d = \frac{m}{V} = \frac{0.275\,g}{0.158\,L} = 1.74051\,g/L,\ d = \frac{P\mathcal{M}}{RT}\quad \text{Rearrange to solve for } \mathcal{M}.$$

$$\mathcal{M} = \frac{dRT}{P} = \frac{1.74051\frac{g}{\cancel{L}} \times 0.08206\frac{\cancel{L}\cdot\cancel{atm}}{K\cdot mol} \times 298\,K}{0.731579\,\cancel{atm}} = 58.2\,g/mol$$

$$\#C = \frac{\mathcal{M} \times 0.8266\,g\,C}{12.01\frac{g\,C}{mol\,C}} = \frac{58.2\frac{g\,\cancel{HC}}{mol\,HC} \times \frac{0.8266\,g\,\cancel{C}}{1\,g\,\cancel{HC}}}{12.01\frac{g\,\cancel{C}}{mol\,C}} = 4.00\frac{mol\,C}{mol\,HC}$$

$$\#H = \frac{\mathcal{M} \times 0.1734\,g\,H}{1.0008\frac{g\,H}{mol\,H}} = \frac{58.2\frac{g\,\cancel{HC}}{mol\,HC} \times \frac{0.1734\,g\,\cancel{H}}{1\,g\,\cancel{HC}}}{1.008\frac{g\,\cancel{H}}{mol\,H}} = 10.0\frac{mol\,H}{mol\,HC}\ \text{Formula is } C_4H_{10} \text{ or butane.}$$

An alternative method would be to calculate the empirical formula (C_2H_5) and then use the ideal gas law to calculate the number of moles of gas (0.00473 mole). The molar mass can be calculated by dividing the mass (0.275 g) by the number of moles. Divide the molar mass by the molar mass (58.2 g/mol) of the empirical formula (29.07 g/mol), giving a multiplier of 2 for the empirical formula.

Check: The answer came up with integer numbers of C and H atoms in the formula and a molecular weight (58) that is reasonable for a gas.

10.101 **Given:** $m_{NiO} = 24.78$ g, $T = 40.0\,°C$, and $P_{Total} = 745$ mmHg **Find:** V_{O_2}
Conceptual Plan: $T \rightarrow P_{H_2O}$ then $P_{Total}, P_{H_2O} \rightarrow P_{O_2}$ then mmHg $\rightarrow$ atm and $°C \rightarrow K$

$$\text{Table 10.3} \qquad P_{Total} = P_{H_2O} + P_{O_2} \qquad \frac{1\,atm}{760\,mmHg} \qquad K = °C + 273.15$$

and $m_{NiO} \rightarrow n_{NiO} \rightarrow n_{O_2}$ **then** $P, V, T \rightarrow n_{O_2}$

$$\frac{1\,mol\,NiO}{74.69\,g\,NiO} \quad \frac{1\,mol\,O_2}{2\,mol\,NiO} \qquad PV = nRT$$

Solution: Table 10.3 states that $P_{H_2O} = 55.40$ mmHg at $40\,°C$ $P_{Total} = P_{H_2O} + P_{O_2}$ Rearrange to solve for P_{O_2}.

$$P_{O_2} = P_{Total} - P_{H_2O} = 745\,mmHg - 55.40\,mmHg = 689.6\,mmHg$$

$$P_{O_2} = 689.6\,\cancel{mmHg} \times \frac{1\,atm}{760\,\cancel{mmHg}} = 0.907368\,atm \quad T = 40.0\,°C + 273.15 = 313.2\,K$$

$$24.78\,g\,\cancel{NiO} \times \frac{1\,mol\,\cancel{NiO}}{74.69\,g\,\cancel{NiO}} \times \frac{1\,mol\,O_2}{2\,mol\,\cancel{NiO}} = 0.1658857\,mol\,O_2 \qquad PV = nRT$$

Rearrange to solve for V. $V_{O_2} = \dfrac{nRT}{P} = \dfrac{0.1658857\,\cancel{mol} \times 0.08206\frac{L\cdot\cancel{atm}}{\cancel{mol}\cdot K} \times 313.2\,K}{0.907368\,\cancel{atm}} = 4.70\,L$

Check: The units (L) are correct. The magnitude of the answer (5) makes sense because we have much less than 0.5 mole of NiO; so we get less than a mole of oxygen. Thus, we expect a volume much less than 22 L.

10.103 **Given:** HCl, K_2S to H_2S, $V_{H_2S} = 42.9$ mL, $P_{H_2S} = 752$ mmHg, and $T = 25.8\,°C$ **Find:** m_{K_2S}
Conceptual Plan: Read description of reaction and convert words to equation, then °C $\rightarrow$ K

$$K = °C + 273.15$$

and mmHg $\rightarrow$ atm and mL $\rightarrow$ L then $P, V, T \rightarrow n_{H_2S} \rightarrow n_{K_2S} \rightarrow m_{K_2S}$

$$\frac{1\ atm}{760\ mmHg} \qquad \frac{1\ L}{1000\ mL} \qquad PV = nRT \qquad \frac{1\ mol\ K_2S}{1\ mol\ H_2S} \quad \frac{1\ mol\ K_2S}{110.27\ g\ K_2S}$$

Solution: $2\,HCl(aq) + K_2S(s) \rightarrow H_2S(g) + 2\,KCl(aq)$

$$T = 25.8\,°C + 273.15 = 299.0\ K,\ P_{H_2S} = 752\ \cancel{mmHg} \times \frac{1\ atm}{760\ \cancel{mmHg}} = 0.989474\ atm$$

$$V_{H_2S} = 42.9\ \cancel{mL} \times \frac{1\ L}{1000\ \cancel{mL}} = 0.0429\ L \quad PV = nRT \quad \text{Rearrange to solve for } n_{H_2S}.$$

$$n_{H_2S} = \frac{PV}{RT} = \frac{0.989474\ \cancel{atm} \times 0.0429\ \cancel{L}}{0.08206\ \dfrac{\cancel{L} \cdot \cancel{atm}}{mol \cdot \cancel{K}} \times 299.0\ \cancel{K}} = 0.00173005\ mol$$

$$0.00173005\ \cancel{mol\ H_2S} \times \frac{1\ \cancel{mol\ K_2S}}{1\ \cancel{mol\ H_2S}} \times \frac{110.27\ g\ K_2S}{1\ \cancel{mol\ K_2S}} = 0.191\ g\ K_2S$$

Check: The units (g) are correct. The magnitude of the answer (0.2) makes sense because such a small volume of gas is generated.

10.105 **Given:** $T = 22\,°C$, $P = 1.02$ atm, and $m = 11.83$ g **Find:** V_{Total}
Conceptual Plan: °C $\rightarrow$ K and $m_{(NH_4)_2CO_3} \rightarrow n_{(NH_4)_2CO_3} \rightarrow n_{Gas}$ then $P, n, T \rightarrow V$

$$K = °C + 273.15 \qquad \frac{1\ mol\ (NH_4)_2CO_3}{96.09\ g\ (NH_4)_2CO_3} \quad \frac{(2 + 1 + 1 = 4)\ mol\ gas}{1\ mol\ (NH_4)_2CO_3} \qquad PV = nRT$$

Solution: $T = 22\,°C + 273.15 = 295\ K$

$$11.83\ \cancel{g\ (NH_4)_2CO_3} \times \frac{1\ \cancel{mol\ (NH_4)_2CO_3}}{96.09\ \cancel{g\ (NH_4)_2CO_3}} \times \frac{4\ mol\ gas}{1\ \cancel{mol\ (NH_4)_2CO_3}} = 0.492455\ mol\ gas$$

$PV = nRT$ Rearrange to solve for V_{Gas}.

$$V_{Gas} = \frac{nRT}{P} = \frac{0.492455\ \cancel{mol\ gas} \times 0.08206\ \dfrac{L \cdot \cancel{atm}}{\cancel{mol} \cdot \cancel{K}} \times 295\ \cancel{K}}{1.02\ \cancel{atm}} = 11.7\ L$$

Check: The units (L) are correct. The magnitude of the answer (12) makes sense because about half a mole of gas is generated.

10.107 **Given:** He and air; $V = 855$ mL, $P = 125$ psi, $T = 25\,°C$, $\mathcal{M}_{air} = 28.8$ g/mol **Find:** $\Delta m = m_{air} - m_{He}$
Conceptual Plan: mL $\rightarrow$ L and psi $\rightarrow$ atm and °C $\rightarrow$ K then $P, T, \mathcal{M} \rightarrow d$

$$\frac{1\ L}{1000\ mL} \qquad \frac{1\ atm}{14.7\ psi} \qquad K = °C + 273.15 \qquad d = \frac{P\mathcal{M}}{RT}$$

then $d, V \rightarrow m$ then $m_{air}, m_{He} \rightarrow \Delta m$

$$d = \frac{m}{V} \qquad \qquad \Delta = m_{air} - m_{He}$$

Solution: $V = 855\ \cancel{mL} \times \dfrac{1\ L}{1000\ \cancel{mL}} = 0.855\ L,\ P = 125\ \cancel{psi} \times \dfrac{1\ atm}{14.7\ \cancel{psi}} = 8.50340\ atm$

$$T = 25\,°C + 273.15 = 298\ K,\ d_{air} = \frac{P\mathcal{M}}{RT} = \frac{8.50340\ \cancel{atm} \times 28.8\ \dfrac{g\ air}{\cancel{mol\ air}}}{0.08206\ \dfrac{\cancel{L} \cdot \cancel{atm}}{\cancel{K} \cdot \cancel{mol}} \times 298\ \cancel{K}} = 10.0147\frac{g\ air}{L},\ d = \frac{m}{V}$$

Rearrange to solve for m. $m = dV$

$$m_{air} = 10.0147 \frac{\text{g air}}{\cancel{L}} \times 0.855 \cancel{L} = 8.56657 \text{ g air}$$

$$d_{He} = \frac{P\mathcal{M}}{RT} = \frac{8.50340 \cancel{\text{atm}} \times 4.03 \frac{\text{g He}}{\cancel{\text{mol He}}}}{0.08206 \frac{\cancel{L} \cdot \cancel{\text{atm}}}{\cancel{K} \cdot \cancel{\text{mol}}} \times 298 \cancel{K}} = 1.40136 \frac{\text{g He}}{L}$$

$$m_{He} = 1.40136 \frac{\text{g He}}{\cancel{L}} \times 0.855 \cancel{L} = 1.19816 \text{ g He}$$

$$\Delta m = m_{air} - m_{He} = 8.56657 \text{ g air} - 1.19816 \text{ g He} = 7.37 \text{ g}$$

Check: The units (g) are correct. We expect the difference to be less than the difference in the molecular weights because we have less than a mole of gas.

10.109 **Given:** $\text{flow}_{Total} = 335 \text{ L/s}$, $P_{NO} = 22.4 \text{ torr}$, $T_{NO} = 955 \text{ K}$, $P_{NH_3} = 755 \text{ torr}$, $T_{NO} = 298 \text{ K}$, and NH_3 purity $= 65.2\%$ **Find:** flow_{NH_3}
Conceptual Plan: torr $\rightarrow$ atm then $P_{NO}, V_{NO}/s, T_{NO} \rightarrow n_{NO}/s \rightarrow n_{NH_3}/s$ **(pure)**

$$\frac{1 \text{ atm}}{760 \text{ torr}} \qquad PV = nRT \quad \frac{4 \text{ mol } NH_3}{4 \text{ mol } NO}$$

then n_{NH_3}/s **(pure)** $\rightarrow n_{NH_3}/s$ **(impure) then** n_{NH_3}/s **(impure)**, $P_{NH_3}, T_{NH_3} \rightarrow V_{NH_3}/s$

$$\frac{100 \text{ mol } NH_3 \text{ impure}}{65.2 \text{ mol } NH_3 \text{ pure}} \qquad PV = nRT$$

Solution: $P_{NO} = 22.4 \cancel{\text{torr}} \times \frac{1 \text{ atm}}{760 \cancel{\text{torr}}} = 0.0294737 \text{ atm}$, $P_{NH_3} = 755 \cancel{\text{torr}} \times \frac{1 \text{ atm}}{760 \cancel{\text{torr}}} = 0.993421 \text{ atm}$

$PV = nRT$ Rearrange to solve for n_{NO}. Note that we can substitute V/s for V and get n/s as a result.

$$\frac{n_{NO}}{s} = \frac{PV}{RT} = \frac{0.0294737 \cancel{\text{atm}} \times 335 \cancel{L}/s}{0.08206 \frac{\cancel{L} \cdot \cancel{\text{atm}}}{\text{mol} \cdot \cancel{K}} \times 955 \cancel{K}} = 0.125992 \frac{\text{mol NO}}{s}$$

$$0.125992 \frac{\cancel{\text{mol NO}}}{s} \times \frac{4 \cancel{\text{mol } NH_3}}{4 \cancel{\text{mol NO}}} \times \frac{100 \text{ mol } NH_3 \text{ impure}}{65.2 \cancel{\text{mol } NH_3 \text{ pure}}} = 0.193240 \frac{\text{mol } NH_3 \text{ impure}}{s} \qquad PV = nRT$$

Rearrange to solve for V_{NH_3}. Note that we can substitute n/s for n and get V/s as a result.

$$\frac{V_{NH_3}}{s} = \frac{nRT}{P} = \frac{0.193240 \frac{\cancel{\text{mol } NH_3 \text{ impure}}}{s} \times 0.08206 \times \frac{\cancel{L} \cdot \cancel{\text{atm}}}{\cancel{\text{mol}} \cdot \cancel{K}} \times 298 \cancel{K}}{0.993421 \cancel{\text{atm}}} = 4.76 \frac{L}{s} \text{ impure } NH_3$$

Check: The units (L/s) are correct. The magnitude of the answer (5) makes sense because we expect it to be less than that for the NO. The NO is at a very low concentration and a high temperature. When this converts to a much higher pressure and lower temperature, it goes down significantly even though the ammonia is impure. From a practical standpoint, you would like a low flow rate to make it economical.

10.111 **Given:** $l = 30.0 \text{ cm}$, $w = 20.0 \text{ cm}$, $h = 15.0 \text{ cm}$, 14.7 psi **Find:** force (lb)
Conceptual Plan: $l, w, h \rightarrow$ Surface Area, $SA(\text{cm}^2) \rightarrow$ Surface Area$(\text{in}^2) \rightarrow$ Force

$$SA = 2(lh) + 2(wh) + 2(lw) \qquad \frac{(1 \text{ in})^2}{(2.54 \text{ cm})^2} \qquad \frac{14.7 \text{ lb}}{1 \text{ in}^2}$$

Solution: $SA = 2(lh) + 2(wh) + 2(lw) = 2(30.0 \text{ cm} \times 15.0 \text{ cm}) + 2(20.0 \text{ cm} \times 15.0 \text{ cm})$
$+ 2(30.0 \text{ cm} \times 20.0 \text{ cm}) = 2700 \text{ cm}^2$

$$2700 \cancel{\text{cm}^2} \times \frac{(1 \text{ in})^2}{(2.54 \cancel{\text{cm}})^2} = 418.50 \text{ in}^2, 418.50 \cancel{\text{in}^2} \times \frac{14.7 \text{ lb}}{1 \cancel{\text{in}^2}} = 6150 \text{ lb}.$$ The can would be crushed.

Check: The units (lb) are correct. The magnitude of the answer (6150) is not unreasonable because there is a large surface area.

10.113 **Given:** $V_1 = 160.0 \text{ L}$, $P_1 = 1855 \text{ psi}$, 3.5 L/balloon, $P_2 = 1.0 \text{ atm} = 14.7 \text{ psi}$, and $T = 298 \text{ K}$ **Find:** # balloons
Conceptual Plan: $V_1, P_1, P_2 \rightarrow V_2$ then L $\rightarrow$ #balloons

$$P_1 V_1 = P_2 V_2 \qquad \frac{1 \text{ balloon}}{3.5 \text{ L}}$$

Solution: $P_1V_1 = P_2V_2$ Rearrange to solve for V_2. $V_2 = \dfrac{P_1}{P_2}V_1 = \dfrac{1855 \text{ psi}}{14.7 \text{ psi}} \times 160.0 \text{ L} = 20\underline{1}90.5 \text{ L}$

$20\underline{1}90.5 \text{ L} \times \dfrac{1 \text{ balloon}}{3.5 \text{ L}} = 5\underline{7}68 \text{ balloons} = 5800 \text{ balloons}$

Check: The units (balloons) are correct. The magnitude of the answer (5800) is reasonable because a store does not have to buy a new helium tank very often.

10.115   **Given:** $r_1 = 2.5$ cm, $P_1 = 4.00$ atm, $T = 298$ K, and $P_2 = 1.00$ atm   **Find:** r_2
Conceptual Plan: $r_1 \rightarrow V_1 V_1, P_1, P_2 \rightarrow V_2$ then $V_2 \rightarrow r_2$

$$V = \frac{4}{3}\pi r^3 \qquad P_1 V_1 = P_2 V_2 \qquad V = \frac{4}{3}\pi r^3$$

Solution: $V = \dfrac{4}{3}\pi\, r^3 = \dfrac{4}{3} \times \pi \times (2.5 \text{ cm})^3 = 65.\underline{4}50 \text{ cm}^3$ $P_1 V_1 = P_2 V_2$ Rearrange to solve for V_2.

$V_2 = \dfrac{P_1}{P_2}V_1 = \dfrac{4.00 \text{ atm}}{1.00 \text{ atm}} \times 65.450 \text{ cm}^3 = 2\underline{6}1.80 \text{ cm}^3, V = \dfrac{4}{3}\pi\, r^3$

Rearrange to solve for r. $r = \sqrt[3]{\dfrac{3V}{4\pi}} = \sqrt[3]{\dfrac{3 \times 2\underline{6}1.80 \text{ cm}^3}{4 \times \pi}} = 4.0 \text{ cm}$

Check: The units (cm) are correct. The magnitude of the answer (4) is reasonable because the bubble will expand as the pressure is decreased.

10.117   **Given:** 2.0 mol CO : 1.0 mol O_2, $V = 2.45$ L, $P_1 = 745$ torr, $P_2 = 552$ torr, and $T = 552\,^\circ$C
Find: % reacted
Conceptual Plan: From $PV = nRT$, **we know that** $P \propto n$; **looking at the chemical reaction, we see that** $2 + 1 = 3$ **moles of gas gets converted to 2 moles of gas. If all of the gas reacts,** $P_2 = 2/3\,P_1$. **Calculate** $-\Delta P$ **for 100% reacted and for actual case. Then calculate % reacted.**

$$-\Delta P_{100\% \text{ reacted}} = P_1 - \frac{2}{3}P_1 \qquad -\Delta P_{\text{actual}} = P_1 - P_2 \qquad \% \text{ reacted} = \frac{\Delta P_{\text{actual}}}{\Delta P_{100\% \text{ reacted}}} \times 100\%$$

Solution: $-\Delta P_{100\% \text{ reacted}} = P_1 - \dfrac{2}{3}P_1 = 745 \text{ torr} - \dfrac{2}{3}745 \text{ torr} = 24\underline{8}.333 \text{ torr}$

$-\Delta P_{\text{actual}} = P_1 - P_2 = 745 \text{ torr} - 552 \text{ torr} = 193 \text{ torr}$

$\% \text{ reacted} = \dfrac{\Delta P_{\text{actual}}}{\Delta P_{100\% \text{ reacted}}} \times 100\% = \dfrac{193 \text{ torr}}{248.333 \text{ torr}} \times 100\% = 77.7\%$

Check: The units (%) are correct. The magnitude of the answer (78) makes sense because the pressure dropped most of the way to the pressure if all of the reactants had reacted $(2/3(745 \text{ torr}) = 497 \text{ torr})$. **Note: This problem can be solved in many ways, including calculating the moles of reactants and products using** $PV = nRT$.

10.119   **Given:** $P_{1,\text{total}} = 2.2$ atm $= CO + O_2$, $P_{2,\text{total}} = 1.9$ atm $= CO + O_2 + CO_2$, $V = 1.0$ L, and $T = 1.0 \times 10^3$ K
Find: mass CO_2 made
Conceptual Plan:
$P_{1,\text{total}} = 2.2 \text{ atm} = P_{1,\text{CO}} + P_{1,O_2}, P_{2,\text{total}} = 1.9 \text{ atm} = P_{2,\text{CO}} + P_{2,O_2} + P_{2,\text{CO}_2}$
Let x = amount of P_{O_2} reacted. From stoichiometry: $P_{2,\text{CO}} = P_{1,\text{CO}} - 2x, P_{2,O_2} = P_{1,O_2} - x$,
$P_{2,\text{CO}_2} = 2x$. Thus, $P_{2,\text{total}} = 1.9 \text{ atm} = P_{1,\text{CO}} - 2x + P_{1,O_2} - x + 2x = P_{1,\text{total}} - x$. Using the initial
conditions: $1.9 \text{ atm} = 2.2 \text{ atm} - x$. So $x = 0.3 \text{ atm}$, and because $2x = P_{2,\text{CO}_2} = 0.6 \text{ atm}$,
$P, V, T \rightarrow n \rightarrow$ g.

$$PV = nRT \qquad \frac{44.01 \text{ g}}{1 \text{ mol}}$$

Solution: $PV = nRT$ Rearrange to solve for n.

$n = \dfrac{PV}{RT} = \dfrac{0.6 \text{ atm} \times 1.0 \text{ L}}{0.08206 \dfrac{\text{L} \cdot \text{atm}}{\text{mol} \cdot \text{K}} \times 1000 \text{ K}} = 0.007\underline{3}117 \text{ mol}$

$0.007\underline{3}117 \text{ mol} \times \dfrac{44.01 \text{ g}}{1 \text{ mol}} = 0.3\underline{2}1788 \text{ g } CO_2 = 0.3 \text{ g } CO_2$

Check: The units (g) are correct. The magnitude of the answer (0.3) makes sense because we have such a small volume at a very high temperature and such a small pressure. This leads us to expect a very small number of moles.

10.121   **Given:** $h_1 = 22.6$ m, $T_1 = 22\,°C$, and $h_2 = 23.8$ m   **Find:** T_2

Conceptual Plan: $°C \rightarrow$ **K because** $V_{cylinder} \propto h$ **we do not need to know** r **to use** $V_1, T_1, T_2 \rightarrow V_2$

$$K = °C + 273.15 \qquad\qquad V = \pi r^2 h \qquad\qquad\qquad \frac{V_1}{T_1} = \frac{V_2}{T_2}$$

Solution: $T_1 = 22\,°C + 273.15 = 295$ K, $\dfrac{V_1}{T_1} = \dfrac{V_2}{T_2}$ Rearrange to solve for T_2.

$$T_2 = T_1 \times \frac{V_2}{V_1} = T_1 \times \frac{\pi r^2 h_2}{\pi r^2 h_1} = 295 \text{ K} \times \frac{23.8 \text{ m}}{22.6 \text{ m}} = 311 \text{ K}$$

Check: The units (K) are correct. We expect the temperature to increase because the volume increased.

10.123   **Given:** He, $V = 0.35$ L, $P_{max} = 88$ atm, and $T = 299$ K   **Find:** m_{He}

Conceptual Plan: $P, V, T \rightarrow n$ **then mol** $\rightarrow m$

$$PV = nRT \qquad\qquad \mathcal{M}$$

Solution: $PV = nRT$ Rearrange to solve for n.

$$n_{He} = \frac{PV}{RT} = \frac{88 \text{ atm} \times 0.35 \text{ L}}{0.08206 \dfrac{\text{L} \cdot \text{atm}}{\text{mol} \cdot \text{K}} \times 299 \text{ K}} = 1.2553 \text{ mol}, \; 1.2553 \text{ mol} \times \frac{4.003 \text{ g}}{1 \text{ mol}} = 5.0 \text{ g He}$$

Check: The units (g) are correct. The magnitude of the answer (5) makes sense because the high pressure and the low volume cancel out (remember 22 L/mol at STP). So we expect ~1 mol; so ~4 g.

10.125   **Given:** 15.0 mL HBr in 1.0 minute and 20.3 mL unknown hydrocarbon gas in 1.0 minute   **Find:** formula of unknown gas

Conceptual Plan: Because these gases are under the same conditions, $V \alpha n$, V, time $\rightarrow$ **Rate then**

$$\text{Rate} = \frac{V}{time}$$

$\mathcal{M}_{HBr}, \textbf{Rate}_{HBr}, \textbf{Rate}_{Unk} \rightarrow \mathcal{M}_{Unk}$

$$\frac{\text{Rate}_{HBr}}{\text{Rate}_{Unk}} = \sqrt{\frac{\mathcal{M}_{Unk}}{\mathcal{M}_{HBr}}}$$

Solution: $\text{Rate}_{HBr} = \dfrac{V}{time} = \dfrac{15.0 \text{ mL}}{1.0 \text{ min}} = 15.0\dfrac{\text{mL}}{\text{min}}$, $\text{Rate}_{Unk} = \dfrac{V}{time} = \dfrac{20.3 \text{ mL}}{1.0 \text{ min}} = 20.3\dfrac{\text{mL}}{\text{min}}$,

$\dfrac{\text{Rate}_{HBr}}{\text{Rate}_{Unk}} = \sqrt{\dfrac{\mathcal{M}_{Unk}}{\mathcal{M}_{HBr}}}$ Rearrange to solve for $\mathcal{M}_{Unk}$.

$$\mathcal{M}_{Unk} = \mathcal{M}_{HBr}\left(\frac{\text{Rate}_{HBr}}{\text{Rate}_{Unk}}\right)^2 = 80.91\frac{\text{g}}{\text{mol}} \times \left(\frac{15.0\frac{\text{mL}}{\text{min}}}{20.3\frac{\text{mL}}{\text{min}}}\right)^2 = 44.2\frac{\text{g}}{\text{mol}} \text{ The formula is } C_3H_8, \text{ or propane, because}$$

we know the molecule contains only H and C and this combination of atoms leads to a molar mass of 44.2 g/mol.

Check: The units (g/mol) are correct. The magnitude of the answer ($<$ HBr) makes sense because the unknown diffused faster and so must be lighter.

10.127   **Given:** 0.583 g neon, $V = 8.00 \times 10^2 \text{ cm}^3$, $P_{Total} = 1.17$ atm, and $T = 295$ K   **Find:** m_{Ar}

Conceptual Plan: $m_{Ne} \rightarrow n_{Ne}$ **and mL** $\rightarrow$ **L then** $n_{Ne}, V, T \rightarrow P_{Ne}$ **then** $P_{Ne}, P_{Total} \rightarrow P_{Ar}$ **then**

$$\frac{1 \text{ mol}}{20.18 \text{ g}} \qquad\qquad \frac{1 \text{ L}}{1000 \text{ mL}} \qquad\qquad PV = nRT \qquad\qquad P_{Total} = P_{Ne} + P_{Ar}$$

$P_{Ar}, V, T \rightarrow n_{Ar} \rightarrow m_{Ar}$

$$PV = nRT \quad \frac{39.95 \text{ g}}{1 \text{ mol}}$$

Solution: $0.583 \text{ g Ne} \times \dfrac{1 \text{ mol Ne}}{20.18 \text{ g Ne}} = 0.02888999 \text{ mol Ne} \quad 8.00 \times 10^2 \text{ mL} \times \dfrac{1 \text{ L}}{1000 \text{ mL}} = 0.800 \text{ L}$

$PV = nRT$ Rearrange to solve for P.

$$P = \frac{nRT}{V} = \frac{0.02888999 \text{ mol} \times 0.08206 \frac{L \cdot atm}{mol \cdot K} \times 295 \text{ K}}{0.800 \text{ L}} = 0.874200 \text{ atm Ne}$$

$P_{Total} = P_{Ne} + P_{Ar}$ Rearrange to solve for P_{Ar}. $P_{Ar} = P_{Total} - P_{Ne} = 1.17 \text{ atm} - 0.874200 \text{ atm} = 0.295800 \text{ atm}$

$PV = nRT$ Rearrange to solve for n. $n_{Ar} = \frac{PV}{RT} = \frac{0.295800 \text{ atm} \times 0.800 \text{ L}}{0.08206 \frac{L \cdot atm}{mol \cdot K} \times 295 \text{ K}} = 0.00977539 \text{ mol Ar}$

$$0.00977539 \text{ mol Ar} \times \frac{39.95 \text{ g Ar}}{1 \text{ mol Ar}} = 0.390527 \text{ g Ar} = 0.39 \text{ g Ar}$$

Check: The units (g) are correct. The magnitude of the answer (0.4) makes sense because the pressure and volume are small.

10.129   **Given:** 75.2% by mass nitrogen + 24.8% by mass krypton and $P_{Total} = 745$ mmHg   **Find:** P_{Kr}
Solution: Assume that 100 g total, so we have 75.2 g N_2 and 24.8 g Kr. Converting these masses to moles,

$$75.2 \text{ g } N_2 \times \frac{1 \text{ mol } N_2}{28.02 \text{ g } N_2} = 2.683797 \text{ mol } N_2 \text{ and } 24.8 \text{ g Kr} \times \frac{1 \text{ mol Kr}}{83.80 \text{ g Kr}} = 0.2959427 \text{ mol Kr}$$

$$P_{Kr} = \chi_{Kr} P_{Total} = \frac{0.2959427 \text{ mol Kr}}{2.683797 \text{ mol } N_2 + 0.2959427 \text{ mol Kr}} \, 745 \text{ mmHg Kr} = 74.0 \text{ mmHg Kr}$$

Check: The units (mmHg) are correct. The magnitude of the answer (74) makes sense because the mixture is mostly nitrogen by mass and this dominance is magnified because the molar mass of krypton is larger than the molar mass of nitrogen.

Challenge Problems

10.131   **Given:** $2 \, NH_3(g) \rightarrow N_2(g) + 3 \, H_2(g)$; $N_2H_4(g) \rightarrow N_2(g) + 2 \, H_2(g)$; initially $P = 0.50$ atm, $T = 300$ K; finally $P = 4.5$ atm, $T = 1200$ K   **Find:** percent N_2H_4 initially
Conceptual Plan: $P_{initial}, T_{initial}, T_{final} \rightarrow P_{final}$ **then determine change in moles of gas**

$$\frac{P_{initial}}{T_{initial}} = \frac{P_{final}}{T_{final}} \qquad \frac{2 \text{ atm added gas}}{2 \text{ atm } NH_3 \text{ reacted}} \text{ and } \frac{2 \text{ atm added gas}}{1 \text{ atm } N_2H_4 \text{ reacted}}$$

$P_1, P_2 \rightarrow \Delta P$ **write expression for** ΔP**, then solve for** P_{1, N_2H_4} **and** P_{1, NH_3} **finally** $P_{1, N_2H_4}, P_{1, NH_3} \rightarrow \% \, N_2H_4$

$$\Delta P = P_2 - P_1 \qquad \Delta P = P_{1, NH_3} \frac{3 \text{ atm added gas}}{2 \text{ atm reacted}} + P_{1, N_2H_4} \frac{2 \text{ atm added gas}}{1 \text{ atm reacted}} \text{ where } P_{1,1200 \text{ K}} = P_{1,NH_3} + P_{1,N_2H_4} \qquad \% \, N_2H_4 = \frac{P_{N_2H_4}}{P_{N_2H_4} + P_{NH_3}} \times 100\%$$

Solution: $\frac{P_{initial}}{T_{initial}} = \frac{P_{final}}{T_{final}}$ Rearrange to solve for P_{final}. $P_2 = P_1 \times \frac{T_2}{T_1} = 0.50 \text{ atm} \times \frac{1200 \text{ K}}{300 \text{ K}} = 2.0 \text{ atm}$ if no reaction occurred. The increase in pressure from 2.0 atm to 4.5 atm (after the reaction occurs) is due to the decomposition reactions.

$\Delta P = P_{final} - P_{initial} = 4.5 \text{ atm} - 2.0 \text{ atm} = 2.5 \text{ atm}$, and $P_{1,1200 \text{ K}} = 2.0 \text{ atm} = P_{1,NH_3} + P_{1,N_2H_4}$ or $P_{1,NH_3} = 2.0 \text{ atm} - P_{1,N_2H_4}$

Substitute this into $\Delta P = P_{1,NH_3} \frac{2 \text{ atm added gas}}{2 \text{ atm reacted}} + P_{1,N_2H_4} \frac{2 \text{ atm added gas}}{1 \text{ atm reacted}}$ and solve for P_{1, N_2H_4}.

$$\Delta P = 2.5 \text{ atm} = (2.0 \text{ atm} - P_{1,N_2H_4}) \frac{2 \text{ atm added gas}}{2 \text{ atm reacted}} + P_{1,N_2H_4} \frac{2 \text{ atm added gas}}{1 \text{ atm reacted}} \rightarrow$$

$P_{1, N_2H_4} = 2.5 \text{ atm} - 2.0 \text{ atm} = 0.5 \text{ atm}$ and $P_{NH_3} = 2.0 \text{ atm} - 0.5 \text{ atm} = 1.5 \text{ atm}$; finally

$$\% \, N_2H_4 = \frac{P_{N_2H_4}}{P_{N_2H_4} + P_{NH_3}} \times 100\% = \frac{0.5 \text{ atm}}{0.5 \text{ atm} + 1.5 \text{ atm}} \times 100\% = 25\% \, N_2H_4 = 30\% \, N_2H_4$$

Check: The units (%) are correct. The magnitude of the answer (30) makes sense because if it were all N_2H_4, the final pressure would have been 6 atm. Because we are closer to the initial pressure than this maximum pressure, less than half of the gas is N_2H_4.

10.133   **Given:** $2 \, CO_2(g) \rightarrow 2 \, CO(g) + O_2(g)$; initially $P = 10.0$ atm, $T = 701$ K; finally $P = 22.5$ atm, $T = 1401$ K
Find: mole percent decomposed

Conceptual Plan: $P_1, T_1, T_2 \rightarrow P_2$ **then determine change in moles of gas**

$$\frac{P_1}{T_1} = \frac{P_2}{T_2}$$

$$\frac{1 \text{ atm added gas}}{2 \text{ atm } CO_2 \text{ reacted}}$$

$P_1, P_2 \rightarrow \Delta P$ **write expression for** ΔP, **then solve for** $P_{CO_2 \text{reacted}}$ **finally**

$$\Delta P = P_2 - P_1 \qquad \Delta P = P_{CO_2 \text{ reacted}} \frac{1 \text{ atm added gas}}{2 \text{ atm } CO_2 \text{ reacted}}$$

$P_2, P_{CO_2 \text{ reacted}} \rightarrow$ **% CO_2 decomposed**

$$\% CO_2 \text{ decomposed} = \frac{P_{CO_2 \text{ reacted}}}{P_2} \times 100\%$$

Solution: $\dfrac{P_1}{T_1} = \dfrac{P_2}{T_2}$ Rearrange to solve for P_1. $P_2 = P_1 \times \dfrac{T_2}{T_1} = 10.0 \text{ atm} \times \dfrac{1401 \text{ K}}{701 \text{ K}} = 19.\underline{9}85735 \text{ atm}$

$\Delta P = P_2 - P_1 = 22.5 \text{ atm} - 19.985735 \text{ atm} = 2.\underline{5}14265 \text{ atm}$

$\Delta P = P_{CO_2 \text{ reacted}} \dfrac{1 \text{ atm added gas}}{2 \text{ atm } CO_2 \text{ reacted}}$ or the pressure increases 1 atm for each 2 atm of gas decomposed, so

$5.\underline{0}2853 \text{ atm}$ decomposes and then

$\% CO_2 \text{ decomposed} = \dfrac{P_{CO_2 \text{ reacted}}}{P_2} \times 100\% = \dfrac{5.02853 \text{ atm}}{19.985735 \text{ atm}} \times 100\% = 25.1606\% CO_2 \text{ decomposed} =$

$25\% CO_2$ decomposed

Check: The units (%) are correct. The magnitude of the answer (25) makes sense because if all of the gas had decomposed, the final pressure would have been 40 atm. Because we are much closer to the initial pressure than this maximum pressure, much less than half of the gas had decomposed.

10.135 **Given:** CH_4: $V = 155 \text{ mL}$ at STP; O_2: $V = 885 \text{ mL}$ at STP; NO: $V = 55.5 \text{ mL}$ at STP; mixed in a flask: $V = 2.0 \text{ L}$, $T = 275 \text{ K}$, and 90.0% of limiting reagent used. **Find:** Ps of all components and P_{Total}

Conceptual Plan: CH_4: $mL \rightarrow L \rightarrow n_{CO_4} \rightarrow n_{CO_2}$ and

$$\frac{1 \text{ L}}{1000 \text{ mL}} \quad \frac{1 \text{ mol}}{22.414 \text{ L}} \quad \frac{1 \text{ mol } CO_2}{5 \text{ mol NO}}$$

O_2: $mL \rightarrow L \rightarrow n_{O_2} \rightarrow n_{CO_2}$ and NO: $mL \rightarrow L \rightarrow n_{NO} \rightarrow n_{CO_2}$

$$\frac{1 \text{ L}}{1000 \text{ mL}} \quad \frac{1 \text{ mol}}{22.414 \text{ L}} \quad \frac{1 \text{ mol } CO_2}{5 \text{ mol } O_2} \qquad \frac{1 \text{ L}}{1000 \text{ mL}} \quad \frac{1 \text{ mol}}{22.414 \text{ L}} \quad \frac{1 \text{ mol } CO_2}{5 \text{ mol NO}}$$

the smallest yield determines the limiting reagent, then initial $n_{NO} \rightarrow$ **reacted** $n_{NO} \rightarrow$ **final** n_{NO}

NO is the limiting reagent   90.0% $0.100 \times$ initial mol_{NO}

reacted $n_{NO} \rightarrow$ **reacted** n_{CH_4} **then initial** n_{CH_4}, **reacted** $n_{CH_4} \rightarrow$ **final** n_{CH_4} **then**

$$\frac{1 \text{ mol } CH_4}{5 \text{ mol NO}} \qquad\qquad \text{initial mol}_{CH_4} - \text{reacted mol}_{CH_4} = \text{final mol}_{CH_4}$$

final n_{CH_4}, $V, T \rightarrow$ **final** P_{CH_4} **and reacted** $n_{NO} \rightarrow$ **reacted** n_{O_2} **then**

$$PV = nRT \qquad\qquad \frac{5 \text{ mol } O_2}{5 \text{ mol NO}}$$

initial n_{O_2}, **reacted** $n_{O_2} \rightarrow$ **final** n_{O_2} **then final** n_{O_2}, $V, T \rightarrow$ **final** P_{O_2} **and**

$$\text{initial mol}_{O_2} - \text{reacted mol}_{O_2} = \text{final mol}_{O_2} \qquad\qquad PV = nRT$$

final n_{NO}, $V, T \rightarrow$ **final** P_{NO} **and theoretical** n_{CO_2} **from NO** $\rightarrow$ **final** n_{CO_2}

$$PV = nRT \qquad\qquad 90.0\%$$

final n_{CO_2}, $V, T \rightarrow P_{CO_2}$ **then final** $n_{CO_2} \rightarrow n_{H_2O}$, $V, T \rightarrow P_{H_2O}$ **and**

$$PV = nRT \qquad\qquad \frac{1 \text{ mol } H_2O}{1 \text{ mol } CO_2} \qquad PV = nRT$$

final $n_{CO_2} \rightarrow n_{NO_2}$ **then** n_{NO_2}, $V, T \rightarrow P_{NO_2}$ **and final** $n_{CO_2} \rightarrow n_{OH}$ **then**

$$\frac{1 \text{ mol } NO_2}{1 \text{ mol } CO_2} \qquad\qquad PV = nRT \qquad\qquad \frac{2 \text{ mol OH}}{1 \text{ mol } CO_2}$$

$n_{OH}, V, T \rightarrow P_{OH}$ finally $P_{CH_4}, P_{O_2}, P_{NO}, P_{CO_2}, P_{H_2O}, P_{NO_2}, P_{OH} \rightarrow P_{T\,total}$

$$PV = nRT \qquad\qquad P_{Total} = \Sigma P$$

Solution: CH_4: $155 \text{ mL} \times \dfrac{1\,L}{1000\,mL} \times \dfrac{1 \text{ mol } CH_4}{22.414\,L} \times \dfrac{1 \text{ mol } CO_2}{1 \text{ mol } CH_4} = 0.00691532 \text{ mol } CO_2$

O_2: $885 \text{ mL} \times \dfrac{1\,L}{1000\,mL} \times \dfrac{1 \text{ mol } O_2}{22.414\,L} = 0.0394843 \text{ mol } O_2 \times \dfrac{1 \text{ mol } CO_2}{5 \text{ mol } O_2} = 0.00789686 \text{ mol } CO_2$

NO: $55.5 \text{ mL} \times \dfrac{1\,L}{1000\,mL} \times \dfrac{1 \text{ mol } NO}{22.414\,L} \times \dfrac{1 \text{ mol } CO_2}{5 \text{ mol } NO} = 0.000495226 \text{ mol } CO_2$

$0.000495226 \text{ mol } CO_2$ is the smallest yield, so NO is the limiting reagent.

$55.5 \text{ mL} \times \dfrac{1\,L}{1000\,mL} \times \dfrac{1 \text{ mol } NO}{22.414\,L} = 0.00247613 \text{ mol } NO$

reacted mol NO $= 0.900 \times$ mol NO $= 0.900 \times 0.00247613$ mol NO $= 0.00222852$ mol NO

unreacted mol NO $= 0.100 \times$ mol NO $= 0.100 \times 0.00247613$ mol NO $= 0.000247613$ mol NO

$$P = \frac{nRT}{V} = \frac{0.000247613 \text{ mol} \times 0.08206 \dfrac{L \cdot atm}{mol \cdot K} \times 275 \text{ K}}{2.0\,L} = 0.00279 \text{ atm NO remaining}$$

$0.00222852 \text{ mol } NO \times \dfrac{1 \text{ mol } CH_4}{5 \text{ mol } NO} = 0.000445704 \text{ mol } CH_4 \text{ reacted}$

$0.00691532 \text{ mol } CH_4 - 0.000445704 \text{ mol } CH_4 \text{ reacted} = 0.00646962 \text{ mol } CH_4 \text{ unreacted then } PV = nRT$

Rearrange to solve for P. $P = \dfrac{nRT}{V} = \dfrac{0.00646962 \text{ mol} \times 0.08206 \dfrac{L \cdot atm}{mol \cdot K} \times 275 \text{ K}}{2.0\,L} = 0.0730 \text{ atm } CH_4 \text{ remaining}$

$0.00222852 \text{ mol } NO \times \dfrac{5 \text{ mol } O_2}{5 \text{ mol } NO} = 0.00222852 \text{ mol } O_2 \text{ reacted}$

$0.0394843 \text{ mol } O_2 - 0.00222852 \text{ mol } O_2 \text{ reacted} = 0.0372558 \text{ mol } O_2 \text{ unreacted}$

$$P = \frac{nRT}{V} = \frac{0.0372558 \text{ mol} \times 0.08206 \dfrac{L \cdot atm}{mol \cdot K} \times 275 \text{ K}}{2.0\,L} = 0.420 \text{ atm } O_2 \text{ remaining}$$

$0.00222852 \text{ mol } NO \times \dfrac{1 \text{ mol } CO_2}{5 \text{ mol } NO} = 0.000445704 \text{ mol } CO_2$

$$P = \frac{nRT}{V} = \frac{0.000445704 \text{ mol} \times 0.08206 \dfrac{L \cdot atm}{mol \cdot K} \times 275 \text{ K}}{2.0\,L} = 0.00503 \text{ atm } CO_2 \text{ produced}$$

$0.00222852 \text{ mol } NO \times \dfrac{1 \text{ mol } H_2O}{5 \text{ mol } NO} = 0.000445704 \text{ mol } H_2O$

$$P = \frac{nRT}{V} = \frac{0.000445704 \text{ mol} \times 0.08206 \dfrac{L \cdot atm}{mol \cdot K} \times 275 \text{ K}}{2.0\,L} = 0.00503 \text{ atm } H_2O \text{ produced}$$

$0.00222852 \text{ mol } NO \times \dfrac{5 \text{ mol } NO_2}{5 \text{ mol } NO} = 0.00222852 \text{ mol } NO_2$

$$P = \frac{nRT}{V} = \frac{0.00222852 \text{ mol} \times 0.08206 \dfrac{L \cdot atm}{mol \cdot K} \times 275 \text{ K}}{2.0\,L} = 0.0251 \text{ atm } NO_2 \text{ produced}$$

$0.00222852 \text{ mol } NO \times \dfrac{2 \text{ mol } OH}{5 \text{ mol } NO} = 0.000891408 \text{ mol } OH$

$$P = \frac{nRT}{V} = \frac{0.000891408 \text{ mol} \times 0.08206 \dfrac{L \cdot atm}{mol \cdot K} \times 275 \text{ K}}{2.0\,L} = 0.0101 \text{ atm } OH \text{ produced}$$

$P_{Total} = \displaystyle\sum P$

$$= 0.0730 \text{ atm} + 0.420 \text{ atm} + 0.00279 \text{ atm} + 0.00503 \text{ atm} + 0.00503 \text{ atm} + 0.0251 \text{ atm} + 0.0101 \text{ atm}$$
$$= 0.541 \text{ atm}$$

Check: The units (atm) are correct. The magnitude of the answers is reasonable. The limiting reagent has the lowest pressure. The product pressures are in line with the ratios of the stoichiometric coefficients.

10.137 **Given:** $P_{CH_4} + P_{C_2H_4} = 0.53$ atm and $P_{CO_2} + P_{H_2O} = 2.2$ atm **Find:** χ_{CH_4}
Conceptual Plan: Write balanced reactions to determine change in moles of gas for CH_4 and C_2H_6.

$$2\,CH_4(g) + 4\,O_2(g) \rightarrow 4\,H_2O(g) + 2\,CO_2(g) \text{ and } 2\,C_2H_6(g) + 7\,O_2(g) \rightarrow 6\,H_2O(g) + 4\,CO_2(g); \text{ thus, } \frac{6 \text{ mol gases}}{2 \text{ mol } CH_4} \quad \frac{10 \text{ mol gases}}{2 \text{ mol } C_2H_6}$$

Write expression for final pressure, substituting in data given $\rightarrow \chi_{CH_4}$.

$$\chi_{CH_4} = \frac{n_{CH_4}}{n_{CH_4} + n_{C_2H_6}} \text{ and } \chi_{C_2H_6} = 1 - \chi_{CH_4}$$

$$P_{CH_4} = \chi_{CH_4} P_{Total} \qquad P_{C_2H_6} = \chi_{C_2H_6} P_{Total} \qquad P_{Final} = \left(\chi_{CH_4} P_{Total} \times \frac{6 \text{ mol gases}}{2 \text{ mol } CH_4} \right) + \left((1 - \chi_{CH_4}) P_{Total} \times \frac{10 \text{ mol gases}}{2 \text{ mol } C_2H_6} \right)$$

Solution:

$$P_{Final} = \left(\chi_{CH_4} \times 0.53 \text{ atm} \times \frac{6 \text{ mol gases}}{2 \text{ mol } CH_4} \right) + \left((1 - \chi_{CH_4}) \times 0.53 \text{ atm} \times \frac{10 \text{ mol gases}}{2 \text{ mol } C_2H_6} \right) = 2.2 \text{ atm}$$

Substitute as above for $\chi_{C_2H_6}$, then solve for $\chi_{CH_4} = 0.42$.

Check: The units (none) are correct. The magnitude of the answer (0.42) makes sense because if it had been all methane, the final pressure would have been 1.59 atm and if it had been all ethane, the final pressure would have been 2.65 atm. Because we are closer to the latter pressure, we expect the mole fraction of methane to be less than 0.5.

Conceptual Problems

10.139 Because the passengers have more mass than the balloon, they have more momentum than the balloon. The passengers will continue to travel in their original direction longer. The car is slowing, so the relative position of the passengers is to move forward and the balloon to move backward. The opposite happens upon acceleration.

10.141 B is the limiting reactant (2.0 L of B requires 1.0 L A to completely react). The final container will have 0.5 L A and 2.0 L C, so the final volume will be 2.5 L. The change will be $[(2.5 \text{ L}/3.5 \text{ L}) \times 100\%] - 100\% = -29\%$.

10.143 (a) False—All gases have the same average kinetic energy at the same temperature.
(b) False—The gases will have the same partial pressures because we have the same number of moles of each.
(c) False—The average velocity of the B molecules will be less than that of the A molecules because the Bs are heavier.
(d) True—Because B molecules are heavier, they will contribute more to the density ($d = m/V$).

10.145 When the volume of a gas is cut in half, the pressure doubles. When the temperature of a gas in kelvins doubles, the pressure doubles. The net effect is that the pressure increases by a factor of four.

10.147 Because the velocity is inversely proportional to the molar mass, the tails on the helium are $\sqrt{20/4} = {\sim}2.2$ times as long as those for neon and $\sqrt{84/4} = {\sim}4.6$ times as long as those for krypton.

smallest particles = He

medium-sized particles = Ne

largest = Kr

Questions for Group Work

10.149   Ideal gas law: $PV = nRT$.
Rearranging, we can get: $P = nRT/V$; $V = nRT/P$; $n = PV/RT$; and $T = PV/nR$.

10.151   All of the gases are in the same container, so they have the same temperature and the same average kinetic energy. If there are equal amounts (in moles) of He, Ar, and Kr, then the mass ratios are 4.003 : 39.95 : 83.80 for He : Ar : Kr, respectively. This is roughly a 1 : 10 : 20 mass ratio for He : Ar : Kr, respectively. The average velocities will be inversely proportional to the square root of the molar masses. Thus, helium will have the highest average velocity; argon's average velocity will be about a third of helium's average velocity; and krypton's average velocity will be between a fifth and a quarter of helium's average velocity. All three gasses have a common number of particles, temperature, and average kinetic energy. All of the other properties will be different.

Data Interpretation and Analysis

10.153   (a)   There is an inverse relationship between the concentration of nitrogen dioxide and the concentration of ozone.

(c)   37 ppmv − 8 ppmv = 29 ppmv. Since $\dfrac{mg}{m^3} = ppmv \times \dfrac{\mathcal{M}}{0.08206 \, \dfrac{L \cdot atm}{K \cdot mol} \times T}$

$$\frac{mg}{m^3} = 29 \, \cancel{ppbv} \times \frac{1 \, ppmv}{10^3 \, \cancel{ppbv}} \times \frac{48.00 \, \cancel{g/mol}}{0.08206 \, \dfrac{L \cdot atm}{K \cdot \cancel{mol}} \times 298.15 \, K} = 0.05\underline{6}9018 \, \frac{mg}{m^3}. \text{ So in } 1.00 \, m^3 \text{ of air, there}$$

is 0.05$\underline{6}$9018 mg O_3 or 5.$\underline{6}$9018 $\times$ 10^{-5} g O_3 or 5.$\underline{6}$9018 $\times$ 10^{-5} $\cancel{g \, O_3}$ $\times \dfrac{1 \, mol \, O_3}{48.00 \, \cancel{g \, O_3}} = 1.2 \times 10^{-6} \, mol \, O_3$.

(e)   Adding up the chemical equations in the model, 1 mole of O_3 consumed for each mole of NO_2 produced. The model is consistent with the result in part d.

11 Liquids, Solids, and Intermolecular Forces

Review Questions

11.1 Ethanol and dimethyl ether are isomers—they have the same chemical formula, C_2H_6O, but are different compounds. In ethanol, the oxygen atom is bonded to a carbon atom and a hydrogen atom. In dimethyl ether, the oxygen atom is bonded to two carbon atoms. The structures of these two molecules are different, and structure determines properties.

11.3 The main properties of liquids are that liquids have much higher densities in comparison to gases and generally have lower densities in comparison to solids; liquids have an indefinite shape and assume the shape of their container; liquids have a definite volume; and liquids are not easily compressed.

11.5 Solids may be crystalline, in which case the atoms or molecules that compose them are arranged in a well-ordered three-dimensional array, or they may be amorphous, in which case the atoms or molecules that compose them have no long-range order.

11.7 Because the most molecular motion occurs in the gas phase and the least molecular motion occurs in the solid phase (atoms are pushed closer together), a substance will be converted from a solid to a liquid and then to a gas as the temperature increases. The strength of the intermolecular interactions is least in the gas phase because there are large distances between particles and they are moving very fast. Intermolecular forces are stronger in liquids and solids, where molecules are "touching" one another. The strength of the interactions in the condensed phases determines at what temperature the substance will melt and boil.

11.9 Intermolecular forces, even the strongest ones, are generally much weaker than bonding forces. The reason for the relative weakness of intermolecular forces compared to bonding forces is related to Coulomb's law $\left(E = \dfrac{1}{4\pi\varepsilon_o} \dfrac{q_1 q_2}{r} \right)$. Bonding forces are the result of large charges (the charges on protons and electrons, q_1 and q_2) interacting at very close distances (r). Intermolecular forces are the result of smaller charges (as we will see in the following discussion) interacting at greater distances.

11.11 The dipole–dipole force exists in all molecules that are polar. Polar molecules have permanent dipoles that interact with the permanent dipoles of neighboring molecules. The positive end of one permanent dipole is attracted to the negative end of another; this attraction is the dipole–dipole force.

11.13 The hydrogen bond is a sort of super dipole–dipole force. Polar molecules containing hydrogen atoms bonded directly to fluorine, oxygen, or nitrogen exhibit an intermolecular force called hydrogen bonding. The large electronegativity difference between hydrogen and these electronegative elements means that the H atoms will have fairly large partial positive charges ($\delta+$), while the F, O, or N atoms will have fairly large partial negative charges ($\delta-$). In addition, because these atoms are all quite small, they can approach one another very closely. The result is a strong attraction between the hydrogen in each of these molecules and the F, O, or N on its neighbors, an attraction called a hydrogen bond.

11.15 Surface tension is the tendency of liquids to minimize their surface area. Molecules at the surface have relatively fewer neighbors with which to interact because there are no molecules above the surface. Consequently, molecules at the surface are inherently less stable—they have higher potential energy—than those in the interior. To increase the surface area of the liquid, some molecules from the interior must be moved to the surface, a process requiring energy. The surface tension of a liquid is the energy required to increase the surface area by a unit amount. Surface tension decreases with decreasing intermolecular forces.

11.17   Capillary action is the ability of a liquid to flow against gravity up a narrow tube. Capillary action results from a combination of two forces: the attraction between molecules in a liquid, called cohesive forces, and the attraction between these molecules and the surface of the tube, called adhesive forces. The adhesive forces cause the liquid to spread out over the surface of the tube, while the cohesive forces cause the liquid to stay together. If the adhesive forces are greater than the cohesive forces (as is the case for water in a glass tube), the attraction to the surface draws the liquid up the tube while the cohesive forces pull along those molecules that are not in direct contact with the tube walls. The liquid rises up the tube until the force of gravity balances the capillary action—the thinner the tube, the higher the rise. If the adhesive forces are smaller than the cohesive forces (as is the case for liquid mercury), the liquid does not rise up the tube at all (and, in fact, will drop to a level below the level of the surrounding liquid).

11.19   The molecules that leave the liquid are at the high end of the energy curve—the most energetic. If no additional heat enters the liquid, the average energy of the entire collection of molecules goes down—much as the class average on an exam goes down when the highest-scoring students are eliminated. So vaporization is an endothermic process; it takes energy to vaporize the molecules in a liquid. Also, vaporization requires overcoming the intermolecular forces that hold liquids together. Because energy must be absorbed to pull the molecules apart, the process is endothermic. Condensation is the opposite process, so it must be exothermic. Also, gas particles have more energy than those in the liquid. It is the least energetic of these that condense, adding energy to the liquid.

11.21   The heat of vaporization (ΔH_{vap}) is the amount of heat required to vaporize 1 mole of a liquid to a gas. The heat of vaporization of a liquid can be used to calculate the amount of heat energy required to vaporize a given mass of the liquid (or the amount of heat given off by the condensation of a given mass of liquid) and can be used to compare the volatility of two substances.

11.23   When a system in dynamic equilibrium is disturbed, the system responds so as to minimize the disturbance and return to a state of equilibrium.

11.25   The boiling point of a liquid is the temperature at which its vapor pressure equals the external pressure. The normal boiling point of a liquid is the temperature at which its vapor pressure equals 1 atm.

11.27   As the temperature rises, more liquid vaporizes and the pressure within the container increases. As more and more gas is forced into the same amount of space, the density of the gas becomes higher and higher. At the same time, the increasing temperature causes the density of the liquid to become lower and lower. At the critical temperature, the meniscus between the liquid and gas disappears, and the gas and liquid phases commingle to form a supercritical fluid.

11.29   Fusion, or melting, is the phase transition from solid to liquid. The term fusion is used for melting because if you heat several crystals of a solid, they will fuse into a continuous liquid upon melting. Fusion is endothermic because solids have less kinetic energy than liquids, so energy must be added to a solid to get it to melt.

11.31   There are two horizontal lines (i.e., heat is added, but the temperature stays constant) in the heating curve because there are two endothermic phase changes. The heat that is added is used to change the phase from solid to liquid or from liquid to gas and, therefore, there is no rise in temperature.

11.33   A phase diagram is a map of the phase of a substance as a function of pressure (on the *y*-axis) and temperature (on the *x*-axis).

When a line is crossed in a phase diagram, it means that a phase transition has occurred.

Problem by Topic

Intermolecular Forces

11.35   (a)   dispersion forces
        (b)   dispersion forces, dipole–dipole forces, and hydrogen bonding
        (c)   dispersion forces and dipole–dipole forces
        (d)   dispersion forces

11.37   (a)   dispersion forces and dipole–dipole forces
        (b)   dispersion forces, dipole–dipole forces, and hydrogen bonding
        (c)   dispersion forces
        (d)   dispersion forces

11.39   (a) CH_4 < (b) CH_3CH_3 < (c) CH_3CH_2Cl < (d) CH_3CH_2OH. The first two molecules only exhibit dispersion forces, so the boiling point increases with increasing molar mass. The third molecule also exhibits dipole–dipole forces, which are stronger than dispersion forces. The last molecule exhibits hydrogen bonding. Because these are by far the strongest intermolecular forces in this group, the last molecule has the highest boiling point.

11.41   (a) CH_3OH has the higher boiling point because it exhibits hydrogen bonding.
        (b) CH_3CH_2OH has the higher boiling point because it exhibits hydrogen bonding.
        (c) CH_3CH_3 has the higher boiling point because it has the larger molar mass.

11.43   (a) Br_2 has the higher vapor pressure because it has the smaller molar mass.
        (b) H_2S has the higher vapor pressure because it does not exhibit hydrogen bonding.
        (c) PH_3 has the higher vapor pressure because it does not exhibit hydrogen bonding.

11.45   (a)   This will not form a homogeneous solution because one is polar and one is nonpolar.
        (b)   This will form a homogeneous solution. There will be ion–dipole interactions between the K^+ and Cl^- ions and the water molecules. There will also be dispersion forces, dipole–dipole forces, and hydrogen bonding between the water molecules.
        (c)   This will form a homogeneous solution. Dispersion forces will be present among all of the molecules.
        (d)   This will form a homogeneous solution. There will be dispersion forces, dipole–dipole forces, and hydrogen bonding among all of the molecules.

Surface Tension, Viscosity, and Capillary Action

11.47   Water will have the higher surface tension because it exhibits hydrogen bonding, a strong intermolecular force. Acetone cannot form hydrogen bonds.

11.49   Compound A will have the higher viscosity because it can interact with other molecules along the entire molecule. The more branched isomer has a smaller surface area, allowing for fewer interactions. Also, the molecule is very flexible, and the molecules can get tangled with each other.

11.51   In a clean glass tube, the water can generate strong adhesive interactions with the glass (due to the dipoles at the surface of the glass). Water experiences adhesive forces with glass that are stronger than its cohesive forces, causing it to climb the surface of a glass tube. When grease or oil coats the glass, this interferes with the formation of these adhesive interactions with the glass because oils are nonpolar and cannot interact strongly with the dipoles in the water. Without experiencing these strong intermolecular forces with oil, the water's cohesive forces will be greater and water will be drawn away from the surface of the tube.

Vaporization and Vapor Pressure

11.53   The water in the 12-cm-diameter beaker will evaporate more quickly because there is more surface area for the molecules to evaporate from. The vapor pressure will be the same in the two containers because the vapor pressure is the pressure of the gas when it is in dynamic equilibrium with the liquid (evaporation rate = condensation rate). The vapor pressure is dependent only on the substance and the temperature. The 12-cm-diameter container will reach this dynamic equilibrium faster.

11.55   The boiling point and higher heat of vaporization of oil are much higher than those of water, so it will not vaporize as quickly as the water. The evaporation of water cools your skin because evaporation is an endothermic process.

11.57   **Given:** 915 kJ from candy bar, water $d = 1.00$ g/mL   **Find:** $L(H_2O)$ vaporized at 100.0 °C
      Other: $\Delta H_{vap}^\circ = 40.7$ kJ/mol
      Conceptual Plan: $q \rightarrow$ mol $H_2O \rightarrow$ g $H_2O \rightarrow$ mL $H_2O \rightarrow$ L H_2O

$$\frac{1\ mol}{40.7\ kJ} \qquad \frac{18.02\ g}{1\ mol} \qquad \frac{1.00\ mL}{1.00\ g} \qquad \frac{1\ L}{1000\ mL}$$

      Solution: $915\ kJ \times \dfrac{1\ mol}{40.7\ kJ} \times \dfrac{18.02\ g}{1\ mol} \times \dfrac{1.00\ mL}{1\ g} \times \dfrac{1\ L}{1000\ mL} = 0.405\ L\ H_2O$

Check: The units (L) are correct. The magnitude of the answer (< 1 L) makes physical sense because we are vaporizing about 22 moles of water.

11.59   **Given:** 0.95 g water condenses on iron block 75.0 g at $T_i = 22\,°C$ **Find:** T_f (iron block)
Other: $\Delta H°_{vap} = 44.0$ kJ/mol; $C_{Fe} = 0.449$ J/g·°C from text
Conceptual Plan: g $H_2O \rightarrow$ mol $H_2O \rightarrow q_{H_2O}$ (kJ) $\rightarrow q_{H_2O}$ (J) $\rightarrow q_{Fe}$ then $q_{Fe}, m_{Fe}, T_i \rightarrow T_f$

$$\frac{1\text{ mol}}{18.02\text{ g}} \qquad \frac{-44.0\text{ kJ}}{1\text{ mol}} \qquad \frac{1000\text{ J}}{1\text{ kJ}} \qquad -q_{H_2O} = q_{Fe} \qquad q = mC_s(T_f - T_i)$$

Solution: $0.95\ \cancel{g} \times \dfrac{1\ \cancel{mol}}{18.02\ \cancel{g}} \times \dfrac{-44.0\ \cancel{kJ}}{1\ \cancel{mol}} \times \dfrac{1000\text{ J}}{1\ \cancel{kJ}} = -2\underline{3}19.64$ J then $-q_{H_2O} = q_{Fe} = 2\underline{3}19.64$ J then

$q = mC_s(T_f - T_i)$.   Rearrange to solve for T_f.

$$T_f = \frac{mC_s T_i + q}{mC_s} = \frac{\left(75.0\ \cancel{g} \times 0.449\ \dfrac{\text{J}}{\cancel{g}\cdot\cancel{°C}} \times 22\ \cancel{°C}\right) + 2\underline{3}19.64\text{ J}}{75.0\ \cancel{g} \times 0.449\ \dfrac{\text{J}}{\cancel{g}\cdot°C}} = 91\,°C$$

Check: The units (°C) are correct. The temperature rose, which is consistent with heat being added to the block. The magnitude of the answer (91 °C) makes physical sense because even though we have $\sim\frac{1}{20}$ of a mole, the energy involved in condensation is very large.

11.61   **Given:**

Temperature (K)	Vapor Pressure (torr)
200	65.3
210	134.3
220	255.7
230	456.0
235	597.0

Find: $\Delta H°_{vap}(NH_3)$ and normal boiling point

Conceptual Plan: To find the heat of vaporization, use Excel or similar software to make a plot of the natural log of vapor pressure (ln *P*) as a function of the inverse of the temperature in K (1/*T*). Then fit the points to a line and determine the slope of the line. Because the *slope* $= -\Delta H_{vap}/R$**, we find the heat of vaporization as follows:**
slope $= -\Delta H_{vap}/R \rightarrow \Delta H_{vap} = -$ *slope* $\times R$ **then J** $\rightarrow$ **kJ**

$$\frac{1\text{ kJ}}{1000\text{ J}}$$

For the normal boiling point, use the equation of the best-fit line, substitute 760 torr for the pressure, and calculate the temperature.
Solution: Data was plotted in Excel.
The slope of the best-fitting line is -2969.9 K.

$$\Delta H_{vap} = -slope \times R = -(-2\underline{9}69.9\ \cancel{K}) \times \frac{8.314\text{ J}}{\cancel{K}\text{ mol}} = \frac{2.4\underline{6}917 \times 10^4\text{ J}}{\text{mol}} \times \frac{1\text{ kJ}}{1000\ \cancel{J}} = 24.7\ \frac{\text{kJ}}{\text{mol}}$$

$$\ln P = -2\underline{9}69.9\ K\left(\frac{1}{T}\right) + 19.03\underline{6} \rightarrow$$

$$\ln 760 = -2\underline{9}69.9\ K\left(\frac{1}{T}\right) + 19.03\underline{6} \rightarrow$$

$$2\underline{9}69.9\ K\left(\frac{1}{T}\right) = 19.03\underline{6} - 6.63\underline{3}32 \rightarrow$$

$$T = \frac{2969.9\ K}{12.40\underline{2}68} = 239\ K$$

Check: The units (kJ/mol) are correct. The magnitude of the answer (25 kJ/mol) is consistent with other values in the text.

11.63   **Given:** ethanol, $\Delta H°_{vap} = 38.56$ kJ/mol; normal boiling point $= 78.4\,°C$ **Find:** $P_{Ethanol}$ at 15 °C
Conceptual Plan: °C $\rightarrow$ K and kJ $\rightarrow$ J then $\Delta H°_{vap}, T_1, P_1, T_2 \rightarrow P_2$

$$K = °C + 273.15 \qquad \frac{1000\text{ J}}{1\text{ kJ}} \qquad\qquad \ln\frac{P_2}{P_1} = \frac{-\Delta H_{vap}}{R}\left(\frac{1}{T_2} - \frac{1}{T_1}\right)$$

Solution: $T_1 = 78.4\,°C + 273.15 = 351.6\ K$ $T_2 = 15\,°C + 273.15 = 288\ K$

$$\frac{38.56 \text{ kJ}}{1 \text{ mol}} \times \frac{1000 \text{ J}}{1 \text{ kJ}} = 3.856 \times 10^4 \frac{\text{J}}{\text{mol}}; \ P_1 = 760 \text{ torr}; \quad \ln \frac{P_2}{P_1} = \frac{-\Delta H_{\text{vap}}}{R}\left(\frac{1}{T_2} - \frac{1}{T_1}\right)$$

Substitute values in equation. $\ln \dfrac{P_2}{760 \text{ torr}} = \dfrac{-3.856 \times 10^4 \frac{\text{J}}{\text{mol}}}{8.314 \frac{\text{J}}{\text{K} \cdot \text{mol}}}\left(\dfrac{1}{288 \text{ K}} - \dfrac{1}{351.6 \text{ K}}\right) = -2.91302 \rightarrow$

$$\frac{P_2}{760 \text{ torr}} = e^{-2.91302} = 0.054311 \rightarrow P_2 = 0.054311 \times 760 \text{ torr} = 41 \text{ torr}$$

Check: The units (torr) are correct. Because 15 °C is significantly below the boiling point, we expect the answer to be much less than 760 torr.

Sublimation and Fusion

11.65   **Given:** 65.8 g water freezes   **Find:** energy released   **Other:** $\Delta H_{\text{fus}}^\circ = 6.02$ kJ/mol from text
Conceptual Plan: g $H_2O \rightarrow$ mol $H_2O \rightarrow q_{H_2O}$ (kJ) $\rightarrow q_{H_2O}$ (J)

$$\frac{1 \text{ mol}}{18.02 \text{ g}} \qquad \frac{-6.02 \text{ kJ}}{1 \text{ mol}} \qquad \frac{1000 \text{ J}}{1 \text{ kJ}}$$

Solution: $65.8 \text{ g} \times \dfrac{1 \text{ mol}}{18.02 \text{ g}} \times \dfrac{-6.02 \text{ kJ}}{1 \text{ mol}} \times \dfrac{1000 \text{ J}}{1 \text{ kJ}} = -21{,}982 \text{ J} = 2.20 \times 10^4 \text{ J}$ so 2.20×10^4 J released

or 22.0 kJ released

Check: The units (J) are correct. The magnitude (22,000 J) makes sense because we are freezing about 3 moles of water. Freezing is exothermic, so heat is released.

11.67   **Given:** 8.5 g ice; 255 g water   **Find:** ΔT of water   **Other:** $\Delta H_{\text{fus}}^\circ = 6.0$ kJ/mol; $C_{\text{s, water}} = 4.18$ J/g · °C from text
Conceptual Plan: The first step is to calculate how much heat is removed from the water to melt the ice.
$q_{\text{ice}} = -q_{\text{water}}$ so g ice $\rightarrow$ mol ice $\rightarrow q_{\text{fus}}$(kJ) $\rightarrow q_{\text{fus}}$(J) $\rightarrow q_{\text{water}}$(J) then $q, m, C_{\text{s}} \rightarrow \Delta T_1$

$$\frac{1 \text{ mol}}{18.02 \text{ g}} \qquad \frac{6.0 \text{ kJ}}{1 \text{ mol}} \qquad \frac{1000 \text{ J}}{1 \text{ kJ}} \qquad q_{\text{water}} = -q_{\text{ice}} \qquad q = mC_{\text{s}}\Delta T_1$$

Now we have slightly cooled water (at a temperature of T_1) in contact with 0.0 °C water, and we can calculate a second temperature drop of the water due to mixing of the water that was ice with the water that was initially room temperature; so $q_{\text{ice}} = -q_{\text{water}}$ with $m, C_{\text{s}} \rightarrow \Delta T_2$ with $\Delta T_1, \Delta T_2 \rightarrow \Delta T_{\text{Total}}$.

$$q = mC_{\text{s}}\Delta T_2 \ \text{then set} \ q_{\text{ice}} = -q_{\text{water}} \qquad \Delta T_{\text{Total}} = \Delta T_1 + \Delta T_2$$

Solution: $8.5 \text{ g} \times \dfrac{1 \text{ mol}}{18.02 \text{ g}} \times \dfrac{6.0 \text{ kJ}}{1 \text{ mol}} \times \dfrac{1000 \text{ J}}{1 \text{ kJ}} = 2.83176 \times 10^3$ J, $q_{\text{water}} = -q_{\text{ice}} = -2.83176 \times 10^3$ J

$q_{\text{water}} = mC_{\text{s}}\Delta T$ Rearrange to solve for ΔT. $\Delta T_1 = \dfrac{q}{mC_{\text{s}}} = \dfrac{-2.83176 \times 10^3 \text{ J}}{255 \text{ g} \times 4.18 \frac{\text{J}}{\text{g} \cdot {}^\circ\text{C}}} = -2.6567 \ {}^\circ\text{C}$

$q = mC_{\text{s}}\Delta T$ Substitute values and set $q_{\text{ice}} = -q_{\text{H}_2\text{O}}$.

$$q_{\text{ice}} = m_{\text{ice}}C_{\text{s, ice}}(T_{\text{f}} - T_{\text{ice, i}}) = 8.5 \text{ g} \times 2.09 \frac{\text{J}}{\text{g} \cdot {}^\circ\text{C}} \times (T_{\text{f}} - 0.0 \ {}^\circ\text{C}) =$$

$$-q_{\text{water}} = -m_{\text{water}}C_{\text{s, water}}\Delta T_{\text{water, 2}} = -255 \text{ g} \times 4.18 \frac{\text{J}}{\text{g} \cdot {}^\circ\text{C}} \times \Delta T_{\text{water, 2}} \rightarrow$$

$17.765 \, T_{\text{f}} = -1065.9 \Delta T_{\text{water, 2}} = -1065.9(T_{\text{f}} - T_{\text{f, 1}})$ Rearrange to solve for T_{f}. $17.765 \, T_{\text{f}} + 1065.9 \, T_{\text{f}} = 1065.9 \, T_{\text{f, 1}} \rightarrow$
$1083.665 \, T_{\text{f}} = 1065.9 \, T_{\text{f, 1}} \rightarrow T_{\text{f}} = 0.983607 \, T_{\text{f, 1}}$ but $\Delta T_1 = (T_{\text{f, 1}} - T_{\text{i, 1}}) = -2.6567 \ {}^\circ\text{C}$, which says that
$T_{\text{f, 1}} = T_{\text{i, 1}} - 2.6567 \ {}^\circ\text{C}$ and $\Delta T_{\text{Total}} = (T_{\text{f}} - T_{\text{i, 1}})$; so
$\Delta T_{\text{Total}} = 0.983607 \, T_{\text{f, 1}} - T_{\text{i, 1}} = 0.983607(T_{\text{i, 1}} - 2.6567 \ {}^\circ\text{C}) - T_{\text{i, 1}} = -2.6131 \ {}^\circ\text{C} - 0.016393 \, T_{\text{i, 1}}$.

This implies that the larger the initial temperature of the water, the larger the temperature drop. If the initial temperature was 90 °C, the temperature drop would be 4.1 °C. If the initial temperature was 25 °C, the temperature drop would be 3.0 °C. If the initial temperature was 5 °C, the temperature drop would be 2.7 °C. This makes physical sense because the lower the initial temperature of the water, the less kinetic energy it initially has and the smaller the heat transfer from the water to the melted ice will be.

Check: The units (°C) are correct. The temperature drop from the melting of the ice is only 2.7 °C because the mass of the water is so much larger than that of the ice.

11.69   **Given:** 10.0 g ice $T_i = -10.0\,^\circ\text{C}$ to steam at $T_f = 110.0\,^\circ\text{C}$ **Find:** heat required (kJ)

Other: $\Delta H^\circ_{\text{fus}} = 6.02\ \text{kJ/mol}$; $\Delta H^\circ_{\text{vap}} = 40.7\ \text{kJ/mol}$; $C_{\text{s, ice}} = 2.09\ \text{J/g} \cdot {}^\circ\text{C}$; $C_{\text{s, water}} = 4.18\ \text{J/g} \cdot {}^\circ\text{C}$; $C_{\text{s, steam}} = 2.01\ \text{J/g} \cdot {}^\circ\text{C}$

Conceptual Plan: Follow the heating curve in Figure 11.33. $q_{\text{Total}} = q_1 + q_2 + q_3 + q_4 + q_5$ **where** $q_1, q_3,$ **and** q_5 **are heating of a single phase then J** $\rightarrow$ **kJ and** q_2 **and** q_4 **are phase transitions.**

$$q = mC_s(T_f - T_i) \qquad\qquad \frac{1\ \text{kJ}}{1000\ \text{J}} \qquad q = m \times \frac{1\ \text{mol}}{18.02\ \text{g}} \times \frac{\Delta H}{1\ \text{mol}}$$

Solution:

Heating ice from $-10.0\,^\circ\text{C}$ to $0\,^\circ\text{C}$:

$$q_1 = m_{\text{ice}}C_{\text{s, ice}}\left(T_{\text{ice, f}} - T_{\text{ice, i}}\right) = 10.0\ \cancel{\text{g}} \times 2.09\ \frac{\text{J}}{\cancel{\text{g}} \cdot \cancel{{}^\circ\text{C}}} \times \left(0.0\,\cancel{{}^\circ\text{C}} - (-10.0\,\cancel{{}^\circ\text{C}})\right) = 209\ \text{J} \times \frac{1\ \text{kJ}}{1000\ \cancel{\text{J}}} = 0.209\ \text{kJ}$$

Melting the ice—phase change:

$$q_2 = m \times \frac{1\ \text{mol}}{18.02\ \text{g}} \times \frac{\Delta H_{\text{fus}}}{1\ \text{mol}} = 10.0\ \cancel{\text{g}} \times \frac{1\ \cancel{\text{mol}}}{18.02\ \cancel{\text{g}}} \times \frac{6.02\ \text{kJ}}{1\ \cancel{\text{mol}}} = 3.3\underline{4}1\ \text{kJ}$$

Heating water to $100\,^\circ\text{C}$:

$$q_3 = m_{\text{water}}C_{\text{s, water}}\left(T_{\text{water, f}} - T_{\text{water, i}}\right) = 10.0\ \cancel{\text{g}} \times 4.18\ \frac{\text{J}}{\cancel{\text{g}} \cdot \cancel{{}^\circ\text{C}}} \times \left(100.0\,\cancel{{}^\circ\text{C}} - 0.0\,\cancel{{}^\circ\text{C}}\right) = 4180\ \text{J} \times \frac{1\ \text{kJ}}{1000\ \cancel{\text{J}}} = 4.18\ \text{kJ}$$

Converting water to steam—phase change:

$$q_4 = m \times \frac{1\ \text{mol}}{18.02\ \text{g}} \times \frac{\Delta H_{\text{vap}}}{1\ \text{mol}} = 10.0\ \cancel{\text{g}} \times \frac{1\ \cancel{\text{mol}}}{18.02\ \cancel{\text{g}}} \times \frac{40.7\ \text{kJ}}{1\ \cancel{\text{mol}}} = 22.5\underline{8}6\ \text{kJ}$$

Heating steam to $110\,^\circ\text{C}$:

$$q_5 = m_{\text{steam}}C_{\text{s, steam}}\left(T_{\text{steam, f}} - T_{\text{steam, i}}\right) = 10.0\ \cancel{\text{g}} \times 2.01\ \frac{\text{J}}{\cancel{\text{g}} \cdot \cancel{{}^\circ\text{C}}} \times \left(110.0\,\cancel{{}^\circ\text{C}} - 100.0\,\cancel{{}^\circ\text{C}}\right)$$

$$= 201\ \text{J} \times \frac{1\ \text{kJ}}{1000\ \cancel{\text{J}}} = 0.201\ \text{kJ}$$

$$q_{\text{Total}} = q_1 + q_2 + q_3 + q_4 + q_5 = 0.209\ \text{kJ} + 3.3\underline{4}1\ \text{kJ} + 4.18\ \text{kJ} + 22.5\underline{8}6\ \text{kJ} + 0.201\ \text{kJ} = 30.5\ \text{kJ}$$

Check: The units (kJ) are correct. The total amount of heat is dominated by the vaporization step. Because we have less than 1 mole, we expect less than 41 kJ.

Phase Diagrams

11.71   (a)   solid
        (b)   liquid
        (c)   gas
        (d)   supercritical fluid
        (e)   solid/liquid equilibrium or fusion curve
        (f)   liquid/gas equilibrium or vaporization curve
        (g)   solid/liquid/gas equilibrium or triple point

11.73   **Given:** nitrogen; normal boiling point $= 77.3$ K; normal melting point $= 63.1$ K; critical temperature $= 126.2$ K; critical pressure $= 2.55 \times 10^4$ torr; triple point at 63.1 K and 94.0 torr
Find: Sketch phase diagram. Does nitrogen have a stable liquid phase at 1 atm?

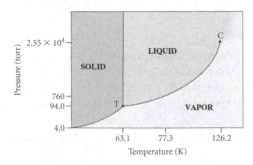

Nitrogen has a stable liquid phase at 1 atm.
Note that the axes are not to scale.

11.75   (a)   0.027 mmHg, the higher of the two triple points

        (b)   The rhombic phase is denser because if we start in the monoclinic phase at 100 °C and increase the pressure, we will cross into the rhombic phase.

The Uniqueness of Water

11.77   Water has a low molar mass (18.02 g/mol), yet it is a liquid at room temperature. Water's high boiling point for its molar mass can be understood by examining the structure of the water molecule. The bent geometry of the water molecule and the highly polar nature of the O—H bonds result in a molecule with a significant dipole moment. Water's two O—H bonds (hydrogen directly bonded to oxygen) allow a water molecule to form very strong hydrogen bonds with four other water molecules, resulting in a relatively high boiling point.

11.79   Water has an exceptionally high specific heat capacity, which has a moderating effect on the climate of coastal cities. Also, its high ΔH_{vap} causes water evaporation and condensation to have a strong effect on temperature. A tremendous amount of heat can be stored in large bodies of water. Heat will be absorbed or released from large bodies of water preferentially over the land around it. In some cities, such as San Francisco, the daily fluctuation in temperature can be less than 10 °C. This same moderating effect occurs over the entire planet, two-thirds of which is covered by water. In other words, without water, the daily temperature fluctuations on our planet might be more like those on Mars, where temperature fluctuations of 63 °C (113 °F) have been measured between early morning and midday.

Cumulative Problems

11.81   The general trend is that melting point increases with increasing molar mass. This is because the electrons of the larger molecules are held more loosely and a stronger dipole moment can be induced more easily. HF is the exception to the rule. It has a relatively high melting point due to strong intermolecular forces due to hydrogen bonding.

11.83   **Given:** $P_{H_2O} = 23.76$ torr at 25 °C; 1.25 g water in 1.5 L container   **Find:** m (H_2O) as liquid

     Conceptual Plan: °C → K and torr → atm then P, V, T → mol (g) → g (g) then g (g), g $(l)_i$ → g $(l)_f$

$$K = °C + 273.15 \qquad \frac{1\ atm}{760\ torr} \qquad PV = nRT \qquad \frac{18.02\ g}{1\ mol} \qquad g\,(l)_f = g(l)_i - g(g)$$

     Solution: $T = 25\ °C + 273.15 = 298$ K $23.76\ \text{torr} \times \dfrac{1\ atm}{760\ \text{torr}} = 0.0312632$ atm then $PV = nRT$

     Rearrange to solve for n. $n = \dfrac{PV}{RT} = \dfrac{0.0312632\ \text{atm} \times 1.5\ \text{L}}{0.08206\ \dfrac{\text{L} \cdot \text{atm}}{\text{K} \cdot \text{mol}} \times 298\ \text{K}} = 0.00191768$ mol in the gas phase then

     $0.00191768\ \text{mol} \times \dfrac{18.02\ g}{1\ \text{mol}} = 0.0345566$ g in the gas phase then

     $g\,(l)_f = g\,(l)_i - g\,(g) = 1.25\ g - 0.0345566\ g = 1.22$ g remaining as liquid. Yes, there is 1.22 g of liquid.

     Check: The units (g) are correct. The magnitude (1.2 g) is expected because very little material is expected to be in the gas phase.

11.85   **Given:** ice: $T_1 = 0\ °C$ exactly, $m = 53.5$ g; water: $T_1 = 75\ °C, m = 115$ g   **Find:** T_f

     Other: $\Delta H°_{fus} = 6.0$ kJ/mol; $C_{water} = 4.18$ J/g · °C

     Conceptual Plan: $q_{ice} = -q_{water}$ so g ice → mol ice → q_{fus}(kJ) → q_{fus}(J) → q_{water}(J) then

$$\frac{1\ mol}{18.02\ g} \qquad \frac{6.02\ kJ}{1\ mol} \qquad \frac{1000\ J}{1\ kJ} \qquad q_{water} = -q_{ice}$$

     q, m, C_s → ΔT then T_i, ΔT → T_2 now we have slightly cooled water in contact with 0.0 °C water

$$q = mC_s\,\Delta T \qquad \Delta T = T_2 - T_i$$

     so $q_{ice} = -q_{water}$ with m, C_s, T_i → T_f

$$q = mC_s(T_f - T_i)\ \text{then set}\ q_{ice} = -q_{water}$$

     Solution: $53.5\ g \times \dfrac{1\ mol}{18.02\ g} \times \dfrac{6.02\ kJ}{1\ mol} \times \dfrac{1000\ J}{1\ kJ} = 1.78828 \times 10^4$ J $q_{water} = -q_{ice} = -1.78828 \times 10^4$ J

     $q_{water} = mC_s\Delta T$ Rearrange to solve for ΔT. $\Delta T = \dfrac{q}{mC_s} = \dfrac{-1.78828 \times 10^4\ J}{115\ g \times 4.18\ \dfrac{J}{g \cdot °C}} = -37.2016\ °C$ then

$\Delta T = T_2 - T_i$ Rearrange to solve for T_2. $T_2 = \Delta T + T_i = -37.\underline{2}016\,°C + 75\,°C = 37.798\,°C$

$q = mC_s(T_f - T_i)$ Substitute values and set $q_{ice} = -q_{water}$.

$q_{ice} = m_{ice}C_{s,\,ice}(T_f - T_{ice,\,i}) = 53.5\,\cancel{g} \times 4.\underline{18}\,\dfrac{J}{\cancel{g}\cdot°C} \times (T_f - 0.0\,°C) =$

$-q_{water} = -m_{water}\,C_{s,\,water}(T_f - T_{water,\,2}) = -115\,\cancel{g} \times 4.\underline{18}\,\dfrac{J}{\cancel{g}\cdot°C} \times (T_f - 3\underline{7}.798\,°C)$

Rearrange to solve for T_f.

$53.5T_f = -115(T_f - 37.798\,°C) \rightarrow 53.5\,T_f = -115\,T_f + 43\underline{4}6.8\,°C \rightarrow -43\underline{4}6.8\,°C = -16\underline{8}.5\,T_f$

$\rightarrow T_f = \dfrac{-43\underline{4}6.8\,°C}{-16\underline{8}.5} = 2\underline{5}.8\,°C = 26\,°C$

Check: The units (°C) are correct. The temperature is between the two initial temperatures. Because the ice mass is about half the water mass, we are not surprised that the temperature is closer to the original ice temperature.

11.87   **Given:** 1 mole of methanol   **Find:** Draw a heating curve beginning at 170 K and ending at 350 K
Other: melting point = 176 K, boiling point = 338 K, ΔH_{fus} = 2.2 kJ/mol, ΔH_{vap} = 35.2 kJ/mol,
$C_{s,solid}$ = 105 J/mol · K, $C_{s,liquid}$ = 81.3 J/mol · K, $C_{s,gas}$ = 48 J/mol · K
Conceptual Plan:
Calculate the temperature range of each phase:

$T_{Starting},\,T_{Melting} \rightarrow \Delta T_{Solid}$ **and** $T_{Melting},\,T_{Boiling} \rightarrow \Delta T_{Liquid}$ **and** $T_{Boiling},\,T_{Ending} \rightarrow \Delta T_{Gas}$

$\quad\quad \Delta T_{Solid} = T_{Melting} - T_{Starting} \quad\quad\quad \Delta T_{Liquid} = T_{Boiling} - T_{Melting} \quad\quad\quad \Delta T_{Gas} = T_{Ending} - T_{Boiling}$

Because there is 1 mole of methanol, the heat for each phase change is simply the ΔH for that phase change.
Calculate the heat required for each step:

$\Delta T_{Solid},\,C_{s,solid} \rightarrow q_{Solid}$ **and** $\Delta T_{Liquid},\,C_{s,liquid} \rightarrow q_{Liquid}$ **and** $\Delta T_{Gas},\,C_{s,gas} \rightarrow q_{Gas}$

$\quad\quad q = C_s \times \Delta T \quad\quad\quad\quad\quad\quad q = C_s \times \Delta T \quad\quad\quad\quad\quad\quad q = C_s \times \Delta T$

Finally, plot each of the segments.
Solution:

$\Delta T_{Solid} = T_{Melting} - T_{Starting} = 176\,K - 170\,K = 6\,K \quad \Delta T_{Liquid} = T_{Boiling} - T_{Melting} = 338\,K - 176\,K = 162\,K$
and $\Delta T_{Gas} = T_{Ending} - T_{Boiling} = 350\,K - 338\,K = 12\,K$

$q_{fus} = \Delta H_{fus} = 2.2\,kJ/mol$ and $q_{vap} = \Delta H_{vap} = 35.2\,kJ/mol$

$q_{Solid} = C_{s,Solid} \times \Delta T_{Solid} = 105\dfrac{J}{mol \cdot K} \times 6\,K = \underline{6}30\,J/mol = 0.\underline{6}3\,kJ/mol$

$q_{Liquid} = C_{s,Liquid} \times \Delta T_{Liquid} = 81.3\dfrac{J}{mol \cdot K} \times 162\,K = 13\underline{1}70.6\,J/mol = 13.\underline{1}706\,kJ/mol$ and

$q_{Gas} = C_{s,Gas} \times \Delta T_{Gas} = 48\dfrac{J}{mol \cdot K} \times 12\,K = 5\underline{7}6\,J/mol = 0.5\underline{7}6\,kJ/mol$

Check: The heating curve has a shape that is similar to that of water.

11.89 **Given:** home: 6.0 m $\times$ 10.0 m $\times$ 2.2 m; $T = 30\,°C$, $P_{H_2O} = 85\%$ of $P^\circ_{H_2O}$ **Find:** $m(H_2O)$ removed
Other: $P^\circ_{H_2O} = 31.8$ mmHg from text
Conceptual Plan: $l, w, h \to V(m^3) \to V(cm^3) \to V(L)$ and $P^\circ_{H_2O} \to P_{H_2O}(mmHg) \to P_{H_2O}(atm)$ and

$$V = lwh \qquad \frac{(100\text{ cm})^3}{(1\text{ m})^3} \qquad \frac{1\text{ L}}{1000\text{ cm}^3} \qquad\qquad P_{H_2O} = 0.85\,P^\circ_{H_2O} \qquad\qquad \frac{1\text{ atm}}{760\text{ mmHg}}$$

$°C \to K$ then $P, V, T \to$ mol $H_2O \to$ g H_2O

$$K = °C + 273.15 \qquad PV = nRT \qquad \frac{18.02\text{ g}}{1\text{ mol}}$$

Solution: $V = lwh = 6.0\text{ m} \times 10.0\text{ m} \times 2.2\text{ m} = 132\text{ m}^3 \times \dfrac{(100\text{ cm})^3}{(1\text{ m})^3} \times \dfrac{1\text{ L}}{1000\text{ cm}^3} = 1.32 \times 10^5\text{ L}$

$P_{H_2O} = 0.85\,P^\circ_{H_2O} = 0.85 \times 31.8\text{ mmHg} \times \dfrac{1\text{ atm}}{760\text{ mmHg}} = 0.035566\text{ atm} \quad T = 30\,°C + 273.15 = 303\text{ K}$

then $PV = nRT$ Rearrange to solve for n.

$$n = \frac{PV}{RT} = \frac{0.035566\text{ atm} \times 1.32 \times 10^5\text{ L}}{0.08206\,\dfrac{\text{L} \cdot \text{atm}}{\text{K} \cdot \text{mol}} \times 303\text{ K}} = 188.81\text{ mol} \text{ then } 188.81\text{ mol} \times \frac{18.02\text{ g}}{1\text{ mol}} = 3400\text{ g to remove}$$

Check: The units (g) are correct. The magnitude of the answer (3400 g) makes sense because the volume of the house is so large. We are removing almost 200 moles of water.

Challenge Problems

11.91 $P_{Total} = P_A + P_B$. Looking at the data, we see that as the relative amount of A increases, the pressure decreases because the vapor pressure of A is less than the vapor pressure of B. Because the solution is almost all A, we expect a vapor pressure just above 24 mmHg. In Chapter 14, we will more fully discuss this relationship and learn that

$$P_A = \chi_A P^\circ_A, \text{ where } \chi_A = \frac{\text{mol A}}{\text{total mol} = \text{mol A} + \text{mol B}}, \chi_B = 1 - \chi_A, \text{ and } P_B = \chi_B P^\circ_B. \text{ So}$$

$$P_{Total} = P_A + P_B = \chi_A P^\circ_A + (1 - \chi_A)P^\circ_B = \frac{5\text{ mol}}{5\text{ mol} + 1\text{ mol}} \times 24\text{ mmHg} + \left(1 - \frac{5\text{ mol}}{5\text{ mol} + 1\text{ mol}}\right) \times 36\text{ mmHg}$$

$$= 26\text{ mmHg}$$

11.93 Decreasing the pressure will decrease the temperature of liquid nitrogen. Because the nitrogen is boiling, its temperature must be constant at a given pressure. As the pressure decreases, the boiling point decreases, as does the temperature. Remember that vaporization is an endothermic process; so as the nitrogen vaporizes, it removes heat from the liquid, dropping its temperature. If the pressure drops below the pressure of the triple point, the phase change will shift from vaporization to sublimation and the liquid nitrogen will become solid.

11.95 **Given:** 1.00 L water, $T_i = 298$ K, $T_f = 373$ K-vapor; $P_{CH_4} = 1.00$ atm **Find:** $V(CH_4)$
Other: $\Delta H^\circ_{comb}(CH_4) = 890.4$ kJ/mol; $C_{water} = 75.2$ J/mol $\cdot$ K; $\Delta H^\circ_{vap}(H_2O) = 40.7$ kJ/mol, $d = 1.00$ g/mL
Conceptual Plan: $L \to mL \to g \to$ mol then heat liquid water: $n, C_s, T_i, T_f \to q_{1,\,water}(J)$

$$\frac{1000\text{ mL}}{1\text{ L}} \quad \frac{1.00\text{ g}}{1.00\text{ mL}} \quad \frac{1\text{ mol}}{18.02\text{ g}} \qquad q = mC_s(T_f - T_i)$$

vaporize water: $n_{water}, \Delta H^\circ_{vap} \to q_{2,\,water}(J)$ then calculate total heat $q_{1,\,water}, q_{2,\,water} \to q_{water}(J)$ then

$$q = n\Delta H \qquad\qquad\qquad q_{1,\,water} + q_{2,\,water} = q_{water}$$

$q_{water}(J) \to -q_{CH_4,\,comb}(J) \to n_{CH_4}$ finally $n_{CH_4}, P, T \to V$

$$q_{water} = -q_{CH_4,\,comb} \qquad q = n\Delta H \qquad\qquad PV = nRT$$

Solution: $1.00\text{ L} \times \dfrac{1000\text{ mL}}{1\text{ L}} \times \dfrac{1.00\text{ g}}{1.00\text{ mL}} \times \dfrac{1\text{ mol}}{18.02\text{ g}} = 55.5247\text{ mol } H_2O$

$q_{1,\,water} = n_{water}C_{water}(T_f - T_i) = 55.5247\text{ mol} \times 75.2\,\dfrac{J}{\text{mol} \cdot \text{K}} \times (373\text{ K} - 298\text{ K}) = 3.1315931 \times 10^5\text{ J}$

$= 313.15931\text{ kJ}$

$$q_{2,\,water} = n\Delta H = 55.5247\ \cancel{mol} \times 40.7\frac{kJ}{\cancel{mol}} = 2.259855 \times 10^3\ kJ$$

$$q_{1,\,water} + q_{2,\,water} = q_{water} = 313.15931\ kJ + 2.259855 \times 10^3\ kJ = 2.57301431 \times 10^3\ kJ$$

$$q_{water} = -q_{CH_4,\,comb} = 2.57301431 \times 10^3\ kJ \text{ then } q_{CH_4,\,comb} = n\Delta H \quad \text{Rearrange to solve for } n.$$

$$n_{CH_4} = \frac{q_{CH_4}}{\Delta H_{CH_4,\,comb}} = \frac{-2.57301431 \times 10^3\ kJ}{-890.4\dfrac{kJ}{mol}} = 2.8897286\ mol \text{ then } PV = nRT$$

Rearrange to solve for V. $V = \dfrac{nRT}{P} = \dfrac{2.8897286\ \cancel{mol} \times 0.08206\dfrac{L \cdot \cancel{atm}}{\cancel{mol} \cdot \cancel{K}} \times 298\ \cancel{K}}{1.00\ \cancel{atm}} = 70.665076\ L = 70.7\ L$

Check: The units (L) are correct. The volume (71 L) is reasonable because we are using about 3 moles of methane.

Conceptual Problems

11.97   The melting of an ice cube in a glass of water will not raise or lower the level of the liquid in the glass as long as the ice is floating in the liquid. This is because the ice will displace a volume of water based on its mass. By the same logic, melting floating icebergs will not raise the ocean levels (assuming that the dissolved solids content, and thus the density, will not change when the icebergs melt). Dissolving ice formations that are supported by land will raise the ocean levels, just as pouring more water into a glass will raise the liquid level in the glass.

11.99   Substance A will have the larger change in vapor pressure with the same temperature change. To understand this, consider the Clausius–Clapeyron equation: $\ln\dfrac{P_2}{P_1} = \dfrac{-\Delta H_{vap}}{R}\left(\dfrac{1}{T_2} - \dfrac{1}{T_1}\right)$. If we use the same temperatures, we see that $\dfrac{P_2}{P_1}\alpha\,e^{-\Delta H_{vap}}$. So the smaller the heat of vaporization, the larger the final vapor pressure. We can also consider that the lower the heat of vaporization is, the easier the substance is converted from a liquid to a gas. This again leads to substance A having the larger change in vapor pressure.

11.101   The liquid segment will have the least steep slope because it takes the most kJ/mol to raise the temperature of the phase.

11.103   The heat of fusion of a substance is always smaller than the heat of vaporization because the number of interactions between particles that are broken is less in fusion than in vaporization. When we melt a solid, the particles have increased mobility but are still strongly interacting with other liquid particles. In vaporization, all of the interactions between particles must be broken (gas particles have essentially no intermolecular interactions), and the particles must absorb enough energy to move much more rapidly.

11.105   The oxygen atoms would be attracted to the $-CH_2$ groups through dipole–dipole forces.

Questions for Group Work

11.107   1-Hexnaol exhibits hydrogen bonding — it has the highest boiling point. 2-Hexanone is polar, but cannot exhibit hydrogen bonding — it has the next highest boiling point. Heptane is nonpolar — it has the lowest boiling point.

11.109   Referring to Figure 11.33, we find that the amount of heat needed to melt 1 mole of water is 6.02 kJ and the amount of heat needed to boil 1 mole of water is 40.7 kJ — it takes a lot more heat to boil water than to melt ice because all of the intermolecular interactions need to be completely broken when boiling but not to melt.
Referring to Figure 11.33, the amount of heat needed to warm 1 mole of ice at 10 °C is 0.38 kJ (0.941 kJ $\times$ (10 °C/25 °C)), the amount of heat needed to warm 1 mole of water at 10 °C is 0.75 kJ (7.52 kJ $\times$ (10 °C/100 °C)), and the amount of heat needed to warm 1 mole of steam at 10 °C is 0.36 kJ (0.905 kJ $\times$ (10 °C/25 °C)). It takes more heat to warm water than to warm ice or steam because there is more motion and disorder in the liquid and gas phases.

Data Interpretation and Analysis

11.111   (a)   The molecular mass of a compound is a factor to be considered in comparing boiling points of compounds exhibiting about the same intermolecular forces. However, for compounds having strong intermolecular forces, the intermolecular forces can outweigh the contribution of molecular mass. For the H_2S, H_2Se, H_2Te series, all three compounds have the same types of intermolecular forces; the boiling point is correlated with molecular mass. Water exhibits hydrogen bonding, which is a very strong intermolecular force, so its boiling point is higher than the other compounds in the series that do not exhibit hydrogen bonding.

(c)   Liquid water exhibits hydrogen bonding, dipole–dipole forces, and dispersion forces. H_2S, H_2Se, and H_2Te all exhibit dipole–dipole forces and London dispersion forces. While water has the highest dipole moment, it has the lowest polarizability. Thus, dispersion forces are stronger in H_2S, H_2Se, and H_2Te compared to H_2O. Only water has hydrogen bonding and it is the strongest intermolecular force. Hydrogen bonding is responsible for the unusually high boiling point of water.

12 Crystalline Solids and Modern Materials

Review Questions

12.1   Graphene is one-atom-thick graphite. It is the thinnest and strongest known material. It conducts heat and electricity, it is transparent, and it is completely impermeable to all substances, including helium.

12.3   A crystalline lattice is the regular arrangements of atoms within a crystalline solid. The crystalline lattice can be represented by a small collection of atoms, ions, or molecules—a fundamental building block called the unit cell. When the unit cell is repeated over and over—like tiles in a floor or the pattern in a wallpaper design, but in three dimensions—the entire lattice can be reproduced.

12.5   Atoms in a simple cubic cell structure have a coordination number of 6, an edge length of $2r$, and one atom in the unit cell. Atoms in a body-centered cubic cell structure have a coordination number of 8, an edge length of $4r/\sqrt{3}$, and two atoms in the unit cell. Atoms in a face-centered cubic cell structure have a coordination number of 12, an edge length of $2\sqrt{2}r$, and four atoms in the unit cell.

12.7   The three types of solids are molecular solids, ionic solids, and atomic solids. Molecular solids are those solids whose composite units are molecules. The lattice sites in a crystalline molecular solid are, therefore, occupied by molecules. Ice (solid H_2O) and dry ice (solid CO_2) are examples of molecular solids. Molecular solids are held together by the kinds of intermolecular forces—dispersion forces, dipole–dipole forces, and hydrogen bonding. Ionic solids are those solids whose composite units are ions. Table salt (NaCl) and calcium fluoride (CaF_2) are good examples of ionic solids. Ionic solids are held together by the coulombic interactions that occur between the cations and anions occupying the lattice sites in the crystal, which is an ionic bond. Atomic solids are those solids whose composite units are individual atoms. Atomic solids can themselves be divided into three categories—nonbonding atomic solids, metallic atomic solids, and network covalent atomic solids—each held together by a different kind of force. Nonbonding atomic solids, which include only the noble gases in their solid form, are held together by relatively weak dispersion forces. Metallic atomic solids, such as iron and gold, are held together by metallic bonds, which in the simplest model are represented by the interaction of metal cations with a sea of electrons that surrounds them. Network covalent atomic solids, such as diamond, graphite, and silicon dioxide, are held together by covalent bonds.

12.9   Nonbonding atomic solids, which include only the noble gases in their solid form, are held together by relatively weak dispersion forces. Metallic atomic solids, such as iron and gold, are held together by metallic bonds, which in the simplest model are represented by the interaction of metal cations with a sea of electrons around them. Network covalent atomic solids, such as diamond, graphite, and silicon dioxide, are held together by covalent bonds.

12.11   The coordination number of the unit cell for an ionic compound represents the number of close cation–anion interactions. Because these interactions lower potential energy, the crystal structure of a particular ionic compound will be the one that maximizes the coordination number while accommodating both charge neutrality (each unit cell must be neutral) and the different sizes of the cations and anions that compose the particular compound. In general, the more similar the radii of the cation and the anion, the higher the possible coordination number.

12.13   The fluorite (CaF_2) structure is shown in Figure 12.15. You can visualize this structure as calcium cations occupying the lattice sites of a face-centered cubic structure, with the larger fluoride anions occupying all eight of the tetrahedral holes located directly beneath each corner atom. Each unit cell contains four calcium cations $[(8 \times 1/8) + (6 \times \frac{1}{2}) = 4]$ and eight fluoride anions (each of the eight fluoride anions is completely contained within the unit cell), resulting in a cation-to-anion ratio of 1:2, as in the formula of the compound.

12.15   Silicates are extended arrays of silicon and oxygen. They are the most common network covalent atomic solids. Quartz is a crystalline silicate where each Si atom is in a tetrahedron surrounded by four O atoms, and each O atom is a bridge connecting the corners of two tetrahedrons.

12.17   Portland cement is a powdered mixture consisting mostly of limestone ($CaCO_3$) and silica (SiO_2), with smaller amounts of alumina (Al_2O_3), iron(III) oxide (Fe_2O_3), and gypsum ($CaSO_4 \cdot 2H_2O$). Unlike clays, which lose water upon setting, Portland cement reacts with water as it hardens.

12.19   Silica melts when heated above 1500 °C. After melting, if cooled quickly, silica does not crystallize back into the quartz structure. Instead, the Si atoms and O atoms form a randomly ordered amorphous structure called a glass.

12.21   In band theory, bands are the combination atomic orbitals of the atoms within a solid crystal that form orbitals that are not localized on individual atoms, but are delocalized over the entire crystal. In band theory, electrons become mobile when they make a transition from the highest occupied molecular orbital into higher-energy empty molecular orbitals. For this reason, the occupied molecular orbitals are often called the valence band and the unoccupied orbitals are called the conduction band.

12.23   Doped semiconductors contain minute amounts of impurities that result in additional electrons in the conduction band or electron "holes" in the valence band. For example, silicon is a group 4A semiconductor. Its valence electrons just fill its valence band. When silicon is doped with phosphorus, a group 5A element with five valence electrons, its conductivity increases. The phosphorus atoms are incorporated into the silicon crystal structure, but each phosphorus atom brings with it one additional electron. Because the valence band is completely full, the additional electrons must go into the conduction band. These electrons are then mobile and can conduct electric current. This type of semiconductor is called an n-type semiconductor because the charge carriers are negatively charged electrons in the conduction band. Silicon can also be doped with a group 3A element, such as gallium, which has only three valence electrons. When gallium is incorporated into the silicon crystal structure, it results in electron "holes," empty molecular orbitals in the valence band. The presence of holes also allows for the movement of electric current because electrons in the valence band can move between holes. In this way, the holes move in the opposite direction of the electrons. This type of semiconductor is called a p-type semiconductor because the hole acts as a positive charge.

12.25   Polymers are long, chainlike molecules composed of repeating units called monomers. Copolymers are long, chainlike molecules composed of two different monomers.

Types of Solids and Their Structures

12.27   **Given:** X-ray with $\lambda = 154$ pm, maximum reflection angle of $\theta = 28.3°$; assume $n = 1$
Find: distance between layers
Conceptual Plan: $\lambda, \theta, n \rightarrow d$

$$n\lambda = 2\,d\sin\theta$$

Solution: $n\lambda = 2\,d\sin\theta$. Rearrange to solve for d. $d = \dfrac{n\,\lambda}{2\sin\theta} = \dfrac{1 \times 154 \text{ pm}}{2\sin 28.3°} = 162$ pm

Check: The units (pm) are correct. The magnitude (162 pm) makes sense because $n = 1$ and the sin is always < 1. The number is consistent with interatomic distances.

12.29   (a)   8 corner atoms $\times$ (1/8 atom/unit cell) = 1 atom/unit cell
(b)   8 corner atoms $\times$ (1/8 atom/unit cell) + 1 atom in center = (1 + 1) atoms/unit cell = 2 atoms/unit cell
(c)   8 corner atoms $\times$ (1/8 atom/unit cell) + 6 face-centered atoms $\times$ (1/2 atom/unit cell)
         = (1 + 3) atoms/unit cell = 4 atoms/unit cell

12.31   **Given:** body-centered cubic structure   **Find:** packing efficiency (fraction filled)
Conceptual Plan: Calculate the volume of the atoms and the volume of the unit cell then $V_{atoms}, V_{unit \, cell} \rightarrow \% \, V_{filled}$

$$\dfrac{2 \text{ atoms}}{\text{unit cell}} \quad V_{atom} = \dfrac{\frac{4}{3}\pi r^3}{\text{atom}} \quad V_{unit \, cell} = l^3 \quad l = \dfrac{4r}{\sqrt{3}} \qquad \% \, V_{filled} = \dfrac{V_{atoms}}{V_{unit \, cell}} \times 100\%$$

Solution: $V_{atoms} = \dfrac{2 \text{ atoms}}{\text{unit cell}} \times \dfrac{\frac{4}{3}\pi r^3}{\text{atoms}} = \dfrac{\frac{8}{3}\pi r^3}{\text{unit cell}}; \; V_{unit\ cell} = l^3 = \left(\dfrac{4r}{\sqrt{3}}\right)^3$ so

$$\%V_{filled} = \dfrac{V_{atoms}}{V_{unit\ cell}} \times 100\% = \dfrac{\frac{8}{3}\pi r^3}{\left(\dfrac{4r}{\sqrt{3}}\right)^3} \times 100\% = \dfrac{\sqrt{3}\pi}{8} \times 100\% = 68.017\%.$$

Check: The units (%) are correct. The magnitude (68%) is consistent with what is stated in the text.

12.33   **Given:** platinum; face-centered cubic structure; $r = 139$ pm   **Find:** edge length of unit cell and density (g/cm^3)
Conceptual Plan: $r \rightarrow l$ and $l \rightarrow V(pm^3) \rightarrow V(cm^3)$ and $\mathcal{M}$, FCC structure $\rightarrow m$ then $m, V \rightarrow d$

$$l = 2\sqrt{2}\,r \quad V = l^3 \quad \dfrac{(1\ cm)^3}{(10^{10}\ pm)^3} \quad\quad m = \dfrac{4\ atoms}{unit\ cell} \times \dfrac{\mathcal{M}}{N_A} \quad\quad d = m/V$$

Solution: $l = 2\sqrt{2}\,r = 2\sqrt{2} \times 139$ pm $= 393.151$ pm $= 393$ pm and

$V = l^3 = (393.151\ \text{pm})^3 \times \dfrac{(1\ cm)^3}{(10^{10}\ pm)^3} = 6.07684 \times 10^{-23}\ cm^3$ and

$m = \dfrac{4\ atoms}{unit\ cell} \times \dfrac{\mathcal{M}}{N_A} = \dfrac{4\ atoms}{unit\ cell} \times \dfrac{195.09\ g}{1\ mol} \times \dfrac{1\ mol}{6.022 \times 10^{23}\ atoms} = 1.295849 \times 10^{-21} \dfrac{g}{unit\ cell}$ then

$d = \dfrac{m}{V} = \dfrac{1.295849 \times 10^{-21} \dfrac{g}{unit\ cell}}{6.07684 \times 10^{-23} \dfrac{cm^3}{unit\ cell}} = 21.3 \dfrac{g}{cm^3}$

Check: The units (pm and g/cm^3) are correct. The magnitude (393 pm) makes sense because it must be larger than the radius of an atom. The magnitude (21 g/cm^3) is consistent for Pt from Chapter 1.

12.35   **Given:** rhodium; face-centered cubic structure; $d = 12.41$ g/cm^3 **Find:** $r(Rh)$
Conceptual Plan: $\mathcal{M}$, FCC structure $\rightarrow m$ then $m, V \rightarrow d$ then $V(cm^3) \rightarrow l(cm) \rightarrow l(pm)$ then $l \rightarrow r$

$$m = \dfrac{4\ atoms}{unit\ cell} \times \dfrac{\mathcal{M}}{N_A} \quad d = m/V \quad V = l^3 \quad \dfrac{10^{10}\ pm}{1\ cm} \quad l = 2\sqrt{2}\,r$$

Solution: $m = \dfrac{4\ atoms}{unit\ cell} \times \dfrac{\mathcal{M}}{N_A} = \dfrac{4\ atoms}{unit\ cell} \times \dfrac{102.905\ g}{1\ mol} \times \dfrac{1\ mol}{6.022 \times 10^{23}\ atoms} = 6.835271 \times 10^{-22} \dfrac{g}{unit\ cell}$

then $d = \dfrac{m}{V}$ Rearrange to solve for V. $V = \dfrac{m}{d} = \dfrac{6.835271 \times 10^{-22} \dfrac{g}{unit\ cell}}{12.41 \dfrac{g}{cm^3}} = 5.507873 \times 10^{-23} \dfrac{cm^3}{unit\ cell}$

then $V = l^3$ Rearrange to solve for l.

$l = \sqrt[3]{V} = \sqrt[3]{5.507873 \times 10^{-23}\ cm^3} = 3.804831 \times 10^{-8}\ cm \times \dfrac{10^{10}\ pm}{1\ cm} = 380.4831$ pm then $l = 2\sqrt{2}\,r$

Rearrange to solve for r. $r = \dfrac{l}{2\sqrt{2}} = \dfrac{380.4831\ \text{pm}}{2\sqrt{2}} = 134.5$ pm

Check: The units (pm) are correct. The magnitude (135 pm) is consistent with an atomic diameter.

12.37   **Given:** polonium, simple cubic structure $d = 9.3$ g/cm^3; $r = 167$ pm; $\mathcal{M} = 209$ g/mol   **Find:** estimate N_A
Conceptual Plan: $r \rightarrow l$ and $l \rightarrow V(pm^3) \rightarrow V(cm^3)$ then $d, V \rightarrow m$ then $\mathcal{M}$, SC structure $\rightarrow m$

$$l = 2r \quad V = l^3 \quad \dfrac{(1\ cm)^3}{(10^{10}\ pm)^3} \quad\quad d = m/V \quad\quad m = \dfrac{1\ atom}{unit\ cell} \times \dfrac{\mathcal{M}}{N_A}$$

Solution: $l = 2r = 2 \times 167$ pm $= 334$ pm and $V = l^3 = (334\ \text{pm})^3 \times \dfrac{(1\ cm)^3}{(10^{10}\ pm)^3} = 3.72597 \times 10^{-23}\ cm^3$ then

$$d = \frac{m}{V} \quad \text{Rearrange to solve for } m. \ m = dV = 9.3 \, \frac{g}{cm^3} \times \frac{3.72597 \times 10^{-23} \, cm^3}{\text{unit cell}} = 3.46515 \times 10^{-22} \, \frac{g}{\text{unit cell}}$$

then $m = \dfrac{1 \text{ atom}}{\text{unit cell}} \times \dfrac{\mathcal{M}}{N_A}$ Rearrange to solve for N_A.

$$N_A = \frac{1 \text{ atom}}{\text{unit cell}} \times \frac{\mathcal{M}}{m} = \frac{1 \text{ atom}}{\text{unit cell}} \times \frac{209 \text{ g}}{1 \text{ mol}} \times \frac{1 \text{ unit cell}}{3.46515 \times 10^{-22} \text{ g}} = 6.0 \times 10^{23} \, \frac{\text{atoms}}{\text{mol}}$$

Check: The units (atoms/mol) are correct. The magnitude (6×10^{23} atoms/mol) is consistent with Avogadro's number.

12.39　(a)　atomic because argon (Ar) is an atom

(b)　molecular because water (H_2O) is a molecule

(c)　ionic because potassium oxide (K_2O) is an ionic solid

(d)　atomic because iron (Fe) is an atom

12.41　LiCl has the highest melting point because it is the only ionic solid in the group. The other three solids are held together by intermolecular forces, while LiCl is held together by stronger coulombic interactions between the cations and anions of the crystal lattice.

12.43　(a)　TiO_2 because it is an ionic solid

(b)　$SiCl_4$ because it has a higher molar mass and, therefore, has stronger dispersion forces

(c)　Xe because it has a higher molar mass and, therefore, has stronger dispersion forces

(d)　CaO because the ions have greater charge and, therefore, stronger dipole–dipole interactions

12.45　The Ti atoms occupy the corner positions and the center of the unit cell: 8 corner atoms × (1/8 atom/unit cell) +1 atom in center = (1 + 1) Ti atoms/unit cell = 2 Ti atoms/unit cell. The O atoms occupy four positions on the top and bottom faces and two positions inside the unit cell: 4 face-centered atoms × (1/2 atom/unit cell) +2 atoms in the interior = (2 + 2) O atoms/unit cell = 4 O atoms/unit cell. Therefore, there are 2 Ti atoms/unit cell and 4 O atoms/unit cell; so the ratio Ti:O is 2:4, or 1:2. The formula for the compound is TiO_2.

12.47　In CsCl: The Cs atoms occupy the center of the unit cell: 1 atom in center = 1 Cs atom/unit cell. The Cl atoms occupy corner positions of the unit cell: 8 corner atoms × (1/8 atom/unit cell) = 1 Cl atom/unit cell. Therefore, there are 1 Cl atom/unit cell and 1 Cl atom/unit cell, so the ratio Cs:Cl is 1:1. The formula for the compound is CsCl, as expected.

In $BaCl_2$: The Ba atoms occupy the corner positions and the face-centered positions of the unit cell: 8 corner atoms × (1/8 atom/unit cell) + 6 face-centered atoms × (1/2 atom/unit cell) = (1 + 3) Ba atoms/unit cell = 4 Ba atoms/unit cell. The Cl atoms occupy eight positions inside the unit cell: 8 Cl atoms/unit cell. Therefore, there are 4 Ba atoms/unit cell and 8 Cl atoms/unit cell, so the ratio Ba:Cl is 4:8, or 1:2. The formula for the compound is $BaCl_2$, as expected.

12.49　Cesium chloride does not have any of the listed crystal structures. Barium chloride has the fluorite crystal structure.

12.51　If all of the anions are removed, we have a face-centered cubic structure.

Ceramics, Cement, and Glass

12.53　(a)　B_4C is a component of a nonoxide ceramic.

(b)　Mg_2SiO_4 is a component of a silicate ceramic.

(c)　$MoSi_2$ is a component of a nonoxide ceramic.

12.55　Boric oxide, B_2O_3, is commonly used to reduce thermal shock cracking and shattering in glass.

12.57　Ca_3SiO_5: The oxidation state of Ca = +2, the oxidation state of O = −2. The oxidation state of Si is deduced from the formula since the sum of the oxidation states must equal the overall charge. 2 (Ca ox state) + (Si ox state) + 5(O ox state) = 0; 3(+2) + (Si ox state) + 5(−2) = 0, so Si ox state = +4.

Semiconductors and Band Theory

12.59   (a)   Zn should have little or no band gap because it is the only metal in the group.

12.61   **Given:** 11.2 g lithium crystal   **Find:** number of molecular orbitals in the conduction band
       Conceptual Plan: g $\rightarrow$ mol $\rightarrow$ Li$_N$ $\rightarrow$ number of valence electrons $\rightarrow$ number of molecular orbitals

$$\frac{1\ mol}{6.94\ g} \qquad \frac{6.022 \times 10^{23}\ atoms}{1\ mol} \qquad \frac{1\ 2s\ electron}{1\ Li\ atom} \qquad\qquad \frac{1\ molecular\ orbital}{1\ 2s\ electron}$$

       Solution:

$$11.2\ g\ Li \times \frac{1\ mol\ Li}{6.94\ g\ Li} \times \frac{6.022 \times 10^{23}\ Li\ atoms}{1\ mol\ Li} \times \frac{1\ 2s\ electron}{1\ Li\ atoms} \times \frac{1\ molecular\ orbital}{1\ 2s\ electron} = 9.72 \times 10^{23}\ molecular\ orbitals$$

       Check: The units (number of molecular orbitals) are correct. The magnitude (10^{23}) is expected since there are so many orbitals because we have over a mole of atoms. Metals can conduct electricity because of these large numbers of orbitals.

12.63   Looking at Table 12.1, a band gap of 6.9 eV is an insulator.

12.65   (a)   p-type semiconductor: Ge is group 4A, and Ga is group 3A; so the Ga will generate electron "holes."
        (b)   n-type semiconductor: Si is group 4A, and As is group 5A; so the As will add electrons to the conduction band.

12.67   **Given:** frequency (v) = 4.29×10^{14} Hz   **Find:** Compare energy to 1.11 eV
       Conceptual Plan: v $\rightarrow$ E(J) $\rightarrow$ E(eV) then compare to 1.1 eV

$$E = hv \quad h = 6.626 \times 10^{-34}\ J\cdot s \qquad \frac{1\ eV}{1.602 \times 10^{-19}\ J}$$

       Solution: $E = hv = (6.626 \times 10^{-34}\ J\cdot s)\left(\dfrac{4.29 \times 10^{14}}{s}\right)\left(\dfrac{1\ eV}{1.602 \times 10^{-19}\ J}\right) = 1.77$ eV Since this is larger than 1.11 eV, this red photon has enough energy.

       Check: The units of the answer (eV) are correct. The magnitude of the answer is reasonable since we are talking about the energy of one photon of visible light.

Polymers

12.69   Because the monomer is $\begin{array}{c} F \\ \diagdown \\ C \end{array}\!\!=\!\!\begin{array}{c} F \\ \diagup \\ C \end{array}$ F, as the polymer forms, the double bond breaks to link with another monomer.

       The structure is

12.71   In a condensation polymer, atoms are eliminated. Here, water is eliminated. The structure is

12.73   In addition, polymer formation, a carbon—carbon pi bond is broken in each monomer, and a carbon—carbon single bond is formed. Thus, the monomer has a carbon—carbon triple bond and 2 carbons – acetylene.

$$H-C\equiv C-H$$

12.75 In condensation polymerization, a small molecule (typically water) is generated when the two monomers react. If water is removed, the other monomer is formaldehyde, CH_2O. The commercial manufacture typically uses phosgene, CCl_2O.

Cumulative Problems

12.77 CsCl has a higher melting point than AgI because of its higher coordination number. In CsCl, one anion bonds to eight cations (and vice versa), while in AgI, one anion bonds only to four cations.

12.79 (a) Atoms are connected across the face diagonal (c), so $c = 4r$.
(b) From the Pythagorean Theorem, $c^2 = a^2 + b^2$; from part (a), $c = 4r$; and for a cubic structure, $a = l, b = l$. So $(4r)^2 = l^2 + l^2 \rightarrow 16r^2 = 2l^2 \rightarrow 8r^2 = l^2 \rightarrow l = \sqrt{8r^2} \rightarrow l = 2\sqrt{2}r$.

12.81 **Given:** diamond, $V(\text{unit cell}) = 0.0454 \text{ nm}^3$; $d = 3.52 \text{ g/cm}^3$ **Find:** number of carbon atoms/unit cell
Conceptual Plan: $V(\text{nm}^3) \rightarrow V(\text{cm}^3)$ then $d, V \rightarrow m \rightarrow \text{mol} \rightarrow \text{atoms}$

$$\frac{(1 \text{ cm})^3}{(10^7 \text{ nm})^3} \qquad d = m/V \quad \frac{1 \text{ mol}}{12.01 \text{ g}} \quad \frac{6.022 \times 10^{23} \text{ atoms}}{1 \text{ mol}}$$

Solution: $0.0454 \text{ nm}^3 \times \dfrac{(1 \text{ cm})^3}{(10^7 \text{ nm})^3} = 4.54 \times 10^{-23} \text{ cm}^3$ then $d = \dfrac{m}{V}$ Rearrange to solve for m.

$m = dV = 3.52 \dfrac{g}{cm^3} \times 4.54 \times 10^{-23} \text{ cm}^3 = 1.59808 \times 10^{-22} \text{ g}$ then

$$\frac{1.59808 \times 10^{-22} \text{ g}}{\text{unit cell}} \times \frac{1 \text{ mol}}{12.01 \text{ g}} \times \frac{6.022 \times 10^{23} \text{ atoms}}{1 \text{ mol}} = 8.01 \frac{\text{C atoms}}{\text{unit cell}} = 8 \frac{\text{C atoms}}{\text{unit cell}}$$

Check: The units (atoms) are correct. The magnitude (8 atoms) makes sense because it is a fairly small number and our answer is within calculation error of an integer.

12.83 **Given:** metal, $d = 7.8748 \text{ g/cm}^3$; $l = 0.28664 \text{ nm}$, body-centered cubic lattice **Find:** $\mathcal{M}$
Conceptual Plan: $l \rightarrow V(\text{nm}^3) \rightarrow V(\text{cm}^3)$ then $d, V \rightarrow m$ then m, FCC structure $\rightarrow \mathcal{M}$

$$V = l^3 \qquad \frac{(1 \text{ cm})^3}{(10^7 \text{ nm})^3} \qquad d = m/V \qquad m = \frac{2 \text{ atoms}}{\text{unit cell}} \times \frac{\mathcal{M}}{N_A}$$

Solution: $V = l^3 = (0.28664 \text{ nm})^3 = 0.02355105602 \text{ nm}^3 \times \dfrac{(1 \text{ cm})^3}{(10^7 \text{ nm})^3} = 2.355105602 \times 10^{-23} \text{ cm}^3$ then $d = \dfrac{m}{V}$

Rearrange to solve for m. $m = dV = 7.8748 \dfrac{g}{cm^3} \times 2.355105602 \times 10^{-23} \text{ cm}^3 = 1.854598559 \times 10^{-22} \text{ g}$

then $m = \dfrac{2 \text{ atoms}}{\text{unit cell}} \times \dfrac{\mathcal{M}}{N_A}$ Rearrange to solve for $\mathcal{M}$.

$\mathcal{M} = \dfrac{\text{unit cell}}{2 \text{ atoms}} \times N_A \times m = \dfrac{\text{unit cell}}{2 \text{ atoms}} \times \dfrac{6.022 \times 10^{23} \text{ atoms}}{1 \text{ mol}} \times \dfrac{1.854598559 \times 10^{-22} \text{ g}}{\text{unit cell}} = 55.842 \dfrac{g}{\text{mol}}$

$= 55.84 \dfrac{g}{\text{mol}}$ iron

Check: The units (g/mol) are correct. The magnitude (55.8 g/mol) makes sense because it is a reasonable atomic mass for a metal and it is close to iron.

12.85 **Given:** KCl, rock salt structure **Find:** density (g/cm^3) **Other:** $r(K^+) = 133 \text{ pm}$; $r(Cl^-) = 181 \text{ pm}$ from Chapter 3
Conceptual Plan: Rock salt structure is a face-centered cubic structure with anions at the lattice points and cations in the holes between lattice sites $\rightarrow$ assume $r = r(Cl^-)$, but $\mathcal{M} = \mathcal{M}(KCl)$

$r(K^+), r(Cl^-) \rightarrow l$ and $l \rightarrow V(pm^3) \rightarrow V(cm^3)$, and FCC structure $\rightarrow m$ then $m, V \rightarrow d$

from Figure 12.13 $l = 2r(Cl^-) + 2r(K^+)$ $V = l^3 \dfrac{(1\,cm)^3}{(10^{10}\,pm)^3}$ $m = \dfrac{4\,\text{formula units}}{\text{unit cell}} \times \dfrac{M}{N_A}$ $d = m/V$

Solution: $l = 2r(Cl^-) + 2r(K^+) = 2(181\,pm) + 2(133\,pm) = 628\,pm$ and

$$V = l^3 = (628\,\text{pm})^3 \times \frac{(1\,cm)^3}{(10^{10}\,\text{pm})^3} = 2.47673 \times 10^{-22}\,cm^3 \text{ and}$$

$$m = \frac{4\,\text{formula units}}{\text{unit cell}} \times \frac{M}{N_A} = \frac{4\,\text{formula units}}{\text{unit cell}} \times \frac{74.55\,g}{1\,mol} \times \frac{1\,mol}{6.022 \times 10^{23}\,\text{formula units}}$$

$$= 4.951843 \times 10^{-22} \frac{g}{\text{unit cell}}$$

$$\text{then } d = \frac{m}{V} = \frac{4.951843 \times 10^{-22}\dfrac{g}{\text{unit cell}}}{2.47673 \times 10^{-22}\dfrac{cm^3}{\text{unit cell}}} = 1.99935\frac{g}{cm^3} = 2.00\frac{g}{cm^3}$$

Check: The units (g/cm^3) are correct. The magnitude ($2\,g/cm^3$) is reasonable for a salt density. The published value is $1.98\,g/cm^3$. This method of estimating the density gives a value that is close to the experimentally measured density.

12.87   **Given:** cubic closest packing structure = cube with touching spheres of radius = r on alternating corners of a cube
Find: body diagonal of cube and radius of tetrahedral hole
Solution: The cell edge length = l and $l^2 + l^2 = (2r)^2 \rightarrow 2l^2 = 4r^2 \rightarrow l^2 = 2r^2$. So, $l = \sqrt{2}r$. Because body diagonal = BD is the hypotenuse of the right triangle formed by the face diagonal and the cell edge, we have $(BD)^2 = l^2 + (2r)^2 = 2r^2 + 4r^2 = 6r^2 \rightarrow BD = \sqrt{6}r$. The radius of the tetrahedral hole = r_T is half the body diagonal minus the radius of the sphere, or

$$r_T = \frac{BD}{2} - r = \frac{\sqrt{6}r}{2} - r = \left(\frac{\sqrt{6}}{2} - 1\right)r = \left(\frac{\sqrt{6} - 2}{2}\right)r = \left(\frac{\sqrt{3}\sqrt{2} - \sqrt{2}\sqrt{2}}{\sqrt{2}\sqrt{2}}\right)r$$

$$= \left(\frac{\sqrt{3} - \sqrt{2}}{\sqrt{2}}\right)r \approx 0.22474r$$

12.89   This transition cannot be used because the energy per photon is too small to have a wavelength short enough to diffract.

Challenge Problems

12.91   (a)   In addition, polymer formation, a carbon—carbon pi bond is broken in each monomer, and a carbon—carbon single bond is formed. Thus, the monomer has a carbon—carbon double bond and 3 carbons – propylene.

(b)   Different relative orientation of the methyl groups in the monomer can result in different relative orientation of the methyl groups in the polymer, which can change the ability of the polymer to crystallize.

12.93   The Rb atoms will occupy the octahedral sites because Rb is a fairly large atom.

Conceptual Problems

12.95   The liquid must be cooled quickly, so it does not have time to crystallize.

12.97   Both structures are three-dimensional structures, where the atoms have a tetrahedral coordination. Both can be viewed as interpenetrating face-centered cubic lattices. In the diamond structure, all of the atoms are the same. In the zinc blende structure, there are two different atoms (a cation and an anion).

12.99   (d) All of these changes are expected to result in an increase in semiconductor conductivity.

Questions for Group Work

12.101   Answers will vary. For example, pairs of students forming covalent bonds by grasping hands may represent a molecular solid, and pairs of students standing in a regular array would represent the structure of the solid.

12.103   Answers will vary. For example, one student may say: "This material is a randomly ordered amorphous structure. One particular example, with the formula SiO_2 is called fused silica. It is transparent to visible and ultraviolet light." The answer would be "glass."

Data Interpretation and Analysis

12.105   (a)   The composition at point A is 20% Cr and 80% Ni, which melts at 1420 °C.
           (c)   The composition at point B is 98% Cr and 2% Ni and has a body-centered cubic structure.

13 Solutions

Review Questions

13.1   A solution is a homogeneous mixture of two or more substances. A solution has at least two compo-
        nents. The majority component is usually called the solvent, and the minority component is usually
        called the solute.

13.3   Ideal gases do not interact with each other in any way (i.e., there are no significant forces between their
        constituent particles). When the two gases mix, their potential energy remains unchanged; so this does
        not drive the mixing. The tendency to mix is related, rather, to a concept called entropy. Entropy is a
        measure of energy randomization or energy dispersal in a system. Recall that a gas at any temperature
        above 0 K has kinetic energy due to the motion of its atoms. When the gases are separated, their kinetic
        energies are also confined to those regions. However, when the gases mix, the kinetic energy of each
        gas becomes spread out, or dispersed, over a larger volume. Therefore, the mixture of the two gases has
        greater energy dispersal, or greater entropy, than the separated components. The pervasive tendency
        for all kinds of energy to spread out, or disperse, whenever they are not restrained from doing so is the
        reason two ideal gases mix.

13.5   Whether two substances will spontaneously mix to form a homogeneous solution is dependent upon a
        number of different types of intermolecular forces, including dispersion forces, dipole–dipole forces,
        hydrogen bonding, and ion–dipole forces.

13.7   The expression "like dissolves like" means that similar kinds of solvents dissolve similar kinds of sol-
        utes. Polar solvents, such as water, dissolve many polar or ionic solutes, and nonpolar solvents, such as
        hexane, dissolve many nonpolar solutes.

13.9   The heat of hydration is the enthalpy change that occurs when 1 mol of gaseous solute ions is dis-
        solved in water. In aqueous solutions, $\Delta H_{\text{solvent}}$ and ΔH_{mix} can be combined into a single term called
        the heat of hydration ($\Delta H_{\text{hydration}}$). Because the ion–dipole interactions that occur between a dis-
        solved ion and the surrounding water molecules are much stronger than the hydrogen bonds in water,
        $\Delta H_{\text{hydration}}$ is always largely negative (exothermic) for ionic compounds. Using the heat of hydration,
        we can write the enthalpy of solution as a sum of just two terms, one endothermic and one exother-
        mic: $\Delta H_{\text{soln}} = \Delta H_{\text{solute}} + \Delta H_{\text{solvent}} + \Delta H_{\text{mix}} = \Delta H_{\text{solute}} + \Delta H_{\text{hydration}} = $ endothermic(+) term +
        exothermic(−) term. For ionic compounds, ΔH_{solute}, the energy required to separate the solute into its
        constituent particles, is simply the negative of the solute's lattice energy ($\Delta H_{\text{solute}} = -\Delta H_{\text{lattice}}$), dis-
        cussed in Section 9.4. For ionic aqueous solutions then, the overall enthalpy of solution depends on the
        relative magnitudes of ΔH_{solute} and $\Delta H_{\text{hydration}}$, with three possible scenarios (in each case, we refer to
        the magnitude or absolute value of ΔH): (1) If $\Delta H_{\text{solute}} < \Delta H_{\text{hydration}}$, the amount of energy required to
        separate the solute into its constituent ions is less than the energy given off when the ions are hydrated.
        ΔH_{soln} is, therefore, negative, the solution process is exothermic, and the solution feels warm to the touch.
        (2) If $\Delta H_{\text{solute}} > \Delta H_{\text{hydration}}$, the amount of energy required to separate the solute into its constituent
        ions is greater than the energy given off when the ions are hydrated. ΔH_{soln} is, therefore, positive,
        the solution process is endothermic (if a solution forms at all), and the resulting solution feels cool to
        the touch. (3) If $\Delta H_{\text{solute}} \approx \Delta H_{\text{hydration}}$, the amount of energy required to separate the solute into its
        constituent ions is about equal to the energy given off when the ions are hydrated. ΔH_{soln} is, therefore,

approximately zero, the solution process is neither appreciably exothermic nor appreciably endothermic, and there is no noticeable change in temperature.

13.11   Although there are exceptions, the solubility of most solids in water increases with increasing temperature. Recrystallization is a common technique to purify a solid. In this technique, the solid is put into water (or some other solvent) at an elevated temperature. Enough solid is added to the solvent to create a saturated solution at the elevated temperature. As the solution cools, it becomes supersaturated and the excess solid begins to come out of solution. If the solution cools slowly, the solid forms crystals as it comes out of solution. The crystalline structure tends to reject impurities, resulting in a purer solid. The solvent is chosen so that the solubility of the impurities is high at the lower temperature, reducing the tendency to co-precipitate.

13.13   The higher the pressure of a gas above a liquid, the more soluble the gas in the liquid. In a sealed can of soda pop, for example, the carbon dioxide is maintained in solution by a high pressure of carbon dioxide within the can. When the can is opened, this pressure is released and the solubility of carbon dioxide decreases, resulting in bubbling.

13.15   The common ways of reporting solution concentration include the following:

(a)   Molarity: $M = \dfrac{\text{amount solute (moles)}}{\text{volume solution (L)}}$

(b)   Molality: $m = \dfrac{\text{amount solute (moles)}}{\text{mass solvent (kg)}}$

(c)   Parts by mass: $\dfrac{\text{mass solute}}{\text{mass solution}} \times$ multiplication factor, where percent by mass (%) factor $= 100$, parts per million by mass (ppm) factor $= 10^6$, and parts per billion by mass (ppb) factor $= 10^9$

(d)   Parts by volume (%, ppm, ppb): $\dfrac{\text{volume solute}}{\text{volume solution}} \times$ multiplication factor, where the same factors as in parts by mass are used

(e)   Mole fraction: $\chi = \dfrac{\text{amount solute (in moles)}}{\text{total amount of solute and solvent (in moles)}}$

(f)   Mole percent: mol % $= \chi \times 100\%$

13.17   The vapor pressure of the solution is lower than the vapor pressure of the pure solvent. The simplest explanation for the lowering of the vapor pressure of a solution relative to that of the pure solvent is related to the concept of dynamic equilibrium. In dynamic equilibrium, the rate of vaporization is equal to the rate of condensation. When a nonvolatile solute is added, however, the solute particles interfere with the ability of the solvent particles to vaporize because they occupy some of the surface area formerly occupied by the solvent. The rate of vaporization is, therefore, diminished compared to that of the pure solvent. The change in the rate of vaporization creates an imbalance in the rates; the rate of condensation is now greater than the rate of vaporization. The net effect is that some of the molecules that were in the gas phase condense into the liquid. As they condense, the reduced number of molecules in the gas phase causes the rate of condensation to decrease. Eventually, the two rates become equal again, but only after the concentration of molecules in the gas phase has decreased, which means a lower vapor pressure for the solution compared to the pure solvent.

13.19   An ideal solution follows Raoult's law at all concentrations for both the solute and the solvent. A nonideal solution exhibits deviations from Raoult's law in the vapor pressure of a component as the mole fraction of this component decreases from 1 (i.e., the pure component).

13.21   A nonvolatile solute lowers the vapor pressure of a solution relative to that of the pure solvent. The vapor pressure lowering occurs at all temperatures, which shifts the vaporization curve in the phase diagram. This means that the temperature must be raised above the pure solvent normal boiling point for the solution vapor pressures to be raised to 1 atm. This shift also results in a lowering of where the vapor pressure curve intersects the solid–gas curve. The net effect is that the solution has a lower melting point and a higher boiling point than the pure solvent.

13.23   Osmosis is defined as the flow of solvent from a solution of lower solute concentration to one of higher solute concentration through a semipermeable membrane—a membrane that selectively allows some substances, but not others, to pass through. The osmotic pressure is the pressure required to stop the osmotic flow and is given by the following equation: $\Pi = MRT$.

Problems by Topic

Solubility

13.25   (a)   hexane, toluene, or CCl_4; dispersion forces
        (b)   water, methanol, acetone; dispersion, dipole–dipole, hydrogen bonding (except for acetone)
        (c)   hexane, toluene, or CCl_4; dispersion forces
        (d)   water, acetone, methanol, ethanol; dispersion, ion–dipole

13.27 $HOCH_2CH_2CH_2OH$ would be more soluble in water because it has —OH groups on both ends of the molecule; so it
        can hydrogen bond on both ends.

13.29   (a)   water; dispersion, dipole–dipole, hydrogen bonding
        (b)   hexane; dispersion forces
        (c)   water; dispersion, dipole–dipole
        (d)   water; dispersion, dipole–dipole, hydrogen bonding

Energetics of Solution Formation

13.31   (a)   endothermic
        (b)   The lattice energy is greater in magnitude than is the heat of hydration.
        (c)

        (d)   The solution forms because chemical systems tend toward greater entropy.

13.33   **Given:** $AgNO_3$: lattice energy $= -820.$ kJ/mol, $\Delta H_{soln} = +22.6$ kJ/mol   **Find:** $\Delta H_{hydration}$
        Conceptual Plan: lattice energy, $\Delta H_{soln} \to \Delta H_{hydration}$

$$\Delta H_{soln} = \Delta H_{solute} + \Delta H_{hydration} \text{ where } \Delta H_{solute} = -\Delta H_{lattice}$$

        Solution: $\Delta H_{soln} = \Delta H_{solute} + \Delta H_{hydration}$ where $\Delta H_{solute} = -\Delta H_{lattice}$ so $\Delta H_{hydration} = \Delta H_{soln} + \Delta H_{lattice}$
        $\Delta H_{hydration} = 22.6$ kJ/mol $- 820.$ kJ/mol $= -797$ kJ/mol

        Check: The units (kJ/mol) are correct. The magnitude of the answer (-800 kJ/mol) makes physical sense because
        the lattice energy is so negative; thus, it dominates the calculation.

13.35   **Given:** LiI: lattice energy $= -7.3 \times 10^2$ kJ/mol, $\Delta H_{hydration} = -793$ kJ/mol; 15.0 g LiI
        Find: ΔH_{soln} and heat evolved
        Conceptual Plan: lattice energy, $\Delta H_{hydration} \to \Delta H_{soln}$ and g $\to$ mol then mol, $\Delta H_{soln} \to q$

$$\Delta H_{soln} = \Delta H_{solute} + \Delta H_{hydration} \text{ where } \Delta H_{solute} = -\Delta H_{lattice} \qquad \frac{1 \text{ mol}}{133.84 \text{ g}} \qquad q = n\Delta H_{soln}$$

        Solution: $\Delta H_{soln} = \Delta H_{solute} + \Delta H_{hydration}$ where $\Delta H_{solute} = -\Delta H_{lattice}$ so $\Delta H_{soln} = \Delta H_{hydration} - \Delta H_{lattice}$

        $\Delta H_{soln} = -793$ kJ/mol $- (-7\underline{3}0$ kJ/mol$) = -\underline{6}3$ kJ/mol $= -6.3 \times 10^1$ kJ/mol and

        $15.0 \text{ g} \times \dfrac{1 \text{ mol}}{133.84 \text{ g}} = 0.11\underline{2}074$ mol then

$$q = n\Delta H_{soln} = 0.11\underline{2}074 \text{ mol} \times -6.3 \times 10^1 \frac{\text{kJ}}{\text{mol}} = -7.1 \text{ kJ or } 7.0 \text{ kJ released}$$

Check: The units (kJ/mol and kJ) are correct. The magnitude of the answer (-60 kJ/mol) makes physical sense because the lattice energy and the heat of hydration are about the same. The magnitude of the heat (7 kJ) makes physical sense because 15 g is much less than a mole; thus, the amount of heat released is going to be small.

Solution Equilibrium and Factors Affecting Solubility

13.37   The solution is unsaturated because we are dissolving 25 g of NaCl per 100 g of water and the solubility from the figure is ~35 g NaCl per 100 g of water at 25 °C.

13.39   At 40 °C, the solution has 45 g of KNO_3 per 100 g of water, and it can contain up to 63 g of KNO_3 per 100 g of water. At 0 °C, the solubility from the figure is ~14 g KNO_3 per 100 g of water; so ~31 g KNO_3 per 100 g of water will precipitate out of solution.

13.41   Because the solubility of gases decreases as the temperature increases, boiling will cause dissolved oxygen to be removed from the solution.

13.43   Henry's law says that as pressure increases, nitrogen will more easily dissolve in blood. To reverse this process, divers should ascend to lower pressures.

13.45   **Given:** room temperature, 80.0 L aquarium, $P_{Total} = 1.0$ atm; $\chi_{N_2} = 0.78$ **Find:** m (N_2)
Other: $k_H(N_2) = 6.1 \times 10^{-4}$ M/L at 25 °C
Conceptual Plan: $P_{Total}, \chi_{N_2} \rightarrow P_{N_2}$ then $P_{N_2}, k_H(N_2) \rightarrow S_{N_2}$ then $L \rightarrow$ mol $\rightarrow$ g

$$P_{N_2} = \chi_{N_2} P_{Total} \qquad S_{N_2} = k_H(N_2)P_{N_2} \quad M = \frac{\text{amount solute (moles)}}{\text{volume solution (L)}} \frac{28.02 \text{ g } N_2}{1 \text{ mol } N_2}$$

Solution: $P_{N_2} = \chi_{N_2} P_{Total} = 0.78 \times 1.0 \text{ atm} = 0.78 \text{ atm}$ then

$$S_{N_2} = k_H(N_2)P_{N_2} = 6.1 \times 10^{-4} \frac{M}{\text{atm}} \times 0.78 \text{ atm} = 4.\underline{7}58 \times 10^{-4} \text{ M then}$$

$$80.0 \text{ L} \times 4.\underline{7}58 \times 10^{-4} \frac{\text{mol}}{\text{L}} \times \frac{28.02 \text{ g}}{1 \text{ mol}} = 1.1 \text{ g}$$

Check: The units (g) are correct. The magnitude of the answer (1 g) seems reasonable because we have 80 L of water and expect much less than a mole of nitrogen.

Concentrations of Solutions

13.47   **Given:** NaCl and water; 112 g NaCl in 1.00 L solution   **Find:** M, m, and mass percent   **Other:** $d = 1.08$ g/mL
Conceptual Plan: $g_{NaCl} \rightarrow$ mol and $L \rightarrow$ mL $\rightarrow g_{soln}$ and $g_{soln}, g_{NaCl} \rightarrow g_{H_2O} \rightarrow kg_{H_2O}$ then

$$\frac{1 \text{ mol NaCl}}{58.44 \text{ g NaCl}} \qquad \frac{1000 \text{ mL}}{1 \text{ L}} \frac{1.08 \text{ g}}{1 \text{ mL}} \qquad g_{H_2O} = g_{soln} - g_{NaCl} \qquad \frac{1 \text{ kg}}{1000 \text{ g}}$$

mol, $V \rightarrow$ M and mol, $kg_{H_2O} \rightarrow m$ and $g_{soln}, g_{NaCl} \rightarrow$ mass percent

$$M = \frac{\text{amount solute (moles)}}{\text{volume solution (L)}} \quad m = \frac{\text{amount solute (moles)}}{\text{mass solvent (kg)}} \quad \text{mass percent} = \frac{\text{mass solute}}{\text{mass solution}} \times 100\%$$

Solution: $112 \text{ g NaCl} \times \dfrac{1 \text{ mol NaCl}}{58.44 \text{ g NaCl}} = 1.9164956 \text{ mol NaCl and}$

$$1.00 \text{ L} \times \frac{1000 \text{ mL}}{1 \text{ L}} \times \frac{1.08 \text{ g}}{1 \text{ mL}} = 1080 \text{ g soln and}$$

$$g_{H_2O} = g_{soln} - g_{NaCl} = 1080 \text{ g} - 112 \text{ g} = 96\underline{8} \text{ g } H_2O \times \frac{1 \text{ kg}}{1000 \text{ g}} = 0.968 \text{ kg } H_2O \text{ then}$$

$$M = \frac{\text{amount solute (moles)}}{\text{volume solution (L)}} = \frac{1.9164956 \text{ mol NaCl}}{1.00 \text{ L soln}} = 1.92 \text{ M and}$$

$$m = \frac{\text{amount solute (moles)}}{\text{mass solvent (kg)}} = \frac{1.9164956 \text{ mol NaCl}}{0.968 \text{ kg H}_2\text{O}} = 2.0 \, m \text{ and}$$

$$\text{mass percent} = \frac{\text{mass solute}}{\text{mass solution}} \times 100\% = \frac{112 \text{ g NaCl}}{1080 \text{ g soln}} \times 100\% = 10.4\% \text{ by mass}$$

Check: The units (M, m, and percent by mass) are correct. The magnitude of the answer (2 M) seems reasonable because we have 112 g NaCl, which is a couple of moles and we have 1 L. The magnitude of the answer (2 m) seems reasonable because it is a little higher than the molarity, which we expect because we only use the solvent weight in the denominator. The magnitude of the answer (10%) seems reasonable because we have 112 g NaCl and just over 1000 g of solution.

13.49   **Given:** initial solution: 50.0 mL of 5.00 M KI; final solution contains 3.05 g KI in 25.0 mL
        Find: final volume of diluted solution
        Conceptual Plan: final solution: $g_{KI} \rightarrow$ mol and mL $\rightarrow$ L then mol, $V \rightarrow M_2$ then $M_1, V_1, M_2 \rightarrow V_2$

$$\frac{1 \text{ mol KI}}{166.00 \text{ g KI}} \qquad \frac{1 \text{ L}}{1000 \text{ mL}} \qquad M = \frac{\text{amount solute (moles)}}{\text{volume solution (L)}} \qquad M_1V_1 = M_2V_2$$

        Solution: $3.05 \text{ g KI} \times \dfrac{1 \text{ mol KI}}{166.00 \text{ g KI}} = 0.01837283 \text{ mol KI}$ and $25.0 \text{ mL} \times \dfrac{1 \text{ L}}{1000 \text{ mL}} = 0.0250 \text{ L}$

        then $M = \dfrac{\text{amount solute (moles)}}{\text{volume solution (L)}} = \dfrac{0.01837283 \text{ mol KI}}{0.0250 \text{ L soln}} = 0.7349132 \text{ M}$ then $M_1V_1 = M_2V_2$

        Rearrange to solve for V_2. $V_2 = \dfrac{M_1}{M_2} \times V_1 = \dfrac{5.00 \text{ M}}{0.7349132 \text{ M}} \times 50.0 \text{ mL} = 340. \text{ mL diluted volume}$

Check: The units (mL) are correct. The magnitude of the answer (340 mL) seems reasonable because we are starting with a concentration of 5 M and ending with a concentration of less than 1 M.

13.51   **Given:** $AgNO_3$ and water; 3.4% Ag by mass, 4.8 L solution   **Find:** m (Ag)   **Other:** $d = 1.01 \text{ g/mL}$
        Conceptual Plan: $L \rightarrow mL \rightarrow g_{soln} \rightarrow g_{Ag}$

$$\frac{1000 \text{ mL}}{1 \text{ L}} \quad \frac{1.01 \text{ g}}{1 \text{ mL}} \quad \frac{3.4 \text{ g Ag}}{100 \text{ g soln}}$$

        Solution: $4.8 \text{ L} \times \dfrac{1000 \text{ mL}}{1 \text{ L}} \times \dfrac{1.01 \text{ g}}{1 \text{ mL}} = 4848 \text{ g soln}$ then

        $4848 \text{ g soln} \times \dfrac{3.4 \text{ g Ag}}{100 \text{ g soln}} = 160 \text{ g Ag} = 1.6 \times 10^2 \text{ g Ag}$

Check: The units (g) are correct. The magnitude of the answer (160 g) seems reasonable because we have almost 5000 g of solution.

13.53   **Given:** Ca^{2+} and water; 0.0085% Ca^{2+} by mass, 1.2 g Ca   **Find:** m (water)
        Conceptual Plan: $g_{Ca} \rightarrow g_{soln} \rightarrow g_{H_2O}$

$$\frac{100 \text{ g soln}}{0.0085 \text{ g Ca}} \qquad g_{H_2O} = g_{soln} - g_{Ca}$$

        Solution: $1.2 \text{ g Ca} \times \dfrac{100 \text{ g soln}}{0.0085 \text{ g Ca}} = 14118 \text{ g soln}$ then

        $g_{H_2O} = g_{soln} - g_{Ca} = 14118 \text{ g} - 1.2 \text{ g} = 1.4 \times 10^4 \text{ g water}$

Check: The units (g) are correct. The magnitude of the answer (10^4 g) seems reasonable because we have such a low concentration of Ca.

13.55   **Given:** concentrated HNO_3: 70.3% HNO_3 by mass, $d = 1.41 \text{ g/mL}$; final solution: 1.15 L of 0.100 M HNO_3
        Find: describe final solution preparation
        Conceptual Plan: $M_2, V_2 \rightarrow mol_{HNO_3} \rightarrow g_{HNO_3} \rightarrow g_{conc \, acid} \rightarrow mL_{conc \, acid}$ then describe method

$$mol = MV \qquad \frac{63.02 \text{ g HNO}_3}{1 \text{ mol HNO}_3} \quad \frac{100 \text{ g conc acid}}{70.3 \text{ g HNO}_3} \quad \frac{1 \text{ mL}}{1.41 \text{ g}}$$

        Solution: $mol = MV = 0.100 \dfrac{\text{mol HNO}_3}{1 \text{ L soln}} \times 1.15 \text{ L soln} = 0.115 \text{ mol HNO}_3$ then

        $0.115 \text{ mol HNO}_3 \times \dfrac{63.02 \text{ g HNO}_3}{1 \text{ mol HNO}_3} \times \dfrac{100 \text{ g conc acid}}{70.3 \text{ g HNO}_3} \times \dfrac{1 \text{ mL conc acid}}{1.41 \text{ g conc acid}} = 7.31 \text{ mL conc acid}$

Prepare the solution by putting about 1.00 L of distilled water in a container. Carefully pour in the 7.31 mL of the concentrated acid, mix the solution, and allow it to cool. Finally, add enough water to generate the total volume of solution (1.15 L). It is important to add acid to water, not the reverse, because such a large amount of heat is released upon mixing.

Check: The units (mL) are correct. The magnitude of the answer (7 mL) seems reasonable because we are starting with such a concentrated solution and diluting it to a low concentration.

13.57   (a)   **Given:** 1.00×10^2 mL of 0.500 M KCl   **Find:** describe final solution preparation
          Conceptual Plan: $\text{mL} \rightarrow \text{L}$ then $M, V \rightarrow \text{mol}_{\text{KCl}} \rightarrow \text{g}_{\text{KCl}}$ **then describe method**

$$\frac{1\ \text{L}}{1000\ \text{mL}} \qquad \text{mol} = MV \qquad \frac{74.55\ \text{g KCl}}{1\ \text{mol KCl}}$$

          Solution: $1.00 \times 10^2\ \text{mL} \times \dfrac{1\ \text{L}}{1000\ \text{mL}} = 0.100\ \text{L}$

$$\text{mol} = MV = 0.500\ \frac{\text{mol KCl}}{1\ \text{L soln}} \times 0.100\ \text{L soln} = 0.0500\ \text{mol KCl}$$

          then $0.0500\ \text{mol KCl} \times \dfrac{74.55\ \text{g KCl}}{1\ \text{mol KCl}} = 3.73\ \text{g KCl}$

Prepare the solution by carefully adding 3.73 g KCl to a 100 mL volumetric flask. Add ~75 mL of distilled water and agitate the solution until the salt dissolves completely. Finally, add enough water to generate a total volume of solution (add water to the mark on the flask).

Check: The units (g) are correct. The magnitude of the answer (4 g) seems reasonable because we are making a small volume of solution and the formula weight of KCl is ~75 g/mol.

   (b)   **Given:** 1.00×10^2 g of 0.500 m KCl   **Find:** describe final solution preparation
          Conceptual Plan:
          $m \rightarrow \text{mol}_{\text{KCl}}/1\ \text{kg solvent} \rightarrow \text{g}_{\text{KCl}}/1\ \text{kg solvent}$ then $\text{g}_{\text{KCl}}/1\ \text{kg solvent}, \text{g}_{\text{soln}} \rightarrow \text{g}_{\text{KCl}}, \text{g}_{\text{H}_2\text{O}}$

$$m = \frac{\text{amount solute (moles)}}{\text{mass solvent (kg)}} \quad \frac{74.55\ \text{g KCl}}{1\ \text{mol KCl}} \qquad\qquad\qquad g_{\text{soln}} = g_{\text{KCl}} + g_{\text{H}_2\text{O}}$$

          then describe method

          Solution: $m = \dfrac{\text{amount solute (moles)}}{\text{mass solvent (kg)}}$ so $0.500\ m = \dfrac{0.500\ \text{mol KCl}}{1\ \text{kg H}_2\text{O}}$ so

$$\frac{0.500\ \text{mol KCl}}{1\ \text{kg H}_2\text{O}} \times \frac{74.55\ \text{g KCl}}{1\ \text{mol KCl}} = \frac{37.28\ \text{g KCl}}{1000\ \text{g H}_2\text{O}} \quad g_{\text{soln}} = g_{\text{KCl}} + g_{\text{H}_2\text{O}} \text{ so } g_{\text{soln}} - g_{\text{KCl}} = g_{\text{H}_2\text{O}}$$

          Substitute into ratio. $\dfrac{37.28\ \text{g KCl}}{1037.28\ \text{g solution}} = \dfrac{x\ \text{g KCl}}{100\ \text{g solution}}$. Multiply both sides by 100 g solution.

$$\frac{37.28\ \text{g KCl}\ (100\ \text{g soln})}{1037.28\ \text{g soln}} = 3.59\ \text{g}$$

$$g_{\text{H}_2\text{O}} = g_{\text{soln}} - g_{\text{KCl}} = 100.\ \text{g} - 3.59\ \text{g} = 96\ \text{g H}_2\text{O}$$

Prepare the solution by carefully adding 3.59 g KCl to a container with 96 g of distilled water and agitate the solution until the salt dissolves completely.

Check: The units (g) are correct. The magnitude of the answer (3.6 g) seems reasonable because we are making a small volume of solution and the formula weight of KCl is ~75 g/mol.

   (c)   **Given:** 1.00×10^2 g of 5.0% KCl by mass   **Find:** describe final solution preparation
          Conceptual Plan: $g_{\text{soln}} \rightarrow g_{\text{KCl}}$ then $g_{\text{KCl}}, g_{\text{soln}} \rightarrow g_{\text{H}_2\text{O}}$

$$\frac{5.0\ \text{g KCl}}{100\ \text{g soln}} \qquad\qquad g_{\text{soln}} = g_{\text{KCl}} + g_{\text{H}_2\text{O}}$$

          then describe method

          Solution: $1.00 \times 10^2\ \text{g soln} \times \dfrac{5.0\ \text{g KCl}}{100\ \text{g soln}} = 5.0\ \text{g KCl}$ then $g_{\text{soln}} = g_{\text{KCl}} + g_{\text{H}_2\text{O}}$

          So $g_{\text{H}_2\text{O}} = g_{\text{soln}} - g_{\text{KCl}} = 100.\ \text{g} - 5.0\ \text{g} = 95\ \text{g H}_2\text{O}$

Prepare the solution by carefully adding 5.0 g KCl to a container with 95 g of distilled water and agitate the solution until the salt dissolves completely.

Check: The units (g) are correct. The magnitude of the answer (5 g) seems reasonable because we are making a small volume of solution and the solution is 5% by mass KCl.

13.59   (a)   **Given:** 28.4 g of glucose $(C_6H_{12}O_6)$ in 355 g water; final volume = 378 mL   **Find:** molarity
            Conceptual Plan: mL → L and $g_{C_6H_{12}O_6} → mol_{C_6H_{12}O_6}$ then $mol_{C_6H_{12}O_6}$, $V → M$

$$\frac{1\ L}{1000\ mL} \qquad \frac{1\ mol\ C_6H_{12}O_6}{180.16\ g\ C_6H_{12}O_6} \qquad M = \frac{amount\ solute\ (moles)}{volume\ solution\ (L)}$$

Solution: $378\ \cancel{mL} \times \dfrac{1\ L}{1000\ \cancel{mL}} = 0.378\ L$ and

$$28.4\ \cancel{g\ C_6H_{12}O_6} \times \frac{1\ mol\ C_6H_{12}O_6}{180.16\ \cancel{g\ C_6H_{12}O_6}} = 0.15\underline{7}638\ mol\ C_6H_{12}O_6$$

$$M = \frac{amount\ solute\ (moles)}{volume\ solution\ (L)} = \frac{0.15\underline{7}638\ mol\ C_6H_{12}O_6}{0.378\ L} = 0.417\ M$$

Check: The units (M) are correct. The magnitude of the answer (0.4 M) seems reasonable because we have 1/8 mole in about 1/3 L.

(b)   **Given:** 28.4 g of glucose $(C_6H_{12}O_6)$ in 355 g water; final volume = 378 mL   **Find:** molality
      Conceptual Plan: $g_{H_2O} → kg_{H_2O}$ and $g_{C_6H_{12}O_6} → mol_{C_6H_{12}O_6}$ then $mol_{C_6H_{12}O_6}$, $kg_{H_2O} → m$

$$\frac{1\ kg}{1000\ g} \qquad \frac{1\ mol\ C_6H_{12}O_6}{180.16\ g\ C_6H_{12}O_6} \qquad m = \frac{amount\ solute\ (moles)}{mass\ solvent\ (kg)}$$

Solution: $355\ \cancel{g} \times \dfrac{1\ kg}{1000\ \cancel{g}} = 0.355\ kg$ and

$$28.4\ \cancel{g\ C_6H_{12}O_6} \times \frac{1\ mol\ C_6H_{12}O_6}{180.16\ \cancel{g\ C_6H_{12}O_6}} = 0.15\underline{7}638\ mol\ C_6H_{12}O_6$$

$$m = \frac{amount\ solute\ (moles)}{mass\ solvent\ (kg)} = \frac{0.15\underline{7}638\ mol\ C_6H_{12}O_6}{0.355\ kg} = 0.444\ m$$

Check: The units (m) are correct. The magnitude of the answer (0.4 m) seems reasonable because we have 1/8 mole in about 1/3 kg.

(c)   **Given:** 28.4 g of glucose $(C_6H_{12}O_6)$ in 355 g water; final volume = 378 mL   **Find:** percent by mass
      Conceptual Plan: $g_{C_6H_{12}O_6}$, $g_{H_2O} → g_{soln}$ then $g_{C_6H_{12}O_6}$, $g_{soln} →$ percent by mass

$$g_{soln} = g_{C_6H_{12}O_6} + g_{H_2O} \qquad mass\ percent = \frac{mass\ solute}{mass\ solution} \times 100\%$$

Solution: $g_{soln} = g_{C_6H_{12}O_6} + g_{H_2O} = 28.4\ g + 355\ g = 38\underline{3}.4\ g\ soln$ then

$$mass\ percent = \frac{mass\ solute}{mass\ solution} \times 100\% = \frac{28.4\ g\ C_6H_{12}O_6}{38\underline{3}.4\ g\ soln} \times 100\% = 7.41\%\ by\ mass$$

Check: The units (percent by mass) are correct. The magnitude of the answer (7%) seems reasonable because we are dissolving 28 g in 355 g.

(d)   **Given:** 28.4 g of glucose $(C_6H_{12}O_6)$ in 355 g water; final volume = 378 mL   **Find:** mole fraction
      Conceptual Plan: $g_{C_6H_{12}O_6} → mol_{C_6H_{12}O_6}$ and $g_{H_2O} → mol_{H_2O}$ then $mol_{C_6H_{12}O_6}$, $mol_{H_2O} → \chi_{C_6H_{12}O_6}$

$$\frac{1\ mol\ C_6H_{12}O_6}{180.16\ g\ C_6H_{12}O_6} \qquad \frac{1\ mol\ H_2O}{18.02\ g\ H_2O} \qquad \chi = \frac{amount\ solute\ (in\ moles)}{total\ amount\ of\ solute\ and\ solvent\ (in\ moles)}$$

Solution: $28.4\ \cancel{g\ C_6H_{12}O_6} \times \dfrac{1\ mol\ C_6H_{12}O_6}{180.16\ \cancel{g\ C_6H_{12}O_6}} = 0.15\underline{7}638\ mol\ C_6H_{12}O_6$ and

$$355\ \cancel{g\ H_2O} \times \frac{1\ mol\ H_2O}{18.02\ \cancel{g\ H_2O}} = 19.\underline{7}003\ mol\ H_2O\ then$$

$$\chi = \frac{amount\ solute\ (in\ moles)}{total\ amount\ of\ solute\ and\ solvent\ (in\ moles)} = \frac{0.15\underline{7}638\ \cancel{mol}}{0.15\underline{7}638\ \cancel{mol} + 19.\underline{7}003\ \cancel{mol}} = 0.00794$$

Check: The units (none) are correct. The magnitude of the answer (0.008) seems reasonable because we have many more grams of water and water has a much lower molecular weight.

(e)   **Given:** 28.4 g of glucose $(C_6H_{12}O_6)$ in 355 g water; final volume = 378 mL   **Find:** mole percent
      Conceptual Plan: use answer from part (d) then $\chi_{C_6H_{12}O_6} →$ mole percent

$$\chi \times 100\%$$

Solution: mole percent $= \chi \times 100\% = 0.00794 \times 100\% = 0.794$ mole percent

Check: The units (mol %) are correct. The magnitude of the answer (0.8) seems reasonable because we have many more grams of water, water has a much lower molecular weight than glucose, and we are increasing the answer from part (d) by a factor of 100.

13.61 **Given:** 3.0% H_2O_2 by mass, $d = 1.01$ g/mL **Find:** molarity
Conceptual Plan:
Assume exactly 100 g of solution; $g_{solution} \rightarrow g_{H_2O_2} \rightarrow mol_{H_2O_2}$ and $g_{solution} \rightarrow mL_{solution} \rightarrow L_{solution}$

$$\frac{3.0 \text{ g } H_2O_2}{100 \text{ g solution}} \quad \frac{1 \text{ mol } H_2O_2}{34.02 \text{ g } H_2O_2} \qquad \frac{1 \text{ mL}}{1.01 \text{ g}} \quad \frac{1 \text{ L}}{1000 \text{ mL}}$$

then $mol_{H_2O_2}, L_{solution} \rightarrow M$

$$M = \frac{\text{amount solute (moles)}}{\text{volume solution (L)}}$$

Solution: $100 \text{ g solution} \times \dfrac{3.0 \text{ g } H_2O_2}{100 \text{ g solution}} \times \dfrac{1 \text{ mol } H_2O_2}{34.02 \text{ g } H_2O_2} = 0.08\underline{8}1834 \text{ mol } H_2O_2$ and

$100 \text{ g solution} \times \dfrac{1 \text{ mL solution}}{1.01 \text{ g solution}} \times \dfrac{1 \text{ L solution}}{1000 \text{ mL solution}} = 0.099\underline{0}099 \text{ L solution}$ then

$M = \dfrac{\text{amount solute (moles)}}{\text{volume solution (L)}} = \dfrac{0.08\underline{8}1834 \text{ mol } H_2O_2}{0.099\underline{0}099 \text{ L solution}} = 0.89 \text{ M } H_2O_2$

Check: The units (M) are correct. The magnitude of the answer (1) seems reasonable because we are starting with a low concentration solution and pure water is ~55.5 M.

13.63 **Given:** 36% HCl by mass **Find:** molality and mole fraction
Conceptual Plan: Assume exactly 100 g of solution; $g_{solution} \rightarrow g_{HCl} \rightarrow mol_{HCl}$ and $g_{HCl}, g_{solution} \rightarrow$

$$\frac{36 \text{ g HCl}}{100 \text{ g solution}} \quad \frac{1 \text{ mol HCl}}{36.46 \text{ g HCl}} \qquad g_{soln} = g_{HCl} + g_{H_2O}$$

$g_{solvent} \rightarrow kg_{solvent}$ then $mol_{HCl}, kg_{solvent} \rightarrow m$ and $g_{solvent} \rightarrow mol_{solvent}$ then $mol_{HCl}, mol_{solvent} \rightarrow \chi_{HCl}$

$$\frac{1 \text{ kg}}{1000 \text{ g}} \qquad m = \frac{\text{amount solute (moles)}}{\text{mass solvent (kg)}} \quad \frac{1 \text{ mol } H_2O}{18.02 \text{ g } H_2O} \qquad \chi = \frac{\text{amount solute (in moles)}}{\text{total amount of solute and solvent (in moles)}}$$

Solution: $100 \text{ g solution} \times \dfrac{36 \text{ g HCl}}{100 \text{ g solution}} = 36 \text{ g HCl} \times \dfrac{1 \text{ mol HCl}}{36.46 \text{ g HCl}} = 0.98\underline{7}383 \text{ mol HCl}$ and

$g_{soln} = g_{HCl} + g_{H_2O}$ **Rearrange to solve for** $g_{solvent}$. $g_{H_2O} = g_{soln} - g_{HCl} = 100 \text{ g} - 36 \text{ g} = 64 \text{ g } H_2O$

$64 \text{ g } H_2O \times \dfrac{1 \text{ kg } H_2O}{1000 \text{ g } H_2O} = 0.064 \text{ kg } H_2O$ then

$m = \dfrac{\text{amount solute (moles)}}{\text{mass solvent (kg)}} = \dfrac{0.98\underline{7}383 \text{ mol HCl}}{0.064 \text{ kg}} = 15 \, m \text{ HCl}$ and

$64 \text{ g } H_2O \times \dfrac{1 \text{ mol } H_2O}{18.02 \text{ g } H_2O} = 3.\underline{5}5161 \text{ mol } H_2O$ then

$\chi = \dfrac{\text{amount solute (in moles)}}{\text{total amount of solute and solvent (in moles)}} = \dfrac{0.98\underline{7}383 \text{ mol}}{0.98\underline{7}383 \text{ mol} + 3.\underline{5}5161 \text{ mol}} = 0.22$

Check: The units (m and unitless) are correct. The magnitude of the answers (15 and 0.2) seems reasonable because we are starting with a high concentration solution and the molar mass of water is much less than that of HCl.

Vapor Pressure of Solutions

13.65 The level has decreased more in the beaker filled with pure water. The dissolved salt in the seawater decreases the vapor pressure and subsequently lowers the rate of vaporization.

13.67 **Given:** 24.5 g of glycerin ($C_3H_8O_3$) in 135 mL water at 30 °C; $P^{\circ}_{H_2O} = 31.8$ torr **Find:** P_{H_2O}
Other: $d(H_2O) = 1.00$ g/mL; glycerin is not an ionic solid
Conceptual Plan: $g_{C_3H_8O_3} \rightarrow mol_{C_3H_8O_3}$ and mL $H_2O \rightarrow g_{H_2O} \rightarrow mol_{H_2O}$ then $mol_{C_3H_8O_3}, mol_{H_2O} \rightarrow \chi_{H_2O}$

$$\frac{1 \text{ mol } C_3H_8O_3}{92.09 \text{ g } C_3H_8O_3} \qquad \frac{1.00 \text{ g}}{1 \text{ mL}} \quad \frac{1 \text{ mol } H_2O}{18.02 \text{ g } H_2O} \qquad \chi = \frac{\text{amount solvent (in moles)}}{\text{total amount of solute and solvent (in moles)}}$$

then $\chi_{H_2O}, P^{\circ}_{H_2O} \rightarrow P_{H_2O}$

$$P_{solution} = \chi_{solvent} P^{\circ}_{solvent}$$

Solution: $24.5 \text{ g C}_3\text{H}_8\text{O}_3 \times \dfrac{1 \text{ mol C}_3\text{H}_8\text{O}_3}{92.09 \text{ g C}_3\text{H}_8\text{O}_3} = 0.2660441 \text{ mol C}_3\text{H}_8\text{O}_3$ and

$135 \text{ mL} \times \dfrac{1.00 \text{ g}}{1 \text{ mL}} \times \dfrac{1 \text{ mol H}_2\text{O}}{18.02 \text{ g H}_2\text{O}} = 7.491676 \text{ mol H}_2\text{O}$ then

$\chi = \dfrac{\text{amount solvent (in moles)}}{\text{total amount of solute and solvent (in moles)}} = \dfrac{7.491676 \text{ mol}}{0.2660441 \text{ mol} + 7.491676 \text{ mol}} = 0.9657059$ then

$P_{solution} = \chi_{solvent} P^{\circ}_{solvent} = 0.9657059 \times 31.8 \text{ torr} = 30.7 \text{ torr}$

Check: The units (torr) are correct. The magnitude of the answer (31 torr) seems reasonable because it is a drop from the pure vapor pressure. Very few moles of glycerin are added, so the pressure will not drop much.

13.69   **Given:** 50.0 g of heptane (C_7H_{16}) and 50.0 g of octane (C_8H_{18}) at 25 °C; $P^{\circ}_{C_7H_{16}} = 45.8$ torr; $P^{\circ}_{C_8H_{18}} = 10.9$ torr

(a)   **Find:** $P_{C_7H_{16}}, P_{C_8H_{18}}$

Conceptual Plan: $\text{g}_{C_7H_{16}} \rightarrow \text{mol}_{C_7H_{16}}$ and $\text{g}_{C_8H_{18}} \rightarrow \text{mol}_{C_8H_{18}}$ then $\text{mol}_{C_7H_{16}}$,

$$\dfrac{1 \text{ mol C}_7\text{H}_{16}}{100.20 \text{ g C}_7\text{H}_{16}} \quad \dfrac{1 \text{ mol C}_8\text{H}_{18}}{114.22 \text{ g C}_8\text{H}_{18}} \qquad \chi_{C_7H_{16}} = \dfrac{\text{amount C}_7\text{H}_{16} \text{ (in moles)}}{\text{total amount (in moles)}}$$

$\text{mol}_{C_8H_{18}} \rightarrow \chi_{C_7H_{16}}, \chi_{C_8H_{18}}$ then $\chi_{C_7H_{16}}, P^{\circ}_{C_7H_{16}} \rightarrow P_{C_7H_{16}}$ and $\chi_{C_8H_{18}}, P^{\circ}_{C_8H_{18}} \rightarrow P_{C_8H_{18}}$

$$\chi_{C_8H_{18}} = 1 - \chi_{C_7H_{16}} \qquad\qquad P_{C_7H_{16}} = \chi_{C_7H_{16}} P^{\circ}_{C_7H_{16}} \qquad\qquad P_{C_8H_{18}} = \chi_{C_8H_{18}} P^{\circ}_{C_8H_{18}}$$

Solution: $50.0 \text{ g C}_7\text{H}_{16} \times \dfrac{1 \text{ mol C}_7\text{H}_{16}}{100.20 \text{ g C}_7\text{H}_{16}} = 0.499002 \text{ mol C}_7\text{H}_{16}$ and

$50.0 \text{ g C}_8\text{H}_{18} \times \dfrac{1 \text{ mol C}_8\text{H}_{18}}{114.22 \text{ g C}_8\text{H}_{18}} = 0.437752 \text{ mol C}_8\text{H}_{18}$ then

$\chi_{C_7H_{16}} = \dfrac{\text{amount C}_7\text{H}_{16} \text{ (in moles)}}{\text{total amount (in moles)}} = \dfrac{0.499002 \text{ mol}}{0.499002 \text{ mol} + 0.437752 \text{ mol}} = 0.532693$ and

$\chi_{C_8H_{18}} = 1 - \chi_{C_7H_{16}} = 1 - 0.532693 = 0.467307$ then

$P_{C_7H_{16}} = \chi_{C_7H_{16}} P^{\circ}_{C_7H_{16}} = 0.532693 \times 45.8 \text{ torr} = 24.4 \text{ torr}$ and

$P_{C_8H_{18}} = \chi_{C_8H_{18}} P^{\circ}_{C_8H_{18}} = 0.467307 \times 10.9 \text{ torr} = 5.09 \text{ torr}$

Check: The units (torr) are correct. The magnitude of the answers (24 and 5 torr) seems reasonable because we expect a drop in half from the pure vapor pressures because we have roughly a 50:50 mole ratio of the two components.

(b)   **Find:** P_{Total}

Conceptual Plan: $P_{C_7H_{16}}, P_{C_8H_{18}} \rightarrow P_{Total}$

$$P_{Total} = P_{C_7H_{16}} + P_{C_8H_{18}}$$

Solution: $P_{Total} = P_{C_7H_{16}} + P_{C_8H_{18}} = 24.4 \text{ torr} + 5.09 \text{ torr} = 29.5 \text{ torr}$

Check: The units (torr) are correct. The magnitude of the answer (30 torr) seems reasonable considering the two pressures.

(c)   **Find:** mass percent composition of the gas phase

Conceptual Plan: because $n \propto P$ and we are calculating a mass percent, which is a ratio of masses, we can simply convert 1 torr to 1 mole so

$P_{C_7H_{16}}, P_{C_8H_{18}} \rightarrow n_{C_7H_{16}}, n_{C_8H_{18}}$ then $\text{mol}_{C_7H_{16}} \rightarrow \text{g}_{C_7H_{16}}$ and $\text{mol}_{C_8H_{18}} \rightarrow \text{g}_{C_8H_{18}}$

$$\dfrac{100.20 \text{ g C}_7\text{H}_{16}}{1 \text{ mol C}_7\text{H}_{16}} \qquad\qquad \dfrac{114.22 \text{ g C}_8\text{H}_{18}}{1 \text{ mol C}_8\text{H}_{18}}$$

then $\text{g}_{C_7H_{16}}, \text{g}_{C_8H_{18}} \rightarrow$ **mass percents**

$$\text{mass percent} = \dfrac{\text{mass solute}}{\text{mass solution}} \times 100\%$$

Solution: so $n_{C_7H_{16}} = 24.4 \text{ mol}$ and $n_{C_8H_{18}} = 5.09 \text{ mol}$ then

$24.4 \text{ mol C}_7\text{H}_{16} \times \dfrac{100.20 \text{ g C}_7\text{H}_{16}}{1 \text{ mol C}_7\text{H}_{16}} = 2444.88 \text{ g C}_7\text{H}_{16}$ and

$5.09 \text{ mol C}_8\text{H}_{18} \times \dfrac{114.22 \text{ g C}_8\text{H}_{18}}{1 \text{ mol C}_8\text{H}_{18}} = 581.380 \text{ g C}_8\text{H}_{18}$ then

$$\text{mass percent} = \frac{\text{mass solute}}{\text{mass solution}} \times 100\% = \frac{2444.88 \text{ g } C_7H_{16}}{2444.88 \text{ g } C_7H_{16} + 581.380 \text{ g } C_8H_{18}} \times 100\% =$$

80.8% by mass C_7H_{16}

then $100\% - 80.8\% = 19.2\%$ by mass C_8H_{18}

Check: The units (mol %) are correct. The magnitude of the answers (81% and 19%) seems reasonable considering the two pressures.

(d)   The two mass percents are different because the vapor is richer in the more volatile component (the lighter molecule).

13.71   **Given:** 4.08 g of chloroform $(CHCl_3)$ and 9.29 g of acetone (CH_3COCH_3); at 35 °C, $P^\circ_{CHCl_3} = 295$ torr; $P^\circ_{CH_3COCH_3} = 332$ torr; assume ideal behavior; $P_{Total\ measured} = 312$ torr
Find: P_{CHCl_3}, $P_{CH_3COCH_3}$, P_{Total}, and if the solution is ideal.
Conceptual Plan: $g_{CHCl_3} \rightarrow mol_{CHCl_3}$ and $g_{CH_3COCH_3} \rightarrow mol_{CH_3COCH_3}$ then

$$\frac{1 \text{ mol } CHCl_3}{119.37 \text{ g } CHCl_3} \qquad \frac{1 \text{ mol } CH_3COCH_3}{58.08 \text{ g } CH_3COCH_3}$$

mol_{CHCl_3}, $mol_{CH_3COCH_3} \rightarrow \chi_{CHCl_3}$, $\chi_{CH_3COCH_3}$ then χ_{CHCl_3}, $P^\circ_{CHCl_3} \rightarrow P_{CHCl_3}$ and

$$\chi_{CHCl_3} = \frac{\text{amount } CHCl_3 \text{ (in moles)}}{\text{total amount (in moles)}} \qquad \chi_{CH_3COCH_3} = 1 - \chi_{CHCl_3} \quad P_{CHCl_3} = \chi_{CHCl_3} P^\circ_{CHCl_3}$$

$\chi_{CH_3COCH_3}$, $P^\circ_{CH_3COCH_3} \rightarrow P_{CH_3COCH_3}$ then P_{CHCl_3}, $P_{CH_3COCH_3} \rightarrow P_{Total}$ then compare values

$$P_{CH_3COCH_3} = \chi_{CH_3COCH_3} P^\circ_{CH_3COCH_3} \qquad\qquad P_{Total} = P_{CHCl_3} + P_{CH_3COCH_{33}}$$

Solution:

$$4.08 \text{ g } CHCl_3 \times \frac{1 \text{ mol } CHCl_3}{119.37 \text{ g } CHCl_3} = 0.0341794 \text{ mol } CHCl_3 \text{ and}$$

$$9.29 \text{ g } CH_3COCH_3 \times \frac{1 \text{ mol } CH_3COCH_3}{58.08 \text{ g } CH_3COCH_3} = 0.159952 \text{ mol } CH_3COCH_3 \text{ then}$$

$$\chi_{CHCl_3} = \frac{\text{amount } CHCl_3 \text{ (in moles)}}{\text{total amount (in moles)}} = \frac{0.0341766 \text{ mol}}{0.0341794 \text{ mol} + 0.159952 \text{ mol}} = 0.176051 \text{ and}$$

$\chi_{CH_3COCH_3} = 1 - \chi_{CHCl_3} = 1 - 0.176051 = 0.823949$ then
$P_{CHCl_3} = \chi_{CHCl_3} P^\circ_{CHCl_3} = 0.176051 \times 295 \text{ torr} = 51.9 \text{ torr}$ and
$P_{CH_3COCH_3} = \chi_{CH_3COCH_3} P^\circ_{CH_3COCH_3} = 0.823949 \times 332 \text{ torr} = 274 \text{ torr}$ then
$P_{Total} = P_{CHCl_3} + P_{CH_3COCH_3} = 51.9 \text{ torr} + 274 \text{ torr} = 326 \text{ torr}$
Because 326 torr $\neq$ 312 torr, the solution is not behaving ideally. The chloroform–acetone interactions are stronger than the chloroform–chloroform and acetone–acetone interactions.

Check: The units (torr) are correct. The magnitude of the answer (326) seems reasonable because each is a fraction of the pure vapor pressure. We are not surprised that the solution is not ideal because the types of bonds in the two molecules are very different.

Freezing Point Depression, Boiling Point Elevation, and Osmosis

13.73   **Given:** 55.8 g of glucose $(C_6H_{12}O_6)$ in 455 g water   **Find:** T_f and T_b **Other:** $K_f = 1.86 \text{ °C}/m$; $K_b = 0.512 \text{ °C}/m$;
Conceptual Plan: $g_{H_2O} \rightarrow kg_{H_2O}$ and $g_{C_6H_{12}O_6} \rightarrow mol_{C_6H_{12}O_6}$ then $mol_{C_6H_{12}O_6}$, $kg_{H_2O} \rightarrow m$

$$\frac{1 \text{ kg}}{1000 \text{ g}} \qquad \frac{1 \text{ mol } C_6H_{12}O_6}{180.16 \text{ g } C_6H_{12}O_6} \qquad m = \frac{\text{amount solute (moles)}}{\text{mass solvent (kg)}}$$

$m, K_f \rightarrow \Delta T_f \rightarrow T_f$ and $m, K_b \rightarrow \Delta T_b \rightarrow T_b$
$\Delta T_f = K_f \times m \quad T_f = T_f^\circ - \Delta T_f \quad \Delta T_b = K_b \times m \quad \Delta T_b = T_b - T_b^\circ$

Solution: $455 \text{ g} \times \dfrac{1 \text{ kg}}{1000 \text{ g}} = 0.455 \text{ kg}$ and $55.8 \text{ g } C_6H_{12}O_6 \times \dfrac{1 \text{ mol } C_6H_{12}O_6}{180.16 \text{ g } C_6H_{12}O_6} = 0.309725 \text{ mol } C_6H_{12}O_6$ then

$$m = \frac{\text{amount solute (moles)}}{\text{mass solvent (kg)}} = \frac{0.309725 \text{ mol } C_6H_{12}O_6}{0.455 \text{ kg}} = 0.680714 \ m \text{ then}$$

$\Delta T_f = K_f \times m = 1.86 \dfrac{\text{°C}}{m} \times 0.680714 \ m = 1.27 \text{ °C}$ then $T_f = T_f^\circ - \Delta T_f = 0.00 \text{ °C} - 1.27 \text{ °C} = -1.27 \text{ °C}$ and

$\Delta T_b = K_b \times m = 0.512 \dfrac{\text{°C}}{m} \times 0.680714 \ m = 0.349 \text{ °C}$ and $\Delta T_b = T_b - T_b^\circ$ so

$T_b = T_b^\circ + \Delta T_b = 100.000\,^\circ\text{C} + 0.349\,^\circ\text{C} = 100.349\,^\circ\text{C}$

Check: The units ($^\circ$C) are correct. The magnitude of the answers (100.3) seems reasonable because the molality is $\sim 2/3$. The shift in boiling point is less than the shift in freezing point because the constant for boiling is smaller than the constant for freezing.

13.75   **Given:** 10.0 g of naphthalene ($C_{10}H_8$) in 100.0 mL benzene (C_6H_6)   **Find:** T_f and T_b
           Other: d(benzene) $= 0.877\ \text{g/cm}^3$; $K_f = 5.12\,^\circ\text{C}/m$; $K_b = 2.53\,^\circ\text{C}/m$; $T_f^\circ = 5.5\,^\circ\text{C}$; $T_b^\circ = 80.1\,^\circ\text{C}$
           Conceptual Plan: $\text{mL}_{C_6H_6} \rightarrow \text{g}_{C_6H_6} \rightarrow \text{kg}_{C_6H_6}$ and $\text{g}_{C_{10}H_8} \rightarrow \text{mol}_{C_{10}H_8}$ then $\text{mol}_{C_{10}H_8}, \text{kg}_{C_6H_6} \rightarrow m$

$$\frac{0.877\ \text{g}}{1\ \text{cm}^3} \qquad \frac{1\ \text{kg}}{1000\ \text{g}} \qquad\qquad \frac{1\ \text{mol}\ C_{10}H_8}{128.16\ \text{g}\ C_{10}H_8} \qquad\qquad m = \frac{\text{amount solute (moles)}}{\text{mass solvent (kg)}}$$

           $m, K_f \rightarrow \Delta T_f \rightarrow T_f$ and $m, K_b \rightarrow \Delta T_b \rightarrow T_b$

$$\Delta T_f = K_f \times m \quad T_f = T_f^\circ - \Delta T_f \quad \Delta T_b = K_b \times m \quad \Delta T_b = T_b - T_b^\circ$$

           Solution: $100.0\ \text{cm}^3 \times \dfrac{0.877\ \text{g}}{1\ \text{cm}^3} \times \dfrac{1\ \text{kg}}{1000\ \text{g}} = 0.0877\ \text{kg}$ and

$$10.0\ \text{g}\ C_{10}H_8 \times \frac{1\ \text{mol}\ C_{10}H_8}{128.16\ \text{g}\ C_{10}H_8} = 0.07\underline{8}02747\ \text{mol}\ C_{10}H_8 \text{ then}$$

$$m = \frac{\text{amount solute (moles)}}{\text{mass solvent (kg)}} = \frac{0.07\underline{8}02747\ \text{mol}\ C_{10}H_8}{0.0877\ \text{kg}} = 0.88\underline{9}7089\ m \text{ then}$$

$$\Delta T_f = K_f \times m = 5.12\frac{^\circ\text{C}}{m} \times 0.88\underline{9}7089\ m = 4.56\,^\circ\text{C} \text{ then } T_f = T_f^\circ - \Delta T_f = 5.5\,^\circ\text{C} - 4.56\,^\circ\text{C} = 0.\underline{9}4\,^\circ\text{C} = 0.9\,^\circ\text{C}$$

           and $\Delta T_b = K_b \times m = 2.53\dfrac{^\circ\text{C}}{m} \times 0.88\underline{9}7089\ m = 2.25\,^\circ\text{C}$ and $\Delta T_b = T_b - T_b^\circ$ so

$$T_b = T_b^\circ + \Delta T_b = 80.1\,^\circ\text{C} + 2.25\,^\circ\text{C} = 82.\underline{3}5\,^\circ\text{C} = 82.4\,^\circ\text{C}$$

           Check: The units ($^\circ$C) are correct. The magnitude of the answers (82.4) seems reasonable because the molality is almost 1. Because the constants are larger for benzene than for water, we expect larger temperature shifts. The shift in boiling point is less than the shift in freezing point because the constant is smaller for freezing.

13.77   **Given:** 17.5 g of unknown nonelectrolyte in 100.0 g water, $T_f = -1.8\,^\circ\text{C}$ **Find:** $\mathcal{M}$
           Other: $K_f = 1.86\,^\circ\text{C}/m$
           Conceptual Plan: $\text{g}_{H_2O} \rightarrow \text{kg}_{H_2O}$ and $T_f \rightarrow \Delta T_f$ then $\Delta T_f, K_f \rightarrow m$ then $m, \text{kg}_{H_2O} \rightarrow \text{mol}_{\text{Unk}}$

$$\frac{1\ \text{kg}}{1000\ \text{g}} \qquad\qquad T_f = T_f^\circ - \Delta T_f \qquad\qquad \Delta T_f = K_f \times m \qquad\qquad m = \frac{\text{amount solute (moles)}}{\text{mass solvent (kg)}}$$

           then $\text{g}_{\text{Unk}}, \text{mol}_{\text{Unk}} \rightarrow \mathcal{M}$

$$\mathcal{M} = \frac{\text{g}_{\text{Unk}}}{\text{mol}_{\text{Unk}}}$$

           Solution: $100.0\ \text{g} \times \dfrac{1\ \text{kg}}{1000\ \text{g}} = 0.1000\ \text{kg}$ and $T_f = T_f^\circ - \Delta T_f$ so

$$\Delta T_f = T_f^\circ - T_f = 0.00\,^\circ\text{C} - (-1.8\,^\circ\text{C}) = +1.8\,^\circ\text{C} \quad \Delta T_f = K_f \times m \quad \text{Rearrange to solve for } m.$$

$$m = \frac{\Delta T_f}{K_f} = \frac{1.8\,^\circ\text{C}}{1.86\dfrac{^\circ\text{C}}{m}} = 0.96\underline{7}74\ m \text{ then } m = \frac{\text{amount solute (moles)}}{\text{mass solvent (kg)}} \text{ so}$$

$$\text{mol}_{\text{Unk}} = m_{\text{Unk}} \times \text{kg}_{H_2O} = 0.96\underline{7}74\ \frac{\text{mol}_{\text{Unk}}}{\text{kg}} \times 0.1000\ \text{kg} = 0.096\underline{7}74\ \text{mol}_{\text{Unk}} \text{ then}$$

$$\mathcal{M} = \frac{\text{g}_{\text{Unk}}}{\text{mol}_{\text{Unk}}} = \frac{17.5\ \text{g}}{0.096\underline{7}74\ \text{mol}} = 180\frac{\text{g}}{\text{mol}} = 1.8 \times 10^2\frac{\text{g}}{\text{mol}}$$

           Check: The units (g/mol) are correct. The magnitude of the answer (180 g/mol) seems reasonable because the molality is ~ 0.1 and we have ~ 18 g. It is a reasonable molecular weight for a solid or liquid.

13.79   **Given:** 24.6 g of glycerin ($C_3H_8O_3$) in 250.0 mL of solution at 298 K   **Find:** Π
           Conceptual Plan: $\text{mL}_{\text{soln}} \rightarrow L_{\text{soln}}$ and $\text{g}_{C_3H_8O_3} \rightarrow \text{mol}_{C_3H_8O_3}$ then $\text{mol}_{C_3H_8O_3}, L_{\text{soln}} \rightarrow \text{M}$ then

$$\frac{1\ \text{L}}{1000\ \text{mL}} \qquad\qquad \frac{1\ \text{mol}\ C_3H_8O_3}{92.09\ \text{g}\ C_3H_8O_3} \qquad\qquad \text{M} = \frac{\text{amount solute (moles)}}{\text{volume solution (L)}}$$

M, T → Π

$\Pi = MRT$

Solution:

$$250.0 \text{ mL} \times \frac{1 \text{ L}}{1000 \text{ mL}} = 0.2500 \text{ L} \text{ and } 24.6 \text{ g } C_3H_8O_3 \times \frac{1 \text{ mol } C_3H_8O_3}{92.09 \text{ g } C_3H_8O_3} = 0.267130 \text{ mol } C_3H_8O_3 \text{ then}$$

$$M = \frac{\text{amount solute (moles)}}{\text{volume solution (L)}} = \frac{0.267130 \text{ mol } C_3H_8O_3}{0.2500 \text{ L}} = 1.06852 \text{ M then}$$

$$\Pi = MRT = 1.06852 \frac{\text{mol}}{\text{L}} \times 0.08206 \frac{\text{L} \cdot \text{atm}}{\text{K} \cdot \text{mol}} \times 298 \text{ K} = 26.1 \text{ atm}$$

Check: The units (atm) are correct. The magnitude of the answer (26 atm) seems reasonable because the molarity is ~1.

13.81   **Given:** 27.55 mg unknown protein in 25.0 mL solution; $\Pi = 3.22$ torr at 25 °C   **Find:** $\mathcal{M}_{\text{unknown protein}}$

Conceptual Plan: °C → K and torr → atm then Π, T → M then mL$_{\text{soln}}$ → L$_{\text{soln}}$ then

$$K = °C + 273.15 \qquad \frac{1 \text{ atm}}{760 \text{ torr}} \qquad \Pi = MRT \qquad \frac{1 \text{ L}}{1000 \text{ mL}}$$

L$_{\text{soln}}$, M → mol$_{\text{unknown protein}}$ and mg → g then g$_{\text{unknown protein}}$, mol$_{\text{unknown protein}}$ → $\mathcal{M}_{\text{unknown protein}}$

$$M = \frac{\text{amount solute (moles)}}{\text{volume solution (L)}} \qquad \frac{1 \text{ g}}{1000 \text{ mg}} \qquad \mathcal{M} = \frac{g_{\text{unknown protein}}}{\text{mol}_{\text{unknown protein}}}$$

Solution: $25 °C + 273.15 = 298$ K and $3.22 \text{ torr} \times \frac{1 \text{ atm}}{760 \text{ torr}} = 0.00423684 \text{ atm}$ then solve $\Pi = MRT$ for M.

$$M = \frac{\Pi}{RT} = \frac{0.00423684 \text{ atm}}{0.08206 \frac{\text{L} \cdot \text{atm}}{\text{K} \cdot \text{mol}} \times 298 \text{ K}} = 1.73258 \times 10^{-4} \frac{\text{mol}}{\text{L}} \text{ then } 25.0 \text{ mL} \times \frac{1 \text{ L}}{1000 \text{ mL}} = 0.0250 \text{ L then}$$

$$M = \frac{\text{amount solute (moles)}}{\text{volume solution (L)}} \qquad \text{Rearrange to solve for mol}_{\text{unknown protein}}.$$

$$\text{mol}_{\text{unknown protein}} = M \times L = 1.73258 \times 10^{-4} \frac{\text{mol}}{\text{L}} \times 0.0250 \text{ L} = 4.33146 \times 10^{-6} \text{ mol and}$$

$$27.55 \text{ mg} \times \frac{1 \text{ g}}{1000 \text{ mg}} = 0.02755 \text{ g then } \mathcal{M} = \frac{g_{\text{unknown protein}}}{\text{mol}_{\text{unknown protein}}} = \frac{0.02755 \text{ g}}{4.33146 \times 10^{-6} \text{ mol}} = 6.36 \times 10^3 \frac{\text{g}}{\text{mol}}$$

Check: The units (g/mol) are correct. The magnitude of the answer (6400 g/mol) seems reasonable for a large biological molecule. A small amount of material is put into 0.025 L; so the concentration is very small, and the molecular weight is large.

13.83   (a)   **Given:** 0.100 m of K_2S, completely dissociated   **Find:** T_f, T_b

Other: $K_f = 1.86 °C/m$; $K_b = 0.512 °C/m$

Conceptual Plan: $m, i, K_f \rightarrow \Delta T_f$ then $\Delta T_f \rightarrow T_f$ and $m, i, K_b \rightarrow \Delta T_b$ then $\Delta T_b \rightarrow T_b$

$$\Delta T_f = K_f \times i \times m_{i=3} \qquad T_f = T_f° - \Delta T_f \qquad \Delta T_b = K_b \times i \times m_{i=3} \qquad T_b = T_b° + \Delta T_b$$

Solution: $\Delta T_f = K_f \times i \times m = 1.86 \frac{°C}{m} \times 3 \times 0.100 \, m = 0.558 °C$ then

$$T_f = T_f° - \Delta T_f = 0.000 °C - 0.558 °C = -0.558 °C \text{ and}$$

$$\Delta T_b = K_b \times i \times m = 0.512 \frac{°C}{m} \times 3 \times 0.100 \, m = 0.154 °C \text{ then}$$

$$T_b = T_b° + \Delta T_b = 100.000 °C + 0.154 °C = 100.154 °C$$

Check: The units (°C) are correct. The magnitude of the answers (-0.6 °C and 100.2 °C) seems reasonable because the molality of the particles is 0.3. The shift in boiling point is less than the shift in freezing point because the constant for boiling is larger than the constant for freezing.

(b)   **Given:** 21.5 g $CuCl_2$ in 4.50×10^2 g water, completely dissociated   **Find:** T_f, T_b

Other: $K_f = 1.86 °C/m$; $K_b = 0.512 °C/m$

Conceptual Plan: $g_{H_2O} \rightarrow kg_{H_2O}$ and $g_{CuCl_2} \rightarrow mol_{CuCl_2}$ then $mol_{CuCl_2}, kg_{H_2O} \rightarrow m$

$$\frac{1 \text{ kg}}{1000 \text{ g}} \qquad \frac{1 \text{ mol } CuCl_2}{134.45 \text{ g } CuCl_2} \qquad m = \frac{\text{amount solute (moles)}}{\text{mass solvent (kg)}}$$

$m, i, K_f \rightarrow \Delta T_f \rightarrow T_f$ and $m, i, K_b \rightarrow \Delta T_b \rightarrow T_b$

$\Delta T_f = K_f \times i \times m_{i=3}$ $T_f = T_f^\circ - \Delta T_f$ $\Delta T_b = K_b \times i \times m_{i=3}$ $T_b = T_b^\circ + \Delta T_b$

Solution:

$$4.5 \times 10^2 \, \cancel{g} \times \frac{1 \, kg}{1000 \, \cancel{g}} = 0.450 \, kg \text{ and } 21.5 \, \cancel{g \, CuCl_2} \times \frac{1 \, mol \, CuCl_2}{134.45 \, \cancel{g \, CuCl_2}} = 0.15\underline{9}911 \, mol \, CuCl_2 \text{ then}$$

$$m = \frac{\text{amount solute (moles)}}{\text{mass solvent (kg)}} = \frac{0.15\underline{9}911 \, mol \, CuCl_2}{0.450 \, kg} = 0.35\underline{5}358 \, m \text{ then}$$

$$\Delta T_f = K_f \times i \times m = 1.86\frac{^\circ C}{\cancel{m}} \times 3 \times 0.35\underline{5}358 \, \cancel{m} = 1.98 \, ^\circ C \text{ then}$$

$$T_f = T_f^\circ - \Delta T_f = 0.000 \, ^\circ C - 1.98 \, ^\circ C = -1.98 \, ^\circ C \text{ and}$$

$$\Delta T_b = K_b \times i \times m = 0.512\frac{^\circ C}{\cancel{m}} \times 3 \times 0.35\underline{5}358 \, \cancel{m} = 0.546 \, ^\circ C \text{ then}$$

$$T_b = T_b^\circ + \Delta T_b = 100.000 \, ^\circ C + 0.546 \, ^\circ C = 100.546 \, ^\circ C$$

Check: The units $(^\circ C)$ are correct. The magnitude of the answers $(-2 \, ^\circ C$ and $100.5 \, ^\circ C)$ seems reasonable because the molality of the particles is ~ 1. The shift in boiling point is less than the shift in freezing point because the constant for boiling is larger than the constant for freezing.

(c)   **Given:** 5.5% by mass $NaNO_3$, completely dissociated   **Find:** T_f, T_b
Other: $K_f = 1.86 \, ^\circ C/m$; $K_b = 0.512 \, ^\circ C/m$
Conceptual Plan: Percent by mass $\rightarrow g_{NaNO_3}, g_{H_2O}$ **then** $g_{H_2O} \rightarrow kg_{H_2O}$ **and** $g_{NaNO_3} \rightarrow mol_{NaNO_3}$

$$\text{mass percent} = \frac{\text{mass solute}}{\text{mass solution}} \times 100\% \qquad \frac{1 \, kg}{1000 \, g} \qquad \frac{1 \, mol \, NaNO_3}{85.00 \, g \, NaNO_3}$$

then $mol_{NaNO_3}, kg_{H_2O} \rightarrow m$ **then** $m, i, K_f \rightarrow \Delta T_f \rightarrow T_f$ **and** $m, i, K_b \rightarrow \Delta T_b \rightarrow T_b$

$$m = \frac{\text{amount solute (moles)}}{\text{mass solvent (kg)}} \qquad \Delta T_f = K_f \times i \times m_{i=2} \quad T_f = T_f^\circ - \Delta T_f \qquad \Delta T_b = K_b \times i \times m_{i=2} \quad T_b = T_b^\circ + \Delta T_b$$

Solution: mass percent $= \dfrac{\text{mass solute}}{\text{mass solution}} \times 100\%$, so 5.5% by mass $NaNO_3$ means 5.5 g $NaNO_3$ and

$$100.0 \, g - 5.5 \, g = 94.5 \, g \text{ water. Then } 94.5 \, \cancel{g} \times \frac{1 \, kg}{1000 \, \cancel{g}} = 0.0945 \, kg \text{ and}$$

$$5.5 \, \cancel{g \, NaNO_3} \times \frac{1 \, mol \, NaNO_3}{85.00 \, \cancel{g \, NaNO_3}} = 0.06\underline{4}706 \, mol \, NaNO_3 \text{ then}$$

$$m = \frac{\text{amount solute (moles)}}{\text{mass solvent (kg)}} = \frac{0.06\underline{4}706 \, mol \, NaNO_3}{0.0945 \, kg} = 0.6\underline{8}472 \, m \text{ then}$$

$$\Delta T_f = K_f \times i \times m = 1.86\frac{^\circ C}{\cancel{m}} \times 2 \times 0.6\underline{8}472 \, \cancel{m} = 2.5 \, ^\circ C \text{ then}$$

$$T_f = T_f^\circ - \Delta T_f = 0.000 \, ^\circ C - 2.5 \, ^\circ C = -2.5 \, ^\circ C \text{ and}$$

$$\Delta T_b = K_b \times i \times m = 0.512\frac{^\circ C}{\cancel{m}} \times 2 \times 0.6\underline{8}472 \, \cancel{m} = 0.70 \, ^\circ C \text{ then}$$

$$T_b = T_b^\circ + \Delta T_b = 100.000 \, ^\circ C + 0.70 \, ^\circ C = 100.70 \, ^\circ C$$

Check: The units $(^\circ C)$ are correct. The magnitude of the answers $(-2.5 \, ^\circ C$ and $100.7 \, ^\circ C)$ seems reasonable because the molality of the particles is ~ 1. The shift in boiling point is less than the shift in freezing point because the constant for boiling is larger than the constant for freezing.

13.85   **Given:** NaCl complete dissociation; 1.0 L water and $T_f = -10.0 \, ^\circ C$ **Find:** mass of NaCl
Other: $K_f = 1.86 \, ^\circ C/m$; $d(\text{water}) = 1.0 \, g/mL$
Conceptual Plan:
$T_f \rightarrow \Delta T_f$ **then** $\Delta T_f, i, K_f \rightarrow m$ **then** $L_{H_2O} \rightarrow mL_{H_2O} \rightarrow g_{H_2O} \rightarrow kg_{H_2O}$ **then** $m, kg_{H_2O} \rightarrow mol_{NaCl}$

$$T_f = T_f^\circ - \Delta T_f \qquad \Delta T_f = K_f \times i \times m \qquad \frac{1000 \, mL}{1 \, L} \quad \frac{1.0 \, g}{1 \, mL} \quad \frac{1 \, kg}{1000 \, g} \qquad m = \frac{\text{amount solute (moles)}}{\text{mass solvent (kg)}}$$

then $mol_{NaCl} \rightarrow g_{NaCl}$

$$\frac{58.44 \, g \, NaCl}{1 \, mol \, NaCl}$$

Solution: $T_f = T_f^\circ - \Delta T_f$ so $\Delta T_f = T_f^\circ - T_f = 0.0 \, ^\circ C - 10.0 \, ^\circ C = -10.0 \, ^\circ C$ then $\Delta T_f = K_f \times i \times m$

Rearrange to solve for m. Because the salt completely dissolves, $i = 2$.

$$m = \frac{\Delta T_f}{K_f \times i} = \frac{10.0\,°C}{1.86\frac{°C}{m} \times 2} = 2.6\underline{8}8172\,m\ \text{NaCl then}\ 1.0\,L \times \frac{1000\,mL}{1\,L} \times \frac{1.0\,g}{1\,mL} \times \frac{1\,kg}{1000\,g} = 1.0\,kg\ \text{then}$$

$$m = \frac{\text{amount solute (moles)}}{\text{mass solvent (kg)}}\ \text{so}\ \text{mol}_{NaCl} = m \times kg_{H_2O} = 2.6\underline{8}8172\frac{\text{mol NaCl}}{kg_{H_2O}} \times 1.0\,kg_{H_2O} = 2.6\underline{8}8172\ \text{mol NaCl}$$

then $2.6\underline{8}8172\ \text{mol NaCl} \times \frac{58.44\ \text{g NaCl}}{1\ \text{mol NaCl}} = 15\underline{7}.097\ \text{g NaCl} = 160\ \text{g NaCl}$

Check: The units (g) are correct. The magnitude of the answer (160 g) seems reasonable because the temperature change is moderate and NaCl has a low formula mass.

13.87   (a)   **Given:** 0.100 m of $FeCl_3$ **Find:** T_f **Other:** $K_f = 1.86\,°C/m$; $i_{measured} = 3.4$
          Conceptual Plan: $m, i, K_f \rightarrow \Delta T_f$ then $\Delta T_f \rightarrow T_f$

$$\Delta T_f = K_f \times i \times m \quad T_f = T_f° - \Delta T_f$$

          Solution: $\Delta T_f = K_f \times i \times m = 1.86\frac{°C}{m} \times 3.4 \times 0.100\,m = 0.632\,°C$ then

          $T_f = T_f° - \Delta T_f = 0.000\,°C - 0.632\,°C = -0.632\,°C$

          Check: The units (°C) are correct. The magnitude of the answer ($-0.6\,°C$) seems reasonable because the theoretical molality of the particles is 0.4.

  (b)   **Given:** 0.085 M of K_2SO_4 at 298 K   **Find:** Π **Other:** $i_{measured} = 2.6$
          Conceptual Plan: $M, i, T \rightarrow \Pi$

$$\Pi = i \times MRT$$

          Solution: $\Pi = i \times MRT = 2.6 \times 0.085\frac{\text{mol}}{L} \times 0.08206\frac{L \cdot atm}{K \cdot mol} \times 298\,K = 5.4\ atm$

          Check: The units (atm) are correct. The magnitude of the answer (5 atm) seems reasonable because the molarity of particles is $\sim 0.2\,m$.

  (c)   **Given:** 1.22% by mass $MgCl_2$ **Find:** T_b **Other:** $K_b = 0.512\,°C/m$; $i_{measured} = 2.7$
          Conceptual Plan: percent by mass $\rightarrow g_{MgCl_2}, g_{H_2O}$ **then** $g_{H_2O} \rightarrow kg_{H_2O}$ **and** $g_{MgCl_2} \rightarrow mol_{MgCl_2}$ **then**

$$\text{mass percent} = \frac{\text{mass solute}}{\text{mass solution}} \times 100\% \qquad \frac{1\,kg}{1000\,g} \qquad \frac{1\,mol\ MgCl_2}{95.21\,g\ MgCl_2}$$

          $mol_{MgCl_2}, kg_{H_2O} \rightarrow m$ **then** $m, i, K_b \rightarrow \Delta T_b \rightarrow T_b$

$$m = \frac{\text{amount solute (moles)}}{\text{mass solvent (kg)}} \qquad \Delta T_b = K_b \times i \times m \quad T_b = T_b° + \Delta T_b$$

          Solution: mass percent $= \frac{\text{mass solute}}{\text{mass solution}} \times 100\%$ so 1.22% by mass $MgCl_2$ means 1.22 g $MgCl_2$ and

          $100.00\,g - 1.22\,g = 98.78\,g$ water then $98.78\,g \times \frac{1\,kg}{1000\,g} = 0.09878\,kg$ and

          $1.22\,g\ MgCl_2 \times \frac{1\ \text{mol}\ MgCl_2}{95.21\,g\ MgCl_2} = 0.012\underline{8}138\ \text{mol}\ MgCl_2$ then

          $m = \frac{\text{amount solute (moles)}}{\text{mass solvent (kg)}} = \frac{0.012\underline{8}138\ \text{mol}\ MgCl_2}{0.09878\,kg} = 0.12\underline{9}721\,m$ then

          $\Delta T_b = K_b \times i \times m = 0.512\frac{°C}{m} \times 2.7 \times 0.12\underline{9}721\,m = 0.18\,°C$ then

          $T_b = T_b° - \Delta T_b = 100.000\,°C + 0.18\,°C = 100.18\,°C$

          Check: The units (°C) are correct. The magnitude of the answer (100.2 °C) seems reasonable because the molality of the particles is $\sim 1/3$.

13.89   **Given:** 1.2 m MX_2 in water and $T_b = 101.4\,°C$ **Find:** i
       Other: $K_b = 0.512\,°C/m$
       Conceptual Plan: $T_b \rightarrow \Delta T_b$ then $m, \Delta T_b, K_b \rightarrow i$

$$\Delta T_b = T_b - T_b° \qquad \Delta T_b = K_b \times i \times m$$

       Solution: $\Delta T_b = T_b - T_b° = 101.4\,°C - 100.0\,°C = 1.4\,°C$ then $\Delta T_b = K_b \times i \times m$

Rearrange to solve for i. $i = \dfrac{\Delta T_b}{K_b \times m} = \dfrac{1.4\,°C}{0.512\dfrac{°C}{m} \times 1.2\,m} = 2.2\underline{7}865 = 2.3$

Check: The units (none) are correct. The magnitude of the answer (2.3) seems reasonable because the formula of the salt is MX_2, where $i = 3$ if it completely dissociated.

13.91   **Given:** 0.100 M of ionic solution, $\Pi = 8.3$ atm at 25 °C   **Find:** $i_{measured}$
         Conceptual Plan: °C → K then Π, M, $T \rightarrow i$

$$K = °C + 273.15 \qquad \Pi = i \times MRT$$

         Solution: 25 °C + 273.15 = 298 K then $\Pi = i \times MRT$. Rearrange to solve for i.

$$i = \frac{\Pi}{MRT} = \frac{8.3\,\text{atm}}{0.100\dfrac{\text{mol}}{L} \times 0.08206\dfrac{L \cdot \text{atm}}{K \cdot \text{mol}} \times 298\,K} = 3.4$$

         Check: The units (none) are correct. The magnitude of the answer (3) seems reasonable for an ionic solution with a high osmotic pressure.

13.93   **Given:** 5.50 % NaCl by mass in water at 25 °C   **Find:** P_{H_2O} **Other:** $P^\circ_{H_2O} = 23.78$ torr
         Conceptual Plan: % NaCl by mass → g_{NaCl}, g_{H_2O} then $g_{NaCl} \rightarrow mol_{NaCl}$ and

$$\frac{5.50\text{ g NaCl}}{100\text{ g (NaCl} + H_2O)} \qquad \frac{1\text{ mol NaCl}}{58.44\text{ g NaCl}}$$

         $g_{H_2O} \rightarrow mol_{H_2O}$ then mol_{NaCl}, $mol_{H_2O} \rightarrow \chi_{H_2O}$ then χ_{H_2O}, $P^\circ_{H_2O} \rightarrow P_{H_2O}$

$$\frac{1\text{ mol } H_2O}{18.02\text{ g } H_2O} \qquad \chi = \frac{\text{amount solvent (in moles)}}{\text{total amount of solute and solvent particles (in moles)}} \qquad P_{solution} = \chi_{solvent}P^\circ_{solvent}$$

         Solution: $\dfrac{5.50\text{ g NaCl}}{100\text{ g (NaCl} + H_2O)}$ means 5.50 g NaCl and $(100\text{ g} - 5.50\text{ g}) = 94.5$ g H_2O then

$$5.50\text{ g NaCl} \times \frac{1\text{ mol NaCl}}{58.44\text{ g NaCl}} = 0.094\underline{1}136\text{ mol NaCl and }94.5\text{ g } H_2O \times \frac{1\text{ mol } H_2O}{18.02\text{ g } H_2O} = 5.2\underline{4}417\text{ mol } H_2O \text{ then}$$

$$\text{number of moles of solute} = i_{NaCl} \times n_{NaCl} \text{ so } \chi_{solv} = \frac{\text{amount solvent (in moles)}}{\text{total amount solute and solvent particles (in moles)}} =$$

$$\frac{5.2\underline{4}708\text{ mol}}{5.2\underline{4}417\text{ mol} + 2(0.094\underline{1}136\text{ mol})} = 0.96\underline{5}89 \text{ then } P_{soln} = \chi_{solv}P^\circ_{solv} = 0.96\underline{5}89 \times 23.78\text{ torr} = 23.0\text{ torr}$$

         Check: The units (torr) are correct. The magnitude of the answer (23 torr) seems reasonable because it is a drop from the pure vapor pressure. Only a fraction of a mole of NaCl is added, so the pressure will not drop much.

Cumulative Problems

13.95   Chloroform is polar and has stronger solute–solvent interactions than does nonpolar carbon tetrachloride.

13.97   **Given:** $KClO_4$: lattice energy $= -599$ kJ/mol, $\Delta H_{hydration} = -548$ kJ/mol; 10.0 g $KClO_4$ in 100.00 mL solution
         Find: ΔH_{soln} and ΔT **Other:** $C_s = 4.05$ J/g °C; $d = 1.05$ g/mL
         Conceptual Plan: Lattice energy, $\Delta H_{hydration} \rightarrow \Delta H_{soln}$ and g → mol then mol, $\Delta H_{soln} \rightarrow q(kJ) \rightarrow q(J)$

$$\Delta H_{soln} = \Delta H_{solute} + \Delta H_{hydration} \text{ where } \Delta H_{solute} = -\Delta H_{lattice} \quad \frac{1\text{ mol}}{138.55\text{ g}} \qquad q = n\,\Delta H_{soln} \quad \frac{1000\text{ J}}{1\text{ kJ}}$$

         then $mL_{soln} \rightarrow g_{soln}$ then q, g_{soln}, $C_s \rightarrow \Delta T$

$$\frac{1.05\text{ g}}{1\text{ mL}} \qquad q = mC_s\Delta T$$

         Solution: $\Delta H_{soln} = \Delta H_{solute} + \Delta H_{hydration}$ where $\Delta H_{solute} = -\Delta H_{lattice}$ so $\Delta H_{soln} = \Delta H_{hydration} - \Delta H_{lattice}$

$$\Delta H_{soln} = -548\text{ kJ/mol} - (-599\text{ kJ/mol}) = +51\text{ kJ/mol and }10.0\text{ g} \times \frac{1\text{ mol}}{138.55\text{ g}} = 0.072\underline{1}761\text{ mol then}$$

$$q = n\,\Delta H_{soln} = 0.072\underline{1}761\text{ mol} \times 51\frac{kJ}{mol} = +3.6\underline{8}10\text{ kJ} \times \frac{1000\text{ J}}{1\text{ kJ}} = +3\underline{6}81.0\text{ J absorbed then}$$

$$100.0\text{ mL} \times \frac{1.05\text{ g}}{1\text{ mL}} = 105\text{ g. Because heat is absorbed when } KClO_4 \text{ dissolves, the temperature will drop or}$$

$q = -3\underline{6}81.0$ J and $q = mC_s\Delta T$. Rearrange to solve for ΔT.

$$\Delta T = \frac{q}{mC_s} = \frac{-3\underline{6}81.0 \text{ J}}{105 \text{ g} \times 4.05 \frac{\text{J}}{\text{g} \cdot °\text{C}}} = -8.7 °\text{C}$$

Check: The units (kJ/mol and °C) are correct. The magnitude of the answer (51 kJ/mol) makes physical sense because the lattice energy is larger than the heat of hydration. The magnitude of the temperature change $(-9\,°\text{C})$ makes physical sense because heat is absorbed and the heat of solution is fairly small.

13.99 **Given:** argon, 0.0537 L; 25 °C, $P_{Ar} = 1.0$ atm to make 1.0 L saturated solution **Find:** $k_H(Ar)$
Conceptual Plan: °C → K and P_{Ar}, V, T → mol$_{Ar}$ then mol$_{Ar}$, V_{soln}, P_{Ar} → $k_H(Ar)$

$$K = °\text{C} + 273.15 \qquad PV = nRT \qquad S_{Ar} = k_H(Ar)\,P_{Ar} \text{ with } S_{Ar} = \frac{\text{mol}_{Ar}}{\text{L}_{soln}}$$

Solution: $25\,°\text{C} + 273.15 = 298$ K and $PV = nRT$ Rearrange to solve for n.

$$n = \frac{PV}{RT} = \frac{1.0 \text{ atm} \times 0.0537 \text{ L}}{0.08206 \dfrac{\text{L} \cdot \text{atm}}{\text{K} \cdot \text{mol}} \times 298 \text{ K}} = 0.00219597 \text{ mol} \text{ then } S_{Ar} = k_H(Ar)P_{Ar} \text{ with } S_{Ar} = \frac{\text{mol}_{Ar}}{\text{L}_{soln}}$$

Substitute in values and rearrange to solve for k_H.

$$k_H(Ar) = \frac{\text{mol}_{Ar}}{\text{L}_{soln}\,P_{Ar}} = \frac{0.00219597 \text{ mol}}{1.0 \text{ L}_{soln} \times 1.0 \text{ atm}} = 2.2 \times 10^{-3} \frac{\text{M}}{\text{atm}}$$

Check: The units (M/atm) are correct. The magnitude of the answer (10^{-3}) seems reasonable because it is consistent with other values in the text.

13.101 **Given:** 0.0020 ppm by mass Hg = legal limit; 0.0040 ppm by mass Hg = contaminated water; 50.0 mg Hg ingested **Find:** volume of contaminated water
Conceptual Plan: mg$_{Hg}$ → g$_{Hg}$ → g$_{H_2O}$ → mL$_{H_2O}$ → L$_{H_2O}$

$$\frac{1 \text{ g}}{1000 \text{ mg}} \qquad \frac{10^6 \text{ g water}}{0.0040 \text{ g Hg}} \qquad \frac{1 \text{ mL}}{1.00 \text{ g}} \qquad \frac{1 \text{ L}}{1000 \text{ mL}}$$

Solution: $50.0 \text{ mg Hg} \times \dfrac{1 \text{ g Hg}}{1000 \text{ mg Hg}} \times \dfrac{10^6 \text{ g water}}{0.0040 \text{ g Hg}} \times \dfrac{1 \text{ mL water}}{1.00 \text{ g water}} \times \dfrac{1 \text{ L water}}{1000 \text{ mL water}} = 1.3 \times 10^4 \text{ L water}$

Check: The units (L) are correct. The magnitude of the answer $(10^4\,\text{L})$ seems reasonable because the concentration is so low.

13.103 **Given:** 12.5% NaCl by mass in water at 55 °C; 2.5 L vapor **Find:** g$_{H_2O}$ in vapor
Other: $P°_{H_2O} = 118$ torr, $i_{NaCl} = 2.0$ (complete dissociation)
Conceptual Plan: % NaCl by mass → g$_{NaCl}$, g$_{H_2O}$ then g$_{NaCl}$ → mol$_{NaCl}$ and g$_{H_2O}$ → mol$_{H_2O}$

$$\frac{12.5 \text{ g NaCl}}{100 \text{ g (NaCl + H}_2\text{O)}} \qquad \frac{1 \text{ mol NaCl}}{58.44 \text{ g NaCl}} \qquad \frac{1 \text{ mol H}_2\text{O}}{18.02 \text{ g H}_2\text{O}}$$

then mol$_{NaCl}$, mol$_{H_2O}$ → χ_{NaCl} → χ_{H_2O} then χ_{H_2O}, $P°_{H_2O}$ → P_{H_2O}

$$\chi = \frac{\text{amount solute (in moles)}}{\text{total amount of solute and solvent (in moles)}} \qquad \chi_{H_2O} = 1 - i_{NaCl}\,\chi_{NaCl} \qquad P_{solution} = \chi_{solvent}P°_{solvent}$$

then torr → atm and °C → K P, V, T → mol$_{H_2O}$ → g$_{H_2O}$

$$\frac{1 \text{ atm}}{760 \text{ torr}} \qquad K = °\text{C} + 273.15 \qquad PV = nRT \qquad \frac{18.02 \text{ g H}_2\text{O}}{1 \text{ mol H}_2\text{O}}$$

Solution: $\dfrac{12.5 \text{ g NaCl}}{100 \text{ g (NaCl + H}_2\text{O)}}$ means 12.5 g NaCl and $(100 \text{ g} - 12.5 \text{ g}) = 87.5 \text{ g H}_2\text{O}$ then

$12.5 \text{ g NaCl} \times \dfrac{1 \text{ mol NaCl}}{58.44 \text{ g NaCl}} = 0.21\underline{3}895 \text{ mol NaCl}$ and $87.5 \text{ g H}_2\text{O} \times \dfrac{1 \text{ mol H}_2\text{O}}{18.02 \text{ g H}_2\text{O}} = 4.8\underline{5}841 \text{ mol H}_2\text{O}$

then $\chi = \dfrac{\text{amount solute (in moles)}}{\text{total amount of solute and solvent (in moles)}} = \dfrac{0.21\underline{3}895 \text{ mol}}{0.21\underline{3}895 \text{ mol} + 4.8\underline{5}841 \text{ mol}} = 0.042\underline{1}692$

then $\chi_{H_2O} = 1 - i_{NaCl}\chi_{NaCl} = 1 - (2.0 \times 0.042\underline{1}492) = 0.91\underline{5}662$ then

$P_{solution} = \chi_{solvent}P°_{solvent} = 0.91\underline{5}662 \times 118 \text{ torr} = 10\underline{8}.048 \text{ torr H}_2\text{O}$ then

$10\underline{8}.048 \text{ torr H}_2\text{O} \times \dfrac{1 \text{ atm}}{760 \text{ torr}} = 0.14\underline{2}168 \text{ atm}$

and $55\,°\text{C} + 273.15 = 328$ K then $PV = nRT$. Rearrange to solve for n.

$$n = \frac{PV}{RT} = \frac{0.142168 \text{ atm} \times 2.5 \text{ L}}{0.08206 \frac{\text{L} \cdot \text{atm}}{\text{K} \cdot \text{mol}} \times 328 \text{ K}} = 0.013205 \text{ mol then}$$

$$0.013205 \text{ mol H}_2\text{O} \times \frac{18.02 \text{ g H}_2\text{O}}{1 \text{ mol H}_2\text{O}} = 0.24 \text{ g H}_2\text{O}$$

Check: The units (g) are correct. The magnitude of the answer (0.2 g) seems reasonable because there is very little mass in a vapor.

13.105   **Given:** $T_b = 106.5\,°\text{C}$ aqueous solution   **Find:** T_f **Other:** $K_f = 1.86\,°\text{C}/m$; $K_b = 0.512\,°\text{C}/m$
Conceptual Plan: $T_b \rightarrow \Delta T_b$ then $\Delta T_b, K_b \rightarrow m$ then $m, K_f \rightarrow \Delta T_f \rightarrow T_f$

$$T_b = T_b° + \Delta T_b \qquad \Delta T_b = K_b \times m \qquad \Delta T_f = K_f \times m \quad T_f = T_f° - \Delta T_f$$

Solution: $T_b = T_b° + \Delta T_b$ so $\Delta T_b = T_b - T_b° = 106.5\,°\text{C} - 100.0\,°\text{C} = 6.5\,°\text{C}$ then $\Delta T_b = K_b \times m$

Rearrange to solve for m. $m = \dfrac{\Delta T_b}{K_b} = \dfrac{6.5\,°\text{C}}{0.512\dfrac{°\text{C}}{m}} = 12.695\, m$ then

$$\Delta T_f = K_f \times m = 1.86\frac{°\text{C}}{m} \times 12.695\, m = 23.6\,°\text{C} \text{ then } T_f = T_f° - \Delta T_f = 0.000\,°\text{C} - 23.6\,°\text{C} = -24\,°\text{C}$$

Check: The units (°C) are correct. The magnitude of the answer ($-24\,°\text{C}$) seems reasonable because the shift in boiling point is less than the shift in freezing point because the constant for boiling is smaller than the constant for freezing.

13.107   (a)   **Given:** 0.90% NaCl by mass per volume; isotonic aqueous solution at 25 °C; KCl; $i = 1.9$
Find: % KCl by mass per volume
Conceptual Plan: Isotonic solutions will have the same number of particles. Because i is the same,

$$\frac{1 \text{ mol KCl}}{1 \text{ mol NaCl}}$$

the new % mass per volume will be the mass ratio of the two salts.

$$\text{percent by mass per volume} = \frac{\text{mass solute}}{V} \times 100\% \quad \frac{1 \text{ mol NaCl}}{58.44 \text{ g NaCl}} \text{ and } \frac{74.55 \text{ g KCl}}{1 \text{ mol KCl}}$$

Solution: percent by mass per volume $= \dfrac{\text{mass solute}}{V} \times 100\%$

$$= \frac{0.0090 \text{ g NaCl}}{V} \times \frac{1 \text{ mol NaCl}}{58.44 \text{ g NaCl}} \times \frac{1 \text{ mol KCl}}{1 \text{ mol NaCl}} \times \frac{74.55 \text{ g KCl}}{1 \text{ mol KCl}} \times 100\%$$

$$= 1.1\% \text{ KCl by mass per volume}$$

Check: The units (% KCl by mass per volume) are correct. The magnitude of the answer (1.1%) seems reasonable because the molar mass of KCl is larger than the molar mass of NaCl.

(b)   **Given:** 0.90% NaCl by mass per volume; isotonic aqueous solution at 25 °C; NaBr; $i = 1.9$
Find: % NaBr by mass per volume
Conceptual Plan: Isotonic solutions will have the same number of particles. Because i is the same,

$$\frac{1 \text{ mol NaBr}}{1 \text{ mol NaCl}}$$

the new % mass per volume will be the mass ratio of the two salts.

$$\text{percent by mass per volume} = \frac{\text{mass solute}}{V} \times 100\% \quad \frac{1 \text{ mol NaCl}}{58.44 \text{ g NaCl}} \text{ and } \frac{102.89 \text{ g NaBr}}{1 \text{ mol NaBr}}$$

Solution: percent by mass per volume $= \dfrac{\text{mass solute}}{V} \times 100\%$

$$= \frac{0.0090 \text{ g NaCl}}{V} \times \frac{1 \text{ mol NaCl}}{58.44 \text{ g NaCl}} \times \frac{1 \text{ mol NaBr}}{1 \text{ mol NaCl}} \times \frac{102.89 \text{ g NaBr}}{1 \text{ mol NaBr}} \times 100\%$$

$$= 1.6\% \text{ NaBr by mass per volume}$$

Check: The units (% NaBr by mass per volume) are correct. The magnitude of the answer (1.6%) seems reasonable because the molar mass of NaBr is larger than the molar mass of NaCl.

(c)   **Given:** 0.90% NaCl by mass per volume; isotonic aqueous solution at 25 °C; glucose ($C_6H_{12}O_6$); $i = 1.9$
Find: % glucose by mass per volume

Conceptual Plan: Isotonic solutions will have the same number of particles. Because glucose is a nonelectrolyte, the i is not the same; then use the mass ratio of the two compounds.

$$\frac{1.9 \text{ mol } C_6H_{12}O_6}{1 \text{ mol NaCl}} \qquad \text{percent by mass per volume} = \frac{\text{mass solute}}{V} \times 100\% \qquad \frac{1 \text{ mol NaCl}}{58.44 \text{ g NaCl}} \text{ and } \frac{180.16 \text{ g } C_6H_{12}O_6}{1 \text{ mol } C_6H_{12}O_6}$$

Solution: percent by mass per volume $= \dfrac{\text{mass solute}}{V} \times 100\% =$

$$\frac{0.0090 \text{ g NaCl}}{V} \times \frac{1 \text{ mol NaCl}}{58.44 \text{ g NaCl}} \times \frac{1.9 \text{ mol } C_6H_{12}O_6}{1 \text{ mol NaCl}} \times \frac{180.16 \text{ g } C_6H_{12}O_6}{1 \text{ mol } C_6H_{12}O_6} \times 100\% =$$

5.3% $C_6H_{12}O_6$ by mass per volume

Check: The units (% $C_6H_{12}O_6$ by mass per volume) are correct. The magnitude of the answer (5.3%) seems reasonable because the molar mass of $C_6H_{12}O_6$ is larger than the molar mass of NaCl and we need more moles of $C_6H_{12}O_6$ because it is a nonelectrolyte.

13.109   **Given:** 4.5701 g of $MgCl_2$ and 43.238 g water, $P_{soln} = 0.3624$ atm, $P^\circ_{soln} = 0.3804$ atm at 348.0 K
Find: $i_{measured}$
Conceptual Plan: $g_{MgCl_2} \rightarrow$ mol$_{MgCl_2}$ and $g_{H_2O} \rightarrow$ mol$_{H_2O}$ then $P_{soln}, P^\circ_{soln}, \rightarrow \chi_{MgCl_2}$

$$\frac{1 \text{ mol } MgCl_2}{95.21 \text{ g } MgCl_2} \qquad \frac{1 \text{ mol } H_2O}{18.02 \text{ g } H_2O} \qquad P_{Soln} = (1 - \chi_{MgCl_2})P^\circ_{H_2O}$$

then mol$_{MgCl_2}$, mol$_{H_2O}$, $\chi_{MgCl_2} \rightarrow i$

$$\chi_{MgCl_2} = \frac{i(\text{moles } MgCl_2)}{\text{moles } H_2O + i(\text{moles } MgCl_2)}$$

Solution: 4.5701 g $MgCl_2 \times \dfrac{1 \text{ mol } MgCl_2}{95.21 \text{ g } MgCl_2} = 0.04800021$ mol $MgCl_2$ and

43.238 g $H_2O \times \dfrac{1 \text{ mol } H_2O}{18.02 \text{ g } H_2O} = 2.3994450$ mol H_2O then $P_{soln} = (1 - \chi_{MgCl_2})P^\circ_{H_2O}$ so

$$\chi_{MgCl_2} = 1 - \frac{P_{soln}}{P^\circ_{H_2O}} = 1 - \frac{0.3624 \text{ atm}}{0.3804 \text{ atm}} = 0.04731861. \text{ Solve for } i.$$

$$i(0.04800021) = 0.04731861(2.3994450 + i(0.04800021)) \rightarrow$$

$$i(0.04800021 - 0.002271303) = 0.1135384 \rightarrow i = \frac{0.1135384}{0.04572891} = 2.483$$

Check: The units (none) are correct. The magnitude of the answer (2.5) seems reasonable for $MgCl_2$ because we expect i to be 3 if it completely dissociates. Because Mg is small and doubly charged, we expect a significant drop from 3.

13.111   **Given:** $T_b = 375.3$ K aqueous solution   **Find:** P_{H_2O} **Other:** $P^\circ_{H_2O} = 0.2467$ atm; $K_b = 0.512\ ^\circ C/m$
Conceptual Plan: $T_b \rightarrow \Delta T_b$ then $\Delta T_b, K_b \rightarrow m$ assume 1 kg water $kg_{H_2O} \rightarrow mol_{H_2O}$ then

$$T_b = T^\circ_b + \Delta T_b \qquad \Delta T_b = K_b \times m \qquad \frac{1 \text{ mol } H_2O}{18.02 \text{ g } H_2O}$$

$m \rightarrow$ mol$_{solute}$ then mol$_{H_2O}$, mol$_{solute} \rightarrow \chi_{H_2O}$ then $\chi_{H_2O}, P^\circ_{H_2O} \rightarrow P_{H_2O}$

$$m = \frac{\text{amount solute (moles)}}{\text{mass solvent (kg)}} \qquad \chi_{H_2O} = \frac{\text{moles } H_2O}{\text{moles } H_2O + \text{moles solute}} \qquad P_{H_2O} = \chi_{H_2O}P^\circ_{H_2O}$$

Solution: $T_b = T^\circ_b + \Delta T_b$ so $\Delta T_b = T_b - T^\circ_b = 375.3$ K $- 373.15$ K $= 2.2$ K $= 2.2\ ^\circ C$ then

$\Delta T_b = K_b \times m$ Rearrange to solve for m. $m = \dfrac{\Delta T_b}{K_b} = \dfrac{2.2\ ^\circ C}{0.512\ \frac{^\circ C}{m}} = 4.296875\ m$ then

$$1000 \text{ g } H_2O \times \frac{1 \text{ mol } H_2O}{18.02 \text{ g } H_2O} = 55.49390 \text{ mol } H_2O \text{ then}$$

$$m = \frac{\text{amount solute (moles)}}{\text{mass solvent (kg)}} = \frac{x \text{ mol}}{1 \text{ kg}} = 4.296875\ m \text{ so } x = 4.296875 \text{ mol then}$$

$$\chi_{H_2O} = \frac{\text{moles } H_2O}{\text{moles } H_2O + \text{moles solute}} = \frac{55.49390 \text{ mol}}{55.49390 \text{ mol} + 4.296875 \text{ mol}} = 0.9281348$$

then $P_{H_2O} = \chi_{H_2O}P^\circ_{H_2O} = 0.9281348 \times 0.2467$ atm $= 0.229$ atm

Check: The units (atm) are correct. The magnitude of the answer (0.229 atm) seems reasonable because the mole fraction is lowered by $\sim 7\%$.

13.113 **Given:** equal masses of carbon tetrachloride (CCl_4) and chloroform ($CHCl_3$) at 316 K; $P^{\circ}_{CCl_4} = 0.354$ atm; $P^{\circ}_{CHCl_3} = 0.526$ atm **Find:** χ_{CCl_4}, χ_{CHCl_3} in vapor and P_{CHCl_3} in flask of condensed vapor

Conceptual Plan: Assume 100 grams of each $g_{CCl_4} \rightarrow mol_{CCl_4}$ **and** $g_{CHCl_3} \rightarrow mol_{CHCl_3}$ **then**

$$\frac{1\ mol\ CCl_4}{153.81\ g\ CCl_4} \qquad \frac{1\ mol\ CHCl_3}{119.37\ g\ CHCl_3}$$

mol_{CCl_4}, $mol_{CHCl_3} \rightarrow \chi_{CCl_4}$, χ_{CHCl_3} **then** χ_{CCl_4}, $P^{\circ}_{CCl_4} \rightarrow P_{CCl_4}$ **and** χ_{CHCl_3}, $P^{\circ}_{CHCl_3} \rightarrow P_{CHCl_3}$ **then**

$$\chi_{CCl_4} = \frac{amount\ CCl_4\ (in\ moles)}{total\ amount\ (in\ moles)} \quad \chi_{CHCl_3} = 1 - \chi_{CCl_4} \qquad P_{CCl_4} = \chi_{CCl_4} P^{\circ}_{CCl_4} \qquad P_{CHCl_3} = \chi_{CHCl_3} P^{\circ}_{CHCl_3}$$

$P^{\circ}_{CCl_4}$, $P_{CHCl_3} \rightarrow P_{Total}$ **then because** $n \propto P$ **and we are calculating a mass percent, which is a ratio of masses,**

$$P_{Total} = P_{CCl_4} + P_{CHCl_3}$$

we can simply convert 1 atm to 1 mole so P_{CCl_4}, $P_{CHCl_3} \rightarrow n_{CHCl_3}$, n_{CHCl_3} **then**

$$\chi_{CCl_4} = \frac{amount\ CCl_4\ (in\ moles)}{total\ amount\ (in\ moles)}$$

mol_{CCl_4}, $mol_{CHCl_3} \rightarrow \chi_{CCl_4}$, χ_{CHCl_3} **then for the second vapor** χ_{CHCl_3}, $P^{\circ}_{CHCl_3} \rightarrow P_{CHCl_3}$

$$\chi_{CHCl_3} = 1 - \chi_{CCl_4} \qquad P_{CHCl_3} = \chi_{CHCl_3} P^{\circ}_{CHCl_3}$$

Solution: $100.00\ g\ CCl_4 \times \dfrac{1\ mol\ CCl_4}{153.81\ g\ CCl_4} = 0.65015279\ mol\ CCl_4$ and

$100.00\ g\ CHCl_3 \times \dfrac{1\ mol\ CHCl_3}{119.37\ g\ CHCl_3} = 0.83773142\ mol\ CHCl_3$ then

$$\chi_{CCl_4} = \frac{amount\ CCl_4\ (in\ moles)}{total\ amount\ (in\ moles)} = \frac{0.65015279\ mol}{0.65015279\ mol\ +\ 0.83773142\ mol} = 0.43696464\ and$$

$\chi_{CHCl_3} = 1 - \chi_{CCl_4} = 1 - 0.43696464 = 0.56303536$ then

$P_{CCl_4} = \chi_{CCl_4} P^{\circ}_{CCl_4} = 0.43696464 \times 0.354\ atm = 0.154685\ atm$ and

$P_{CHCl_3} = \chi_{CHCl_3} P^{\circ}_{CHCl_3} = 0.56303536 \times 0.526\ atm = 0.296157\ atm$ then

$P_{Total} = P_{CCl_4} + P_{CHCl_3} = 0.154685\ atm + 0.296157\ atm = 0.450841\ atm$ then

$mol_{CCl_4} = 0.154687\ mol$ and $mol_{CHCl_3} = 0.296157\ mol$ then

$$\chi_{CCl_4} = \frac{amount\ CCl_4\ (in\ moles)}{total\ amount\ (in\ moles)} = \frac{0.154685\ mol}{0.154685\ mol\ +\ 0.296157\ mol} = 0.343102 = 0.343\ in\ the\ first\ vapor\ and$$

$\chi_{CHCl_3} = 1 - \chi_{CCl_4} = 1 - 0.343102 = 0.656898 = 0.657$ in the first vapor; then in the second vapor,

$P_{CHCl_3} = \chi_{CHCl_3} P^{\circ}_{CHCl_3} = 0.656898 \times 0.526\ atm = 0.345528\ atm = 0.346\ atm$

Check: The units (none and atm) are correct. The magnitude of the answers seems reasonable because we expect the lighter component to be found preferentially in the vapor phase. This effect is magnified in the second vapor.

13.115 **Given:** 49.0% H_2SO_4 by mass, $d = 1.39$ g/cm^3, 25.0 mL diluted to 99.8 cm^3 **Find:** molarity

Conceptual Plan: Initial $mL_{solution} \rightarrow g_{solution} \rightarrow g_{H_2SO_4} \rightarrow mol_{H_2SO_4}$ **and final** $mL_{solution} \rightarrow L_{solution}$

$$\frac{1.39\ g}{1\ mL} \quad \frac{49.0\ g\ H_2SO_4}{100\ g\ solution} \quad \frac{1\ mol\ H_2SO_4}{98.08\ g\ H_2SO_4} \qquad \frac{1\ L}{1000\ mL}$$

then $mol_{H_2SO_4}$, $L_{solution} \rightarrow M$

$$M = \frac{amount\ solute\ (moles)}{volume\ solution\ (L)}$$

Solution:

$$25.0\ mL\ solution \times \frac{1.39\ g\ solution}{1\ mL\ solution} \times \frac{49.0\ g\ H_2SO_4}{100\ g\ solution} \times \frac{1\ mol\ H_2SO_4}{98.08\ g\ H_2SO_4} = 0.173608\ mol\ H_2SO_4\ and$$

$$99.8\ mL\ solution \times \frac{1\ L\ solution}{1000\ mL\ solution} = 0.0998\ L\ solution\ then$$

$$M = \frac{amount\ solute\ (moles)}{volume\ solution\ (L)} = \frac{0.173608\ mol\ H_2SO_4}{0.0998\ L\ solution} = 1.74\ M\ H_2SO_4$$

Check: The units (M) are correct. The magnitude of the answer (1.74 M) seems reasonable because the solutions is $\sim 1/6$ sulfuric acid.

13.117   **Given:** 10.05 g of unknown compound in 50.0 g water, $T_f = -3.16\,°C$, mass percent composition of the compound is
60.97% C and 11.94% H; the rest is O   **Find:** molecular formula
Other: $K_f = 1.86\,°C/m;\ d = 1.00\ g/mL$
Conceptual Plan: $g_{H_2O} \rightarrow kg_{H_2O}$ and $T_f \rightarrow \Delta T_f$ then $\Delta T_f, K_f \rightarrow m$ then $m, kg_{H_2O} \rightarrow mol_{Unk}$

$$\frac{1\ kg}{1000\ g} \qquad T_f = T_f^° - \Delta T_f \qquad \Delta T_f = K_f \times m \qquad m = \frac{amount\ solute\ (moles)}{mass\ solvent\ (kg)}$$

then $g_{Unk}, mol_{Unk} \rightarrow M \rightarrow g_C, g_H, g_O \rightarrow mol_C, mol_H, mol_O \rightarrow$ **molecular formula**

$$\mathcal{M} = \frac{g_{Unk}}{mol_{Unk}} \text{ mass percents} \qquad \frac{1\ mol\ C}{12.01\ g\ C} \quad \frac{1\ mol\ H}{1.008\ g\ H} \quad \frac{1\ mol\ O}{16.00\ g\ O}$$

Solution: $50.0\ g \times \dfrac{1\ kg}{1000\ g} = 0.0500\ kg$ and $T_f = T_f^° - \Delta T_f$ so

$\Delta T_f = T_f^° - T_f = 0.00\,°C - (-3.16\,°C) = +3.16\,°C$ $\Delta T_f = K_f \times m.$ Rearrange to solve for m.

$$m = \frac{\Delta T_f}{K_f} = \frac{3.16\,°C}{1.86\dfrac{°C}{m}} = 1.6\underline{9}892\ m \text{ then } m = \frac{amount\ solute\ (moles)}{mass\ solvent\ (kg)} \text{ so}$$

$$mol_{Unk} = m_{Unk} \times kg_{H_2O} = 1.6\underline{9}892\ \frac{mol\ Unk}{kg} \times 0.0500\ kg = 0.08\underline{4}94600\ mol\ Unk \text{ then}$$

$$\mathcal{M} = \frac{g_{Unk}}{mol_{Unk}} = \frac{10.05\ g}{0.08\underline{4}94600\ mol} = 118.\underline{3}105\frac{g}{mol} \text{ then}$$

$$\frac{118.\underline{3}105\ g\ Unk}{1\ mol\ Unk} \times \frac{60.97\ g\ C}{100\ g\ Unk} \times \frac{1\ mol\ C}{12.01\ g\ C} = \frac{6.01\ mol\ C}{1\ mol\ Unk}$$

$$\frac{118.\underline{3}105\ g\ Unk}{1\ mol\ Unk} \times \frac{11.94\ g\ H}{100\ g\ Unk} \times \frac{1\ mol\ H}{1.008\ g\ H} = \frac{14.0\ mol\ H}{1\ mol\ Unk} \text{ and}$$

$$\frac{118.\underline{3}105\ g\ Unk}{1\ mol\ Unk} \times \frac{(100 - (60.97 + 11.94))\ g\ O}{100\ g\ Unk} \times \frac{1\ mol\ O}{16.00\ g\ O} = \frac{2.00\ mol\ O}{1\ mol\ Unk}$$

So the molecular formula is $C_6H_{14}O_2$.

Check: The units (formula) are correct. The magnitude of the answer (formula with ~118 g/mol) seems reasonable
because the molality is ~1.7 and we have ~10 g. It is a reasonable molecular weight for a solid or liquid. The formula
does have the correct molar mass.

13.119   **Given:** 100.0 mL solution 13.5% by mass NaCl, $d = 1.12\ g/mL;\ T_b = 104.4\,°C$ **Find:** g NaCl or water to add
Other: $K_b = 0.512\,°C/m;\ i_{measured} = 1.8$
Conceptual Plan: $T_b \rightarrow \Delta T_b$ then $\Delta T_b, i, K_b \rightarrow m$ then $mL_{solution} \rightarrow g_{solution} \rightarrow g_{NaCl} \rightarrow mol_{NaCl}$ then

$$\Delta T_b = T_b - T_b^° \qquad \Delta T_b = K_b \times i \times m \qquad \frac{1.12\ g\ solution}{1\ mL\ solution} \quad \frac{13.5\ g\ NaCl}{100\ g\ solution} \quad \frac{1\ mol\ NaCl}{58.44\ g\ NaCl}$$

$m, mol_{NaCl} \rightarrow kg_{H_2O} \rightarrow g_{H_2O}$ and $g_{solution}, g_{NaCl} \rightarrow g_{H_2O}$ then compare the initial and final g_{H_2O} then

$$m = \frac{amount\ solute\ (moles)}{mass\ solvent\ (kg)} \quad \frac{1000\ g}{1\ kg} \qquad g_{solution} = g_{NaCl} + g_{H_2O}$$

**calculate the total NaCl in final solution by scaling up the amount from the initial solution. Then calculate the
difference between the needed and starting amounts of NaCl.**
Solution: $\Delta T_b = T_b - T_b^° = 104.4\,°C - 100.0\,°C = 4.4\,°C$ then $\Delta T_b = K_b \times i \times m$

Rearrange to solve for m. $m = \dfrac{\Delta T_b}{K_b\ i} = \dfrac{4.4\,°C}{0.512\dfrac{°C}{m} \times 1.8} = 4.\underline{7}74306\ m$ NaCl then

$$100.0\ mL\ solution \times \frac{1.12\ g\ solution}{1\ mL\ solution} = 112\ g\ solution \qquad 112\ g\ solution \times \frac{13.5\ g\ NaCl}{100\ g\ solution} = 15.\underline{1}2\ g\ NaCl$$

$$15.\underline{1}2\ NaCl \times \frac{1\ mol\ NaCl}{58.44\ g\ NaCl} = 0.25\underline{8}7269\ mol\ NaCl$$

then $m = \dfrac{amount\ solute\ (moles)}{mass\ solvent\ (kg)}$ Rearrange to solve for kg_{H_2O}.

$$kg_{H_2O} = \frac{mol_{NaCl}}{m} = \frac{0.2587269 \text{ mol NaCl}}{\dfrac{4.774306 \text{ mol NaCl}}{1 \text{ kg}_{H_2O}}} = 0.0541915 \text{ kg}_{H_2O} \times \frac{1000 \text{ g}_{H_2O}}{1 \text{ kg}_{H_2O}} = 54.1915 \text{ g}_{H_2O} \text{ in final solution}$$

then $g_{solution} = g_{NaCl} + g_{H_2O} = 112 \text{ g solution} - 15.\underline{1}2 \text{ g NaCl} = 96.88 \text{ g } H_2O$ in initial solution. Comparing the initial and final solutions, there is a lot more water in the initial solution; so NaCl needs to be added.

In the solution with a boiling point of 104.4 °C, $\dfrac{15.\underline{1}2 \text{ g NaCl}}{54.1915 \text{ g } H_2O} = \dfrac{x \text{ g NaCl}}{96.88 \text{ g } H_2O}$. Solve for x g NaCl.

$$x \text{ g NaCl} = \frac{15.\underline{1}2 \text{ g NaCl}}{54.1915 \text{ g } H_2O} \times 96.88 \text{ g } H_2O = 27.031 \text{ g NaCl; so the amount to be added is}$$

$27.\underline{0}31 \text{ g NaCl} - 15.\underline{1}2 \text{ g NaCl} = 11.911 \text{ g NaCl} = 12 \text{ g NaCl}$

Check: The units (g) are correct. The magnitude of the answer (12 g) seems reasonable because there is approximately twice as much water in the initial solution as is desired; so the NaCl amount needs to be approximately doubled.

Challenge Problems

13.121   **Given:** N_2: $k_H(N_2) = 6.1 \times 10^{-4} \text{ M/L at } 25 \text{ °C}$; 14.6 mg/L at 50 °C and 1.00 atm $P_{N_2} = 0.78 \text{ atm}$
O_2: $k_H(O_2) = 1.3 \times 10^{-3} \text{ M/L at } 25 \text{ °C}$; 27.8 mg/L at 50 °C and 1.00 atm; $P_{O_2} = 0.21 \text{ atm}$; and 1.5 L water
Find: $V(N_2)$ and $V(O_2)$
Conceptual Plan: at 25 °C : $P_{Total}, \chi_{N_2} \rightarrow P_{N_2}$ then $P_{N_2}, k_H(N_2) \rightarrow S_{N_2}$ then $L \rightarrow mol$

$$\qquad\qquad P_{N_2} = \chi_{N_2} P_{Total} \qquad\qquad S_{N_2} = k_H(N_2) P_{N_2} \qquad S_{N_2}$$

at 50 °C : $L \rightarrow mL \rightarrow mg \rightarrow g \rightarrow mol$ then $mol_{25 \text{ °C}}, mol_{25 \text{ °C}} \rightarrow mol_{removed}$ then °C $\rightarrow$ K

$$\quad \frac{1000 \text{ mL}}{1 \text{ L}} \quad \frac{14.6 \text{ mg}}{1 \text{ L}} \quad \frac{1 \text{ g}}{1000 \text{ mg}} \quad \frac{1 \text{ mol}}{28.02 \text{ g}} \qquad\qquad mol_{removed} = mol_{25 \text{ °C}} - mol_{50 \text{ °C}} \qquad K = \text{°C} + 273.15$$

then $P, n, T \rightarrow V$

$$PV = nRT$$

at 25 °C : $P_{Total}, \chi_{O_2} \rightarrow P_{O_2}$ then $P_{O_2}, k_H(O_2) \rightarrow S_{O_2}$ then $L \rightarrow mol$

$$\qquad\qquad P_{O_2} = \chi_{O_2} P_{Total} \qquad\qquad S_{O_2} = k_H(O_2) P_{O_2} \qquad S_{O_2}$$

at 50 °C : $L \rightarrow mL \rightarrow mg \rightarrow g \rightarrow mol$ then $mol_{25 \text{ °C}}, mol_{25 \text{ °C}} \rightarrow mol_{removed}$ then °C $\rightarrow$ K

$$\quad \frac{1000 \text{ mL}}{1 \text{ L}} \quad \frac{27.8 \text{ mg}}{1 \text{ L}} \quad \frac{1 \text{ g}}{1000 \text{ mg}} \quad \frac{1 \text{ mol}}{32.00 \text{ g}} \qquad\qquad mol_{removed} = mol_{25 \text{ °C}} - mol_{50 \text{ °C}} \qquad K = \text{°C} + 273.15$$

then $P, n, T \rightarrow V$

$$PV = nRT$$

Solution: at 25 °C: $P_{N_2} = \chi_{N_2} P_{Total} = 0.78 \times 1.0 \text{ atm} = 0.78 \text{ atm}$ then

$$S_{N_2} = k_H(N_2) P_{N_2} = 6.1 \times 10^{-4} \frac{M}{atm} \times 0.78 \text{ atm} = 4.\underline{7}58 \times 10^{-4} \text{ M then}$$

$$1.5 \text{ L} \times 4.\underline{7}58 \times 10^{-4} \frac{mol}{L} = 0.00071370 \text{ mol}$$

$$\text{at 50 °C: } 1.5 \text{ L} \times \frac{14.6 \text{ mg}}{1 \text{ L} \cdot atm} \times 0.78 \text{ atm} \times \frac{1 \text{ g}}{1000 \text{ mg}} \times \frac{1 \text{ mol}}{28.02 \text{ g}} = 0.00060964 \text{ mol then}$$

$mol_{removed} = mol_{25 \text{ °C}} - mol_{50 \text{ °C}} = 0.000713\underline{7}0 \text{ mol} - 0.000609\underline{6}4 \text{ mol} = 1.\underline{0}41 \times 10^{-4} \text{ mol } N_2$
then 50 °C + 273.15 = 323 K then $PV = nRT$. Rearrange to solve for V.

$$V = \frac{nRT}{P} = \frac{1.\underline{0}41 \times 10^{-4} \text{ mol} \times 0.08206 \dfrac{L \cdot atm}{K \cdot mol} \times 323 \text{ K}}{1.00 \text{ atm}} = 0.002\underline{7}592 \text{ L } N_2$$

at 25 °C: $P_{O_2} = \chi_{O_2} P_{Total} = 0.21 \times 1.0 \text{ atm} = 0.21 \text{ atm}$ then

$$S_{O_2} = k_H(O_2) P_{O_2} = 1.3 \times 10^{-3} \frac{M}{atm} \times 0.21 \text{ atm} = 2.\underline{7}3 \times 10^{-4} \text{ M then}$$

$$1.5 \text{ L} \times 2.\underline{7}3 \times 10^{-4} \frac{mol}{L} = 0.0004\underline{0}95 \text{ mol}$$

$$\text{at 50 °C: } 1.5 \text{ L} \times \frac{27.8 \text{ mg}}{1 \text{ L} \cdot atm} \times 0.21 \text{ atm} \times \frac{1 \text{ g}}{1000 \text{ mg}} \times \frac{1 \text{ mol}}{32.00 \text{ g}} = 0.00027366 \text{ mol then}$$

$\text{mol}_{\text{removed}} = \text{mol}_{25\,°C} - \text{mol}_{50\,°C} = 0.0004095 \text{ mol} - 0.00027366 \text{ mol} = 1.358 \times 10^{-4} \text{ mol O}_2$
then $50\,°C + 273.15 = 323 \text{ K}$ then $PV = nRT$. Rearrange to solve for V.

$$V = \frac{nRT}{P} = \frac{1.358 \times 10^{-4} \text{ mol} \times 0.08206 \dfrac{L \cdot atm}{K \cdot mol} \times 323 \text{ K}}{1.00 \text{ atm}} = 0.0035994 \text{ L O}_2 \text{ finally}$$

$V_{\text{Total}} = V_{N_2} + V_{O_2} = 0.0027592 \text{ L} + 0.0035994 \text{ L} = 0.0064 \text{ L}$

Check: The units (L) are correct. The magnitude of the answer (0.006 L) seems reasonable because we have so little dissolved gas at room temperature and most is still soluble at 50 °C.

13.123   **Given:** 1.10 g glucose ($C_6H_{12}O_6$) and sucrose ($C_{12}H_{22}O_{11}$) mixture in 25.0 mL solution and $\Pi = 3.78$ atm at 298 K
Find: percent composition of mixture
Conceptual Plan: $\Pi, T \rightarrow M$ then $\text{mL}_{\text{soln}} \rightarrow L_{\text{soln}}$ then $L_{\text{soln}}, M \rightarrow \text{mol}_{\text{mixture}}$ then

$$\Pi = MRT \qquad \frac{1 \text{ L}}{1000 \text{ mL}} \qquad M = \frac{\text{amount solute (moles)}}{\text{volume solution (L)}}$$

$\text{mol}_{\text{mixture}}, g_{\text{mixture}} \rightarrow \text{mol}_{C_6H_{12}O_6}, \text{mol}_{C_{12}H_{22}O_{11}}$ then

$g_{\text{mixture}} = \text{mol } C_6H_{12}O_6 \times \dfrac{180.16 \text{ g } C_6H_{12}O_6}{1 \text{ mol } C_6H_{12}O_6} + \text{mol } C_{12}H_{22}O_{11} \times \dfrac{342.30 \text{ g } C_{12}H_{22}O_{11}}{1 \text{ mol } C_{12}H_{22}O_{11}}$ *with* $\text{mol}_{\text{mixture}} = \text{mol}_{C_6H_{12}O_6} + \text{mol}_{C_{12}H_{22}O_{11}}$

$\text{mol}_{C_6H_{12}O_6} \rightarrow g_{C_6H_{12}O_6}$ **and** $\text{mol}_{C_{12}H_{22}O_{11}} \rightarrow g_{C_{12}H_{22}O_{11}}$ **and** $g_{C_6H_{12}O_6}, g_{C_{12}H_{22}O_{11}} \rightarrow$ **mass percents**

$\dfrac{180.16 \text{ g } C_6H_{12}O_6}{1 \text{ mol } C_6H_{12}O_6} \qquad \dfrac{342.30 \text{ g } C_{12}H_{22}O_{11}}{1 \text{ mol } C_{12}H_{22}O_{11}} \qquad \text{mass percent} = \dfrac{\text{mass solute}}{\text{mass solution}} \times 100\%$

Solution: $\Pi = MRT$. Rearrange to solve for M.

$$M = \frac{\Pi}{RT} = \frac{3.78 \text{ atm}}{0.08206 \dfrac{L \cdot atm}{K \cdot mol} \times 298 \text{ K}} = 0.154577 \frac{\text{mol mixture}}{L} \text{ then } 25.0 \text{ mL} \times \frac{1 \text{ L}}{1000 \text{ mL}} = 0.0250 \text{ L}$$

then $M = \dfrac{\text{amount solute (moles)}}{\text{volume solution (L)}}$ so

$\text{mol}_{\text{mixture}} = M \times L_{\text{soln}} = 0.154577 \dfrac{\text{mol mixture}}{L} \times 0.0250 \text{ L} = 0.00386442 \text{ mol mixture}$ then

$g_{\text{mixture}} = \text{mol } C_6H_{12}O_6 \times \dfrac{180.16 \text{ g } C_6H_{12}O_6}{1 \text{ mol } C_6H_{12}O_6} + \text{mol } C_{12}H_{22}O_{11} \times \dfrac{342.30 \text{ g } C_{12}H_{22}O_{11}}{1 \text{ mol } C_{12}H_{22}O_{11}}$ with

$\text{mol}_{\text{mixture}} = \text{mol}_{C_6H_{12}O_6} + \text{mol}_{C_{12}H_{22}O_{11}}$ so

$1.10 \text{ g} = \text{mol } C_6H_{12}O_6 \times \dfrac{180.16 \text{ g } C_6H_{12}O_6}{1 \text{ mol } C_6H_{12}O_6} + (0.00386442 \text{ mol} - \text{mol } C_6H_{12}O_6) \times \dfrac{342.30 \text{ g } C_{12}H_{22}O_{11}}{1 \text{ mol } C_{12}H_{22}O_{11}} \rightarrow$

$1.10 = 180.16 \times \text{mol } C_6H_{12}O_6 + 1.32228 - 342.30 \times \text{mol } C_6H_{12}O_6 \rightarrow 162.14 \, x \text{ mol } C_6H_{12}O_6 = 0.22228 \rightarrow$

$x \text{ mol } C_6H_{12}O_6 = \dfrac{0.22228}{162.14} = 0.00137091 \text{ mol } C_6H_{12}O_6$ then

$\text{mol}_{C_{12}H_{22}O_{11}} = \text{mol}_{\text{mixture}} - \text{mol}_{C_6H_{12}O_6} = 0.00386442 \text{ mol} - 0.00137091 \text{ mol}$
$= 0.0024935 \text{ mol } C_{12}H_{22}O_{11}$ then

$0.00137091 \text{ mol } C_6H_{12}O_6 \times \dfrac{180.16 \text{ g } C_6H_{12}O_6}{1 \text{ mol } C_6H_{12}O_6} = 0.24698 \text{ g } C_6H_{12}O_6$ and

$0.0024935 \text{ mol } C_{12}H_{22}O_{11} \times \dfrac{342.30 \text{ g } C_{12}H_{22}O_{11}}{1 \text{ mol } C_{12}H_{22}O_{11}} = 0.85353 \text{ g } C_{12}H_{22}O_{11}$ finally

$\text{mass percent} = \dfrac{\text{mass solute}}{\text{mass solution}} \times 100\% = \dfrac{0.24698 \text{ g } C_6H_{12}O_6}{0.24698 \text{ g } C_6H_{12}O_6 + 0.85353 \text{ g } C_{12}H_{22}O_{11}} \times 100\%$

$= 22.44\% \text{ C}_6H_{12}O_6 \text{ by mass and } 100.00\% - 22.44\% = 78\% \text{ C}_{12}H_{22}O_{11} \text{ by mass}$

Check: The units (% by mass) are correct. We expect the percent by $C_6H_{12}O_6$ to be larger than that for $C_{12}H_{22}O_{11}$ because the $g_{\text{mixture}}/\text{mol}_{\text{mixture}} = 285$ g/mol, which is closer to $C_{12}H_{22}O_{11}$ than $C_6H_{12}O_6$ and the molar mass of $C_{12}H_{22}O_{11}$ is larger than the molar mass of $C_6H_{12}O_6$. In addition, and most definitively, the masses obtained for sucrose and glucose sum to 1.1 g, the initial amount of solid dissolved.

13.125   **Given:** isopropyl alcohol $((CH_3)_2CHOH)$ and propyl alcohol $(CH_3CH_2CH_2OH)$ at 313 K; solution 2/3 by mass isopropyl alcohol $P_{2/3} = 0.110$ atm; solution 1/3 by mass isopropyl alcohol $P_{1/3} = 0.089$ atm
          Find: P°_{iso} and P°_{pro} and explain why they are different
          Conceptual Plan: Because these are isomers, they have the same molar mass and so the fraction by mass is the same as the mole fraction so mole fractions, P_{soln}s $\rightarrow P^\circ$s

$$\chi_{iso} = \frac{\text{amount iso (in moles)}}{\text{total amount (in moles)}} \qquad \chi_{pro} = 1 - \chi_{iso} \qquad P_{iso} = \chi_{iso}P^\circ_{iso} \qquad P_{pro} = \chi_{pro}P^\circ_{pro} \text{ and } P_{soln} = P_{iso} + P_{pro}$$

          Solution:
          Solution 1: $\chi_{iso} = 2/3$ and $\chi_{iso} = 1/3$ $P_{soln} = P_{iso} + P_{pro}$ so 0.110 atm $= 2/3P^\circ_{iso} + 1/3P^\circ_{pro}$
          Solution 2: $\chi_{iso} = 1/3$ and $\chi_{iso} = 2/3$ $P_{soln} = P_{iso} + P_{pro}$ so 0.089 atm $= 1/3P^\circ_{iso} + 2/3P^\circ_{pro}$ We now have two equations and two unknowns and a number of ways to solve this. One way is to rearrange the first equation for P°_{iso} and substitute into the other equation. Thus, $P^\circ_{iso} = 3/2(0.110 \text{ atm} - 1/3P^\circ_{pro})$ and

$$0.089 \text{ atm} = \frac{1}{3}\frac{3}{2}(0.110 \text{ atm} - 1/3 P^\circ_{pro}) + \frac{2}{3}P^\circ_{pro} \rightarrow 0.089 \text{ atm} = 0.0550 \text{ atm} - \frac{1}{6}P^\circ_{pro} + \frac{2}{3}P^\circ_{pro} \rightarrow$$

$$\frac{1}{2}P^\circ_{pro} = 0.03\underline{4}0 \text{ atm} \rightarrow P^\circ_{pro} = 0.06\underline{8}0 \text{ atm} = 0.068 \text{ atm and then}$$

$$P^\circ_{iso} = 3/2(0.110 \text{ atm} - 1/3P^\circ_{pro}) = 3/2(0.110 \text{ atm} - 1/3(0.06\underline{8}0 \text{ atm})) = 0.131 \text{ atm}$$

          The major intermolecular attractions are between the OH groups. The OH group at the end of the chain in propyl alcohol is more accessible than the one in the middle of the chain in isopropyl alcohol. In addition, the molecular shape of propyl alcohol is a straight chain of carbon atoms, while that of isopropyl alcohol has a branched chain and is more like a ball. The contact area between two ball-like objects is smaller than that of two chain-like objects. The smaller contact area in isopropyl alcohol means that the molecules do not attract each other as strongly as do those of propyl alcohol. As a result of both of these factors, the vapor pressure of isopropyl alcohol is higher.

          Check: The units (atm) are correct. The magnitude of the answers seems reasonable because both solution partial pressures are ~0.1 atm.

13.127   **Given:** 0.1000 m H_2SO_4 solution; complete dissociation to H^+ and HSO_4^-; limited dissociation to SO_4^{2-}; $T_f = 272.76$ K   **Find:** $m(SO_4^{2-})$ **Other:** $K_f = 1.86$ K/m
          Conceptual Plan: $T_f \rightarrow \Delta T_f$ then $m, \Delta T_f, K_f \rightarrow i \rightarrow m(SO_4^{2-})$

$$T_f = T_f^\circ - \Delta T_f \qquad\qquad \Delta T_f = K_f im \quad m(SO_4^{2-}) = m\, H_2SO_4\,(i - 2.0)/2$$

          Solution: $T_f = T_f^\circ - \Delta T_f$ Rearrange to solve for ΔT_f. So
          $\Delta T_f = T_f^\circ - T_f = 273.15 \text{ K} - 272.76 \text{ K} = -0.39$ K then $\Delta T_f = K_f im$. Rearrange to solve for i.

$$i = \frac{\Delta T_f}{K_f m} = \frac{0.39 \text{ K}}{1.86\dfrac{K}{m} \times 0.1000\, m} = 2.09\underline{6}8.$$

          Remember that when H_2SO_4 completely dissociates, two particles are formed (H^+ and HSO_4^-); so $i = 2$. When HSO_4^- dissociates, two particles are generated (SO_4^{2-} and H^+).

$$m(SO_4^{2-}) = m\, H_2SO_4\,\frac{i - 2}{2} = (0.1000\, m)\frac{2.09\underline{6}8 - 2}{2} = 0.004\underline{8}4\, m = 0.005\, m\, SO_4^{2-}$$

          Check: The units (m) are correct. The magnitude of the answer (0.005 m) seems reasonable because not much of the HSO_4^- dissociates.

13.129   **Given:** $Na_2CO_3 + NaHCO_3 = 11.60$ g in 1.00 L; treat 300.0 cm^3 of solution with HNO_3 and collect 0.940 L CO_2 at 298 K and 0.972 atm   **Find:** $M(Na_2CO_3)$ and $M(NaHCO_3)$
          Conceptual Plan: $P, V, T \rightarrow n$ in 300.0 cm^3 then n in 300.0 cm^3 $\rightarrow n$ in 1.00 L then

$$PV = nRT \qquad\qquad\qquad\qquad \text{take ratio of volumes}$$

          set up equations for the total mass and the total moles and solve. Then calculate concentrations.

$$g_{Na_2CO_3} + g_{NaHCO_3} = 11.60 \text{ g}\frac{1 \text{ mol } Na_2CO_3}{105.99 \text{ g } Na_2CO_3}\frac{1 \text{ mol } NaHCO_3}{84.01 \text{ g } NaHCO_3} \quad n_{Na_2CO_3} + n_{NaHCO_3} = n$$

          Solution: $PV = nRT$. Rearrange to solve for n. $n = \dfrac{PV}{RT}$

$$n_{CO_2} = \frac{0.972 \text{ atm} \times 0.940 \text{ L}}{0.08206\dfrac{L \cdot atm}{mol \cdot K} \times 298 \text{ K}} = 0.0373\underline{6}340 \text{ mol } CO_2 \text{ in 300.0 cm}^3 \text{ then take ratio of moles to volume to get the}$$

          moles in 1.00 L.

$$\frac{0.03736340 \text{ mol } CO_2}{300.0 \text{ cm}^3} \times 1000 \text{ cm}^3 = 0.1245447 \text{ mol } CO_2 \text{ in } 1.00 \text{ L because 1 mole of } CO_2 \text{ is generated for each}$$

mole of carbonate. So $n = n_{Na_2CO_3} + n_{NaHCO_3} = 0.1245447 \text{ mol}$ and $g_{Na_2CO_3} + g_{NaHCO_3} = 11.60 \text{ g}$ or

$g_{Na_2CO_3} = 11.60 \text{ g} - g_{NaHCO_3}$ using molar masses and substituting

$$n_{Na_2CO_3} + n_{NaHCO_3} = 0.1245447 \text{ mol} = g_{Na_2CO_3} \times \frac{1 \text{ mol } Na_2CO_3}{105.99 \text{ g } Na_2CO_3} + g_{NaHCO_3} \times \frac{1 \text{ mol } NaHCO_3}{84.01 \text{ g } NaHCO_3} =$$

$$(11.60 \text{ g} - g_{NaHCO_3}) \times \frac{1 \text{ mol } Na_2CO_3}{105.99 \text{ g } Na_2CO_3} + g_{NaHCO_3} \times \frac{1 \text{ mol } NaHCO_3}{84.01 \text{ g } NaHCO_3}$$

$$\rightarrow 0.1245447 \text{ mol} = 0.10944429 \text{ mol} - g_{NaHCO_3} \times 0.009434852 \frac{\text{mol}}{\text{g}} + g_{NaHCO_3} 0.011903345 \frac{\text{mol}}{\text{g}}$$

$$\rightarrow 0.01510041 \text{ mol} = g_{NaHCO_3} \times 0.002468493 \frac{\text{mol}}{\text{g}} \rightarrow$$

$$g_{NaHCO_3} = \frac{0.01510041 \text{ mol}}{0.002468493 \frac{\text{mol}}{\text{g}}} = 6.117259 \text{ g } NaHCO_3 = 6.1 \text{ g } NaHCO_3 \text{ and then } g_{Na_2CO_3} =$$

$11.60 \text{ g} - g_{NaHCO_3} = 11.60 \text{ g} - 6.117259 \text{ g } NaHCO_3 = 5.48274 \text{ g } Na_2CO_3 = 5.5 \text{ g } Na_2CO_3 \text{ then M} = \text{mol/L}$

$$\frac{6.117259 \text{ g } NaHCO_3 \times \frac{1 \text{ mol } NaHCO_3}{84.01 \text{ g } NaHCO_3}}{1.00 \text{ L}} = 0.073 \text{ M } NaHCO_3 \text{ and}$$

$$\frac{5.48274 \text{ g } Na_2CO_3 \times \frac{1 \text{ mol } Na_2CO_3}{105.99 \text{ g } Na_2CO_3}}{1.00 \text{ L}} = 0.052 \text{ M } Na_2CO_3$$

Check: The units (M and M) are correct. The magnitude of the answers (0.073 M and 0.052 M) makes sense because the number of moles of CO_2 is small (0.12 M). The balance of the two components makes sense because if it were all Na_2CO_3, the number of moles would have been 0.109 mole (11.6/105.99) and if it were all $NaHCO_3$, the number of moles would have been 0.138 mole (11.6/84.01)—the actual number of moles is roughly in the middle of these two values.

Conceptual Problems

13.131   The warm coolant water should not be put directly in the river without being cooled because it will raise the temperature of the water. When water is warmed, there is less dissolved oxygen in the water; this will be detrimental to aquatic life that depends on the dissolved oxygen.

13.133   (e)   NaCl. If all of the substances have the same cost per kilogram, we need to determine which substance will generate the largest number of particles per kilogram (or gram). $HOCH_2CH_2OH$ generates 1 mol particle/62.07 g, NaCl generates 2 mol particles/58.44 g, KCl generates 2 mol particles/74.55 g, $MgCl_2$ generates 3 mol particles/95.22 g, and $SrCl_2$ generates 3 particles/158.52 g. So NaCl will generate 1 mole of particles for each 29 g.

Questions for Group Work

13.135   1-Propanol contains three carbon atoms and an alcohol functional group (—OH). The end with the carbon chain is nonpolar and so is soluble in nonpolar solvents, like hexane. The alcohol group exhibits hydrogen bonding, and so it can interact well with water. Since hexane is nonpolar, it cannot interact very well with water (a very polar molecule with strong hydrogen bonding), so hexane has little solubility in water.

13.137

	Increasing Temperature	Increasing Pressure
Solubility of gas in water	Decreases	Increases
Solubility of a solid in water	Generally increases	No effect

13.139   **Given:** 0.165 m sucrose **Other:** $K_f = 1.86\,°C/m$; $K_b = 0.512\,°C/m$ **Find:** T_f and T_b

Conceptual Plan: $m, K_f \rightarrow \Delta T_f \rightarrow T_f$ and $m, K_b \rightarrow \Delta T_b \rightarrow T_b$

$$\Delta T_f = K_f m \quad T_f = T_f° - \Delta T_f \qquad \Delta T_b = K_b m \quad \Delta T_b = T_b - T_b°$$

Solution: $\Delta T_f = K_f m = 1.86\,\dfrac{°C}{m} \times 0.165\,m = 0.307\,°C$ then $T_f = T_f° - \Delta T_f = 0.000\,°C - 0.307\,°C = -0.307\,°C$

and $\Delta T_b = K_b m = 0.512\,\dfrac{°C}{m} \times 0.165\,m = 0.0845\,°C$ and $\Delta T_b = T_b - T_b°$ so

$T_b = T_b° + \Delta T_b = 100.000\,°C + 0.0845\,°C = 100.0845\,°C$

Check: The units (°C) are correct. The magnitudes of the answers seem reasonable since the molality is $\sim 0.2\,m$. The shift in boiling point is less than the shift in freezing point because the constant is smaller for freezing.

In order to boil this sucrose solution, the solution needs to be raised to a slightly higher temperature than for pure water. In addition, the heat capacity of a sucrose solution is slightly higher than pure water. This results in a slightly increased time to boil the sucrose solution.

The syrup will freeze in a typical freezer because the freezing point has been dropped by less than 1 °C.

Data Interpretation and Analysis

13.140   (a)   The higher salinity levels are found east of North and South America, east of the Arabian Peninsula. The lower salinity levels are found near the North and South Poles. In general, warmer waters have higher salinity levels.

(c)   Highest salinity 38 g NaCl/1000 g seawater, so 962 g water/1000 g seawater = 0.962 kg water/ 1 kg seawater

$38\,g\,NaCl \times \dfrac{1\,mol\,NaCl}{58.44\,g\,NaCl} = 0.6\underline{5}024\,mol\,NaCl$, assume $\quad i = 1.9$

$\Delta T_f = i \times m \times K_f = 1.9 \times \dfrac{0.65024\,mol\,NaCl}{0.962\,kg\,water} \times \dfrac{1.86\,°C}{m} = 2.\underline{3}887\,°C;\ T_f = -2.4\,°C.$

Lowest salinity 33 g NaCl/1000 g seawater, so 967 g water/1000 g seawater = 0.967 kg water/ 1 kg seawater

$33\,g\,NaCl \times \dfrac{1\,mol\,NaCl}{58.44\,g\,NaCl} = 0.5\underline{6}468\,mol\,NaCl$, assume $\quad i = 1.9$

$\Delta T_f = i \times m \times K_f = 1.9 \times \dfrac{0.56468\,mol\,NaCl}{0.967\,kg\,water} \times \dfrac{1.86\,°C}{m} = 2.\underline{0}637\,°C;\ T_f = -2.1\,°C.$

14 Chemical Kinetics

Review Questions

14.1 Unlike mammals, which actively regulate their body temperature through metabolic activity, lizards are ectotherms—their body temperature depends on their surroundings. When splashed with cold water, a lizard's body gets colder. The drop in body temperature immobilizes the lizard because its movement depends on chemical reactions that occur within its muscles, and the rates of those reactions—how fast they occur—are highly sensitive to temperature. In other words, when the temperature drops, the reactions that produce movement occur more slowly; therefore, the movement itself slows down. Cold reptiles are lethargic, unable to move very quickly. For this reason, reptiles try to maintain their body temperature in a narrow range by moving between sun and shade.

14.3 Since reactions occur as a result of collisions between particles, the more particles are in a reaction vessel, the higher the concentration of particles, the more collisions are possible. The more collisions that occur, the faster the reaction rate.

14.5 The rate of a chemical reaction is measured as a change in the amounts of reactants or products (usually in terms of concentration) divided by the change in time. Typical units are molarity per second (M/s), molarity per minute (M/min), and molarity per year (M/yr), depending on how fast the reaction proceeds.

14.7 The average rate of the reaction can be calculated for any time interval as

$$\text{Rate} = -\frac{1}{a}\frac{[A]_{t_2} - [A]_{t_1}}{t_2 - t_1} = -\frac{1}{b}\frac{[B]_{t_2} - [B]_{t_1}}{t_2 - t_1} = \frac{1}{c}\frac{[C]_{t_2} - [C]_{t_1}}{t_2 - t_1} = \frac{1}{d}\frac{[D]_{t_2} - [D]_{t_1}}{t_2 - t_1} \text{ for the}$$

chemical reaction $a\text{A} + b\text{B} \rightarrow c\text{C} + d\text{D}$. The instantaneous rate of the reaction is the rate at any one point in time, represented by the instantaneous slope of the plot of concentration versus time at that point. We can obtain the instantaneous rate from the slope of the tangent to this curve at the point of interest.

14.9 The reaction order cannot be determined by the stoichiometry of the reaction. It can only be determined by running controlled experiments where the concentrations of the reactants are varied and the reaction rates are measured and analyzed.

14.11 The rate law shows the relationship between the rate of a reaction and the concentrations of the reactants. The integrated rate law for a chemical reaction is a relationship between the concentration of a reactant and time.

14.13 The half-life ($t_{1/2}$) of a reaction is the time required for the concentration of a reactant to fall to one-half of its initial value. For a zero-order reaction, $t_{1/2} = \dfrac{[A]_0}{2k}$. For a first-order reaction, $t_{1/2} = \dfrac{0.693}{k}$. For a second-order reaction, $t_{1/2} = \dfrac{1}{k[A]_0}$.

14.15 The modern form of the Arrhenius equation, which relates the rate constant (k) and the temperature in kelvin (T), is as follows: $k = A\,e^{-E_a/RT}$, where R is the gas constant ($8.314\text{ J/mol} \cdot \text{K}$), A is a constant called the frequency factor (or the pre-exponential factor), and E_a is called the activation energy (or activation barrier). The frequency factor is the number of times the reactants approach the activation barrier per unit time. The exponential factor $(-E_a/RT)$ is the fraction of approaches that are successful in surmounting the activation barrier and forming products. The exponential factor increases with increasing temperature, but decreases with an increasing value for the activation energy. As the temperature increases, the number of collisions increases and the number of molecules having enough thermal energy to surmount the activation barrier increases. At any given temperature, a sample of molecules will have a distribution of energies, as shown in Figure 14.14. Under common circumstances, only a small fraction of the molecules have enough energy to make it over the activation barrier. Because of the shape of the energy distribution curve, however, a small change in temperature results in a large difference in the number of molecules having enough energy to surmount the activation barrier.

14.17   The orientation factor says that if two molecules are to react with each other, they must collide in such a way that allows the necessary bonds to break and form. The small orientation factor indicates that the orientational requirements for this reaction are fairly stringent—the molecules must be aligned in a very specific way for the reaction to occur. When two molecules with sufficient energy and the correct orientation collide, something unique happens: The electrons on one of the atoms or molecules are attracted to the nuclei of the other; some bonds begin to weaken, and other bonds begin to form. If all goes well, the reactants go through the transition state and are transformed into the products and a chemical reaction occurs.

14.19   An elementary step is a single step in a reaction mechanism. Elementary steps cannot be broken down into simpler steps—they occur as they are written. Elementary steps are characterized by their molecularity, the number of reactant particles involved in the step. The molecularity of the three most common types of elementary steps are as follows: unimolecular—$A \rightarrow$ products and Rate $= k[A]$; bimolecular—$A + A \rightarrow$ products and Rate $= k[A]^2$; and bimolecular—$A + B \rightarrow$ products and Rate $= k[A][B]$. Elementary steps in which three reactant particles collide, called termolecular steps, are very rare because the probability of three particles simultaneously colliding is small.

14.21   Reaction intermediates are species that are formed in one step of a mechanism and consumed in another step. An intermediate is not found in the balanced equation for the overall reaction but plays a key role in the mechanism.

14.23   In homogeneous catalysis, the catalyst exists in the same phase as the reactants. In heterogeneous catalysis, the catalyst exists in a phase different from that of the reactants. The most common type of heterogeneous catalyst is a solid catalyst.

14.25   Enzymes are biological catalysts that increase the rates of biochemical reactions. Enzymes are large protein molecules with complex three-dimensional structures. Within that structure is a specific area called the active site. The properties and shape of the active site are just right to bind the reactant molecule, usually called the substrate. The substrate fits into the active site in a manner that is analogous to a key fitting into a lock.

Reaction Rates

14.27   (a) $$\text{Rate} = -\frac{1}{2}\frac{\Delta[\text{HBr}]}{\Delta t} = \frac{\Delta[\text{H}_2]}{\Delta t} = \frac{\Delta[\text{Br}_2]}{\Delta t}$$

(b)   **Given:** first 25.0 s; 0.600 M to 0.512 M   **Find:** average rate
Conceptual Plan: $t_1, t_2, [\text{HBr}]_1, [\text{HBr}]_2 \rightarrow$ **average rate**

$$\text{Rate} = -\frac{1}{2}\frac{\Delta[\text{HBr}]}{\Delta t}$$

Solution:

$$\text{Rate} = -\frac{1}{2}\frac{[\text{HBr}]_{t_2} - [\text{HBr}]_{t_1}}{t_2 - t_1} = -\frac{1}{2}\frac{0.512\text{ M} - 0.600\text{ M}}{25.0\text{ s} - 0.0\text{ s}} = 1.\underline{76} \times 10^{-3}\text{ M}\cdot\text{s}^{-1} = 1.8 \times 10^{-3}\text{ M}\cdot\text{s}^{-1}$$

Check: The units $(\text{M}\cdot\text{s}^{-1})$ are correct. The magnitude of the answer $(10^{-3}\text{ M}\cdot\text{s}^{-1})$ makes physical sense because rates are always positive and we are not changing the concentration much in 25 s.

(c)   **Given:** 1.50 L vessel, first 15.0 s of reaction, and part (b) data   **Find:** mol_{Br_2} formed
Conceptual Plan: Average rate, $t_1, t_2, \rightarrow \Delta[\text{Br}_2]$ then $\Delta[\text{Br}_2], \text{L} \rightarrow \text{mol}_{\text{Br}_2}$ formed

$$\text{Rate} = \frac{\Delta[\text{Br}_2]}{\Delta t} \qquad\qquad M = \frac{\text{mol}_{\text{Br}_2}}{\text{L}}$$

Solution: Rate $= 1.\underline{7}6 \times 10^{-3} \text{ M} \cdot \text{s}^{-1} = \dfrac{\Delta[\text{Br}_2]}{\Delta t} = \dfrac{\Delta[\text{Br}_2]}{15.0 \text{ s} - 0.0 \text{ s}}$ Rearrange to solve for $\Delta[\text{Br}_2]$.

$\Delta[\text{Br}_2] = 1.\underline{7}6 \times 10^{-3} \dfrac{\text{M}}{\text{s}} \times 15.0 \text{ s} = 0.02\underline{6}4 \text{ M}$ then $\text{M} = \dfrac{\text{mol}_{\text{Br}_2}}{\text{L}}$ Rearrange to solve for mol_{Br_2}.

$0.02\underline{6}4 \dfrac{\text{mol Br}_2}{\text{L}} \times 1.50 \text{ L} = 0.040 \text{ mol Br}_2$

Check: The units (mol) are correct. The magnitude of the answer (0.04 mol) makes physical sense because the time is shorter than in part (b) and we need to divide by 2 and multiply by 1.5 because of the stoichiometric coefficient difference and the volume of the vessel, respectively.

14.29 (a) Rate $= -\dfrac{1}{2}\dfrac{\Delta[\text{A}]}{\Delta t} = -\dfrac{\Delta[\text{B}]}{\Delta t} = \dfrac{1}{3}\dfrac{\Delta[\text{C}]}{\Delta t}$

(b) **Given:** $\dfrac{\Delta[\text{A}]}{\Delta t} = -0.100 \text{ M/s}$ **Find:** $\dfrac{\Delta[\text{B}]}{\Delta t}$ and $\dfrac{\Delta[\text{C}]}{\Delta t}$

Conceptual Plan: $\dfrac{\Delta[\text{A}]}{\Delta t} \rightarrow \dfrac{\Delta[\text{B}]}{\Delta t}$ and $\dfrac{\Delta[\text{C}]}{\Delta t}$

$$\text{Rate} = -\dfrac{1}{2}\dfrac{\Delta[\text{A}]}{\Delta t} = -\dfrac{\Delta[\text{B}]}{\Delta t} = \dfrac{1}{3}\dfrac{\Delta[\text{C}]}{\Delta t}$$

Solution: Rate $= -\dfrac{1}{2}\dfrac{\Delta[\text{A}]}{\Delta t} = -\dfrac{\Delta[\text{B}]}{\Delta t} = \dfrac{1}{3}\dfrac{\Delta[\text{C}]}{\Delta t}$. Substitute value and solve for the two desired

values. $-\dfrac{1}{2}\dfrac{-0.100 \text{ M}}{\text{s}} = -\dfrac{\Delta[\text{B}]}{\Delta t}$ so $\dfrac{\Delta[\text{B}]}{\Delta t} = -0.0500 \text{ M} \cdot \text{s}^{-1}$ and $-\dfrac{1}{2}\dfrac{-0.100 \text{ M}}{\text{s}} = \dfrac{1}{3}\dfrac{\Delta[\text{C}]}{\Delta t}$

so $\dfrac{\Delta[\text{C}]}{\Delta t} = 0.150 \text{ M} \cdot \text{s}^{-1}$

Check: The units $(\text{M} \cdot \text{s}^{-1})$ are correct. The magnitude of the answer $(-0.05 \text{ M} \cdot \text{s}^{-1})$ makes physical sense because fewer moles of B are reacting for every mole of A and the change in concentration with time is negative because this is a reactant. The magnitude of the answer $(0.15 \text{ M} \cdot \text{s}^{-1})$ makes physical sense because more moles of C are being formed for every mole of A reacting and the change in concentration with time is positive because this is a product.

14.31 **Given:** $\text{Cl}_2(g) + 3\text{F}_2(g) \rightarrow 2\text{ClF}_3(g)$; $\Delta[\text{Cl}_2]/\Delta t = -0.012 \text{ M/s}$ **Find:** $\Delta[\text{F}_2]/\Delta t$; $\Delta[\text{ClF}_3]/\Delta t$; and Rate
Conceptual Plan: Write the expression for the rate with respect to each species then

$$\text{Rate} = -\dfrac{\Delta[\text{Cl}_2]}{\Delta t} = -\dfrac{1}{3}\dfrac{\Delta[\text{F}_2]}{\Delta t} = \dfrac{1}{2}\dfrac{\Delta[\text{ClF}_3]}{\Delta t}$$

rate expression, $\dfrac{\Delta[\text{Cl}_2]}{\Delta t} \rightarrow \dfrac{\Delta[\text{F}_2]}{\Delta t}$; $\dfrac{\Delta[\text{ClF}_3]}{\Delta t}$; **Rate**

$$\text{Rate} = -\dfrac{\Delta[\text{Cl}_2]}{\Delta t} = -\dfrac{1}{3}\dfrac{\Delta[\text{F}_2]}{\Delta t} = \dfrac{1}{2}\dfrac{\Delta[\text{ClF}_3]}{\Delta t}$$

Solution: Rate $= -\dfrac{\Delta[\text{Cl}_2]}{\Delta t} = -\dfrac{1}{3}\dfrac{\Delta[\text{F}_2]}{\Delta t} = \dfrac{1}{2}\dfrac{\Delta[\text{ClF}_3]}{\Delta t}$ so

$-\dfrac{\Delta[\text{Cl}_2]}{\Delta t} = -\dfrac{1}{3}\dfrac{\Delta[\text{F}_2]}{\Delta t}$ Rearrange to solve for $\dfrac{\Delta[\text{F}_2]}{\Delta t}$. $\dfrac{\Delta[\text{F}_2]}{\Delta t} = 3\dfrac{\Delta[\text{Cl}_2]}{\Delta t} = 3(-0.012 \text{ M} \cdot \text{s}^{-1}) = -0.036 \text{ M} \cdot \text{s}^{-1}$

and $-\dfrac{\Delta[\text{Cl}_2]}{\Delta t} = \dfrac{1}{2}\dfrac{\Delta[\text{ClF}_3]}{\Delta t}$ Rearrange to solve for $\dfrac{\Delta[\text{ClF}_3]}{\Delta t}$.

$\dfrac{\Delta[\text{ClF}_3]}{\Delta t} = -2\dfrac{\Delta[\text{Cl}_2]}{\Delta t} = -2(-0.012 \text{ M} \cdot \text{s}^{-1}) = 0.024 \text{ M} \cdot \text{s}^{-1}$

Check: The units $(\text{M} \cdot \text{s}^{-1})$ are correct. The magnitude of the answers $(-0.036 \text{ M} \cdot \text{s}^{-1}$ and $0.024 \text{ M} \cdot \text{s}^{-1})$ makes physical sense because F_2 is being used at three times the rate of Cl_2, ClF_3 is being formed at two times the rate of Cl_2 disappearance, and Cl_2 has a stoichiometric coefficient of 1.

14.33 (a) **Given:** $[\text{C}_4\text{H}_8]$ versus time data **Find:** average rate between 0 and 10 s and between 40 and 50 s
Conceptual Plan: $t_1, t_2, [\text{C}_4\text{H}_8]_1, [\text{C}_4\text{H}_8]_2 \rightarrow$ **average rate**

$$\text{Rate} = -\dfrac{\Delta[\text{C}_4\text{H}_8]}{\Delta t}$$

Solution: For 0 to 10 s, Rate $= -\dfrac{[C_4H_8]_{t_2} - [C_4H_8]_{t_1}}{t_2 - t_1} = -\dfrac{0.913\,M - 1.000\,M}{10.\,s - 0.\,s} = 8.7 \times 10^{-3}\,M \cdot s^{-1}$ and

for 40 to 50 s, Rate $= -\dfrac{[C_4H_8]_{t_2} - [C_4H_8]_{t_1}}{t_2 - t_1} = -\dfrac{0.637\,M - 0.697\,M}{50.\,s - 40.\,s} = 6.0 \times 10^{-3}\,M \cdot s^{-1}$

Check: The units $(M \cdot s^{-1})$ are correct. The magnitude of the answers $(10^{-3}\,M \cdot s^{-1})$ makes physical sense because rates are always positive and we are not changing the concentration much in 10 s. Also, reactions slow as they proceed because the concentration of the reactants is decreasing.

(b)   **Given:** $[C_4H_8]$ versus time data   **Find:** $\dfrac{\Delta[C_2H_4]}{\Delta t}$ between 20 and 30 s

Conceptual Plan: $t_1, t_2, [C_4H_8]_1, [C_4H_8]_2 \to \dfrac{\Delta[C_2H_4]}{\Delta t}$

$$\text{Rate} = -\frac{\Delta[C_4H_8]}{\Delta t} = \frac{1}{2}\frac{\Delta[C_2H_4]}{\Delta t}$$

Solution: Rate $= -\dfrac{[C_4H_8]_{t_2} - [C_4H_8]_{t_1}}{t_2 - t_1} = -\dfrac{0.763\,M - 0.835\,M}{30.\,s - 20.\,s} = 7.2 \times 10^{-3}\,M \cdot s^{-1} = \dfrac{1}{2}\dfrac{\Delta[C_2H_4]}{\Delta t}$

Rearrange to solve for $\dfrac{\Delta[C_2H_4]}{\Delta t}$. So $\dfrac{\Delta[C_2H_4]}{\Delta t} = 2(7.2 \times 10^{-3}\,M \cdot s^{-1}) = 1.4 \times 10^{-2}\,M \cdot s^{-1}$

Check: The units $(M \cdot s^{-1})$ are correct. The magnitude of the answer $(10^{-2}\,M \cdot s^{-1})$ makes physical sense because the rate of product formation is always positive and we are not changing the concentration much in 10 s. The rate of change of the product is faster than the decline of the reactant because of the stoichiometric coefficients.

14.35   (a)   **Given:** $[Br_2]$ versus time plot
Find: (i) average rate between 0 and 25 s; (ii) instantaneous rate at 25 s; (iii) instantaneous rate of HBr formation at 50 s
Conceptual Plan: (i) $t_1, t_2, [Br_2]_1, [Br_2]_2 \to$ **average rate then**

$$\text{Rate} = -\frac{\Delta[Br_2]}{\Delta t}$$

(ii) **draw tangent at 25 s and determine slope** $\to$ **instantaneous rate then**

$$\text{Rate} = -\frac{\Delta[Br_2]}{\Delta t}$$

(iii) **draw tangent at 50 s and determine slope** $\to$ **instantaneous rate** $\to$ $\dfrac{\Delta[HBr]}{\Delta t}$

$$\text{Rate} = -\frac{\Delta[Br_2]}{\Delta t} \qquad \text{Rate} = \frac{1}{2}\frac{\Delta[HBr]}{\Delta t}$$

Solution:

(i) Rate $= -\dfrac{[Br_2]_{t_2} - [Br_2]_{t_1}}{t_2 - t_1} = -\dfrac{0.75\,M - 1.00\,M}{25\,s - 0.\,s} = 1.0 \times 10^{-2}\,M \cdot s^{-1}$ and (ii) at 25 s:

Slope $= \dfrac{\Delta y}{\Delta x} = \dfrac{0.68\,M - 0.85\,M}{35\,s - 15\,s} = -8.5 \times 10^{-3}\,M \cdot s^{-1}$

because the slope $= \dfrac{\Delta[Br_2]}{\Delta t}$ and Rate $= -\dfrac{\Delta[Br_2]}{\Delta t}$,

then Rate $= -(-8.5 \times 10^{-3}\,M \cdot s^{-1}) = 8.5 \times 10^{-3}\,M \cdot s^{-1}$

(iii) at 50 s:

Slope $= \dfrac{\Delta y}{\Delta x} = \dfrac{0.53\,M - 0.66\,M}{60.\,s - 40.\,s} = -6.5 \times 10^{-3}\,M \cdot s^{-1}$

because the slope $= \dfrac{\Delta[Br_2]}{\Delta t}$ and

Rate $= -\dfrac{\Delta[Br_2]}{\Delta t} = \dfrac{1}{2}\dfrac{\Delta[HBr]}{\Delta t}$ then

$\dfrac{\Delta[HBr]}{\Delta t} = -2\dfrac{\Delta[Br_2]}{\Delta t} = -2(-6.5 \times 10^{-3}\,M \cdot s^{-1}) = 1.3 \times 10^{-2}\,M \cdot s^{-1}$

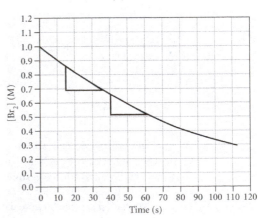

Check: The units $(M \cdot s^{-1})$ are correct. The magnitude of the first answer is larger than that of the second answer because the rate is slowing down and the first answer includes the initial portion of the data. The magnitude of the answers $(10^{-3} \, M \cdot s^{-1})$ makes physical sense because rates are always positive and we are not changing the concentration much.

(b)   **Given:** $[Br_2]$ versus time data and $[HBr]_0 = 0 \, M$ **Find:** plot [HBr] with time

Conceptual Plan: Because $Rate = -\dfrac{\Delta[Br_2]}{\Delta t} = \dfrac{1}{2}\dfrac{\Delta[HBr]}{\Delta t}$. **The rate of change of [HBr] will be twice that of $[Br_2]$. The plot will start at the origin.**

Solution:

Check: The units (M versus s) are correct. The plot makes sense because it has the same general shape of the original plot except that we are increasing instead of decreasing our concentration axis by a factor of two (to account for the difference in stoichiometric coefficients).

The Rate Law and Reaction Orders

14.37   (a)   **Given:** Rate versus [A] plot   **Find:** reaction order
Conceptual Plan: Look at shape of plot and match to possibilities.
Solution: The plot is a linear plot, so Rate α [A] or the reaction is first order.

Check: The order of the reaction is a common reaction order.

(b)   **Given:** part (a)   **Find:** sketch plot of [A] versus time
Conceptual Plan: Using the result from part (a), shape plot of [A] versus time should be curved with [A] decreasing. Use 1.0 M as initial concentration.
Solution:

Rate $= k$ [A]
plugging in any of the values from the graph
$0.002 \, M \cdot s^{-1} = k \, [0.2 \, M]$
$k = \dfrac{0.002 \, M \cdot s^{-1}}{0.2 \, M}$
$k = 0.01 \, s^{-1}$

Check: The plot has a shape that matches the one in the text for first-order plots.

(c)   **Given:** part (a)   **Find:** write a rate law and estimate k
Conceptual Plan: Using result from part (a), the slope of the plot is the rate constant.

Solution: Slope $= \dfrac{\Delta y}{\Delta x} = \dfrac{0.010\dfrac{M}{s} - 0.00\dfrac{M}{s}}{1.0 \, M - 0.0 \, M} = 0.010 \, s^{-1}$ so Rate $= k[A]^1$ or Rate $= k[A]$ or

Rate $= 0.010 \text{ s}^{-1} [A]$

Check: The units (s^{-1}) are correct. The magnitude of the answer (10^{-2} s^{-1}) makes physical sense because of the rate and concentration data. Remember that concentration is in units of M; so plugging the rate constant into the equation has the units of the rate as $M \cdot \text{s}^{-1}$, which is correct.

14.39   **Given:** reaction order: (a) first order, (b) second order, and (c) zero order   **Find:** units of k

Conceptual Plan: Using rate law, rearrange to solve for k.

$$\text{Rate} = k[A]^n, \text{ where } n = \text{reaction order}$$

Solution: For all cases, rate has units of $M \cdot \text{s}^{-1}$ and $[A]$ has units of M.

(a)   Rate $= k[A]^1 = k[A]$ so $k = \dfrac{\text{Rate}}{[A]} = \dfrac{\frac{M}{\text{s}}}{M} = \text{s}^{-1}$

(b)   Rate $= k[A]^2$ so $k = \dfrac{\text{Rate}}{[A]^2} = \dfrac{\frac{M}{\text{s}}}{M \cdot M} = M^{-1} \cdot \text{s}^{-1}$

(c)   Rate $= k[A]^0 = k = M \cdot \text{s}^{-1}$

Check: The units $(\text{s}^{-1}, M^{-1} \cdot \text{s}^{-1}, \text{ and } M \cdot \text{s}^{-1})$ are correct. The units for k change with the reaction order so that the units on the rate remain as $M \cdot \text{s}^{-1}$.

14.41   **Given:** A, B, and C react to form products. Reaction is first order in A, second order in B, and zero order in C. **Find:** (a) rate law; (b) overall order of reaction; (c) factor change in rate if [A] doubled; (d) factor change in rate if [B] doubled; (e) factor change in rate if [C] doubled; and (f) factor change in rate if [A], [B], and [C] doubled

Conceptual Plan:

(a)   **Using general rate law form, substitute values for orders.**

$$\text{Rate} = k[A]^m[B]^n[C]^p, \text{ where } m, n, \text{ and } p = \text{reaction orders}$$

(b)   **Using rate law in part (a), add up all reaction orders.**

$$overall\ reaction\ order = m + n + p$$

(c)   **Through (f), using rate law from part (a), substitute concentration changes.**

$$\frac{\text{Rate 2}}{\text{Rate 1}} = \frac{k[A]_2^1[B]_2^2}{k[A]_1^1[B]_1^2}$$

Solution:

(a) $m = 1, n = 2,$ and $p = 0$ so Rate $= k[A]^1[B]^2[C]^0$ or Rate $= k[A][B]^2$

(b) $overall\ reaction\ order = m + n + p = 1 + 2 + 0 = 3$, so it is a third-order reaction overall

(c) $\dfrac{\text{Rate 2}}{\text{Rate 1}} = \dfrac{k[A]_2^1[B]_2^2}{k[A]_1^1[B]_1^2}$ and $[A]_2 = 2[A]_1, [B]_2 = [B]_1, [C]_2 = [C]_1,$ so $\dfrac{\text{Rate 2}}{\text{Rate 1}} = \dfrac{k(2[A]_1)^1[B]_1^2}{k[A]_1^1[B]_1^2} = 2$

     so the reaction rate doubles (factor of 2)

(d) $\dfrac{\text{Rate 2}}{\text{Rate 1}} = \dfrac{k[A]_2^1[B]_2^2}{k[A]_1^1[B]_1^2}$ and $[A]_2 = [A]_1, [B]_2 = 2[B]_1, [C]_2 = [C]_1,$ so $\dfrac{\text{Rate 2}}{\text{Rate 1}} = \dfrac{k[A]_1(2[B]_1)^2}{k[A]_1^1[B]_1^2} = 2^2 = 4$

     so the reaction rate quadruples (factor of 4)

(e) $\dfrac{\text{Rate 2}}{\text{Rate 1}} = \dfrac{k[A]_2^1[B]_2^2}{k[A]_1^1[B]_1^2}$ and $[A]_2 = [A]_1, [B]_2 = [B]_1, [C]_2 = 2[C]_1,$ so $\dfrac{\text{Rate 2}}{\text{Rate 1}} = \dfrac{k[A]_1[B]_1^2}{k[A]_1[B]_1^2} = 1$

     so the reaction rate is unchanged (factor of 1)

(f) $\dfrac{\text{Rate 2}}{\text{Rate 1}} = \dfrac{k[A]_2^1[B]_2^2}{k[A]_1^1[B]_1^2}$ and $[A]_2 = 2[A]_1, [B]_2 = 2[B]_1, [C]_2 = 2[C]_1,$ so

     $\dfrac{\text{Rate 2}}{\text{Rate 1}} = \dfrac{k(2[A]_1)^1(2[B]_1)^2}{k[A]_1[B]_1^2} = 2 \times 2^2 = 8$ so the reaction rate goes up by a factor of 8

Check: The units (none) are correct. The rate law is consistent with the orders given, and the overall order is larger than any of the individual orders. The factors are consistent with the reaction orders. The larger the order, the larger the factor. When all concentrations are changed, the rate changes the most. If a reactant is not in the rate law, then changing its concentration has no effect on the reaction rate.

14.43 **Given:** table of [A] versus initial rate   **Find:** rate law and k

Conceptual Plan: Using general rate law form, compare rate ratios to determine reaction order.

$$\frac{\text{Rate 2}}{\text{Rate 1}} = \frac{k[A]_2^n}{k[A]_1^n}$$

Then use one of the concentration/initial rate pairs to determine k.

$$\text{Rate} = k[A]^n$$

Solution: $\dfrac{\text{Rate 2}}{\text{Rate 1}} = \dfrac{k[A]_2^n}{k[A]_1^n}$ Comparing the first two sets of data, $\dfrac{0.210\ \cancel{M/s}}{0.053\ \cancel{M/s}} = \dfrac{k(0.200\ \text{M})^n}{k(0.100\ \text{M})^n}$ and $3.9623 = 2^n$

so $n = 2$ If we compare the first and the last data sets, $\dfrac{0.473\ \cancel{M/s}}{0.053\ \cancel{M/s}} = \dfrac{k(0.300\ \text{M})^n}{k(0.100\ \text{M})^n}$ and $8.9245 = 3^n$ so $n = 2$

This second comparison is not necessary, but it increases our confidence in the reaction order. So Rate $= k[A]^2$. Selecting the second data set and rearranging the rate equation,

$$k = \frac{\text{Rate}}{[A]^2} = \frac{0.210\dfrac{\text{M}}{\text{s}}}{(0.200\ \text{M})^2} = 5.25\ \text{M}^{-1}\cdot\text{s}^{-1},\ \text{so Rate} = 5.25\ \text{M}^{-1}\cdot\text{s}^{-1}[A]^2$$

Check: The units (none and $\text{M}^{-1}\cdot\text{s}^{-1}$) are correct. The rate law is a common form. The rate is changing more rapidly than the concentration, so second order is consistent. The rate constant is consistent with the units necessary to get rate as M/s, and the magnitude is reasonable because we have a second-order reaction.

14.45 **Given:** table of $[NO_2]$ and $[F_2]$ versus initial rate   **Find:** rate law, k, and overall order

Conceptual Plan: Using general rate law form, compare rate ratios to determine reaction order of each reactant. Be sure to choose data that changes only one concentration at a time.

$$\frac{\text{Rate 2}}{\text{Rate 1}} = \frac{k[NO_2]_2^m\,[F_2]_2^n}{k[NO_2]_1^m\,[F_2]_1^n}$$

Then use one of the concentration/initial rate pairs to determine k.

$$\text{Rate} = k[NO_2]^m\,[F_2]^n$$

Solution: $\dfrac{\text{Rate 2}}{\text{Rate 1}} = \dfrac{k[NO_2]_2^m\,[F_2]_2^n}{k[NO_2]_1^m\,[F_2]_1^n}$ Comparing the first two sets of data,

$\dfrac{0.051\ \cancel{M/s}}{0.026\ \cancel{M/s}} = \dfrac{k(0.200\ \text{M})^m\,(\cancel{0.100\ \text{M}})^n}{k(0.100\ \text{M})^m\,(\cancel{0.100\ \text{M}})^n}$ and $1.9615 = 2^m$; so $m = 1$. If we compare the second and third

data sets, $\dfrac{0.103\ \cancel{M/s}}{0.051\ \cancel{M/s}} = \dfrac{k(\cancel{0.200\ \text{M}})^m\,(0.200\ \text{M})^n}{k(\cancel{0.200\ \text{M}})^m\,(0.100\ \text{M})^n}$ and $2 = 2^n$, $n = 1$. Other comparisons can be made, but they are

not necessary. They should reinforce these values of the reaction orders. So Rate $= k[NO_2][F_2]$. Selecting the last

data set and rearranging the rate equation, $k = \dfrac{\text{Rate}}{[NO_2][F_2]} = \dfrac{0.411\dfrac{\text{M}}{\text{s}}}{(0.400\ \text{M})(0.400\ \text{M})} = 2.57\ \text{M}^{-1}\cdot\text{s}^{-1}$; so

Rate $= 2.57\ \text{M}^{-1}\cdot\text{s}^{-1}\,[NO_2][F_2]$, and the reaction is second order overall.

Check: The units (none and $\text{M}^{-1}\cdot\text{s}^{-1}$) are correct. The rate law is a common form. The rate is changing as rapidly as each concentration is changing, which is consistent with first order in each reactant. The rate constant is consistent with the units necessary to get rate as M/s, and the magnitude is reasonable because we have a second-order reaction.

The Integrated Rate Law and Half-Life

14.47 (a) The reaction is zero order. Because the slope of the plot is independent of the concentration, there is no dependence of the concentration of the reactant in the rate law.

(b) The reaction is first order. The expression for the half-life of a first-order reaction is $t_{1/2} = \dfrac{0.693}{k}$, which is independent of the reactant concentration.

(c) The reaction is second order. The integrated rate expression for a second-order reaction is $\dfrac{1}{[A]_t} = kt + \dfrac{1}{[A]_0}$, which is linear when the inverse of the concentration is plotted versus time.

14.49   **Given:** table of [AB] versus time   **Find:** reaction order, k, and [AB] at 25 s
Conceptual Plan: Look at the data and see if any common reaction orders can be eliminated. If the data does not show an equal concentration drop with time, zero order can be eliminated. Look for changes in the half-life (compare time for concentration to drop to one-half of any value). If the half-life is not constant, the first order can be eliminated. If the half-life is getting longer as the concentration drops, this might suggest second order. Plot the data as indicated by the appropriate rate law. Determine k from the slope of the plot. Finally, calculate the [AB] at 25 s by using the appropriate integrated rate expression.
Solution: By the preceding logic, we can eliminate both the zero-order and first-order reactions. (Alternatively, you could make all three plots and only one should be linear.) This suggests that we should have a second-order reaction. Plot 1/[AB] versus time.

Because $\dfrac{1}{[\text{AB}]_t} = kt + \dfrac{1}{[\text{AB}]_0}$, the slope will be the rate constant. The slope can be determined by measuring $\Delta y / \Delta x$ on the plot or by using functions such as "add trendline" in Excel. Thus, the rate constant is 0.0225 $\text{M}^{-1}\cdot\text{s}^{-1}$, and the rate law is Rate = 0.0225 $\text{M}^{-1}\cdot\text{s}^{-1}[\text{AB}]^2$.

Finally, use $\dfrac{1}{[\text{AB}]_t} = kt + \dfrac{1}{[\text{AB}]_0}$; substitute the values of $[\text{AB}]_0$, 25 s, and k; and rearrange to solve

for [AB] at 25 s. $[\text{AB}]_t = \dfrac{1}{kt + \dfrac{1}{[\text{AB}]_0}} = \dfrac{1}{(0.0225\ \text{M}^{-1}\cdot\text{s}^{-1})(25\ \text{s}) + \left(\dfrac{1}{0.950\ \text{M}}\right)} = 0.619\ \text{M}$

Check: The units (none, $\text{M}^{-1}\cdot\text{s}^{-1}$, and M) are correct. The rate law is a common form. The plot was linear, confirming second-order kinetics. The rate constant is consistent with the units necessary to get the rate as M/s, and the magnitude is reasonable because we have a second-order reaction. The [AB] at 25 s is between the values at 0 s and 50 s.

14.51   **Given:** table of $[\text{C}_4\text{H}_8]$ versus time   **Find:** reaction order, k, and reaction rate when $[\text{C}_4\text{H}_8] = 0.25\ \text{M}$
Conceptual Plan: Look at the data and see if any common reaction orders can be eliminated. If the data does not show an equal concentration drop with time, zero order can be eliminated. Look for changes in half-life (compare time for concentration to drop to one-half of any value). If the half-life is not constant, the first order can be eliminated. If the half-life is getting longer as the concentration drops, this might suggest second order. Plot the data as indicated by the appropriate rate law. Determine k from the slope of the plot. Finally, calculate the reaction rate when $[\text{C}_4\text{H}_8] = 0.25\ \text{M}$ by using the rate law.

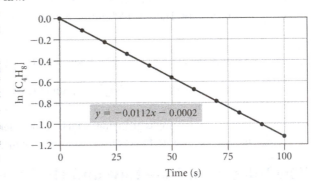

Solution: By the preceding logic, we can see that the reaction is most likely first order. It takes about 60 s for any concentration to be cut to half its value. Plot $\ln[\text{C}_4\text{H}_8]$ versus time. Because $\ln[\text{A}]_t = -kt + \ln[\text{A}]_0$, the negative of the slope will be the rate constant. The slope can be determined by measuring $\Delta y / \Delta x$ on the plot or by using functions such as "add trendline" in Excel. Thus, the rate constant is 0.0112 s^{-1}, and the rate law is Rate = 0.0112 $\text{s}^{-1}[\text{C}_4\text{H}_8]$. Finally, use Rate = 0.0112 $\text{s}^{-1}[\text{C}_4\text{H}_8]$ and substitute the values of $[\text{C}_4\text{H}_8]$: Rate = 0.0112 $\text{s}^{-1}[0.25\ \text{M}] = 2.8 \times 10^{-3}\ \text{M}\cdot\text{s}^{-1}$

Check: The units (none, s^{-1}, and $\text{M}\cdot\text{s}^{-1}$) are correct. The rate law is a common form. The plot was linear, confirming first-order kinetics. The rate constant is consistent with the units necessary to get rate as M/s, and the magnitude is reasonable because we have a first-order reaction. The rate when $[\text{C}_4\text{H}_8] = 0.25\ \text{M}$ is consistent with the average rate using 90 s and 100 s.

14.53   **Given:** plot of ln[A] versus time has slope = $-0.0045/\text{s}$; $[\text{A}]_0 = 0.250\ \text{M}$ **Find:** (a) k, (b) rate law, (c) $t_{1/2}$, and (d) [A] after 225 s
Conceptual Plan:
(a)   A plot of ln[A] versus time is linear for a first-order reaction. Using $\ln[\text{A}]_t = -kt + \ln[\text{A}]_0$, the rate constant is the negative of the slope.

(b)   **Rate law is first order. Add rate constant from part (a).**

(c)   **For a first-order reaction, $t_{1/2} = \dfrac{0.693}{k}$. Substitute k from part (a).**

(d)   **Use the integrated rate law, $\ln[A]_t = -kt + \ln[A]_0$, and substitute k and the initial concentration.**

Solution:

(a)   Because the rate constant is the negative of the slope, $k = 4.5 \times 10^{-3}\,\text{s}^{-1}$.

(b)   Because the reaction is first order, $\text{Rate} = k[A] = 4.5 \times 10^{-3}\,\text{s}^{-1}\,[A]$.

(c) $t_{1/2} = \dfrac{0.693}{k} = \dfrac{0.693}{0.0045/\text{s}} = 1.5 \times 10^2\,\text{s}$

(d) $\ln[A]_t = -kt + \ln[A]_0$ and substitute k and the initial concentration. So
$\ln[A]_t = -(0.0045/\text{s})(225\,\text{s}) + \ln 0.250\,\text{M} = -2.39879$ and $[A]_{250\,\text{s}} = e^{-2.39879} = 0.0908\,\text{M}$

Check: The units (s^{-1}, none, s, and M) are correct. The rate law is a common form. The rate constant is consistent with value of the slope. The half-life is consistent with a small value of k. The concentration at 225 s is consistent with being between one and two half-lives.

14.55   **Given:** decomposition of SO_2Cl_2, first order; $k = 1.42 \times 10^{-4}\,\text{s}^{-1}$
Find: (a) $t_{1/2}$, (b) t to decrease to 25% of $[SO_2Cl_2]_0$, (c) t to 0.78 M when $[SO_2Cl_2]_0 = 1.00$ M, and
(d) $[SO_2Cl_2]$ after 2.00×10^2 s and 5.00×10^2 s when $[SO_2Cl_2]_0 = 0.150$ M
Conceptual Plan:

(a) $k \rightarrow t_{1/2}$

$t_{1/2} = \dfrac{0.693}{k}$

(b) $[SO_2Cl_2]_0,\ 25\%\text{ of }[SO_2Cl_2]_0,\ k \rightarrow t$

$\ln[A]_t = -kt + \ln[A]_0$

(c) $[SO_2Cl_2]_0,\ [SO_2Cl_2]_t,\ k \rightarrow t$

$\ln[A]_t = -kt + \ln[A]_0$

(d) $[SO_2Cl_2]_0,\ t,\ k \rightarrow [SO_2Cl_2]_t$

$\ln[A]_t = -kt + \ln[A]_0$

Solution:

(a) $t_{1/2} = \dfrac{0.693}{k} = \dfrac{0.693}{1.42 \times 10^{-4}\,\text{s}^{-1}} = 4.88 \times 10^3\,\text{s}$

(b) $[SO_2Cl_2]_t = 0.25\,[SO_2Cl_2]_0$ Because $\ln[SO_2Cl_2]_t = -kt + \ln[SO_2Cl_2]_0$, rearrange to solve for t.

$t = -\dfrac{1}{k}\ln\dfrac{[SO_2Cl_2]_t}{[SO_2Cl_2]_0} = -\dfrac{1}{1.42 \times 10^{-4}\,\text{s}^{-1}}\ln\dfrac{0.25\,[SO_2Cl_2]_0}{[SO_2Cl_2]_0} = 9.8 \times 10^3\,\text{s}$

(c) $[SO_2Cl_2]_t = 0.78$ M; $[SO_2Cl_2]_0 = 1.00$ M   Because $\ln[SO_2Cl_2]_t = -kt + \ln[SO_2Cl_2]_0$, rearrange to solve
for t. $t = -\dfrac{1}{k}\ln\dfrac{[SO_2Cl_2]_t}{[SO_2Cl_2]_0} = -\dfrac{1}{1.42 \times 10^{-4}\,\text{s}^{-1}}\ln\dfrac{0.78\,\text{M}}{1.00\,\text{M}} = 1.7 \times 10^3\,\text{s}$

(d) $[SO_2Cl_2]_0 = 0.150$ M and $t = 2.00 \times 10^2$ s
$\ln[SO_2Cl_2]_t = -(1.42 \times 10^{-4}\,\text{s}^{-1})(2.00 \times 10^2\,\text{s}) + \ln 0.150\,\text{M} = -1.92552 \rightarrow$
$[SO_2Cl_2]_t = e^{-1.92552} = 0.146$ M
$[SO_2Cl_2]_0 = 0.150$ M and $t = 5.00 \times 10^2$ s
$\ln[SO_2Cl_2]_t = -(1.42 \times 10^{-4}\,\text{s}^{-1})(5.00 \times 10^2\,\text{s}) + \ln 0.150\,\text{M} = -1.96812 \rightarrow$
$[SO_2Cl_2]_t = e^{-1.96812} = 0.140$ M

Check: The units (s, s, s, and M) are correct. The rate law is a common form. The half-life is consistent with a small value of k. The time to 25% is consistent with two half-lives. The time to 0.78 M is consistent with being less than one half-life. The final concentrations are consistent with the time being less than one half-life.

14.57   **Given:** $t_{1/2}$ for radioactive decay of U-238 $= 4.5$ billion years and independent of $[\text{U-238}]_0$
Find: t to decrease by 10%; number U-238 atoms today, when 1.5×10^8 atoms formed 13.8 billion years ago
Conceptual Plan: $t_{1/2}$ **independent of concentration implies first-order kinetics,** $t_{1/2} \rightarrow k$ **then**

$t_{1/2} = \dfrac{0.693}{k}$

90% of $[\text{U-238}]_0$, $k \rightarrow t$ and $[\text{U-238}]_0$, t, $k \rightarrow [\text{U-238}]_t$

$$\ln[A]_t = -kt + \ln[A]_0 \qquad \ln[A]_t = -kt + \ln[A]_0$$

Solution: $t_{1/2} = \dfrac{0.693}{k}$ Rearrange to solve for k. $k = \dfrac{0.693}{t_{1/2}} = \dfrac{0.693}{4.5 \times 10^9 \text{ yr}} = 1.\underline{5}4 \times 10^{-10} \text{ yr}^{-1}$ then

$[\text{U-238}]_t = 0.10 \, [\text{U-238}]_0$ Because $\ln[\text{U-238}]_t = -kt + \ln[\text{U-238}]_0$, rearrange to solve for t.

$$t = -\frac{1}{k}\ln\frac{[\text{U-238}]_t}{[\text{U-238}]_0} = -\frac{1}{1.\underline{5}4 \times 10^{-10} \text{ yr}^{-1}}\ln\frac{0.90 \, \cancel{[\text{U-238}]_0}}{\cancel{[\text{U-238}]_0}} = 6.8 \times 10^8 \text{ yr}$$

and $[\text{U-238}]_0 = 1.15 \times 10^{18}$ atoms; $t = 13.8 \times 10^9$ yr

$\ln[\text{U-238}]_t = -kt + \ln[\text{U-238}]_0 = -(1.\underline{5}4 \times 10^{-10}\,\cancel{\text{yr}^{-1}})(13.8 \times 10^9 \,\cancel{\text{yr}})$

$+\ln(1.5 \times 10^{18}\text{ atoms}) = 39.726797 \rightarrow$

$[\text{U-238}]_t = e^{39.726797} = 1.8 \times 10^{17}$ atoms

Check: The units (yr and atoms) are correct. The time to 10% decay is consistent with less than one half-life. The final concentration is consistent with the time being about three half-lives.

The Effect of Temperature and the Collision Model

14.59

14.61 **Given:** activation energy $= 56.8 \text{ kJ/mol}$, frequency factor $= 1.5 \times 10^{11}/\text{s}$, $25\,°\text{C}$ **Find:** rate constant

Conceptual Plan: $°\text{C} \rightarrow \text{K}$ and $\text{kJ/mol} \rightarrow \text{J/mol}$ then $E_a, T, A \rightarrow k$

$$\text{K} = °\text{C} + 273.15 \qquad \frac{1000 \text{ J}}{1 \text{ kJ}} \qquad k = Ae^{-E_a/RT}$$

Solution: $T = 25\,°\text{C} + 273.15 = 298 \text{ K}$ and $\dfrac{56.8 \text{ kJ}}{\text{mol}} \times \dfrac{1000 \text{ J}}{1 \text{ kJ}} = 5.68 \times 10^4 \dfrac{\text{J}}{\text{mol}}$ then

$$k = Ae^{-E_a/RT} = (1.5 \times 10^{11} \text{ s}^{-1})e^{\dfrac{-5.68 \times 10^4\,\frac{\text{J}}{\cancel{\text{mol}}}}{\left(8.314\,\frac{\text{J}}{\text{K}\cdot\cancel{\text{mol}}}\right)298\,\text{K}}} = 17 \text{ s}^{-1}$$

Check: The units (s^{-1}) are correct. The rate constant is consistent with a large activation energy and a large frequency factor.

14.63 **Given:** plot of $\ln k$ versus $1/T$ (in K) is linear with a slope of -7445 K **Find:** E_a

Conceptual Plan: Because $\ln k = \dfrac{-E_a}{R}\left(\dfrac{1}{T}\right) + \ln A$ plot of $\ln k$ versus $1/T$ will have a *slope* $= -E_a/R$.

Solution: Because the *slope* $= -7445 \text{ K} = -E_a/R$,

$$E_a = -(\text{slope})R = -(-7445 \text{ K})\left(8.314\frac{\text{J}}{\text{K}\cdot\text{mol}}\right)\left(\frac{1\text{kJ}}{1000 \text{ J}}\right) = 61.90\frac{\text{kJ}}{\text{mol}}$$

Check: The units (kJ/mol) are correct. The activation energy is typical for many reactions.

14.65   **Given:** table of rate constant versus T **Find:** E_a and A

Conceptual Plan: Because $\ln k = \dfrac{-E_a}{R}\left(\dfrac{1}{T}\right) + \ln A$, a plot of $\ln k$ versus $1/T$ will have a *slope* $= -E_a/R$

and an *intercept* $= \ln A$.

Solution: The slope can be determined by measuring $\Delta y/\Delta x$ on the plot or by using functions such as "add trendline" in Excel.

Because the *slope* $= -30189 \text{ K} = -E_a/R$,

$E_a = -(slope)R =$

$-(-30189 \text{ K})\left(8.314\dfrac{J}{K \cdot mol}\right)\left(\dfrac{1 \text{ kJ}}{1000 \text{ J}}\right)$

$= 251\dfrac{\text{kJ}}{\text{mol}}$ and *intercept* $= 27.399 = \ln A$ then

$A = e^{intercept} = e^{27.399} = 7.93 \times 10^{11} \text{s}^{-1}$.

Check: The units (kJ/mol and s^{-1}) are correct. The plot was linear, confirming Arrhenius behavior. The activation and frequency factor are typical for many reactions.

14.67   **Given:** table of rate constant versus T **Find:** E_a and A

Conceptual Plan: Because $\ln k = \dfrac{-E_a}{R}\left(\dfrac{1}{T}\right) + \ln A$, a plot of $\ln k$ versus $1/T$ will have a *slope* $= -E_a/R$ and

an *intercept* $= \ln A$.

Solution: The slope can be determined by measuring $\Delta y/\Delta x$ on the plot or by using functions such as "add trendline" in Excel.

Because the *slope* $= -2767.2 \text{ K} = -E_a/R$,

$E_a = -(slope)R$

$= -(-2767.2 \text{ K})\left(8.314\dfrac{J}{K \cdot mol}\right)\left(\dfrac{1 \text{ kJ}}{1000 \text{ J}}\right)$

$= 23.0\dfrac{\text{kJ}}{\text{mol}}$ and *intercept* $= 25.112 = \ln A$ then

$A = e^{intercept} = e^{25.112} = 8.05 \times 10^{10} \text{ s}^{-1}$.

Check: The units (kJ/mol and s^{-1}) are correct. The plot was linear, confirming Arrhenius behavior. The activation and frequency factor are typical for many reactions.

14.69   **Given:** rate constant $= 0.0117/\text{s}$ at 400.0 K and 0.689/s at 450.0 K   **Find:** (a) E_a and (b) rate constant at 425 K

Conceptual Plan:

(a) $k_1, T_1, k_2, T_2, \rightarrow E_a$ then J/mol $\rightarrow$ kJ/mol

$$\ln\left(\frac{k_2}{k_1}\right) = \frac{E_a}{R}\left(\frac{1}{T_1} - \frac{1}{T_2}\right) \qquad \frac{1 \text{ kJ}}{1000 \text{ J}}$$

(b) $E_a, k_1, T_1, T_2 \rightarrow k_2$

$$\ln\left(\frac{k_2}{k_1}\right) = \frac{E_a}{R}\left(\frac{1}{T_1} - \frac{1}{T_2}\right)$$

Solution:

(a) $\ln\left(\dfrac{k_2}{k_1}\right) = \dfrac{E_a}{R}\left(\dfrac{1}{T_1} - \dfrac{1}{T_2}\right)$ Rearrange to solve for E_a.

$$E_a = \frac{R \ln\left(\dfrac{k_2}{k_1}\right)}{\left(\dfrac{1}{T_1} - \dfrac{1}{T_2}\right)} = \frac{8.314 \dfrac{J}{K \cdot mol} \ln\left(\dfrac{0.689 \text{ s}^{-1}}{0.0117 \text{ s}^{-1}}\right)}{\left(\dfrac{1}{400.0 \text{ K}} - \dfrac{1}{450.0 \text{ K}}\right)} = 1.22 \times 10^5 \frac{J}{mol} \times \frac{1 \text{ kJ}}{1000 \text{ J}} = 122\frac{\text{kJ}}{\text{mol}}$$

(b) $\ln\left(\dfrac{k_2}{k_1}\right) = \dfrac{E_a}{R}\left(\dfrac{1}{T_1} - \dfrac{1}{T_2}\right)$ with $k_1, = 0.0117/\text{s}, T_1 = 400.0 \text{ K}, T_2 = 425 \text{ K}$ Rearrange to solve for k_2.

$$\ln k_2 = \frac{E_a}{R}\left(\frac{1}{T_1} - \frac{1}{T_2}\right) + \ln k_1 = \frac{1.22 \times 10^5 \frac{J}{mol}}{8.314 \frac{J}{K \cdot mol}}\left(\frac{1}{400.0\,K} - \frac{1}{425\,K}\right) + \ln 0.0117\ s^{-1} = -2.2902 \rightarrow$$

$$k_2 = e^{-2.2902} = 0.101\ s^{-1}$$

Check: The units (kJ/mol and s^{-1}) are correct. The activation energy is typical for a reaction. The rate constant at 425 K is between the values given at 400 K and 450 K.

14.71   **Given:** rate constant doubles from 10.0 °C to 20.0 °C   **Find:** E_a

        Conceptual Plan: °C $\rightarrow$ K then $k_1,\ T_1,\ k_2,\ T_2 \rightarrow E_a$ then J/mol $\rightarrow$ kJ/mol

$$K = °C + 273.15 \qquad \ln\left(\frac{k_2}{k_1}\right) = \frac{E_a}{R}\left(\frac{1}{T_1} - \frac{1}{T_2}\right) \qquad \frac{1\,kJ}{1000\,J}$$

        Solution: $T_1 = 10.0\ °C + 273.15 = 283.2$ K and $T_2 = 20.0\ °C + 273.15 = 293.2$ K and $k_2 = 2\,k_1$ then

$$\ln\left(\frac{k_2}{k_1}\right) = \frac{E_a}{R}\left(\frac{1}{T_1} - \frac{1}{T_2}\right). \text{ Rearrange to solve for } E_a.$$

$$E_a = \frac{R \ln\left(\frac{k_2}{k_1}\right)}{\left(\frac{1}{T_1} - \frac{1}{T_2}\right)} = \frac{8.314\,\frac{J}{K \cdot mol}\ \ln\left(\frac{2\,k_1}{k_1}\right)}{\left(\frac{1}{283.2\,K} - \frac{1}{293.2\,K}\right)} = 4.7851 \times 10^4\,\frac{J}{mol} \times \frac{1\,kJ}{1000\,J} = 47.85\,\frac{kJ}{mol}$$

        Check: The units (kJ/mol) are correct. The activation energy is typical for a reaction.

14.73   Reaction (a) would have the faster rate because the orientation factor, p, would be larger for this reaction because the reactants are symmetrical.

Reaction Mechanisms

14.75   The overall reaction is the sum of the elementary steps in the proposed mechanism. Because the first reaction is the slow step, it is the rate-determining step. Using this first step to determine the rate law, Rate $= k_1[AB]^2$. Because this is the observed rate law, this mechanism is consistent with the experimental data.

14.77   (a)   The overall reaction is the sum of the steps in the mechanism:

$$Cl_2(g) \underset{k_2}{\overset{k_1}{\rightleftharpoons}} 2\,\cancel{Cl}(g)$$
$$\cancel{Cl}(g) + CHCl_3(g) \xrightarrow{k_3} HCl(g) + \cancel{CCl_3}(g)$$
$$\cancel{Cl}(g) + \cancel{CCl_3}(g) \xrightarrow{k_4} CCl_4(g)$$
$$\overline{Cl_2(g) + CHCl_3(g) \longrightarrow HCl(g) + CCl_4(g)}$$

        (b)   The intermediates are the species that are generated by one step and consumed by other steps. These are a $Cl(g)$ and $CCl_3(g)$.

        (c)   Because the second step is the rate-determining step, Rate $= k_3[Cl][CHCl_3]$. Because Cl is an intermediate, its concentration cannot appear in the rate law. Using the fast equilibrium in the first step, we see that $k_1[Cl_2] = k_2[Cl]^2$ or $[Cl] = \sqrt{\frac{k_1}{k_2}[Cl_2]}$. Substituting this into the first rate expression, we get

$$\text{Rate} = k_3\sqrt{\frac{k_1}{k_2}}[Cl_2]^{1/2}[CHCl_3]. \text{ Simplifying this expression, we see that Rate} = k[Cl_2]^{1/2}[CHCl_3].$$

Catalysis

14.79   Heterogeneous catalysts require a large surface area because catalysis can only happen at the active sites on the surface. A greater surface area means greater opportunity for the substrate to react, which results in a speedier reaction.

14.81   Assume rate ratio $\propto k$ ratio (because concentration terms will cancel each other) and $k = Ae^{-E_a/RT}$. $T = 25\,°C + 273.15 = 298$ K, $E_{a_1} = 1.25 \times 10^5$ J/mol, and $E_{a_2} = 5.5 \times 10^4$ J/mol. Ratio of rates will be

$$\frac{k_2}{k_1} = \frac{Ae^{-E_{a_2}/RT}}{Ae^{-E_{a_1}/RT}} = \frac{e^{\frac{-5.5 \times 10^4\,\frac{J}{mol}}{\left(8.314\,\frac{J}{K \cdot mol}\right)298\,K}}}{e^{\frac{-1.25 \times 10^5\,\frac{J}{mol}}{\left(8.314\,\frac{J}{K \cdot mol}\right)298\,K}}} = \frac{e^{-22.199}}{e^{-50.453}} = 10^{12}$$

Note that no additional significant figures are quoted since the exponent has its last significant digit in the 1's place.

Cumulative Problems

14.83   **Given:** table of $[CH_3CN]$ versus time   **Find:** (a) reaction order, k; (b) $t_{1/2}$; and (c) t for 90% conversion
Conceptual Plan: (a) and (b) Look at the data and see if any common reaction orders can be eliminated. If the data does not show an equal concentration drop with time, zero order can be eliminated. Look for changes in the half-life (compare time for concentration to drop to one-half of any value). If the half-life is not constant, the first order can be eliminated. If the half-life is getting longer as the concentration drops, this might suggest second order. Plot the data as indicated by the appropriate rate law, or if it is first order and there is an obvious half-life in the data, a plot is not necessary. Determine k from the slope of the plot (or using the half-life equation for first order). (c) Finally, calculate the time to 90% conversion using the appropriate integrated rate equation.
Solution: (a) and (b) By the preceding logic, we can see that the reaction is first order. It takes 15.0 h for the concentration to be cut in half for any concentration (1.000 M to 0.501 M, 0.794 M to 0.398 M, and 0.631 M to 0.316 M), so $t_{1/2} = 15.0$ h. Then use $t_{1/2} = \dfrac{0.693}{k}$ and rearrange to solve for k.

$$k = \frac{0.693}{t_{1/2}} = \frac{0.693}{15.0\,h} = 0.0462\,h^{-1}$$

(c) $[CH_3CN]_t = 0.10\,[CH_3CN]_0$ Because $\ln[CH_3CN]_t = -kt + \ln[CH_3CN]_0$, rearrange to solve for t.

$$t = -\frac{1}{k}\ln\frac{[CH_3CN]_t}{[CH_3CN]_0} = -\frac{1}{0.0462\,h^{-1}}\ln\frac{0.10[CH_3CN]_0}{[CH_3CN]_0} = 49.8\,h$$

Check: The units (none, h^{-1}, h, and h) are correct. The rate law is a common form. The data showed a constant half-life very clearly. The rate constant is consistent with the units necessary to get rate as M/s, and the magnitude is reasonable because we have a first-order reaction. The time to 90% conversion is consistent with a time between three and four half-lives.

14.85   **Given:** Rate $= k\dfrac{[A][C]^2}{[B]^{1/2}} = 0.0115$ M/s at certain initial concentrations of A, B, and C; double A and C concentration and triple B concentration   **Find:** reaction rate
Conceptual Plan: $[A]_1, [B]_1, [C]_1$, Rate 1, $[A]_2, [B]_2, [C]_2 \rightarrow$ **Rate 2**

$$\frac{\text{Rate 2}}{\text{Rate 1}} = \frac{k\dfrac{[A]_2[C]_2^2}{[B]_2^{1/2}}}{k\dfrac{[A]_1[C]_1^2}{[B]_1^{1/2}}}$$

Solution: $\dfrac{\text{Rate 2}}{\text{Rate 1}} = \dfrac{k\dfrac{[A]_2[C]_2^2}{[B]_2^{1/2}}}{k\dfrac{[A]_1[C]_1^2}{[B]_1^{1/2}}}$ Rearrange to solve for Rate 2. $\text{Rate 2} = \dfrac{k\dfrac{[A]_2[C]_2^2}{[B]_2^{1/2}}}{k\dfrac{[A]_1[C]_1^2}{[B]_1^{1/2}}}\text{Rate 1}$

$[A]_2 = 2[A]_1$, $[B]_2 = 3[B]_1$, $[C]_2 = 2[C]_1$, and Rate 1 $= 0.0115$ M/s so

$$\text{Rate 2} = \frac{k\dfrac{2\,\cancel{[A]_1}(2\,\cancel{[C]_1})^2}{(3\,\cancel{[B]_1})^{1/2}}}{k\dfrac{\cancel{[A]_1}\,\cancel{[C]_1^2}}{\cancel{[B]_1^{1/2}}}}\,0.0115\frac{M}{s} = \frac{2^3}{3^{1/2}}\,0.0115\frac{M}{s} = 0.0531\frac{M}{s}$$

Check: The units $(M \cdot s^{-1})$ are correct. They should increase because we have a factor of eight (2^3) divided by the square root of three (1.73).

14.87   **Given:** table of P_{Total} versus time   **Find:** rate law, k, and P_{Total} at 2.00×10^4 s

Conceptual Plan: Because 2 moles of gas are generated for each mole of CH₃CHO decomposed, this $P_{CH_3CHO} = P^\circ_{Total} - (P_{Total} - P^\circ_{Total})$. Look at the data and see if any common reaction orders can be eliminated. If the data does not show an equal P_{Total} rise (or P_{CH_3CHO} drop) with time, zero order can be eliminated. There is not enough data to look for changes in the half-life (compare time for P_{CH_3CHO} to drop to one-half of any value). It does appear that the half-life is getting longer, so the first order can be eliminated. Plot the data as indicated by the appropriate rate law. Determine k from the slope of the plot. Finally, calculate the P_{CH_3CHO} at 2.00×10^4 s using the appropriate integrated rate expression and convert this to P_{Total} using the reaction stoichiometry.

Solution: Calculate $P_{CH_3CHO} = P^\circ_{Total} - (P_{Total} - P^\circ_{Total})$.

Time(s)	P_{Total}(atm)	P_{CH_3CHO}(atm)
0	0.22	0.22
1000	0.24	0.20
3000	0.27	0.17
7000	0.31	0.13

By the preceding logic, we can eliminate both the zero-order and the first-order reactions. (Alternatively, you could make all three plots, and only one should be linear.) This suggests that we should have a second-order reaction. Plot $1/P_{CH_3CHO}$ versus time. Because $\dfrac{1}{P_{CH_3CHO}} = kt + \dfrac{1}{P^\circ_{CH_3CHO}}$, the slope will be the rate constant. The slope can

be determined by measuring $\Delta y / \Delta x$ on the plot or by using functions such as "add trendline" in Excel. Thus, the rate constant is 4.5×10^{-4} atm$^{-1} \cdot$ s^{-1}, and the rate law is Rate $= 4.5 \times 10^{-4}$ atm$^{-1} \cdot$ s$^{-1} P_{CH_3CHO}$.

Finally, use $\dfrac{1}{P_{CH_3CHO}} = kt + \dfrac{1}{P^\circ_{CH_3CHO}}$; substitute the values of $P^\circ_{CH_3CHO}$, 2.00×10^4 s, and k and rearrange to solve for $P^\circ_{CH_3CHO}$ at 2.00×10^4 s.

$$P_{CH_3CHO} = \cfrac{1}{kt + \cfrac{1}{P^\circ_{CH_3CHO}}} =$$

$$\cfrac{1}{(4.5 \times 10^{-4}\,\text{atm}^{-1} \cdot \cancel{s^{-1}})(2.00 \times 10^4\,\cancel{s}) + \left(\cfrac{1}{0.22\,\text{atm}}\right)} = 0.07\underline{3}8255\,\text{atm} = 0.074\,\text{atm}$$

Finally, from the first equation in the solution, $P_{Total} = 2P^\circ_{Total} - P_{CH_3CHO} = 2(0.22\,\text{atm}) - 0.0738255\,\text{atm} = 0.36\underline{6}175\,\text{atm} = 0.37\,\text{atm}$.

Check: The units (none, atm$^{-1} \cdot$ s^{-1}, and atm) are correct. The rate law is a common form. The plot was linear, confirming second-order kinetics. The rate constant is consistent with the units necessary to get rate as atm/s, and the magnitude is reasonable because we have a second-order reaction. The P_{Total} at 2.00×10^4 s is consistent with the changes we see in the data table through 7000 s.

14.89   **Given:** N₂O₅ decomposes to NO₂ and O₂, first order in $[N_2O_5]$; $t_{1/2} = 2.81$ h at 25 °C; $V = 1.5$ L, $P^\circ_{N_2O_5} = 745$ torr

Find: P_{O_2} after 215 minutes

Conceptual Plan: Write a balanced reaction. Then $t_{1/2} \rightarrow k$ then °C $\rightarrow$ K and torr $\rightarrow$ atm then

$$N_2O_5 \rightarrow 2\,NO_2 + \tfrac{1}{2}O_2 \qquad t_{1/2} = \frac{0.693}{k} \qquad K = {}^\circ C + 273.15 \qquad \frac{1\,\text{atm}}{760\,\text{torr}}$$

$P^\circ_{N_2O_5}, V, T \rightarrow n/V$ then min $\rightarrow$ h then $[N_2O_5]_0, t, k \rightarrow [N_2O_5]_t$ then $[N_2O_5]_0, [N_2O_5]_t \rightarrow [O_2]_t$

$$PV = nRT \qquad \frac{1\text{ h}}{60\text{ min}} \qquad \ln[A]_t = -kt + \ln[A]_0 \qquad [O_2]_t = ([N_2O_5]_0 - [N_2O_5]_t) \times \frac{1/2\text{ mol }O_2}{1\text{ mol }N_2O_5}$$

then $[O_2]_t, V, T \rightarrow P^\circ_{O_2}$ and finally atm $\rightarrow$ torr

$$PV = nRT \qquad \frac{760\text{ torr}}{1\text{ atm}}$$

Solution: $t_{1/2} = \dfrac{0.693}{k}$ and rearrange to solve for k. $k = \dfrac{0.693}{t_{1/2}} = \dfrac{0.693}{2.81\text{ h}} = 0.246619\text{ h}^{-1}$ then

$T = 25\,°C + 273.15 = 298\text{ K}$. $745\text{ torr} \times \dfrac{1\text{ atm}}{760\text{ torr}} = 0.980263\text{ atm}$ then $PV = nRT$ Rearrange to solve for n/V.

$\dfrac{n}{V} = \dfrac{P}{RT} = \dfrac{0.980263\text{ atm}}{0.08206\dfrac{L \cdot atm}{K \cdot mol} \times 298\text{ K}} = 0.0400862\text{ M}$ then $215\text{ min} \times \dfrac{1\text{ h}}{60\text{ min}} = 3.58333\text{ h}$

Because $\ln[N_2O_5]_t = -kt + \ln[N_2O_5]_0 = -(0.246619\text{ h}^{-1})(3.58333\text{ h}) + \ln(0.0400862\text{ M}) = -4.10044 \rightarrow$

$[N_2O_5]_t = e^{-4.10044} = 0.0165654\text{ M}$ then

$$[O_2]_t = ([N_2O_5]_0 - [N_2O_5]_t) \times \frac{1/2\text{ mol }O_2}{1\text{ mol }N_2O_5} = \left(0.0400862\frac{\text{mol }N_2O_5}{L} - 0.0165654\frac{\text{mol }N_2O_5}{L}\right) \times \frac{1/2\text{ mol }O_2}{1\text{ mol }N_2O_5}$$

$= 0.0117604\text{ M }O_2$

then finally $PV = nRT$ and rearrange to solve for P

$$P = \frac{n}{V}RT = 0.0117604\frac{\text{mol}}{L} \times 0.08206\frac{L\text{ atm}}{K \cdot mol} \times 298\text{ K} = 0.287587\text{ atm} \times \frac{760\text{ torr}}{1\text{ atm}} = 219\text{ torr}$$

Check: The units (torr) are correct. The pressure is reasonable because it must be less than one-half of the original pressure.

14.91   **Given:** I_2 formation from I atoms, second order in I; $k = 1.5 \times 10^{10}\text{ M}^{-1} \cdot s^{-1}$, $[I]_0 = 0.0100\text{ M}$
        Find: t to decrease by 95%
        Conceptual Plan: $[I]_0, [I]_t, k \rightarrow t$

$$\frac{1}{[A]_t} = kt + \frac{1}{[A]_0}$$

Solution: $[I]_t = 0.05[I]_0 = 0.05 \times 0.0100\text{ M} = 0.0005\text{ M}$ $\dfrac{1}{[I]_t} = kt + \dfrac{1}{[I]_0}$ Rearrange to solve for t.

$$t = \frac{1}{k}\left(\frac{1}{[I]_t} - \frac{1}{[I]_0}\right) = \frac{1}{(1.5 \times 10^{10}\text{ M}^{-1} \cdot s^{-1})}\left(\frac{1}{0.0005\text{ M}} - \frac{1}{0.0100\text{ M}}\right) = 1.267 \times 10^{-7}\text{ s} = 1 \times 10^{-7}\text{ s}$$

Check: The units (s) are correct. We expect the time to be extremely small because the rate constant is so large.

14.93   **Given:** $AB(aq) \rightarrow A(g) + B(g)$; $k = 0.0118\text{ M}^{-1} \cdot s^{-1}$; 250.0 mL of 0.100 M AB; collect gas over water
        $T = 25.0\,°C$, $P_{Total} = 755.1\text{ mmHg}$, and $V = 200.0\text{ mL}$; $P^\circ_{H_2O} = 23.8\text{ mmHg}$ **Find:** t
        Conceptual Plan: $P_{Total}, P_{H_2O} \rightarrow P_A + P_B$ then mmHg $\rightarrow$ atm and mL $\rightarrow$ L

$$P_{Total} = P_{H_2O} + P_A + P_B \qquad \frac{1\text{ atm}}{760\text{ mmHg}} \qquad \frac{1\text{ L}}{1000\text{ mL}}$$

and $°C \rightarrow K$ $P, V, T \rightarrow n_{A+B} \rightarrow \Delta n_{AB}$ then $[AB]_0, V_{AB}, \Delta n_{AB} \rightarrow [AB]$ then $k, [AB] \rightarrow t$

$$K = °C + 273.15 \quad PV = nRT \quad \Delta n_{AB} = \tfrac{1}{2}n_{A+B} \qquad [AB] = [AB]_0 - \frac{\Delta n_{AB}}{V_{AB} \times \dfrac{1\text{ L}}{1000\text{ mL}}} \quad \frac{1}{[AB]_t} = kt + \frac{1}{[AB]_0}$$

Solution: $P_{Total} = P_{H_2O} + P_A + P_B$ Rearrange to solve for $P_A + P_B$. $P_A + P_B = P_{Total} - P_{H_2O} = 755.1\text{ mmHg}$
$-23.8\text{ mmHg} = 731.3\text{ mmHg}$

$P_A + P_B = 731.3\text{ mmHg} \times \dfrac{1\text{ atm}}{760\text{ mmHg}} = 0.96223684\text{ atm}$ $V = 200.0\text{ mL} \times \dfrac{1\text{ L}}{1000\text{ mL}} = 0.2000\text{ L}$

$T = 25.0\,°C + 273.15 = 298.2\text{ K}$ $PV = nRT$ Rearrange to solve for n. $n = \dfrac{PV}{RT}$

$$n_{A+B} = \frac{0.96223684\text{ atm} \times 0.2000\text{ L}}{0.08206\dfrac{L \cdot atm}{mol \cdot K} \times 298.2\text{ K}} = 0.0078645309\text{ mol A + B}$$ Because 1 mole each of A and B is generated for

each mole of AB reacting, $\Delta n_{AB} = \tfrac{1}{2}n_{A+B} = \tfrac{1}{2}(0.0078645309\text{ mol A + B}) = 0.0039322655\text{ mol AB}$ then

$$[AB] = [AB]_0 - \frac{\Delta n_{AB}}{V_{AB} \times \frac{1\,L}{1000\,mL}} = 0.100\,M - \frac{0.0039322655\,\text{mol AB}}{250.0\,mL \times \frac{1\,L}{1000\,mL}} = 0.100\,M - 0.015729062\,M$$

$$= 0.0842709\,M$$

Because $\dfrac{1}{[AB]_t} = kt + \dfrac{1}{[AB]_0}$, rearrange to solve for t.

$$t = \frac{1}{k}\left(\frac{1}{[XY]_t} - \frac{1}{[XY]_0}\right) = \frac{1}{(0.0118\,M^{-1}\cdot s^{-1})}\left(\frac{1}{0.0842709\,M} - \frac{1}{0.100\,M}\right) = 158.1773\,s = 160\,s$$

Check: The units (s) are correct. The magnitude of the answer (160 s) makes sense because the rate constant is $0.0118\ M^{-1}\,s^{-1}$ and a small volume of gas is generated.

14.95   (a)   There are two elementary steps in the reaction mechanism because there are two peaks in the reaction progress diagram.

         (b)

         (c)   The first step is the rate-limiting step because it has the higher activation energy.

         (d)   The overall reaction is exothermic because the products are at a lower energy than the reactants.

14.97   **Given:** n-butane desorption from single crystal aluminum oxide, first order; $k = 0.128\ s^{-1}$ at 150 K; initially completely covered

         Find: (a) $t_{1/2}$; (b) t for 25% and for 50% to desorb; (c) fraction remaining after 10 s and 20 s

         Conceptual Plan: (a) $k \rightarrow t_{1/2}$ (b) $[C_4H_{10}]_0, [C_4H_{10}]_t, k \rightarrow t$ (c) $[C_4H_{10}]_0, t, k \rightarrow [C_4H_{10}]_t$

$$t_{1/2} = \frac{0.693}{k} \qquad\qquad \ln[A]_t = -kt + \ln[A]_0 \qquad \ln[A]_t = -kt + \ln[A]_0$$

         Solution:

         (a) $t_{1/2} = \dfrac{0.693}{k} = \dfrac{0.693}{0.128\ s^{-1}} = 5.41\ s$

         (b) $\ln[C_4H_{10}]_t = -kt + \ln[C_4H_{10}]_0$ Rearrange to solve for t. For 25% desorbed, $[C_4H_{10}]_t = 0.75\,[C_4H_{10}]_0$

                and $t = -\dfrac{1}{k}\ln\dfrac{[C_4H_{10}]_t}{[C_4H_{10}]_0} = -\dfrac{1}{0.128\ s^{-1}}\ln\dfrac{0.75[C_4H_{10}]_0}{[C_4H_{10}]_0} = 2.2\ s$. For 50% desorbed,

                $[C_4H_{10}]_t = 0.50[C_4H_{10}]_0$ and $t = -\dfrac{1}{k}\ln\dfrac{[C_4H_{10}]_t}{[C_4H_{10}]_0} = -\dfrac{1}{0.128\ s^{-1}}\ln\dfrac{0.50\,[C_4H_{10}]_0}{[C_4H_{10}]_0} = 5.4\ s$

         (c)   For 10 s, $\ln[C_4H_{10}]_t = -kt + \ln[C_4H_{10}]_0 = -(0.128\ s^{-1})(10\ s) + \ln(1.00) = -1.28 \rightarrow$

                $[C_4H_{10}]_t = e^{-1.28} = 0.28 =$ fraction covered

                For 20 s, $\ln[C_4H_{10}]_t = -kt + \ln[C_4H_{10}]_0 = -(0.128\ s^{-1})(20\ s) + \ln(1.00) = -2.56 \rightarrow$

                $[C_4H_{10}]_t = e^{-2.56} = 0.077 =$ fraction covered

                Check: The units (s, s, s, none, and none) are correct. The half-life is reasonable considering the size of the rate constant. The time to 25% desorbed is less than one half-life. The time to 50% desorbed is the half-life. The fraction at 10 s is consistent with about two half-lives. The fraction covered at 20 s is consistent with about four half-lives.

14.99   **(a)**   **Given:** table of rate constant versus T **Find:** E_a and A

Conceptual Plan: First, convert temperature data into kelvin ($°C + 273.15 = K$). Because

$$\ln k = \frac{-E_a}{R}\left(\frac{1}{T}\right) + \ln A, \text{ a plot of } \ln k \text{ versus } 1/T \text{ will have a } slope = -E_a/R \text{ and an } intercept = \ln A.$$

Solution: The slope can be determined by measuring $\Delta y/\Delta x$ on the plot or by using functions such as "add trendline" in Excel. Because the *slope* $= -10759$ K $= -E_a/R$,
$E_a = -(slope)R$

$$= -(-10759\,K)\left(8.314\frac{J}{K \cdot mol}\right)\left(\frac{1\,kJ}{1000\,J}\right)$$

$$= 89.5\frac{kJ}{mol}$$

and *intercept* $= 26.769 = \ln A$ then
$A = e^{intercept} = e^{26.769} = 4.22 \times 10^{11}\,s^{-1}$

Check: The units (kJ/mol and s^{-1}) are correct. The plot was linear, confirming Arrhenius behavior. The activation and frequency factor are typical for many reactions.

(b)   **Given:** part (a) results   **Find:** k at 15 °C
Conceptual Plan: °C → K then T, E_a, A → k

$$°C + 273.15 = K \qquad \ln k = \frac{-E_a}{R}\left(\frac{1}{T}\right) + \ln A$$

Solution: $15\,°C + 273.15 = 288$ K then

$$\ln k = \frac{-E_a}{R}\left(\frac{1}{T}\right) + \ln A = \frac{-89.5\frac{kJ}{mol} \times \frac{1000\,J}{1\,kJ}}{8.314\frac{J}{K \cdot mol}}\left(\frac{1}{288\,K}\right) + \ln(4.22 \times 10^{11}\,s^{-1}) = -10.610 \rightarrow$$

$k = e^{-10.610} = 2.5 \times 10^{-5}\,M^{-1} \cdot s^{-1}$

Check: The units $(M^{-1} \cdot s^{-1})$ are correct. The value of the rate constant is less than the value at 25 °C.

(c)   **Given:** part (a) results, 0.155 M C_2H_5Br and 0.250 M OH^- at 75 °C   **Find:** initial reaction rate
Conceptual Plan: °C → K then T, E_a, A → k then k, $[C_2H_5Br]$, $[OH^-]$ → initial reaction rate

$$°C + 273.15 = K \qquad \ln k = \frac{-E_a}{R}\left(\frac{1}{T}\right) + A \qquad\qquad Rate = k[C_2H_5Br][OH^-]$$

Solution: $75\,°C + 273.15 = 348$ K then

$$\ln k = \frac{-E_a}{R}\left(\frac{1}{T}\right) + \ln A = \frac{-89.5\frac{kJ}{mol} \times \frac{1000\,J}{1\,kJ}}{8.314\frac{J}{K \cdot mol}}\left(\frac{1}{348\,K}\right) + \ln(4.22 \times 10^{11}\,s^{-1}) = -4.1656 \rightarrow$$

$k = e^{-4.1656} = 1.5521 \times 10^{-2}\,M^{-1} \cdot s^{-1}$
Rate $= k[C_2H_5Br][OH^-] = (1.5521 \times 10^{-2}\,M^{-1} \cdot s^{-1})(0.155\,M)(0.250\,M) = 6.0 \times 10^{-4}\,M \cdot s^{-1}$

Check: The units $(M \cdot s^{-1})$ are correct. The value of the rate is reasonable considering the value of the rate constant (larger than in the table) and the fact that the concentrations are less than 1 M.

14.101   **(a)**   No, because the activation energy is zero. This means that the rate constant ($k = Ae^{-E_a/RT}$) will be independent of temperature.

      (b)   No bond is broken, and the two radicals (CH_3) attract each other.

      (c)   Formation of diatomic gases from atomic gases

14.103   **Given:** $t_{1/2}$ for radioactive decay of C-14 $= 5730$ years; bone has 19.5% C-14 in living bone
Find: age of bone
Conceptual Plan: Radioactive decay implies first-order kinetics, $t_{1/2} \rightarrow k$ **then** 19.5% **of** $[\text{C-14}]_0, k \rightarrow t$

$$t_{1/2} = \frac{0.693}{k} \qquad \qquad \ln[A]_t = -kt + \ln[A]_0$$

Solution: $t_{1/2} = \dfrac{0.693}{k}$ Rearrange to solve for k. $k = \dfrac{0.693}{t_{1/2}} = \dfrac{0.693}{5730 \text{ yr}} = 1.2\underline{0}942 \times 10^{-4} \text{ yr}^{-1}$ then

$[\text{C-14}]_t = 0.195\,[\text{C-14}]_0$ Because $\ln[\text{C-14}]_t = -kt + \ln[\text{C-14}]_0$, rearrange to solve for t.

$$t = -\frac{1}{k}\ln\frac{[\text{C-14}]_t}{[\text{C-14}]_0} = -\frac{1}{1.2\underline{0}942 \times 10^{-4}\text{ yr}^{-1}}\ln\frac{0.195\,\cancel{[\text{C-14}]_0}}{\cancel{[\text{C-14}]_0}} = 1.35 \times 10^4 \text{ yr}$$

Check: The units (yr) are correct. The time to 19.5% decay is consistent with the time being between two and three half-lives.

14.105   (a)   For each, check that all steps sum to overall reaction and that the predicted rate law is consistent with experimental data (Rate $= k[\text{H}_2][\text{I}_2]$).
For the first mechanism, the single step is the overall reaction. The rate law is determined by the stoichiometry; so Rate $= k[\text{H}_2][\text{I}_2]$, and the mechanism is valid.
For the second mechanism, the overall reaction is the sum of the steps in the mechanism:

$$\text{I}_2(g) \underset{k_2}{\overset{k_1}{\rightleftharpoons}} 2\,\cancel{\text{I}(g)}$$
$$\text{H}_2(g) + 2\,\cancel{\text{I}(g)} \xrightarrow{k_3} 2\,\text{HI}(g) \qquad \text{So the sum matches the overall reaction.}$$

$$\overline{\text{H}_2(g) + \text{I}_2(g) \longrightarrow 2\,\text{HI}(g)}$$

Because the second step is the rate-determining step, Rate $= k_3[\text{H}_2][\text{I}]^2$. Because I is an intermediate, its concentration cannot appear in the rate law. Using the fast equilibrium in the first step, we see that

$k_1[\text{I}_2] = k_2[\text{I}]^2$ or $[\text{I}]^2 = \dfrac{k_1}{k_2}[\text{I}_2]$. Substituting this into the first rate expression, we get

Rate $= k_3\dfrac{k_1}{k_2}[\text{H}_2][\text{I}_2]$, which can be simplified to rate $= k[\text{H}_2][\text{I}_2]$ and the mechanism is valid.

(b)   To distinguish between mechanisms, you could look for the buildup of I(g), the intermediate in the second mechanism.

14.107   The steps in the mechanism are as follows:

$$\text{Br}_2(g) \underset{k_{-1}}{\overset{k_1}{\rightleftharpoons}} 2\,\text{Br}(g)$$
$$\cancel{\text{Br}(g)} + \text{H}_2(g) \xrightarrow{k_2} \text{HBr}(g) + \cancel{\text{H}(g)}$$

$$\overline{\cancel{\text{H}(g)} + \text{Br}_2(g) \xrightarrow{k_3} \text{HBr}(g) + \cancel{\text{Br}(g)}}$$

Because the second step is the rate-determining step, Rate $= k_2[\text{H}_2][\text{Br}]$. Because Br is an intermediate, its concentration cannot appear in the rate law. Using the fast equilibrium in the first step, we see that

$k_1[\text{Br}_2] = k_{-1}[\text{Br}]^2$ or $[\text{Br}] = \sqrt{\dfrac{k_1}{k_{-1}}}[\text{Br}_2]^{1/2}$. Substituting this into the first rate expression, we get

Rate $= k_2\sqrt{\dfrac{k_1}{k_{-1}}}[\text{H}_2][\text{Br}_2]^{1/2}$. The rate law is 3/2 order overall.

14.109   (a)   For a zero-order reaction, the rate is independent of the concentration. If the first half goes in the first 100 minutes, the second half will go in the second 100 minutes. This means that none, or 0%, will be left at 200 minutes.
(b)   For a first-order reaction, the half-life is independent of concentration. This means that if half of the reactant decomposes in the first 100 minutes, half of this (or another 25% of the original amount) will decompose in the second 100 minutes. This means that at 200 minutes, 50% + 25% = 75% has decomposed or 25% remains.
(c)   For a second-order reaction, $t_{1/2} = \dfrac{1}{k[\text{A}]_0} = 100$ min and the integrated rate expression is $\dfrac{1}{[\text{A}]_t} = kt + \dfrac{1}{[\text{A}]_0}$.

We can rearrange the first expression to solve for k as $k = \dfrac{1}{100 \text{ min } [\text{A}]_0}$. Substituting this and 200 minutes

into the integrated rate expression, we get $\dfrac{1}{[\text{A}]_t} = \dfrac{200 \cancel{\text{ min}}}{100 \cancel{\text{ min}} [\text{A}]_0} + \dfrac{1}{[\text{A}]_0} \rightarrow \dfrac{1}{[\text{A}]_t} = \dfrac{3}{[\text{A}]_0} \rightarrow \dfrac{[\text{A}]_t}{[\text{A}]_0} = \dfrac{1}{3}$ or 33% remains.

14.111   Using the energy diagram shown and Hess's law, we can see that the activation energy for the decomposition is equal to the activation energy for the formation reaction plus the heat of formation of 2 moles of HI, or $E_{a, \text{formation}} = E_{a, \text{decomposition}} + 2\Delta H_f^\circ(\text{HI})$. So
$E_{a, \text{formation}} = 185 \text{ kJ} + 2 \text{ mol}(-5.65 \text{ kJ/mol}) = 174 \text{ kJ}$.

Check: Because the reaction is endothermic, we expect the activation energy in the reverse direction to be less than that in the forward direction.

Note: Energy axis is not to scale.

Challenge Problems

14.113   (a)   Because the rate-determining step involves the collision of two molecules, the expected reaction order would be second order.

(b)   The proposed mechanism is

$$\text{CH}_3\text{NC} + \text{CH}_3\text{NC} \underset{k_2}{\overset{k_1}{\rightleftharpoons}} \text{CH}_3\text{NC*} + \text{CH}_3\text{NC} \qquad \text{(fast)}$$

$$\text{CH}_3\text{NC*} \overset{k_3}{\longrightarrow} \text{CH}_3\text{CN} \qquad \text{(slow)}$$

$$\overline{\text{CH}_3\text{NC} \longrightarrow \text{CH}_3\text{CN}} \qquad \text{So the sum matches the overall reaction.}$$

$\text{CH}_3\text{NC*}$ is the activated molecule. Because the second step is the rate-determining step, Rate $= k_3[\text{CH}_3\text{NC*}]$. Because $\text{CH}_3\text{NC*}$ is an intermediate, its concentration cannot appear in the rate law. Using the fast equilibrium in the first step, we see that $k_1[\text{CH}_3\text{NC}]^2 = k_2[\text{CH}_3\text{NC*}][\text{CH}_3\text{NC}]$ or $[\text{CH}_3\text{NC*}] = \dfrac{k_1}{k_2}[\text{CH}_3\text{NC}]$. Substituting this into the first rate expression, we get Rate $= k_3\dfrac{k_1}{k_2}[\text{CH}_3\text{NC}]$, which simplifies to Rate $= k[\text{CH}_3\text{NC}]$. This matches the experimental observation of first order, and the mechanism is valid.

14.115   Rate $= k[\text{A}]^2$ and Rate $= -\dfrac{d[\text{A}]}{dt}$, so $\dfrac{d[\text{A}]}{dt} = -k[\text{A}]^2$. Moving the A terms to the left and the t and constants to the right, we have $\dfrac{d[\text{A}]}{[\text{A}]^2} = -k \, dt$. Integrating, we get $\displaystyle\int_{[\text{A}]_0}^{[\text{A}]} \dfrac{d[\text{A}]}{[\text{A}]^2} = -\int_0^t k \, dt$. When we evaluate this integral,

$-[\text{A}]^{-1}\Big|_{[\text{A}]_0}^{[\text{A}]} = -kt\Big|_0^t \rightarrow -[\text{A}]_t^{-1} - (-[\text{A}]_0^{-1}) = -kt \rightarrow [\text{A}]_t^{-1} = kt + [\text{A}]_0^{-1}$ or $\dfrac{1}{[\text{A}]_t} = kt + \dfrac{1}{[\text{A}]_0}$, which is the desired integrated rate law.

14.117   For this mechanism, the overall reaction is the sum of the steps in the mechanism:

$$\text{Cl}_2(g) \underset{k_2}{\overset{k_1}{\rightleftharpoons}} 2\,\text{Cl}(g)$$

$$\text{Cl}(g) + \text{CO}(g) \underset{k_4}{\overset{k_3}{\rightleftharpoons}} \text{ClCO}(g)$$

$$\text{ClCO}(g) + \text{Cl}_2(g) \overset{k_5}{\longrightarrow} \text{Cl}_2\text{CO}(g) + \text{Cl}(g)$$

$$\overline{\text{CO}(g) + 2\,\text{Cl}_2(g) \rightarrow \text{Cl}_2\text{CO}(g) + 2\,\text{Cl}(g)}$$

No overall reaction is given. Because the third step is the rate-determining step, Rate $= k_5[\text{ClCO}][\text{Cl}_2]$. Because ClCO is an intermediate, its concentration cannot appear in the rate law. Using the fast equilibrium in the second step, we see that $k_3[\text{Cl}][\text{CO}] = k_4[\text{ClCO}]$ or $[\text{ClCO}] = \dfrac{k_3}{k_4}[\text{Cl}][\text{CO}]$. Substituting this expression into the first rate expression, we get Rate $= k_5\dfrac{k_3}{k_4}[\text{Cl}][\text{CO}][\text{Cl}_2]$. Because Cl is an intermediate, its concentration cannot appear in the rate law. Using the fast equilibrium in the first step, we see that $k_1[\text{Cl}_2] = k_2[\text{Cl}]^2$ or $[\text{Cl}] = \sqrt{\dfrac{k_1}{k_2}}[\text{Cl}_2]$. Substituting this expression into last rate expression, we get Rate $= k_5 \dfrac{k_3}{k_4}\sqrt{\dfrac{k_1}{k_2}}[Cl_2][CO][Cl_2] = k_5 \dfrac{k_3}{k_4}\sqrt{\dfrac{k_1}{k_2}}[CO][Cl_2]^{3/2}$.

Simplifying this expression, we see that Rate $= k[CO][Cl_2]^{3/2}$.

14.119 For the elementary reaction $2\,NOCl(g) \underset{k_{-1}}{\overset{k_1}{\rightleftharpoons}} 2\,NO(g) + Cl_2(g)$, we see that $k_1[NOCl]^2 = k_{-1}[NO]^2[Cl_2]$. For each mole of NOCl that reacts, 1 mole of NO and 0.5 mole of Cl_2 are generated. Because before any reaction only NOCl is present, $[NO] = 2[Cl_2]$. Substituting this into the first expression, we get $k_1[NOCl]^2 = k_{-1}(2[Cl_2])^2[Cl_2] \to k_1[NOCl]^2 = 4k_{-1}[Cl_2]^3$. Rearranging and substituting the specific values into this expression, we get

$$[Cl_2] = \sqrt[3]{\dfrac{k_1}{4k_{-1}}}[NOCl]^{2/3} = \sqrt[3]{\dfrac{7.8\times 10^{-2}\,\dfrac{L^2}{mol^2\cdot s}}{4\left(4.7\times 10^2\,\dfrac{L^2}{mol\cdot s}\right)}}\left(0.12\,\dfrac{mol}{L}\right)^{2/3} = 0.00842235\,\dfrac{mol}{L}\,Cl_2$$

$= 0.0084\ \text{M}\ Cl_2$ and $[Cl_2] = 2[NO] = 2\left(0.00842235\,\dfrac{mol}{L}\right) = 0.0168445\,\dfrac{mol}{L}NO = 0.017\ \text{M NO}$

Conceptual Problems

14.121 Reactant concentrations drop more quickly for first-order reactions than for second-order reactions, so reaction A must be second order. A plot of 1/[A] versus time will be linear $\left(\dfrac{1}{[A]_t} = kt + \dfrac{1}{[A]_0}\right)$. Reaction B is first order. A plot of ln[A] versus time will be linear $(\ln[A]_t = -kt + \ln[A]_0)$.

Questions for Group Work

14.123 It is very tempting to think that the same descriptor that relates the concentration change to the rate change is the order of the reaction. This statement is incorrect because the reaction order is first order. For a first-order reaction, Rate $= k[A]^1$, so doubling the concentration of A doubles the reaction rate. For a second-order reaction, Rate $= k[A]^2$, so doubling the concentration of A quadruples the reaction rate ($2^2 = 4$).

Data Interpretation and Analysis

14.125 (a) Since the reaction rate doubles when the concentration of each of the reactants is doubled, the reaction is first order in both methane and ozone.

(c) Since $\ln k = \dfrac{-E_a}{R}\left(\dfrac{1}{T}\right) + \ln A = \dfrac{-63.98288\,\dfrac{kJ}{mol}}{\left(8.314\,\dfrac{J}{K\cdot mol}\right)\left(\dfrac{1kJ}{1000\,J}\right)}\left(\dfrac{1}{273K}\right) + 18.902$

$= -28.18974 + 18.902 = -9.28774$

then $k = e^{-9.28774} = 9.25517\times 10^{-5}\,M^{-1}\cdot s^{-1}$. Convert reactant concentrations from ppm to molar:

$$[CH_4] = 1.8\ \text{ppm}\ CH_4 = \dfrac{1.8\ L\,CH_4}{10^6\ \text{L air}}\times\dfrac{1\ mol\ CH_4}{22.4\ L\,CH_4} = 8.0357\times 10^{-8}\ \text{M}\ CH_4$$

and

$$[O_3] = 5.0\ \text{ppm}\ O_3 = \dfrac{5.0\ L\,O_3}{10^6\ \text{L air}}\times\dfrac{1\ mol\ O_3}{22.4\ L\,O_3} = 2.2321\times 10^{-7}\ \text{M}\ O_3.$$

Rate $= k[CH_4][O_3] = (9.25517\times 10^{-5}\,M^{-1}\cdot s^{-1})(8.0357\times 10^{-8}\ \text{M}\ CH_4)(2.2321\times 10^{-7}\,\text{M}\ O_3)$

$$= 1.66008\times 10^{-18}\,M/s = 1.7\times 10^{-18}\,M/s.$$

15 Chemical Equilibrium

Review Questions

15.1   Like adult hemoglobin, fetal hemoglobin is in equilibrium with oxygen. However, the equilibrium constant for fetal hemoglobin is larger than the equilibrium constant for adult hemoglobin, meaning that the reaction tends to go farther in the direction of the product. Consequently, fetal hemoglobin loads oxygen at a lower oxygen concentration than does adult hemoglobin. In the placenta, fetal blood flows in close proximity to maternal blood, without the two ever mixing. Because of the different equilibrium constants, the maternal hemoglobin unloads oxygen that the fetal hemoglobin then binds and carries into its own circulatory system. Nature has thus evolved a chemical system through which the mother's hemoglobin can in effect hand off oxygen to the hemoglobin of the fetus.

15.3   The general expression of the equilibrium constant is $K = \frac{[C]^c[D]^d}{[A]^a[B]^b}$.

15.5   If you reverse the equation, invert the equilibrium constant. $K_{\text{reverse}} = \dfrac{1}{K_{\text{forward}}}$

       If you multiply the coefficients in the equation by a factor, raise the equilibrium constant to the same factor. That is, if you multiply the reaction by n, raise the equilibrium constant K to the n. $K' = K^n$

15.7 K_c is the equilibrium constant with respect to concentration in molarity. K_p is the equilibrium constant with respect to partial pressures in atmospheres. The two can be related as follows: $K_p = K_c(RT)^{\Delta n}$.

15.9   The concentration of a solid does not change because a solid does not expand to fill its container. Its concentration, therefore, depends only on its density, which is constant as long as some solid is present. Consequently, solids are not included in the equilibrium expression. Similarly, the concentration of a pure liquid does not change, so liquids are also excluded from the equilibrium expression.

15.11   When we know the initial concentrations of the reactants and products and the equilibrium concentration of one reactant or product, the other equilibrium concentrations can be deduced from the stoichiometry of the reaction. From the initial and equilibrium concentration of the one reactant, we can find the change in concentration for that reactant. Using the stoichiometry of the reaction, we can determine the equilibrium concentrations of the other reactants and products. We generally use an ICE table (I = Initial C = Change E = Equilibrium) to keep track of the changes.

15.13   Standard states are defined as 1 M concentration, or 1 atm pressure. So the value of K_c and K_p when the reactants and products are in their standard states is 1.

15.15   To solve a problem given initial concentrations and the equilibrium constant, you would prepare an ICE table, calculate Q, compare Q and K_c, predict the direction of the reaction, represent the change with x, sum the table, determine the equilibrium values, put the equilibrium values in the equilibrium expression, and solve for x. Determine the reactant and product concentration.

15.17   According to Le Châtelier's principle, when a chemical system at equilibrium is disturbed, the system shifts in a direction that minimizes the disturbance.

15.19   If a chemical system is at equilibrium:
- decreasing the volume causes the reaction to shift in the direction that has the fewer moles of gas particles.
- increasing the volume causes the reaction to shift in the direction that has the greater number of moles of gas particles.
- if a reaction has an equal number of moles of gas on both sides of the chemical equation, then a change in volume produces no effect on the equilibrium.
- adding an inert gas to the mixture at a fixed volume has no effect on the equilibrium.

Problems by Topic
Equilibrium and the Equilibrium Constant

15.21   The equilibrium constant is defined as the concentrations of the products raised to their stoichiometric coefficients divided by the concentrations of the reactants raised to their stoichiometric coefficients.

(a) $K = \dfrac{[SbCl_3][Cl_2]}{[SbCl_5]}$

(b) $K = \dfrac{[NO]^2[Br_2]}{[BrNO]^2}$

(c) $K = \dfrac{[CS_2][H_2]^4}{[CH_4][H_2S]^2}$

(d) $K = \dfrac{[CO_2]^2}{[CO]^2[O_2]}$

15.23   With an equilibrium constant of 1.4×10^{-5}, the value of the equilibrium constant is small; therefore, the concentration of reactants will be greater than the concentration of products. This is independent of the initial concentration of the reactants and products.

15.25   (i)   has 10 H_2 and 10 I_2
               (ii)   has 7 H_2 and 7 I_2 and 6 HI
               (iii)   has 5 H_2 and 5 I_2 and 10 HI
               (iv)   has 4 H_2 and 4 I_2 and 12 HI
               (v)   has 3 H_2 and 3 I_2 and 14 HI
               (vi)   has 3 H_2 and 3 I_2 and 14 HI

(a)   Concentrations of (v) and (vi) are the same, so the system reached equilibrium at (v).

(b)   If a catalyst was added to the system, the system would reach the conditions at (v) sooner because a catalyst speeds up the reaction but does not change the equilibrium conditions.

(c)   The final figure (vi) would have the same amount of reactants and products because a catalyst speeds up the reaction but does not change the equilibrium concentrations.

15.27   (a)   If you reverse the reaction, invert the equilibrium constant. So $K' = \dfrac{1}{K_p} = \dfrac{1}{2.26 \times 10^4} = 4.42 \times 10^{-5}$.

                 The reactants will be favored.

(b)   If you multiply the coefficients in the equation by a factor, raise the equilibrium constant to the same factor. So $K' = (K_p)^{1/2} = (2.26 \times 10^4)^{1/2} = 1.50 \times 10^2$. The products will be favored.

(c)   Begin with the reverse of the reaction and invert the equilibrium constant.

$$K_{reverse} = \frac{1}{K_p} = \frac{1}{2.26 \times 10^4} = 4.42 \times 10^{-5}$$

Then multiply the reaction by 2 and raise the value of $K_{reverse}$ to the second power.

$K' = (K_{reverse})^2 = (4.42 \times 10^{-5})^2 = 1.95 \times 10^{-9}$. The reactants will be favored.

15.29 To find the equilibrium constant for reaction 3, you need to combine reactions 1 and 2 to get reaction 3. Begin by reversing reaction 2; then multiply reaction 1 by 2 and add the two new reactions. When you add reactions, you multiply the values of K.

$$N_2(g) + O_2(g) \rightleftharpoons 2\,\cancel{NO}(g) \qquad\qquad K_1 = \frac{1}{K_p} = \frac{1}{2.1 \times 10^{30}} = 4.\underline{7}6 \times 10^{-31}$$

$$2\,\cancel{NO}(g) + Br_2(g) \rightleftharpoons 2\,NOBr(g) \qquad\qquad K_2 = (K_p)^2 = (5.3)^2 = 28.\underline{0}9$$

$$N_2(g) + O_2(g) + Br_2(g) \rightleftharpoons 2\,NOBr(g) \qquad\qquad K_3 = K_1 K_2 = (4.\underline{7}6 \times 10^{-31})(28.\underline{0}9) = 1.3 \times 10^{-29}$$

K_p, K_c, and Heterogeneous Equilibria

15.31 (a) **Given:** $K_p = 6.26 \times 10^{-22}$ $T = 298\,K$ **Find:** K_c
Conceptual Plan: $K_p \rightarrow K_c$

$$K_p = K_c(RT)^{\Delta n}$$

Solution: $\Delta n = $ mol product gas $-$ mol reactant gas $= 2 - 1 = 1$

$$K_c = \frac{K_p}{(RT)^{\Delta n}} = \frac{6.26 \times 10^{-22}}{\left(0.08206 \dfrac{L \cdot atm}{mol \cdot K} \times 298\,K\right)^1} = 2.56 \times 10^{-23}$$

Check: Substitute into the equation and confirm that you get the original value of K_p.

$$K_p = K_c(RT)^{\Delta n} = (2.56 \times 10^{-23})\left(0.08206 \dfrac{L \cdot atm}{mol \cdot K} \times 298\right)^1 = 6.26 \times 10^{-22}$$

(b) **Given:** $K_p = 7.7 \times 10^{24}$ $T = 298\,K$ **Find:** K_c
Conceptual Plan: $K_p \rightarrow K_c$

$$K_p = K_c(RT)^{\Delta n}$$

Solution: $\Delta n = $ mol product gas $-$ mol reactant gas $= 4 - 2 = 2$

$$K_c = \frac{K_p}{(RT)^{\Delta n}} = \frac{7.7 \times 10^{24}}{\left(0.08206 \dfrac{L \cdot atm}{mol \cdot K} \times 298\,K\right)^2} = 1.3 \times 10^{22}$$

Check: Substitute into the equation and confirm that you get the original value of K_p.

$$K_p = K_c(RT)^{\Delta n} = (1.3 \times 10^{22})\left(0.08206 \dfrac{L \cdot atm}{mol \cdot K} \times 298\right)^2 = 7.7 \times 10^{24}$$

(c) **Given:** $K_p = 81.9$ $T = 298\,K$ **Find:** K_c
Conceptual Plan: $K_p \rightarrow K_c$

$$K_p = K_c(RT)^{\Delta n}$$

Solution: $\Delta n = $ mol product gas $-$ mol reactant gas $= 2 - 2 = 0$

$$K_c = \frac{K_p}{(RT)^{\Delta n}} = \frac{81.9}{\left(0.08206 \dfrac{L \cdot atm}{mol \cdot K} \times 298\,K\right)^0} = 81.9$$

Check: Substitute into the equation and confirm that you get the original value of K_p.

$$K_p = K_c(RT)^{\Delta n} = (81.9)\left(0.08206 \dfrac{L \cdot atm}{mol \cdot K} \times 298\right)^0 = 81.9$$

15.33 (a) Because H_2O is a liquid, it is omitted from the equilibrium expression. $K_{eq} = \dfrac{[HCO_3^-][OH^-]}{[CO_3^{2-}]}$

(b) Because $KClO_3$ and KCl are both solids, they are omitted from the equilibrium expression. $K_{eq} = [O_2]^3$

(c)   Because H_2O is a liquid, it is omitted from the equilibrium expression. $K_{eq} = \dfrac{[H_3O^+][F^-]}{[HF]}$

(d)   Because H_2O is a liquid, it is omitted from the equilibrium expression. $K_{eq} = \dfrac{[NH_4^+][OH^-]}{[NH_3]}$

Relating the Equilibrium Constant to Equilibrium Concentrations and Equilibrium Partial Pressures

15.35   **Given:** at equilibrium: $[CO] = 0.105$ M, $[H_2] = 0.114$ M, $[CH_3OH] = 0.185$ M   **Find:** K_c
         Conceptual Plan: Balanced reaction → equilibrium expression → K_c

         Solution: $K_c = \dfrac{[CH_3OH]}{[CO][H_2]^2} = \dfrac{(0.185)}{(0.105)(0.114)^2} = 136$

         Check: The answer is reasonable because the concentration of products is greater than the concentration of reactants and the equilibrium constant should be greater than 1.

15.37   At 500 K: **Given:** at equilibrium: $[N_2] = 0.115$ M, $[H_2] = 0.105$ M, and $[NH_3] = 0.439$ M   **Find:** K_c
         Conceptual Plan: Balanced reaction → equilibrium expression → K_c

         Solution: $K_c = \dfrac{[NH_3]^2}{[N_2][H_2]^3} = \dfrac{(0.439)^2}{(0.115)(0.105)^3} = 1.45 \times 10^3$

         Check: The value is reasonable because the concentration of products is greater than the concentration of reactants.

         At 575 K: **Given:** at equilibrium: $[N_2] = 0.110$ M, $[NH_3] = 0.128$ M, $K_c = 9.6$ **Find:** $[H_2]$
         Conceptual Plan: Balanced reaction → equilibrium expression → $[H_2]$

         Solution: $K_c = \dfrac{[NH_3]^2}{[N_2][H_2]^3}$ $9.6 = \dfrac{(0.128)^2}{(0.110)(x)^3}$ $x = 0.249 = \underline{0.25}$

         Check: Plug the value for x back into the equilibrium expression and check the value.

         $$9.6 = \dfrac{(0.128)^2}{(0.110)(0.249)^3}$$

         At 775 K: **Given:** at equilibrium: $[N_2] = 0.120$ M, $[H_2] = 0.140$ M, $K_c = 0.0584$ **Find:** $[NH_3]$
         Conceptual Plan: Balanced reaction → equilibrium expression → $[NH_3]$

         Solution: $K_c = \dfrac{[NH_3]^2}{[N_2][H_2]^3}$ $0.0584 = \dfrac{(x)^2}{(0.120)(0.140)^3}$ $x = 0.00439$

         Check: Plug the value for x back into the equilibrium expression and check the value.

         $$0.0584 = \dfrac{(0.00439)^2}{(0.120)(0.140)^3}$$

15.39   **Given:** $P_{NO} = 108$ torr; $P_{Br_2} = 126$ torr, $K_p = 28.4$ **Find:** P_{NOBr}
         Conceptual Plan: torr → atm and then balanced reaction → equilibrium expression → P_{NOBr}

         $$\dfrac{1\ atm}{760\ torr}$$

         Solution: $P_{NO} = 108\ \cancel{torr} \times \dfrac{1\ atm}{760\ \cancel{torr}} = 0.14\underline{2}1$ atm $P_{Br_2} = 126\ \cancel{torr} \times \dfrac{1\ atm}{760\ \cancel{torr}} = 0.16\underline{5}8$ atm

         $K_p = \dfrac{P_{NOBr}^2}{P_{NO}^2 P_{Br_2}}$ $28.4 = \dfrac{x^2}{(0.14\underline{2}1)^2(0.16\underline{5}8)}$ $x = 0.308$ atm $= 234$ torr

         Check: Plug the value for x back into the equilibrium expression and check the value.

         $$28.3 = \dfrac{(0.308)^2}{(0.14\underline{2}1)^2(0.16\underline{5}8)}$$

15.41   **Given:** $P_{A,\,initial} = 1.32$ atm; $P_{A,\,eq} = 0.25$ atm   **Find:** K_p
              Conceptual Plan: 1. **Prepare ICE table.**
                                       2. **Calculate pressure change for known value.**
                                       3. **Calculate pressure changes for other reactants/products.**
                                       4. **Determine equilibrium pressures.**
                                       5. **Write the equilibrium expression and determine K_p.**
              Solution: $A(g) \rightleftharpoons 2B(g)$

	P_A	P_B
Initial	1.32 atm	0.00
Change	$-x$	$+2x$
Equil	0.25 atm	$2x$

$$x = 1.32 - 0.25 = 1.07 \text{ atm}$$
$$2x = 2(1.07 \text{ atm}) = 2.14 \text{ atm} = P_B$$
$$K_p = \frac{P_B^2}{P_A} = \frac{(2.14)^2}{(0.25)} = 18.32 = \underline{18}$$

Check: The units (none) are correct. The magnitude is reasonable (18) since there are significant concentrations of both reactants and products.

15.43   **Given:** $[Fe^{3+}]_{initial} = 1.0 \times 10^{-3}$ M; $[SCN^-]_{initial} = 8.0 \times 10^{-4}$ M; $[FeSCN^{2+}]_{eq} = 1.7 \times 10^{-4}$ M   **Find:** K_c
              Conceptual Plan: 1. **Prepare ICE table.**
                                       2. **Calculate concentration change for known value.**
                                       3. **Calculate concentration changes for other reactants/products.**
                                       4. **Determine equilibrium concentration.**
                                       5. **Write the equilibrium expression and determine K_c.**
              Solution: $Fe^{3+}(aq) + SCN^-(aq) \rightleftharpoons FeSCN^{3+}(aq)$

	$[Fe^{3+}]$	$[SCN^-]$	$[FeSCN^{3+}]$
Initial	1.0×10^{-3}	8.0×10^{-4}	0.00
Change	-1.7×10^{-4}	-1.7×10^{-4}	$+1.7 \times 10^{-4}$
Equil	8.3×10^{-4}	6.3×10^{-4}	1.7×10^{-4}

$$K_c = \frac{[FeSCN^{2+}]}{[Fe^{3+}][SCN^-]} = \frac{(1.7 \times 10^{-4})}{(8.3 \times 10^{-4})(6.3 \times 10^{-4})} = 3.3 \times 10^2$$

Check: The units (none) are correct. The magnitude is reasonable (330) since there are significant concentrations of both reactants and products.

15.45   **Given:** 3.67 L flask, 0.763 g H_2 initial, 96.9 g I_2 initial, 90.4 g HI equilibrium   **Find:** K_c
              Conceptual Plan: g $\rightarrow$ mol $\rightarrow$ M and then

$$n = \frac{g}{\text{molar mass}} \qquad M = \frac{n}{V}$$

                                       1. **Prepare ICE table.**
                                       2. **Calculate concentration change for known value.**
                                       3. **Calculate concentration changes for other reactants/products.**
                                       4. **Determine equilibrium concentration.**
                                       5. **Write the equilibrium expression and determine K_c.**

Solution: $0.763 \text{ g } H_2 \times \dfrac{1 \text{ mol } H_2}{2.016 \text{ g } H_2} = 0.3785 \text{ mol } H_2$ then $\dfrac{0.3785 \text{ mol } H_2}{3.67 \text{ L}} = 0.103 \text{ M}$ This is an initial concentration.

$96.9 \text{ g } I_2 \times \dfrac{1 \text{ mol } I_2}{253.80 \text{ g } I_2} = 0.3818 \text{ mol } I_2$ then $\dfrac{0.3818 \text{ mol } I_2}{3.67 \text{ L}} = 0.104 \text{ M}$ This is an initial concentration.

$90.4 \text{ g HI} \times \dfrac{1 \text{ mol HI}}{127.91 \text{ g HI}} = 0.7068 \text{ mol HI}$ then $\dfrac{0.7068 \text{ mol HI}}{3.67 \text{ L}} = 0.193 \text{ M}$ This is an equilibrium concentration.

$$H_2(g) \quad + \quad I_2(g) \rightleftharpoons 2\,HI(g)$$

	$[H_2]$	$[I_2]$	$[HI]$
Initial	0.103	0.104	0.00
Change	-0.0965	-0.0965	$+0.193$
Equil	0.0065	0.0075	0.193

Because HI gained 0.193 M, H_2 and I_2 had to lose $0.193/2 = 0.0965$ M from the stoichiometry of the balanced reaction.

$$K_c = \frac{[HI]^2}{[H_2][I_2]} = \frac{(0.193)^2}{(0.0065)(0.0075)} = 764$$

Check: The units (none) are correct. The magnitude is reasonable (764) since there are higher concentrations of the products than the reactants.

The Reaction Quotient and Reaction Direction

15.47   **Given:** $K_c = 8.5 \times 10^{-3}$; $[NH_3] = 0.166$ M; $[H_2S] = 0.166$ M   **Find:** whether solid will form or decompose
Conceptual Plan: Calculate $Q \rightarrow$ compare $Q \rightarrow$ and K_c
Solution: $Q = [NH_3][H_2S] = (0.166)(0.166) = 0.0276$
$Q = 0.0276$ and $K_c = 8.5 \times 10^{-3}$ so $Q > K_c$ and the reaction will shift to the left; so more solid will form.

15.49   **Given:** 6.55 g Ag_2SO_4, 1.5 L solution, $K_c = 1.1 \times 10^{-5}$ **Find:** whether more solid will dissolve
Conceptual Plan: g $Ag_2SO_4 \rightarrow$ mol $Ag_2SO_4 \rightarrow [Ag_2SO_4] \rightarrow [Ag^+], [SO_4^{2-}] \rightarrow$ calculate Q and compare to K_c

$$\frac{1\text{ mol } Ag_2SO_4}{311.80\text{ g}} \qquad [\,] = \frac{\text{mol } Ag_2SO_4}{\text{vol solution}} \qquad\qquad Q = [Ag^+]^2[SO_4^{2-}]$$

Solution: $6.55 \text{ g } Ag_2SO_4 \left(\dfrac{1\text{ mol } Ag_2SO_4}{311.80 \text{ g } Ag_2SO_4} \right) = 0.0210$ mol Ag_2SO_4

$$\frac{0.0210 \text{ mol } Ag_2SO_4}{1.5 \text{ L solution}} = 0.0140 \text{ M } Ag_2SO_4$$

$[Ag^+] = 2[Ag_2SO_4] = 2(0.0140 \text{ M}) = 0.0280 \text{ M}$; $[SO_4^{2-}] = [Ag_2SO_4] = 0.0140 \text{ M}$
$Q = [Ag^+]^2[SO_4^{2-}] = (0.0280)^2(0.0140) = 1.1 \times 10^{-5}$
$Q = K_c$, so the system is at equilibrium and is a saturated solution. Therefore, if more solid is added, it will not dissolve.

Finding Equilibrium Concentrations from Initial Concentrations and the Equilibrium Constant

15.51   (a)   **Given:** $[A] = 1.0$ M, $[B] = 0.0$, $K_c = 4.0$; $a = 1, b = 1$ **Find:** $[A]$, $[B]$ at equilibrium
Conceptual Plan: Prepare an ICE table, calculate Q, compare Q and K_c, predict the direction of the reaction, represent the change with x, sum the table, determine the equilibrium values, put the equilibrium values in the equilibrium expression, and solve for x. Determine [A] and [B].
Solution: $A(g) \rightleftharpoons B(g)$

	$[A]$	$[B]$
Initial	1.0	0.00
Change	$-x$	$+x$
Equil	$1.0 - x$	x

$$Q = \frac{[B]}{[A]} = \frac{0}{1.0} = 0 \quad Q < K \quad \text{Therefore, the reaction will proceed to the right by } x.$$

$$K_c = \frac{[B]}{[A]} = \frac{(x)}{(1.0 - x)} = 4.0; \quad x = 0.80$$

$[A] = 1 - 0.80 = 0.20$ M; $[B] = 0.80$ M

Check: Plug the values into the equilibrium expression: $K_c = \dfrac{0.80}{0.20} = 4.0$

(b) **Given:** $[A] = 1.0\,M$, $[B] = 0.0$, $K_c = 4.0$; $a = 2, b = 2$ **Find:** $[A]$, $[B]$ at equilibrium
Conceptual Plan: Prepare an ICE table, calculate Q, compare Q and K_c, predict the direction of the reaction, represent the change with x, sum the table, determine the equilibrium values, put the equilibrium values in the equilibrium expression, and solve for x. Determine [A] and [B].
Solution: $\quad 2\,A(g) \rightleftharpoons 2\,B(g)$

	$[A]$	$[B]$
Initial	1.0	0.00
Change	$-2x$	$+2x$
Equil	$1.0 - 2x$	$2x$

$$Q = \frac{[B]^2}{[A]^2} = \frac{0}{1.0} = 0 \quad Q < K \quad \text{Therefore, the reaction will proceed to the right by } x.$$

$$K_c = \frac{[B]^2}{[A]^2} = \frac{(2x)^2}{(1.0 - 2x)^2} = 4.0$$

$$\sqrt{\frac{(2x)^2}{(1.0 - 2x)^2}} = \sqrt{4.0} \quad \text{Take the square root of this expression. } x = 0.33$$

$$[A] = 1 - 2(0.33) = 0.34\,M; \quad [B] = 2(0.33) = 0.66\,M$$

Check: Plug the values into the equilibrium expression: $K_c = \dfrac{(0.66)^2}{(0.34)^2} = 3.8.$

(c) **Given:** $[A] = 1.0\,M$, $[B] = 0.0$, $K_c = 4.0$; $a = 1, b = 2$ **Find:** $[A]$, $[B]$ at equilibrium
Conceptual Plan: Prepare an ICE table, calculate Q, compare Q and K_c, predict the direction of the reaction, represent the change with x, sum the table, determine the equilibrium values, put the equilibrium values in the equilibrium expression, and solve for x. Determine [A] and [B].
Solution: $\quad A(g) \rightleftharpoons 2\,B(g)$

	$[A]$	$[B]$
Initial	1.0	0.00
Change	$-x$	$+2x$
Equil	$1.0 - x$	$2x$

$$Q = \frac{[B]^2}{[A]} = \frac{0}{1.0} = 0 \quad Q < K \quad \text{Therefore, the reaction will proceed to the right by } x.$$

$$K_c = \frac{[B]^2}{[A]} = \frac{(2x)^2}{(1.0 - x)} = 4.0 \quad 4x^2 + 4x - 4 = 0 \quad \text{Solve using the quadratic equation, found in Appendix I.}$$

$$x = -1.6 \quad \text{or } x = 0.62; \text{ therefore, } x = 0.62.$$

$$[A] = 1.0 - 0.62 = 0.38\,M \quad [B] = 2x = 2(0.62) = 1.2\,M$$

Check: Plug the values into the equilibrium expression: $K_c = \dfrac{(1.2)^2}{0.38} = 3.8.$

15.53 **Given:** $[N_2O_4] = 0.0500\,M$, $[NO_2] = 0.0$, $K_c = 0.513$ **Find:** $[N_2O_4]$, $[NO_2]$ at equilibrium
Conceptual Plan: Prepare an ICE table, calculate Q, compare Q and K_c, predict the direction of the reaction, represent the change with x, sum the table, determine the equilibrium values, put the equilibrium values in the equilibrium expression, and solve for x. Determine $[N_2O_4]$ and $[NO_2]$.
Solution:

$$N_2O_4(g) \rightleftharpoons 2\,NO_2(g)$$

	$[N_2O_4]$	$[NO_2]$
Initial	0.0500	0.00
Change	$-x$	$+2x$
Equil	$0.0500 - x$	$2x$

$$Q = \frac{[NO_2]^2}{[N_2O_4]} = \frac{0}{0.0500} = 0 \qquad Q < K \quad \text{Therefore, the reaction will proceed to the right by } x.$$

$$K_c = \frac{[NO_2]^2}{[N_2O_4]} = \frac{(2x)^2}{(0.0500 - x)} = 0.513 \quad 4x^2 + 0.513x - 0.02565 = 0$$

$$\frac{-b \pm \sqrt{b^2 - 4ac}}{2a} = \frac{-0.513 \pm \sqrt{(0.513)^2 - 4(4)(-0.02565)}}{2(4)} = \frac{-0.513 \pm \sqrt{0.6735}}{2(4)}$$

$x = -0.1667$ or $x = 0.0385$ Therefore, $x = 0.0385$.

$[N_2O_4] = 0.0500 - 0.0385 = 0.0115 \, M$ $[NO_2] = 2x = 2(0.0385) = 0.0770 \, M$

Check: Plug the values into the equilibrium expression: $K_c = \dfrac{(0.0770)^2}{0.0115} = 0.516$.

15.55   **Given:** $[CO] = 0.20 \, M$, $[CO_2] = 0.0$, $K_c = 4.0 \times 10^3$ **Find:** $[CO_2]$ at equilibrium

Conceptual Plan: Prepare an ICE table, calculate Q, compare Q and K_c, predict the direction of the reaction, represent the change with x, sum the table, determine the equilibrium values, put the equilibrium values in the equilibrium expression, and solve for x. Determine $[CO]$, $[Cl_2]$, and $[COCl_2]$.

Solution:

$$NiO(s) + CO(g) \rightleftharpoons Ni(s) + CO_2(g)$$

	$[CO]$	$[Cl_2]$
Initial	0.20	0.0
Change	$-x$	$+x$
Equil	$0.20 - x$	x

$$Q = \frac{[CO_2]}{[CO]} = \frac{0}{(0.20)} = 0 \qquad Q < K \quad \text{Therefore, the reaction will proceed to the right by } x.$$

$$K_c = \frac{[CO_2]}{[CO]} = \frac{x}{(0.20 - x)} = 4.0 \times 10^3 \qquad 4.0 \times 10^3(0.20 - x) = x$$

$$x = 0.199$$

$$[CO_2] = 0.20 \, M$$

Check: Because the equilibrium constant is so large, the reaction goes essentially to completion; therefore, it is reasonable that the concentration of the product is 0.20 M.

15.57   **Given:** $[HC_2H_3O_2] = 0.210 \, M$, $[H_3O^+] = 0.0$, $[C_2H_3O_2^-] = 0.0$, $K_c = 1.8 \times 10^{-5}$

Find: $[HC_2H_3O_2]$, $[H_2O^+]$, $[C_2H_3O_2^-]$ at equilibrium

Conceptual Plan: Prepare an ICE table, calculate Q, compare Q and K_c, predict the direction of the reaction, represent the change with x, sum the table, determine the equilibrium values, put the equilibrium values in the equilibrium expression, and solve for x. Determine $[CO]$, $[Cl_2]$, and $[COCl_2]$.

Solution:

$$HC_2H_3O_2(aq) + H_2O(l) \rightleftharpoons H_3O^+(aq) + C_2H_3O_2^-(aq)$$

	$HC_2H_3O_2$	$[H_2O]$	$[H_3O^+]$	$[C_2H_3O_2^-]$
Initial	0.210		0.0	0.0
Change	$-x$		$+x$	$+x$
Equil	$0.210 - x$		x	x

$$Q = \frac{[H_3O^+][C_2H_3O_2^-]}{[HC_2H_3O_2]} = \frac{0}{(0.210)} = 0 \qquad Q < K \quad \text{Therefore, the reaction will proceed to the right by } x.$$

$$K_c = \frac{[H_3O^+][C_2H_3O_2^-]}{[HC_2H_3O_2]} = \frac{(x)(x)}{(0.210 - x)} = 1.8 \times 10^{-5}$$

Assume that x is small compared to 0.210.

$$x^2 = 0.210(1.8 \times 10^{-5})$$

$x = 0.00194$ Check assumption: $\dfrac{0.00194}{0.210} \times 100 = 0.92\%$; assumption is valid.

$[H_3O^+] = [C_2H_3O_2^-] = 0.00194\ M$

$[HC_2H_3O_2] = 0.210 - 0.00194 = 0.2081 = 0.208\ M$

Check: Plug the values into the equilibrium expression: $K_c = \dfrac{(0.00194)(0.00194)}{(0.208)} = 1.81 \times 10^{-5}$.

The answer is the same to two significant figures with the true value, so the answers are valid.

15.59   **Given:** $P_{Br_2} = 755$ torr, $P_{Cl_2} = 735$ torr, $P_{BrCl} = 0.0$, $K_p = 1.11 \times 10^{-4}$ **Find:** P_{BrCl} at equilibrium
Conceptual Plan: Torr → atm and then prepare an ICE table, calculate Q, compare Q and K_c, predict the direction of the reaction, represent the change with x, sum the table, determine the equilibrium values, put the equilibrium values in the equilibrium expression, and solve for x. Determine P_{BrCl}.

Solution: $P_{Br_2} = 755\ \text{torr} \times \dfrac{1\ \text{atm}}{760\ \text{torr}} = 0.9934\ \text{atm}$ $P_{Cl_2} = 735\ \text{torr} \times \dfrac{1\ \text{atm}}{760\ \text{torr}} = 0.9671\ \text{atm}$

$$Br_2(g) \ + \ Cl_2(g) \rightleftharpoons 2\,BrCl(g)$$

	P_{Br_2}	P_{Cl_2}	P_{BrCl}
Initial	0.9934	0.9671	0.0
Change	$-x$	$-x$	$+2x$
Equil	$0.9934 - x$	$0.9671 - x$	$2x$

$Q = \dfrac{P_{BrCl}^2}{P_{Br_2}P_{Cl_2}} = \dfrac{0}{(0.9934)(0.9671)} = 0$ $Q < K$ Therefore, the reaction will proceed to the right by x.

$K_p = \dfrac{P_{BrCl}^2}{P_{Br_2}P_{Cl_2}} = \dfrac{(2x)^2}{(0.9934 - x)(0.9671 - x)} = 1.11 \times 10^{-4}$

Assume that x is small compared to 0.9934 and 0.9671.

$\dfrac{(2x)^2}{(0.9934)(0.9671)} = 1.11 \times 10^{-4}$ $4x^2 = 1.066 \times 10^{-4}$

$x = 0.00516\ \text{atm} = 3.92\ \text{torr}$

$P_{BrCl} = 2x = 2(3.92\ \text{torr}) = 7.84\ \text{torr}$

Check: Plug the values into the equilibrium expression:

$K_c = \dfrac{(2(0.00516))^2}{(0.9934 - 0.00516)(0.9671 - 0.00516)} = 1.120 \times 10^{-4} = 1.12 \times 10^{-4}$.

This is within 0.01×10^{-4} of the true value; therefore, the answers are valid.

15.61   (a)   **Given:** $[A] = 1.0\ M$, $[B] = [C] = 0.0$, $K_c = 1.0$ **Find:** $[A], [B], [C]$ at equilibrium
Conceptual Plan: Prepare an ICE table, calculate Q, compare Q and K_c, predict the direction of the reaction, represent the change with x, sum the table, determine the equilibrium values, put the equilibrium values in the equilibrium expression, and solve for x. Determine $[A], [B],$ and $[C]$.
Solution:

$$A(g) \rightleftharpoons B(g) \ + \ C(g)$$

	$[A]$	$[B]$	$[C]$
Initial	1.0	0.0	0.0
Change	$-x$	$+x$	$+x$
Equil	$1.0 - x$	x	x

$Q = \dfrac{[B][C]}{[A]} = \dfrac{0}{(1.0)} = 0$ $Q < K$ Therefore, the reaction will proceed to the right by x.

$K_c = \dfrac{[B][C]}{[A]} = \dfrac{(x)(x)}{(1.0 - x)} = 1.0$

$x^2 = 1.0(1.0 - x)$

$x^2 + x - 1 = 0$

$$\frac{-b \pm \sqrt{b^2 - 4ac}}{2a} = \frac{-1 \pm \sqrt{1^2 - 4(1)(-1)}}{2(1)}$$

$x = 0.6\underline{1}8$ or $x = -1.618$ Therefore, $x = 0.6\underline{1}8$.

$[B] = [C] = x = 0.6\underline{1}8 = 0.62 \text{ M}$

$[A] = 1.0 - 0.6\underline{1}8 = 0.382 = 0.38 \text{ M}$

Check: Plug the values into the equilibrium expression:

$K_c = \dfrac{(0.62)(0.62)}{(0.38)} = 1.01 = 1.0$, which is the equilibrium constant; so the values are correct.

(b)   **Given:** $[A] = 1.0 \text{ M}$, $[B] = [C] = 0.0$, $K_c = 0.010$ **Find:** $[A], [B], [C]$ at equilibrium
Conceptual Plan: Prepare an ICE table, calculate Q, compare Q and K_c, predict the direction of the reaction, represent the change with x, sum the table, determine the equilibrium values, put the equilibrium values in the equilibrium expression, and solve for x. Determine $[A], [B],$ and $[C]$.

Solution:

$$A(g) \rightleftharpoons B(g) + C(g)$$

	$[A]$	$[B]$	$[C]$
Initial	1.0	0.0	0.0
Change	$-x$	$+x$	$+x$
Equil	$1.0 - x$	x	x

$Q = \dfrac{[B][C]}{[A]} = \dfrac{0}{(1.0)} = 0$ $Q < K$ Therefore, the reaction will proceed to the right by x.

$K_c = \dfrac{[B][C]}{[A]} = \dfrac{(x)(x)}{(1.0 - x)} = 0.010$

$x^2 = 0.010(1.0 - x)$

$x^2 + 0.010x - 0.010 = 0$

$$\frac{-b \pm \sqrt{b^2 - 4ac}}{2a} = \frac{-(0.010) \pm \sqrt{(0.010)^2 - 4(1)(-0.010)}}{2(1)}$$

$x = 0.09\underline{5}12$ or $x = -0.1051$ Therefore, $x = 0.09\underline{5}12$.

$[B] = [C] = x = 0.09\underline{5}12 = 0.095 \text{ M}$

$[A] = 1.0 - 0.09\underline{5}12 = 0.90488 = 0.90 \text{ M}$

Check: Plug the values into the equilibrium expression:

$K_c = \dfrac{(0.095)(0.095)}{(0.90)} = 0.01003 = 0.010$, which is the equilibrium constant; so the values are correct.

(c)   **Given:** $[A] = 1.0 \text{ M}$, $[B] = [C] = 0.0$, $K_c = 1.0 \times 10^{-5}$ **Find:** $[A], [B], [C]$ at equilibrium
Conceptual Plan: Prepare an ICE table, calculate Q, compare Q and K_c, predict the direction of the reaction, represent the change with x, sum the table, determine the equilibrium values, put the equilibrium values in the equilibrium expression, and solve for x. Determine $[A], [B],$ and $[C]$.

Solution:

$$A(g) \rightleftharpoons B(g) + C(g)$$

	$[A]$	$[B]$	$[C]$
Initial	1.0	0.0	0.0
Change	$-x$	$+x$	$+x$
Equil	$1.0 - x$	x	x

$Q = \dfrac{[B][C]}{[A]} = \dfrac{0}{(1.0)} = 0$ $Q < K$ Therefore, the reaction will proceed to the right by x.

$K_c = \dfrac{[B][C]}{[A]} = \dfrac{(x)(x)}{(1.0 - x)} = 1.0 \times 10^{-5}$

Assume that x is small compared to 1.0.

$$x^2 = 1.0(1.0 \times 10^{-5})$$

$x = 0.003\underline{1}6$ Check assumption: $\dfrac{0.003\underline{1}6}{1.0} \times 100\% = 0.32\%$; assumption is valid.

$[B] = [C] = x = 0.003\underline{1}6 = 0.0032\,\text{M}$

$[A] = 1.0 - 0.003\underline{1}6 = 0.9968 = 1.0\,\text{M}$

Check: Plug the values into the equilibrium expression:

$$K_c = \frac{(0.0032)(0.0032)}{(1.0)} = 1.024 \times 10^{-5} = 1.0 \times 10^{-5},$$

which is the equilibrium constant; so the values are correct.

Le Châtelier's Principle

15.63   **Given:** $CO(g) + Cl_2(g) \rightleftharpoons COCl_2(g)$ at equilibrium   **Find:** What is the effect of each of the following?

(a) $COCl_2$ is added to the reaction mixture. Adding $COCl_2$ increases the concentration of $COCl_2$ and causes the reaction to shift to the left.

(b) Cl_2 is added to the reaction mixture. Adding Cl_2 increases the concentration of Cl_2 and causes the reaction to shift to the right.

(c) $COCl_2$ is removed from the reaction mixture. Removing the $COCl_2$ decreases the concentration of $COCl_2$ and causes the reaction to shift to the right.

15.65   **Given:** $2\,KClO_3(s) \rightleftharpoons 2\,KCl(s) + 3\,O_2(g)$ at equilibrium   **Find:** What is the effect of each of the following?

(a) O_2 is removed from the reaction mixture. Removing the O_2 decreases the concentration of O_2 and causes the reaction to shift to the right.

(b)   KCl is added to the reaction mixture. Adding KCl does not cause any change in the reaction. KCl is a solid, and the concentration remains constant; so the addition of more solid does not change the equilibrium concentration.

(c) $KClO_3$ is added to the reaction mixture. Adding $KClO_3$ does not cause any change in the reaction. $KClO_3$ is a solid, and the concentration remains constant; so the addition of more solid does not change the equilibrium concentration.

(d) O_2 is added to the reaction mixture. Adding O_2 increases the concentration of O_2 and causes the reaction to shift to the left.

15.67   (a)   **Given:** $I_2(g) \rightleftharpoons 2\,I(g)$ at equilibrium   **Find:** the effect of increasing the volume

The chemical equation has 2 moles of gas on the right and 1 mole of gas on the left. Increasing the volume of the reaction mixture decreases the pressure and causes the reaction to shift to the right (toward the side with more moles of gas particles).

(b)   **Given:** $2\,H_2S(g) \rightleftharpoons 2\,H_2(g) + S_2(g)$ **Find:** the effect of decreasing the volume

The chemical equation has 3 moles of gas on the right and 2 moles of gas on the left. Decreasing the volume of the reaction mixture increases the pressure and causes the reaction to shift to the left (toward the side with fewer moles of gas particles).

(c)   **Given:** $I_2(g) + Cl_2(g) \rightleftharpoons 2\,ICl(g)$ **Find:** the effect of decreasing the volume

The chemical equation has 2 moles of gas on the right and 2 moles of gas on the left. Decreasing the volume of the reaction mixture increases the pressure but causes no shift in the reaction because the moles are equal on both sides.

15.69   **Given:** $C(s) + CO_2(g) \rightleftharpoons 2\,CO(g)$ is endothermic.   **Find:** the effect of increasing the temperature

Because the reaction is endothermic, we can think of the heat as a reactant. Increasing the temperature is equivalent to adding a reactant, causing the reaction to shift to the right. This will cause an increase in the concentration of products and a decrease in the concentration of reactants; therefore, the value of K will increase.

Find: the effect of decreasing the temperature

Because the reaction is endothermic, we can think of the heat as a reactant. Decreasing the temperature is equivalent to removing a reactant, causing the reaction to shift to the left. This will cause a decrease in the concentration of products and an increase in the concentration of reactants; therefore, the value of K will decrease.

15.71   **Given:** $C(s) + 2\,H_2(g) \rightleftharpoons CH_4(g)$ is exothermic.   **Find:** Determine which will favor CH_4.

(a)   Adding more C to the reaction mixture does *not* favor CH_4. Adding C does not cause any change in the reaction. C is a solid, and the concentration remains constant; so the addition of more solid does not change the equilibrium concentration.

(b)   Adding more H_2 to the reaction mixture favors CH_4. Adding H_2 increases the concentration of H_2, causing the reaction to shift to the right.

(c)   Raising the temperature of the reaction mixture does *not* favor CH_4. Because the reaction is exothermic, we can think of heat as a product. Raising the temperature is equivalent to adding a product, causing the reaction to shift to the left.

(d)   Lowering the volume of the reaction mixture favors CH_4. The chemical equation has 1 mole of gas on the right and 2 moles of gas on the left. Decreasing the volume of the reaction mixture increases the pressure and causes the reaction to shift to the right (toward the side with fewer moles of gas particles).

(e)   Adding a catalyst to the reaction mixture does *not* favor CH_4. A catalyst added to the reaction mixture only speeds up the reaction; it does not change the equilibrium concentration.

(f)   Adding neon gas to the reaction mixture does *not* favor CH_4. Adding an inert gas to a reaction mixture at a fixed volume has no effect on the equilibrium.

Cumulative Problems

15.73   (a)   To find the value of K for the new equation, combine the two given equations to yield the new equation. Reverse equation 1 and use $1/K_1$, then add to equation 2. To find K for equation 3, use $(1/K_1)(K_2)$.

$$HbO_2(aq) \rightleftharpoons \cancel{Hb(aq)} + O_2(aq) \qquad K_1 = 1/1.8$$
$$\cancel{Hb(aq)} + CO(aq) \rightleftharpoons HbCO(aq) \qquad K_2 = 306$$

$$\overline{HbO_2(aq) + CO(aq) \rightleftharpoons HbCO(aq) + O_2(aq) \quad K_3 = K_1K_2 = (1/1.8)(306) = 170}$$

(b)   **Given:** $O_2 = 20\%, CO = 0.10\%$ **Find:** the ratio $\dfrac{[HbCO]}{[HbO_2]}$

Conceptual Plan: Determine the equilibrium expression and then determine $\dfrac{[HbCO]}{[HbO_2]}$.

Solution: $K = \dfrac{[HbCO][O_2]}{[HbO_2][CO]}$ $170 = \dfrac{[HbCO](20.0)}{[HbO_2](0.10)}$ then $\dfrac{[HbCO]}{[HbO_2]} = 170\left(\dfrac{0.10}{20.0}\right) = \dfrac{0.85}{1.0}$

Because the ratio is almost 1:1, 0.10% CO will replace about 50% of the O_2 in the blood. The CO blocks the uptake of O_2 by the blood and is therefore highly toxic.

15.75   (a)   **Given:** 4.45 g CO_2, 10.0 L, 1200 K, 2.00 g C, $K_p = 5.78$ **Find:** total pressure

Conceptual Plan: g $CO_2 \rightarrow$ mol CO_2 and g $C \rightarrow$ mol C and then determine limiting reactant

$$\dfrac{1\ mol\ CO_2}{44.01\ g\ CO_2} \qquad\qquad \dfrac{1\ mol\ C}{12.01\ g\ C}$$

and then mol $CO_2 \rightarrow P_{CO_2}$. Prepare an ICE table, represent the change with x, sum the table,

$$PV = nRT$$

determine the equilibrium values, put the equilibrium values in the equilibrium expression, and solve for x.

Solution: $CO_2(g) + C(s) \rightleftharpoons 2\,CO(g)$

$$4.45\ \cancel{g\ CO_2} \times \dfrac{1\ mol}{44.01\ \cancel{g\ CO_2}} = 0.10\underline{1}1\ mol\ CO_2 \quad 2.00\ \cancel{g\ C} \times \dfrac{1\ mol}{12.01\ \cancel{g\ C}} = 0.16\underline{6}5\ mol\ C$$

Because the stoichiometry is 1:1, the CO_2 is the limiting reactant.

$$P_{CO_2} = \dfrac{(0.10\underline{1}1\ \cancel{mol})\left(\dfrac{0.08206\ \cancel{L}\cdot atm}{\cancel{mol}\cdot\cancel{K}}\right)(1200\ \cancel{K})}{10.0\ \cancel{L}} = 0.99\underline{5}6\ atm$$

$$CO_2(g) + C(s) \rightleftharpoons 2\,CO(g)$$

	P_{CO_2}	P_{CO}
Initial	0.996	0.0
Change	$-x$	$+2x$
Equil	$0.996 - x$	$2x$

The reaction will proceed to the right by x.

$$K_p = \frac{P_{CO}^2}{P_{CO_2}} = \frac{(2x)^2}{(0.996 - x)} = 5.78. \text{ Solve using the quadratic equation, found in Appendix I.}$$

$$x = 0.678 \text{ atm}$$

$$P_{CO_2} = 0.996 \text{ atm} - 0.678 \text{ atm} = 0.318 \text{ atm} \quad P_{CO} = 2(0.678 \text{ atm}) = 1.356 \text{ atm}$$

$$P_{total} = 1.67 \text{ atm}$$

Check: Plug the values for the partial pressure into the equilibrium expression. $\dfrac{(1.356)^2}{0.318} = 5.78$, which is the value of the equilibrium constant.

(b) **Given:** 4.45 g CO_2, 10.0 L, 1200 K, 0.50 g C, $K_p = 5.78$ **Find:** total pressure

Conceptual Plan: g $CO_2 \rightarrow$ mol CO_2 and g C $\rightarrow$ mol C and then determine limiting reactant

$$\frac{1 \text{ mol } CO_2}{44.01 \text{ g } CO_2} \qquad \frac{1 \text{ mol C}}{12.01 \text{ g C}}$$

and then mol $CO_2 \rightarrow P_{CO_2}$ and mol C $\rightarrow$ mol CO $\rightarrow P_{CO}$

$$PV = nRT \qquad\qquad PV = nRT$$

Solution: $CO_2(g) + C(s) \rightleftharpoons 2\,CO(g)$

$$4.45 \text{ g } CO_2 \times \frac{1 \text{ mol}}{44.01 \text{ g } CO_2} = 0.1011 \text{ mol } CO_2 \qquad 0.50 \text{ g C} \times \frac{1 \text{ mol}}{12.01 \text{ g C}} = 0.0416 \text{ mol C}$$

Because the stoichiometry is 1:1, the C is the limiting reactant; therefore, the moles of CO formed will be determined from the reaction, not the equilibrium.

$$0.0416 \text{ mol C} \times \frac{2 \text{ mol CO}}{1 \text{ mol C}} = 0.0832 \text{ mol CO}$$

$$CO_2(g) + C(s) \rightleftharpoons 2\,CO(g)$$

	CO_2	C	CO
Initial	0.1011	0.0416	0.0
Change	-0.0416	-0.0416	$+2(0.0416)$
Equil	0.0595	0	0.0832

$$P_{CO_2} = \frac{(0.0595 \text{ mol})\left(\dfrac{0.08206 \text{ L} \cdot \text{atm}}{\text{mol} \cdot \text{K}}\right)(1200 \text{ K})}{10.0 \text{ L}} = 0.586 \text{ atm}$$

$$P_{CO} = \frac{(0.0832 \text{ mol})\left(\dfrac{0.08206 \text{ L} \cdot \text{atm}}{\text{mol} \cdot \text{K}}\right)(1200 \text{ K})}{10.0 \text{ L}} = 0.819 \text{ atm}$$

$$P_{total} = 0.586 + 0.819 = 1.405 = 1.41 \text{ atm}$$

Check: The pressure is less than the equilibrium pressure, which is reasonable because the C was the limiting reactant.

15.77 **Given:** $V = 10.0$ L, $T = 650$ K, 1.0 g MgO, $P_{CO_2} = 0.026$ atm, $K_p = 0.0260$
Find: mass $MgCO_3$ when volume is 0.100 L
Conceptual Plan: $P(10.0 \text{ L}) \rightarrow P(0.100 \text{ L})$. Prepare an ICE table, represent the change with x, sum the table, determine the equilibrium value, put the equilibrium values in the equilibrium expression, and solve for x. Then determine moles CO_2, the limiting reactant, and the mass of $MgCO_3$ formed.

$$P_1V_1 = P_2V_2 \quad PV = nRT$$

Solution:

$P_1V_1 = P_2V_2$ $(0.0260 \text{ atm})(10.0 \text{ L}) = (x)(0.100 \text{ L})$ $x = 2.60 \text{ atm}$

$$MgCO_3(s) \rightleftharpoons MgO_3(s) + CO_2(g)$$

		2.60
Initial		2.60
Change		$-x$
Equil		$2.60 - x$

$K_p = P_{CO_2} = 0.026 = 2.60 - x$ $x = 2.579 \text{ atm}$

$$n_{CO_2} = \frac{(2.579 \text{ atm})10.0 \text{ L}}{\left(\dfrac{0.08206 \text{ L} \cdot \text{atm}}{\text{mol} \cdot \text{K}}\right)(650 \text{ K})} = 0.4835 \text{ mol } CO_2 \quad 1.0 \text{ g MgO} \times \frac{1 \text{ mol}}{40.31 \text{ g MgO}} = 0.0248 \text{ mol MgO}$$

Therefore, MgO is the limiting reactant and produces 0.0248 mol $MgCO_3$.

$$0.0248 \text{ mol MgCO}_3 \times \frac{84.32 \text{ g } MgCO_3}{1 \text{ mol MgCO}_3} = 2.1 \text{ g } MgCO_3$$

Check: The units (g) are correct. The magnitude (2) is reasonable since the pressure and volume are small. Much less than 1 mole is anticipated.

15.79   **Given:** $C_2H_4(g) + Cl_2(g) \rightleftharpoons C_2H_4Cl_2(g)$ is exothermic   **Find:** Which of the following will maximize $C_2H_4Cl_2$?

     (a)   Increasing the reaction volume will not maximize $C_2H_4Cl_2$. The chemical equation has 1 mole of gas on the right and 2 moles of gas on the left. Increasing the volume of the reaction mixture decreases the pressure and causes the reaction to shift to the left (toward the side with more moles of gas particles).

     (b)   Removing $C_2H_4Cl_2$ as it forms will maximize $C_2H_4Cl_2$. Removing the $C_2H_4Cl_2$ will decrease the concentration of $C_2H_4Cl_2$ and will cause the reaction to shift to the right, producing more $C_2H_4Cl_2$.

     (c)   Lowering the reaction temperature will maximize $C_2H_4Cl_2$. The reaction is exothermic, so we can think of heat as a product. Lowering the temperature will cause the reaction to shift to the right, producing more $C_2H_4Cl_2$.

     (d)   Adding Cl_2 will maximize $C_2H_4Cl_2$. Adding Cl_2 increases the concentration of Cl_2, so the reaction shifts to the right, which will produce more $C_2H_4Cl_2$.

15.81   **Given:** reaction 1 at equilibrium: $P_{H_2} = 0.958 \text{ atm}$, $P_{I_2} = 0.877 \text{ atm}$, $P_{HI} = 0.020 \text{ atm}$; reaction 2: $P_{H_2} = P_{I_2} = 0.621 \text{ atm}$, $P_{HI} = 0.101 \text{ atm}$ **Find:** Is reaction 2 at equilibrium? If not, what is the P_{HI} at equilibrium? **Conceptual Plan: Use equilibrium partial pressures to determine K_p. Use K_p to determine whether reaction 2 is at equilibrium. Prepare an ICE table, calculate Q, compare Q and K_p, predict the direction of the reaction, represent the change with x, sum the table, determine the equilibrium values, put the equilibrium values in the equilibrium expression, and solve for x. Determine P_{HI}.**

     Solution: $H_2(g) + I_2(g) \rightleftharpoons 2 HI(g)$

Reaction 1:	P_{H_2}	P_{I_2}	P_{HI}
Equil	0.958	0.877	0.020

$$K_p = \frac{P_{HI}^2}{P_{H_2}P_{I_2}} = \frac{(0.020)^2}{(0.958)(0.877)} = 4.7610 \times 10^{-4}$$

$$Q = \frac{P_{HI}^2}{P_{H_2}P_{I_2}} = \frac{(0.101)^2}{(0.621)(0.621)} = 0.0264 \qquad Q > K, \text{ so the reaction shifts to the left.}$$

$$H_2(g) \; + \; I_2(g) \rightleftharpoons 2 HI(g)$$

Reaction 2:	P_{H_2}	P_{I_2}	P_{HI}
Initial	0.621	0.621	0.101
Change	$+x$	$+x$	$-2x$
Equil	$0.621 + x$	$0.621 + x$	$0.101 - 2x$

$$K_p = \frac{P_{HI}^2}{P_{H_2}P_{I_2}} = \frac{(0.101 - 2x)^2}{(0.621 + x)(0.621 + x)} = 4.7610 \times 10^{-4}$$

$$\sqrt{\frac{(0.101 - 2x)^2}{(0.621 + x)(0.621 + x)}} = \sqrt{4.7\underline{6}10 \times 10^{-4}}$$

$$\frac{(0.101 - 2x)}{(0.621 + x)} = 2.1\underline{8}2 \times 10^{-2}$$

$x = 0.04325 = 0.0433$

$P_{H_2} = P_{I_2} = 0.621 + x = 0.621 + 0.0433 = 0.664$ atm

$P_{HI} = 0.101 - 2x = 0.101 - 2(0.0433) = 0.0144$ atm

Check: Plug the values into the equilibrium expression:

$$K_p = \frac{(0.0144)^2}{(0.664)^2} = 4.703 \times 10^{-4} = 4.70 \times 10^{-4};$$ this value is close to the original equilibrium constant.

15.83   **Given:** 200.0 L container; 1.27 kg N_2; 0.310 kg H_2; 725 K; $K_p = 5.3 \times 10^{-5}$ **Find:** mass in g of NH_3 and % yield
Conceptual Plan: $K_p \rightarrow K_c$ and then kg $\rightarrow$ g $\rightarrow$ mol $\rightarrow$ M and then prepare an ICE table. Represent

$$K_p = K_c(RT)^{\Delta n} \qquad \frac{1000\text{ g}}{1\text{ kg}} \quad \frac{\text{g}}{\text{molar mass}} \quad \frac{\text{mol}}{\text{vol}}$$

the change with x, sum the table, determine the equilibrium values, put the equilibrium values in the equilibrium expression, and solve for x. Determine [NH_3]. Then M $\rightarrow$ mol $\rightarrow$ g and then determine

$$\text{M} \times \text{vol} \quad \text{mol} \times \text{molar mass}$$

theoretical yield $NH_3 \rightarrow$ % yield.

$$\text{determine limiting reactant } \% \text{ yield} = \frac{\text{actual yield}}{\text{theoretical yield}} \times 100\%$$

Solution: $K_p = K_c(RT)^{\Delta n}$ $K_c = \dfrac{K_p}{(RT)^{\Delta n}} = \dfrac{5.3 \times 10^{-5}}{\left(\left(0.08206\,\dfrac{\text{L} \cdot \text{atm}}{\text{mol} \cdot \text{K}}\right)(725\text{ K})\right)^{-2}} = 0.1\underline{8}76$

$n_{N_2} = 1.27\text{ kg }N_2 \times \dfrac{1000\text{ g}}{1\text{ kg}} \times \dfrac{1\text{ mol }N_2}{28.02\text{ g }N_2} = 45.\underline{3}25\text{ mol }N_2$ $[N_2] = \dfrac{45.325\text{ mol}}{200.0\text{ L}} = 0.22\underline{6}63\text{ M}$

$n_{H_2} = 0.310\text{ kg }H_2 \times \dfrac{1000\text{ g}}{1\text{ kg}} \times \dfrac{1\text{ mol }H_2}{2.016\text{ g }H_2} = 153.\underline{7}7\text{ mol }H_2$ $[H_2] = \dfrac{153.77\text{ mol}}{200.0\text{ L}} = 0.76\underline{8}85\text{ M}$

$$N_2(g) \quad + \quad 3\,H_2(g) \rightleftharpoons 2\,NH_3(g)$$

Reaction 1:	$[N_2]$	$[H_2]$	$[NH_3]$
Initial	0.2266	0.7689	0.0
Change	$-x$	$-3x$	$+2x$
Equil	$0.2266 - x$	$0.7689 - 3x$	$2x$

Reaction shifts to the right.

$$K_c = \frac{[NH_3]^2}{[N_2][H_2]^3} = \frac{(2x)^2}{(0.2266 - x)(0.7689 - 3x)^3} = 0.1876$$

Assume that x is small compared to 0.2268 and $3x$ is small compared to 0.7689.

$$\frac{(2x)^2}{(0.2266)(0.7689)^3} = 0.1877 \qquad x = 0.06951$$

Check assumptions: $\dfrac{0.06951}{0.2268} \times 100\% = 30.6$, which is not valid, and $\dfrac{3(0.06951)}{0.7689} \times 100\% = 27.1\%$.

Use method of successive substitution to solve for x. This yields $x = 0.0460$.
$[NH_3] = 2x = 2(0.0460) = 0.0920$ M

Check: Plug the values into the equilibrium expression:

$$K_c = \frac{(0.0920)^2}{(0.2266 - 0.0460)((0.7689 - 3(0.0460))^3} = 0.1866;$$

this value is close to the original equilibrium constant.

Determine grams NH_3: $\dfrac{0.0922 \text{ mol NH}_3}{L} \times 200.0 \ L \times \dfrac{17.03 \text{ g NH}_3}{\text{mol NH}_3} = 3\underline{1}3.8 \text{ g} = 3.1 \times 10^2 \text{ g}$

Determine the theoretical yield and the limiting reactant:

$1.27 \text{ kg N}_2 \times \dfrac{1000 \text{ g}}{1 \text{ kg}} \times \dfrac{1 \text{ mol N}_2}{28.02 \text{ g N}_2} \times \dfrac{2 \text{ mol NH}_3}{1 \text{ mol N}_2} \times \dfrac{17.03 \text{ g NH}_3}{\text{mol NH}_3} = 1544 \text{ g NH}_3$

$0.310 \text{ kg H}_2 \times \dfrac{1000 \text{ g}}{1 \text{ kg}} \times \dfrac{1 \text{ mol H}_2}{2.016 \text{ g H}_2} \times \dfrac{2 \text{ mol NH}_3}{3 \text{ mol H}_2} \times \dfrac{17.03 \text{ g NH}_3}{\text{mol NH}_3} = 1745 \text{ g NH}_3$

N_2 produces the least amount of NH_3; therefore, it is the limiting reactant, and the theoretical yield is $1.54 \times 10^3 \text{ g NH}_3$.

% yield $= \dfrac{3.1 \times 10^2 \text{ g}}{1.54 \times 10^3 \text{ g}} \times 100\% = 20.\%$

15.85   **Given:** at equilibrium: $P_{CO} = 0.30$ atm; $P_{Cl_2} = 0.10$ atm; $P_{COCl_2} = 0.60$ atm, add 0.40 atm Cl_2
              Find: P_{CO} when system returns to equilibrium
              Conceptual Plan: Use equilibrium partial pressures to determine K_p. For the new conditions, prepare an ICE table, represent the change with x, sum the table, determine the equilibrium values, put the equilibrium values in the equilibrium expression, and solve for x. Determine P_{CO}.
              Solution: $CO(g) + Cl_2(g) \rightleftharpoons COCl_2(g)$

Condition 1:	P_{CO}	P_{Cl_2}	P_{COCl_2}
	0.30	0.10	0.60

$K_p = \dfrac{P_{COCl_2}}{P_{CO}P_{Cl_2}} = \dfrac{(0.60)}{(0.30)(0.10)} = 20$

$$CO(g) \quad + \quad Cl_2(g) \rightleftharpoons COCl_2(g)$$

Condition 2:	P_{CO}	P_{Cl_2}	P_{COCl_2}
Initial	0.30	0.10 + 0.40	0.60
Change	$-x$	$-x$	$+x$
Equil	$0.30 - x$	$0.50 - x$	$0.60 + x$

Reaction shifts to the right because the concentration of Cl_2 was increased.

$K_p = \dfrac{P_{COCl_2}}{P_{CO}P_{Cl_2}} = \dfrac{(0.60 + x)}{(0.30 - x)(0.50 - x)} = 20$

$20x^2 - 17x + 2.4 = 0$

$\dfrac{-b \pm \sqrt{b^2 - 4ac}}{2a} = \dfrac{-(-17) \pm \sqrt{(-17)^2 - 4(20)(2.4)}}{2(20)}$

$x = 0.67$ or 0.18 so $x = 0.18$
$P_{CO} = 0.30 - 0.18 = 0.12$ atm; $P_{Cl_2} = 0.50 - 0.18 = 0.32$; $P_{COCl_2} = 0.60 + 0.18 = 0.78$ atm

              Check: Plug the values into the equilibrium expression:

$K_p = \dfrac{(0.78)}{(0.12)(0.32)} = 20.3 = 20.$; this is the same as the original equilibrium constant.

15.87   **Given:** $K_p = 0.76$; P_{total} at equilibrium $= 1.0$ atm   **Find:** $P_{initial}$ CCl_4
              Conceptual Plan: Prepare an ICE table, represent the P_{CCl_4} with A and the change with x, sum the table, determine the equilibrium values, use the total pressure, and solve for A in terms of x. Determine partial pressure of each at equilibrium, use the equilibrium expression to determine x, and determine A.

Solution: $CCl_4(g) \rightleftharpoons C(s) + 2Cl_2(g)$

	P_{CCl_4}	P_C	P_{Cl_2}
Initial	A	constant	0.00
Change	$-x$		$+2x$
Equil	$A - x$		$2x$

$P_{total} = P_{CCl_4} + P_{Cl_2}$ $1.0 = A - x + 2x$ $A = 1.0 - x$

$P_{CCl_4} = (A - x) = (1.0 - x) - x = 1.0 - 2x;$ $P_{Cl_2} = 2x$

$$K_p = \frac{P_{Cl_2}^2}{P_{CCl_4}} = \frac{(2x)^2}{(1.0 - 2x)} = 0.76$$

$4x^2 + 1.52x - 0.76 = 0$ Solve the quadratic equation. $x = 0.285$ or -0.665 so $x = 0.2\underline{8}5$

$A = 1.0 - x = 1.0 - 0.285 = 0.715 = 0.72$ atm

Check: Plug the values into the equilibrium expression:

$$K_p = \frac{P_{Cl_2}^2}{P_{CCl_4}} = \frac{(2x)^2}{(A - x)} = \frac{(2(0.285))^2}{(0.715 - 0.285)} = 0.756 = 0.76;$$ the original equilibrium is constant.

15.89   **Given:** $V = 0.654$ L, $T = 1000$ K, $K_p = 3.9 \times 10^{-2}$ **Find:** mass CaO at equilibrium

Conceptual Plan: $K_p \rightarrow P_{CO_2} \rightarrow n_{(CO_2)} \rightarrow n_{(CaO)} \rightarrow g$

$\qquad\qquad\qquad\qquad PV = nRT \quad$ stoichiometry $g = n(\text{molar mass})$

Solution: Because $CaCO_3$ and CaO are solids, they are not included in the equilibrium expression.

$$K_p = P_{CO_2} = 3.9 \times 10^{-2}; \quad n = \frac{PV}{RT} = \frac{(3.9 \times 10^{-2} \text{ atm})(0.654 \text{ L})}{\left(0.08206 \dfrac{\text{L} \cdot \text{atm}}{\text{mol} \cdot \text{K}}\right)(1000 \text{ K})} = 3.\underline{1}08 \times 10^{-4} \text{ mol } CO_2$$

$$3.\underline{1}08 \times 10^{-4} \text{ mol } CO_2 \times \frac{1 \text{ mol CaO}}{1 \text{ mol } CO_2} \times \frac{56.08 \text{ g CaO}}{1 \text{ mol CaO}} = 0.0174 \text{ g} = 0.0174 \text{ g CaO}$$

Check: The small value of K would give a small amount of products, so we would not expect a large mass of CaO to form.

15.91   **Given:** $K_p = 3.10$, initial $P_{CO} = 215$ torr, $P_{Cl_2} = 245$ torr   **Find:** mole fraction $COCl_2$

Conceptual Plan: P in torr $\rightarrow P$ in atm. **Prepare an ICE table, represent the change with x, sum the table,**

$\qquad\qquad\qquad\qquad \dfrac{1 \text{ atm}}{760 \text{ torr}}$

determine the equilibrium values, use the total pressure, and solve for mole fraction.

$\qquad\qquad\qquad\qquad\qquad\qquad\qquad\qquad\qquad\qquad\qquad\qquad \dfrac{P_{COCl_2}}{P_{total}}$

Solution: $P_{CO} = (215 \text{ torr})\left(\dfrac{1 \text{ atm}}{760 \text{ torr}}\right) = 0.28\underline{2}9$ atm $P_{Cl_2} = (245 \text{ torr})\left(\dfrac{1 \text{ atm}}{760 \text{ torr}}\right) = 0.32\underline{2}4$ atm

$$CO(g) + Cl_2(g) \rightleftharpoons COCl_2(g)$$

Initial	0.28\underline{2}9	0.32\underline{2}4	0
Change	$-x$	$-x$	$+x$
Equil	$0.28\underline{2}9 - x$	$0.32\underline{2}4 - x$	x

$Q < K$, so the reaction shifts to the right.

$$K_p = \frac{P_{COCl_2}}{P_{CO} P_{Cl_2}} = \frac{x}{(0.28\underline{2}9 - x)(0.32\underline{2}4 - x)} = 3.10 \text{ so } 3.10x^2 - 2.87634x + 0.2827 = 0 \text{ Solve using the quadratic}$$

equation, found in Appendix I. $x = 0.81\underline{6}1$ or $0.11\underline{1}7$ so $x = 0.11\underline{1}7$ P is proportional to n under like conditions so the mole fraction can be determined using the pressure values.

$P_{CO} = 0.28\underline{2}9 - 0.11\underline{1}7 = 0.17\underline{1}2$ $P_{Cl_2} = 0.32\underline{2}4 - 0.11\underline{1}7 = 0.21\underline{0}7$ $P_{COCl_2} = 0.11\underline{1}7$

$$\text{mole fraction } COCl_2 = \frac{P_{COCl_2}}{P_{CO} + P_{Cl_2} + P_{COCl_2}} = \frac{0.11\underline{1}7}{0.17\underline{1}2 + 0.21\underline{0}7 + 0.11\underline{1}7} = 0.22\underline{6}3 = 0.226$$

Check: Plug the equilibrium pressures into the equilibrium expression:

$$K_p = \frac{0.11\underline{17}}{(0.17\underline{12})(0.21\underline{07})} = 3.0966 = 3.10, \text{ which is the equilibrium constant; so the answer is reasonable.}$$

Challenge Problems

15.93 (a) **Given:** $P_{NO} = 522$ torr, $P_{O_2} = 421$ torr; at equilibrium, $P_{total} = 748$ torr   **Find:** K_p
Conceptual Plan: Prepare an ICE table, represent the change with x, sum the table, determine the equilibrium values, use the total pressure, and solve for x. torr → atm → K_p
Solution: $2\,NO(g) + O_2(g) \rightleftharpoons 2\,NO_2(g)$

	P_{NO}	P_{O_2}	P_{NO_2}
Initial	522 torr	421 torr	0.00
Change	$-2x$	$-x$	$+2x$
Equil	$522 - 2x$	$421 - x$	$2x$

$P_{total} = P_{NO} + P_{O_2} + P_{NO_2}$ $748 = 522 - 2x + (421 - x) + 2x$
$x = 195$ torr $P_{NO} = (522 - 2(195)) = 132$ torr
$P_{O_2} = (421 - 195) = 226$ torr $P_{NO_2} = 2(195) = 390$ torr

$$P_{NO} = 132 \text{ torr} \times \frac{1 \text{ atm}}{760 \text{ torr}} = 0.17\underline{37} \text{ atm;} \qquad P_{O_2} = 226 \text{ torr} \times \frac{1 \text{ atm}}{760 \text{ torr}} = 0.29\underline{74} \text{ atm}$$

$$P_{NO_2} = 390 \text{ torr} \times \frac{1 \text{ atm}}{760 \text{ torr}} = 0.51\underline{32} \text{ atm}$$

$$K_p = \frac{P_{NO_2}^2}{P_{NO}^2 P_{O_2}} = \frac{(0.51\underline{32})^2}{(0.17\underline{37})^2(0.29\underline{74})} = 29.35 = 29.3$$

(b) **Given:** $P_{NO} = 255$ torr, $P_{O_2} = 185$ torr, $K_p = 29.3$ **Find:** equilibrium P_{NO_2}
Conceptual Plan:
torr → atm and then prepare an ICE table. Represent the change with x, sum the table,
$$\frac{1 \text{ atm}}{760 \text{ torr}}$$
determine the equilibrium values, put the equilibrium values in the equilibrium expression, and solve for x. Determine P_{NO_2}.

Solution: $P_{NO} = 255 \text{ torr} \times \dfrac{1 \text{ atm}}{760 \text{ torr}} = 0.33\underline{55} \text{ atm}$ $\qquad P_{O_2} = 185 \text{ torr} \times \dfrac{1 \text{ atm}}{760 \text{ torr}} = 0.24\underline{34} \text{ atm}$

$2\,NO(g) + O_2(g) \rightleftharpoons 2\,NO_2(g)$

	P_{NO}	P_{O_2}	P_{NO_2}
Initial	0.3355	0.2434	0.00
Change	$-2x$	$-x$	$+2x$
Equil	$0.3355 - 2x$	$0.2434 - x$	$2x$

$$K_p = \frac{P_{NO_2}^2}{P_{NO}^2 P_{O_2}} = \frac{(2x)^2}{(0.3355 - 2x)^2(0.2434 - x)} = 29.3$$

$-117.2x^3 + 63.847x^2 - 12.867x + 0.80282 = 0$. Solve using successive approximations or a cubic equation calculator found on the Internet.
$x = 0.11\underline{13}$ $P_{NO_2} = 2x = 2(0.11\underline{13}) = 0.22\underline{26}$ atm
$P_{NO} = (0.3355 - 2(0.1113)) = 0.1129$ $P_{O_2} = (0.2434 - 0.1113) = 0.1321$

$$0.22\underline{26} \text{ atm} \times \frac{760 \text{ torr}}{1 \text{ atm}} = 169.2 \text{ torr} = 169 \text{ torr}$$

Check: Plug the values into the equilibrium expression:

$$K_p = \frac{(0.22\underline{26})^2}{(0.1129)^2(0.1321)} = 29.428 = 29.4; \text{ this is within 0.1 of the original equilibrium constant.}$$

15.95   **Given:** P_{NOCl} at equilibrium $= 115$ torr; $K_p = 0.27$, $T = 700$ K   **Find:** initial pressure NO, Cl_2
          Conceptual Plan: torr → atm and then prepare an ICE table. Represent the change with x, sum the table,

$$\frac{1 \text{ atm}}{760 \text{ torr}}$$

determine the equilibrium values, put the equilibrium values in the equilibrium expression, and determine initial pressure.

Solution: $115 \text{ torr} \times \dfrac{1 \text{ atm}}{760 \text{ torr}} = 0.151\underline{3} \text{ atm}$

$$2 \, NO(g) \; + \; Cl_2(g) \rightleftharpoons 2 \, NOCl(g)$$

	P_{NO}	P_{Cl_2}	P_{NOCl}
Initial	A	A	0.00
Change	$-2x$	$-x$	$+2x$
Equil	$A - 2x$	$A - x$	0.151
	$A - 0.151$	$A - 0.0756$	

Let $A =$ initial pressure of NO and Cl_2.

$2x = 0.151$; $x = 0.0756$

$$K_p = \frac{P^2_{NOCl}}{P^2_{NO} \, P_{Cl_2}} = \frac{(0.151)^2}{(A - 0.151)^2(A - 0.0756)} = 0.27$$

$0.27A^3 - 0.1019A^2 + 0.01231A - 0.023266 = 0$. Solve using successive approximations or a cubic equation calculator found on the Internet.
$A = 0.566$
$P_{NO} = P_{Cl_2} = A = 0.566 \text{ atm} = 430 \text{ torr}$

Check: Plug the values into the equilibrium expression:

$$K_p = \frac{P^2_{NOCl}}{P^2_{NO} \, P_{Cl_2}} = \frac{(0.151)^2}{(0.566 - 0.151)^2(0.566 - 0.0756)} = 0.2699 = 0.27;$$

this is the same as the original equilibrium constant.

15.97   **Given:** $P = 0.750$ atm, density $= 0.520$ g/L, $T = 337\,°C$ **Find:** K_c
          Conceptual Plan: Prepare an ICE table, represent the P_{CCl_4} with A and the change with x, sum the table, determine the equilibrium values, use the total pressure, and solve for A in terms of x. Determine partial pressure of each at equilibrium in terms of x, use the density to determine the apparent molar mass, and use

$$d = \frac{P\mathcal{M}}{RT}$$

the mole fraction (in terms of P) and the molar mass of each gas to determine x.

$$\chi_A = \frac{P_A}{P_{total}}$$

Solution: $2 \, NO_2(g) \rightleftharpoons 2 \, NO(g) + O_2(g)$

	P_{NO_2}	P_{NO}	P_{O_2}
Initial	A	0.00	0.00
Change	$-2x$	$+2x$	$+x$
Equil	$A - 2x$	$2x$	x

$P_{total} = P_{NO_2} + P_{NO} + P_{O_2}$ $0.750 = A - 2x + 2x + x$ $A = 0.750 - x$
$P_{NO_2} = (A - 2x) = (0.750 - x) - 2x = (0.750 - 3x);$ $P_{NO} = (2x);$ $P_{O_2} = x$

$$d = \frac{P\mathcal{M}}{RT} \qquad \mathcal{M} = \frac{dRT}{P} = \frac{\left(0.520 \, \dfrac{g}{L}\right)\left(0.08206 \, \dfrac{L \cdot atm}{mol \cdot K}\right)(610 \text{ K})}{0.750 \text{ atm}} = 34.71 \text{ g/mol}$$

$$\mathcal{M} = \chi_{NO_2}\mathcal{M}_{NO_2} + \chi_{NO}\mathcal{M}_{NO} + \chi_{O_2}\mathcal{M}_{O_2} = \frac{P_{NO_2}}{P_{total}}\mathcal{M}_{NO_2} + \frac{P_{NO}}{P_{total}}\mathcal{M}_{NO} + \frac{P_{O_2}}{P_{total}}\mathcal{M}_{O_2}$$

$$P_{total}\mathcal{M} = P_{NO_2}\mathcal{M}_{NO_2} + P_{NO}\mathcal{M}_{NO} + P_{O_2}\mathcal{M}_{O_2}$$

$(0.750)(34.7) = (0.750 - 3x)(46.0) + 2x(30.0) + x(32.0)$
$x = 0.184$

$$K_p = \frac{P_{NO}^2 P_{O_2}}{P_{NO_2}^2} = \frac{(2x)^2(x)}{(0.750 - 3x)} = \frac{(2(0.184))^2(0.184)}{(0.750 - 3(0.184))^2} = 0.635\underline{6}$$

$$K_p = K_c(RT)^{\Delta n} \quad 0.635\underline{6} = K_c\left(\left(0.08206 \frac{\text{L} \cdot \text{atm}}{\text{mol} \cdot \text{K}}\right)(610\,\cancel{K})\right)^1$$

$$K_c = 1.27 \times 10^{-2}$$

Check: The units (none) are correct. The magnitude (0.01) is reasonable since there are higher concentrations of the products than the reactants.

15.99 **Given:** $P_{total} = 3.0$ atm, mole fraction $O_2 = 0.12$, $T = 600$ K **Find:** K_p
Conceptual Plan: mole fraction $\rightarrow P_{O_2} \rightarrow P_{SO_2} \rightarrow P_{SO_3} \rightarrow K_p$

$$\text{mole fraction} = \frac{P_{O_2}}{P_{total}} \qquad \frac{2 P_{SO_2}}{P_{O_2}} \qquad P_{SO_3} = P_{total} - P_{SO_2} - P_{O_2} \qquad K_p = \frac{P_{SO_2}^2 P_{O_2}}{P_{SO_3}^2}$$

Solution: $P_{O_2} = (0.12)(3.0) = 0.36$ atm $\quad P_{SO_2} = 2P_{O_2} = 2(0.36 \text{ atm}) = 0.72$ atm

$$P_{SO_3} = 3.0 - 0.36 - 0.72 = 1.\underline{9}2 \quad K_p = \frac{P_{SO_2}^2 P_{O_2}}{P_{SO_3}^2} = \frac{(0.72)^2(0.36)}{(1.\underline{9}2)^2} = 0.050\underline{6} = 5.1 \times 10^{-2}$$

Check: The units (none) are correct. The magnitude (0.05) is reasonable since there are higher concentrations of the products than the reactants.

Conceptual Problems

15.101 Yes, the direction will depend on the volume. If the initial moles of A and B are equal, the initial concentrations of A and B are equal regardless of the volume. Because $K_c = \dfrac{[B]^2}{[A]} = 1$, if the volume is such that the $[A] = [B] < 1.0$, then $Q < K$ and the reaction goes to the right to reach equilibrium. However, if the volume is such that the $[A] = [B] > 1.0$, then $Q > K$ and the reaction goes to the left to reach equilibrium.

15.103 An examination of the data shows that when $P_A = 1.0$, $P_B = 1.0$; therefore, $K_p = \dfrac{P_B^b}{P_A^a} = \dfrac{(1.0)^b}{(1.0)^a} = 1.0$. Therefore, the value of the numerator and denominator must be equal. We see from the data that $P_B = \sqrt{P_A}$, so $P_B^2 = P_A$. Because the stoichiometric coefficients become exponents in the equilibrium expression, $a = 1$ and $b = 2$.

Questions for Group Work

15.105 To find the value of K for the new equation, combine the three given equations to yield the new equation. Use equation a as is, reverse and halve equation b, and reverse and halve equation c. To find K for new reaction, use $K_a/(K_b K_c)^{1/2}$.

$$C(s) + \cancel{O_2(g)} \rightleftharpoons \cancel{CO_2(g)} \qquad\qquad K = 1.363 \times 10^{69}$$

$$H_2O(g) \rightleftharpoons H_2(g) + \tfrac{1}{2}\cancel{O_2(g)} \qquad K = \sqrt{\frac{1}{K_b}} = \sqrt{\frac{1}{1.389 \times 10^{80}}} = 8.48\underline{4}942 \times 10^{-41}$$

$$\cancel{CO_2(g)} \rightleftharpoons CO(g) + \tfrac{1}{2}\cancel{O_2(g)} \qquad K = \sqrt{\frac{1}{K_c}} = \sqrt{\frac{1}{1.477 \times 10^{90}}} = 8.22\underline{8}293 \times 10^{-46}$$

$$\overline{C(s) + H_2O(g) \rightleftharpoons CO(g) + H_2(g) \qquad K = (1.363 \times 10^{69}) \times (8.48\underline{4}942 \times 10^{-41}) \times (8.22\underline{8}293 \times 10^{-46})}$$

$$= 9.516 \times 10^{-17}$$

15.107 For the first reaction, A $\rightleftharpoons$ B, $K = \dfrac{[B]}{[A]}$. For the second reaction, C $\rightleftharpoons$ 2 D, $K = \dfrac{[D]^2}{[C]} \neq \dfrac{[D]}{[C]}$. Since K is the same as the ratio of products to reactants for the first reaction, it is independent of the starting concentrations. Since K is not the same as the ratio of products to reactants for the second reaction, the ratio of products to reactants is dependent on the starting concentrations. The equilibrium constant value is independent of the starting concentrations.

15.109   (a)   When a concentration of a reactant is increased, the denominator of Q increases and the reaction must shift to the right.

       (b)   When a solid product is added, the value of Q will not change (as long as the volume is not changed) because solids do not appear in the equilibrium expression.

       (c)   When the volume is decreased, the reaction will shift to the side of the reaction that has the smaller number of moles of gas. If the number of moles of gas is the same on both sides of the reaction, the reaction will not shift as the volume is changed.

       (d)   Heat can be considered a reactant in an endothermic reaction and a product in an exothermic reaction. As the temperature is increased, an endothermic reaction will shift to the right. As the temperature is increased, an exothermic reaction will shift to the left.

Data Interpretation and Analysis

15.110   (a) $CaSiO_3(s) + CO_2(g) \rightleftharpoons CaCO_3(s) + SiO_2(s)$ (System 1) $K_P = \dfrac{1}{P_{CO_2}} = \dfrac{1}{91 \text{ atm}} = 0.011$

              $MgSiO_3(s) + CaCO_3(s) + SiO_2(s) \rightleftharpoons CaMgSi_2O_6(s) + CO_2(g)$ (System 2)

              $K_P = P_{CO_2} = 91 \text{ atm} = 91$

       (c)   The partial pressure of carbon dioxide in part a (91 atm) is a little lower than the midpoint partial pressure of carbon dioxide of System 1 in part b, so System 1 is most likely responsible for the carbon dioxide on the surface of Venus.

16 Acids and Bases

Review Questions

16.1 There are two problems with this treatment. First, the neutralization reaction between a strong acid and a strong base is highly exothermic, so Batman's treatment is likely to have produced a significant amount of heat, leading to potential burns for Black Mask. Second, strong bases are themselves caustic to the skin, so Batman's base has the potential to cause additional burns as well. The standard treatment for spilling concentrated acid on the skin is to rinse with large amounts of water for an extended period of time. The water dilutes the acid and washes it away.

16.3 A carboxylic acid is an organic acid that contains the following group of atoms: $H{-}O{-}\overset{\overset{\displaystyle O}{\|}}{C}{-}$. Carboxylic acids are often found in substances derived from living organisms. Examples include citric acid, malic acid, and acetic acid.

16.5 The hydronium ion is H_3O^+. In water, H^+ ions always associate with H_2O molecules to form hydronium ions and other associated species with the general formula $H(H_2O)_n{}^+$.

16.7 According to Huheey, "The differences between the various acid–base concepts are not concerned with which is right, but which is most convenient to use in a particular situation." There is no single correct definition; we use the definition that is best for a particular situation.

16.9 A conjugate acid–base pair consists of two substances related to each other by the transfer of a proton. In the reaction:

$NH_3(aq) + H_2O(l) \rightleftharpoons NH_4{}^+(aq) + OH^-(aq)$, the conjugate pairs are $NH_3/NH_4{}^+$ and H_2O/OH^-.

16.11 For an $H{-}Y$ binary acid, the factors affecting the ease with which this hydrogen will be donated are the polarity of the bond and the strength of the bond.

16.13 A diprotic acid contains two ionizable protons. An example is H_2SO_4.

A triprotic acid contains three ionizable protons. An example is H_3PO_4.

16.15 The autoionization of water is $H_2O(l) + H_2O(l) \rightleftharpoons H_3O^+(aq) + OH^-(aq)$ and has the ion product of water $K_w = [H_3O^+][OH^-]$.

It can also be written $H_2O(l) \rightleftharpoons H^+(aq) + OH^-(aq)$, and the ion product of water is $K_w = [H^+][OH^-]$. At 25 °C, $K_w = 1.0 \times 10^{-14}$.

16.17 We define pH as $pH = -\log[H_3O^+]$. An acidic solution has a pH < 7, a basic solution has a pH > 7, and a neutral solution has a pH $= 7$.

16.19 We can neglect the contribution of the autoionization of water to the H_3O^+ concentration in a solution of a strong or weak acid because it is negligible compared to the concentration of H_3O^+ ions generated from the dissociation of the acid. Even the weakest of the weak acids has an ionization constant that is four orders of magnitude larger than that of water.

16.21   The percent ionization is defined as percent ionization $= \dfrac{\text{concentration of ionized acid}}{\text{initial concentration of acid}} \times 100$. The percent ionization of a weak acid decreases with the increasing concentration of the acid.

16.23 $B(aq) + H_2O(l) \rightleftharpoons BH^+(aq) + OH^-(aq)$

16.25   At 25 °C, the product of K_a for an acid and K_b for its conjugate base is $K_a \times K_b = K_w = 1.0 \times 10^{-14}$.

16.27   For most polyprotic acids, K_{a_1} is much larger than K_{a_2}. Therefore, the amount of H_3O^+ contributed by the first ionization step is much larger than that contributed by the second or third ionization step. In addition, the production of H_3O^+ by the first step inhibits additional production of H_3O^+ by the second step because of Le Châtelier's principle.

16.29   A Lewis acid is an electron pair acceptor. A Lewis base is an electron pair donor.

Problems by Topic

The Nature and Definitions of Acids and Bases

16.31   (a)   acid $HNO_3(aq) \rightarrow H^+(aq) + NO_3^-(aq)$
        (b)   acid $NH_4^+(aq) \rightarrow H^+(aq) + NH_3(aq)$
        (c)   base $KOH(aq) \rightarrow K^+(aq) + OH^-(aq)$
        (d)   acid $HC_2H_3O_2(aq) \rightarrow H^+(aq) + C_2H_3O_2^-(aq)$

16.33   (a)   Because H_2CO_3 donates a proton to H_2O, it is the acid. After H_2CO_3 donates the proton, it becomes HCO_3^-, the conjugate base. Because H_2O accepts a proton, it is the base. After H_2O accepts the proton, it becomes H_3O^+, the conjugate acid.
        (b)   Because H_2O donates a proton to NH_3, it is the acid. After H_2O donates the proton, it becomes OH^-, the conjugate base. Because NH_3 accepts a proton, it is the base. After NH_3 accepts the proton, it becomes NH_4^+, the conjugate acid.
        (c)   Because HNO_3 donates a proton to H_2O, it is the acid. After HNO_3 donates the proton, it becomes NO_3^-, the conjugate base. Because H_2O accepts a proton, it is the base. After H_2O accepts the proton, it becomes H_3O^+, the conjugate acid.
        (d)   Because H_2O donates a proton to C_5H_5N, it is the acid. After H_2O donates the proton, it becomes OH^-, the conjugate base. Because C_5H_5N accepts a proton, it is the base. After C_5H_5N accepts the proton, it becomes $C_5H_5NH^+$, the conjugate acid.

16.35   (a) Cl^- $HCl(aq) + H_2O(l) \rightarrow H_3O^+(aq) + Cl^-(aq)$
        (b) HSO_3^- $H_2SO_3(aq) + H_2O(l) \rightleftharpoons H_3O^+(aq) + HSO_3^-(aq)$
        (c) CHO_2^- $HCHO_2(aq) + H_2O(l) \rightleftharpoons H_3O^+(aq) + CHO_2^-(aq)$
        (d) F^- $HF(aq) + H_2O(l) \rightleftharpoons H_3O^+(aq) + F^-(aq)$

16.37 $H_2PO_4^-(aq) + H_2O(l) \rightleftharpoons H_3O^+(aq) + HPO_4^{2-}(aq)$
        $H_2PO_4^-(aq) + H_2O(l) \rightleftharpoons H_3PO_4(aq) + OH^-(aq)$

Molecular Structure and Acid Strength

16.39   (a)   HCl is the stronger acid. HCl is the weaker bond; therefore, it is more acidic.
        (b)   HF is the stronger acid. F is more electronegative than is O, so the bond is more polar and more acidic.
        (c) H_2Se is the stronger acid. The H—Se bond is weaker; therefore, it is more acidic.

16.41   (a) H_2SO_4 is the stronger acid because it has more oxygen atoms.
        (b) $HClO_2$ is the stronger acid because it has more oxygen atoms.
        (c)   HClO is the stronger acid because Cl is more electronegative than is Br.
        (d) CCl_3COOH is the stronger acid because Cl is more electronegative than is H.

16.43 S^{2-} is the stronger base. Base strength is determined from the corresponding acid. The weaker the acid, the stronger the base. H_2S is the weaker acid because it has a stronger bond.

Acid Strength and K_a

16.45   (a) HNO_3 is a strong acid.
         (b) HCl is a strong acid.
         (c) HBr is a strong acid.
         (d) H_2SO_3 is a weak acid. $H_2SO_3(aq) + H_2O(l) \rightleftharpoons H_3O^+(aq) + HSO_3^-(aq)$

$$K_a = \frac{[H_3O^+][HSO_3^-]}{[H_2SO_3]}$$

16.47   The strength of an acid is determined by how much it ionizes and how much undissociated acid is left over. The larger the fraction of the acid that ionized, the stronger the acid.
         (a)   contains no HA, 10 H^+, and 10 A^-
         (b)   contains 3 HA, 3 H^+, and 7 A^-
         (c)   contains 9 HA, 1 H^+, and 1 A^-
         So solution a $>$ solution b $>$ solution c.

Autoionization of Water and pH

16.49   (a)   **Given:** $K_w = 1.0 \times 10^{-14}$, $[H_3O^+] = 1.2 \times 10^{-8}$ M   **Find:** $[OH^-]$
                Conceptual Plan: $[H_3O^+] \rightarrow [OH^-]$

                                $K_w = 1.0 \times 10^{-14} = [H_3O^+][OH^-]$

                Solution:
                $K_w = 1.0 \times 10^{-14} = (1.2 \times 10^{-8})[OH^-]$
                $[OH^-] = 8.3 \times 10^{-7}$ M
                $[OH^-] > [H_3O^+]$, so the solution is basic.
                Check: The units (M) are correct. The magnitude (8×10^{-7} M) makes sense because $[H^+] < 10^{-7}$ M.

         (b)   **Given:** $K_w = 1.0 \times 10^{-14}$, $[H_3O^+] = 8.5 \times 10^{-5}$ M   **Find:** $[OH^-]$
                Conceptual Plan: $[H_3O^+] \rightarrow [OH^-]$

                                $K_w = 1.0 \times 10^{-14} = [H_3O^+][OH^-]$

                Solution:
                $K_w = 1.0 \times 10^{-14} = (8.5 \times 10^{-5})[OH^-]$
                $[OH^-] = 1.2 \times 10^{-10}$ M
                $[H_3O^+] > [OH^-]$, so the solution is acidic.
                Check: The units (M) are correct. The magnitude (1×10^{-10} M) makes sense because $[H^+] > 10^{-7}$ M.

         (c)   **Given:** $K_w = 1.0 \times 10^{-14}$, $[H_3O^+] = 3.5 \times 10^{-2}$ M   **Find:** $[OH^-]$
                Conceptual Plan: $[H_3O^+] \rightarrow [OH^-]$

                                $K_w = 1.0 \times 10^{-14} = [H_3O^+][OH^-]$

                Solution:
                $K_w = 1.0 \times 10^{-14} = (3.5 \times 10^{-2})[OH^-]$
                $[OH^-] = 2.9 \times 10^{-13}$ M
                $[H_3O^+] > [OH^-]$, so the solution is acidic.
                Check: The units (M) are correct. The magnitude (3×10^{-13} M) makes sense because $[H^+] > 10^{-7}$ M.

16.51   (a)   **Given:** $[H_3O^+] = 1.7 \times 10^{-8}$ M   **Find:** pH and pOH
                Conceptual Plan: $[H_3O^+] \rightarrow pH \rightarrow pOH$

                                $pH = -\log[H_3O^+]$ pH + pOH = 14

                Solution: $pH = -\log(1.7 \times 10^{-8}) = 7.77$ $pOH = 14.00 - 7.77 = 6.23$
                $pH > 7$, so the solution is basic.
                Check: The units (none and none) are correct. When $[H^+] < 10^{-7}$ M the pH > 7, pOH < 7.

(b)   **Given:** $[H_3O^+] = 1. \times 10^{-7}$ M   **Find:** pH and pOH
Conceptual Plan: $[H_3O^+] \rightarrow pH \rightarrow pOH$

$$pH = -\log[H_3O^+] \quad pH + pOH = 14$$

Solution: $pH = -\log(1.0 \times 10^{-7}) = 7.00 \qquad pOH = 14.00 - 7.00 = 7.00$
pH = 7, so the solution is neutral.

Check: The units (none and none) are correct. When $[H^+] = 1 \times 10^{-7}$ M the pH = 7 = pOH.

(c)   **Given:** $[H_3O^+] = 2.2 \times 10^{-6}$ M   **Find:** pH and pOH
Conceptual Plan: $[H_3O^+] \rightarrow pH \rightarrow pOH$

$$pH = -\log[H_3O^+] \quad pH + pOH = 14$$

Solution: $pH = -\log(2.2 \times 10^{-6}) = 5.66 \qquad pOH = 14.00 - 5.66 = 8.34$
pH < 7, so the solution is acidic.

Check: The units (none and none) are correct. When $[H^+] > 10^{-7}$ M the pH < 7, pOH > 7.

16.53 $pH = -\log[H_3O^+] \quad K_w = 1.0 \times 10^{-14} = [H_3O^+][OH^-]$

$[H_3O^+]$	$[OH^-]$	pH	Acidic or basic
7.1×10^{-4} M	1.4×10^{-11} M	**3.15**	acidic
$\mathbf{3.7 \times 10^{-9}}$ **M**	2.7×10^{-6} M	8.43	basic
8×10^{-12} M	1×10^{-3} M	**11.1**	basic
6.3×10^{-4} M	$\mathbf{1.6 \times 10^{-11}}$ **M**	3.21	acidic

$$[H_3O^+] = 10^{-3.15} = 7.1 \times 10^{-4} \text{ M} \qquad [OH^-] = \frac{1.0 \times 10^{-14}}{7.1 \times 10^{-4}} = 1.4 \times 10^{-11} \text{ M}$$

$$[OH^-] = \frac{1.0 \times 10^{-14}}{3.7 \times 10^{-9}} = 2.7 \times 10^{-6} \text{ M} \qquad pH = -\log(3.7 \times 10^{-9}) = 8.43$$

$$[H_3O^+] = 10^{-11.1} = 8 \times 10^{-12} \text{ M} \qquad [OH^-] = \frac{1.0 \times 10^{-14}}{8 \times 10^{-12}} = 1 \times 10^{-3} \text{ M}$$

$$[H_3O^+] = \frac{1.0 \times 10^{-14}}{1.6 \times 10^{-11}} = 6.3 \times 10^{-4} \text{ M} \qquad pH = -\log(6.2 \times 10^{-4}) = 3.21$$

Check: When pH < 7, $[H^+] > 10^{-7}$ M and $[OH^-] < 10^{-7}$ M. When pH > 7, $[H^+] < 10^{-7}$ M and $[OH^-] > 10^{-7}$ M.

16.55   **Given:** $K_w = 2.4 \times 10^{-14}$ at 37 °C   **Find:** $[H_3O^+]$, pH
Conceptual Plan: $K_w \rightarrow [H_3O^+] \rightarrow pH$

$$K_w = [H_3O^+][OH^-] \quad pH = -\log[H_3O^+]$$

Solution: $H_2O(l) + H_2O(l) \rightleftharpoons H_3O^+(aq) + OH^-(aq)$

$$K_w = [H_3O^+][OH^-]$$
$$[H_3O^+] = [OH^-] = \sqrt{K_w} = \sqrt{2.4 \times 10^{-14}} = 1.5 \times 10^{-7} \text{ M}$$
$$pH = -\log[H_3O^+] = -\log(1.5 \times 10^{-7}) = 6.82$$

Check: The value of K_w increased, indicating more products formed; so the H_3O^+ increases and the pH decreases from the values at 25 °C.

16.57   (a)   **Given:** $[H_3O^+] = 0.044$ M   **Find:** pH
Conceptual Plan: $[H_3O^+] \rightarrow pH$

$$pH = -\log[H_3O^+]$$

Solution: $pH = -\log(0.044) = 1.3\underline{5}7 = 1.36$

(b)   **Given:** $[H_3O^+] = 0.045$ M   **Find:** pH
Conceptual Plan: $[H_3O^+] \rightarrow pH$

$$pH = -\log[H_3O^+]$$

Solution: $pH = -\log(0.045) = 1.3\underline{4}7 = 1.35$

(c)   **Given:** $[H_3O^+] = 0.046\,M$ **Find:** pH
Conceptual Plan: $[\mathbf{H_3O^+}] \rightarrow \mathbf{pH}$

$$pH = -\log[H_3O^+]$$

Solution: $pH = -\log(0.046) = 1.337 = 1.34$

If the pH of the solution did not carry as many decimal places as the significant digits to the right of the decimal point, you would not be able to distinguish a difference in the pH of solutions (b) and (c). Also, solution (a) would have a pH that was too high.

Acid Solutions

16.59   (a)   **Given:** 0.25 M HCl (strong acid)   **Find:** $[H_3O^+], [OH^-], pH$
Conceptual Plan: $[\mathbf{HCl}] \rightarrow [\mathbf{H_3O^+}] \rightarrow \mathbf{pH}$ and then $[\mathbf{H_3O^+}] \rightarrow [\mathbf{OH^-}]$

$$[HCl] = [H_3O^+] \quad pH = -\log[H_3O^+] \qquad [H_3O^+][OH^-] = 1.0 \times 10^{-14}$$

Solution: $0.25\,M\,HCl = 0.25\,M\,H_3O^+ \quad pH = -\log(0.25) = 0.60$
$$[OH^-] = 1.0 \times 10^{-14}/0.25\,M = 4.0 \times 10^{-14}$$

Check: HCl is a strong acid with a relatively high concentration, so we expect the pH to be low and the $[OH^-]$ to be small.

(b)   **Given:** 0.015 M HNO_3 (strong acid)   **Find:** $[H_3O^+], [OH^-], pH$
Conceptual Plan: $[\mathbf{HNO_3}] \rightarrow [\mathbf{H_3O^+}] \rightarrow \mathbf{pH}$ and then $[\mathbf{H_3O^+}] \rightarrow [\mathbf{OH^-}]$

$$[HNO_3] = [H_3O^+] \quad pH = -\log[H_3O^+] \qquad [H_3O^+][OH^-] = 1.0 \times 10^{-14}$$

Solution: $0.015\,M\,HNO_3 = 0.015\,M\,H_3O^+ \quad pH = -\log(0.015) = 1.82$
$$[OH^-] = 1.0 \times 10^{-14}/0.015\,M = 6.7 \times 10^{-13}$$

Check: HNO_3 is a strong acid, so we expect the pH to be low and the $[OH^-]$ to be small.

(c)   **Given:** 0.052 M HBr and 0.015 M HNO_3 (strong acids)   **Find:** $[H_3O^+], [OH^-], pH$
Conceptual Plan: $[\mathbf{HBr}] + [\mathbf{HNO_3}] \rightarrow [\mathbf{H_3O^+}] \rightarrow \mathbf{pH}$ and then $[\mathbf{H_3O^+}] \rightarrow [\mathbf{OH^-}]$

$$[HBr]+[HNO_3] = [H_3O^+] \quad pH = -\log[H_3O^+] \qquad [H_3O^+][OH^-] = 1.0 \times 10^{-14}$$

Solution: $0.052\,M\,HBr = 0.052\,M\,H_3O^+$ and $0.020\,M\,HNO_3 = 0.020\,M\,H_3O^+$
$$\text{Total } H_3O^+ = 0.072\,M + 0.020\,M = 0.072\,M; pH = -\log(0.072) = 1.14$$
$$[OH^-] = 1.0 \times 10^{-14}/0.072\,M = 1.4 \times 10^{-13}\,M$$

Check: HBr and HNO_3 are both strong acids that completely dissociate. This gives a relatively high concentration, so we expect the pH to be low and the $[OH^-]$ to be small.

(d)   **Given:** $HNO_3 = 0.655\%$ by mass, $d_{solution} = 1.01\,g/mL$ **Find:** $[H_3O^+], [OH^-], pH$
Conceptual Plan:
% mass $HNO_3 \rightarrow$ g $HNO_3 \rightarrow$ mol HNO_3 and then g soln $\rightarrow$ mL soln $\rightarrow$ L soln $\rightarrow$ M HNO_3

$$\frac{\%}{100} \qquad \frac{1\text{ mol }HNO_3}{63.018\text{ g }HNO_3} \qquad\qquad \frac{1.01\text{ g soln}}{1\text{ mL soln}} \quad \frac{1000\text{ mL soln}}{1\text{ L soln}} \quad \frac{\text{mol }HNO_3}{1\text{ L soln}}$$

$\rightarrow$ M $\mathbf{H_3O^+} \rightarrow$ pH and then $[\mathbf{H_3O^+}] \rightarrow [\mathbf{OH^-}]$

$$[HNO_3] = [H_3O^+] \qquad pH = -\log[H_3O^+] \quad [H_3O^+][OH^-] = 1.0 \times 10^{-14}$$

Solution: $\dfrac{0.655\text{ g }HNO_3}{100\text{ g soln}} \times \dfrac{1\text{ mol }HNO_3}{63.018\text{ g }HNO_3} \times \dfrac{1.01\text{ g soln}}{1\text{ mL soln}} \times \dfrac{1000\text{ mL soln}}{1\text{ L soln}} = 0.105\,M\,HNO_3$

$0.105\,M\,HNO_3 = 0.105\,M\,H_3O^+ \quad pH = -\log(0.105) = 0.979$
$$[OH^-] = 1.00 \times 10^{-14}/0.105\,M = 9.52 \times 10^{-14}\,M$$

Check: HNO_3 is a strong acid that completely dissociates. This gives a relatively high concentration, so we expect the pH to be low and the $[OH^-]$ to be small.

16.61   (a)   **Given:** $pH = 1.25, 0.250\,L$ **Find:** g HI
Conceptual Plan: $\mathbf{pH} \rightarrow [\mathbf{H_3O^+}] \rightarrow [\mathbf{HI}] \rightarrow \mathbf{mol\ HI} \rightarrow \mathbf{g\ HI}$

$$pH = -\log[H_3O^+] \quad [H_3O^+] = [HI] \quad mol = MV \quad g = \frac{127.9\text{ g HI}}{\text{mol HI}}$$

Solution: $[H_3O^+] = 10^{-1.25} = 0.056 \text{ M} = [HI] \dfrac{0.056 \text{ mol HI}}{1 \text{ L}} \times 0.250 \text{ L} \times \dfrac{127.9 \text{ g HI}}{1 \text{ mol HI}} = 1.8 \text{ g HI}$

(b)   **Given:** pH $= 1.75, 0.250$ L   **Find:** g HI

     Conceptual Plan: pH $\rightarrow [H_3O^+] \rightarrow [HI] \rightarrow$ mol HI $\rightarrow$ g HI

$$pH = -\log[H_3O^+] \quad [H_3O^+] = [HI] \quad \text{mol} = MV \quad g = \dfrac{127.9 \text{ g HI}}{\text{mol HI}}$$

     Solution: $[H_3O^+] = 10^{-1.75} = 0.01\underline{7}8 \text{ M} = [HI] \dfrac{0.01\underline{7}8 \text{ mol HI}}{1 \text{ L}} \times 0.250 \text{ L} \times \dfrac{127.9 \text{ g HI}}{1 \text{ mol HI}} = 0.57 \text{ g HI}$

(c)   **Given:** pH $= 2.85, 0.250$ L   **Find:** g HI

     Conceptual Plan: pH $\rightarrow [H_3O^+] \rightarrow [HI] \rightarrow$ mol HI $\rightarrow$ g HI

$$pH = -\log[H_3O^+] \quad [H_3O^+] = [HI] \quad \text{mol} = MV \quad g = \dfrac{127.9 \text{ g HI}}{\text{mol HI}}$$

     Solution: $[H_3O^+] = 10^{-2.85} = 0.0014 \text{ M} = [HI] \dfrac{0.0014 \text{ mol HI}}{1 \text{ L}} \times 0.250 \text{ L} \times \dfrac{127.9 \text{ g HI}}{1 \text{ mol HI}} = 0.045 \text{ g HI}$

     Check: The units (g) are correct. The magnitudes of the answers make sense since the volume is small (<1 L). The higher the pH, the lower the amount of HI.

16.63   **Given:** 224 mL HCl, 27.2 °C, 1.02 atm, 1.5 L solution   **Find:** pH

     Conceptual Plan: vol HCl $\rightarrow$ mol HCl $\rightarrow [HCl] \rightarrow [H_3O^+] \rightarrow$ pH

$$PV = nRT \qquad M = \dfrac{\text{mol HCl}}{\text{L soln}} \quad [HCl] = [H_3O^+] \quad pH = -\log[H_3O^+]$$

     Solution: $n = \dfrac{(1.02 \text{ atm})(224 \text{ mL})\left(\dfrac{1 \text{ L}}{1000 \text{ mL}}\right)}{\left(\dfrac{0.08206 \text{ L} \cdot \text{atm}}{\text{mol} \cdot \text{K}}\right)((27.2 + 273.15) \text{ K})} = 0.009\underline{2}7 \text{ mol}$

$$[HCl] = \dfrac{0.009\underline{2}7 \text{ mol}}{1.5 \text{ L}} = 0.006\underline{1}8 \text{ M} = [H_3O^+]; pH = -\log(0.006\underline{1}8) = 2.21$$

     Check: The units (none) are correct. The magnitude of the answer makes sense since HCl is an acid and we will have a low pH.

16.65   **Given:** 0.100 M benzoic acid, $K_a = 6.5 \times 10^{-5}$ **Find:** $[H_3O^+]$, pH

     Conceptual Plan: Write a balanced reaction. Prepare an ICE table, represent the change with x, sum the table, determine the equilibrium values, put the equilibrium values in the equilibrium expression, and solve for x. Determine $[H_3O^+]$ and pH.

     Solution: $\quad HC_7H_5O_2(aq) + H_2O(l) \rightleftharpoons H_3O^+(aq) + C_7H_5O_2^-(aq)$

Initial	0.100	0.0	0.0
Change	$-x$	$+x$	$+x$
Equil	$0.100 - x$	x	x

$$K_a = \dfrac{[H_3O^+][C_7H_5O_2^-]}{[HC_7H_5O_2]} = \dfrac{(x)(x)}{(0.100 - x)} = 6.5 \times 10^{-5}$$

Assume that x is small compared to 0.100.

$$x^2 = (6.5 \times 10^{-5})(0.100); \quad x = 2.5 \times 10^{-3} \text{ M} = [H_3O^+]$$

Check assumption: $\dfrac{2.5 \times 10^{-3}}{0.100} \times 100\% = 2.5\%$; assumption is valid.

$$pH = -\log(2.5 \times 10^{-3}) = 2.60$$

     Check: The units (M and none) are correct. The magnitude of the pH (3) makes sense since we expect a pH $> -\log(0.100) = 1.00$ when we have a weak acid.

16.67   (a)   **Given:** 0.500 M HNO_2, $K_a = 4.6 \times 10^{-4}$ **Find:** pH

Conceptual Plan: Write a balanced reaction. Prepare an ICE table, represent the change with x, sum the table, determine the equilibrium values, put the equilibrium values in the equilibrium expression, and solve for x. Determine $[H_3O^+]$ and pH.

Solution: $HNO_2(aq) + H_2O(l) \rightleftharpoons H_3O^+(aq) + NO_2^-(aq)$

	HNO_2		H_3O^+	NO_2^-
Initial	0.500		0.0	0.0
Change	$-x$		$+x$	$+x$
Equil	$0.500 - x$		x	x

$$K_a = \frac{[H_3O^+][NO_2^-]}{[HNO_2]} = \frac{(x)(x)}{(0.500 - x)} = 4.6 \times 10^{-4}$$

Assume that x is small compared to 0.500.

$$x^2 = (4.6 \times 10^{-4})(0.500); \quad x = 0.015 \text{ M} = [H_3O^+]$$

Check assumption: $\dfrac{0.015}{0.500} \times 100\% = 3.0\%$; assumption is valid.

$$pH = -\log(0.015) = 1.82$$

Check: The units (none) are correct. The magnitude of the answer (2) makes sense since we expect a pH $> -\log(0.500) = 0.301$ when we have a weak acid.

(b)   **Given:** 0.100 M HNO_2, $K_a = 4.6 \times 10^{-4}$ **Find:** pH

Conceptual Plan: Write a balanced reaction. Prepare an ICE table, represent the change with x, sum the table, determine the equilibrium values, put the equilibrium values in the equilibrium expression, and solve for x. Determine $[H_3O^+]$ and pH.

Solution: $HNO_2(aq) + H_2O(l) \rightleftharpoons H_3O^+(aq) + NO_2^-(aq)$

	HNO_2		H_3O^+	NO_2^-
Initial	0.100		0.0	0.0
Change	$-x$		$+x$	$+x$
Equil	$0.100 - x$		x	x

$$K_a = \frac{[H_3O^+][NO_2^-]}{[HNO_2]} = \frac{(x)(x)}{(0.100 - x)} = 4.6 \times 10^{-4}$$

Assume that x is small compared to 0.100.

$$x^2 = (4.6 \times 10^{-4})(0.100) \quad x = 0.0068 \text{ M} = [H_3O^+]$$

Check assumption: $\dfrac{0.0068}{0.100} \times 100\% = 6.8\%$; assumption is not valid; solve using the quadratic equation.

$$x^2 = (4.6 \times 10^{-4})(0.100 - x); \quad x^2 + 4.6 \times 10^{-4}x - 4.6 \times 10^{-5} = 0$$

$$x = 0.00656$$

$$pH = -\log(0.00656) = 2.18$$

Check: The units (none) are correct. The magnitude of the answer (2) makes sense since we expect a pH $> -\log(0.100) = 1.00$ when we have a weak acid.

(c)   **Given:** 0.0100 M HNO_2, $K_a = 4.6 \times 10^{-4}$ **Find:** pH

Conceptual Plan: Write a balanced reaction. Prepare an ICE table, represent the change with x, sum the table, determine the equilibrium values, put the equilibrium values in the equilibrium expression, and solve for x. Determine $[H_3O^+]$ and pH.

Solution: $HNO_2(aq) + H_2O(l) \rightleftharpoons H_3O^+(aq) + NO_2^-(aq)$

	HNO_2		H_3O^+	NO_2^-
Initial	0.0100		0.0	0.0
Change	$-x$		$+x$	$+x$
Equil	$0.0100 - x$		x	x

$$K_a = \frac{[H_3O^+][NO_2^-]}{[HNO_2]} = \frac{(x)(x)}{(0.0100 - x)} = 4.6 \times 10^{-4}$$

Assume that x is small compared to 0.100.

$$x^2 = (4.6 \times 10^{-4})(0.0100); \quad x = 0.0021 \text{ M} = [H_3O^+]$$

Check assumption: $\frac{0.0021}{0.0100} \times 100\% = 21\%$; assumption is not valid; solve using the quadratic equation.

$$x^2 = (4.6 \times 10^{-4})(0.0100 - x) \quad x^2 + 4.6 \times 10^{-4}x - 4.6 \times 10^{-6} = 0$$

$$x = 0.0019$$

$$pH = -\log(0.0019) = 2.72$$

Check: The units (none) are correct. The magnitude of the answer (3) makes sense since we expect a pH $> -\log(0.0100) = 2.00$ when we have a weak acid.

16.69 **Given:** 15.0 mL glacial acetic, $d = 1.05$ g/mL, dilute to 1.50 L, $K_a = 1.8 \times 10^{-5}$ **Find:** pH
Conceptual Plan: mL acetic acid → g acetic acid → mol acetic acid → M and then write a balanced reaction.

$$\frac{1.05 \text{ g}}{1 \text{ mL}} \qquad \frac{1 \text{ mol acetic acid}}{60.05 \text{ g}} \qquad M = \frac{\text{mol}}{L}$$

Prepare an ICE table, represent the change with x, sum the table, determine the equilibrium values, put the equilibrium values in the equilibrium expression, and solve for x. Determine $[H_3O^+]$ and pH.

Solution: $15.0 \text{ mL} \times \dfrac{1.05 \text{ g}}{1 \text{ mL}} \times \dfrac{1 \text{ mol}}{60.05 \text{ g}} \times \dfrac{1}{1.50 \text{ L}} = 0.1749 \text{ M}$

	$HC_2H_3O_2(aq) + H_2O(l) \rightleftharpoons H_3O^+(aq) + C_2H_3O_2^-(aq)$		
Initial	0.1749	0.0	0.0
Change	$-x$	$+x$	$+x$
Equil	$0.1749 - x$	x	x

$$K_a = \frac{[H_3O^+][C_2H_3O_2^-]}{[HC_2H_3O_2]} = \frac{(x)(x)}{(0.1749 - x)} = 1.8 \times 10^{-5}$$

Assume that x is small compared to 0.1749.

$$x^2 = (1.8 \times 10^{-5})(0.1749); \quad x = 0.00177 \text{ M} = [H_3O^+]$$

Check assumption: $\dfrac{0.00177}{0.1749} \times 100\% = 1.0\%$; assumption is valid.

$$pH = -\log(0.00177) = 2.75$$

Check: The units (none) are correct. The magnitude of the answer (3) makes sense since we have an acid concentration <1 M and we have a weak acid.

16.71 **Given:** 0.185 M HA, pH = 2.95 **Find:** K_a
Conceptual Plan: pH → $[H_3O^+]$ and then write a balanced reaction. Prepare an ICE table, calculate equilibrium concentrations, and plug into the equilibrium expression to solve for K_a.

Solution: $[H_3O^+] = 10^{-2.95} = 0.00112 \text{ M} = [A^-]$

	$HA(aq) + H_2O(l) \rightleftharpoons H_3O^+(aq) + A^-(aq)$		
Initial	0.185	0.0	0.0
Change	$-x$	$+x$	$+x$
Equil	$0.185 - 0.00112$	0.00112	0.00112

$$K_a = \frac{[H_3O^+][A^-]}{[HA]} = \frac{(0.00112)(0.00112)}{(0.185 - 0.00112)} = 6.82 \times 10^{-6}$$

Check: The units (none) are correct. The magnitude of the answer makes sense since we expect a pH $> -\log(0.185) = 0.733$. Thus, we have a weak acid with a small K.

16.73   **Given:** 0.125 M HCN, $K_a = 4.9 \times 10^{-10}$ **Find:** % ionization
Conceptual Plan: Write a balanced reaction. Prepare an ICE table, represent the change with x, sum the table, determine the equilibrium values, put the equilibrium values in the equilibrium expression, solve for x, and then $x \rightarrow$ % ionization.

$$\% \text{ ionization} = \frac{x}{[\text{HCN}]_{\text{original}}} \times 100\%$$

Solution: $\text{HCN}(aq) + \text{H}_2\text{O}(l) \rightleftharpoons \text{H}_3\text{O}^+(aq) + \text{CN}^-(aq)$

	HCN	H₃O⁺	CN⁻
Initial	0.125	0.0	0.0
Change	$-x$	$+x$	$+x$
Equil	$0.125 - x$	x	x

$$K_a = \frac{[\text{H}_3\text{O}^+][\text{CN}^-]}{[\text{HCN}]} = \frac{(x)(x)}{(0.125 - x)} = 4.9 \times 10^{-10}$$

Assume that x is small compared to 0.125.

$$x^2 = (4.9 \times 10^{-10})(0.125); \quad x = 7.83 \times 10^{-6}$$

$$\% \text{ ionization} = \frac{7.83 \times 10^{-6}}{0.125} \times 100\% = 0.0063\% \text{ ionized}$$

Check: The units (%) are correct. The magnitude of the answer makes sense since we have a very weak acid (very small K).

16.75   (a)   **Given:** 1.00 M HC₂H₃O₂, $K_a = 1.8 \times 10^{-5}$ **Find:** % ionization
Conceptual Plan: Write a balanced reaction. Prepare an ICE table, represent the change with x, sum the table, determine the equilibrium values, put the equilibrium values in the equilibrium expression, solve for x, and then $x \rightarrow$ % ionization.

$$\% \text{ ionization} = \frac{x}{[\text{HC}_2\text{H}_3\text{O}_2]_{\text{original}}} \times 100\%$$

Solution: $\text{HC}_2\text{H}_3\text{O}_2(aq) + \text{H}_2\text{O}(l) \rightleftharpoons \text{H}_3\text{O}^+(aq) + \text{C}_2\text{H}_3\text{O}_2^-(aq)$

	HC₂H₃O₂	H₃O⁺	C₂H₃O₂⁻
Initial	1.00	0.0	0.0
Change	$-x$	$+x$	$+x$
Equil	$1.00 - x$	x	x

$$K_a = \frac{[\text{H}_3\text{O}^+][\text{C}_2\text{H}_3\text{O}_2^-]}{[\text{HC}_2\text{H}_3\text{O}_2]} = \frac{(x)(x)}{(1.00 - x)} = 1.8 \times 10^{-5}$$

Assume that x is small compared to 1.00.

$$x^2 = (1.8 \times 10^{-5})(1.00); \quad x = 0.00424$$

$$\% \text{ ionization} = \frac{0.00424}{1.00} \times 100\% = 0.42\% \text{ ionized}$$

Check: The units (%) are correct. The magnitude of the answer makes sense since we have a weak acid (small K).

(b)   **Given:** 0.500 M HC₂H₃O₂, $K_a = 1.8 \times 10^{-5}$ **Find:** % ionization
Conceptual Plan: Write a balanced reaction. Prepare an ICE table, represent the change with x, sum the table, determine the equilibrium values, put the equilibrium values in the equilibrium expression, solve for x, and then $x \rightarrow$ % ionization.

$$\% \text{ ionization} = \frac{x}{[\text{HC}_2\text{H}_3\text{O}_2]_{\text{original}}} \times 100\%$$

Solution: $\text{HC}_2\text{H}_3\text{O}_2(aq) + \text{H}_2\text{O}(l) \rightleftharpoons \text{H}_3\text{O}^+(aq) + \text{C}_2\text{H}_3\text{O}_2^-(aq)$

	HC₂H₃O₂	H₃O⁺	C₂H₃O₂⁻
Initial	0.500	0.0	0.0
Change	$-x$	$+x$	$+x$
Equil	$0.500 - x$	x	x

$$K_a = \frac{[H_3O^+][C_2H_3O_2^-]}{[HC_2H_3O_2]} = \frac{(x)(x)}{(0.500 - x)} = 1.8 \times 10^{-5}$$

Assume that x is small compared to 0.500.

$$x^2 = (1.8 \times 10^{-5})(0.500) \quad x = 0.00300$$

$$\% \text{ ionization} = \frac{0.00300}{0.500} \times 100\% = 0.60\% \text{ ionized}$$

Check: The units (%) are correct. The magnitude of the answer makes sense since we have a weak acid (small K).

(c) **Given:** 0.100 M $HC_2H_3O_2$, $K_a = 1.8 \times 10^{-5}$ **Find:** % ionization
Conceptual Plan: Write a balanced reaction. Prepare an ICE table, represent the change with x, sum the table, determine the equilibrium values, put the equilibrium values in the equilibrium expression, solve for x, and then $x \rightarrow$ % ionization.

$$\% \text{ ionization} = \frac{x}{[HC_2H_3O_2]_{original}} \times 100\%$$

Solution: $HC_2H_3O_2(aq) + H_2O(l) \rightleftharpoons H_3O^+(aq) + C_2H_3O_2^-(aq)$

Initial	0.100	0.0	0.0
Change	$-x$	$+x$	$+x$
Equil	$0.100 - x$	x	x

$$K_a = \frac{[H_3O^+][C_2H_3O_2^-]}{[HC_2H_3O_2]} = \frac{(x)(x)}{(0.100 - x)} = 1.8 \times 10^{-5}$$

Assume that x is small compared to 0.100.

$$x^2 = (1.8 \times 10^{-5})(0.100) \quad x = 0.00134$$

$$\% \text{ ionization} = \frac{0.00134}{0.100} \times 100\% = 1.3\% \text{ ionized}$$

Check: The units (%) are correct. The magnitude of the answer makes sense since we have a weak acid (small K).

(d) **Given:** 0.0500 M $HC_2H_3O_2$, $K_a = 1.8 \times 10^{-5}$ **Find:** % ionization
Conceptual Plan: Write a balanced reaction. Prepare an ICE table, represent the change with x, sum the table, determine the equilibrium values, put the equilibrium values in the equilibrium expression, solve for x, and then $x \rightarrow$ % ionization.

$$\% \text{ ionization} = \frac{x}{[HC_2H_3O_2]_{original}} \times 100\%$$

Solution: $HC_2H_3O_2(aq) + H_2O(l) \rightleftharpoons H_3O^+(aq) + C_2H_3O_2^-(aq)$

Initial	0.0500	0.0	0.0
Change	$-x$	$+x$	$+x$
Equil	$0.0500 - x$	x	x

$$K_a = \frac{[H_3O^+][C_2H_3O_2^-]}{[HC_2H_3O_2]} = \frac{(x)(x)}{(0.0500 - x)} = 1.8 \times 10^{-5}$$

Assume that x is small compared to 0.0500.

$$x^2 = (1.8 \times 10^{-5})(0.0500); \quad x = 9.49 \times 10^{-4}$$

$$\% \text{ ionization} = \frac{9.49 \times 10^{-4}}{0.0500} \times 100\% = 1.9\% \text{ ionized}$$

Check: The units (%) are correct. The magnitude of the answer makes sense since we have a weak acid (small K).

16.77   Given: 0.148 M HA, 1.55% dissociation   **Find:** K_a
            Conceptual Plan: $M \rightarrow [H_3O^+] \rightarrow K_a$ **and then write a balanced reaction, determine equilibrium concentration, and plug into the equilibrium expression.**
            Solution: $(0.148 \text{ M HA})(0.0155) = 0.002294;$ $[H_3O^+] = [A^-]$

$$HA(aq) + H_2O(l) \rightleftharpoons H_3O^+(aq) + A^-(aq)$$

Initial	0.148	0.0	0.0
Change	$-x$	$+x$	$+x$
Equil	$0.148 - 0.002294$	0.002294	0.002294

$$K_a = \frac{[H_3O^+][A^-]}{[HA]} = \frac{(0.002294)(0.002294)}{(0.148 - 0.002294)} = 3.61 \times 10^{-5}$$

Check: The units (none) are correct. The magnitude of the answer makes sense since we have a low amount of the acid that dissociates. Thus, we have a weak acid with a small K.

16.79   (a)   **Given:** 0.250 M HF, $K_a = 3.5 \times 10^{-4}$ **Find:** pH, % dissociation
            Conceptual Plan: Write a balanced reaction. Prepare an ICE table, represent the change with x, sum the table, determine the equilibrium values, put the equilibrium values in the equilibrium expression, solve for x, and then $x \rightarrow$ % ionization.

$$\% \text{ ionization} = \frac{x}{[HF]_{original}} \times 100\%$$

Solution: $HF(aq) + H_2O(l) \rightleftharpoons H_3O^+(aq) + F^-(aq)$

Initial	0.250	0.0	0.0
Change	$-x$	$+x$	$+x$
Equil	$0.250 - x$	x	x

$$K_a = \frac{[H_3O^+][F^-]}{[HF]} = \frac{(x)(x)}{(0.250 - x)} = 3.5 \times 10^{-4}$$

Assume that x is small compared to 0.250.
$$x^2 = (3.5 \times 10^{-4})(0.250) \quad x = 0.00935 \text{ M} = [H_3O^+]$$
$$\frac{0.00935}{0.250} \times 100\% = 3.7\%$$
$$pH = -\log(0.00935) = 2.03$$

Check: The units (% and none) are correct. The magnitude of the answer makes sense. We have a weak acid with a small K, so we expect a pH $> -\log(0.250) = 0.602$ and a small percent dissociation.

        (b)   **Given:** 0.100 M HF, $K_a = 3.5 \times 10^{-4}$ **Find:** pH, % dissociation
            Conceptual Plan: Write a balanced reaction. Prepare an ICE table, represent the change with x, sum the table, determine the equilibrium values, put the equilibrium values in the equilibrium expression, solve for x, and then $x \rightarrow$ % ionization.

$$\% \text{ ionization} = \frac{x}{[HF]_{original}} \times 100\%$$

Solution: $HF(aq) + H_2O(l) \rightleftharpoons H_3O^+(aq) + F^-(aq)$

Initial	0.100	0.0	0.0
Change	$-x$	$+x$	$+x$
Equil	$0.100 - x$	x	x

$$K_a = \frac{[H_3O^+][F^-]}{[HF]} = \frac{(x)(x)}{(0.100 - x)} = 3.5 \times 10^{-4}$$

Assume that x is small compared to 0.100.
$$x^2 = (3.5 \times 10^{-4})(0.100); \quad x = 0.00592 \text{ M} = [H_3O^+]$$
$$\frac{0.00592}{0.100} \times 100\% = 5.9\%$$

$x > 5.0\%$ Therefore, the assumption is invalid; solve using the quadratic equation.

$x^2 + 3.5 \times 10^{-4}x - 3.5 \times 10^{-5} = 0$

$x = 0.00574$ or $- 0.00609$

$pH = -\log(0.00574) = 2.24$

$\% \text{ dissociation} = \dfrac{0.00574}{0.100} \times 100\% = 5.7\%$

Check: The units (% and none) are correct. The magnitude of the answer makes sense. We have a weak acid with a small K, so we expect a pH $> -\log(0.100) = 1.000$ and a small percent dissociation.

(c) **Given:** 0.050 M HF, $K_a = 3.5 \times 10^{-4}$ **Find:** pH, % dissociation

Conceptual Plan: Write a balanced reaction. Prepare an ICE table, represent the change with x, sum the table, determine the equilibrium values, put the equilibrium values in the equilibrium expression, solve for x, and then $x \rightarrow$ % ionization.

$$\% \text{ ionization} = \dfrac{x}{[\text{HF}]_{\text{original}}} \times 100\%$$

Solution: $\text{HF}(aq) + \text{H}_2\text{O}(l) \rightleftharpoons \text{H}_3\text{O}^+(aq) + \text{F}^-(aq)$

	HF		H₃O⁺	F⁻
Initial	0.050		0.0	0.0
Change	$-x$		$+x$	$+x$
Equil	$0.050 - x$		x	x

$$K_a = \dfrac{[\text{H}_3\text{O}^+][\text{F}^-]}{[\text{HF}]} = \dfrac{(x)(x)}{(0.050 - x)} = 3.5 \times 10^{-4}$$

Assume that x is small compared to 0.050.

$x^2 = (3.5 \times 10^{-4})(0.050);$ $x = 0.00418 \text{ M} = [\text{H}_3\text{O}^+]$

$\dfrac{0.00418}{0.050} \times 100\% = 8.4\%$

The assumption is invalid; solve using the quadratic equation.

$x^2 + 3.5 \times 10^{-4}\, x - 1.75 \times 10^{-5} = 0$

$x = 0.00401$ or -0.00436

$pH = -\log(0.00401) = 2.40$

$\% \text{ dissociation} = \dfrac{0.00401}{0.050} \times 100\% = 8.0\%$

Check: The units (% and none) are correct. The magnitude of the answer makes sense. We have a weak acid with a small K, so we expect a pH $> -\log(0.050) = 1.30$ and a small percent dissociation.

16.81 (a) **Given:** 0.115 M HBr (strong acid), 0.125 M HCHO₂ (weak acid) **Find:** pH
Conceptual Plan: Because the mixture is a strong acid and a weak acid, the strong acid will dominate. Use the concentration of the strong acid to determine $[\text{H}_3\text{O}^+]$ and then pH.
Solution: 0.115 M HBr = 0.115 M $[\text{H}_3\text{O}^+]$ pH $= -\log(0.115) = 0.939$

(b) **Given:** 0.150 M HNO₂ (weak acid), 0.085 M HNO₃ (strong acid) **Find:** pH
Conceptual Plan: Because the mixture is a strong acid and a weak acid, the strong acid will dominate. Use the concentration of the strong acid to determine $[\text{H}_3\text{O}^+]$ and then pH.
Solution: 0.085 M HNO₃ = 0.085 M $[\text{H}_3\text{O}^+]$ pH $= -\log(0.085) = 1.07$

(c) **Given:** 0.185 M HCHO₂, $K_a = 1.8 \times 10^{-4}$; 0.225 M HC₂H₃O₂, $K_a = 1.8 \times 10^{-5}$ **Find:** pH
Conceptual Plan: Because the mixture is a weak acid and a weak acid and the K values are only 10^1 apart, you need to find $[\text{H}_3\text{O}^+]$ from each reaction. Write a balanced reaction. Prepare an ICE table, represent the change with x, sum the table, determine the equilibrium values, put the equilibrium values in the equilibrium expression, and solve for x.
Solution: $\text{HCHO}_2(aq) + \text{H}_2\text{O}(l) \rightleftharpoons \text{H}_3\text{O}^+(aq) + \text{CHO}_2^-(aq)$

	HCHO₂		H₃O⁺	CHO₂⁻
Initial	0.185		0.0	0.0
Change	$-x$		$+x$	$+x$
Equil	$0.185 - x$		x	x

$$K_a = \frac{[H_3O^+][CHO_2^-]}{[HCHO_2]} = \frac{(x)(x)}{(0.185 - x)} = 1.8 \times 10^{-4}$$

Assume that x is small compared to 0.100.

$$x^2 = (1.8 \times 10^{-4})(0.185); \qquad x = 0.005\underline{7}706$$

$$HC_2H_3O_2(aq) + H_2O(l) \rightleftharpoons H_3O^+(aq) + C_2H_3O_2^-(aq)$$

Initial	0.225	0.005\underline{7}706	0.0
Change	$-x$	$+x$	$+x$
Equil	$0.225 - x$	$0.005\underline{7}706 + x$	x

$$K_a = \frac{[H_3O^+][C_2H_3O_2^-]}{[HC_2H_3O_2]} = \frac{(0.005\underline{7}706 + x)(x)}{(0.225 - x)} = 1.8 \times 10^{-5}$$

Assume that x is small compared to 0.225.

$$x^2 + 0.005\underline{7}706x - 4.05 \times 10^{-6} = 0; \qquad x = 0.006\underline{3}250$$

$$[H_3O^+] = 0.005\underline{7}706 + x = 0.005\underline{7}706 + 0.006\underline{3}250 = 0.006\underline{4}031 \text{ M}$$

$$pH = -\log(0.006\underline{4}031) = 2.19$$

Check: The units (none) are correct. The magnitude of the answer makes sense. We have weak acids, so we expect a pH $> -\log(0.185 + 0.225) = 0.387$.

(d)   **Given:** 0.050 M $HC_2H_3O_2$, $K_a = 1.8 \times 10^{-5}$; 0.050 M HCN, $K_a = 4.9 \times 10^{-10}$ **Find:** pH
Conceptual Plan: Because the values of K are more than 10^1 apart, the acid with the larger K will dominate the reaction. Write a balanced reaction. Prepare an ICE table, represent the change with x, sum the table, determine the equilibrium values, put the equilibrium values in the equilibrium expression, and solve for x.
Solution: $HC_2H_3O_2(aq) + H_2O(l) \rightleftharpoons H_3O^+(aq) + C_2H_3O_2^-(aq)$

Initial	0.0500	0.0	0.0
Change	$-x$	$+x$	$+x$
Equil	$0.0500 - x$	x	x

$$K_a = \frac{[H_3O^+][C_2H_3O_2^-]}{[HC_2H_3O_2]} = \frac{(x)(x)}{(0.0500 - x)} = 1.8 \times 10^{-5}$$

Assume that x is small compared to 0.0500.

$$x^2 = (1.8 \times 10^{-5})(0.0500) \quad x = 9.\underline{4}9 \times 10^{-4}$$

$$pH = -\log(9.\underline{4}9 \times 10^{-4}) = 3.02$$

Check: The units (none) are correct. The magnitude of the answer makes sense. We have weak acids, so we expect a pH $> -\log(0.050 + 0.050) = 1.000$.

Base Solutions

16.83   (a)   **Given:** 0.15 M NaOH   **Find:** $[OH^-]$, $[H_3O^+]$, pH, pOH
Conceptual Plan: $[NaOH] \rightarrow [OH^-] \rightarrow [H_3O^+] \rightarrow pH \rightarrow pOH$

$$K_w = [H_3O^+][OH^-] \quad pH = -\log[H_3O^+] \quad pH + pOH = 14$$

Solution: $[OH^-] = [NaOH] = 0.15 \text{ M}$

$$[H_3O^+] = \frac{K_w}{[OH^-]} = \frac{1.0 \times 10^{-14}}{0.15 \text{ M}} = 6.7 \times 10^{-14} \text{ M}$$

$$pH = -\log(6.7 \times 10^{-14}) = 13.17$$
$$pOH = 14.00 - 13.17 = 0.83$$

(b)   **Given:** 1.5×10^{-3} M $Ca(OH)_2$ **Find:** $[OH^-]$, $[H_3O^+]$, pH, pOH
Conceptual Plan: $[Ca(OH)_2] \rightarrow [OH^-] \rightarrow [H_3O^+] \rightarrow pH \rightarrow pOH$

$$K_w = [H_3O^+][OH^-] \quad pH = -\log[H_3O^+] \quad pH + pOH = 14$$

Solution: $[OH^-] = 2[Ca(OH)_2] = 2(1.5 \times 10^{-3}) = 0.0030\ M$

$$[H_3O^+] = \frac{K_w}{[OH^-]} = \frac{1.0 \times 10^{-14}}{0.0030\ M} = 3.\underline{33} \times 10^{-12}\ M$$

$pH = -\log(3.\underline{33} \times 10^{-12}) = 11.48$
$pOH = 14.00 - 11.48 = 2.52$

(c) **Given:** $4.8 \times 10^{-4}\ M\ Sr(OH)_2$ **Find:** $[OH^-]$, $[H_3O^+]$, pH, pOH
Conceptual Plan: $[Sr(OH)_2] \rightarrow [OH^-] \rightarrow [H_3O^+] \rightarrow pH \rightarrow pOH$

$$K_w = [H_3O^+][OH^-] \quad pH = -\log[H_3O^+] \quad pH + pOH = 14$$

Solution: $[OH^+] = [Sr(OH)_2] = 2(4.8 \times 10^{-4}) = 9.6 \times 10^{-4}\ M$

$$[H_3O^-] = \frac{K_w}{[OH^-]} = \frac{1.0 \times 10^{-14}}{9.6 \times 10^{-4}\ M} = 1.\underline{04} \times 10^{-11}\ M$$

$pH = -\log(1.\underline{04} \times 10^{-11}) = 10.98$
$pOH = 14.00 - 10.98 = 3.02$

Check: The units (M, M, none, and none) are correct. All of these solutions are bases, so $[OH^-] > 10^{-7}\ M$, $[H^+] < 10^{-7}\ M$, pH > 7, and pOH < 7. The higher the base concentration, the lower the $[H^+]$, the higher the pH, and the lower the pOH.

(d) **Given:** $8.7 \times 10^{-5}\ M\ KOH$ **Find:** $[OH^-]$, $[H_3O^+]$, pH, pOH
Conceptual Plan: $[KOH] \rightarrow [OH^-] \rightarrow [H_3O^+] \rightarrow pH \rightarrow pOH$

$$K_w = [H_3O^+][OH^-] \quad pH = -\log[H_3O^+] \quad pH + pOH = 14$$

Solution: $[OH^-] = [KOH] = 8.7 \times 10^{-5}\ M$

$$[H_3O^+] = \frac{K_w}{[OH^-]} = \frac{1.0 \times 10^{-14}}{8.7 \times 10^{-5}\ M} = 1.\underline{1} \times 10^{-10}\ M$$

$pH = -\log(1.\underline{1} \times 10^{-10}) = 9.96$
$pOH = 14.00 - 9.96 = 4.04$

Check: The units (M, M, none, and none) are correct. All of these solutions are bases, so $[OH^-] > 10^{-7}\ M$, $[H^+] < 10^{-7}\ M$, pH > 7, and pOH < 7. The higher the base concentration, the lower the $[H^+]$, the higher the pH, and the lower the pOH.

16.85 **Given:** 3.85% KOH by mass, $d = 1.01\ g/mL$ **Find:** pH
Conceptual Plan:
% mass → g KOH → mol KOH and mass soln → mL soln → L soln → M KOH → $[OH^+]$

$$\frac{1\ mol\ KOH}{56.11\ g\ KOH} \qquad \frac{1.01\ g\ soln}{1\ mL\ soln} \quad \frac{1000\ mL\ soln}{1\ L\ soln} \quad \frac{mol\ KOH}{L\ soln}$$

→ pOH → pH

$$pOH = -\log[OH^-] \quad pH + pOH = 14$$

Solution: $\dfrac{3.85\ g\ \cancel{KOH}}{100.0\ g\ \cancel{soln}} \times \dfrac{1\ mol\ KOH}{56.11\ g\ \cancel{KOH}} \times \dfrac{1.01\ g\ \cancel{soln}}{1\ \cancel{mL\ soln}} \times \dfrac{1000\ \cancel{mL\ soln}}{1\ L\ soln} = 0.\underline{6930}\ M\ KOH$

$[OH^-] = [KOH] = 0.\underline{6930}\ M \quad pOH = -\log(0.\underline{6930}) = 0.159; \quad pH = 14.000 - 0.159 = 13.841$

Check: The units (none) are correct. This is a solution, so $[OH^-] > 10^{-7}\ M$, pH > 7, and pOH < 7. The concentration is fairly high, so we expect a high pH.

16.87 **Given:** 3.55 L, pH = 12.4; 0.855 M KOH **Find:** Vol
Conceptual Plan: $pH \rightarrow [H_3O^+] \rightarrow [OH^-]$ and then $V_1M_1 = V_2M_2$

$$[H_3O^+] = 10^{-pH} \quad 1.0 \times 10^{-14} = [H_3O^+][OH^-] \quad V_1M_1 = V_2M_2$$

Solution: $[H_3O^+] = 10^{-12.4} = 3.98 \times 10^{-13} \quad 1.0 \times 10^{-14} = 3.98 \times 10^{-13}[OH^-]$
$[OH^-] = 0.02\underline{513}\ M$
$V_1M_1 = V_2M_2 \quad V_1(0.855\ M) = (3.55\ L)(0.0251\ M) \quad V_1 = 0.104\ L$

Check: The units (L) are correct. The magnitude of the answer (0.1 L) makes sense since the concentration of the original solution is fairly high.

16.89   (a) $NH_3(aq) + H_2O(l) \rightleftharpoons NH_4^+(aq) + OH^-(aq)$ $K_b = \dfrac{[NH_4^+][OH^-]}{[NH_3]}$

        (b) $HCO_3^-(aq) + H_2O(l) \rightleftharpoons H_2CO_3(aq) + OH^-(aq)$ $K_b = \dfrac{[H_2CO_3][OH^-]}{[HCO_3^-]}$

        (c) $CH_3NH_2(aq) + H_2O(l) \rightleftharpoons CH_3NH_3^+(aq) + OH^-(aq)$ $K_b = \dfrac{[CH_3NH_3^+][OH^-]}{[CH_3NH_2]}$

16.91   **Given:** 0.15 M NH_3, $K_b = 1.76 \times 10^{-5}$ **Find:** $[OH^-]$, pH, pOH
        Conceptual Plan: Write a balanced reaction. Prepare an ICE table, represent the change with x, sum the table, determine the equilibrium values, put the equilibrium values in the equilibrium expression, and solve for x.
        $x = [OH^-] \rightarrow [pOH] \rightarrow pH$
        $pOH = -\log[OH^-]$ $pH + pOH = 14$

        Solution: $NH_3(aq) + H_2O(l) \rightleftharpoons NH_4^+(aq) + OH^-(aq)$

Initial	0.15	0.0	0.0
Change	$-x$	$+x$	$+x$
Equil	$0.15 - x$	x	x

        $K_b = \dfrac{[NH_4^+][OH^-]}{[NH_3]} = \dfrac{(x)(x)}{(0.15 - x)} = 1.76 \times 10^{-5}$

        Assume that x is small.
        $x^2 = (1.76 \times 10^{-5})(0.15)$ $x = [OH^-] = 0.00162$ M
        $pOH = -\log(0.00162) = 2.79$
        $pH = 14.00 - 2.79 = 11.21$
        Check: The units (M, none, and none) are correct. The magnitude of the pH makes sense. We have a weak base, so we expect a pH $< 14 - \log(0.15) = 13.18$; $[OH^-] > 10^{-7}$ M, and pOH > 0.82.

16.93   **Given:** $pK_b = 10.4$, 455 mg/L caffeine   **Find:** pH
        Conceptual Plan: $pK_b \rightarrow K_b$ and then mg/L $\rightarrow$ g/L $\rightarrow$ mol/L and then write a balanced reaction. Prepare an ICE table, represent the change with x, sum the table, determine the equilibrium values, put the equilibrium values in the equilibrium expression, and solve for x.
        $x = [OH^-] \rightarrow [pOH] \rightarrow pH$
        $pOH = -\log[OH^-]$ $pH + pOH = 14$

        Solution: $K_b = 10^{-10.4} = 3.98 \times 10^{-11}$

        $\dfrac{455 \text{ mg caffeine}}{1 \text{ L soln}} \times \dfrac{1 \text{ g caffeine}}{1000 \text{ mg caffeine}} \times \dfrac{1 \text{ mol caffeine}}{194.20 \text{ g}} = 0.002343$ M caffeine

        $C_8H_{10}N_4O_2(aq) + H_2O(l) \rightleftharpoons HC_8H_{10}N_4O_2^+(aq) + OH^-(aq)$

Initial	0.002343	0.0	0.0
Change	$-x$	$+x$	$+x$
Equil	$0.002343 - x$	x	x

        $K_b = \dfrac{[HC_8H_{10}N_4O_2^+][OH^-]}{[C_8H_{10}N_4O_2]} = \dfrac{(x)(x)}{(0.002343 - x)} = 3.98 \times 10^{-11}$

        Assume that x is small.
        $x^2 = (3.98 \times 10^{-11})(0.002343)$ $x = [OH^-] = 3.05 \times 10^{-7}$ M

        $\dfrac{3.05 \times 10^{-7} \text{ M}}{0.002343} \times 100\% = 0.013\%$; assumption is valid.

        $pOH = -\log(3.05 \times 10^{-7}) = 6.5$
        $pH = 14.00 - 6.5 = 7.5$

Check: The units (none) are correct. We expect a pH just above neutral because we have a very low concentration of a weak base.

16.95   **Given:** 0.150 M morphine, pH = 10.5   **Find:** K_b
Conceptual Plan:
pH → pOH → [OH⁻] and then write a balanced equation, prepare an ICE table, and determine

$$pH + pOH = 14 \quad pOH = -\log[OH^-]$$

equilibrium concentrations → K_b.
Solution: $pOH = 14.0 - 10.5 = 3.5;$ $[OH^-] = 10^{-3.5} = 3.16 \times 10^{-4} = [\text{Hmorphine}^+]$

morphine(aq) + H$_2$O(l) ⇌ Hmorphine$^+(aq)$ + OH$^-(aq)$		
Initial	0.150	0.0	0.0
Change	$-x$	$+x$	$+x$
Equil	$0.150 - x$	3.16×10^{-4}	3.16×10^{-4}

$$K_b = \frac{[\text{Hmorphine}^+][OH^-]}{[\text{morphine}]} = \frac{(3.16 \times 10^{-4})(3.16 \times 10^{-4})}{(0.150 - 3.16 \times 10^{-4})} = 6.67 \times 10^{-7} = 7 \times 10^{-7}$$

Check: The units (none) are correct. The magnitude of the K makes sense since the pH is much less than $14 - \log(0.150) = 13.176$.

Acid–Base Properties of Ions and Salts

16.97   (a)   pH-neutral: Br$^-$ is the conjugate base of a strong acid; therefore, it is pH-neutral.
(b)   weak base: ClO$^-$ is the conjugate base of a weak acid; therefore, it is a weak base.
ClO$^-(aq)$ + H$_2$O(l) ⇌ HClO(aq) + OH$^-(aq)$
(c)   weak base: CN$^-$ is the conjugate base of a weak acid; therefore, it is a weak base.
CN$^-(aq)$ + H$_2$O(l) ⇌ HCN(aq) + OH$^-(aq)$
(d)   pH-neutral: Cl$^-$ is the conjugate base of a strong acid; therefore, it is pH-neutral.

16.99   **Given:** $[F^-] = 0.140$ M, $K_a(HF) = 6.8 \times 10^{-4}$ **Find:** $[OH^-]$, pH
Conceptual Plan: Determine K_b. Write a balanced reaction. Prepare an ICE table, represent the change

$$K_b = \frac{K_w}{K_a}$$

with x, sum the table, determine the equilibrium values, put the equilibrium values in the equilibrium expression, and solve for x. Determine $[OH^-]$ → pOH → pH.

$$pOH = -\log[OH^-] \quad pH + pOH = 14$$

Solution:

F$^-(aq)$ + H$_2$O(l) ⇌ HF(aq) + OH$^-(aq)$		
Initial	0.140	0.0	0.0
Change	$-x$	$+x$	$+x$
Equil	$0.140 - x$	x	x

$$K_b = \frac{K_w}{K_a} = \frac{1 \times 10^{-14}}{6.8 \times 10^{-4}} = 1.47 \times 10^{-11} = \frac{(x)(x)}{(0.140 - x)}$$

Assume that x is small.
$x = 1.43 \times 10^{-6}$ M $= [OH^-]$ $pOH = -\log(1.43 \times 10^{-6}) = 5.843$
pH = $14.00 - 5.843 = 8.16$

Check: The units (M and none) are correct. We expect $[OH^-] > (1 \times 10^{-14}/0.140) = 7 \times 10^{-14}$ M and $7 > pH > -\log(0.140) = 0.854$.

16.101   (a)   weak acid: NH$_4^+$ is the conjugate acid of a weak base; therefore, it is a weak acid.
NH$_4^+(aq)$ + H$_2$O(l) ⇌ H$_3$O$^+(aq)$ + NH$_3(aq)$
(b)   pH-neutral: Na$^+$ is the counterion of a strong base; therefore, it is pH-neutral.
(c)   weak acid: The Co^{3+} cation is a small, highly charged metal cation; therefore, it is a weak acid.
Co(H$_2$O)$_6^{3+}(aq)$ + H$_2$O(l) ⇌ Co(H$_2$O)$_5$(OH)$^{2+}(aq)$ + H$_3$O$^+(aq)$

(d)   weak acid: $CH_2NH_3^+$ is the conjugate acid of a weak base; therefore, it is a weak acid.

$$CH_2NH_3^+(aq) + H_2O(l) \rightleftharpoons H_3O^+(aq) + CH_2NH_2(aq)$$

16.103   (a)   acidic: $FeCl_3$ Fe^{3+} is a small, highly charged metal cation; therefore, it is acidic. Cl^- is the conjugate base of a strong acid; therefore, it is pH-neutral.

(b)   basic: NaF Na^+ is the counterion of a strong base; therefore, it is pH-neutral. F^- is the conjugate base of a weak acid; therefore, it is basic.

(c)   pH-neutral: $CaBr_2$ Ca^{2+} is the counterion of a strong base; therefore, it is pH-neutral. Br^- is the conjugate base of a strong acid; therefore, it is pH-neutral.

(d)   acidic: NH_4Br NH_4^+ is the conjugate acid of a weak base; therefore, it is acidic. Br^- is the conjugate base of a strong acid; therefore, it is pH-neutral.

(e)   acidic: $C_6H_5NH_3NO_2$ $C_6H_5NH_3^+$ is the conjugate acid of a weak base; therefore, it is acidic. NO_2^- is the conjugate base of a weak acid; therefore, it is basic. To determine pH, compare K values.

$$K_a(C_6H_5NH_3^+) = \frac{1.0 \times 10^{-14}}{3.9 \times 10^{-10}} = 2.6 \times 10^{-5} \quad K_b(NO_2^-) = \frac{1.0 \times 10^{-14}}{4.6 \times 10^{-4}} = 2.2 \times 10^{-11}$$

$K_a > K_b$; therefore, the solution is acidic.

16.105   **Conceptual Plan:** Identify each species and determine whether it is acidic, basic, or neutral.

$NaCl$ pH-neutral: Na^+ is the counterion of a strong base; therefore, it is pH-neutral. Cl^- is the conjugate base of a strong acid; therefore, it is pH-neutral.

NH_4Cl acidic: NH_4^+ is the conjugate acid of a weak base; therefore, it is acidic. Cl^- is the conjugate base of a strong acid; therefore, it is pH-neutral.

$NaHCO_3$ basic: Na^+ is the counterion of a strong base; therefore, it is pH-neutral. HCO_3^- is the conjugate base of a weak acid; therefore, it is basic.

NH_4ClO_2 acidic: NH_4^+ is the conjugate acid of a weak base; therefore, it is acidic. ClO_2^- is the conjugate base of a weak acid; therefore, it is basic. $K_a(NH_4^+) = 5.6 \times 10^{-10}$ $K_b(ClO_2^-) = 9.1 \times 10^{-13}$ $K_a > K_b$; therefore, the solution is acidic.

$NaOH$ strong base

Increasing acidity: $NaOH < NaHCO_3 < NaCl < NH_4ClO_2 < NH_4Cl$

16.107   (a)   **Given:** 0.10 M NH_4Cl **Find:** pH

Conceptual Plan: Identify each species and determine which will contribute to pH. Write a balanced reaction. Prepare an ICE table, represent the change with x, sum the table, determine the equilibrium values, put the equilibrium values in the equilibrium expression, and solve for x. Determine $[H_3O^+] \rightarrow$ pH.

Solution: NH_4^+ is the conjugate acid of a weak base; therefore, it is acidic. Cl^- is the conjugate base of a strong acid; therefore, it is pH-neutral.

$$NH_4^+(aq) + H_2O(l) \rightleftharpoons NH_3(aq) + H_3O^+(aq)$$

	NH_4^+	NH_3	H_3O^+
Initial	0.10	0.0	0.0
Change	$-x$	$+x$	$+x$
Equil	$0.10 - x$	x	x

$$K_a = \frac{K_w}{K_b} = \frac{1.0 \times 10^{-14}}{1.76 \times 10^{-5}} = 5.\underline{6}8 \times 10^{-10} = \frac{(x)(x)}{(0.10 - x)}$$

Assume that x is small.

$x = 7.\underline{5}4 \times 10^{-6}$ M $= [H_3O^+]$ pH $= -\log(7.\underline{5}4 \times 10^{-6}) = 5.12$

(b)   **Given:** 0.10 M $NaC_2H_3O_2$ **Find:** pH

Conceptual Plan: Identify each species and determine which will contribute to pH. Write a balanced reaction. Prepare an ICE table, represent the change with x, sum the table, determine the equilibrium values, put the equilibrium values in the equilibrium expression, and solve for x. Determine $[OH^-] \rightarrow$ pOH $\rightarrow$ pH.

$\text{pOH} = -\log[OH^-]$ $\text{pH} + \text{pOH} = 14$

Solution: Na^+ is the counterion of a strong base; therefore, it is pH-neutral. $C_2H_3O_2^-$ is the conjugate base of a weak acid; therefore, it is basic.

$$C_2H_3O_2^-(aq) + H_2O(l) \rightleftharpoons HC_2H_3O_2(aq) + OH^-(aq)$$

Initial	0.10	0.0	0.0
Change	$-x$	$+x$	$+x$
Equil	$0.10 - x$	x	x

$$K_b = \frac{K_w}{K_a} = \frac{1.0 \times 10^{-14}}{1.8 \times 10^{-5}} = 5.56 \times 10^{-10} = \frac{(x)(x)}{(0.10 - x)}$$

Assume that x is small.

$x = 7.46 \times 10^{-6} \text{ M} = [OH^-]$ $pOH = -\log(7.46 \times 10^{-6}) = 5.13$
$pH = 14.00 - 5.13 = 8.87$

(c) **Given:** 0.10 M NaCl **Find:** pH
Conceptual Plan: Identify each species and determine which will contribute to pH.
Solution: Na^+ is the counterion of a strong base; therefore, it is pH-neutral. Cl^- is the conjugate base of a strong acid; therefore, it is pH-neutral.
$pH = 7.0$

16.109   **Given:** 0.15 M KF **Find:** concentration of all species
Conceptual Plan: Identify each species and determine which will contribute to pH. Write a balanced reaction. Prepare an ICE table, represent the change with x, sum the table, determine the equilibrium values, put the equilibrium values in the equilibrium expression, and solve for x. Then $[OH^-] \rightarrow [H_3O^+]$.

$$K_w = [H_3O^+][OH^-]$$

Solution: K^+ is the counterion of a strong base; therefore, it is pH-neutral. F^- is the conjugate base of a weak acid; therefore, it is basic.

$$F^-(aq) + H_2O(l) \rightleftharpoons HF(aq) + OH^-(aq)$$

Initial	0.15	0.0	0.0
Change	$-x$	$+x$	$+x$
Equil	$0.15 - x$	x	x

$$K_b = \frac{K_w}{K_a} = \frac{1.0 \times 10^{-14}}{6.8 \times 10^{-4}} = \frac{(x)(x)}{(0.15 - x)} = 1.47 \times 10^{-11}$$

Assume that x is small.

$x = 1.485 \times 10^{-6} \text{ M} = [OH^-] = [HF];$ $[H_3O^+] = \dfrac{K_w}{[OH^-]} = \dfrac{1.0 \times 10^{-14}}{1.485 \times 10^{-6}} = 6.7 \times 10^{-9} \text{ M}$

$[K^+] = 0.15 \text{ M}$
$[F^-] = (0.15 - 1.485 \times 10^{-6}) = 0.15 \text{ M}$
$[HF] = 1.5 \times 10^{-6} \text{ M}$
$[OH^-] = 1.5 \times 10^{-6} \text{ M}$
$[H_3O^+] = 6.7 \times 10^{-9} \text{ M}$

Check: The units (Ms) are correct. We expect $[K^+]$ and $[F^-]$ to be at or near 0.15 M and all other concentrations to be very low.

16.111   (a) F^- is a stronger base than is Cl^-.
F^- is the conjugate base of HF (a weak acid); Cl^- is the conjugate base of HCl (a strong acid); the weaker the acid, the stronger the conjugate base.

(b) NO_2^- is a stronger base than is NO_3^-.
NO_2^- is the conjugate base of HNO_2 (a weak acid); NO_3^- is the conjugate base of HNO_3 (a strong acid); the weaker the acid, the stronger the conjugate base.

(c) ClO^- is a stronger base than is F^-.
F^- is the conjugate base of HF ($K_a = 3.5 \times 10^{-4}$); ClO^- is the conjugate base of HClO ($K_a = 2.9 \times 10^{-8}$); HClO is the weaker acid, so ClO^- will be the stronger base—the weaker the acid, the stronger the conjugate base.

Polyprotic Acids

16.113 $H_3PO_4(aq) + H_2O(l) \rightleftharpoons H_3O^+(aq) + H_2PO_4^-(aq)$ $K_{a_1} = \dfrac{[H_3O^+][H_2PO_4^-]}{[H_3PO_4]}$

$H_2PO_4^-(aq) + H_2O(l) \rightleftharpoons H_3O^+(aq) + HPO_4^{2-}(aq)$ $K_{a_2} = \dfrac{[H_3O^+][HPO_4^{2-}]}{[H_2PO_4^-]}$

$HPO_4^{2-}(aq) + H_2O(l) \rightleftharpoons H_3O^+(aq) + PO_4^{3-}(aq)$ $K_{a_3} = \dfrac{[H_3O^+][PO_4^{3-}]}{[HPO_4^{2-}]}$

16.115 **(a)** **Given:** 0.350 M H_3PO_4 $K_{a_1} = 7.5 \times 10^{-3}, K_{a_2} = 6.2 \times 10^{-8}$ **Find:** $[H_3O^+]$, pH
Conceptual Plan: K_{a_1} is much larger than K_{a_2}, so use K_{a_1} to calculate $[H_3O^+]$. **Write a balanced reaction. Prepare an ICE table, represent the change with x, sum the table, determine the equilibrium values, put the equilibrium values in the equilibrium expression, and solve for x.**
Solution: $H_3PO_4(aq) + H_2O(l) \rightleftharpoons H_3O^+(aq) + H_2PO_4^-(aq)$

	H_3PO_4		H_3O^+	$H_2PO_4^-$
Initial	0.350		0.0	0.0
Change	$-x$		$+x$	$+x$
Equil	$0.350 - x$		x	x

$K_{a_1} = \dfrac{[H_3O^+][H_2PO_4^-]}{[H_3PO_4]} = \dfrac{(x)(x)}{(0.350 - x)} = 7.5 \times 10^{-3}$

Assume that x is small compared to 0.350.

$x^2 = (7.5 \times 10^{-3})(0.350); \quad x = 0.0505 \text{ M} = [H_3O^+]$

Check assumption: $\dfrac{0.0505}{0.350} \times 100\% = 14.4\%$ assumption is not valid, solve using the quadratic equation.

$x^2 + 7.5 \times 10^{-3}x - 0.002625 = 0; \quad x = 0.04\underline{7}62 \text{ M} = [H_3O^+]$
$\text{pH} = -\log(0.04762) = 1.32$

Check: The units (none) are correct. We expect pH slightly above $-\log((0.350 \times 0.0075)^{1/2}) = 1.29$.

(b) **Given:** 0.350 M $H_2C_2O_4$ $K_{a_1} = 6.0 \times 10^{-2}, K_{a_2} = 6.0 \times 10^{-5}$ **Find:** $[H_3O^+]$, pH
Conceptual Plan: K_{a_1} is much larger than K_{a_2}, so use K_{a_1} to calculate $[H_3O^+]$. **Write a balanced reaction. Prepare an ICE table, represent the change with x, sum the table, determine the equilibrium values, put the equilibrium values in the equilibrium expression, and solve for x.**
Solution: $H_2C_2O_4(aq) + H_2O(l) \rightleftharpoons H_3O^+(aq) + HC_2O_4^-(aq)$

	$H_2C_2O_4$		H_3O^+	$HC_2O_4^-$
Initial	0.350		0.0	0.0
Change	$-x$		$+x$	$+x$
Equil	$0.350 - x$		x	x

$K_{a_1} = \dfrac{[H_3O^+][HC_2O_4^-]}{[H_2C_2O_4]} = \dfrac{(x)(x)}{(0.350 - x)} = 6.0 \times 10^{-2}$

$x^2 + 6.0 \times 10^{-2}x - 0.021 = 0 \quad x = 0.1\underline{1}79 = 0.12 \text{ M } [H_3O^+]$
$\text{pH} = -\log(0.1179) = 0.93$

Check: The units (none) are correct. We expect pH slightly above $-\log((0.350 \times 0.060)^{1/2}) = 0.84$.

16.117 **Given:** 0.500 M H_2SO_3 $K_{a_1} = 1.6 \times 10^{-2}, K_{a_2} = 6.4 \times 10^{-8}$ **Find:** concentration of all species
Conceptual Plan: K_{a_1} is much larger than K_{a_2}, so use K_{a_1} to calculate $[H_3O^+]$. **Write a balanced reaction. Prepare an ICE table, represent the change with x, sum the table, determine the equilibrium values, put the equilibrium values in the equilibrium expression, and solve for x. Repeat this process by preparing a second ICE table using these values and representing the change with y.**
Solution: $H_2SO_3(aq) + H_2O(l) \rightleftharpoons H_3O^+(aq) + HSO_3^-(aq)$

	H_2SO_3		H_3O^+	HSO_3^-
Initial	0.500		0.0	0.0
Change	$-x$		$+x$	$+x$
Equil	$0.500 - x$		x	x

$$K_{a_1} = \frac{[H_3O^+][HSO_3^-]}{[H_2SO_3]} = \frac{(x)(x)}{(0.500 - x)} = 1.6 \times 10^{-2}$$

$$x^2 + 1.6 \times 10^{-2}x - 0.0080 = 0; \quad x = 0.08\underline{1}8 = 0.082 \text{ M} = [H_3O^+] = [HSO_3^-]$$

Use the values from reaction 1 in reaction 2.

$$HSO_3^-(aq) + H_2O(l) \rightleftharpoons H_3O^+(aq) + SO_3^{2-}(aq)$$

Initial	0.0818	0.0818	0.0
Change	$-y$	$+y$	$+y$
Equil	$0.0818 - y$	$0.0818 + y$	y

$$K_{a_1} = \frac{[H_3O^+][SO_3^{2-}]}{[HSO_3^-]} = \frac{(0.0818 + y)(y)}{(0.0818 - y)} = 6.4 \times 10^{-8}$$

Assume that y is small. $y = 6.4 \times 10^{-8}$

$[H_2SO_3] = 0.500 - 0.0818 = 0.418$ M

$[HSO_3^-] = x = 0.0818 = 0.082$ M

$[SO_3^{2-}] = y = 6.4 \times 10^{-8}$ M

$[H_3O^+] = x + y = 0.0818 \text{ M} + 6.4 \times 10^{-8} \text{ M} = 0.082$ M

$$[OH^-] = \frac{K_w}{[H_3O^+]} = \frac{1.0 \times 10^{-14}}{0.0818} = 1.2 \times 10^{-13} \text{ M}$$

Check: The units (Ms) are correct. We expect $[H_2SO_3]$ to be the highest at just under 0.500 M and $[SO_3^{2-}]$ to be much lower than $[HSO_3^-]$ and near K_{a2}. We expect $[H_3O^+]$ near $(0.500 \times 0.016)^{1/2} = 0.089$ M.

16.119   (a)   **Given:** $[H_2SO_4] = 0.50$ M $K_{a_2} = 0.012$ **Find:** $[H_3O^+]$, pH

Conceptual Plan: The first ionization step is strong. Use K_{a_2} and reaction 2. Write a balanced reaction. Prepare an ICE table, represent the change with x, sum the table, determine the equilibrium values, put the equilibrium values in the equilibrium expression, and solve for x.

Solution: $H_2SO_4(aq) + H_2O(l) \rightarrow H_3O^+(aq) + HSO_4^-(aq)$ strong

$$[H_3O^+] = [HSO_4^-] = 0.50 \text{ M}$$

$$HSO_4^-(aq) + H_2O(l) \rightleftharpoons H_3O^+(aq) + SO_4^{2-}(aq)$$

Initial	0.50	0.50	0.0
Change	$-x$	$+x$	$+x$
Equil	$0.500 - x$	$0.50 + x$	x

$$K_{a_2} = \frac{[H_3O^+][SO_4^{2-}]}{[HSO_4^-]} = \frac{(0.50 + x)(x)}{(0.50 - x)} = 0.012$$

$$x^2 + 0.512x - 0.006 = 0; \quad x = 0.01\underline{1}5 \text{ M} = [H_3O^+] \text{ from second ionization step}$$

$[H_3O^+] = 0.50 + 0.012 = 0.51$ M

$pH = -\log(0.51) = 0.29$

(b)   **Given:** $[H_2SO_4] = 0.10$ M $K_{a_2} = 0.012$ **Find:** $[H_3O^+]$, pH

Conceptual Plan: The first ionization step is strong. Use K_{a_2} and reaction 2. Write a balanced reaction. Prepare an ICE table, represent the change with x, sum the table, determine the equilibrium values, put the equilibrium values in the equilibrium expression, and solve for x.

Solution: $H_2SO_4(aq) + H_2O(l) \rightarrow H_3O^+(aq) + HSO_4^-(aq)$ strong

$$[H_3O^+] = [HSO_4^-] = 0.10 \text{ M}$$

$$HSO_4^-(aq) + H_2O(l) \rightleftharpoons H_3O^+(aq) + SO_4^{2-}(aq)$$

Initial	0.10	0.10	0.0
Change	$-x$	$+x$	$+x$
Equil	$0.10 - x$	$0.10 + x$	x

$$K_{a_2} = \frac{[H_3O^+][SO_4^{2-}]}{[HSO_4^-]} = \frac{(0.10 + x)(x)}{(0.10 - x)} = 0.012$$

$x^2 + 0.112x - 0.0012 = 0; \quad x = 0.009848 \text{ M}$

$\dfrac{0.009848}{0.10} \times 100\% = 9.8\%$ contribution of second ionization step is not negligible.

$[H_3O^+] = 0.10 + 0.009848 = 0.109848 \text{ M} = 0.11 \text{ M}$
$\text{pH} = -\log(0.11) = 0.96$

(c)   **Given:** $[H_2SO_4] = 0.050 \text{ M}$ $K_{a_2} = 0.012$ **Find:** $[H_3O^+]$, pH
     Conceptual Plan: The first ionization step is strong. Use K_{a_2} and reaction 2. Write a balanced reaction. Prepare an ICE table, represent the change with x, sum the table, determine the equilibrium values, put the equilibrium values in the equilibrium expression, and solve for x.

     Solution: $H_2SO_4(aq) + H_2O(l) \rightarrow H_3O^+(aq) + HSO_4^-(aq)$ strong

     $[H_3O^+] = [HSO_4^-] = 0.050 \text{ M}$

	$HSO_4^-(aq) + H_2O(l) \rightleftharpoons$	$H_3O^+(aq) +$	$SO_4^{2-}(aq)$
Initial	0.050	0.050	0.0
Change	$-x$	$+x$	$+x$
Equil	$0.050 - x$	$0.050 + x$	x

     $K_{a_2} = \dfrac{[H_3O^+][SO_4^{2-}]}{[HSO_4^-]} = \dfrac{(0.050 + x)(x)}{(0.050 - x)} = 0.012$

     $x^2 + 0.062\,x - 0.0006 = 0; \quad x = 0.008509$

     $\dfrac{0.008509}{0.05} \times 100\% = 17\%$ contribution of second ionization step is not negligible.

     $[H_3O^+] = 0.050 + 0.0085 = 0.0585 = 0.059 \text{ M}$ $\text{pH} = -\log(0.059) = 1.23$

     Check: The units (M and none) are correct. We expect pH to be just below $-\log[H_2SO_4]$ and $[H^+]$ to be just above $[H_2SO_4]$. The lower the initial acid concentration, the bigger the differences from these values.

Lewis Acids and Bases

16.121   (a)   Lewis acid: Fe^{3+} has an empty d orbital and can accept a lone pair of electrons.
        (b)   Lewis acid: BH_3 has an empty p orbital to accept a lone pair of electrons.
        (c)   Lewis base: NH_3 has a lone pair of electrons to donate.
        (d)   Lewis base: F^- has a lone pair of electrons to donate.

16.123   (a) Fe^{3+} accepts an electron pair from H_2O; so Fe^{3+} is the Lewis acid, and H_2O is the Lewis base.
        (b) Zn^{2+} accepts an electron pair from NH_3; so Zn^{2+} is the Lewis acid, and NH_3 is the Lewis base.
        (c)   The empty p orbital on B accepts an electron pair from $(CH_3)_3N$; so BF_3 is the Lewis acid, and $(CH_3)_3N$ is the Lewis base.

Cumulative Problems

16.125   (a)   weak acid: The beaker contains 10 HF molecules, 2 H_3O^+ ions, and 2 F^- ions. Because both the molecules and the ions exist in solution, the acid is a weak acid.
        (b)   strong acid: The beaker contains 12 H_3O^+ ions and 12 I^- ions. Because the molecules are completely ionized in solution, the acid is a strong acid.
        (c)   weak acid: The beaker contains 10 $HCHO_2$ molecules, 2 H_3O^+ ions, and 2 CHO_2^- ions. Because both the molecules and the ions exist in solution, the acid is a weak acid.
        (d)   strong acid: The beaker contains 12 H_3O^+ ions and 12 NO_3^- ions. Because the molecules are completely ionized in solution, the acid is a strong acid.

16.127 $HbH^+(aq) + O_2(aq) \rightleftharpoons HbO_2(aq) + H^+(aq)$
     Using Le Châtelier's principle, we see that if the $[H^+]$ increases, the reaction will shift left, and if the $[H^+]$ decreases, the reaction will shift right. So if the pH of blood is too acidic (low pH; $[H^+]$ increased), the reaction will shift to the left. This will cause less of the HbO_2 in the blood and decrease the oxygen-carrying capacity of the hemoglobin in the blood.

16.129   **Given:** 4.00×10^2 mg $Mg(OH)_2$, 2.00×10^2 mL HCl solution, pH $= 1.3$ **Find:** volume neutralized, % neutralized
Conceptual Plan:

mg $Mg(OH)_2 \rightarrow$ g $Mg(OH)_2 \rightarrow$ mol $Mg(OH)_2$ and then pH $\rightarrow [H_3O^+]$ and then mol $Mg(OH)_2$

$$\frac{1 \text{ g Mg(OH)}_2}{1000 \text{ mg}} \qquad \frac{1 \text{ mol Mg(OH)}_2}{58.326 \text{ g}} \qquad\qquad pH = -\log[H_3O^+]$$

mol $OH^- \rightarrow$ mol $H_3O^+ \rightarrow$ vol $H_3O^+ \rightarrow$ % neutralized

$$\frac{2 \text{ mol OH}^-}{1 \text{ mol Mg(OH)}_2} \quad \frac{1 \text{ mol H}_3\text{O}^+}{1 \text{ mol OH}^-} \quad \frac{\text{mol H}_3\text{O}^+}{M(\text{H}_3\text{O}^+)} \quad \frac{\text{vol HCl neutralized}}{\text{total vol HCl}} \times 100$$

Solution: $[H_3O^+] = 10^{-1.3} = 0.05012$ M $= 0.05$ M

$$4.00 \times 10^2 \text{ mg Mg(OH)}_2 \times \frac{1 \text{ g Mg(OH)}_2}{1000 \text{ mg Mg(OH)}_2} \times \frac{1 \text{ mol Mg(OH)}_2}{58.326 \text{ g Mg(OH)}_2} \times \frac{2 \text{ mol OH}^-}{1 \text{ mol Mg(OH)}_2}$$

$$\times \frac{1 \text{ mol H}_3\text{O}^+}{1 \text{ mol OH}^-} \times \frac{1 \text{ L}}{0.05012 \text{ mol H}_3\text{O}^+} \times \frac{1000 \text{ mL}}{1 \text{ L}} = 273.8 \text{ mL} = 274 \text{ mL neutralized}$$

The stomach contains 2.00×10^2 mL HCl at pH $= 1.3$, and 4.00×10^2 mg will neutralize 274 mL of pH 1.3 HCl; so all of the stomach acid will be neutralized.

16.131   **Given:** pH of Great Lakes acid rain $= 4.5$, West Coast $= 5.4$ **Find:** $[H_3O^+]$ and ratio of Great Lakes/West Coast
Conceptual Plan: pH $\rightarrow [H_3O^+]$ and then ratio of $[H_3O^+]$ Great Lakes to West Coast

$$pH = -\log[H_3O^+]$$

Solution: Great Lakes: $[H_3O^+] = 10^{-4.5} = 3.16 \times 10^{-5}$ M   West Coast: $[H_3O^+] = 10^{-5.4} = 3.98 \times 10^{-6}$ M

$$\frac{\text{Great Lakes}}{\text{West Coast}} = \frac{3.16 \times 10^{-5} \text{ M}}{3.98 \times 10^{-6} \text{ M}} = 7.94 = 8 \text{ times more acidic}$$

Check: The answers seem reasonable in light of the fact that pH is a log scale and the difference in pH is almost <1 pH unit.

16.133   **Given:** 6.5×10^2 mg aspirin, 8 oz water, $pK_a = 3.5$ **Find:** pH of solution
Conceptual Plan:

mg aspirin $\rightarrow$ g aspirin $\rightarrow$ mol aspirin and ounces $\rightarrow$ quart $\rightarrow$ L and then [aspirin] and pK_a

$$\frac{1 \text{ g aspirin}}{1000 \text{ mg}} \quad \frac{1 \text{ mol aspirin}}{180.15 \text{ g}} \qquad\qquad \frac{32 \text{ oz}}{1 \text{ qt}} \quad \frac{1.0567 \text{ qt}}{1 \text{ L}} \qquad \frac{\text{mol aspirin}}{\text{L soln}} \quad pK_a = -\log K_a$$

$\rightarrow K_a$. Write a balanced reaction. Prepare an ICE table, represent the change with x, sum the table, determine the equilibrium values, put the equilibrium values in the equilibrium expression, and solve for x. Determine $[H_3O^+] \rightarrow$ pH.

$$pH = -\log[H_3O^+]$$

Solution: $\dfrac{6.5 \times 10^2 \text{ mg aspirin}}{8 \text{ oz}} \times \dfrac{1 \text{ g aspirin}}{1000 \text{ mg aspirin}} \times \dfrac{1 \text{ mol aspirin}}{180.15 \text{ g}} \times \dfrac{32 \text{ oz}}{1 \text{ qt}} \times \dfrac{1.0567 \text{ qt}}{1 \text{ L}} = 0.0153$ M

$$K_a = 10^{-3.5} = 3.16 \times 10^{-4}$$

	aspirin(aq) + H$_2$O(l) $\rightleftharpoons$	H$_3$O$^+$(aq) +	aspirin$^-$(aq)
Initial	0.0153	0.0	0.0
Change	$-x$	$+x$	$+x$
Equil	$0.0153 - x$	x	x

$$K_a = \frac{[H_3O^+][\text{aspirin}^-]}{[\text{aspirin}]} = \frac{(x)(x)}{(0.0153 - x)} = 3.16 \times 10^{-4}$$

$x^2 + 3.16 \times 10^{-4}x - 4.83 \times 10^{-6} = 0$;   Solve the quadratic equation. $x = 2.04 \times 10^{-3}$ M $= [H_3O^+]$

$pH = -\log(2.04 \times 10^{-3}) = 2.69 = 2.7$

Check: The units (none) are correct. The answer seems reasonable since aspirin is an acid and patients are warned of potential stomach problems issuing from taking aspirin.

16.135 (a)   **Given:** 0.0100 M $HClO_4$ **Find:** pH
        Conceptual Plan: $[HClO_4] \rightarrow [H_3O^+] \rightarrow pH$

$$[HClO_4] = [H_3O^+] \qquad pH = -\log[H_3O^+]$$

Solution: $HClO_4$ is a strong acid, so 0.0100 M $HClO_4$ = 0.0100 M H_3O^+ pH $= -\log(0.0100) = 2.000$

Check: The units (none) are correct. The answer is reasonable since perchloric acid is a strong acid.

(b)   **Given:** 0.115 M $HClO_2$, $K_a = 1.1 \times 10^{-2}$ **Find:** pH
        Conceptual Plan: Write a balanced reaction. Prepare an ICE table, represent the change with x, sum the table, determine the equilibrium values, put the equilibrium values in the equilibrium expression, and solve for x. Determine $[H_3O^+]$ and pH.

Solution: $HClO_2(aq) + H_2O(l) \rightleftharpoons H_3O^+(aq) + ClO_2^-(aq)$

		H_3O^+	ClO_2^-
Initial	0.115	0.0	0.0
Change	$-x$	$+x$	$+x$
Equil	$0.115 - x$	x	x

$$K_a = \frac{[H_3O^+][ClO_2^-]}{[HClO_2]} = \frac{(x)(x)}{(0.115 - x)} = 1.1 \times 10^{-2}$$

Assume that x is small compared to 0.115.

$$x^2 = (1.1 \times 10^{-2})(0.115); \qquad x = 0.035\underline{6} \text{ M} = [H_3O^+]$$

Check assumption: $\dfrac{0.035\underline{6}}{0.115} \times 100\% = 31.0\%$; assumption is not valid; solve using the quadratic equation.

$$x^2 + 1.1 \times 10^{-2}x - 0.001265 = 0; \quad \text{Solve the quadratic equation.}$$

$x = 0.030\underline{4}9 \text{ or } -0.0415$

pH $= -\log(0.030\underline{4}9) = 1.52$

Check: The units (none) are correct. We expect pH near $-\log((0.115 \times 0.011)^{1/2}) = 1.49$.

(c)   **Given:** 0.045 M $Sr(OH)_2$ **Find:** pH
        Conceptual Plan: $[Sr(OH)_2] \rightarrow [OH^-] \rightarrow [H_3O^+] \rightarrow pH$

$$K_w = [H_3O^+][OH^-] \quad pH = -\log[H_3O^+] \quad pH + pOH = 14$$

Solution: $Sr(OH)_2$ is a strong base, so $[OH^-] = 2[Sr(OH)_2] = 2(0.045) = 0.090$ M

$$[H_3O^+] = \frac{K_w}{[OH^-]} = \frac{1.0 \times 10^{-14}}{0.090 \text{ M}} = 1.\underline{1}1 \times 10^{-13} \text{ M}$$

pH $= -\log(1.\underline{1}1 \times 10^{-13}) = 12.95$

Check: The units (none) are correct. The answer is reasonable since strontium hydroxide is a strong base.

(d)   **Given:** 0.0852 KCN, $K_a(HCN) = 4.9 \times 10^{-10}$ **Find:** pH
        Conceptual Plan: Identify each species and determine which will contribute to pH. Write a balanced reaction. Prepare an ICE table, represent the change with x, sum the table, determine the equilibrium values, put the equilibrium values in the equilibrium expression, and solve for x. Determine $[OH^-] \rightarrow pOH \rightarrow pH$.

$$pOH = -\log[OH^-] \quad pH + pOH = 14$$

Solution: K^+ is the counterion of a strong base; therefore, it is pH-neutral. CN^- is the conjugate base of a weak acid; therefore, it is basic.

$$CN^-(aq) + H_2O(l) \rightleftharpoons HCN(aq) + OH^-(aq)$$

		HCN	OH^-
Initial	0.0852	0.0	0.0
Change	$-x$	$+x$	$+x$
Equil	$0.0852 - x$	x	x

$$K_b = \frac{K_w}{K_a} = \frac{1.0 \times 10^{-14}}{4.9 \times 10^{-10}} = 2.0\underline{4} \times 10^{-5} = \frac{(x)(x)}{(0.0852 - x)}$$

Assume that x is small.

$x = 1.32 \times 10^{-3}\,\text{M} = [\text{OH}^-]$; $\text{pOH} = -\log(1.32 \times 10^{-3}) = 2.88$

$\text{pH} = 14.00 - 2.88 = 11.12$

Check: The units (none) are correct. We expect pH significantly less than $14 - \log(0.0852) = 12.930$.

(e) **Given:** 0.155 NH_4Cl, $K_b\,(\text{NH}_3) = 1.76 \times 10^{-5}$ **Find:** pH

Conceptual Plan: Identify each species and determine which will contribute to pH. Write a balanced reaction. Prepare an ICE table, represent the change with x, sum the table, determine the equilibrium values, put the equilibrium values in the equilibrium expression, and solve for x. Determine $[\text{H}_3\text{O}^+] \to \text{pH}$.

Solution: NH_4^+ is the conjugate acid of a weak base; therefore, it is acidic. Cl^- is the conjugate base of a strong acid; therefore, it is pH-neutral.

$$\text{NH}_4^+(aq) + \text{H}_2\text{O}(l) \rightleftharpoons \text{NH}_3(aq) + \text{H}_3\text{O}^+(aq)$$

	NH_4^+	NH_3	H_3O^+
Initial	0.155	0.0	0.0
Change	$-x$	$+x$	$+x$
Equil	$0.155 - x$	x	x

$$K_a = \frac{K_w}{K_b} = \frac{1 \times 10^{-14}}{1.76 \times 10^{-5}} = 5.682 \times 10^{-10} = \frac{(x)(x)}{(0.155 - x)}$$

Assume that x is small.

$x = 9.385 \times 10^{-6}\,\text{M} = [\text{H}_3\text{O}^+]$; $\text{pH} = -\log(9.385 \times 10^{-6}) = 5.028$

Check: The units (none) are correct. We expect pH significantly above $\log(0.155) = 0.810$.

16.137 (a) **Given:** 0.0550 M HI (strong acid), 0.00850 M HF (weak acid) **Find:** pH

Conceptual Plan: Because the mixture is a strong acid and a weak acid, the strong acid will dominate. Use the concentration of the strong acid to determine $[\text{H}_3\text{O}^+]$ and then pH.

Solution: 0.0550 M HI $= 0.0550$ M $[\text{H}_3\text{O}^+]$ pH $= -\log(0.0550) = 1.260$

Check: The units (none) are correct. The answer is reasonable since hydroiodic acid is a strong acid and will dominate the calculation of pH.

(b) **Given:** 0.112 M NaCl (salt), 0.0953 M KF (salt) **Find:** pH

Conceptual Plan: Identify each species and determine which will contribute to pH. Write a balanced reaction. Prepare an ICE table, represent the change with x, sum the table, determine the equilibrium values, put the equilibrium values in the equilibrium expression, and solve for x. Determine $[\text{H}_3\text{O}^+] \to \text{pH}$.

Solution: Na^+ is the counterion of a strong base; therefore, it is pH-neutral. Cl^- is the conjugate base of a strong acid; therefore, it is pH-neutral. K^+ is the counterion of a strong base; therefore, it is pH-neutral. F^- is the conjugate base of a weak acid. Therefore, it will produce a basic solution.

$$\text{F}^-(aq) + \text{H}_2\text{O}(l) \rightleftharpoons \text{HF}(aq) + \text{OH}^-(aq)$$

	F^-	HF	OH^-
Initial	0.0953	0.0	0.0
Change	$-x$	$+x$	$+x$
Equil	$0.0953 - x$	x	x

$$K_b = \frac{K_w}{K_a} = \frac{1.0 \times 10^{-14}}{3.5 \times 10^{-4}} = \frac{(x)(x)}{(0.0953 - x)}$$

Assume that x is small.

$x = 1.650 \times 10^{-6}\,\text{M} = [\text{OH}^-]$; $\text{pOH} = -\log(1.650 \times 10^{-6}) = 5.78$

$\text{pH} = 14.00 - 5.78 = 8.22$

Check: The units (none) are correct. The answer is reasonable since NaCl will not affect the pH and KF will make the solution slightly basic.

(c) **Given:** 0.132 M NH_4Cl (salt), 0.150 M HNO_3 (strong acid) **Find:** pH

Conceptual Plan: Because the mixture is a strong acid and a salt, the strong acid will dominate. Use the concentration of the strong acid to determine $[\text{H}_3\text{O}^+]$ and then pH.

Solution: 0.150 M $\text{HNO}_3 = 0.150$ M $[\text{H}_3\text{O}^+]$ pH $= -\log(0.150) = 0.824$

Check: The units (none) are correct. The answer is reasonable since nitric acid is a strong acid and will dominate the calculation of pH.

(d)   **Given:** 0.0887 M NaC$_7$H$_5$O$_2$ (salt), 0.225 M KBr (salt)

Conceptual Plan: Identify each species and determine which will contribute to pH. Write a balanced reaction. Prepare an ICE table, represent the change with x, sum the table, determine the equilibrium values, put the equilibrium values in the equilibrium expression, and solve for x. Determine $[H_3O^+] \rightarrow$ pH.

Solution: Na$^+$ is the counterion of a strong base; therefore, it is pH-neutral. C$_7$H$_5$O$_2^-$ is the conjugate base of a weak acid. Therefore, it will produce a basic solution. K$^+$ is the counterion of a strong base; therefore, it is pH-neutral. Cl$^-$ is the conjugate base of a strong acid; therefore, it is pH-neutral.

$$C_7H_5O_2^-(aq) + H_2O(l) \rightleftharpoons HC_7H_5O_2(aq) + OH^-(aq)$$

	C$_7$H$_5$O$_2^-$	HC$_7$H$_5$O$_2$	OH$^-$
Initial	0.0887	0.0	0.0
Change	$-x$	$+x$	$+x$
Equil	$0.0887 - x$	x	x

$$K_b = \frac{K_w}{K_a} = \frac{1.0 \times 10^{-14}}{6.5 \times 10^{-5}} = \frac{(x)(x)}{(0.0887 - x)}$$

Assume that x is small.

$$x = 3.694 \times 10^{-6}\,M = [OH^-]; \quad pOH = -\log(3.694 \times 10^{-6}) = 5.43$$
$$pH = 14.00 - 5.43 = 8.57$$

Check: The units (none) are correct. The answer is reasonable since KBr will not affect the pH and NaC$_7$H$_5$O$_2$ will make the solution slightly basic.

(e)   **Given:** 0.0450 M HCl (strong acid), 0.0225 M HNO$_3$ (strong acid)   **Find:** pH

Conceptual Plan: Because the mixture is a strong acid and a strong acid, $[H_3O^+]$ is the sum of the concentration of both acids, and then determine pH.

Solution: 0.0450 M HCl = 0.0450 $[H_3O^+]$, 0.0225 M HNO$_3$ = 0.0225 M $[H_3O^+]$

$[H_3O^+] = 0.0450 + 0.0225 = 0.0675$ M; pH $= -\log(0.0675) = 1.171$

Check: The units (none) are correct. The answer is reasonable since hydrochloric acid and nitric acid are strong acids. The two acid concentrations determine the pH.

16.139   (a)   sodium cyanide = NaCN; nitric acid = HNO$_3$

$$H^+(aq) + CN^-(aq) \rightleftharpoons HCN(aq)$$

          (b)   ammonium chloride = NH$_4$Cl; sodium hydroxide = NaOH

$$NH_4^+(aq) + OH^-(aq) \rightleftharpoons NH_3(aq) + H_2O(l)$$

          (c)   sodium cyanide = NaCN; ammonium bromide = NH$_4$Br

$$NH_4^+(aq) + CN^-(aq) \rightleftharpoons NH_3(aq) + HCN(aq)$$

          (d)   potassium hydrogen sulfate = KHSO$_4$; lithium acetate = LiC$_2$H$_3$O$_2$

$$HSO_4^-(aq) + C_2H_3O_2^-(aq) \rightleftharpoons SO_4^{2-}(aq) + HC_2H_3O_2(aq)$$

          (e)   sodium hypochlorite = NaClO; ammonia = NH$_3$

              No reaction; both are bases.

16.141   **Given:** 1.0 M urea, pH = 7.050   **Find:** K_a Hurea$^+$

Conceptual Plan: pH $\rightarrow$ pOH $\rightarrow [OH^-] \rightarrow K_b$(urea) $\rightarrow K_a$(Hurea$^+$). Write a balanced reaction.

$$pH + pOH = 14 \qquad pOH = -\log[OH^-]$$

Prepare an ICE table, represent the change with x, sum the table, determine the equilibrium values, put the equilibrium values in the equilibrium expression, and determine K_b.

Solution: pOH = 14.000 − 7.050 = 6.950; $[OH^-] = 10^{-6.950} = 1.122 \times 10^{-7}$ M

$$urea(aq) + H_2O(l) \rightleftharpoons Hurea^+(aq) + OH^-(aq)$$

	urea	Hurea$^+$	OH$^-$
Initial	1.0	0.0	0.0
Change	$-x$	$+x$	$+x$
Equil	$1.0 - 1.122 \times 10^{-7}$	1.122×10^{-7}	1.122×10^{-7}

$$K_b = \frac{[Hurea^+][OH^-]}{[urea]} = \frac{(1.122 \times 10^{-7})(1.122 \times 10^{-7})}{(1.0 - 1.122 \times 10^{-7})} = 1.2589 \times 10^{-14}$$

$$K_a = \frac{K_w}{K_b} = \frac{1.00 \times 10^{-14}}{1.2589 \times 10^{-14}} = 0.7943 = 0.794$$

Check: The units (none) are correct. The answer is reasonable since the pH of a 1 M solution is close to neutral.

16.143   **Given:** $Ca(Lact)_2$, $[Ca^{2+}] = 0.26$ M, pH $= 8.40$ **Find:** K_a lactic acid
Conceptual Plan: $[Ca^{2+}] \rightarrow [Lact^-]$; determine K_b lactate ion. Prepare an ICE table, represent the

$$\frac{2 \text{ mol lactate ion}}{1 \text{ mol } Ca^{2+}}$$

change with x, sum the table, and determine the equilibrium constant. $K_b \rightarrow K_a$

$$K_a = \frac{K_w}{K_b}$$

Solution: 0.26 M $Ca^{2+} \left(\dfrac{2 \text{ mol Lact}^-}{1 \text{ mol } Ca^{2+}} \right) = 0.52$ M $Lact^-$

	$Lact^-(aq) + H_2O(l) \rightleftharpoons$	$H Lact(aq) +$	$OH^-(aq)$
Initial	0.52	0	0
Change	$-x$	$+x$	$+x$
Equil	$0.52 - x$	x	x

pH $= 8.40$ $[H_3O^+] = 10^{-8.40} = 4.0 \times 10^{-9}$; $[OH^-] = \dfrac{1.0 \times 10^{-14}}{4.0 \times 10^{-9}} = 2.5 \times 10^{-6}$ M $= x$

$K_b = \dfrac{[H Lact][OH^-]}{[Lact^-]} = \dfrac{(2.5 \times 10^{-6})(2.5 \times 10^{-6})}{(0.52 - 2.5 \times 10^{-6})} = 1.2 \times 10^{-11}$

$K_a = \dfrac{K_w}{K_b} = \dfrac{1.0 \times 10^{-14}}{1.2 \times 10^{-11}} = 8.3 \times 10^{-4}$

Check: The units (none) are correct. The answer is reasonable since lactic acid is a weak acid.

Challenge Problems

16.145   The calculation is incorrect because it neglects the contribution from the autoionization of water. HI is a strong acid, so $[H_3O^+]$ from HI $= 1.0 \times 10^{-7}$.

	$H_2O(l) + H_2O(l) \rightleftharpoons$	$H_3O^+(aq) +$	$OH^-(aq)$
Initial		1×10^{-7}	0.0
Change	$-x$	$+x$	$+x$
Equil		$1 \times 10^{-7} + x$	x

$K_w = [H_3O^+][OH^-] = 1.0 \times 10^{-14}$

$(1 \times 10^{-7} + x)(x) = 1.0 \times 10^{-14}$ $x^2 + 1 \times 10^{-7}x - 1.0 \times 10^{-14} = 0$

$x = 6.18 \times 10^{-8}$

$[H_3O^+] = (1 \times 10^{-7} + x) = (1 \times 10^{-7} + 6.18 \times 10^{-8}) = 1.618 \times 10^{-7}$ M

pH $= -\log(1.618 \times 10^{-7}) = 6.79$

16.147   **Given:** 0.00115 M HCl, 0.0100 M $HClO_2$, $K_a = 1.1 \times 10^{-2}$ **Find:** pH
Conceptual Plan: Use HCl to determine $[H_3O^+]$. Use $[H_3O^+]$ and $HClO_2$ to determine dissociation of $HClO_2$. Write a balanced reaction, prepare an ICE table, calculate equilibrium concentrations, and plug into the equilibrium expression.

Solution: 0.00115 M HCl $= 0.00115$ M H_3O^+

	$HClO_2(aq) + H_2O(l) \rightleftharpoons$	$H_3O^+(aq) +$	$ClO_2^-(aq)$
Initial	0.0100	0.00115	0.0
Change	$-x$	$+x$	$+x$
Equil	$0.01000 - x$	$0.00115 + x$	x

$K_a = \dfrac{[H_3O^+][ClO_2^-]}{[HClO_2]} = \dfrac{(0.00115 + x)(x)}{(0.0100 - x)} = 1.1 \times 10^{-2}$

$x^2 + 0.01215x - 1.1 \times 10^{-4} = 0$; $x = 0.006045$

$[H_3O^+] = 0.00115 + 0.006045 = 0.007195$

pH $= -\log(0.007195) = 2.14$

Check: The units (none) are correct. The answer is reasonable since hydrochloric acid is a strong acid and chlorous acid is a fairly strong weak acid.

16.149   **Given:** 1.0 M HA, $K_a = 1.0 \times 10^{-8}$, $K = 4.0$ for reaction 2   **Find:** $[H^+]$, $[A^-]$, $[HA_2^-]$

Conceptual Plan: Combine reaction 1 and reaction 2; then determine the equilibrium expression and the value of K. Prepare an ICE table, calculate equilibrium concentrations, and plug into the equilibrium expression.

Solution:

$$HA(aq) \rightleftharpoons H^+(aq) + A^-(aq) \qquad K = 1.0 \times 10^{-8}$$
$$HA(aq) + A^-(aq) \rightleftharpoons HA_2^-(aq) \qquad K = 4.0$$

$$2\,HA(aq) \rightleftharpoons H^+(aq) + HA_2^-(aq) \qquad K = 4.0 \times 10^{-8}$$

	$2\,HA$	H^+	HA_2^-
Initial	1.0	0.0	0.0
Change	$-2x$	$+x$	$+x$
Equil	$1.0 - 2x$	x	x

$$K = \frac{[H^+][HA_2^-]}{[HA]^2} = 4.0 \times 10^{-8} = \frac{(x)(x)}{(1.0 - 2x)^2}$$

Take the square root of both sides of the equation. $x = 2.0 \times 10^{-4}\,M = [H^+]$

	$HA(aq) \rightleftharpoons$	$H^+(aq)$ +	$A^-(aq)$	$K = 1.0 \times 10^{-8}$
Initial	1.0	2.0×10^{-4}	0.0	
Change	$-y$	$+y$	$+y$	
Equil	$1.0 - y$	$2.0 \times 10^{-4} + y$	y	

$$K = \frac{[H^+][A^-]}{[HA]} = 1.0 \times 10^{-8} = \frac{(2.0 \times 10^{-4} + y)(y)}{(1.0 - y)}$$

Assume that y is small compared to 1.0.

$y^2 + 2.0 \times 10^{-4}y - 1.0 \times 10^{-8} = 0$; $y = 4.1\underline{4} \times 10^{-5}$

$[H^+] = x + y = 2.0 \times 10^{-4}\,M + 4.1\underline{4} \times 10^{-5}\,M = 2.1 \times 10^{-4}\,M$

$[A^-] = y = 4.1 \times 10^{-5}\,M$

$[HA_2^-] = x = 2.0 \times 10^{-4}\,M$

Check: The units (Ms) are correct. The values are reasonable since the value of K for the second reaction is greater than 1.

16.151   **Given:** 0.200 mol NH_4CN, 1.00 L, $K_b\,(NH_3) = 1.76 \times 10^{-5}$, $K_a\,(HCN) = 4.9 \times 10^{-10}$ **Find:** pH

Conceptual Plan: $K_b\,NH_3 \rightarrow K_a\,NH_4^+$; $K_a\,HCN \rightarrow K_b\,CN^-$ **Prepare an ICE table, represent the**

$$K_a K_b = K_w$$

change with x, sum the table, and determine the equilibrium conditions.

Solution: $K_a(NH_4^+) = \dfrac{1.0 \times 10^{-14}}{1.76 \times 10^{-5}} = 5.68 \times 10^{-10}$ $K_b(CN^-) = \dfrac{1.0 \times 10^{-14}}{4.9 \times 10^{-10}} = 2.0 \times 10^{-5}$

0.200 M NH_4CN = 0.200 M NH_4^+ and 0.200 M CN^-

Because the value of K for CN^- is greater than K for NH_4^+, the CN^- reaction will be larger and the solution will be basic.

	$CN^-(aq) + H_2O(l) \rightleftharpoons$	$HCN(aq) +$	$OH^-(aq)$
Initial	0.200	0	0
Change	$-x$	$+x$	$+x$
Equil	$0.200 - x$	x	x

$K_b = \dfrac{[HCN][OH^-]}{[CN^-]}$; $2.0 \times 10^{-5} = \dfrac{(x)(x)}{(0.200 - x)}$ Solve using the quadratic equation.

$x = 0.0020\,M = [OH^-]$

$$[H_3O^+] = \frac{1.0 \times 10^{-14}}{0.0020} = 5.0 \times 10^{-12}; \quad pH = -\log(5.0 \times 10^{-12}) = 11.30$$

Check: The units (none) are correct. The value is reasonable since the value of K for CN^- is greater than the K for NH_4^+.

16.153   **Given:** mixture Na_2CO_3 and $NaHCO_3 = 82.2$ g, 1.0 L, pH = 9.95, $K_a (H_2CO_3) = K_{a_1} = 4.3 \times 10^{-7}$, $K_{a_2} = 5.6 \times 10^{-11}$ **Find:** mass $NaHCO_3$

Conceptual Plan: Let $x = $ g $NaHCO_3$, $y = $ g $Na_2CO_3 \rightarrow$ mol $NaHCO_3$, $Na_2CO_3 \rightarrow [NaHCO_3], [Na_2CO_3]$

$$\frac{1 \text{ mol } NaHCO_3}{84.01 \text{ g}} \qquad \frac{1 \text{ mol } Na_2CO_3}{105.99 \text{ g}} \qquad M = \frac{mol}{L}$$

$\rightarrow [HCO_3^-], [CO_3^{2-}], K_a (HCO_3^-) \rightarrow K_b(CO_3^{2-})$. **Prepare an ICE table, represent the change with z,**

$$\frac{HCO_3^-}{NaHCO_3} \quad \frac{CO_3^{-2}}{Na_2CO_3} \qquad K_b = \frac{K_w}{K_a}$$

sum the table, and determine the equilibrium conditions.

Solution: Let $x = $ g $NaHCO_3$ and $y = $ g Na_2CO_3 $x + y = 82.2$, so $y = 82.2 - x$

$$\text{mol } NaHCO_3 = x \text{ g } NaHCO_3 \left(\frac{1 \text{ mol}}{84.01 \text{ g } NaHCO_3} \right) = \frac{x}{84.01}$$

$$[HCO_3^-] = \left(\frac{\frac{x}{84.01} \text{ mol } NaHCO_3}{1 \text{ L}} \right) \frac{1 \text{ mol } HCO_3^-}{1 \text{ mol } NaHCO_3} = \frac{x}{84.01} \text{M } HCO_3^-$$

$$\text{mol } Na_2CO_3 = 82.2 - x \text{ g } Na_2CO_3 \left(\frac{1 \text{ mol}}{105.99 \text{ g } Na_2CO_3} \right) = \frac{82.2 - x}{105.99} Na_2CO_3$$

$$[CO_3^{-2}] = \left(\frac{\frac{82.2 - x}{105.99} \text{ mol } Na_2CO_3}{1 \text{ L}} \right) \frac{1 \text{ mol } CO_3^{2-}}{1 \text{ mol } Na_2CO_3} = \frac{82.2 - x}{105.99} \text{ M } CO_3^{2-}$$

Because the pH of the solution is basic, it is the hydrolysis of CO_3^{2-} that dominates in the solution.

$$K_b(CO_3^{2-}) = \frac{1.0 \times 10^{-14}}{5.6 \times 10^{-11}} = 1.79 \times 10^{-4} \text{ and } [H_3O^+] = 10^{-9.95} = 1.12 \times 10^{-10} \text{ M}; \quad [OH^-] = \frac{1.0 \times 10^{-14}}{1.12 \times 10^{-10}}$$

$$= 8.93 \times 10^{-5} \text{ M}$$

	$CO^{2-}(aq) + H_2O(l) \rightleftharpoons$	$HCO_3^-(aq)$	+	$OH^-(aq)$
Initial	$\frac{82.2 - x}{105.99}$	$\frac{x}{84.01}$		
Change	$-z$	$+z$		
Equil	$\left(\frac{82.2 - x}{105.99}\right) - z$	$\left(\frac{x}{84.01}\right) + z$		8.93×10^{-5}

$$K_b(CO_3^{2-}) = 1.79 \times 10^{-4} = \frac{[HCO_3^-][OH^-]}{[CO_3^{2-}]} = \frac{\left(\left(\frac{x}{84.01}\right) - z\right)(8.93 \times 10^{-5})}{\left(\frac{82.2 - x}{105.99}\right) - z} \qquad \text{Assume that } z \text{ is small.}$$

$$\frac{1.79 \times 10^{-4}}{8.93 \times 10^{-5}} = \frac{\left(\frac{x}{84.01}\right)}{\left(\frac{82.2 - x}{105.99}\right)}; \quad x = 50.38 \text{ g} = \underline{50.4 \text{ g } NaHCO_3}$$

Conceptual Problems

16.155   Solution (b) would be most acidic.

(a)   0.0100 M HCl (strong acid) and 0.0100 M KOH (strong base) because the concentrations are equal, the acid and base will completely neutralize each other, and the resulting solution will be pH-neutral.

(b)   0.0100 M HF (weak acid) and 0.0100 M KBr (salt). K_a (HF) $= 3.5 \times 10^{-4}$. The weak acid will produce an acidic solution. K^+ is the counterion of a strong base and is pH-neutral. Br^- is the conjugate base of a strong acid and is pH-neutral.

(c)   0.0100 M NH_4Cl (salt) and 0.100 M CH_3NH_3Br. $K_b(NH_3) = 1.8 \times 10^{-5}$, $K_b(CH_3NH_2) = 4.4 \times 10^{-4}$. NH_4^+ is the conjugate acid of a weak base, and $CH_3NH_3^+$ is the conjugate acid of a weak base. Cl^- and Br^- are the conjugate bases of strong acids and will be pH-neutral. K_a of $NH_4^+ = 5.68 \times 10^{-10}$, K_a of $CH_3NH_3^+ = 2.27 \times 10^{-11}$. Since NH_4^+ has a higher K_a, it will have a higher concentration of $[H_3O^+]$ ions and, therefore, will be more acidic. The solution will be acidic. However, because the K_a for the conjugate acids in this solution is smaller K_a for HF, the solution will be acidic, but not as acidic as HF.

(d)   0.100 M NaCN (salt) and 0.100 M $CaCl_2$. Na^+ and Ca^{2+} ion are the counterions of a strong base; therefore, they are pH-neutral. Cl^- is the conjugate base of a strong acid and is pH-neutral. CN^- is the conjugate base of a weak acid and will produce a basic solution.

16.157 $CH_3COOH < CH_2ClCOOH < CHCl_2COOH < CCl_3COOH$

Because Cl is more electronegative than is H, as you add Cl, you increase the number of electronegative atoms, which pulls the electron density away from the O—H group, polarizing the O—H bond, making it more acidic.

Questions for Group Work

16.159   Acids have the following general properties: a sour taste, the ability to dissolve many metals, the ability to turn blue litmus paper red, and the ability to neutralize bases.

Bases have the following general properties: a bitter taste, a slippery feel, the ability to turn red litmus paper blue, and the ability to neutralize acids.

16.161   (a)   A strong acid completely ionizes in solution, and the pH will equal the $-\log$ of the acid concentration. An example is HCl. A weak acid only partially ionizes in solution, and the pH will be greater than the $-\log$ of the acid concentration. An example is HF. Table 15.3 lists the strong acids. If the acid is not in this table, it is a weak acid.

(b)   For a strong acid, HA, pH $= -\log[HA]$.

(c)   For a weak acid, write a balanced reaction for the ionization of the acid. Prepare an ICE table; represent the change with x; sum the table and determine the equilibrium values; put the equilibrium values in the equilibrium expression and solve for x. Finally, determine $[H_3O^+]$ and pH $(-\log[H_3O^+])$.

(d)   Since $K_a \times K_b = K_w$, $K_b = K_w/K_a$. The larger the K_a the smaller the K_b.

(e)   Since $K_w = [H_3O^+] \times [OH^-]$, $[H_3O^+] = K_w/[OH^-]$. The larger the $[OH^-]$, the smaller the $[H_3O^+]$.

16.163   An Arrhenius acid is a substance that produces H^+ in aqueous solution. An example is HCl.

A Brønsted–Lowry base is a proton (H^+) acceptor. Examples are H_2O and F^-.

A Lewis acid is an electron pair acceptor. An example is BF_3.

Data Interpretation and Analysis

16.164   (a)   According to the table, 63 ppm SO_2 (63 mg SO_2/L) is needed. If the initial concentration is 22 mg SO_2/L, the free SO_2 needs to increase by 41 mg SO_2/L. Since there is 225 L of wine,

$$225 \text{ L wine} \times \frac{41 \text{ mg } SO_2}{1 \text{ L wine}} \times \frac{1 \text{ g } SO_2}{1000 \text{ mg } SO_2} = 9.225 \text{ g } SO_2 = 9.2 \text{ g } SO_2.$$

(c) $S_2O_5^{2-} + H_2O \rightleftharpoons 2HSO_3^-$

(e)   For part a, $9.225 \text{ g } SO_2 \times \dfrac{100 \text{ g } K_2S_2O_5}{57.6286431 \text{ g } SO_2} = 16.008 \text{ g } K_2S_2O_5 = 16 \text{ g } K_2S_2O_5.$

For part b, $8.55 \text{ g } SO_2 \times \dfrac{100 \text{ g } K_2S_2O_5}{57.6286431 \text{ g } SO_2} = 14.836 \text{ g } K_2S_2O_5 = 15 \text{ g } K_2S_2O_5.$

17 Aqueous Ionic Equilibrium

Review Questions

17.1 The pH range of human blood is between 7.36 and 7.42. This nearly constant blood pH is maintained by buffers that are chemical systems that resist pH changes, neutralizing an added acid or base. An important buffer system in blood is a mixture of carbonic acid (H_2CO_3) and bicarbonate ion (HCO_3^-).

17.3 The common ion effect occurs when a solution contains two substances $(HC_2H_3O_2$ and $NaC_2H_3O_2)$ that share a common ion $(C_2H_3O_2^-)$. The presence of the $C_2H_3O_2^-(aq)$ ion causes the acid to ionize even less than it normally would, resulting in a less acidic solution (higher pH). This effect is an example of Le Châtelier's principle shifting an equilibrium because of the addition (or removal) of the common ion from the solution.

17.5 When the concentrations of the conjugate acid and base components of a buffer system are equal, the pH is equal to the pK_a of the weak acid of the buffer system. When more of the acid component is present, the pH becomes more acidic (pH drops). When more of the base component is present, the pH becomes more basic (pH rises). The pH in both cases can be calculated using the Henderson–Hasselbalch equation.

17.7 To find the pH of this solution, determine which component is the acid and which is the base, and substitute their concentrations into the Henderson–Hasselbalch equation. The pK_a is the negative of the log of the equilibrium constant of the acid dissociation reaction where the weak acid component and water are the reactants and the conjugate base and H_3O^+ are the products. At the end, confirm that the "x is small" approximation is valid by calculating the $[H_3O^+]$ from the pH. Because H_3O^+ is formed by ionization of the acid, the calculated $[H_3O^+]$ has to be less than 0.05 (or 5%) of the initial concentration of the acid for the "x is small" approximation to be valid.

17.9 The relative concentrations of the acid and conjugate base should not differ by more than a factor of 10 for a buffer to be reasonably effective. Using the Henderson–Hasselbalch equation, this means that the pH should be within one pH unit of the weak acid's pK_a.

17.11 The titration of weak acid by a strong base always has a basic equivalence point because at the equivalence point, all of the acid has been converted into its conjugate base, resulting in a weakly basic solution.

17.13
   (a) The initial pH of the solution is simply the pH of the strong acid. Because strong acids completely dissociate, the concentration of H_3O^+ is the concentration of the strong acid and $pH = -\log[H_3O^+]$.

   (b) Before the equivalence point, H_3O^+ is in excess. Calculate the $[H_3O^+]$ by subtracting the number of moles of added OH^- from the initial number of moles of H_3O^+ and dividing by the *total* volume. Then convert to pH using $-\log[H_3O^+]$.

   (c) At the equivalence point, neither reactant is in excess and the $pH = 7.00$.

   (d) Beyond the equivalence point, OH^- is in excess. Calculate the $[OH^-]$ by subtracting the initial number of moles of H_3O^+ from the number of moles of added OH^- and dividing by the *total* volume. Then convert to pH using $-\log[H_3O^+]$.

17.15   When a polyprotic acid is titrated with a strong base and if K_{a_1} and K_{a_2} are sufficiently different, the pH curve will have two equivalence points because the two acidic protons will be titrated sequentially. The titration of the first acidic proton will be completed before the titration of the second acidic proton.

17.17   The end point is the point when the indicator changes color in an acid–base titration. The equivalence point is when stoichiometrically equivalent amounts of acid and base have reacted. With the correct indicator, the end point of the titration will occur at the equivalence point.

17.19   The solubility product constant (K_{sp}) is the equilibrium expression for a chemical equation representing the dissolution of an ionic compound. The expression of the solubility product constant of A_mX_n is $K_{sp} = [A^{n+}]^m [X^{m-}]^n$.

17.21   In accordance with Le Châtelier's principle, the presence of a common ion in solution causes the equilibrium to shift to the left (compared to its position with pure water as the solvent), which means that less of the ionic compound dissolves. Thus, the solubility of an ionic compound is lower in a solution containing a common ion than in pure water. The exact value of the solubility can be calculated by working an equilibrium problem in which the concentration of the common ion is accounted for in the initial conditions. The molar solubility of a compound, A_mX_n, in a solution with an initial concentration of $A^{n+} = [A^{n+}]_0$ and an initial concentration of $X^{m-} = [X^{m-}]_0$ can be computed directly from K_{sp} by solving for S in the expression $K_{sp} = ([A^{n+}]_0 + mS)^m ([X^{m-}]_0 + nS)^n$.

17.23 Q is the reaction quotient, the product of the concentrations of the ionic components raised to their stoichiometric coefficients, and K is the product of the concentrations of the ionic components raised to their stoichiometric coefficients at equilibrium. A solution contains an ionic compound: if $Q < K_{sp}$, the solution is unsaturated. More of the solid ionic compound can dissolve in the solution: if $Q = K_{sp}$, the solution is saturated, the solution is holding the equilibrium amount of the dissolved ions, and additional solid will not dissolve in the solution. If $Q > K_{sp}$, the solution is supersaturated and, under most circumstances, the excess solid will precipitate out.

Problems by Topic

The Common Ion Effect and Buffers

17.25   The only solution in which HNO_2 will ionize less is (d) 0.10 M $NaNO_2$. It is the only solution that generates a common ion NO_2^- with nitrous acid.

17.27   (a)   **Given:** 0.20 M $HCHO_2$ and 0.15 M $NaCHO_2$ **Find:** pH   **Other:** $K_a (HCHO_2) = 1.8 \times 10^{-4}$

            Conceptual Plan: M $NaCHO_2 \rightarrow$ M CHO_2^- then M $HCHO_2$, M $CHO_2^- \rightarrow [H_3O^+] \rightarrow$ pH

                $NaCHO_2(aq) \rightarrow Na^+(aq) + CHO_2^-(aq)$ ICE table   pH $= -\log[H_3O^+]$

            Solution: Because 1 CHO_2^- ion is generated for each $NaCHO_2$, $[CHO_2^-] = 0.15$ M CHO_2^-.

                $HCHO_2(aq) + H_2O(l) \rightleftharpoons H_3O^+(aq) + CHO_2^-(aq)$

	$[HCHO_2]$		$[H_3O^+]$	$[CHO_2^-]$
Initial	0.20		≈ 0.00	0.15
Change	$-x$		$+x$	$+x$
Equil	$0.20 - x$		$+x$	$0.15 + x$

        $K_a = \dfrac{[H_3O^+][CHO_2^-]}{[HCHO_2]} = 1.8 \times 10^{-4} = \dfrac{x(0.15 + x)}{0.20 - x}$ Assume that x is small ($x << 0.15 < 0.20$), so

        $\dfrac{x(0.15 + \cancel{x})}{0.20 - \cancel{x}} = 1.8 \times 10^{-4} = \dfrac{x(0.15)}{0.20}$ and $x = 2.4 \times 10^{-4}$ M $= [H_3O^+]$. Confirm that the more stringent

        assumption is valid.

        $\dfrac{2.4 \times 10^{-4}}{0.15} \times 100\% = 0.16\%$, so the assumption is valid. Finally,

        pH $= -\log[H_3O^+] = -\log(2.4 \times 10^{-4}) = 3.62$.

        Check: The units (none) are correct. The magnitude of the answer makes physical sense because pH should be greater than $-\log(0.20) = 0.70$ since this is a weak acid and there is a common ion effect.

(b) **Given:** 0.16 M NH_3 and 0.22 M NH_4Cl **Find:** pH **Other:** K_b (NH_3) $= 1.79 \times 10^{-5}$
Conceptual Plan: M $NH_4Cl \rightarrow$ M NH_4^+ then M NH_3, M $NH_4^+ \rightarrow [OH^-] \rightarrow [H_3O^+] \rightarrow$ pH

$$NH_4Cl(aq) \rightarrow NH_4^+(aq) + Cl^-(aq) \qquad \text{ICE table} \qquad K_w = [H_3O^+][OH^-] \qquad pH = -\log[H_3O^+]$$

Solution: Because 1 NH_4^+ ion is generated for each NH_4Cl, $[NH_4^+] = 0.22$ M NH_4^+.

$$NH_3(aq) + H_2O(l) \rightleftharpoons NH_4^+(aq) + OH^-(aq)$$

	$[NH_3]$	$[NH_4^+]$	$[OH^-]$
Initial	0.16	0.22	≈ 0.00
Change	$-x$	$+x$	$+x$
Equil	$0.16 - x$	$0.22 + x$	$+x$

$$K_b = \frac{[NH_4^+][OH^-]}{[NH_3]} = 1.79 \times 10^{-5} = \frac{(0.22 + x)x}{0.16 - x}$$

Assume that x is small ($x \ll 0.16 < 0.22$), so $\dfrac{(0.22 + \cancel{x})x}{0.16 - \cancel{x}} = 1.79 \times 10^{-5} = \dfrac{(0.22)x}{0.16}$ and

$x = 1.\underline{3}0182 \times 10^{-5}$ M $= [OH^-]$. Confirm that the more stringent assumption is valid.

$$\frac{1.30182 \times 10^{-5}}{0.16} \times 100\% = 8.1 \times 10^{-3}\%, \text{ so the assumption is valid.}$$

$$K_w = [H_3O^+][OH^-] \text{ so } [H_3O^+] = \frac{K_w}{[OH^-]} = \frac{1.0 \times 10^{-14}}{1.30182 \times 10^{-5}} = 7.\underline{6}816 \times 10^{-10} \text{ M}$$

Finally, pH $= -\log[H_3O^+] = -\log(7.6816 \times 10^{-10}) = 9.11$.

Check: The units (none) are correct. The magnitude of the answer makes physical sense because pH should be less than $14 + \log(0.16) = 13.2$ since this is a weak base and there is a common ion effect.

17.29 **Given:** 0.15 M $HC_7H_5O_2$ in pure water and in 0.10 M $NaC_7H_5O_2$
Find: % ionization in both solutions **Other:** $K_a(HC_7H_5O_2) = 6.5 \times 10^{-5}$
Conceptual Plan: Pure water: M $HC_7H_5O_2 \rightarrow [H_3O^+] \rightarrow$ % ionization then in $NaC_7H_5O_2$ solution:

$$\text{ICE table} \qquad \% \text{ ionization} = \frac{[H_3O^+]_{equil}}{[HC_7H_5O_2]_0} \times 100\%$$

M $NaC_7H_5O_2 \rightarrow$ M $C_7H_5O_2^-$ then M $HC_7H_5O_2$, M $C_7H_5O_2^- \rightarrow [H_3O^+] \rightarrow$ % ionization

$$NaC_7H_5O_2(aq) \rightarrow Na^+(aq) + C_7H_5O_2^-(aq) \qquad \text{ICE table} \qquad \% \text{ ionization} = \frac{[H_3O^-]_{equil}}{[HC_7H_5O_2]_0} \times 100\%$$

Solution: in pure water:

$$HC_7H_5O_2(aq) + H_2O(l) \rightleftharpoons H_3O^+(aq) + C_7H_5O_2^-(aq)$$

	$[HC_7H_5O_2]$	$[H_3O^+]$	$[C_7H_5O_2^-]$
Initial	0.15	≈ 0.00	0.00
Change	$-x$	$+x$	$+x$
Equil	$0.15 - x$	$+x$	$+x$

$$K_a = \frac{[H_3O^+][C_7H_5O_2^-]}{[HC_7H_5O_2]} = 6.5 \times 10^{-5} = \frac{x^2}{0.15 - x}$$

Assume that x is small ($x \ll 0.10$), so $\dfrac{x^2}{0.15 - \cancel{x}} = 6.5 \times 10^{-5} = \dfrac{x^2}{0.15}$ and $x = 3.\underline{1}225 \times 10^{-3}$ M $= [H_3O^+]$. Then

$$\% \text{ ionization} = \frac{[H_3O^+]_{equil}}{[HC_7H_5O_2]_0} \times 100\% = \frac{3.\underline{1}225 \times 10^{-3}}{0.15} \times 100\% = 2.1\%, \text{ which also confirms that the}$$

assumption is valid (because it is less than 5%). In $NaC_7H_5O_2$ solution: because one $C_7H_5O_2^-$ ion is generated for each $NaC_7H_5O_2$, $[C_7H_5O_2^-] = 0.10$ M $C_7H_5O_2^-$.

$$HC_7H_5O_2(aq) + H_2O(l) \rightleftharpoons H_3O^+(aq) + C_7H_5O_2^-(aq)$$

	$[HC_7H_5O_2]$		$[H_3O^+]$	$[C_7H_5O_2^-]$
Initial	0.15		≈ 0.00	0.10
Change	$-x$		$+x$	$+x$
Equil	$0.15 - x$		$+x$	$0.10 + x$

$$K_a = \frac{[H_3O^+][C_7H_5O_2^-]}{[HC_7H_5O_2]} = 6.5 \times 10^{-5} = \frac{x(0.10 + x)}{0.15 - x} \quad \text{Assume that } x \text{ is small } (x \ll 0.10 < 0.15), \text{ so}$$

$$\frac{x(0.10 + \cancel{x})}{0.15 - \cancel{x}} = 6.5 \times 10^{-5} = \frac{x(0.10)}{0.15} \text{ and } x = 9.\underline{7}5 \times 10^{-5}\,M = [H_3O^+]. \text{ Then}$$

$$\% \text{ ionization} = \frac{[H_3O^+]_{equil}}{[HC_7H_5O_2]_0} \times 100\% = \frac{9.\underline{7}5 \times 10^{-5}}{0.15} \times 100\% = 0.065\%, \text{ which also confirms that the}$$

assumption is valid (because it is less than 5%). The percent ionization in the sodium benzoate solution is less than that in pure water because of the common ion effect. An increase in one of the products (benzoate ion) shifts the equilibrium to the left, so less acid dissociates.

Check: The units (%) are correct. The magnitude of the answer makes physical sense because the acid is weak; so the percent ionization is low. With a common ion present, the percent ionization decreases.

17.31   (a)   **Given:** 0.15 M HF   **Find:** pH   **Other:** $K_a(\text{HF}) = 3.5 \times 10^{-4}$
          Conceptual Plan: $\text{M HF} \rightarrow [H_3O^+] \rightarrow pH$
                         ICE table $pH = -\log[H_3O^+]$

          Solution:

$$\text{HF}(aq) + H_2O(l) \rightleftharpoons H_3O^+(aq) + F^-(aq)$$

	$[HF]$		$[H_3O^+]$	$[F^-]$
Initial	0.15		≈ 0.00	0.00
Change	$-x$		$+x$	$+x$
Equil	$0.15 - x$		$+x$	$+x$

$$K_a = \frac{[H_3O^+][F^-]}{[HF]} = 3.5 \times 10^{-4} = \frac{x^2}{0.15 - x} \quad \text{Assume that } x \text{ is small } (x \ll 0.15), \text{ so}$$

$$\frac{x^2}{0.15 - \cancel{x}} = 3.5 \times 10^{-4} = \frac{x^2}{0.15} \text{ and } x = 7.\underline{2}457 \times 10^{-3}\,M = [H_3O^+]. \text{ Confirm that the}$$

$$\text{assumption is valid. } \frac{7.\underline{2}457 \times 10^{-3}}{0.15} \times 100\% = 4.8\% < 5\%, \text{ so the assumption is valid.}$$

Finally, $pH = -\log[H_3O^+] = -\log(7.\underline{2}457 \times 10^{-3}) = 2.14.$

Check: The units (none) are correct. The magnitude of the answer makes physical sense because the pH should be greater than $-\log(0.15) = 0.82$ since this is a weak acid.

   (b)   **Given:** 0.15 M NaF   **Find:** pH   **Other:** $K_a(\text{HF}) = 3.5 \times 10^{-4}$
          Conceptual Plan: $\text{M NaF} \rightarrow \text{M F}^- \text{ and } K_a \rightarrow K_b \text{ then M F}^- \rightarrow [OH^-] \rightarrow [H_3O^+] \rightarrow pH$
             $\text{NaF}(aq) \rightarrow \text{Na}^+(aq) + F^-(aq)$ $K_w = K_a K_b$ ICE table $K_w = [H_3O^+][OH^-]$ $pH = -\log[H_3O^+]$

          Solution:
          Because one F^- ion is generated for each NaF, $[F^-] = 0.15$ M F^-. Because $K_w = K_a K_b$, rearrange to solve for K_b.

$$K_b = \frac{K_w}{K_a} = \frac{1.0 \times 10^{-14}}{3.5 \times 10^{-4}} = 2.\underline{8}571 \times 10^{-11}$$

$$F^-(aq) + H_2O(l) \rightleftharpoons \text{HF}(aq) + OH^-(aq)$$

	$[F^-]$		$[HF]$	$[OH^-]$
Initial	0.15		0.00	≈ 0.00
Change	$-x$		$+x$	$+x$
Equil	$0.15 - x$		$+x$	$+x$

$$K_b = \frac{[HF][OH^-]}{[F^-]} = 2.\underline{8}571 \times 10^{-11} = \frac{x^2}{0.15 - x}$$

Assume that x is small ($x \ll 0.15$), so $\dfrac{x^2}{0.15 - \cancel{x}} = 2.\underline{8}571 \times 10^{-11} = \dfrac{x^2}{0.15}$ and

$x = 2.\underline{0}702 \times 10^{-6} \, M = [OH^-]$.

Confirm that the assumption is valid. $\dfrac{2.\underline{0}702 \times 10^{-6}}{0.15} \times 100\% = 0.0014\% < 5\%$, so the assumption is valid.

$K_w = [H_3O^+][OH^-]$, so $[H_3O^+] = \dfrac{K_w}{[OH^-]} = \dfrac{1.0 \times 10^{-14}}{2.0702 \times 10^{-6}} = 4.\underline{8}305 \times 10^{-9} \, M$.

Finally, $pH = -\log[H_3O^+] = -\log(4.\underline{8}305 \times 10^{-9}) = 8.32$.

Check: The units (none) are correct. The magnitude of the answer makes physical sense because the pH should be slightly basic because the fluoride ion is a very weak base.

(c) **Given:** 0.15 M HF and 0.15 M NaF   **Find:** pH   **Other:** $K_a(HF) = 3.5 \times 10^{-4}$
Conceptual Plan: M NaF $\rightarrow$ M F$^-$ then M HF, M F$^-$ $\rightarrow$ [H$_3$O$^+$] $\rightarrow$ pH

$$NaF(aq) \rightarrow Na^+(aq) + F^-(aq) \qquad\qquad \text{ICE table} \quad pH = -\log[H_3O^+]$$

Solution: Because one F$^-$ ion is generated for each NaF, $[F^-] = 0.15$ M F$^-$.

$$HF(aq) + H_2O(l) \rightleftharpoons H_3O^+(aq) + F^-(aq)$$

	[HF]	[H$_3$O$^+$]	[F$^-$]
Initial	0.15	≈ 0.00	0.15
Change	$-x$	$+x$	$+x$
Equil	$0.15 - x$	$+x$	$0.15 + x$

$$K_a = \frac{[H_3O^+][F^-]}{[HF]} = 3.5 \times 10^{-4} = \frac{x(0.15 + x)}{0.15 - x}$$

Assume that x is small ($x \ll 0.15$), so $\dfrac{x(0.15 + \cancel{x})}{0.15 - \cancel{x}} = 3.5 \times 10^{-4} = \dfrac{x(\cancel{0.15})}{\cancel{0.15}}$ and

$x = 3.5 \times 10^{-4} \, M = [H_3O^+]$.

Confirm that the assumption is valid. $\dfrac{3.5 \times 10^{-4}}{0.15} \times 100\% = 0.23\% < 5\%$, so the assumption is valid.

Finally, $pH = -\log[H_3O^+] = -\log(3.5 \times 10^{-4}) = 3.46$.

Check: The units (none) are correct. The magnitude of the answer makes physical sense since the pH should be greater than that in part (a) (2.14) because of the common ion effect suppressing the dissociation of the weak acid.

17.33   When an acid (such as HCl) is added, it reacts with the conjugate base of the buffer system as follows: HCl + NaC$_2$H$_3$O$_2$ $\rightarrow$ HC$_2$H$_3$O$_2$ + NaCl. When a base (such as NaOH) is added, it reacts with the weak acid of the buffer system as follows: NaOH + HC$_2$H$_3$O$_2$ $\rightarrow$ H$_2$O + NaC$_2$H$_3$O$_2$. The reaction generates the other buffer system component.

17.35   (a) **Given:** 0.15 M HCHO$_2$ and 0.10 M NaCHO$_2$ **Find:** pH   **Other:** $K_a(HCHO_2) = 1.8 \times 10^{-4}$
Conceptual Plan: Identify acid and base components then M NaCHO$_2$ $\rightarrow$ M CHO$_2^-$ then

$$\text{acid} = HCHO_2 \text{ base} = CHO_2^- \qquad\qquad NaCHO_2(aq) \rightarrow Na^+(aq) + CHO_2^-(aq)$$

K_a, M HCHO$_2$, M CHO$_2^-$ $\rightarrow$ pH

$$pH = pK_a + \log\frac{[\text{base}]}{[\text{acid}]}$$

Solution: Acid = HCHO$_2$, so $[\text{acid}] = [HCHO_2] = 0.15$ M. Base = CHO$_2^-$. Because one CHO$_2^-$ ion is generated for each NaCHO$_2$, $[CHO_2^-] = 0.10$ M CHO$_2^- = [\text{base}]$. Then

$$pH = pK_a + \log\frac{[\text{base}]}{[\text{acid}]} = -\log(1.8 \times 10^{-4}) + \log\frac{0.10 \, M}{0.15 \, M} = 3.57.$$

Note that to use the Henderson–Hasselbalch equation, the assumption that x is small must be valid. This was confirmed in Problem 17.27.

Check: The units (none) are correct. The magnitude of the answer makes physical sense because the pH should be less than the pK_a of the acid because there is more acid than base. The answer agrees with Problem 17.27.

(b) **Given:** 0.12 M NH_3 and 0.18 M NH_4Cl **Find:** pH   **Other:** $K_b(NH_3) = 1.76 \times 10^{-5}$
Conceptual Plan: Identify acid and base components then M $NH_4Cl \rightarrow$ M NH_4^+ and $K_b \rightarrow pK_b \rightarrow pK_a$

$$\text{acid} = NH_4^+ \text{ base} = NH_3 \quad NH_4Cl(aq) \rightarrow NH_4^+(aq) + Cl^-(aq) \qquad pK_b = -\log K_b \quad 14 = pK_a + pK_b$$

then pK_a, M NH_3, M $NH_4^+ \rightarrow$ pH

$$pH = pK_a + \log \frac{[\text{base}]}{[\text{acid}]}$$

Solution: Base = NH_3, $[\text{base}] = [NH_3] = 0.12$ M. Acid = NH_4^+. Because one NH_4^+ ion is generated for each NH_4Cl, $[NH_4^+] = 0.18$ M $NH_4^+ = [\text{acid}]$.

Because $K_b(NH_3) = 1.76 \times 10^{-5}$, $pK_b = -\log K_b = -\log(1.76 \times 10^{-5}) = 4.75$. Because $14 = pK_a + pK_b$,

$$pK_a = 14 - pK_b = 14 - 4.75 = 9.25. \text{ Then } pH = pK_a + \log \frac{[\text{base}]}{[\text{acid}]} = 9.25 + \log \frac{0.12\,M}{0.18\,M} = 9.07.$$

Note that to use the Henderson–Hasselbalch equation, the assumption that x is small must be valid. This was confirmed in Problem 17.28.

Check: The units (none) are correct. The magnitude of the answer makes physical sense because the pH should be less than the pK_a of the acid since there is more acid than base. The answer agrees with Problem 17.28 within the error of the value.

17.37   (a)   **Given:** 0.135 M HClO and 0.155 M KClO   **Find:** pH   **Other:** $K_a(\text{HClO}) = 2.9 \times 10^{-8}$
Conceptual Plan: Identify acid and base components then M KClO $\rightarrow$ M ClO^- then

$$\text{acid} = \text{HClO, base} = ClO^- \qquad KClO(aq) \rightarrow K^+(aq) + ClO^-(aq)$$

M HClO, M $ClO^- \rightarrow$ pH

$$pH = pK_a + \log \frac{[\text{base}]}{[\text{acid}]}$$

Solution: Acid = HClO, so $[\text{acid}] = [\text{HClO}] = 0.135$ M. Base = ClO^-. Because 1 ClO^- ion is generated for each KClO, $[ClO^-] = 0.155$ M $ClO^- = [\text{base}]$. Then

$$pH = pK_a + \log \frac{[\text{base}]}{[\text{acid}]} = -\log(2.9 \times 10^{-8}) + \log \frac{0.155\,M}{0.135\,M} = 7.60.$$

Check: The units (none) are correct. The magnitude of the answer makes physical sense because pH should be greater than the pK_a of the acid since there is more base than acid.

(b)   **Given:** 1.05% by mass $C_2H_5NH_2$ and 1.10% by mass $C_2H_5NH_3Br$ **Find:** pH
Other: $K_b(C_2H_5NH_2) = 5.6 \times 10^{-4}$
Conceptual Plan: Assume exactly 100 g of solution. Because both components are in the same solution (i.e., the same final volume of solution), the ratio of the moles of each component is the same as the ratio of the molarity of these components and only the relative number of moles needs to be calculated.

$$g_{\text{solution}} \rightarrow g_{C_2H_5NH_2} \rightarrow \text{mol}_{C_2H_5NH_2} \text{ and } g_{\text{solution}} \rightarrow g_{C_2H_5NH_3Br} \rightarrow \text{mol}_{C_2H_5NH_3Br}$$

$$\frac{1.05\ g\ C_2H_5NH_2}{100\ g\ \text{solution}} \quad \frac{1\ \text{mol}\ C_2H_5NH_2}{45.09\ g\ C_2H_5NH_2} \qquad \frac{1.10\ g\ C_2H_5NH_3Br}{100\ g\ \text{solution}} \quad \frac{1\ \text{mol}\ C_2H_5NH_3Br}{125.99\ g\ C_2H_5NH_3Br}$$

identify acid and base components then M $C_2H_5NH_3Br \rightarrow$ M $C_2H_5NH_3^+$ and

$$\text{acid} = C_2H_5NH_3^+, \text{ base} = C_2H_5NH_2 \quad C_2H_5NH_3Br(aq) \rightarrow C_2H_5NH_3^+(aq) + Br^-(aq)$$

$K_b \rightarrow pK_b \rightarrow pK_a$ then pK_a, M $C_2H_5NH_2$, M $C_2H_5NH_3^+ \rightarrow$ pH

$$pK_b = -\log K_b, \ 14 = pK_a + pK_b \qquad\qquad pH = pK_a + \log \frac{[\text{base}]}{[\text{acid}]}$$

Solution:

$$100.\ g\ \text{solution} \times \frac{1.05\ g\ C_2H_5NH_2}{100\ g\ \text{solution}} \times \frac{1\ \text{mol}\ C_2H_5NH_2}{45.09\ g\ C_2H_5NH_2} = 0.02328676\ \text{mol}\ C_2H_5NH_2 \text{ and}$$

$$100.\ g\ \text{solution} \times \frac{1.10\ g\ C_2H_5NH_3Br}{100\ g\ \text{solution}} \times \frac{1\ \text{mol}\ C_2H_5NH_3Br}{125.99\ g\ C_2H_5NH_3Br} = 0.008730852\ \text{mol}\ C_2H_5NH_3Br$$

Then base = $C_2H_5NH_2$, moles of base = mol $C_2H_5NH_2$ = 0.02328676 mol; acid = $C_2H_5NH_3^+$.
Because 1 $C_2H_5NH_3^+$ ion is generated for each $C_2H_5NH_3Br$, mol $C_2H_5NH_3^+$ = 0.008730852 mol.
Because K_b $(C_2H_5NH_2)$ = 5.6×10^{-4}, pK_b = $-\log K_b$ = $-\log (5.6 \times 10^{-4})$ = 3.25.
Because 14 = pK_a + pK_b, pK_a = 14 $-$ pK_b = 14 $-$ 3.25 = 10.75, then

$$pH = pK_a + \log \frac{\text{mol base}}{\text{mol acid}} = 10.75 + \log \frac{0.02328676 \text{ mol}}{0.008730852 \text{ mol}} = 11.18.$$

Check: The units (none) are correct. The magnitude of the answer makes physical sense because pH should be greater than the pK_a of the acid since there is more base than acid.

(c) **Given:** 10.0 g $HC_2H_3O_2$ and 10.0 g $NaC_2H_3O_2$ in 150.0 mL solution **Find:** pH
Other: K_a $(HC_2H_3O_2)$ = 1.8×10^{-5}
Conceptual Plan: Identify acid and base components then mL $\rightarrow$ L and g $HC_2H_3O_2$ $\rightarrow$ mol $HC_2H_3O_2$

$$\text{acid} = HC_2H_3O_2, \text{base} = C_2H_3O_2^- \qquad \frac{1 \text{ L}}{1000 \text{ mL}} \qquad \frac{1 \text{ mol } HC_2H_3O_2}{60.05 \text{ g } HC_2H_3O_2}$$

then mol $HC_2H_3O_2$, L $\rightarrow$ M $HC_2H_3O_2$ and g $NaC_2H_3O_2$ $\rightarrow$ mol $NaC_2H_3O_2$ then

$$M = \frac{\text{mol}}{L} \qquad \frac{1 \text{ mol } NaC_2H_3O_2}{82.03 \text{ g } NaC_2H_3O_2}$$

mol $NaC_2H_3O_2$, L $\rightarrow$ M $NaC_2H_3O_2$ $\rightarrow$ M $C_2H_3O_2^-$ then M $HC_2H_3O_2$, M $C_2H_3O_2^-$ $\rightarrow$ pH

$$M = \frac{\text{mol}}{L} \qquad NaC_2H_3O_2(aq) \rightarrow Na^+(aq) + C_2H_3O_2^-(aq) \qquad pH = pK_a + \log \frac{[\text{base}]}{[\text{acid}]}$$

Solution: $150.0 \text{ mL} \times \dfrac{1 \text{ L}}{1000 \text{ mL}} = 0.1500 \text{ L}$ and

$$10.0 \text{ g } HC_2H_3O_2 \times \frac{1 \text{ mol } HC_2H_3O_2}{60.05 \text{ g } HC_2H_3O_2} = 0.166528 \text{ mol } HC_2H_3O_2$$

$$\text{then } M = \frac{\text{mol}}{L} = \frac{0.166528 \text{ mol } HC_2H_3O_2}{0.1500 \text{ L}} = 1.11019 \text{ M } HC_2H_3O_2 \text{ and}$$

$$10.0 \text{ g } NaC_2H_3O_2 \times \frac{1 \text{ mol } NaC_2H_3O_2}{82.03 \text{ g } NaC_2H_3O_2} = 0.121906 \text{ mol } NaC_2H_3O_2 \text{ then}$$

$$M = \frac{\text{mol}}{L} = \frac{0.121906 \text{ mol } NaC_2H_3O_2}{0.1500 \text{ L}} = 0.812706 \text{ M } NaC_2H_3O_2$$

Acid = $HC_2H_3O_2$, so [acid] = $[HC_2H_3O_2]$ = 1.11019 M and base = $C_2H_3O_2^-$. Because 1 $C_2H_3O_2^-$ ion is generated for each $NaC_2H_3O_2$, $[C_2H_3O_2^-]$ = 0.812706 M $C_2H_3O_2^-$ = [base]. Then

$$pH = pK_a + \log \frac{[\text{base}]}{[\text{acid}]} = -\log (1.8 \times 10^{-5}) + \log \frac{0.812706 \text{ M}}{1.11019 \text{ M}} = 4.61.$$

Check: The units (none) are correct. The magnitude of the answer makes physical sense because pH should be less than the pK_a of the acid since there is more acid than base.

17.39 (a) **Given:** 50.0 mL of 0.15 M $HCHO_2$ and 75.0 mL of 0.13 M $NaCHO_2$ **Find:** pH
Other: $K_a(HCHO_2)$ = 1.8×10^{-4}
Conceptual Plan: Identify acid and base components then mL $HCHO_2$, mL $NaCHO_2$ $\rightarrow$ total mL then

$$\text{acid} = HCHO_2, \text{base} = CHO_2^- \qquad \text{total mL} = \text{mL } HCHO_2 + \text{mL } NaCHO_2$$

mL $HCHO_2$, M $HCHO_2$, total mL $\rightarrow$ buffer M $HCHO_2$ and

$$M_1V_1 = M_2V_2$$

mL $NaCHO_2$, M $NaCHO_2$, total mL $\rightarrow$ buffer M $NaCHO_2$ $\rightarrow$ buffer M CHO_2^- then

$$M_1V_1 = M_2V_2 \qquad NaCHO_2(aq) \rightarrow Na^+(aq) + CHO_2^-(aq)$$

K_a, M $HCHO_2$, M CHO_2^- $\rightarrow$ pH

$$pH = pK_a + \log \frac{[\text{base}]}{[\text{acid}]}$$

Solution: total mL = mL $HCHO_2$ + mL $NaCHO_2$ = 50.0 mL + 75.0 mL = 125.0 mL
Then because $M_1V_1 = M_2V_2$, rearrange to solve for M_2.

$$M_2 = \frac{M_1V_1}{V_2} = \frac{(0.15 \text{ M})(50.0 \text{ mL})}{125.0 \text{ mL}} = 0.060 \text{ M } HCHO_2 \text{ and}$$

$$M_2 = \frac{M_1 V_1}{V_2} = \frac{(0.13 \text{ M})(75.0 \text{ mL})}{125.0 \text{ mL}} = 0.078 \text{ M NaCHO}_2. \text{ Acid} = \text{HCHO}_2, \text{ so } [\text{acid}] = [\text{HCHO}_2] =$$

0.060 M. Base $= \text{CHO}_2^-$. Because one CHO_2^- ion is generated for each NaCHO_2, $[\text{CHO}_2^-] =$

$0.078 \text{ M CHO}_2^- = [\text{base}]$. Then $\text{pH} = \text{p}K_a + \log \dfrac{[\text{base}]}{[\text{acid}]} = -\log (1.8 \times 10^{-4}) + \log \dfrac{0.078 \text{ M}}{0.060 \text{ M}} = 3.86.$

Check: The units (none) are correct. The magnitude of the answer makes physical sense because the pH should be greater than the pK_a of the acid since there is more base than acid.

(b)   **Given:** 125.0 mL of 0.10 M NH_3 and 250.0 mL of 0.10 M NH_4Cl **Find:** pH
      Other: $K_b(\text{NH}_3) = 1.76 \times 10^{-5}$
      Conceptual Plan: Identify acid and base components then mL NH_3, mL $\text{NH}_4\text{Cl} \rightarrow$ total mL then

$$\text{acid} = \text{NH}_4^+ \text{ base} = \text{NH}_3 \qquad\qquad \text{total mL} = \text{mL NH}_3 + \text{mL NH}_4\text{Cl}$$

mL NH_3, M NH_3, total mL $\rightarrow$ buffer M NH_3 and

$$M_1 V_1 = M_2 V_2$$

mL NH_4Cl, M NH_4Cl, total mL $\rightarrow$ buffer M $\text{NH}_4\text{Cl} \rightarrow$ buffer M NH_4^+ and $K_b \rightarrow \text{p}K_b \rightarrow \text{p}K_a$ then

$$M_1 V_1 = M_2 V_2 \qquad \text{NH}_4\text{Cl}(aq) \rightarrow \text{NH}_4^+(aq) + \text{Cl}^-(aq) \quad \text{p}K_b = -\log K_b \quad 14 = \text{p}K_a + \text{p}K_b$$

pK_a, M NH_3, M $\text{NH}_4^+ \rightarrow$ pH

$$\text{pH} = \text{p}K_a + \log \frac{[\text{base}]}{[\text{acid}]}$$

Solution: total mL $=$ mL $\text{NH}_3 +$ mL $\text{NH}_4\text{Cl} = 125.0$ mL $+ 250.0$ mL $= 375.0$ mL
Then because $M_1 V_1 = M_2 V_2$, rearrange to solve for M_2.

$$M_2 = \frac{M_1 V_1}{V_2} = \frac{(0.10 \text{ M})(125.0 \text{ mL})}{375.0 \text{ mL}} = 0.033333 \text{ M NH}_3 \text{ and}$$

$$M_2 = \frac{M_1 V_1}{V_2} = \frac{(0.10 \text{ M})(250.0 \text{ mL})}{375.0 \text{ mL}} = 0.066667 \text{ M NH}_4\text{Cl}. \text{ Base} = \text{NH}_3, [\text{base}] = [\text{NH}_3] =$$

0.033333 M acid $= \text{NH}_4^+$. Because one NH_4^+ ion is generated for each NH_4Cl, $[\text{NH}_4^+] =$
0.0666667 M $\text{NH}_4^+ = [\text{acid}]$. Because $K_b(\text{NH}_3) = 1.79 \times 10^{-5}$,
$\text{p}K_b = -\log K_b = -\log (1.76 \times 10^{-5}) = 4.75$. Because $14 = \text{p}K_a + \text{p}K_b$,

$$\text{p}K_a = 14 - \text{p}K_b = 14 - 4.75 = 9.25. \text{ Then pH} = \text{p}K_a + \log \frac{[\text{base}]}{[\text{acid}]} = 9.25 + \log \frac{0.033333 \text{ M}}{0.066667 \text{ M}} = 8.95.$$

Check: The units (none) are correct. The magnitude of the answer makes physical sense because the pH should be less than the pK_a of the acid since there is more acid than base.

17.41   **Given:** NaF/HF buffer at pH $= 4.00$ **Find:** $[\text{NaF}]/[\text{HF}]$ **Other:** $K_a(\text{HF}) = 3.5 \times 10^{-4}$
        Conceptual Plan: Identify acid and base components then pH, $K_a \rightarrow [\text{NaF}]/[\text{HF}]$

$$\text{acid} = \text{HF} \text{ base} = \text{F}^- \qquad\qquad \text{pH} = \text{p}K_a + \log \frac{[\text{base}]}{[\text{acid}]}$$

Solution: $\text{pH} = \text{p}K_a + \log \dfrac{[\text{base}]}{[\text{acid}]} = -\log (3.5 \times 10^{-4}) + \log \dfrac{[\text{NaF}]}{[\text{HF}]} = 4.00.$ Solve for $[\text{NaF}]/[\text{HF}]$.

$$\log \frac{[\text{NaF}]}{[\text{HF}]} = 4.00 - 3.46 = 0.54 \rightarrow \frac{[\text{NaF}]}{[\text{HF}]} = 10^{0.54} = 3.5.$$

Check: The units (none) are correct. The magnitude of the answer makes physical sense because the pH is greater than the pK_a of the acid; so there needs to be more base than acid.

17.43   **Given:** 150.0 mL buffer of 0.15 M benzoic acid at pH $= 4.25$ **Find:** mass sodium benzoate
        Other: $K_a(\text{HC}_7\text{H}_5\text{O}_2) = 6.5 \times 10^{-5}$
        Conceptual Plan: Identify acid and base components then pH, K_a, $[\text{HC}_7\text{H}_5\text{O}_2] \rightarrow [\text{NaC}_7\text{H}_5\text{O}_2]$

$$\text{acid} = \text{HC}_7\text{H}_5\text{O}_2 \text{ base} = \text{C}_7\text{H}_5\text{O}_2^- \qquad\qquad \text{pH} = \text{p}K_a + \log \frac{[\text{base}]}{[\text{acid}]}$$

mL $\rightarrow$ L then $[\text{NaC}_7\text{H}_5\text{O}_2]$, L $\rightarrow$ mol $\text{NaC}_7\text{H}_5\text{O}_2 \rightarrow$ g $\text{NaC}_7\text{H}_5\text{O}_2$

$$\frac{1 \text{ L}}{1000 \text{ mL}} \qquad\qquad M = \frac{\text{mol}}{\text{L}} \qquad\qquad \frac{144.10 \text{ g NaC}_7\text{H}_5\text{O}_2}{1 \text{ mol NaC}_7\text{H}_5\text{O}_2}$$

Solution: $pH = pK_a + \log\dfrac{[\text{base}]}{[\text{acid}]} = -\log(6.5 \times 10^{-5}) + \log\dfrac{[NaC_7H_5O_2]}{0.15 \text{ M}} = 4.25$. Solve for $[NaC_7H_5O_2]$.

$$\log\dfrac{[NaC_7H_5O_2]}{0.15 \text{ M}} = 4.25 - 4.1870866 = 0.06291 \rightarrow \dfrac{[NaC_7H_5O_2]}{0.15 \text{ M}} = 10^{0.06291} = 1.1559 \rightarrow$$

$[NaC_7H_5O_2] = 0.17338 \text{ M}$

Convert to moles using $M = \dfrac{\text{mol}}{L}$.

$$\dfrac{0.17338 \text{ mol } NaC_7H_5O_2}{1 \text{ L}} \times 0.150 \text{ L} = 0.026007 \text{ mol } NaC_7H_5O_2$$

$$0.026007 \text{ mol } NaC_7H_5O_2 \times \dfrac{144.10 \text{ g } NaC_7H_5O_2}{1 \text{ mol } NaC_7H_5O_2} = 3.7 \text{ g } NaC_7H_5O_2$$

Check: The units (g) are correct. The magnitude of the answer makes physical sense because the volume of solution is small and the concentration is low; so much less than a mole is needed.

17.45   (a)   **Given:** 250.0 mL buffer of 0.250 M $HC_2H_3O_2$ and 0.250 M $NaC_2H_3O_2$ **Find:** initial pH
     Other: $K_a(HC_2H_3O_2) = 1.8 \times 10^{-5}$
     Conceptual Plan: Identify acid and base components then M $NaC_2H_3O_2 \rightarrow$ M $C_2H_3O_2^-$ then

         acid = $HC_2H_3O_2$ base = $C_2H_3O_2^-$ $NaC_2H_3O_2(aq) \rightarrow Na^+(aq) + C_2H_3O_2^-(aq)$

     M $HC_2H_3O_2$, M $C_2H_3O_2^- \rightarrow$ pH

$$pH = pK_a + \log\dfrac{[\text{base}]}{[\text{acid}]}$$

     Solution: Acid = $HC_2H_3O_2$, so $[\text{acid}] = [HC_2H_3O_2] = 0.250$ M. Base = $C_2H_3O_2^-$. Because one $C_2H_3O_2^-$ ion is generated for each $NaC_2H_3O_2$, $[C_2H_3O_2^-] = 0.250$ M $C_2H_3O_2^- = [\text{base}]$. Then

$$pH = pK_a + \log\dfrac{[\text{base}]}{[\text{acid}]} = -\log(1.8 \times 10^{-5}) + \log\dfrac{0.250 \text{ M}}{0.250 \text{ M}} = 4.74.$$

     Check: The units (none) are correct. The magnitude of the answer makes physical sense because the pH is equal to the pK_a of the acid since there are equal amounts of acid and base.

   (b)   **Given:** 250.0 mL buffer of 0.250 M $HC_2H_3O_2$ and 0.250 M $NaC_2H_3O_2$, add 0.0050 mol HCl   **Find:** pH
     Other: $K_a(HC_2H_3O_2) = 1.8 \times 10^{-5}$
     Conceptual Plan: Part I: Stoichiometry:
     mL $\rightarrow$ L then $[NaC_2H_3O_2]$, L $\rightarrow$ mol $NaC_2H_3O_2$ and $[HC_2H_3O_2]$, L $\rightarrow$ mol $HC_2H_3O_2$

       $\dfrac{1 \text{ L}}{1000 \text{ mL}}$ $M = \dfrac{\text{mol}}{L}$ $M = \dfrac{\text{mol}}{L}$

     write balanced equation then

       $HCl + NaC_2H_3O_2 \rightarrow HC_2H_3O_2 + NaCl$

     mol $NaC_2H_3O_2$, mol $HC_2H_3O_2$, mol HCl $\rightarrow$ mol $NaC_2H_3O_2$, mol $HC_2H_3O_2$ then

                               set up stoichiometry table

     Part II: Equilibrium:
     mol $NaC_2H_3O_2$, mol $HC_2H_3O_2$, L, $K_a \rightarrow$ pH

$$pH = pK_a + \log\dfrac{[\text{base}]}{[\text{acid}]}$$

     Solution: $250.0 \text{ mL} \times \dfrac{1 \text{ L}}{1000 \text{ mL}} = 0.2500$ L then

$$\dfrac{0.250 \text{ mol } HC_2H_3O_2}{1 \text{ L}} \times 0.250 \text{ L} = 0.0625 \text{ mol } HC_2H_3O_2 \text{ and}$$

$$\dfrac{0.250 \text{ mol } NaC_2H_3O_2}{1 \text{ L}} \times 0.250 \text{ L} = 0.0625 \text{ mol } NaC_2H_3O_2.$$ Set up a table to track changes:

	$HCl(aq)$	+	$NaC_2H_3O_2(aq)$	$\rightarrow$	$HC_2H_3O_2(aq)$	+	$NaCl(aq)$
Before addition	≈ 0.00 mol		0.0625 mol		0.0625 mol		0.00 mol
Addition	0.0050 mol		—		—		—
After addition	≈ 0.00 mol		0.0575 mol		0.0675 mol		0.0050 mol

Because the amount of HCl is small, there are still significant amounts of both buffer components; so the Henderson–Hasselbalch equation can be used to calculate the new pH.

$$pH = pK_a + \log \frac{[\text{base}]}{[\text{acid}]} = -\log (1.8 \times 10^{-5}) + \log \frac{\dfrac{0.0575 \text{ mol}}{0.250 \text{ L}}}{\dfrac{0.0675 \text{ mol}}{0.250 \text{ L}}} = 4.68$$

Check: The units (none) are correct. The magnitude of the answer makes physical sense because the pH dropped slightly when acid was added.

(c)   **Given:** 250.0 mL buffer of 0.250 M $HC_2H_3O_2$ and 0.250 M $NaC_2H_3O_2$, add 0.0050 mol NaOH
Find: pH   **Other:** $K_a(HC_2H_3O_2) = 1.8 \times 10^{-5}$
Conceptual Plan: Part I: Stoichiometry:
mL → L then [$NaC_2H_3O_2$], L → mol $NaC_2H_3O_2$ and [$HC_2H_3O_2$], L → mol $HC_2H_3O_2$

$$\frac{1 \text{ L}}{1000 \text{ mL}} \qquad\qquad M = \frac{\text{mol}}{\text{L}} \qquad\qquad M = \frac{\text{mol}}{\text{L}}$$

write balanced equation then

$$NaOH + HC_2H_3O_2 \rightarrow H_2O + NaC_2H_3O_2$$

mol $NaC_2H_3O_2$, mol $HC_2H_3O_2$, mol NaOH → mol $NaC_2H_3O_2$, mol $HC_2H_3O_2$ then

set up stoichiometry table

Part II: Equilibrium:
mol $NaC_2H_3O_2$, mol $HC_2H_3O_2$, L, K_a → pH

$$pH = pK_a + \log \frac{[\text{base}]}{[\text{acid}]}$$

Solution: $250.0 \text{ mL} \times \dfrac{1 \text{ L}}{1000 \text{ mL}} = 0.2500 \text{ L}$ then

$$\frac{0.250 \text{ mol } HC_2H_3O_2}{1 \text{ L}} \times 0.2500 \text{ L} = 0.0625 \text{ mol } HC_2H_3O_2 \text{ and}$$

$$\frac{0.250 \text{ mol } NaC_2H_3O_2}{1 \text{ L}} \times 0.2500 \text{ L} = 0.0625 \text{ mol } NaC_2H_3O_2. \text{ Set up a table to track changes:}$$

	$NaOH(aq)$	$+$	$HC_2H_3O_2(aq)$	$\rightarrow$	$NaC_2H_3O_2(aq)$	$+$	$H_2O(l)$
Before addition	≈ 0.00 mol		0.0625 mol		0.0625 mol		—
Addition	0.0050 mol		—		—		—
After addition	≈ 0.00 mol		0.0575 mol		0.0675 mol		—

Because the amount of NaOH is small, there are still significant amounts of both buffer components; so the Henderson–Hasselbalch equation can be used to calculate the new pH.

$$pH = pK_a + \log \frac{[\text{base}]}{[\text{acid}]} = -\log (1.8 \times 10^{-5}) + \log \frac{\dfrac{0.0675 \text{ mol}}{0.2500 \text{ L}}}{\dfrac{0.0575 \text{ mol}}{0.2500 \text{ L}}} = 4.81$$

Check: The units (none) are correct. The magnitude of the answer makes physical sense because the pH rose slightly when base was added.

17.47   (a)   **Given:** 500.0 mL pure water   **Find:** initial pH and pH after adding 0.010 mol HCl
Conceptual Plan: Pure water has a pH of 7.00 then mL → L then mol HCl, L → [H_3O^+] → pH

$$\frac{1 \text{ L}}{1000 \text{ mL}} \qquad\qquad M = \frac{\text{mol}}{\text{L}} \qquad pH = -\log [H_3O^+]$$

Solution: Pure water has a pH of 7.00, so initial pH = 7.00, $500.0 \text{ mL} \times \dfrac{1 \text{ L}}{1000 \text{ mL}} = 0.5000 \text{ L}$,

then $M = \dfrac{\text{mol}}{\text{L}} = \dfrac{0.010 \text{ mol HCl}}{0.5000 \text{ L}} = 0.020 \text{ M HCl}$. Because HCl is a strong acid, it dissociates completely; so

$pH = -\log [H_3O^-] = -\log (0.020) = 1.70$.

Check: The units (none) are correct. The magnitude of the answers makes physical sense because the pH starts neutral and then drops significantly when acid is added and no buffer is present.

(b) **Given:** 500.0 mL buffer of 0.125 M $HC_2H_3O_2$ and 0.115 M $NaC_2H_3O_2$
Find: initial pH and pH after adding 0.010 mol HCl   **Other:** $K_a(HC_2H_3O_2) = 1.8 \times 10^{-5}$
Conceptual Plan: Initial pH:
Identify acid and base components then M $NaC_2H_3O_2 \rightarrow$ M $C_2H_3O_2^-$ then

$$\text{acid} = HC_2H_3O_2, \text{base} = C_2H_3O_2^- \quad NaC_2H_3O_2(aq) \rightarrow Na^+(aq) + C_2H_3O_2^-(aq)$$

M $HC_2H_3O_2$, M $C_2H_3O_2^- \rightarrow$ pH

$$pH = pK_a + \log \frac{[\text{base}]}{[\text{acid}]}$$

pH after HCl addition: Part I: Stoichiometry:
mL $\rightarrow$ L then $[NaC_2H_3O_2]$, L $\rightarrow$ mol $NaC_2H_3O_2$ and $[HC_2H_3O_2]$, L $\rightarrow$ mol $HC_2H_3O_2$

$$\frac{1 \text{ L}}{1000 \text{ mL}} \qquad M = \frac{\text{mol}}{L} \qquad M = \frac{\text{mol}}{L}$$

write balanced equation then

$$HCl + NaC_2H_3O_2 \rightarrow HC_2H_3O_2 + NaCl$$

mol $NaC_2H_3O_2$, mol $HC_2H_3O_2$, mol HCl $\rightarrow$ mol $NaC_2H_3O_2$, mol $HC_2H_3O_2$ then

set up stoichiometry table

Part II: Equilibrium:
mol $NaC_2H_3O_2$, mol $HC_2H_3O_2$, L, $K_a \rightarrow$ pH

$$pH = pK_a + \log \frac{[\text{base}]}{[\text{acid}]}$$

Solution: Initial pH: Acid = $HC_2H_3O_2$, so $[\text{acid}] = [HC_2H_3O_2] = 0.125$ M. Base = $C_2H_3O_2^-$. Because one $C_2H_3O_2^-$ ion is generated for each $NaC_2H_3O_2$, $[C_2H_3O_2^-] = 0.115$ M $C_2H_3O_2^- = [\text{base}]$. Then

$$pH = pK_a + \log \frac{[\text{base}]}{[\text{acid}]} = -\log (1.8 \times 10^{-5}) + \log \frac{0.115 \text{ M}}{0.125 \text{ M}} = 4.71.$$

pH after HCl addition:

$$500.0 \text{ mL} \times \frac{1 \text{ L}}{1000 \text{ mL}} = 0.5000 \text{ L then} \quad \frac{0.125 \text{ mol } HC_2H_3O_2}{1 \text{ L}} \times 0.5000 \text{ L} = 0.0625 \text{ mol } HC_2H_3O_2 \text{ and}$$

$$\frac{0.115 \text{ mol } NaC_2H_3O_2}{1 \text{ L}} \times 0.5000 \text{ L} = 0.0575 \text{ mol } NaC_2H_3O_2. \text{ Set up a table to track changes:}$$

	$HCl(aq)$	+	$NaC_2H_3O_2(aq)$	$\rightarrow$	$HC_2H_3O_2(aq)$	+	$NaCl(aq)$
Before addition	≈ 0.00 mol		0.0575 mol		0.0625 mol		0.00 mol
Addition	0.010 mol		—		—		—
After addition	≈ 0.00 mol		0.0475 mol		0.0725 mol		0.10 mol

Because the amount of HCl is small, there are still significant amounts of both buffer components; so the Henderson–Hasselbalch equation can be used to calculate the new pH.

$$pH = pK_a + \log \frac{[\text{base}]}{[\text{acid}]} = -\log (1.8 \times 10^{-5}) + \log \frac{\frac{0.0475 \text{ mol}}{0.5000 \text{ L}}}{\frac{0.0725 \text{ mol}}{0.5000 \text{ L}}} = 4.56$$

Check: The units (none) are correct. The magnitude of the answers makes physical sense because the pH started below the pK_a of the acid and it dropped slightly when acid was added.

(c) **Given:** 500.0 mL buffer of 0.155 M $C_2H_5NH_2$ and 0.145 M $C_2H_5NH_3Cl$
Find: initial pH and pH after adding 0.010 mol HCl   **Other:** $K_b(C_2H_5NH_2) = 5.6 \times 10^{-4}$
Conceptual Plan: Initial pH:
Identify acid and base components then M $C_2H_5NH_3Cl \rightarrow$ M $C_2H_5NH_3^+$ then

$$\text{acid} = C_2H_5NH_3^+ \text{ base} = C_2H_5NH_2 \quad C_2H_5NH_3Cl(aq) \rightarrow C_2H_5NH_3^+(aq) + Cl^-(aq)$$

and $K_b \rightarrow pK_b \rightarrow pK_a$ then pK_a, M $C_2H_5NH_2$, M $C_2H_5NH_3^+ \rightarrow$ pH

$$pK_b = -\log K_b \quad 14 = pK_a + pK_b \qquad pH = pK_a + \log \frac{[\text{base}]}{[\text{acid}]}$$

pH after HCl addition: Part I: Stoichiometry:

$mL \rightarrow L$ then $[C_2H_5NH_2], L \rightarrow$ mol $C_2H_5NH_2$ and

$$\frac{1\ L}{1000\ mL} \qquad M = \frac{mol}{L}$$

$[C_2H_5NH_3Cl], L \rightarrow$ mol $C_2H_5NH_3Cl$ **write balanced equation then**

$$M = \frac{mol}{L} \qquad HCl + C_2H_5NH_2 \rightarrow C_2H_5NH_3Cl$$

mol $C_2H_5NH_2$, mol $C_2H_5NH_3Cl$, mol HCl $\rightarrow$ mol $C_2H_5NH_2$, mol $C_2H_5NH_3Cl$ then

set up stoichiometry table

Part II: Equilibrium:

mol $C_2H_5NH_2$, mol $C_2H_5NH_3Cl$, L, $K_a \rightarrow$ pH

$$pH = pK_a + \log\frac{[base]}{[acid]}$$

Solution: Base $= C_2H_5NH_2$, so $[base] = [C_2H_5NH_2] = 0.155$ M. Acid $= C_2H_5NH_3{}^+$. Because one $C_2H_5NH_3{}^+$ ion is generated for each $C_2H_5NH_3Cl$, $[C_2H_5NH_3{}^+] = 0.145$ M $C_2H_5NH_3{}^+ = [acid]$. Because $K_b(C_2H_5NH_2) = 5.6 \times 10^{-4}$, $pK_b = -\log K_b = -\log(5.6 \times 10^{-4}) = 3.25$. Because $14 = pK_a + pK_b$, $pK_a = 14 - pK_b = 14 - 3.25 = 10.75$ then

$$pH = pK_a + \log\frac{[base]}{[acid]} = 10.75 + \log\frac{0.155\ M}{0.145\ M} = 10.78.$$

pH after HCl addition: $500.0\ \cancel{mL} \times \dfrac{1\ L}{1000\ \cancel{mL}} = 0.5000$ L then

$$\frac{0.155\ mol\ C_2H_5NH_2}{1\ \cancel{L}} \times 0.5000\ \cancel{L} = 0.0775\ mol\ C_2H_5NH_2\ and$$

$$\frac{0.145\ mol\ C_2H_5NH_3Cl}{1\ \cancel{L}} \times 0.5000\ \cancel{L} = 0.0725\ mol\ C_2H_5NH_3Cl.\ \text{Set up a table to track changes:}$$

	HCl(aq)	+	$C_2H_5NH_2(aq)$	$\rightarrow$	$C_2H_5NH_3Cl(aq)$
Before addition	$\approx$0.00 mol		0.0775 mol		0.0725 mol
Addition	0.010 mol		—		—
After addition	$\approx$0.00 mol		0.06$\underline{7}$5 mol		0.08$\underline{2}$5 mol

Because the amount of HCl is small, there are still significant amounts of both buffer components; so the Henderson–Hasselbalch equation can be used to calculate the new pH.

$$pH = pK_a + \log\frac{[base]}{[acid]} = 10.75 + \log\frac{\dfrac{0.06\underline{7}5\ \cancel{mol}}{\cancel{0.5000\ L}}}{\dfrac{0.08\underline{2}5\ \cancel{mol}}{\cancel{0.5000\ L}}} = 10.66$$

Check: The units (none) are correct. The magnitude of the answers makes physical sense because the initial pH should be greater than the pK_a of the acid because there is more base than acid and the pH drops slightly when acid is added.

17.49   **Given:** 350.00 mL 0.150 M HF and 0.150 M NaF buffer
Find: mass NaOH to raise pH to 4.00 and mass NaOH to raise pH to 4.00 with buffer concentrations raised to 0.350 M
Other: $K_a(HF) = 3.5 \times 10^{-4}$
Conceptual Plan: Identify acid and base components. Because $[NaF] = [HF]$, then initial pH $= pK_a$

$$acid = HF,\ base = F^- \qquad\qquad\qquad pH = pK_a$$

final pH, $pK_a \rightarrow [NaF]/[HF]$ and mL $\rightarrow$ L then $[HF], L \rightarrow$ mol HF and $[NaF], L \rightarrow$ mol NaF

$$pH = pK_a + \log\frac{[base]}{[acid]} \qquad \frac{1\ L}{1000\ mL} \qquad M = \frac{mol}{L} \qquad M = \frac{mol}{L}$$

then write balanced equation then

$$NaOH + HF \rightarrow NaF + H_2O$$

mol HF, mol NaF, $[NaF]/[HF] \rightarrow$ mol NaOH $\rightarrow$ g NaOH.

set up stoichiometry table $\qquad \dfrac{40.00\ g\ NaOH}{1\ mol\ NaOH}$

Finally, when the buffer concentrations are raised to 0.350 M, simply multiply the g NaOH by the ratio of concentrations (0.350 M/0.150 M).

Solution: initial pH $= pK_a = -\log(3.5 \times 10^{-4}) = 3.46$ then

$$pH = pK_a + \log \frac{[\text{base}]}{[\text{acid}]} = -\log(3.5 \times 10^{-4}) + \log \frac{[\text{NaF}]}{[\text{HF}]} = 4.00. \text{ Solve for [NaF]/[HF].}$$

$$\log \frac{[\text{NaF}]}{[\text{HF}]} = 4.00 - 3.46 = 0.54 \rightarrow \frac{[\text{NaF}]}{[\text{HF}]} = 10^{0.54} = 3.5.\ 350.0\ \cancel{\text{mL}} \times \frac{1\ \text{L}}{1000\ \cancel{\text{mL}}} = 0.3500\ \text{L then}$$

$$\frac{0.150\ \text{mol HF}}{1\ \cancel{L}} \times 0.3500\ \cancel{L} = 0.0525\ \text{mol HF and } \frac{0.150\ \text{mol NaF}}{1\ \cancel{L}} \times 0.3500\ \cancel{L} = 0.0525\ \text{mol NaF}$$

Set up a table to track changes:

	NaOH(aq)	+	HF(aq)	→	NaF(aq)	+	H$_2$O(aq)
Before addition	$\approx$0.00 mol		0.0525 mol		0.0525 mol		—
Addition	x		—		—		—
After addition	$\approx$0.00 mol		$(0.0525 - x)$ mol		$(0.0525 + x)$ mol		—

Because $\dfrac{[\text{NaF}]}{[\text{HF}]} = 3.5 = \dfrac{(0.0525 + x)\ \cancel{\text{mol}}}{(0.0525 - x)\ \cancel{\text{mol}}}$, solve for x. Note that the ratio of moles is the same as the ratio of concentrations because the volume for both terms is the same. $3.5(0.0525 - x) = (0.0525 + x) \rightarrow$ $0.1\underline{8}375 - 3.5x = 0.0525 + x \rightarrow 0.1\underline{3}125 = 4.5x \rightarrow x = 0.029\underline{1}67\ \text{mol NaOH then}$

$$0.029\underline{1}67\ \cancel{\text{mol NaOH}} \times \frac{40.00\ \text{g NaOH}}{1\ \cancel{\text{mol NaOH}}} = 1.\underline{1}667\ \text{g NaOH} = 1.2\ \text{g NaOH}$$

To scale the amount of NaOH up to a 0.350 M HF and NaF solution, multiply the NaOH mass by the ratio of concentrations. $1.1667\ \text{g NaOH} \times \dfrac{0.350\ \text{M}}{0.150\ \text{M}} = 2.7\ \text{g NaOH}$

Check: The units (g) are correct. The magnitude of the answers makes physical sense because there is much less than a mole of each of the buffer components; so there must be much less than a mole of NaOH. The higher the buffer concentrations, the higher the buffer capacity and the mass of NaOH it can neutralize.

17.51   (a)   Yes, this will be a buffer because NH$_3$ is a weak base and NH$_4^+$ is its conjugate acid. The ratio of base to acid is $0.10/0.15 = 0.67$, so the pH will be within 1 pH unit of the pK_a.

       (b)   No, this will not be a buffer solution because HCl is a strong acid and NaOH is a strong base.

       (c)   Yes, this will be a buffer because HF is a weak acid and the NaOH will convert $(20.0/50.0) \times 100\% = 40\%$ of the acid to its conjugate base.

       (d)   No, this will not be a buffer solution because both components are bases.

       (e)   No, this will not be a buffer solution because both components are bases.

17.53   (a)   **Given:** blood buffer 0.024 M HCO$_3^-$ and 0.0012 M H$_2$CO$_3$, $pK_a = 6.1$ **Find:** initial pH

       Conceptual Plan: Identify acid and base components then M HCO$_3^-$, M H$_2$CO$_3$ → pH

$$\text{acid} = \text{H}_2\text{CO}_3 \quad \text{base} = \text{HCO}_3^- \qquad\qquad pH = pK_a + \log\frac{[\text{base}]}{[\text{acid}]}$$

       Solution: Acid $= \text{H}_2\text{CO}_3$, so $[\text{acid}] = [\text{H}_2\text{CO}_3] = 0.0012\ \text{M.}$ Base $= \text{HCO}_3^-$, so $[\text{base}] = [\text{HCO}_3^-] = 0.024\ \text{M HCO}_3^-$. Then $pH = pK_a + \log\dfrac{[\text{base}]}{[\text{acid}]} = 6.1 + \log\dfrac{0.024\ \text{M}}{0.0012\ \text{M}} = 7.4$.

       Check: The units (none) are correct. The magnitude of the answer makes physical sense because the pH is greater than the pK_a of the acid since there is more base than acid.

       (b)   **Given:** 5.0 L of blood buffer   **Find:** mass HCl to lower pH to 7.0

       Conceptual Plan: Final pH, pK_a → $[\text{HCO}_3^-]/[\text{H}_2\text{CO}_3]$ then $[\text{HCO}_3^-]$, L → mol HCO$_3^-$ and

$$pH = pK_a + \log\frac{[\text{base}]}{[\text{acid}]} \qquad\qquad M = \frac{\text{mol}}{\text{L}}$$

       $[\text{H}_2\text{CO}_3]$, L → mol H$_2$CO$_3$ then write balanced equation then

$$M = \frac{\text{mol}}{\text{L}} \qquad\qquad \text{H}^+ + \text{HCO}_3^- \rightarrow \text{H}_2\text{CO}_3$$

mol HCO$_3^-$, mol H$_2$CO$_3$, [HCO$_3^-$]/[H$_2$CO$_3$] → mol HCl → g HCl

set up stoichiometry table $\dfrac{36.46 \text{ g HCl}}{1 \text{ mol HCl}}$

Solution: $\text{pH} = \text{p}K_a + \log \dfrac{[\text{base}]}{[\text{acid}]} = 6.1 + \log \dfrac{[\text{HCO}_3^-]}{[\text{H}_2\text{CO}_3]} = 7.0$. Solve for $[\text{HCO}_3^-]/[\text{H}_2\text{CO}_3]$.

$\log \dfrac{[\text{HCO}_3^-]}{[\text{H}_2\text{CO}_3]} = 7.0 - 6.1 = 0.9 \rightarrow \dfrac{[\text{HCO}_3^-]}{[\text{H}_2\text{CO}_3]} = 10^{0.9} = 7.9433$. Then

$\dfrac{0.024 \text{ mol HCO}_3^-}{1 \text{ L}} \times 5.0 \text{ L} = 0.12 \text{ mol HCO}_3^-$ and $\dfrac{0.0012 \text{ mol H}_2\text{CO}_3}{1 \text{ L}} \times 5.0 \text{ L} = 0.0060 \text{ mol H}_2\text{CO}_3$.

Because HCl is a strong acid, $[\text{HCl}] = [\text{H}^+]$. Set up a table to track changes:

	H$^+$(aq)	+	HCO$_3^-$(aq)	→	H$_2$CO$_3$(aq)
Before addition	≈ 0.00 mol		0.12 mol		0.0060 mol
Addition	x		—		—
After addition	≈ 0.00 mol		(0.12 − x) mol		(0.0060 + x) mol

Because $\dfrac{[\text{HCO}_3^-]}{[\text{H}_2\text{CO}_3]} = 7.9433 = \dfrac{(0.12 - x) \text{ mol}}{(0.0060 + x) \text{ mol}}$, solve for x. Note that the ratio of moles is the same as the ratio of concentrations because the volume for both terms is the same.

$7.9433(0.0060 + x) = (0.12 - x) \rightarrow 0.0476598 + 7.9433x = 0.12 - x \rightarrow 8.9433x = 0.07234 \rightarrow$

$x = 0.0080888 \text{ mol HCl then } 0.0080888 \text{ mol HCl} \times \dfrac{36.46 \text{ g HCl}}{1 \text{ mol HCl}} = 0.29492 \text{ g HCl} = 0.3 \text{ g HCl}$

Check: The units (g) are correct. The amount of acid needed is small because the concentrations of the buffer components are very low and the buffer starts only 0.4 pH unit above the final pH.

(c)   **Given:** 5.0 L of blood buffer   **Find:** mass NaOH to raise pH to 7.8
Conceptual Plan: Final pH, pK_a → [HCO$_3^-$]/[H$_2$CO$_3$] then [HCO$_3^-$], L → mol HCO$_3^-$ and

$$\text{pH} = \text{p}K_a + \log \frac{[\text{base}]}{[\text{acid}]} \qquad\qquad M = \frac{\text{mol}}{\text{L}}$$

[H$_2$CO$_3$], L → mol H$_2$CO$_3$ then write balanced equation then

$$M = \frac{\text{mol}}{\text{L}} \qquad\qquad \text{OH}^- + \text{H}_2\text{CO}_3 \rightarrow \text{HCO}_3^- + \text{H}_2\text{O}$$

mol HCO$_3^-$, mol H$_2$CO$_3$, [HCO$_3^-$]/[H$_2$CO$_3$] → mol NaOH → g NaOH

set up stoichiometry table $\dfrac{40.00 \text{ g NaOH}}{1 \text{ mol NaOH}}$

Solution: $\text{pH} = \text{p}K_a + \log \dfrac{[\text{base}]}{[\text{acid}]} = 6.1 + \log \dfrac{[\text{HCO}_3^-]}{[\text{H}_2\text{CO}_3]} = 7.8$

Solve for $[\text{HCO}_3^-]/[\text{H}_2\text{CO}_3]$. $\log \dfrac{[\text{HCO}_3^-]}{[\text{H}_2\text{CO}_3]} = 7.8 - 6.1 = 1.7 \rightarrow \dfrac{[\text{HCO}_3^-]}{[\text{H}_2\text{CO}_3]} = 10^{1.7} = 50.11872$. Then

$\dfrac{0.024 \text{ mol HCO}_3^-}{1 \text{ L}} \times 5.0 \text{ L} = 0.12 \text{ mol HCO}_3^-$ and $\dfrac{0.0012 \text{ mol H}_2\text{CO}_3}{1 \text{ L}} \times 5.0 \text{ L} = 0.0060 \text{ mol H}_2\text{CO}_3$.

Because NaOH is a strong base, $[\text{NaOH}] = [\text{OH}^-]$. Set up a table to track changes:

	OH$^-$(aq)	+	H$_2$CO$_3$(aq)	→	HCO$_3^-$(aq)	+	H$_2$O(l)
Before addition	≈ 0.00 mol		0.0060 mol		0.12 mol		—
Addition	x		—		—		
After addition	≈ 0.00 mol		(0.0060 − x) mol		(0.12 + x) mol		—

Because $\dfrac{[\text{HCO}_3^-]}{[\text{H}_2\text{CO}_3]} = 50.11872 = \dfrac{(0.12 + x) \text{ mol}}{(0.0060 - x) \text{ mol}}$, solve for x. Note that the ratio of moles is the same as the ratio of concentrations because the volume for both terms is the same.

$50.11872(0.0060 - x) = (0.12 + x) \rightarrow 0.30071 - 50.11872x = 0.12 + x \rightarrow 51.11872x = 0.18071$
$\rightarrow x = 0.0035351 \text{ mol NaOH then}$

$$0.0035351 \text{ mol NaOH} \times \frac{40.00 \text{ g NaOH}}{1 \text{ mol NaOH}} = 0.14140 \text{ g NaOH} = 0.14 \text{ g NaOH}$$

Check: The units (g) are correct. The amount of base needed is small because the concentrations of the buffer components are very low.

17.55 **Given:** $HC_2H_3O_2/KC_2H_3O_2$, $HClO_2/KClO_2$, NH_3/NH_4Cl, and $HClO/KClO$ potential buffer systems to create buffer at $pH = 7.20$ **Find:** best buffer system and ratio of component masses
Other: $K_a(HC_2H_3O_2) = 1.8 \times 10^{-5}$, $K_a(HClO_2) = 1.8 \times 10^{-4}$, $K_b(NH_3) = 1.76 \times 10^{-5}$, $K_a(HClO) = 2.9 \times 10^{-8}$
Conceptual Plan: Calculate pK_a of all potential buffer acids for the base $K_b \rightarrow pK_b \rightarrow pK_a$ and

$$pK_a = -\log K_a \qquad\qquad pK_b = -\log K_b \quad 14 = pK_a + pK_b$$

choose the pK_a that is closest to 7.20 then pH, $K_a \rightarrow [\text{base}]/[\text{acid}] \rightarrow$ **mass base/mass acid**

$$pH = pK_a + \log \frac{[\text{base}]}{[\text{acid}]} \qquad \frac{\mathcal{M}(\text{base})}{\mathcal{M}(\text{acid})}$$

Solution: For $HC_2H_3O_2/KC_2H_3O_2$: $pK_a = -\log K_a = -\log(1.8 \times 10^{-5}) = 4.74$
For $HClO_2/KClO_2$: $pK_a = -\log K_a = -\log(1.8 \times 10^{-4}) = 3.74$
For NH_3/NH_4Cl: $pK_b = -\log K_b = -\log(1.76 \times 10^{-5}) = 4.75$
Because $14 = pK_a + pK_b$, $pK_a = 14 - pK_b = 14 - 4.75 = 9.25$.
And for $HClO/KClO$: $pK_a = -\log K_a = -\log(2.9 \times 10^{-8}) = 7.54$ So the HClO/KClO buffer system has the pK_a

that is closest to 7.20. So $pH = pK_a + \log \dfrac{[\text{base}]}{[\text{acid}]} = 7.54 + \log \dfrac{[\text{KClO}]}{[\text{HClO}]} = 7.20$. Solve for $[\text{KClO}]/[\text{HClO}]$.

$$\log \frac{[\text{KClO}]}{[\text{HClO}]} = 7.20 - 7.54 = -0.34 \rightarrow \frac{[\text{KClO}]}{[\text{HClO}]} = 10^{-0.34} = 0.457088 \quad \text{Then convert to mass ratio using}$$

$$\frac{\mathcal{M}(\text{base})}{\mathcal{M}(\text{acid})}, 0.457088 \; \frac{\dfrac{\text{KClO mol}}{L}}{\dfrac{\text{HClO mol}}{L}} \times \frac{90.55 \text{ g KClO}}{\text{mol KClO}} \bigg/ \frac{52.46 \text{ g HClO}}{\text{mol HClO}} = 0.79 \; \frac{\text{g KClO}}{\text{g HClO}}.$$

Check: The units (none and g base/g acid) are correct. The buffer system with the K_a closest to 10^{-7} is the best choice. The magnitude of the answer makes physical sense because the buffer needs more acid than base (and this fact is not overcome by the heavier molar mass of the base).

17.57 **Given:** 500.0 mL of 0.100 M HNO_2/0.150 M KNO_2 buffer and (a) 250 mg NaOH, (b) 350 mg KOH, (c) 1.25 g HBr, and (d) 1.35 g HI **Find:** whether buffer capacity is exceeded
Conceptual Plan: mL $\rightarrow$ L then $[HNO_2]$, L $\rightarrow$ mol HNO_2 and $[KNO_2]$, L $\rightarrow$ mol KNO_2

$$\frac{1 \text{ L}}{1000 \text{ mL}} \qquad\qquad M = \frac{\text{mol}}{L} \qquad\qquad M = \frac{\text{mol}}{L}$$

then calculate moles of acid or base to be added to the buffer mg $\rightarrow$ g $\rightarrow$ mol then

$$\frac{1 \text{ g}}{1000 \text{ mg}} \quad \mathcal{M}$$

compare the added amount to the buffer amount of the opposite component. **Ratio of base/acid must be between 0.1 and 10 to maintain the buffer integrity.**

Solution: $500.00 \text{ mL} \times \dfrac{1 \text{ L}}{1000 \text{ mL}} = 0.5000 \text{ L}$ then $\dfrac{0.100 \text{ mol } HNO_2}{1 \text{ L}} \times 0.5000 \text{ L} = 0.0500 \text{ mol } HNO_2$ and

$$\frac{0.150 \text{ mol } KNO_2}{1 \text{ L}} \times 0.5000 \text{ L} = 0.0750 \text{ mol } KNO_2$$

(a) For NaOH: $250 \text{ mg NaOH} \times \dfrac{1 \text{ g NaOH}}{1000 \text{ mg NaOH}} \times \dfrac{1 \text{ mol NaOH}}{40.00 \text{ g NaOH}} = 0.00625 \text{ mol NaOH}$. Because the buffer

contains 0.0500 mol acid, the amount of acid is reduced by $0.00625/0.0500 = 0.125$ and the ratio of base/acid is still between 0.1 and 10. The buffer capacity is not exceeded.

(b) For KOH: $350 \text{ mg KOH} \times \dfrac{1 \text{ g KOH}}{1000 \text{ mg KOH}} \times \dfrac{1 \text{ mol KOH}}{56.11 \text{ g KOH}} = 0.00624 \text{ mol KOH}$. Because the buffer

contains 0.0500 mol acid, the amount of acid is reduced by $0.00624/0.0500 = 0.125$ and the ratio of base/acid is still between 0.1 and 10. The buffer capacity is not exceeded.

(c)   For HBr: $1.25 \text{ g HBr} \times \dfrac{1 \text{ mol HBr}}{80.91 \text{ g HBr}} = 0.0154493 \text{ mol HBr}$. Because the buffer contains 0.0750 mol base,

the amount of acid is reduced by $0.0154/0.0750 = 0.206$ and the ratio of base/acid is still between 0.1 and 10. The buffer capacity is not exceeded.

(d)   For HI: $1.35 \text{ g HI} \times \dfrac{1 \text{ mol HI}}{127.91 \text{ g HI}} = 0.0105543 \text{ mol HI}$. Because the buffer contains 0.0750 mol base, the

amount of acid is reduced by $0.0106/0.0750 = 0.141$ and the ratio of base/acid is still between 0.1 and 10. The buffer capacity is not exceeded.

Titrations, pH Curves, and Indicators

17.59   (i)   The equivalence point of a titration is where the pH rises sharply as base is added. The pH at the equivalence point is the midpoint of the sharp rise at ~50 mL added base. For (a), the pH = ~8, and for (b), the pH = ~7.

       (ii)   Graph (a) represents a weak acid, and graph (b) represents a strong acid. A strong acid titration starts at a lower pH, has a flatter initial region, and has a sharper rise at the equivalence point than does a weak acid. The pH at the equivalence point of a strong acid is neutral, while the pH at the equivalence point of a weak acid is basic.

17.61   **Given:** 20.0 mL 0.200 M KOH and 0.200 M CH_3NH_2 titrated with 0.100 M HI

      (a)   **Find:** volume of base to reach equivalence point

          Conceptual Plan: The answer for both titrations will be the same because the initial concentration and volumes of the bases are the same. Write balanced equation then mL → L then

$$HI + KOH \rightarrow KI + H_2O \text{ and } HI + CH_3NH_2 \rightarrow CH_3NH_3I \qquad \dfrac{1 \text{ L}}{1000 \text{ mL}}$$

          [base], L → mol base then set mol base = mol acid and [HI], mol HI → L HI → mL HI

$$M = \dfrac{\text{mol}}{\text{L}} \qquad \text{balanced equation has 1:1 stoichiometry} \qquad M = \dfrac{\text{mol}}{\text{L}} \quad \dfrac{1000 \text{ mL}}{1 \text{ L}}$$

          Solution: $20.0 \text{ mL base} \times \dfrac{1 \text{ L}}{1000 \text{ mL}} = 0.0200 \text{ L base}$ then

$$\dfrac{0.200 \text{ mol base}}{1 \text{ L}} \times 0.0200 \text{ L} = 0.00400 \text{ mol base}. \text{ So mol base} = 0.00400 \text{ mol} = \text{mol HI then}$$

$$0.00400 \text{ mol HI} \times \dfrac{1 \text{ L HI}}{0.100 \text{ mol HI}} = 0.0400 \text{ L HI} \times \dfrac{1000 \text{ mL}}{1 \text{ L}} = 40.0 \text{ mL HI for both titrations}$$

          Check: The units (mL) are correct. The volume of acid is twice the volume of bases because the concentration of the base is twice that of the acid in each case. The answer for both titrations is the same because the stoichiometry is the same for both titration reactions.

      (b)   The pH at the equivalence point will be neutral for KOH (because it is a strong base), and it will be acidic for CH_3NH_2 (because it is a weak base and will produce a conjugate acid when titrated).

      (c)   The initial pH will be lower for CH_3NH_2 (because it is a weak base and will only partially dissociate) and not raise the pH as high as KOH will (because it is a strong base and so dissociates completely) at the same base concentration.

      (d)   The titration curves will look like the following:

KOH:

CH_3NH_2:

Important features to include are a high initial pH (pH is over 13 for a strong base and is lower for a weak base), flat initial region (very flat for strong base, not as flat for weak base where pH halfway to equivalence point is the pK_b of the base), sharp drop at equivalence point, pH at equivalence point (neutral for strong base and lower for weak base), and then flattening out at low pH.

17.63   (a)   The equivalence point of a titration is where the pH rises sharply as base is added. The volume at the equivalence point is ~30 mL. The pH at the equivalence point is the midpoint of the sharp rise at ~30 mL added base, which is a pH = ~9.

(b)   At 0 mL, the pH is calculated by doing an equilibrium calculation of a weak acid in water (as done in Chapter 16).

(c)   The pH halfway to the equivalence point is equal to the pK_a of the acid, or ~15 mL.

(d)   The pH at the equivalence point, or ~30 mL, is calculated by doing an equilibrium problem with the K_b of the acid. At the equivalence point, all of the acid has been converted to its conjugate base.

(e)   Beyond the equivalence point (30 mL), there is excess base. All of the acid has been converted to its conjugate base, so the pH is calculated by focusing on this excess base concentration.

17.65   **Given:** 35.0 mL of 0.175 M HBr titrated with 0.200 M KOH

(a)   **Find:** initial pH
Conceptual Plan: Because HBr is a strong acid, it will dissociate completely; so initial
$$\text{pH} = -\log[\text{H}_3\text{O}^+] = -\log[\text{HBr}].$$
Solution: $\text{pH} = -\log[\text{HBr}] = -\log 0.175 = 0.757$
Check: The units (none) are correct. The pH is reasonable because the concentration is greater than 0.1 M, and when the acid dissociates completely, the pH becomes less than 1.

(b)   **Find:** volume of base to reach equivalence point
Conceptual Plan: Write balanced equation then mL $\rightarrow$ L **then** [HBr], L $\rightarrow$ mol HBr **then**

$$\text{HBr} + \text{KOH} \rightarrow \text{KBr} + \text{H}_2\text{O} \qquad \frac{1\ \text{L}}{1000\ \text{mL}} \qquad \text{M} = \frac{\text{mol}}{\text{L}}$$

set mol acid (HBr) = mol base (KOH) and [KOH], mol KOH $\rightarrow$ L KOH $\rightarrow$ mL KOH

$$\text{balanced equation has 1:1 stoichiometry} \qquad \text{M} = \frac{\text{mol}}{\text{L}} \qquad \frac{1000\ \text{mL}}{1\ \text{L}}$$

Solution: $35.0\ \text{mL HBr} \times \dfrac{1\ \text{L}}{1000\ \text{mL}} = 0.0350\ \text{L HBr}$ then

$$\frac{0.175\ \text{mol HBr}}{1\ \text{L}} \times 0.0350\ \text{L} = 0.006125\ \text{mol HBr}$$

So mol acid = mol HBr = 0.006125 mol = mol KOH then

$$0.006125\ \text{mol KOH} \times \frac{1\ \text{L}}{0.200\ \text{mol KOH}} = 0.030625\ \text{L KOH} \times \frac{1000\ \text{mL}}{1\ \text{L}} = 30.6\ \text{mL KOH}$$

Check: The units (mL) are correct. The volume of base is a little less than the volume of acid because the concentration of the base is a little greater than that of the acid.

(c)   **Find:** pH after adding 10.0 mL of base
Conceptual Plan: Use calculations from part (b). Then mL $\rightarrow$ L **then** [KOH], L $\rightarrow$ mol KOH **then**

$$\frac{1\ \text{L}}{1000\ \text{mL}} \qquad \text{M} = \frac{\text{mol}}{\text{L}}$$

mol HBr, mol KOH $\rightarrow$ **mol excess HBr and** L HBr, L KOH $\rightarrow$ total L **then**

$$\text{set up stoichiometry table} \qquad \text{L HBr} + \text{L KOH} = \text{total L}$$

mol excess HBr, L $\rightarrow$ [HBr] $\rightarrow$ pH.

$$\text{M} = \frac{\text{mol}}{\text{L}} \qquad \text{pH} = -\log[\text{HBr}]$$

Solution: $10.0\ \text{mL KOH} \times \dfrac{1\ \text{L}}{1000\ \text{mL}} = 0.0100\ \text{L KOH}$ then

$$\frac{0.200\ \text{mol KOH}}{1\ \text{L}} \times 0.0100\ \text{L} = 0.00200\ \text{mol KOH}$$

Because KOH is a strong base, [KOH] = [OH$^-$]. Set up a table to track changes:

	KOH(aq)	+	HBr(aq)	$\rightarrow$	KBr(aq)	+	H$_2$O(l)
Before addition	$\approx$0.00 mol		0.006125 mol		0.00 mol		—
Addition	0.00200 mol		—		—		—
After addition	$\approx$0.00 mol		0.004125 mol		0.00200 mol		—

Then 0.0350 L HBr + 0.0100 L KOH = 0.0450 L total volume.

So mol excess acid = mol HBr = 0.004125 mol in 0.0450 L so

$$[\text{HBr}] = \frac{0.004125 \text{ mol HBr}}{0.0450 \text{ L}} = 0.0916667 \text{ M and}$$

pH = $-\log[\text{HBr}] = -\log 0.0916667 = 1.038$

Check: The units (none) are correct. The pH is a little higher than the initial pH, which is expected because this is a strong acid.

(d)   **Find:** pH at equivalence point

Solution: Because this is a strong acid–strong base titration, the pH at the equivalence point is neutral, or 7.

(e)   **Find:** pH after adding 5.0 mL of base beyond the equivalence point

Conceptual Plan: Use calculations from parts (b) and (c). Then the pH is only dependent on the amount of excess base and the total solution volumes.

mL excess → L excess then [KOH], L excess → mol KOH excess

$$\frac{1 \text{ L}}{1000 \text{ mL}} \qquad\qquad M = \frac{\text{mol}}{\text{L}}$$

then L HBr, L KOH to equivalence point, L KOH excess → total L then

L HBr + L KOH to equivalence point + L KOH excess = total L

mol excess KOH, total L → [KOH] = [OH⁻] → [H₃O⁺] → pH

$$M = \frac{\text{mol}}{\text{L}} \qquad K_w = [\text{H}_3\text{O}^+][\text{OH}^-] \qquad \text{pH} = -\log[\text{H}_3\text{O}^+]$$

Solution: 5.0 mL KOH $\times \dfrac{1 \text{ L}}{1000 \text{ mL}} = 0.0050$ L KOH excess then

$\dfrac{0.200 \text{ mol KOH}}{1 \text{ L}} \times 0.0050 \text{ L} = 0.0010$ mol KOH excess. Then 0.0350 L HBr + 0.0306 L KOH + 0.0050 L

KOH = 0.0706 L total volume. $[\text{KOH excess}] = \dfrac{0.0010 \text{ mol KOH excess}}{0.0706 \text{ L}} = 0.014164$ M KOH excess.

Because KOH is a strong base, $[\text{KOH}]$ excess = $[\text{OH}^-]$. $K_w = [\text{H}_3\text{O}^+][\text{OH}^-]$, so

$$[\text{H}_3\text{O}^+] = \frac{K_w}{[\text{OH}^-]} = \frac{1.0 \times 10^{-14}}{0.014164} = 7.06 \times 10^{-13} \text{ M}.$$

Finally, pH = $-\log[\text{H}_3\text{O}^+] = -\log(7.06 \times 10^{-13}) = 12.15$.

Check: The units (none) are correct. The pH is rising sharply at the equivalence point, so the pH after 5 mL past the equivalence point should be quite basic.

17.67   **Given:** 25.0 mL of 0.115 M RbOH titrated with 0.100 M HCl

(a)   **Find:** initial pH

Conceptual Plan: Because RbOH is a strong base, it will dissociate completely, so

[RbOH] = [OH⁻] → [H₃O⁺] → pH

$$K_w = [\text{H}_3\text{O}^+][\text{OH}^-] \qquad \text{pH} = -\log[\text{H}_3\text{O}^+]$$

Solution: Because RbOH is a strong base, $[\text{RbOH}]$ excess = $[\text{OH}^-]$. $K_w = [\text{H}_3\text{O}^+][\text{OH}^-]$, so

$$[\text{H}_3\text{O}^+] = \frac{K_w}{[\text{OH}^-]} = \frac{1.0 \times 10^{-14}}{0.115} = 8.69565 \times 10^{-14} \text{ M and}$$

pH = $-\log[\text{H}_3\text{O}^+] = -\log(8.69565 \times 10^{-14}) = 13.06$.

Check: The units (none) are correct. The pH is reasonable because the concentration is greater than 0.1 M, and when the base dissociates completely, the pH becomes greater than 13.

(b)　**Find:** volume of acid to reach equivalence point
　　Conceptual Plan: Write balanced equation then mL $\rightarrow$ L then [RbOH], L $\rightarrow$ mol RbOH then

$$HCl + RbOH \rightarrow RbCl + H_2O \qquad \frac{1\,L}{1000\,mL} \qquad M = \frac{mol}{L}$$

　　set mol base (RbOH) $=$ mol acid (HCl) and [HCl], mol HCl $\rightarrow$ L HCl $\rightarrow$ mL HCl

$$\text{balanced equation has 1:1 stoichiometry} \qquad M = \frac{mol}{L} \qquad \frac{1000\,mL}{1\,L}$$

　　Solution: $25.0\ \text{mL RbOH} \times \dfrac{1\,L}{1000\,mL} = 0.0250\ \text{L RbOH}$ then

$$\frac{0.115\ \text{mol RbOH}}{1\,L} \times 0.0250\ L = 0.002875\ \text{mol RbOH. So mol base} = \text{mol RbOH} = 0.002875\ \text{mol} = \text{mol}$$

　　HCl then $0.002875\ \text{mol HCl} \times \dfrac{1\,L}{0.100\ \text{mol HCl}} = 0.02875\ \text{L HCl} \times \dfrac{1000\,mL}{1\,L} = 28.8\ \text{mL HCl.}$

　　Check: The units (mL) are correct. The volume of acid is greater than the volume of base because the concentration of the base is a little greater than that of the acid.

(c)　**Find:** pH after adding 5.0 mL of acid
　　Conceptual Plan: Use calculations from part (b). Then mL $\rightarrow$ L then [HCl], L $\rightarrow$ mol HCl then

$$\frac{1\,L}{1000\,mL} \qquad M = \frac{mol}{L}$$

　　mol RbOH, mol HCl $\rightarrow$ mol excess RbOH and L RbOH, L HCl $\rightarrow$ total L then

$$\text{set up stoichiometry table} \qquad \text{L RbOH} + \text{L HCl} = \text{total L}$$

　　mol excess RbOH, L $\rightarrow$ [RbOH] $=$ [OH$^-$] $\rightarrow$ [H$_3$O$^+$] $\rightarrow$ pH.

$$M = \frac{mol}{L} \qquad K_w = [H_3O^+][OH^-] \qquad pH = -\log\,[H_3O^+]$$

　　Solution: $5.0\ \text{mL HCl} \times \dfrac{1\,L}{1000\,mL} = 0.0050\ \text{L HCl}$ then $\dfrac{0.100\ \text{mol HCl}}{1\,L} \times 0.0050\ L = 0.00050\ \text{mol HCl}$

　　Because HCl is a strong acid, $[\text{HCl}] = [\text{H}_3\text{O}^+]$. Set up a table to track changes:

	HCl(*aq*)	+	RbOH(*aq*)	$\rightarrow$	RbCl(*aq*)	+	H$_2$O(*l*)
Before addition	0.00 mol		0.002875 mol		0.00 mol		—
Addition	0.00050 mol		—		—		—
After addition	$\approx$0.00 mol		0.002375 mol		0.00050 mol		—

　　Then $0.0250\ \text{L RbOH} + 0.0050\ \text{L HCl} = 0.0300\ \text{L total volume. So mol excess base} = \text{mol RbOH} =$

　　0.002375 mol in 0.0300 L, so $[\text{RbOH}] = \dfrac{0.002375\ \text{mol RbOH}}{0.0300\ \text{L}} = 0.0791667$ M. Because RbOH is a strong

　　base, $[\text{RbOH}]$ excess $= [\text{OH}^-]$. $K_w = [\text{H}_3\text{O}^+][\text{OH}^-]$ so $[\text{H}_3\text{O}^+] = \dfrac{K_w}{[\text{OH}^-]} = \dfrac{1.0 \times 10^{-14}}{0.0791667}$

　　$= 1.26316 \times 10^{-13}$ M and pH $= -\log\,[\text{H}_3\text{O}^+] = -\log\,(1.26316 \times 10^{-13}) = 12.90$

　　Check: The units (none) are correct. The pH is a little lower than the initial pH, which is expected because this is a strong base.

(d)　**Find:** pH at equivalence point
　　Solution: Because this is a strong acid–strong base titration, the pH at the equivalence point is neutral, or 7.

(e)　**Find:** pH after adding 5.0 mL of acid beyond the equivalence point
　　Conceptual Plan: Use calculations from parts (b) and (c). Then the pH is only dependent on the amount of excess acid and the total solution volumes. Then mL excess $\rightarrow$ L excess then [HCl], L excess $\rightarrow$ mol HCl excess

$$\frac{1\,L}{1000\,mL} \qquad M = \frac{mol}{L}$$

　　then L RbOH, L HCl to equivalence point, L HCl excess $\rightarrow$ total L then

$$\text{L RbOH} + \text{L HCl to equivalence point} + \text{L HCl excess} = \text{total L}$$

　　mol excess HCl, total L $\rightarrow$ [HCl] $=$ [H$_3$O$^+$] $\rightarrow$ pH.

$$M = \frac{mol}{L} \qquad pH = -\log\,[H_3O^+]$$

Solution: $5.0 \; \cancel{\text{mL HCl}} \times \dfrac{1 \, \text{L}}{1000 \; \cancel{\text{mL}}} = 0.0050 \, \text{L HCl excess}$ then

$\dfrac{0.100 \, \text{mol HCl}}{1 \; \cancel{\text{L}}} \times 0.0050 \; \cancel{\text{L}} = 0.00050 \, \text{mol HCl excess}$. Then $0.0250 \, \text{L RbOH} + 0.0288 \, \text{L HCl} + 0.0050 \, \text{L}$

$\text{HCl} = 0.0588 \, \text{L total volume.} \; [\text{HCl excess}] = \dfrac{0.00050 \, \text{mol HCl excess}}{0.0588 \, \text{L}} = 0.008\underline{5}034 \, \text{M HCl excess}$

Because HCl is a strong acid, $[\text{HCl}]$ excess $= [\text{H}_3\text{O}^+]$.

Finally, $\text{pH} = -\log[\text{H}_3\text{O}^+] = -\log(0.008\underline{5}034) = 2.07$.

Check: The units (none) are correct. The pH is dropping sharply at the equivalence point, so the pH after 5 mL past the equivalence point should be quite acidic.

17.69   **Given:** 20.0 mL of 0.105 M $\text{HC}_2\text{H}_3\text{O}_2$ titrated with 0.125 M NaOH   **Other:** $K_a(\text{HC}_2\text{H}_3\text{O}_2) = 1.8 \times 10^{-5}$

    (a)   **Find:** initial pH

         Conceptual Plan: Because $\text{HC}_2\text{H}_3\text{O}_2$ is a weak acid, set up an equilibrium problem using the initial concentration.

         So M $\text{HC}_2\text{H}_3\text{O}_2 \rightarrow [\text{H}_3\text{O}^+] \rightarrow$ pH

                      ICE table $\text{pH} = -\log[\text{H}_3\text{O}^+]$

         Solution:

$$\text{HC}_3\text{H}_3\text{O}_2(aq) + \text{H}_2\text{O}(l) \rightleftharpoons \text{H}_3\text{O}^+(aq) + \text{C}_2\text{H}_3\text{O}_2^-(aq)$$

	$[\text{HC}_2\text{H}_3\text{O}_2]$	$[\text{H}_3\text{O}^+]$	$[\text{C}_2\text{H}_3\text{O}_2^-]$
Initial	0.105	≈ 0.00	0.00
Change	$-x$	$+x$	$+x$
Equil	$0.105 - x$	$+x$	$+x$

$K_a = \dfrac{[\text{H}_3\text{O}^+][\text{C}_2\text{H}_3\text{O}_2^-]}{[\text{HC}_2\text{H}_3\text{O}_2]} = 1.8 \times 10^{-5} = \dfrac{x^2}{0.105 - x}$ Assume that x is small ($x \ll 0.105$), so

$\dfrac{x^2}{0.105 - \cancel{x}} = 1.8 \times 10^{-5} = \dfrac{x^2}{0.105}$ and $x = 1.\underline{3}748 \times 10^{-3} \, \text{M} = [\text{H}_3\text{O}^+]$. Confirm that the assumption is valid.

$\dfrac{1.\underline{3}748 \times 10^{-3}}{0.105} \times 100\% = 1.3\% < 5\%$, so the assumption is valid. Finally,

$\text{pH} = -\log[\text{H}_3\text{O}^+] = -\log(1.\underline{3}748 \times 10^{-3}) = 2.86$.

         Check: The units (none) are correct. The magnitude of the answer makes physical sense because the pH should be greater than $-\log(0.105) = 0.98$ since this is a weak acid.

    (b)   **Find:** volume of base to reach equivalence point

         Conceptual Plan: Write a balanced equation then mL $\rightarrow$ L then $[\text{HC}_2\text{H}_3\text{O}_2]$, L $\rightarrow$ mol $\text{HC}_2\text{H}_3\text{O}_2$ then

            $\text{HC}_2\text{H}_3\text{O}_2 + \text{NaOH} \rightarrow \text{NaC}_2\text{H}_3\text{O}_2 + \text{H}_2\text{O}$ $\dfrac{1 \, \text{L}}{1000 \, \text{mL}}$ $M = \dfrac{\text{mol}}{\text{L}}$

         set mol acid($\text{HC}_2\text{H}_3\text{O}_2$) = mol base(NaOH) and [NaOH], mol NaOH $\rightarrow$ L NaOH $\rightarrow$ mL NaOH

                 balanced equation has 1:1 stoichiometry $M = \dfrac{\text{mol}}{\text{L}}$ $\dfrac{1000 \, \text{mL}}{1 \, \text{L}}$

         Solution: $20.0 \; \cancel{\text{mL HC}_2\text{H}_3\text{O}_2} \times \dfrac{1 \, \text{L}}{1000 \; \cancel{\text{mL}}} = 0.0200 \, \text{L HC}_2\text{H}_3\text{O}_2$ then

$\dfrac{0.105 \, \text{mol HC}_2\text{H}_3\text{O}_2}{1 \; \cancel{\text{L}}} \times 0.0200 \; \cancel{\text{L}} = 0.00210 \, \text{mol HC}_2\text{H}_3\text{O}_2$

So mol acid $= \text{mol HC}_2\text{H}_3\text{O}_2 = 0.00210 \, \text{mol} = \text{mol NaOH}$ then

$0.00210 \; \cancel{\text{mol NaOH}} \times \dfrac{1 \, \text{L}}{0.125 \; \cancel{\text{mol NaOH}}} = 0.0168 \; \cancel{\text{L NaOH}} \times \dfrac{1000 \, \text{mL}}{1 \; \cancel{\text{L}}} = 16.8 \, \text{mL NaOH}$

         Check: The units (mL) are correct. The volume of base is a little less than the volume of acid because the concentration of the base is a little greater than that of the acid.

(c) **Find:** pH after adding 5.0 mL of base

Conceptual Plan: Use calculations from part (b). Then mL → L then [NaOH], L → mol NaOH then

$$\frac{1\ L}{1000\ mL} \qquad\qquad M = \frac{mol}{L}$$

mol $HC_2H_3O_2$, mol NaOH → mol excess $HC_2H_3O_2$, mol $C_2H_3O_2^-$ and

set up stoichiometry table

L $HC_2H_3O_2$, L NaOH → total L then

L $HC_2H_3O_2$ + L NaOH = total L

mol excess $HC_2H_3O_2$, L → $[HC_2H_3O_2]$ and mol excess $C_2H_3O_2^-$, L → $[C_2H_3O_2^-]$ then

$$M = \frac{mol}{L} \qquad\qquad\qquad M = \frac{mol}{L}$$

M $HC_2H_3O_2$, M $C_2H_3O_2^- \rightarrow [H_3O^+] \rightarrow$ pH.

ICE table $\qquad$ pH = $-\log [H_3O^+]$

Solution: $5.0\ \cancel{mL\ NaOH} \times \dfrac{1\ L}{1000\ \cancel{mL}} = 0.0050\ L\ NaOH$ then

$\dfrac{0.125\ mol\ NaOH}{1\ \cancel{L}} \times 0.0050\ \cancel{L} = 0.000625\ mol\ NaOH$. Set up a table to track changes:

	NaOH(aq)	+	$HC_2H_3O_2(aq)$	→	$NaC_2H_3O_2(aq)$	+	$H_2O(l)$
Before addition	0.00 mol		0.00210 mol		0.00 mol		—
Addition	0.000625 mol		—		—		—
After addition	≈0.00 mol		0.001475 mol		0.000625 mol		—

Then $0.0200\ L\ HC_2H_3O_2 + 0.0050\ L\ NaOH = 0.0250\ L$ total volume. Then

$$[HC_2H_3O_2] = \frac{0.001475\ mol\ HC_2H_3O_2}{0.0250\ L} = 0.0590\ M \text{ and}$$

$$[NaC_2H_3O_2] = \frac{0.000625\ mol\ C_2H_3O_2^-}{0.0250\ L} = 0.025\ M.$$

Because one $C_2H_3O_2^-$ ion is generated for each $NaC_2H_3O_2$, $[C_2H_3O_2^-] = 0.025\ M\ C_2H_3O_2^-$.

$$HC_2H_3O_2(aq) + H_2O(l) \rightleftharpoons H_3O^+(aq) + C_2H_3O_2^-(aq)$$

	$[HC_2H_3O_2]$	$[H_3O^+]$	$[C_2H_3O_2^-]$
Initial	0.0590	≈0.00	0.025
Change	$-x$	$+x$	$+x$
Equil	$0.0590 - x$	$+x$	$0.025 + x$

$$K_a = \frac{[H_3O^+][C_2H_3O_2^-]}{[HC_2H_3O_2]} = 1.8 \times 10^{-5} = \frac{x(0.025 + x)}{0.0590 - x} \quad \text{Assume that } x \text{ is small } (x << 0.025 < 0.0590), \text{ so}$$

$$\frac{x(0.025 + \cancel{x})}{0.0590 - \cancel{x}} = 1.8 \times 10^{-5} = \frac{x(0.025)}{0.0590} \text{ and } x = 4.248 \times 10^{-5}\ M = [H_3O^+]. \text{ Confirm that the assumption}$$

is valid.

$$\frac{4.248 \times 10^{-5}}{0.025} \times 100\% = 0.17\% < 5\%, \text{ so the assumption is valid.}$$

Finally, pH = $-\log [H_3O^+]$ = $-\log (4.248 \times 10^{-5})$ = 4.37.

Check: The units (none) are correct. The pH is a little higher than the initial pH, which is expected because some of the acid has been neutralized.

(d) **Find:** pH at one-half the equivalence point

Conceptual Plan: Because this is a weak acid–strong base titration, the pH at one-half the equivalence point is the pK_a of the weak acid.

Solution: pH = pK_a = $-\log K_a$ = $-\log (1.8 \times 10^{-5})$ = 4.74

Check: The units (none) are correct. Because this is a weak acid–strong base titration, the pH at one-half the equivalence point is the pK_a of the weak acid; so it should be a little below 5.

(e)   **Find:** pH at equivalence point

Conceptual Plan: Use calculations from part (b). Then because all of the weak acid has been converted to its conjugate base, the pH is only dependent on the hydrolysis reaction of the conjugate base. The mol $C_2H_3O_2^- = $ initial mol $HC_2H_3O_2$ and L $HC_2H_3O_2$,

L NaOH to equivalence point $\rightarrow$ total L then

$$\text{L } HC_2H_3O_2 + \text{L NaOH} = \text{total L}$$

mol excess $C_2H_3O_2^-$, L $\rightarrow$ $[C_2H_3O_2^-]$ and $K_a \rightarrow K_b$ then do an equilibrium calculation:

$$M = \frac{\text{mol}}{\text{L}} \qquad\qquad K_w = K_aK_b$$

$[C_2H_3O_2^-], K_b \rightarrow [OH^-] \rightarrow [H_3O^+] \rightarrow$ pH

$$\text{set up ICE table} \quad K_w = [H_3O^+][OH^-] \quad pH = -\log[H_3O^+]$$

Solution: mol $C_2H_3O_2^- = $ initial mol $HC_2H_3O_2 = 0.00210$ mol and total volume $=$ L $HC_2H_3O_2 + $

L NaOH $= 0.020$ L $+ 0.0168$ L $= 0.0368$ L then

$$[C_2H_3O_2^-] = \frac{0.00210 \text{ mol } C_2H_3O_2^-}{0.0368 \text{ L}} = 0.05\underline{7}0652 \text{ M and } K_w = K_aK_b. \text{ Rearrange to solve for } K_b.$$

$$K_b = \frac{K_w}{K_a} = \frac{1.0 \times 10^{-14}}{1.8 \times 10^{-5}} = 5.\underline{5}556 \times 10^{-10}. \text{ Set up an ICE table:}$$

$$C_2H_3O_2^-(aq) + H_2O(l) \rightleftharpoons HC_2H_3O_2(aq) + OH^-(aq)$$

	$[C_2H_3O_2^-]$	$[HC_2H_3O_2]$	$[OH^-]$
Initial	0.0570652	≈ 0.00	≈ 0.00
Change	$-x$	$+x$	$+x$
Equil	$0.05\underline{7}0652 - x$	$+x$	$+x$

$$K_b = \frac{[HC_2H_3O_2][OH^-]}{[C_2H_3O_2^-]} = 5.\underline{5}556 \times 10^{-10} = \frac{x^2}{0.05\underline{7}0652 - x}. \quad \text{Assume that } x \text{ is small } (x \ll 0.057), \text{ so}$$

$$\frac{x^2}{0.05\underline{7}0652 - \cancel{x}} = 5.\underline{5}556 \times 10^{-10} = \frac{x^2}{0.05\underline{7}0652} \text{ and } x = 5.\underline{6}306 \times 10^{-6} \text{ M} = [OH^-].$$

Confirm that the assumption is valid. $\dfrac{5.\underline{6}306 \times 10^{-6}}{0.05\underline{7}0652} \times 100\% = 0.0099\% < 5\%,$ so the assumption is valid.

$$K_w = [H_3O^+][OH^-], \text{ so } [H_3O^+] = \frac{K_w}{[OH^-]} = \frac{1.0 \times 10^{-14}}{5.\underline{6}305 \times 10^{-6}} = 1.\underline{7}760 \times 10^{-9} \text{ M}.$$

Finally, pH $= -\log[H_3O^+] = -\log(1.\underline{7}760 \times 10^{-9}) = 8.75.$

Check: The units (none) are correct. Because this is a weak acid–strong base titration, the pH at the equivalence point is basic.

(f)   **Find:** pH after adding 5.0 mL of base beyond the equivalence point

Conceptual Plan: Use calculations from parts (b) and (c). Then the pH is only dependent on the amount of excess base and the total solution volumes.

mL excess $\rightarrow$ L excess then [NaOH], L excess $\rightarrow$ mol NaOH excess

$$\frac{1 \text{ L}}{1000 \text{ mL}} \qquad\qquad M = \frac{\text{mol}}{\text{L}}$$

then L $HC_2H_3O_2$, L NaOH to equivalence point, L NaOH excess $\rightarrow$ total L then

$$\text{L } HC_2H_3O_2 + \text{L NaOH to equivalence point} + \text{L NaOH excess} = \text{total L}$$

mol excess NaOH, total L $\rightarrow$ [NaOH] $= [OH^-] \rightarrow [H_3O^+] \rightarrow$ pH

$$M = \frac{\text{mol}}{\text{L}} \qquad\qquad K_w = [H_3O^+][OH^-] \quad pH = -\log[H_3O^+]$$

Solution: $5.0 \text{ mL NaOH} \times \dfrac{1 \text{ L}}{1000 \text{ mL}} = 0.0050 \text{ L NaOH excess then}$

$\dfrac{0.125 \text{ mol NaOH}}{1 \text{ L}} \times 0.0050 \text{ L} = 0.000625 \text{ mol NaOH excess. Then } 0.0200 \text{ L } HC_2H_3O_2 + 0.0168 \text{ L NaOH}$

$+ 0.0050 \text{ L NaOH} = 0.0418 \text{ L total volume.}$

$$[\text{NaOH excess}] = \frac{0.000625 \text{ mol NaOH excess}}{0.0418 \text{ L}} = 0.0149522 \text{ M NaOH excess}.$$ Because NaOH is a strong base, $[\text{NaOH}]$ excess $= [\text{OH}^-]$. The strong base overwhelms the weak base, which is insignificant in the calculation. $K_w = [\text{H}_3\text{O}^+][\text{OH}^-]$, so $[\text{H}_3\text{O}^+] = \dfrac{K_w}{[\text{OH}^-]} = \dfrac{1.0 \times 10^{-14}}{0.0149522} = 6.688 \times 10^{-13} \text{ M}.$

Finally, pH $= -\log[\text{H}_3\text{O}^+] = -\log(6.688 \times 10^{-13}) = 12.17.$

Check: The units (none) are correct. The pH is rising sharply at the equivalence point, so the pH after 5 mL past the equivalence point should be quite basic.

17.71 **Given:** 25.0 mL of 0.175 M CH_3NH_2 titrated with 0.150 M HBr   **Other:** $K_b(CH_3NH_2) = 4.4 \times 10^{-4}$

(a) **Find:** initial pH

Conceptual Plan: Because CH_3NH_2 is a weak base, set up an equilibrium problem using the initial concentration, so M CH_3NH_2 → $[OH^-]$ → $[H_3O^+]$ → pH

$$\text{ICE table} \qquad K_w = [\text{H}_3\text{O}^+][\text{OH}^-] \qquad \text{pH} = -\log[\text{H}_3\text{O}^+]$$

Solution:

$$CH_3NH_2(aq) + H_2O(l) \rightleftharpoons CH_3NH_3^+(aq) + OH^-(aq)$$

	$[\text{CH}_3\text{NH}_2]$	$[\text{CH}_3\text{NH}_3^+]$	$[\text{OH}^-]$
Initial	0.175	0.00	≈ 0.00
Change	$-x$	$+x$	$+x$
Equil	$0.175 - x$	$+x$	$+x$

$$K_b = \frac{[\text{CH}_3\text{NH}_3^+][\text{OH}^-]}{[\text{CH}_3\text{NH}_2]} = 4.4 \times 10^{-4} = \frac{x^2}{0.175 - x}$$

Assume that x is small ($x \ll 0.175$), so $\dfrac{x^2}{0.175 - x} = 4.4 \times 10^{-4} = \dfrac{x^2}{0.175}$ and $x = 8.7750 \times 10^{-3}$ M $= [\text{OH}^-]$.

Confirm that the assumption is valid. $\dfrac{8.7750 \times 10^{-3}}{0.175} \times 100\% = 5.0\%$, so the assumption is valid.

$$K_w = [\text{H}_3\text{O}^+][\text{OH}^-], \text{ so } [\text{H}_3\text{O}^+] = \frac{K_w}{[\text{OH}^-]} = \frac{1.0 \times 10^{-14}}{8.7750 \times 10^{-3}} = 1.1396 \times 10^{-12} \text{ M}.$$

Finally, pH $= -\log[\text{H}_3\text{O}^+] = -\log(1.1396 \times 10^{-12}) = 11.94.$

Check: The units (none) are correct. The magnitude of the answer makes physical sense because the pH should be less than $14 + \log(0.175) = 13.2$ since this is a weak base.

(b) **Find:** volume of acid to reach equivalence point

Conceptual Plan: Write a balanced equation, then mL → L then $[CH_3NH_2]$, L → mol CH_3NH_2

$$HBr + CH_3NH_2 \rightarrow CH_3NH_3Br + H_2O \qquad \frac{1 \text{ L}}{1000 \text{ mL}} \qquad M = \frac{\text{mol}}{\text{L}}$$

then set mol base $[CH_3NH_2]$ = mol acid (HBr) and $[HBr]$, mol HBr → L HBr → mL HBr

$$\text{balanced equation has 1:1 stoichiometry} \qquad M = \frac{\text{mol}}{\text{L}} \qquad \frac{1000 \text{ mL}}{1 \text{ L}}$$

Solution: $25.0 \text{ mL CH}_3\text{NH}_2 \times \dfrac{1 \text{ L}}{1000 \text{ mL}} = 0.0250 \text{ L CH}_3\text{NH}_2$ then

$\dfrac{0.175 \text{ mol CH}_3\text{NH}_2}{1 \text{ L}} \times 0.0250 \text{ L} = 0.004375 \text{ mol CH}_3\text{NH}_2.$ So mol base = mol CH_3NH_2 = 0.004375 mol

= mol HBr then $0.004375 \text{ mol HBr} \times \dfrac{1 \text{ L}}{0.150 \text{ mol HBr}} = 0.0291667 \text{ L HBr} \times \dfrac{1000 \text{ mL}}{1 \text{ L}} = 29.2 \text{ mL HBr}.$

Check: The units (mL) are correct. The volume of acid is greater than the volume of base because the concentration of the base is a little greater than that of the acid.

(c) **Find:** pH after adding 5.0 mL of acid

Conceptual Plan: Use calculations from part (b). Then mL → L then $[HBr]$, L → mol HBr

$$\frac{1 \text{ L}}{1000 \text{ mL}} \qquad M = \frac{\text{mol}}{\text{L}}$$

then mol CH_3NH_2, mol HBr → mol excess CH_3NH_2 and L CH_3NH_2, L HBr → total L.

set up stoichiometry table   L CH_3NH_2 + L HBr = total L

Because there are significant concentrations of both the acid and the conjugate base species, this is a buffer solution; so the Henderson–Hasselbalch equation $\left(pH = pK_a + \log \dfrac{[\text{base}]}{[\text{acid}]} \right)$ can be used.

Convert K_b to K_a using $K_w = K_a K_b$. Also note that the ratio of concentrations is the same as the ratio of moles because the volume is the same for both species.

Solution: $5.0 \text{ mL HBr} \times \dfrac{1 \text{ L}}{1000 \text{ mL}} = 0.0050 \text{ L HBr}$ then $\dfrac{0.150 \text{ mol HBr}}{1 \text{ L}} \times 0.0050 \text{ L} = 0.00075 \text{ mol HBr}$

Set up a table to track changes:

	HBr(aq)	+	$CH_3NH_2(aq)$	→	$CH_3NH_3Br(aq)$
Before addition	0.00 mol		0.004375 mol		0.00 mol
Addition	0.00075 mol		—		—
After addition	≈0.00 mol		0.003625 mol		0.000750 mol

then $K_w = K_a K_b$ so

$K_a = \dfrac{K_w}{K_b} = \dfrac{1.0 \times 10^{-14}}{4.4 \times 10^{-4}} = 2.2727 \times 10^{-11}$ M. Then use the Henderson–Hasselbalch equation because the

solution is a buffer. $pH = pK_a + \log \dfrac{[\text{base}]}{[\text{acid}]} = -\log(2.2727 \times 10^{-11}) + \log \dfrac{0.003625}{0.000750} = 11.33$

Check: The units (none) are correct. The pH is a little lower than the last pH, which is expected because some of the base has been neutralized.

(d)   **Find:** pH at one-half the equivalence point

      Conceptual Plan: Because this is a weak base–strong acid titration, the pH at one-half the equivalence point is the pK_a of the conjugate acid of weak base.

      Solution: $pH = pK_a = -\log K_a = -\log(2.2727 \times 10^{-11}) = 10.64$

      Check: The units (none) are correct. Because this is a weak acid–strong base titration, the pH at one-half the equivalence point is the pK_a of the conjugate acid of the weak base; so it should be a little below 11.

(e)   **Find:** pH at equivalence point

      Conceptual Plan: Use previous calculations. Because all of the weak base has been converted to its conjugate acid, the pH is only dependent on the hydrolysis reaction of the conjugate acid. The mol $CH_3NH_3^+ = $ initial mol CH_3NH_2 and

      L CH_3NH_2, L HBr to equivalence point → total L then mol $CH_3NH_3^+$, L → $[CH_3NH_3^+]$

L CH_3NH_2 + L HBr = total L $M = \dfrac{\text{mol}}{\text{L}}$

      then do an equilibrium calculation: $[CH_3NH_3^+]$, K_a → $[H_3O^+]$ → pH

set up ICE table $pH = -\log[H_3O^+]$

      Solution: mol base = mol acid = mol $CH_3NH_3^+$ = 0.004375 mol. Then total volume = L CH_3NH_2 +

      L HBr = 0.0250 L + 0.0292 L = 0.0542 then $[CH_3NH_3^+] = \dfrac{0.004375 \text{ mol } CH_3NH_3^+}{0.0542 \text{ L}} = 0.0807196$ M.

      Set up an ICE table:

$CH_3NH_3^+(aq) + H_2O(l) \rightleftharpoons CH_3NH_2(aq) + H_3O^+(aq)$

	$[CH_3NH_3^+]$	$[CH_3NH_2]$	$[H_3O^+]$
Initial	0.0807196	≈0.00	≈0.00
Change	$-x$	$+x$	$+x$
Equil	$0.0807196 - x$	$+x$	$+x$

$K_a = \dfrac{[CH_3NH_2][H_3O^+]}{[CH_3NH_3^+]} = 2.2727 \times 10^{-11} = \dfrac{x^2}{0.0807196 - x}$ Assume that x is small ($x \ll 0.0807$), so

$\dfrac{x^2}{0.0807196 - x} = 2.2727 \times 10^{-11} = \dfrac{x^2}{0.0807196}$ and $x = 1.3544 \times 10^{-6} = [H_3O^+]$.

Confirm that the assumption is valid. $\dfrac{1.3544 \times 10^{-6}}{0.080\underline{7}106} \times 100\% = 0.0017\% < 5\%$, so the assumption is valid.

Finally, $pH = -\log\left[H_3O^+\right] = -\log\left(1.\underline{3}544 \times 10^{-6}\right) = 5.87$.

Check: The units (none) are correct. Because this is a weak base–strong acid titration, the pH at the equivalence point is acidic.

(f) **Find:** pH after adding 5.0 mL of acid beyond the equivalence point

Conceptual Plan: Use calculations from parts (b) and (c). Then the pH is only dependent on the amount of excess acid and the total solution volumes.

mL excess $\rightarrow$ L excess then [HBr], L excess $\rightarrow$ mol HBr excess

$$\frac{1\ L}{1000\ mL} \qquad\qquad M = \frac{mol}{L}$$

then L CH_3NH_2, L HBr to equivalence point, L HBr excess $\rightarrow$ total L then

$$L\ CH_3NH_2 + L\ HBr\ to\ equivalence\ point + L\ HBr\ excess = total\ L$$

mol excess HBr, total L $\rightarrow$ [HBr] $= \left[H_3O^+\right] \rightarrow$ pH

$$M = \frac{mol}{L} \qquad\qquad pH = -\log\left[H_3O^+\right]$$

Solution: $5.0\ \text{mL HBr} \times \dfrac{1\ L}{1000\ mL} = 0.0050\ L\ HBr\ \text{excess then}$

$\dfrac{0.150\ mol\ HBr}{1\ L} \times 0.0050\ L = 0.00075\ mol\ HBr\ \text{excess. Then } 0.0250\ L\ CH_3NH_2 + 0.0292\ L\ HBr +$

$0.0050\ L\ HBr = 0.0592\ L\ \text{total volume.}$

$$\left[HBr\ \text{excess}\right] = \frac{0.00075\ mol\ HBr\ \text{excess}}{0.0592\ L} = 0.01\underline{2}669\ M\ HBr\ \text{excess}$$

Because HBr is a strong acid, $\left[HBr\right]$ excess $= \left[H_3O^+\right]$. The strong acid overwhelms the weak acid, which is insignificant in the calculation. Finally, $pH = -\log\left[H_3O^+\right] = -\log\left(0.01\underline{2}669\right) = 1.90$.

Check: The units (none) are correct. The pH is dropping sharply at the equivalence point, so the pH after 5 mL past the equivalence point should be quite acidic.

17.73 (i) Acid a is more concentrated because the equivalence point (where sharp pH rise occurs) is at a higher volume of added base.

(ii) Acid b has the larger K_a because the pH at a volume of added base equal to half the equivalence point volume is lower.

17.75 **Given:** 0.229 g unknown monoprotic acid titrated with 0.112 M NaOH and curve

Find: molar mass and pK_a of acid

Conceptual Plan: The equivalence point is where sharp pH rise occurs. The pK_a is the pH at a volume of added base equal to half the equivalence point volume. Then mL NaOH $\rightarrow$ L NaOH

$$\frac{1\ L}{1000\ mL}$$

then [NaOH], L NaOH $\rightarrow$ mol NaOH $=$ mol acid then mol acid, g acid $\rightarrow$ molar mass.

$$M = \frac{mol}{L} \qquad\qquad \frac{g\ acid}{mol\ acid}$$

Solution: The equivalence point is at 25 mL NaOH. The pH at $0.5 \times 25\ mL = 13\ mL$ is $\sim 3 = pK_a$. Then

$$25\ \text{mL NaOH} \times \frac{1\ L}{1000\ mL} = 0.025\ L\ NaOH\ \text{then}$$

$$\frac{0.112\ mol\ NaOH}{1\ L} \times 0.025\ L = 0.0028\ mol\ NaOH = 0.0028\ mol\ acid\ \text{then}$$

$$\text{molar mass} = \frac{0.229\ g\ acid}{0.0028\ mol\ acid} = 82\ g/mol$$

Check: The units (none and g/mol) are correct. The pK_a is consistent with a weak acid. The molar mass is reasonable for an acid (> 1 g/mol).

17.77   **Given:** 20.0 mL of 0.115 M sulfurous acid (H_2SO_3) titrated with 0.1014 M KOH   **Find:** volume of base added

Conceptual Plan: Because this is a diprotic acid, each proton is titrated sequentially. Write balanced equations.

$$H_2SO_3 + OH^- \rightarrow HSO_3^- + H_2O \text{ and } HSO_3^- + OH^- \rightarrow SO_3^{2-} + H_2O$$

Then mL $\rightarrow$ L then $[H_2SO_3]$, L $\rightarrow$ mol H_2SO_3 then set mol base (H_2SO_3) = mol acid (KOH) and

$$\frac{1\,L}{1000\,mL} \qquad\qquad M = \frac{mol}{L} \qquad\qquad \textit{balanced equation has 1:1 stoichiometry (1st equivalence point)}$$

[KOH], mol KOH $\rightarrow$ L KOH $\rightarrow$ mL KOH the volume to the second equivalence point will be

$$M = \frac{mol}{L} \qquad \frac{1000\,mL}{1\,L}$$

double the volume to the first equivalence point.

Solution: $20.0\ \cancel{mL\ H_2SO_3} \times \dfrac{1\,L}{1000\ \cancel{mL}} = 0.0200\ L\ H_2SO_3$ then

$\dfrac{0.115\ mol\ H_2SO_3}{1\ \cancel{L}} \times 0.0200\ \cancel{L} = 0.00230\ mol\ H_2SO_3.$ So mol base = mol H_2SO_3 = 0.00230 mol = mol KOH

then $0.00230\ \cancel{mol\ KOH} \times \dfrac{1\,L}{0.1014\ \cancel{mol\ KOH}} = 0.02268245\ \cancel{L\ KOH} \times \dfrac{1000\ mL}{1\ \cancel{L}} = 22.7\ mL\ KOH$ to first

equivalence point. The volume to the second equivalence point is simply twice this amount, or 45.4 mL, to the second equivalence point.

Check: The units (mL) are correct. The volume of base is greater than the volume of acid because the concentration of the acid is a little greater than that of the base. The volume to the second equivalence point is twice the volume to the first equivalence point.

17.79   The indicator will be in its acid form at an acidic pH, so the color in the HCl sample will be red. The color change will occur over the pH range from pH = pK_a − 1.0 to pH = pK_a + 1.0, so the color will start to change at pH = 5.0 − 1.0 = 4.0 and finish changing by pH = 5.0 + 1.0 = 6.0.

17.81   Because the exact conditions of the titration are not given, a rough calculation will suffice. Recall that at the equivalence point, the moles of acid and base are equal. If it is assumed that the concentrations of the acid and the base are equal, the total volume of the solution will have doubled. Assuming an acid and base concentration of 0.1 M, the conjugate based formed must have a concentration of ~0.05 M. From earlier calculations, it can be seen that

the $K_b = \dfrac{K_w}{K_a} = \dfrac{[OH^-]^2}{0.05}$; thus,

$$[OH^-] = \sqrt{\frac{0.05\,K_w}{K_a}} = \sqrt{\frac{5 \times 10^{-16}}{K_a}} \text{ and the pH} = 14 + \log\sqrt{\frac{5 \times 10^{-16}}{K_a}}.$$

(a)   For HF, the $K_a = 3.5 \times 10^{-4}$; so the above equation approximates the pH at the equivalence point of ~8.0. Looking at Table 17.1, we find that phenol red or *m*-nitrophenol will change at the appropriate pH range.

(b)   For HCl, the pH at the equivalence point is 7 because HCl is a strong acid. Looking at Table 17.1, we find that alizarin, bromthymol blue, *m*-nitrophenol, or phenol red will change at the appropriate pH range.

(c)   For HCN, the $K_a = 4.9 \times 10^{-10}$; so the preceding equation approximates the pH at the equivalence point of ~11.0. Looking at Table 17.1, we see that alizarin yellow R will change at the appropriate pH range.

Solubility Equilibria

17.83   For the dissolution reaction, start with the ionic compound as a solid and put it in equilibrium with the appropriate cation and anion, making sure to include the appropriate stoichiometric coefficients. The K_{sp} expression is the product of the concentrations of the cation and anion concentrations raised to their stoichiometric coefficients.

(a) $BaSO_4(s) \rightleftharpoons Ba^{2+}(aq) + SO_4^{2-}(aq)$ and $K_{sp} = [Ba^{2+}][SO_4^{2-}]$

(b) $PbBr_2(s) \rightleftharpoons Pb^{2+}(aq) + 2\,Br^-(aq)$ and $K_{sp} = [Pb^{2+}][Br^-]^2$

(c) $Ag_2CrO_4(s) \rightleftharpoons 2\,Ag^+(aq) + CrO_4^{2-}(aq)$ and $K_{sp} = [Ag^+]^2[CrO_4^{2-}]$

17.85   **Given:** ionic compound formula and Table 17.2 of K_{sp} values   **Find:** molar solubility (S)
Conceptual Plan: Because the balanced chemical equation for the dissolution of A_mX_n is $A_mX_n(s) \rightleftharpoons mA^{n+}(aq) + nX^{m-}(aq)$, the expression of the solubility product constant of A_mX_n is $K_{sp} = [A^{n+}]^m[X^{m-}]^n$. The molar solubility of a compound, A_mX_n, can be computed directly from K_{sp} by solving for S in the expression $K_{sp} = (mS)^m(nS)^n = m^m n^n S^{m+n}$.
Solution:

(a)   For AgBr, $K_{sp} = 5.35 \times 10^{-13}$, A = Ag^+, $m = 1$, X = Br^-, and $n = 1$; so $K_{sp} = 5.35 \times 10^{-13} = S^2$.

Rearrange to solve for S. $S = \sqrt{5.35 \times 10^{-13}} = 7.31 \times 10^{-7}$ M

(b)   For $Mg(OH)_2$, $K_{sp} = 2.06 \times 10^{-13}$, A = Mg^{2+}, $m = 1$, X = OH^-, and $n = 2$; so $K_{sp} = 2.06 \times 10^{-13} = 2^2 S^3$.

Rearrange to solve for S. $S = \sqrt[3]{\dfrac{2.06 \times 10^{-13}}{4}} = 3.72 \times 10^{-5}$ M

(c)   For CaF_2, $K_{sp} = 1.46 \times 10^{-10}$, A = Ca^{2+}, $m = 1$, X = F^-, and $n = 2$; so $K_{sp} = 1.46 \times 10^{-10} = 2^2 S^3$.

Rearrange to solve for S. $S = \sqrt[3]{\dfrac{1.46 \times 10^{-10}}{4}} = 3.32 \times 10^{-4}$ M

Check: The units (M) are correct. The molar solubilities are much less than 1 and dependent not only on the value of the K_{sp}, but also on the stoichiometry of the ionic compound. The more ions generated, the greater the molar solubility for the same value of the K_{sp}.

17.87   **Given:** ionic compound formula and molar solubility (S)   **Find:** K_{sp}
Conceptual Plan: The expression of the solubility product constant of A_mX_n is $K_{sp} = [A^{n+}]^m[X^{m-}]^n$. The molar solubility of a compound, A_mX_n, can be computed directly from K_{sp} by solving for S in the expression $K_{sp} = (mS)^m(nS)^n = m^m n^n S^{m+n}$.

Solution:

(a)   For MX, $S = 3.27 \times 10^{-11}$ M, A = M^+, $m = 1$, X = X^-, and $n = 1$; so $K_{sp} = S^2 = (3.27 \times 10^{-11})^2 = 1.07 \times 10^{-21}$.

(b)   For PbF_2, $S = 5.63 \times 10^{-3}$ M, A = Pb^{2+}, $m = 1$, X = F^-, and $n = 2$; so $K_{sp} = 2^2 S^3 = 2^2(5.63 \times 10^{-3})^3 = 7.14 \times 10^{-7}$.

(d)   For MgF_2, $S = 2.65 \times 10^{-4}$ M, A = Mg^{2+}, $m = 1$, X = F^-, and $n = 2$; so $K_{sp} = 2^2 S^3 = 2^2(2.65 \times 10^{-4})^3 = 7.44 \times 10^{-11}$.

Check: The units (none) are correct. The K_{sp} values are much less than 1 and dependent not only on the value of the solubility, but also on the stoichiometry of the ionic compound. The more ions generated, the smaller the K_{sp} for the same value of the S.

17.89   **Given:** ionic compound formulas AX and AX_2 and $K_{sp} = 1.5 \times 10^{-5}$ **Find:** higher molar solubility (S)
Conceptual Plan: The expression of the solubility product constant of A_mX_n is $K_{sp} = [A^{n+}]^m[X^{m-}]^n$. The molar solubility of a compound, A_mX_n, can be computed directly from K_{sp} by solving for S in the expression $K_{sp} = (mS)^m(nS)^n = m^m n^n S^{m+n}$.

Solution: For AX, $K_{sp} = 1.5 \times 10^{-5}$, $m = 1$, and $n = 1$; so $K_{sp} = 1.5 \times 10^{-5} = S^2$. Rearrange to solve for S.

$S = \sqrt{1.5 \times 10^{-5}} = 3.9 \times 10^{-3}$ M. For AX_2, $K_{sp} = 1.5 \times 10^{-5}$, $m = 1$, and $n = 2$; so $K_{sp} = 1.5 \times 10^{-5} = 2^2 S^3$.

Rearrange to solve for S. $S = \sqrt[3]{\dfrac{1.5 \times 10^{-5}}{4}} = 1.6 \times 10^{-2}$ M. Because 10^{-2} M > 10^{-3} M, AX_2 has a higher molar solubility.

Check: The units (M) are correct. The more ions generated, the greater the molar solubility for the same value of the K_{sp}.

17.91   **Given:** $Fe(OH)_2$ in 100.0 mL solution   **Find:** grams of $Fe(OH)_2$ **Other:** $K_{sp} = 4.87 \times 10^{-17}$
Conceptual Plan: The expression of the solubility product constant of A_mX_n is $K_{sp} = [A^{n+}]^m[X^{m-}]^n$. The molar solubility of a compound, A_mX_n, can be computed directly from K_{sp} by solving for S in the expression $K_{sp} = (mS)^m(nS)^n = m^m n^n S^{m+n}$. Then solve for S, then mL $\rightarrow$ L then

$$\frac{1\,L}{1000\,mL}$$

$S, L \rightarrow$ mol $Fe(OH)_2 \rightarrow$ g $Fe(OH)_2$.

$$M = \frac{mol}{L} \qquad \frac{89.87\,g\,Fe(OH)_2}{1\,mol\,Fe(OH)_2}$$

Solution: For $Fe(OH)_2$, $K_{sp} = 4.87 \times 10^{-17}$, $A = Fe^{2+}$, $m = 1$, $X = OH^-$, and $n = 2$; so $K_{sp} = 4.87 \times 10^{-17} = 2^2 S^3$.

Rearrange to solve for S. $S = \sqrt[3]{\dfrac{4.87 \times 10^{-17}}{4}} = 2.30050 \times 10^{-6}\,M$. Then $100.0\,\cancel{mL} \times \dfrac{1\,L}{1000\,\cancel{mL}} = 0.1000\,L$

then $\dfrac{2.30050 \times 10^{-6}\,mol\,Fe(OH)_2}{1\,\cancel{L}} \times 0.1000\,\cancel{L} = 2.30050 \times 10^{-7}\,\cancel{mol\,Fe(OH)_2} \times \dfrac{89.87\,g\,Fe(OH)_2}{1\,\cancel{mol\,Fe(OH)_2}}$

$= 2.07 \times 10^{-5}\,g\,Fe(OH)_2$.

Check: The units (g) are correct. The solubility rules from Chapter 4 (most hydroxides are insoluble) suggest that very little $Fe(OH)_2$ will dissolve; so the magnitude of the answer is not surprising.

17.93   **(a)**   **Given:** BaF_2 **Find:** molar solubility (S) in pure water   **Other:** $K_{sp}(BaF_2) = 2.45 \times 10^{-5}$
Conceptual Plan: The expression of the solubility product constant of A_mX_n is $K_{sp} = [A^{n+}]^m[X^{m-}]^n$. The molar solubility of a compound, A_mX_n, can be computed directly from K_{sp} by solving for S in the expression $K_{sp} = (mS)^m(nS)^n = m^m n^n S^{m+n}$.

Solution: For BaF_2, $K_{sp} = 2.45 \times 10^{-5}$, $A = Ba^{2+}$, $m = 1$, $X = F^-$, and $n = 2$; so $K_{sp} = 2.45 \times 10^{-5} = 2^2 S^3$.

Rearrange to solve for S. $S = \sqrt[3]{\dfrac{2.45 \times 10^{-5}}{4}} = 1.83 \times 10^{-2}\,M$

(b)   **Given:** BaF_2 **Find:** molar solubility (S) in 0.10 M $Ba(NO_3)_2$ **Other:** $K_{sp}(BaF_2) = 2.45 \times 10^{-5}$
Conceptual Plan: M $Ba(NO_3)_2 \rightarrow$ M Ba^{2+} then M $Ba^{2+}, K_{sp} \rightarrow S$

$$Ba(NO_3)_2(s) \rightarrow Ba^{2+}(aq) + 2\,NO_3^-(aq) \qquad\qquad \text{ICE table}$$

Solution: Because one Ba^{2+} ion is generated for each $Ba(NO_3)_2$, $[Ba^{2+}] = 0.10\,M$.

$BaF_2(s)$ $\rightleftharpoons$	$Ba^{2+}(aq)$	$+\ 2\,F^-(aq)$
Initial	0.10	0.00
Change	S	$2S$
Equil	$0.10 + S$	$2S$

$K_{sp}(BaF_2) = [Ba^{2+}][F^-]^2 = 2.45 \times 10^{-5} = (0.10 + S)(2S)^2$

Assume that $S << 0.10$, $2.45 \times 10^{-5} = (0.10)(2S)^2$, and $S = 7.83 \times 10^{-3}\,M$. Confirm that the assumption is valid. $\dfrac{7.83 \times 10^{-3}}{0.10} \times 100\% = 7.8\% > 5\%$, so the assumption is not valid. Because expanding the expression will give a third-order polynomial, that is not easily solved directly. Solve by successive approximations. Substitute $S = 7.83 \times 10^{-3}\,M$ for the S term that is part of a sum [i.e., the one in $(0.10 + S)$]. Thus, $2.45 \times 10^{-5} = (0.10 + 7.83 \times 10^{-3})(2S)^2$ and $S = 7.53 \times 10^{-3}\,M$. Substitute this new S value again. Thus, $2.45 \times 10^{-5} = (0.10 + 7.53 \times 10^{-3})(2S)^2$ and $S = 7.55 \times 10^{-3}\,M$. Substitute this new S value again. Thus, $2.45 \times 10^{-5} = (0.10 + 7.55 \times 10^{-3})(2S)^2$ and $S = 7.55 \times 10^{-3}\,M$. So the solution has converged and $S = 7.55 \times 10^{-3}\,M$.

(c)   **Given:** BaF_2 **Find:** molar solubility (S) in 0.15 M NaF   **Other:** $K_{sp}(BaF_2) = 2.45 \times 10^{-5}$
Conceptual Plan: M NaF $\rightarrow$ M F^- then M $F^-, K_{sp} \rightarrow S$

$$NaF(s) \rightarrow Na^+(aq) + F^-(aq) \qquad\qquad \text{ICE table}$$

Solution: Because one F^- ion is generated for each NaF, $[F^-] = 0.15$ M.

$$BaF_2(s) \rightleftharpoons Ba^{2+}(aq) + 2\,F^-(aq)$$

Initial	0.00	0.15
Change	S	2S
Equil	S	0.15 + 2S

$K_{sp}(BaF_2) = [Ba^{2+}][F^-]^2 = 2.45 \times 10^{-5} = (S)(0.15 + 2S)^2$

Because $2S \ll 0.15$, $2.45 \times 10^{-5} = (S)(0.15)^2$ and $S = 1.09 \times 10^{-3}$ M. Confirm that the assumption is valid. $\dfrac{2\,(1.09 \times 10^{-3})}{0.15} \times 100\% = 1.5\% < 5\%$, so the assumption is valid.

Check: The units (M) are correct. The solubility of the BaF_2 decreases in the presence of a common ion. The effect of the anion is greater because the K_{sp} expression has the anion concentration squared.

17.95   **Given:** $Ca(OH)_2$ **Find:** molar solubility (S) in buffers at (a) pH = 4, (b) pH = 7, and (c) pH = 9
Other: $K_{sp}\,(Ca(OH)_2) = 4.68 \times 10^{-6}$
Conceptual Plan: $pH \rightarrow [H_3O^+] \rightarrow [OH^-]$ then $M\ OH^-, K_{sp} \rightarrow S$

$$[H_3O^+] = 10^{-pH} \qquad K_w = [H_3O^+][OH^-] \qquad \text{set up ICE table}$$

Solution:

(a)   pH = 4, so $[H_3O^+] = 10^{-pH} = 10^{-4} = 1 \times 10^{-4}$ M then $K_w = [H_3O^+][OH^-]$ so

$$[OH^-] = \frac{K_w}{[H_3O^+]} = \frac{1.0 \times 10^{-14}}{1 \times 10^{-4}} = 1 \times 10^{-10}\ \text{M then}$$

$$Ca(OH)_2(s) \rightleftharpoons Ca^{2+}(aq) + 2\,OH^-(aq)$$

Initial	0.00	1×10^{-10}
Change	S	—
Equil	S	1×10^{-10}

$K_{sp}(Ca(OH)_2) = [Ca^{2+}][OH^-]^2 = 4.68 \times 10^{-6} = S(1 \times 10^{-10})^2$ and $S = 5 \times 10^{14}$ M

(b)   pH = 7, so $[H_3O^+] = 10^{-pH} = 10^{-7} = 1 \times 10^{-7}$ M then $K_w = [H_3O^+][OH^-]$ so

$$[OH^-] = \frac{K_w}{[H_3O^+]} = \frac{1.0 \times 10^{-14}}{1 \times 10^{-7}} = 1 \times 10^{-7}\ \text{M then}$$

$$Ca(OH)_2(s) \rightleftharpoons Ca^{2+}(aq) + 2\,OH^-(aq)$$

Initial	0.00	1×10^{-7}
Change	S	—
Equil	S	1×10^{-7}

$K_{sp}(Ca(OH)_2) = [Ca^{2+}][OH^-]^2 = 4.68 \times 10^{-6} = S(1 \times 10^{-7})^2$ and $S = 5 \times 10^8$ M

(c)   pH = 9, so $[H_3O^+] = 10^{-pH} = 10^{-9} = 1 \times 10^{-9}$ M then $K_w = [H_3O^+][OH^-]$ so

$$[OH^-] = \frac{K_w}{[H_3O^+]} = \frac{1.0 \times 10^{-14}}{1 \times 10^{-9}} = 1 \times 10^{-5}\ \text{M then}$$

$$Ca(OH)_2(s) \rightleftharpoons Ca^{2+}(aq) + 2\,OH^-(aq)$$

Initial	0.00	1×10^{-5}
Change	S	—
Equil	S	1×10^{-5}

$K_{sp}(Ca(OH)_2) = [Ca^{2+}][OH^-]^2 = 4.68 \times 10^{-6} = S(1 \times 10^{-5})^2$ and $S = 5 \times 10^4$ M

Check: The units (M) are correct. The solubility of the $Ca(OH)_2$ decreases as the pH increases (and the hydroxide ion concentration increases). These molar solubilities are not achievable because the saturation point of pure $Ca(OH)_2$ is ~30 M. The bottom line is that as long as the hydroxide concentration can be controlled with a buffer, the $Ca(OH)_2$ will be very soluble.

17.97 (a) $BaCO_3$ will be more soluble in acidic solutions because CO_3^{2-} is basic. In acidic solutions, it can be converted to HCO_3^- and H_2CO_3. These species are not CO_3^{2-}, so they do not appear in the K_{sp} expression.

(b) CuS will be more soluble in acidic solutions because S^{2-} is basic. In acidic solutions, it can be converted to HS^- and H_2S. These species are not S^{2-}, so they do not appear in the K_{sp} expression.

(c) AgCl will not be more soluble in acidic solutions because Cl^- will not react with acidic solutions because HCl is a strong acid.

(d) PbI_2 will not be more soluble in acidic solutions because I^- will not react with acidic solutions since HI is a strong acid.

Precipitation and Qualitative Analysis

17.99 **Given:** 0.015 M NaF and 0.010 M $Ca(NO_3)_2$ **Find:** Will a precipitate form? If so, identify it.
Other: $K_{sp}(CaF_2) = 1.46 \times 10^{-10}$
Conceptual Plan: Look at all possible combinations and consider the solubility rules from Chapter 4. Salts of alkali metals (Na) are very soluble, so NaF and $NaNO_3$ will be very soluble. Nitrate compounds are very soluble, so $NaNO_3$ will be very soluble. The only possibility for a precipitate is CaF_2. Determine whether a precipitate will form by determining the concentration of the Ca^{2+} and F^- in solution. Then compute the reaction quotient, Q. If $Q > K_{sp}$, a precipitate will form.
Solution: Because the only possible precipitate is CaF_2, calculate the concentrations of Ca^{2+} and F^-. $NaF(s) \rightarrow Na^+(aq) + F^-(aq)$. Because one F^- ion is generated for each NaF, $[F^-] = 0.015$ M.
$Ca(NO_3)_2(s) \rightarrow Ca^{2+}(aq) + 2 NO_3^-(aq)$. Because one Ca^{2+} ion is generated for each $Ca(NO_3)_2$, $[Ca^{2+}] = 0.010$ M.
Then calculate $Q(CaF_2)$, $A = Ca^{2+}$, $m = 1$, $X = F^-$, and $n = 2$. Because $Q = [A^{n+}]^m [X^{m-}]^n$,
$Q(CaF_2) = [Ca^{2+}][F^-]^2 = (0.010)(0.015)^2 = 2.3 \times 10^{-6} > 1.46 \times 10^{-10} = K_{sp}(CaF_2)$; so a precipitate will form.

Check: The units (none) are correct. The solubility of the CaF_2 is low, and the concentration of ions is extremely large compared to the K_{sp}; so a precipitate will form.

17.101 **Given:** 75.0 mL of NaOH with pOH = 2.58 and 125.0 mL of 0.0018 M $MgCl_2$
Find: Will a precipitate form? If so, identify it.   **Other:** $K_{sp}(Mg(OH)_2) = 2.06 \times 10^{-13}$
Conceptual Plan: Look at all possible combinations and consider the solubility rules from Chapter 4. Salts of alkali metals (Na) are very soluble, so NaOH and NaCl will be very soluble. Chloride compounds are generally very soluble, so $MgCl_2$ and NaCl will be very soluble. The only possibility for a precipitate is $Mg(OH)_2$. Determine whether a precipitate will form by determining the concentration of the Mg^{2+} and OH^- in solution. Because pOH, not NaOH concentration, is given, pOH $\rightarrow [OH^-]$ then

$$[OH^-] = 10^{-pOH}$$

mix solutions and calculate diluted concentrations mL NaOH, mL $MgCl_2$ $\rightarrow$ mL total then

$$mL\ NaOH + mL\ MgCl_2 = total\ mL$$

mL, initial M $\rightarrow$ final M then compute the reaction quotient, Q.

$$M_1 V_1 = M_2 V_2$$

If $Q > K_{sp}$, a precipitate will form.
Solution: Because the only possible precipitate is $Mg(OH)_2$, calculate the concentrations of Mg^{2+} and OH^-.
For NaOH at pOH = 2.58, so $[OH^-] = 10^{-pOH} = 10^{-2.58} = 2.63027 \times 10^{-3}$ M and
$MgCl_2(s) \rightarrow Mg^{2+}(aq) + 2 Cl^-(aq)$. Because one Mg^{2+} ion is generated for each $MgCl_2$, $[Mg^{2+}] = 0.0018$ M.
Then total mL = mL NaOH + mL $MgCl_2$ = 75.0 mL + 125.0 mL = 200.0 mL. Then $M_1 V_1 = M_2 V_2$;

rearrange to solve for M_2. $M_2 = M_1 \dfrac{V_1}{V_2} = 2.63027 \times 10^{-3}$ M $OH^- \times \dfrac{75.0\ mL}{200.0\ mL} = 9.8635 \times 10^{-4}$ M OH^- and

$M_2 = M_1 \dfrac{V_1}{V_2} = 0.0018$ M $Mg^+ \times \dfrac{125.0\ mL}{200.0\ mL} = 1.125 \times 10^{-3}$ M Mg^{2+}. Calculate $Q(Mg(OH)_2)$, $A = Mg^{2+}$,

$m = 1$, $X = OH^-$, and $n = 2$. Because $Q = [A^{n+}]^m [X^{m-}]^n$, $Q(Mg(OH)_2) = [Mg^{2+}][OH^-]^2 =$

$(1.125 \times 10^{-3})(9.8635 \times 10^{-4})^2 = 1.1 \times 10^{-9} > 2.06 \times 10^{-13} = K_{sp}(Mg(OH)_2)$; so a precipitate will form.

Check: The units (none) are correct. The solubility of the $Mg(OH)_2$ is low, and the NaOH (a base) is high enough that the product of the concentration of ions is large compared to the K_{sp}; so a precipitate will form.

17.103 **Given:** KOH as precipitation agent in (a) 0.015 M $CaCl_2$, (b) 0.0025 M $Fe(NO_3)_2$, and (c) 0.0018 M $MgBr_2$
Find: concentration of KOH necessary to form a precipitate
Other: $K_{sp}(Ca(OH)_2) = 4.68 \times 10^{-6}$, $K_{sp}(Fe(OH)_2) = 4.87 \times 10^{-17}$, $K_{sp}(Mg(OH)_2) = 2.06 \times 10^{-13}$
Conceptual Plan: The solubility rules from Chapter 4 state that most hydroxides are insoluble, so all precipitates will be hydroxides. Determine the concentration of the cation in solution. Because all metals have an oxidation

state of $+2$ and $[OH^-] = [KOH]$, all of the $K_{sp} = [\text{cation}][KOH]^2$; so $[KOH] = \sqrt{\dfrac{K_{sp}}{[\text{cation}]}}$.

Solution:

(a) $CaCl_2(s) \rightarrow Ca^{2+}(aq) + 2\,Cl^-(aq)$. Because one Ca^{2+} ion is generated for each $CaCl_2$, $[Ca^{2+}] = 0.015$ M.

Then $[KOH] = \sqrt{\dfrac{K_{sp}}{[\text{cation}]}} = \sqrt{\dfrac{4.68 \times 10^{-6}}{0.015}} = 0.018$ M KOH.

(b) $Fe(NO_3)_2(s) \rightarrow Fe^{2+}(aq) + 2\,NO_3^-(aq)$. Because one Fe^{2+} ion is generated for each $Fe(NO_3)_2$, $[Fe^{2+}] = 0.0025$ M.

Then $[KOH] = \sqrt{\dfrac{K_{sp}}{[\text{cation}]}} = \sqrt{\dfrac{4.87 \times 10^{-17}}{0.0025}} = 1.4 \times 10^{-7}$ M KOH.

(c) $MgBr_2(s) \rightarrow Mg^{2+}(aq) + 2\,Br^-(aq)$. Because one Mg^{2+} ion is generated for each $MgBr_2$, $[Mg^{2+}] = 0.0018$ M.

Then $[KOH] = \sqrt{\dfrac{K_{sp}}{[\text{cation}]}} = \sqrt{\dfrac{2.06 \times 10^{-13}}{0.0018}} = 1.1 \times 10^{-5}$ M KOH.

Check: The units (none) are correct. Because all cations have an oxidation state of $+2$, it can be seen that the [KOH] needed to precipitate the hydroxide is lower the smaller the K_{sp}.

17.105 **Given:** solution with 0.010 M Ba^{2+} and 0.020 M Ca^{2+}, add Na_2SO_4 to form precipitates
Find: (a) which ion precipitates first and minimum $[Na_2SO_4]$ needed and (b) [first cation] when second cation precipitates **Other:** $K_{sp}(BaSO_4) = 1.07 \times 10^{-10}$, $K_{sp}(CaSO_4) = 7.10 \times 10^{-5}$
Conceptual Plan: (a) The precipitates that will form are $BaSO_4$ and $CaSO_4$.
Use the equation derived in Problem 17.19 to define K_{sp}. Substitute concentration of cation and solve for

for ionic compound, A_mX_n, $K_{sp} = [A^{n+}]^m[X^{m-}]^n$

concentration of anion to form precipitate. The cation with the lower anion concentration will precipitate first.
(b) **Substitute the higher anion concentration into the K_{sp} expression for the first cation to precipitate and calculate the amount of this first cation to remain in solution.**

Solution:

(a) Derive expression for $K_{sp}(BaSO_4)$, A = Ba^{2+}, $m = 1$, X = SO_4^{2-}, and $n = 1$. Because $K_{sp} = [Ba^{2+}][SO_4^{2-}]$,
$K_{sp}(BaSO_4) = 1.07 \times 10^{-10} = 0.010\,[SO_4^{2-}]$. Solve for $[SO_4^{2-}]$. $[SO_4^{2-}] = 1.07 \times 10^{-8}$ M SO_4^{2-}.
Because $Na_2SO_4(s) \rightarrow 2\,Na^+(aq) + SO_4^{2-}(aq)$. Because one SO_4^{2-} ion is generated for each
Na_2SO_4, $[Na_2SO_4] = 1.1 \times 10^{-8}$ M Na_2SO_4 to precipitate $BaSO_4$. Derive expression for
$K_{sp}(CaSO_4)$, A = Ca^{2+}, $m = 1$, X = SO_4^{2-}, and $n = 1$. Because $K_{sp} = [Ca^{2+}][SO_4^{2-}]$,
$K_{sp}(CaSO_4) = 7.10 \times 10^{-5} = 0.020\,[SO_4^{2-}]$. Solve for $[SO_4^{2-}]$. $[SO_4^{2-}] = 0.0036$ M $SO_4^{2-} = $
0.0036 M $Na_2SO_4 = [Na_2SO_4]$ to precipitate $CaSO_4$. Because 1.1×10^{-8} M $Na_2SO_4 \ll 0.0036$ M
Na_2SO_4, the Ba^{2+} will precipitate first.

(b) Because Ca^{2+} will not precipitate until $[Na_2SO_4] = 0.00355$ M Na_2SO_4, substitute this value into the K_{sp}
expression for $BaSO_4$. So $K_{sp}(BaSO_4) = [Ba^{2+}][SO_4^{2-}] = 1.07 \times 10^{-10} = [Ba^{2+}]\,0.0035$. Solve for
$[Ba^{2+}]$. $[Ba^{2+}] = 3.0 \times 10^{-8}$ M Ba^{2+}.

Check: The units (none, M, and M) are correct. Comparing the two K_{sp} values, it can be seen that the Ba^{2+} will precipitate first because the solubility product is so much lower. Because the K_{sp} value is so low, the concentration of precipitating agent is very low. Because the $CaSO_4$ K_{sp} value is so much higher, the higher $[SO_4^{2-}]$ to precipitate Ca will force the concentration of Ba^{2+} to very low levels.

Complex Ion Equilibria

17.107   **Given:** solution with 1.1×10^{-3} M $Zn(NO_3)_2$ and 0.150 M NH_3 **Find:** $[Zn^{2+}]$ at equilibrium
Other: $K_f(Zn(NH_3)_4^{2+}) = 2.8 \times 10^9$
Conceptual Plan: Write a balanced equation and expression for K_f. Use initial concentrations to set up an ICE table. Because the K_f is so large, assume that reaction essentially goes to completion. Solve for $[Zn^{2+}]$ at equilibrium.
Solution: $Zn(NO_3)_2(s) \rightarrow Zn^{2+}(aq) + 2\,NO_3^-(aq)$. Because one Zn^{2+} ion is generated for each $Zn(NO_3)_2$,
$[Zn^{2+}] = 1.1 \times 10^{-3}$ M. Balanced equation is:

$$Zn^{2+}(aq) \quad + \quad 4\,NH_3(aq) \quad \rightleftharpoons \quad Zn(NH_3)_4^{2+}(aq)$$

	$[Zn^{2+}]$	$[NH_3]$	$[Zn(NH_3)_4^{2+}]$
Initial	1.1×10^{-3}	0.150	0.00
Change	$\approx 1.1 \times 10^{-3}$	$\approx -4(1.1 \times 10^{-3})$	$\approx 1.1 \times 10^{-3}$
Equil	x	0.1456	1.1×10^{-3}

Set up an ICE table with initial concentrations. Because K_f is so large and because initially $[NH_3] > 4\,[Zn^{2+}]$, the reaction essentially goes to completion.

Write the equilibrium expression and solve for x.

$$K_f = \frac{[Zn(NH_3)_4^{2+}]}{[Zn^{2+}][NH_3]^4} = 2.8 \times 10^9 = \frac{1.1 \times 10^{-3}}{x\,(0.1456)^4}.$$ So $x = 8.7 \times 10^{-10}$ M Zn^{2+}. Because x is insignificant compared
to the initial concentration, the assumption is valid.

Check: The units (M) are correct. Because K_f is so large, the reaction essentially goes to completion and $[Zn^{2+}]$ is extremely small.

17.109   **Given:** $FeS(s) + 6\,CN^-(aq) \rightleftharpoons Fe(CN)_6^{4-}(aq) + S^{2-}(aq)$ use K_{sp} and K_f values   **Find:** K
Other: $K_f(Fe(CN)_6^{4-}) = 1.5 \times 10^{35}$, $K_{sp}(FeS) = 3.72 \times 10^{-19}$
Conceptual Plan: Identify the appropriate solid and complex ion. Write balanced equations for dissolving the solid and forming the complex ion. Add these two reactions to get the desired overall reaction. Using the rules from Chapter 15, multiply the individual reaction Ks to get the overall K for the sum of these reactions.
Solution: Identify the solid as FeS and the complex ion as $Fe(CN)_6^{4-}$. Write the individual reactions and add them together.

$$FeS(s) \rightleftharpoons \cancel{Fe^{2+}(aq)} + S^{2-}(aq) \qquad\qquad K_{sp} = 1.5 \times 10^{-19}$$
$$\cancel{Fe^{2+}(aq)} + 6\,CN^-(aq) \rightleftharpoons Fe(CN)_6^{4-}(aq) \qquad K_f = 3.72 \times 10^{35}$$
$$\overline{FeS(s) + 6\,CN^-(aq) \rightleftharpoons Fe(CN)_6^{4-}(aq) + S^{2-}(aq)}$$

Because the overall reaction is the simple sum of the two reactions, the overall reaction $K = K_f \times K_{sp} = (1.5 \times 10^{35}) \times (3.72 \times 10^{-19}) = 5.6 \times 10^{16}$.

Check: The units (none) are correct. Because K_f is so large, it overwhelms the K_{sp} and the overall reaction is spontaneous.

Cumulative Problems

17.111   **Given:** 150.0 mL solution of 2.05 g sodium benzoate and 2.47 g benzoic acid   **Find:** pH
Other: $K_a(HC_7H_5O_2) = 6.5 \times 10^{-5}$
Conceptual Plan: g $NaC_7H_5O_2$ → mol $NaC_7H_5O_2$ and g $HC_7H_5O_2$ → mol $HC_7H_5O_2$

$$\frac{1 \text{ mol } NaC_7H_5O_2}{144.10 \text{ g } NaC_7H_5O_2} \qquad\qquad \frac{1 \text{ mol } HC_7H_5O_2}{122.12 \text{ g } HC_7H_5O_2}$$

Because the two components are in the same solution, the ratio of [base]/[acid] = (mol base)/(mol acid). Then K_a, mol $NaC_7H_5O_2$, mol $HC_7H_5O_2$ → pH.

$$pH = pK_a + \log\frac{[\text{base}]}{[\text{acid}]}$$

Solution: $2.05 \text{ g } \cancel{NaC_7H_5O_2} \times \dfrac{1 \text{ mol } NaC_7H_5O_2}{144.10 \text{ g } \cancel{NaC_7H_5O_2}} = 0.0142262 \text{ mol } NaC_7H_5O_2$ and

$2.47 \text{ g } \cancel{HC_7H_5O_2} \times \dfrac{1 \text{ mol } HC_7H_5O_2}{122.12 \text{ g } \cancel{HC_7H_5O_2}} = 0.0202260 \text{ mol } HC_7H_5O_2$ then

$$pH = pK_a + \log\frac{[\text{base}]}{[\text{acid}]} = pK_a + \log\frac{\text{mol base}}{\text{mol acid}} = -\log\,(6.5 \times 10^{-5}) + \log\frac{0.0142262\;\cancel{\text{mol}}}{0.0202260\;\cancel{\text{mol}}} = 4.03$$

Check: The units (none) are correct. The magnitude of the answer makes physical sense because the pH is a little lower than the pK_a of the acid because there is more acid than base in the buffer solution.

17.113   **Given:** 150.0 mL of 0.25 M $HCHO_2$ and 75.0 ml of 0.20 M NaOH   **Find:** pH
Other: $K_a(HCHO_2) = 1.8 \times 10^{-4}$
Conceptual Plan: In this buffer, the base is generated by converting some of the formic acid to the formate ion.
Part I: Stoichiometry:
mL → L then L, initial $HCHO_2$ M → mol $HCHO_2$ then mL → L then

$$\frac{1\,L}{1000\,mL} \qquad\qquad M = \frac{\text{mol}}{L} \qquad\qquad \frac{1\,L}{1000\,mL}$$

L, initial NaOH M → mol NaOH then write a balanced equation then

$$M = \frac{\text{mol}}{L} \qquad\qquad NaOH + HCHO_2 \rightarrow H_2O + NaCHO_2$$

mol $HCHO_2$, mol NaOH → mol $NaCHO_2$, mol $HCHO_2$ then

set up stoichiometry table

Part II: Equilibrium:
Because the two components are in the same solution, the ratio of [base]/[acid] = (mol base)/(mol acid).
Then K_a, mol $NaCHO_2$, mol $HCHO_2$ → pH.

$$pH = pK_a + \log\frac{[\text{base}]}{[\text{acid}]}$$

Solution: $150.0\;\cancel{mL} \times \dfrac{1\,L}{1000\;\cancel{mL}} = 0.1500\,L$ then

$$0.1500\;L\,\cancel{HCHO_2} \times \frac{0.25\;\text{mol}\;HCHO_2}{1\;L\,\cancel{HCHO_2}} = 0.0375\;\text{mol}\;HCHO_2$$

Then $75.0\;\cancel{mL} \times \dfrac{1\,L}{1000\;\cancel{mL}} = 0.0750\,L$ then $0.0750\;L\,\cancel{NaOH} \times \dfrac{0.20\;\text{mol}\;NaOH}{1\;L\,\cancel{NaOH}} = 0.015\;\text{mol}\;NaOH$

Set up a table to track changes:

	NaOH(aq)	+	$HCHO_2$(aq)	→	$NaCHO_2$(aq)	+	H_2O(l)
Before addition	0.00 mol		0.0375 mol		0.00 mol		—
Addition	0.015 mol		—		—		—
After addition	≈0.00 mol		0.0225 mol		0.015 mol		—

Because the amount of NaOH is small, there are significant amounts of both buffer components; so the Henderson–Hasselbalch equation can be used to calculate the pH.

$$pH = pK_a + \log\frac{[\text{base}]}{[\text{acid}]} = pK_a + \log\frac{\text{mol base}}{\text{mol acid}} = -\log\,(1.8 \times 10^{-4}) + \log\frac{0.015\;\cancel{\text{mol}}}{0.0225\;\cancel{\text{mol}}} = 3.57$$

Check: The units (none) are correct. The magnitude of the answer makes physical sense because the pH is a little lower than the pK_a of the acid since there is more acid than base in the buffer solution.

17.115   **Given:** 1.0 L of buffer of 0.25 mol NH_3 and 0.25 mol NH_4Cl; adjust to pH = 8.75
Find: mass NaOH or HCl   **Other:** $K_b(NH_3) = 1.76 \times 10^{-5}$
Conceptual Plan: To decide which reagent needs to be added to adjust pH, calculate the initial pH. Because the mol NH_3 = mol NH_4Cl, the pH = pK_a so $K_b \rightarrow pK_b \rightarrow pK_a$ then

acid = NH_4^+ base = NH_3 $pK_b = -\log K_b$ $14 = pK_a + pK_b$

final pH, $pK_a \rightarrow [NH_3]/[NH_4^+]$ then $[NH_3]$, L → mol $[NH_3]$ and $[NH_4^+]$, L → mol $[NH_4^+]$

$pH = pK_a + \log\dfrac{[\text{base}]}{[\text{acid}]}$ $M = \dfrac{\text{mol}}{L}$ $M = \dfrac{\text{mol}}{L}$

then write a balanced equation then

$H^+ + NH_3 \rightarrow NH_4^+$

mol NH_3, mol NH_4^+, $[NH_3]/[NH_4^+]$ → mol HCl → g HCl.

set up stoichiometry table $\dfrac{36.46\;\text{g HCl}}{1\;\text{mol HCl}}$

Solution: Because $K_b(NH_3) = 1.76 \times 10^{-5}$, $pK_b = -\log K_b = -\log(1.76 \times 10^{-5}) = 4.75$. Because $14 = pK_a + pK_b$, $pK_a = 14 - pK_b = 14 - 4.75 = 9.25$. Because the desired pH is lower (8.75), HCl (a strong acid) needs to be added. Then $pH = pK_a + \log \dfrac{[\text{base}]}{[\text{acid}]} = 9.25 + \log \dfrac{[NH_3]}{[NH_4^+]} = 8.75$. Solve for $\dfrac{[NH_3]}{[NH_4^+]}$.

$\log \dfrac{[NH_3]}{[NH_4^+]} = 8.75 - 9.25 = -0.50 \rightarrow \dfrac{[NH_3]}{[NH_4^+]} = 10^{-0.50} = 0.3\underline{1}623$ then

$\dfrac{0.25 \text{ mol } NH_3}{1 \, \cancel{L}} \times 1.0 \, \cancel{L} = 0.25 \text{ mol } NH_3$ and

$\dfrac{0.25 \text{ mol } NH_4Cl}{1 \, \cancel{L}} \times 1.0 \, \cancel{L} = 0.25 \text{ mol } NH_4Cl = 0.25 \text{ mol } NH_4^+$. Because HCl is a strong acid, $[\text{HCl}] = [H^+]$.

Set up a table to track changes:

	$H^+(aq)$	$+$	$NH_3(aq)$	$\rightarrow$	$NH_4^+(aq)$
Before addition	≈ 0.00 mol		0.25 mol		0.25 mol
Addition	x		—		—
After addition	≈ 0.00 mol		$(0.25 - x)$ mol		$(0.25 + x)$ mol

Because $\dfrac{[NH_3]}{[NH_4^+]} = 0.3\underline{1}623 = \dfrac{(0.25 - x) \, \cancel{\text{mol}}}{(0.25 + x) \, \cancel{\text{mol}}}$, solve for x. Note that the ratio of moles is the same as the ratio of concentrations because the volume for both terms is the same. $0.3\underline{1}623(0.25 + x) = (0.25 - x) \rightarrow 0.07\underline{9}0575 + 0.3\underline{1}623x = 0.25 - x \rightarrow 1.3\underline{1}623x = 0.1\underline{7}094 \rightarrow x = 0.1\underline{2}987 \text{ mol HCl}$ then

$0.1\underline{2}987 \, \cancel{\text{mol HCl}} \times \dfrac{36.46 \text{ g HCl}}{1 \, \cancel{\text{mol HCl}}} = 4.7 \text{ g HCl}$

Check: The units (g) are correct. The magnitude of the answer makes physical sense because there is much less than a mole of each of the buffer components; so there must be much less than a mole of HCl.

17.117　**(a)**　**Given:** potassium hydrogen phthalate $= KHP = KHC_8H_4O_4$ titration with NaOH　**Find:** balanced equation
Conceptual Plan: The reaction will be a titration of the acid proton, leaving the phthalate ion intact. The K will not be titrated because it is basic.
Solution: $NaOH(aq) + KHC_8H_4O_4(aq) \rightarrow Na^+(aq) + K^+(aq) + C_8H_4O_4^{2-}(aq) + H_2O(l)$
Check: An acid–base reaction generates a salt (soluble here) and water. There is only one acidic proton in KHP.

(b)　**Given:** 0.5527 g KHP titrated with 25.87 mL of NaOH solution　**Find:** [NaOH]
Conceptual Plan:
g KHP → mol KHP → mol NaOH and mL → L then mol NaOH and mL → M NaOH

$\dfrac{1 \text{ mol KHP}}{204.22 \text{ g KHP}}$　1:1 from balanced equation　　$\dfrac{1 \text{ L}}{1000 \text{ mL}}$　　$M = \dfrac{\text{mol}}{\text{L}}$

Solution: $0.5527 \, \cancel{\text{g KHP}} \times \dfrac{1 \text{ mol KHP}}{204.22 \, \cancel{\text{g KHP}}} = 0.00270\underline{6}395 \text{ mol KHP}$; mol KHP = mol acid = mol base =

$0.00270\underline{6}395 \text{ mol NaOH}$ then $25.87 \, \cancel{\text{mL}} \times \dfrac{1 \text{ L}}{1000 \, \cancel{\text{mL}}} = 0.02587 \text{ L}$ then

$[\text{NaOH}] = \dfrac{0.00270\underline{6}395 \text{ mol NaOH}}{0.02587 \text{ L}} = 0.1046 \text{ M NaOH}$

Check: The units (M) are correct. The magnitude of the answer makes physical sense because there is much less than a mole of acid. The magnitude of the moles of acid and base is smaller than the volume of base in liters.

17.119　**Given:** 0.25 mol weak acid with 10.0 mL of 3.00 M KOH diluted to 1.5000 L has pH $= 3.85$　**Find:** pK_a of acid
Conceptual Plan: mL → L then M KOH, L → mol KOH then write a balanced reaction

$\dfrac{1 \text{ L}}{1000 \text{ mL}}$　　　$M = \dfrac{\text{mol}}{\text{L}}$　　　$KOH + HA \rightarrow NaA + H_2O$

added mol KOH, initial mol acid → equil. mol KOH, equil. mol acid then
set up stoichiometry table

equil. mol KOH, equil. mol acid, pH → pK_a

$$pH = pK_a + \log \frac{[\text{base}]}{[\text{acid}]}$$

Solution: $10.00 \text{ mL} \times \dfrac{1 \text{ L}}{1000 \text{ mL}} = 0.01000 \text{ L}$ then $0.01000 \text{ L KOH} \times \dfrac{3.00 \text{ mol KOH}}{1 \text{ L KOH}} = 0.0300 \text{ mol KOH}$

Because KOH is a strong base, $[\text{KOH}] = [\text{OH}^-]$. Set up a table to track changes:

	KOH(aq)	+	HA(aq)	→	KA(aq)	+	H$_2$O(l)
Before addition	≈0.00 mol		0.25 mol		0.00 mol		—
Addition	0.0300 mol		—		—		—
After addition	≈0.00 mol		0.22 mol		0.0300 mol		—

Because the ratio of base to acid is between 0.1 and 10, it is a buffer solution. Note that the ratio of moles is the same as the ratio of concentrations because the volume for both terms is the same.

$$pH = pK_a + \log \frac{[\text{base}]}{[\text{acid}]} = pK_a + \log \frac{0.0300 \text{ mol}}{0.22 \text{ mol}} = 3.85 \text{ Solve for } pK_a.$$

$$pK_a = 3.85 - \log \frac{0.0300 \text{ mol}}{0.22 \text{ mol}} = 4.72$$

Check: The units (none) are correct. The magnitude of the answer makes physical sense because there is more acid than base at equilibrium; so the pK_a is higher than the pH of the solution.

17.121   **Given:** 0.552 g ascorbic acid dissolved in 20.00 mL and titrated with 28.42 mL of 0.1103 M KOH solution; pH = 3.72 when 10.0 mL KOH added   **Find:** molar mass and K_a of acid

Conceptual Plan: mL → L then M KOH, L → mol KOH → mol acid then mol acid, g acid → $\mathcal{M}$

$$\frac{1 \text{ L}}{1000 \text{ mL}} \qquad M = \frac{\text{mol}}{\text{L}} \quad \text{1:1 for monoprotic acid} \qquad \mathcal{M} = \frac{\text{g acid}}{\text{mol acid}}$$

For the second part of the problem:

mL → L then M KOH, L → mol KOH then write a balanced reaction

$$\frac{1 \text{ L}}{1000 \text{ mL}} \qquad M = \frac{\text{mol}}{\text{L}} \qquad KOH + HA \rightarrow KA + H_2O$$

added mol KOH, initial mol acid → equil. mol KOH, equil. mol acid then

set up stoichiometry table

equil. mol KOH, equil. mol acid, pH → pK_a → K_a.

$$pH = pK_a + \log \frac{[\text{base}]}{[\text{acid}]} \qquad pK_a = -\log K_a$$

Solution: $28.42 \text{ mL} \times \dfrac{1 \text{ L}}{1000 \text{ mL}} = 0.02842 \text{ L}$ then

$0.02842 \text{ L KOH} \times \dfrac{0.1103 \text{ mol KOH}}{1 \text{ L KOH}} = 0.003134726 \text{ mol KOH} = \text{mol base} = \text{mol acid} = 0.003134726 \text{ mol}$

ascorbic acid then $M = \dfrac{\text{g acid}}{\text{mol acid}} = \dfrac{0.552 \text{ g acid}}{0.003134726 \text{ mol acid}} = 176 \text{ g/mol}$. For the second part of the problem,

$10.00 \text{ mL} \times \dfrac{1 \text{ L}}{1000 \text{ mL}} = 0.01000 \text{ L}$ then $0.01000 \text{ L KOH} \times \dfrac{0.1103 \text{ mol KOH}}{1 \text{ L KOH}} = 0.001103 \text{ mol KOH}$

Because KOH is a strong base, $[\text{KOH}] = [\text{OH}^-]$. Set up a table to track changes:

	KOH(aq)	+	HA(aq)	→	KA(aq)	+	H$_2$O(l)
Before addition	≈0.00 mol		0.003134726 mol		0.00 mol		—
Addition	0.001103 mol		—		—		—
After addition	≈0.00 mol		0.002031726 mol		0.001103 mol		—

Because the ratio of base to acid is between 0.1 and 10, it is a buffer solution. Note that the ratio of moles is the same as the ratio of concentrations because the volume for both terms is the same.

$$pH = pK_a + \log \frac{[\text{base}]}{[\text{acid}]} = pK_a + \log \frac{0.001103 \text{ mol}}{0.002031726 \text{ mol}} = 3.72. \text{ Solve for } pK_a.$$

$$pK_a = 3.72 - \log \frac{0.001103}{0.002031726} = 3.98529 \text{ and so } pK_a = -\log K_a$$

or $K_a = 10^{-pK_a} = 10^{-3.98529} = 1.0 \times 10^{-4}$.

Check: The units (g/mol and none) are correct. The magnitude of the answer makes physical sense because there is much less than a mole of acid and about half a gram of acid; so the molar mass will be high. The number is reasonable for an acid (must be >20 g/mol—lightest acid is HF). The K_a is reasonable because the pK_a is within 1 unit of the pH when the titration solution is behaving as a buffer.

17.123   **Given:** saturated $CaCO_3$ solution; precipitate 1.00×10^2 mg $CaCO_3$ **Find:** volume of solution evaporated
Other: $K_{sp}(CaCO_3) = 4.96 \times 10^{-9}$
Conceptual Plan: mg $CaCO_3 \rightarrow$ g $CaCO_3 \rightarrow$ mol $CaCO_3$

$$\frac{1 \text{ g } CaCO_3}{1000 \text{ mg } CaCO_3} \qquad \frac{1 \text{ mol } CaCO_3}{100.09 \text{ g } CaCO_3}$$

The expression of the solubility product constant of A_mX_n is $K_{sp} = [A^{n+}]^m [X^{m-}]^n$. The molar solubility of a compound, A_mX_n, can be computed directly from K_{sp} by solving for S in the expression $K_{sp} = (mS)^m (nS)^n = m^n n^n S^{m+n}$. Then mol $CaCO_3$, $S \rightarrow L$.

$$M = \frac{mol}{L}$$

Solution: $1.00 \times 10^2 \text{ mg } CaCO_3 \times \frac{1 \text{ g } CaCO_3}{1000 \text{ mg } CaCO_3} \times \frac{1 \text{ mol } CaCO_3}{100.09 \text{ g } CaCO_3} = 9.99101 \times 10^{-4} \text{ mol } CaCO_3$ then

$K_{sp} = 4.96 \times 10^{-9}$, $A = Ca^{2+}$, $m = 1$, $X = CO_3^{2-}$, and $n = 1$; so $K_{sp} = 4.96 \times 10^{-9} = S^2$. Rearrange to solve for S. $S = \sqrt{4.96 \times 10^{-9}} = 7.04273 \times 10^{-5}$ M. Finally,

$$9.99101 \times 10^{-4} \text{ mol } CaCO_3 \times \frac{1 \text{ L}}{7.04273 \times 10^{-5} \text{ mol } CaCO_3} = 14.2 \text{ L}.$$

Check: The units (L) are correct. The volume should be large because the solubility is low.

17.125   **Given:** $[Ca^{2+}] = 9.2$ mg/dL and $K_{sp}(Ca_2P_2O_7) = 8.64 \times 10^{-13}$ **Find:** $[P_2O_7^{4-}]$ to form precipitate
Conceptual Plan: mg Ca^{2+}/dL $\rightarrow$ g Ca^{2+}/dL $\rightarrow$ mol Ca^{2+}/dL $\rightarrow$ mol Ca^{2+}/L then

$$\frac{1 \text{ g } Ca^{2+}}{1000 \text{ mg } Ca^{2+}} \qquad \frac{1 \text{ mol } Ca^{2+}}{40.08 \text{ g } Ca^{2+}} \qquad \frac{10 \text{ dL}}{1 \text{ L}}$$

write a balanced equation and expression for K_{sp}. Then $[Ca^{2+}]$, $K_{sp} \rightarrow [P_2O_7^{4-}]$.

Solution: $9.2 \frac{\text{mg } Ca^{2+}}{\text{dL}} \times \frac{1 \text{ g } Ca^{2+}}{1000 \text{ mg } Ca^{2+}} \times \frac{1 \text{ mol } Ca^{2+}}{40.08 \text{ g } Ca^{2+}} \times \frac{10 \text{ dL}}{1 \text{ L}} = 2.29541 \times 10^{-3}$ M Ca^{2+} then write

equation $Ca_2P_2O_7(s) \rightarrow 2 Ca^{2+}(aq) + P_2O_7^{4-}(aq)$. So $K_{sp} = [Ca^{2+}]^2 [P_2O_7^{4-}] = 8.64 \times 10^{-13} = (2.29541 \times 10^{-3})^2 [P_2O_7^{4-}]$. Solve for $[P_2O_7^{4-}]$ then $[P_2O_7^{4-}] = 1.6 \times 10^{-7}$ M.

Check: The units (M) are correct. Because K_{sp} is so small and the calcium concentration is relatively high, the diphosphate concentration required is at a very low level.

17.127   **Given:** CuX in 0.150 M NaCN   **Find:** molar solubility (S)
Other: $K_f(Cu(CN)_4^{2-}) = 1.0 \times 10^{25}$, $K_{sp}(CuX) = [Cu^{2+}][X^{2-}] = 1.27 \times 10^{-36}$
Conceptual Plan: Identify the appropriate solid and complex ion. Write balanced equations for dissolving the solid and forming the complex ion. Add these two reactions to get the desired overall reaction. Using the rules from Chapter 15, multiply the individual reaction Ks to get the overall K for the sum of these reactions. Then M NaCN, $K \rightarrow S$.
ICE table
Solution: Identify the solid as MX and the complex ion as $M(CN)_4^{2-}$. Write the individual reactions and add them together.

$CuX(s) \rightleftharpoons Cu^+(aq) + X^{2-}(aq)$	$K_{sp} = 1.27 \times 10^{-36}$
$Cu^{2+}(aq) + 4 CN^-(aq) \rightleftharpoons Cu(CN)_4^{2-}(aq)$	$K_f = 1.0 \times 10^{25}$

$$\overline{Cu(s) + 4 CN^-(aq) \rightleftharpoons Cu(CN)_4^{2-}(aq) + X^{2-}(aq)}$$

Because the overall reaction is the simple sum of the two reactions, the overall reaction $K = K_f K_{sp} = (1.0 \times 10^{25}) \times (1.27 \times 10^{-36}) = 1.27 \times 10^{-11}$. $NaCN(s) \rightarrow Na^+(aq) + CN^-(aq)$. Because one CN^- ion is generated for each NaCN, $[CN^-] = 0.150$ M. Set up an ICE table:

$$Cu(s) \quad + \quad 4\,CN^-(aq) \rightleftharpoons Cu(CN)_4^{2-}(aq) \quad + \quad X^{2-}(aq)$$

	$[CN^-]$	$[Cu(CN)_4^{2-}]$	$[X^{2-}]$
Initial	0.150	0.00	0.00
Change	$-4S$	$+S$	$+S$
Equil	$0.150 - 4S$	$+S$	$+S$

$$K = \frac{[Cu(CN)_4^{2-}][X^{2-}]}{[CN^-]^4} = 1.\underline{2}7 \times 10^{-11} = \frac{S^2}{(0.150 - 4S)^4}$$

Assume that S is small ($4S << 0.150$), so $\dfrac{S^2}{(0.150 - \cancel{4S})^4} = 1.\underline{2}7 \times 10^{-11} = \dfrac{S^2}{(0.150)^4}$ and $S = 8.\underline{0}183 \times 10^{-8} =$

8.0×10^{-8} M. Confirm that the assumption is valid. $\dfrac{4(8.\underline{0}183 \times 10^{-8})}{0.150} \times 100\% = 0.00021\% << 5\%$, so the assumption is valid.

Check: The units (M) are correct. Because K_f is large, the overall K is larger than the original K_{sp} and the solubility of MX increases over that of pure water ($\sqrt{1.27 \times 10^{-36}} = 1.13 \times 10^{-18}$ M).

17.129 **Given:** 100.0 mL of 0.36 M NH_2OH and 50.0 mL of 0.26 M HCl and $K_b(NH_2OH) = 1.10 \times 10^{-8}$ **Find:** pH
Conceptual Plan: Identify acid and base components mL $\rightarrow$ L then $[NH_2OH]$, L $\rightarrow$ mol NH_2OH

$$\text{acid} = NH_3OH^+ \text{ base} = NH_2OH \qquad \frac{1\,L}{1000\,mL} \qquad M = \frac{mol}{L}$$

then mL $\rightarrow$ L then [HCl], L $\rightarrow$ mol HCl then write balanced equation then

$$\frac{1\,L}{1000\,mL} \qquad M = \frac{mol}{L} \qquad HCl + NH_2OH \rightarrow NH_3OHCl$$

mol NH_2OH, mol HCL $\rightarrow$ mol excess NH_2OH, mol NH_3OH^+.

set up stoichiometry table

Because there are significant amounts of both the acid and the conjugate base species, this is a buffer solution;

so the Henderson–Hasselbalch equation $\left(pH = pK_a + \log \dfrac{[base]}{[acid]} \right)$ **can be used. Convert K_b to K_a using**

$K_w = K_a K_b$. **Also note that the ratio of concentrations is the same as the ratio of moles because the volume is the same for both species.**

Solution: $100 \cancel{mL\,NH_2OH} \times \dfrac{1\,L}{1000\,\cancel{mL}} = 0.1\,L\,NH_2OH$ then

$\dfrac{0.36\,mol\,NH_2OH}{1\,\cancel{L}} \times 0.1000\,\cancel{L} = 0.036\,mol\,NH_2OH$. $50.0\,\cancel{mL\,HCl} \times \dfrac{1\,L}{1000\,\cancel{mL}} = 0.0500\,L\,HCl$ then

$\dfrac{0.26\,mol\,HCl}{1\,\cancel{L}} \times 0.0500\,\cancel{L} = 0.013\,mol\,HCl$. Set up a table to track changes:

	$HCl(aq)$	$+$	$NH_2OH(aq)$	$\rightarrow$	$NH_3OHCl(aq)$
Before addition	0.00 mol		0.036 mol		0.00 mol
Addition	0.013 mol		—		—
After addition	$\approx$0.00 mol		0.023 mol		0.013 mol

Then $K_w = K_a K_b$, so $K_a = \dfrac{K_w}{K_b} = \dfrac{1.0 \times 10^{-14}}{1.10 \times 10^{-8}} = 9.\underline{0}909 \times 10^{-7}$ M. Then use the Henderson–Hasselbalch equation because the solution is a buffer. Note that the ratio of moles is the same as the ratio of concentrations because the volume for both terms is the same.

$$pH = pK_a + \log \frac{[base]}{[acid]} = -\log(9.\underline{0}909 \times 10^{-7}) + \log \frac{0.023\,\cancel{mol}}{0.013\,\cancel{mol}} = 6.2\underline{8}918 = 6.29$$

Check: The units (none) are correct. The magnitude of the answer makes physical sense because the pH should be more than the pK_a of the acid since there is more base than acid.

17.131 **Given:** 25.0 mL of NaOH titrated with 19.6 mL of 0.189 M HCl solution; 10.0 mL of H_3PO_4 titrated with 34.9 mL NaOH **Find:** concentration of H_3PO_4 solution
Conceptual Plan: Write the first balanced reaction then mL $\rightarrow$ L then M HCl, L $\rightarrow$ mol HCl $\rightarrow$ mol NaOH

$$HCl + NaOH \rightarrow NaCl + H_2O \qquad \frac{1\,L}{1000\,mL} \qquad M = \frac{mol}{L} \qquad 1{:}1$$

then mL → L then mol NaOH, L → M NaOH then write 2nd balanced reaction then mL → L

$$\frac{1\,L}{1000\,mL}$$ H$_3$PO$_4$ + 3 NaOH → Na$_3$PO$_4$ + 3 H$_2$O $$\frac{1\,L}{1000\,mL}$$

then M NaOH, L → mol NaOH → mol H$_3$PO$_4$ then mL → L then mol H$_3$PO$_4$, L → M H$_3$PO$_4$.

$$M = \frac{mol}{L}$$ 3:1 $$\frac{1\,L}{1000\,mL}$$ $$M = \frac{mol}{L}$$

Solution: In the first titration, $19.6\ mL \times \dfrac{1\,L}{1000\,mL} = 0.0196\ L$ then

$0.0196\ L\,HCl \times \dfrac{0.189\ mol\ HCl}{1\ L\,HCl} = 0.0037044\ mol\ HCl \times \dfrac{1\ mol\ NaOH}{1\ mol\ HCl} = 0.0037044\ mol\ NaOH$ then

$25.0\ mL \times \dfrac{1\,L}{1000\,mL} = 0.0250\ L$ then $\dfrac{0.0037044\ mol\ NaOH}{0.0250\ L\ NaOH} = 0.148176\ M\ NaOH$

In the second titration, $34.9\ mL \times \dfrac{1\,L}{1000\,mL} = 0.0349\ L$ then

$0.0349\ L\,NaOH \times \dfrac{0.148176\ mol\ NaOH}{1\ L\,NaOH} = 0.00517134\ mol\ NaOH \times \dfrac{1\ mol\ H_3PO_4}{3\ mol\ NaOH} = 0.00172378\ mol\ H_3PO_4$

then $10.0\ mL \times \dfrac{1\,L}{1000\,mL} = 0.0100\ L$ then $\dfrac{0.00172378\ mol\ H_3PO_4}{0.0100\ L\ H_3PO_4} = 0.172\ M\ H_3PO_4$

Check: The units (M) are correct. The magnitude of the answer makes physical sense because the concentration of NaOH is a little lower than the HCl (because the volume of NaOH is greater than HCl) and the concentration of H$_3$PO$_4$ is more than the NaOH (because the ratio of the volume of NaOH to volume of H$_3$PO$_4$ is just over 3 and H$_3$PO$_4$ is a triprotic acid).

17.133   **Given:** $(CH_3)_2NH/(CH_3)_2NH_2Cl$ buffer at pH = 10.43   **Find:** relative masses of $(CH_3)_2NH$ and $(CH_3)_2NH_2Cl$
Other: $K_b((CH_3)_2NH) = 5.4 \times 10^{-4}$
Conceptual Plan: $K_b → pK_b → pK_a$ then pH, $pK_a → [(CH_3)_2NH]/[(CH_3)_2NH_2{}^+]$ then

$pK_b = -\log K_b$ $14 = pK_a + pK_b$ acid = $(CH_3)_2NH_2{}^+$ base = $(CH_3)_2NH$ $pH = pK_a + \log\dfrac{[\text{base}]}{[\text{acid}]}$

$[(CH_3)_2NH]/[(CH_3)_2NH_2{}^+] → g(CH_3)_2NH/g(CH_3)_2NH_2{}^+$

$$\frac{45.08\ g\,(CH_3)_2NH}{1\ mol\,(CH_3)_2NH}$$ $$\frac{1\ mol\,(CH_3)_2NH_2Cl}{81.54\ g\,(CH_3)_2NH_2Cl}$$

Solution: Because $K_b((CH_3)_2NH) = 5.4 \times 10^{-4}$, $pK_b = -\log K_b = -\log(5.4 \times 10^{-4}) = 3.27$. Because

$14 = pK_a + pK_b$, $pK_a = 14 - pK_b = 14 - 3.27 = 10.73$. Because $[\text{acid}] = [(CH_3)_2NH_2{}^+] = [(CH_3)_2NH_2Cl]$

and $[\text{base}] = [(CH_3)_2NH]$, $pH = pK_a + \log\dfrac{[\text{base}]}{[\text{acid}]} = 10.73 + \log\dfrac{[(CH_3)_2NH]}{[(CH_3)_2NH_2Cl]} = 10.43$. Solve for

$\dfrac{[(CH_3)_2NH]}{[(CH_3)_2NH_2Cl]}$. $\log\dfrac{[(CH_3)_2NH]}{[(CH_3)_2NH_2Cl]} = 10.43 - 10.73 = -0.30 → \dfrac{[(CH_3)_2NH]}{[(CH_3)_2NH_2Cl]} = 10^{-0.30} = 0.501187$.

Then $\dfrac{0.501187\ mol\,(CH_3)_2NH}{1\ mol\,(CH_3)_2NH_2Cl} \times \dfrac{45.08\ g\,(CH_3)_2NH}{1\ mol\,(CH_3)_2NH} \times \dfrac{1\ mol\,(CH_3)_2NH_2Cl}{81.54\ g\,(CH_3)_2NH_2Cl} = \dfrac{0.277084\ g\,(CH_3)_2NH}{g\,(CH_3)_2NH_2Cl}$

$= \dfrac{0.28\ g\ (CH_3)_2NH}{g\ (CH_3)_2NH_2Cl}$ or $\dfrac{3.6\ g\ (CH_3)_2NH_2Cl}{g\ (CH_3)_2NH}$

Check: The units (g/g) are correct. The magnitude of the answer makes physical sense because there are more moles of acid than base in the buffer (pH < pK_a) and the molar mass of the acid is greater than the molar mass of the base. Thus, the ratio of the mass of the base to the mass of the acid is expected to be less than 1.

17.135   **Given:** $HC_7H_5O_2/C_7H_5O_2Na$ buffer at pH = 4.55, complete dissociation of $C_7H_5O_2Na$, $d = 1.01$ g/mL; $T_f = -2.0\,°C$
Find: $[HC_7H_5O_2]$ and $[C_7H_5O_2Na]$ **Other:** $K_a(HC_7H_5O_2) = 6.5 \times 10^{-5}$, $K_f = 1.86\,°C/m$
Conceptual Plan: $K_a → pK_a$ then pH, $pK_a → [C_7H_5O_2Na]/[HC_7H_5O_2]$ and $T_f → \Delta T_f$ then

$pK_a = -\log K_a$ acid = $HC_7H_5O_2$ base = $C_7H_5O_2{}^-$ $pH = pK_a + \log\dfrac{[\text{base}]}{[\text{acid}]}$ $T_f = T_f° - \Delta T_f$

$\Delta T_f, i, K_f \rightarrow m$ then assume 1 kg water (or 1000 g water)

$$\Delta T_f = K_f \times m \qquad m = \frac{\text{amount solute (moles)}}{\text{mass solvent (kg)}}$$

Assume that $HC_7H_5O_2$ does not dissociate and that $C_7H_5O_2Na$ completely dissociates, then mol particles = mol $HC_7H_5O_2$ + 2(mol $C_7H_5O_2Na$). Use$[C_7H_5O_2Na]/[HC_7H_5O_2]$ and total mol particles = mol $[HC_7H_5O_2]$ + 2(mol $C_7H_5O_2Na$) to solve for mol $HC_7H_5O_2$ and mol $C_7H_5O_2Na$. Then mol $HC_7H_5O_2 \rightarrow$ g $HC_7H_5O_2$ and mol $C_7H_5O_2$ Na $\rightarrow$ g $C_7H_5O_2$ Na then

$$\frac{122.12 \text{ g } HC_7H_5O_2}{1 \text{ mol } HC_7H_5O_2} \qquad\qquad \frac{144.10 \text{ g } C_7H_5O_2Na}{1 \text{ mol } C_7H_5O_2Na}$$

g $HC_7H_5O_2$, g $C_7H_5O_2Na$, g water $\rightarrow$ g solution $\rightarrow$ mL solution $\rightarrow$ L solution then

$$\text{g } HC_7H_5O_2 + \text{g } C_7H_5O_2Na + \text{g water} = \text{g solution} \qquad d = \frac{1 \text{ mL}}{1.01 \text{ g}} \quad \frac{1 \text{ L}}{1000 \text{ mL}}$$

mol $HC_7H_5O_2$, L $\rightarrow$ M $HC_7H_5O_2$ and mol $C_7H_5O_2Na$, L $\rightarrow$ M $C_7H_5O_2Na$.

$$M = \frac{\text{mol}}{L} \qquad\qquad M = \frac{\text{mol}}{L}$$

Solution: Because $K_a(HC_7H_5O_2) = 6.5 \times 10^{-5}$, $pK_a = -\log K_a = -\log(6.5 \times 10^{-5}) = 4.19$. Because $[\text{acid}] = [HC_7H_5O_2]$ and $[\text{base}] = [C_7H_5O_2^-] = [C_7H_5O_2Na]$,

$$pH = pK_a + \log\frac{[\text{base}]}{[\text{acid}]} = 4.19 + \log\frac{[C_7H_5O_2Na]}{[HC_7H_5O_2]} = 4.55. \text{ Solve for } \frac{[C_7H_5O_2Na]}{[HC_7H_5O_2]}.$$

$$\log\frac{[C_7H_5O_2Na]}{[HC_7H_5O_2]} = 4.55 - 4.19 = 0.36 \rightarrow \frac{[C_7H_5O_2Na]}{[HC_7H_5O_2]} = 10^{0.36} = 2.29087. \text{ Then } T_f = T_f^\circ - \Delta T_f, \text{ so}$$

$\Delta T_f = T_f^\circ - T_f = 0.0\,°C - (-2.0\,°C) = 2.0\,°C$. Then $\Delta T_f = K_f \times m$. Rearrange to solve for m.

$$m = \frac{\Delta T_f}{K_f} = \frac{2.0\,°C}{1.86\frac{°C}{m}} = 1.07527 \frac{\text{mol particles}}{\text{kg solvent}}. \text{ Assume 1 kg water, so we have } 1.07527 \text{ mol particles.}$$

Assume that $HC_7H_5O_2$ does not dissociate and that $C_7H_5O_2Na$ completely dissociates; then 1.07527 mol particles = mol $HC_7H_5O_2$ + 2(mol $C_7H_5O_2Na$). Use $[C_7H_5O_2Na]/[HC_7H_5O_2] = 2.29087$ and 1.07527 mol particles = mol $[HC_7H_5O_2]$ + 2(mol $C_7H_5O_2Na$) to solve for mol $HC_7H_5O_2$ and mol $C_7H_5O_2Na$. So mol $C_7H_5O_2Na = 2.29087(\text{mol } HC_7H_5O_2) \rightarrow 1.07527$ mol particles = mol $HC_7H_5O_2$ + 2(2.29087 mol $HC_7H_5O_2$) $\rightarrow$ 1.07527 mol particles = 5.58174 mol $HC_7H_5O_2 \rightarrow$ mol $HC_7H_5O_2 = 0.192641$ mol and mol $C_7H_5O_2Na = 2.29087 \times$ mol $HC_7H_5O_2 = 2.29087 \times 0.192641$ mol $HC_7H_5O_2 = 0.441315$ mol $C_7H_5O_2Na$.

$$0.192641 \text{ mol } HC_7H_5O_2 \times \frac{122.12 \text{ g } HC_7H_5O_2}{1 \text{ mol } HC_7H_5O_2} = 23.5253 \text{ g } HC_7H_5O_2 \text{ and}$$

$$0.441315 \text{ mol } C_7H_5O_2Na \times \frac{144.10 \text{ g } C_7H_5O_2Na}{1 \text{ mol } C_7H_5O_2Na} = 63.5935 \text{ g } C_7H_5O_2Na \text{ then}$$

23.5253 g $HC_7H_5O_2$ + 63.5935 g $C_7H_5O_2Na$ + 1000 g water = 1087.119 g solution then

$$1087.119 \text{ g solution} \times \frac{1 \text{ mL}}{1.01 \text{ g}} \times \frac{1 \text{ L}}{1000 \text{ mL}} = 1.076355 \text{ L. Finally,}$$

$$\frac{0.192641 \text{ mol } HC_7H_5O_2}{1.076355 \text{ L}} = 0.178975 \text{ M } HC_7H_5O_2 = 0.18 \text{ M } HC_7H_5O_2 \text{ and}$$

$$\frac{0.441315 \text{ mol } C_7H_5O_2Na}{1.076355 \text{ L}} = 0.410009 \text{ M } C_7H_5O_2Na = 0.41 \text{ M } C_7H_5O_2Na.$$

Check: The units (M and M) are correct. The magnitude of the answer makes physical sense because there are more moles of base than acid in the buffer $(pH > pK_a)$.

Challenge Problems

17.137   **Given:** 10.0 L of 75 ppm $CaCO_3$ and 55 ppm $MgCO_3$ (by mass)

Find: mass Na_2CO_3 to precipitate 90.0% of ions

Other: $K_{sp}(CaCO_3) = 4.96 \times 10^{-9}$ and $K_{sp}(MgCO_3) = 6.82 \times 10^{-6}$

Conceptual Plan: Assume that the density of water is 1.00 g/mL L water → mL water → g water then

$$\frac{1000 \text{ mL}}{1 \text{ L}} \qquad \frac{1.00 \text{ g water}}{1 \text{ mL}}$$

g water → g CaCO$_3$ → mol CaCO$_3$ → mol Ca^{2+} and g water → g MgCO$_3$ → mol MgCO$_3$ then

$$\frac{75 \text{ g CaCO}_3}{10^6 \text{ g water}} \qquad \frac{1 \text{ mol CaCO}_3}{100.09 \text{ g CaCO}_3} \qquad \frac{1 \text{ mol Ca}^{2+}}{1 \text{ mol CaCO}_3} \qquad \frac{55 \text{ g MgCO}_3}{10^6 \text{ g water}} \qquad \frac{1 \text{ mol MgCO}_3}{84.32 \text{ g MgCO}_3}$$

mol MgCO$_3$ → mol Mg^{2+} then comparing the two K_{sp} values, essentially all of the Ca^{2+} will

$$\frac{1 \text{ mol Mg}^{2+}}{1 \text{ mol MgCO}_3}$$

precipitate before the Mg^{2+} will begin to precipitate. Because 90.0% of the ions are to be precipitates, 10.0% of the ions will be left in solution (all will be Mg^{2+}).

$$(0.100)(\text{mol Ca}^{2+} + \text{mol Mg}^{2+})$$

Calculate the moles of ions remaining in solution. Then mol Mg^{2+}, L → M Mg^{2+}.

$$M = \frac{\text{mol}}{L}$$

The solubility product constant (K_{sp}) is the equilibrium expression for a chemical equation representing the dissolution of an ionic compound. The expression of the solubility product constant of A$_m$X$_n$ is $K_{sp} = [A^{n+}]^m[X^{m-}]^n$. Use this equation with M Mg^{2+} and K_{sp} → M CO$_3^{2-}$ then M CO$_3^{2-}$, L → mol CO$_3^{2-}$

$$\text{for ionic compound, A}_m\text{X}_n, K_{sp} = [A^{n+}]^m[X^{m-}]^n \qquad\qquad M = \frac{\text{mol}}{L}$$

then mol CO$_3^{2-}$ → mol Na$_2$CO$_3$ → g Na$_2$CO$_3$.

$$\frac{1 \text{ mol CO}_3^{2-}}{1 \text{ mol Na}_2\text{CO}_3} \qquad \frac{105.99 \text{ g Na}_2\text{CO}_3}{1 \text{ mol Na}_2\text{CO}_3}$$

Solution: $10.0 \text{ L} \times \dfrac{1000 \text{ mL}}{1 \text{ L}} \times \dfrac{1.00 \text{ g water}}{1 \text{ mL}} = 1.00 \times 10^4$ g water then

$$1.00 \times 10^4 \text{ g water} \times \frac{75 \text{ g CaCO}_3}{10^6 \text{ g water}} \times \frac{1 \text{ mol CaCO}_3}{100.09 \text{ g CaCO}_3} \times \frac{1 \text{ mol Ca}^{2+}}{1 \text{ mol CaCO}_3} = 0.0074933 \text{ mol Ca}^{2+} \text{ and}$$

$$1.00 \times 10^4 \text{ g water} \times \frac{55 \text{ g MgCO}_3}{10^6 \text{ g water}} \times \frac{1 \text{ mol MgCO}_3}{84.32 \text{ g MgCO}_3} \times \frac{1 \text{ mol Mg}^{2+}}{1 \text{ mol MgCO}_3} = 0.0065228 \text{ mol Mg}^{2+} \text{ so the ions}$$

remaining in solution after 90.0% precipitate out $= (0.100)(\text{mol Ca}^{2+} + \text{mol Mg}^{2+})$

$= (0.100)(0.0074933 \text{ mol Ca}^{2+} + 0.0065228 \text{ mol Mg}^{2+}) = 0.00140161$ mol ions

so $\dfrac{0.00140161 \text{ mol Mg}^{2+}}{10.0 \text{ L}} = 0.000140161 \text{ M Mg}^{2+}$. Then $K_{sp} = 6.82 \times 10^{-6}$, A = Mg^{2+}, $m = 1$, X = CO$_3^{2-}$, and

$n = 1$, so $K_{sp} = 6.82 \times 10^{-6} = [\text{Mg}^{2+}][\text{CO}_3^{2-}] = (0.00140161)[\text{CO}_3^{2-}]$. Rearrange to solve for $[\text{CO}_3^{2-}]$. So

$[\text{CO}_3^{2-}] = 0.0486583$ M. Then

$$\frac{0.0486583 \text{ mol CO}_3^{2-}}{1 \text{ L}} \times 10.0 \text{ L} \times \frac{1 \text{ mol Na}_2\text{CO}_3}{1 \text{ mol CO}_3^{2-}} \times \frac{105.99 \text{ g Na}_2\text{CO}_3}{1 \text{ mol Na}_2\text{CO}_3} = 51.6 \text{ g Na}_2\text{CO}_3.$$

Check: The units (g) are correct. The mass is reasonable to put in a washing machine load.

17.139   **Given:** excess Mg(OH)$_2$ in 1.00 L of 1.0 M NH$_4$Cl has pH = 9.00   **Find:** K_{sp}(Mg(OH)$_2$)
Other: K_b(NH$_3$) = 1.76×10^{-5}
Conceptual Plan: M NH$_4$Cl → M NH$_4^+$ and K_b → K_a then final pH → [H$_3$O$^+$] then

$$\text{NH}_4\text{Cl}(aq) \rightarrow \text{NH}_4^+(aq) + \text{Cl}^-(aq) \qquad K_w = K_a K_b \qquad [\text{H}_3\text{O}^+] = 10^{-\text{pH}}$$

M NH$_4^+$, M H$_3$O$^+$, K_a → x. Because x is significant compared to initial M NH$_4^+$, this is a buffer solution.

ICE table

The NH$_4^+$ is neutralized with Mg(OH)$_2$. Because Mg(OH)$_2$(s) ⇌ Mg^{2+}(aq) + 2 OH$^-$(aq), 2 moles of OH$^-$ are generated for each mole of Mg(OH)$_2$ dissolved. Thus, $\frac{1}{2}$(x mol OH$^-$) = mol Mg(OH)$_2$ dissolved in 1.00 L of solution. Because there is 1.00 L solution, mol Mg(OH)$_2$ = [Mg^{2+}] and [H$_3$O$^+$] → [OH$^-$].

$$K_w = [\text{H}_3\text{O}^+][\text{OH}^-]$$

Finally, write an expression for K_{sp}(Mg(OH)$_2$) and substitute values for [Mg^{2+}] and [OH$^-$].
Solution: Because one NH$_4^+$ ion is generated for each NH$_4$Cl, [NH$_4^+$] = 1.0 M NH$_4^+$. Because

$K_w = K_a K_b$, rearrange to solve for K_a. $K_a = \dfrac{K_w}{K_b} = \dfrac{1.0 \times 10^{-14}}{1.76 \times 10^{-5}} = 5.6818 \times 10^{-10}$. Final pH = 9.00, so

$[H_3O^+] = 10^{-pH} = 10^{-9.00} = 1.0 \times 10^{-9}$ M. Set up an ICE table:

$$NH_4^+(aq) + H_2O(l) \rightleftharpoons H_3O^+(aq) + NH_3(aq)$$

	$[NH_4^+]$	$[H_3O^+]$	$[NH_3]$
Initial	1.0	≈ 0.00	0.00
Change	$-x$	$+x$	$+x$
Equil	$1.0 - x$	1×10^{-9}	x

$K_a = \dfrac{[H_3O^+][NH_3]}{[NH_4^+]} = 5.\underline{6}818 \times 10^{-10} = \dfrac{(1.0 \times 10^{-9})x}{1.0 - x}$

Solve for x. $5.\underline{6}818 \times 10^{-10}(1.0 - x) = (1.0 \times 10^{-9})x \rightarrow 5.\underline{6}818 \times 10^{-10} = (1.0 \times 10^{-9} + 5.\underline{6}818 \times 10^{-10})x \rightarrow x = 0.3\underline{6}2318$, so this is a buffer solution. Because there is 1.00 L of solution, 0.362318 mol of NH_4^+ is neutralized with $Mg(OH)_2$. Because $Mg(OH)_2(s) \rightleftharpoons Mg^{2+}(aq) + 2\,OH^-(aq)$, 2 moles of OH^- are generated for each mole of $Mg(OH)_2$ dissolved. Thus, $\frac{1}{2}(0.362318$ mol $OH^-) = 0.181159$ mol $Mg(OH)_2$ was dissolved in 1.00 L of solution. Thus, the $[Mg^{2+}] = 0.181159$ M. Because $K_w = [H_3O^+][OH^-]$,

$[OH^-] = \dfrac{K_w}{[H_3O^+]} = \dfrac{1.0 \times 10^{-14}}{1.0 \times 10^{-9}} = 1.0 \times 10^{-5}$ M then

$K_{sp}(Mg(OH)_2) = [Mg^{2+}][OH^-]^2 = (0.1\underline{8}1159)(1.0 \times 10^{-5})^2 = 1.8 \times 10^{-11}$.

Check: The units (none) are correct. The magnitude of the answer makes physical sense because the concentration of NH_4Cl is high; so it took a significant amount of $Mg(OH)_2$ to raise the pH to 9.00. Note that this number disagrees with the accepted value for the $K_{sp}(Mg(OH)_2)$. This is most likely due to errors in the measurements in this experiment.

17.141 (a)   **Given:** $Au(OH)_3$ in pure water   **Find:** molar solubility (S)   **Other:** $K_{sp} = 5.5 \times 10^{-46}$
Conceptual Plan: Use equations derived in Problems 17.19 and 17.20 and solve for S. Then

$$\text{for ionic compound, } A_mX_n, K_{sp} = [A^{n+}]^m[X^{m-}]^n = m^m n^n S^{m+n}$$

check answer for validity.
Solution: For $Au(OH)_3$, $K_{sp} = 5.5 \times 10^{-46}$, $A = Au^{3+}$, $m = 1$, $X = OH^-$, and $n = 3$; so $K_{sp} = [Au^{3+}][OH^-]^3$

$= 5.5 \times 10^{-46} = 3^3 S^4$. Rearrange to solve for S. $S \sqrt[4]{\dfrac{5.5 \times 10^{-46}}{27}} = 2.1 \times 10^{-12}$ M. This answer

suggests that the $[OH^-] = 3(2.1 \times 10^{-12}$ M$) = 6.3 \times 10^{-12}$ M. This result is lower than what is found in pure water (1.0×10^{-7} M), so substitute this value for $[OH^-]$ and solve for $S = [Au^{3+}]$. So $K_{sp} = [Au^{3+}][OH^-]^3 = 5.5 \times 10^{-46} = S(1.0 \times 10^{-7}$ M$)^3$, and solving for S gives $S = 5.5 \times 10^{-25}$ M.

Check: The units (M) are correct. Because K_{sp} is so small, the autoionization of water must be considered, and the solubility is smaller than what is normally anticipated.

(b)   **Given:** $Au(OH)_3$ in 1.0 M HNO_3 **Find:** molar solubility (S)   **Other:** $K_{sp} = 5.5 \times 10^{-46}$
Conceptual Plan: Because HNO_3 is a strong acid, it will neutralize the gold(III) hydroxide (through the reaction of H^+ with OH^+ to form water (the reverse of the autoionization of water equilibrium). Write balanced equations for dissolving the solid and for the neutralization reaction. Add these two reactions to get the desired overall reaction. Using the rules from Chapter 15, multiply the individual reaction Ks to get the overall K for the sum of these reactions. Then M HNO_3, $K \rightarrow S$.

ICE table

Solution: Identify the solid as $Au(OH)_3$. Write the individual reactions and add them together.

$$Au(OH)_3(s) \rightleftharpoons Au^{3+}(aq) + 3\,\cancel{OH^-(aq)} \qquad K_{sp} = 5.5 \times 10^{-46}$$

$$3\,H^+(aq) + 3\,\cancel{OH^-(aq)} \rightleftharpoons 3\,H_2O(l) \qquad \left(\dfrac{1}{K_w}\right)^3 = \left(\dfrac{1}{1.0 \times 10^{-14}}\right)^3$$

$$Au(OH)_3(s) + 3\,H^+(aq) \rightleftharpoons Au^{3+}(aq) + 3\,H_2O(l)$$

Because the overall reaction is the sum of the dissolution reaction and three times the reverse of the autoionization of water reaction, the overall reaction

$$K = K_{sp}\left(\dfrac{1}{K_w}\right)^3 = (5.5 \times 10^{-46})\left(\dfrac{1}{1.0 \times 10^{-14}}\right)^3 = 5.5 \times 10^{-4} = \dfrac{[Au^{3+}]}{[H^+]^3}; \text{ then because } HNO_3 \text{ is a strong}$$

acid, it will completely dissociate to H^+ and NO_3^-. Set up an ICE table:
$$Au(OH)_3(s) + 3\,H^+(aq) \rightleftharpoons Au^{3+}(aq) + 3\,H_2O(l)$$

	$[H^+]$	$[Au^{3+}]$
Initial	1.0	0.00
Change	$-3S$	$+S$
Equil	$1.0 - 3S$	$+S$

$$K = \frac{[Au^{3+}]}{[H^+]^3} = 5.5 \times 10^{-4} = \frac{S}{(1.0 - 3S)^3}.$$

Assume that S is small ($3S \ll 1.0$), so $\dfrac{S}{(1.0 - \cancel{3S})^3} = 5.5 \times 10^{-4}\,M = \dfrac{S}{(1.0)^3} = S$. Confirm that the

assumption is valid. $\dfrac{3(5.5 \times 10^{-4})}{1.0} \times 100\% = 0.017\% \ll 5\%$, so the assumption is valid.

Check: The units (M) are correct. K is much larger than the original K_{sp}, so the solubility of $Au(OH)_3$ increases over that of pure water.

17.143 **Given:** 1.00 L of 0.100 M $MgCO_3$ **Find:** volume of 0.100 M Na_2CO_3 to precipitate 99% of Mg^{2+} ions
Other: $K_{sp}(MgCO_3) = 6.82 \times 10^{-6}$
Conceptual Plan: Because 99% of the Mg^{2+} ions are to be precipitated, 1% of the ions will be left in solution.

$$(0.01)(0.100\,M\,Mg^{2+})$$

Let x = required volume (in L). Calculate the amount of CO_3^{2-} added and the amount of Mg^{2+} that does not precipitate and remains in solution. Use these to calculate the $[Mg^{2+}]$ and $[CO_3^{2-}]$. The solubility product constant (K_{sp}) is the equilibrium expression for a chemical equation representing the dissolution of an ionic compound. The expression of the solubility product constant of A_mX_n is $K_{sp} = [A^{n+}]^m[X^{m-}]^n$. Substitute these expressions in this equation to $[Mg^{2+}], [CO_3^{2-}], K_{sp} \rightarrow x$.

$$\text{for ionic compound, } A_mX_n, K_{sp} = [A^{n+}]^m[X^{m-}]^n$$

Solution: Because 99% of the Mg^{2+} ions are to be precipitated, 1% of the ions will be left in solution, or $(0.01)(0.100\,M\,Mg^{2+}) = 0.001\,M\,Mg^{2+}$. Let x = required volume (in L). The volume of the solution after precipitation is $(1.00 + x)$. The amount of CO_3^{2-} added $= (0.100M)(x\,L) = 0.100x$ mol CO_3^{2-}. The amount of Mg^{2+} that does not precipitate and remains in solution is $(0.100\,M)(1.00\,L)(0.01) = 1.00 \times 10^{-3}$ mol, and the amount that precipitates $= 0.099$ mol, which is also equal to the amount of CO_3^{2-} used. The amount of CO_3^{2-} remaining in solution is $(0.10x - 0.099)$. Thus, $[Mg^{2+}] = 1.00 \times 10^{-3}mol/(1.00 + x)$ L and $[CO_3^{2-}] = (0.10x - 0.099)$ mol$/(1.00 + x)$ L. Then $K_{sp} = 6.82 \times 10^{-6}$, $A = Mg^{2+}, m = 1, X = CO_3^{2-}$, and

$n = 1$; so $K_{sp} = 6.82 \times 10^{-6} = [Mg^{2+}][CO_3^{2-}] = \dfrac{(1.00 \times 10^{-3})(0.10x - 0.099)}{(1.00 + x)^2}$. Rearrange to solve for x.

$1.00 + 2.00x + x^2 = \dfrac{1.0 \times 10^{-4}x - 9.9 \times 10^{-5}}{6.82 \times 10^{-6}} \rightarrow 0 = x^2 - 12.6628x + 15.5161$. Using quadratic equation,

$x = 1.\underline{3}75\,L = 1.4\,L$.

Check: The units (L) are correct. The necessary concentration is very low, so the volume is fairly large.

17.145 **Given:** 1.0 L solution with 0.10 M $Ba(OH)_2$ and excess $Zn(OH)_2$ **Find:** pH
Other: $K_{sp}(Zn(OH)_2) = 3 \times 10^{-15}$, $K_f(Zn(OH)_4^{2-}) = 2 \times 10^{15}$
Conceptual Plan: Because $[Ba(OH)_2] = 0.10$ M, $[OH^-] = 0.20$ M. Write balanced equations for the solubility of $Zn(OH)_2$ and reaction with excess OH^- and expressions for K_{sp} and K_f. Use initial concentrations to set up an ICE table. Solve for $[OH^-]$ at equilibrium. Then $[OH^-] \rightarrow [H_3O^+] \rightarrow pH$.

$$K_w = [H_3O^+][OH^-] \qquad pH = -\log[H_3O^+]$$

Solution: Write two reactions and combine.

$Zn(OH)_2(s) \rightleftharpoons Zn^{2+}\cancel{(aq)} + \cancel{2}\,OH^-(aq)$ with $K_{sp} = [Zn^{2+}][OH^-]^2 = 3 \times 10^{-15}$

$\cancel{Zn^{2+}(aq)} + \cancel{2}4\,OH^-(aq) \rightleftharpoons Zn(OH)_4^{2-}(aq)$ with $K_f = \dfrac{[Zn(OH)_4^{2-}]}{[Zn^{2+}][OH^-]^4} = 2 \times 10^{15}$

$Zn(OH)_2(s) + 2\,OH^-(aq) \rightleftharpoons Zn(OH)_4^{2-}(aq)$

with $K = K_{sp}K_f = \cancel{[Zn^{2+}]}\cancel{[OH^-]}^2 \dfrac{[Zn(OH)_4^{2-}]}{\cancel{[Zn^{2+}]}\cancel{[OH^-]}^{\cancel{4}2}} = (3 \times 10^{-15})(2 \times 10^{15})$

$K = \dfrac{[Zn(OH)_4^{2-}]}{[OH^-]^2} = 6$. Set up an ICE table with initial concentration and solve for x.

$Zn(OH)_2(s) + 2\,OH^-(aq) \rightleftharpoons Zn(OH)_4^{2-}(aq)$

	$[\text{OH}^-]$	$[\text{Zn(OH)}_4^{2-}]$
Initial	0.20	0.00
Change	$-2x$	x
Equil	$0.20 - 2x$	x

$$K = \frac{[\text{Zn(OH)}_4^{2-}]}{[\text{OH}^-]^2} = \frac{x}{(0.20 - 2x)^2} = 6$$

$x = 6(0.20 - 2x)^2 \rightarrow x = 6(4x^2 - 0.80x + 0.040) \rightarrow x = 24x^2 - 4.8x + 0.24 \rightarrow 0 = 24x^2 - 5.8x + 0.24.$
Using quadratic equation, $x = 0.0530049 \rightarrow [\text{OH}^-] = 0.20 - 2x = 0.20 - 2(0.0530049) = 0.093990$ M OH$^-$.
Then $K_w = [\text{H}_3\text{O}^+][\text{OH}^-]$, so

$$[\text{H}_3\text{O}^+] = \frac{K_w}{[\text{OH}^-]} = \frac{1.0 \times 10^{-14}}{0.093990} = 1.06394 \times 10^{-13} \text{ M. Finally,}$$

$$\text{pH} = -\log[\text{H}_3\text{O}^+] = -\log(1.06394 \times 10^{-13}) = 12.97.$$

Check: The units (none) are correct. Because the pH of the solution before the addition of the Zn(OH)$_2$ is 13.30, the reaction decreases the $[\text{OH}^-]$ and the pH drops.

Conceptual Problems

17.147   If the concentration of the acid is greater than the concentration of the base, the pH will be less than the pK_a. If the concentration of the acid is equal to the concentration of the base, the pH will be equal to the pK_a. If the concentration of the acid is less than the concentration of the base, the pH will be greater than the pK_a.

    (a)   pH $<$ pK_a

    (b)   pH $>$ pK_a

    (c)   pH $=$ pK_a; the OH$^-$ will convert half of the acid to base.

    (d)   pH $>$ pK_a; the OH$^-$ will convert more than half of the acid to base.

17.149   Only (a) is the same for both solutions. The volume to the first equivalence point will be the same because the number of moles of acid is the same. The pH profiles of the two titrations will be different.

17.151   (a)   The solubility will be unchanged because the pH is constant and no common ions are added.

    (b)   The solubility will be less because extra fluoride ions are added, suppressing the solubility of the fluoride ionic compound.

    (c)   The solubility will increase because some of the fluoride ion will be converted to HF; so more of the ionic compound can be dissolved.

Questions for Group Work

17.153   Start with the balanced equation for the ionization of a weak acid: HA (aq) + H$_2$O (l) $\rightleftharpoons$ H$_3$O$^+$ (aq) + A$^-$ (aq)

Write the expression for the equilibrium constant: $K_a = \dfrac{[\text{H}_3\text{O}^+][\text{A}^-]}{[\text{HA}]}$.

Rearrange this expression to solve for [H$_3$O$^+$]: $[\text{H}_3\text{O}^+] = K_a \dfrac{[\text{HA}]}{[\text{A}^-]}$

Take the negative logarithm of this equation: $-\log[\text{H}_3\text{O}^+] = -\log\left(K_a \dfrac{[\text{HA}]}{[\text{A}^-]}\right)$

Since the $-\log(AB) = -\log A - \log B$, rearrange the equation: $-\log[\text{H}_3\text{O}^+] = -\log K_a - \log \dfrac{[\text{HA}]}{[\text{A}^-]}$

Since pH $= -\log[\text{H}_3\text{O}^+]$ and p$K_a = -\log K_a$, substituting into the equation: pH $= \text{p}K_a - \log \dfrac{[\text{HA}]}{[\text{A}^-]}$

Since the $-\log(A/B) = \log B/A$, rearrange the equation: pH $= \text{p}K_a + \log \dfrac{[\text{A}^-]}{[\text{HA}]}$

17.155   (a)   Since this is a solubility equilibrium, the equilibrium constant will be $K_{sp} = 4.96 \times 10^{-9}$.

         (b) $K_{sp} = 4.96 \times 10^{-9} = [Ca^{2+}][CO_3^{2-}] = S^2$ and so $S = 7.04 \times 10^{-5}$ M.

         (c) $200. \text{ mL} \times \dfrac{1 \text{ L}}{1000 \text{ mL}} \times \dfrac{7.04 \times 10^{-5} \text{ mol CaCO}_3}{1 \text{ L}} \times \dfrac{100.09 \text{ g CaCO}_3}{1 \text{ mol CaCO}_3} \times \dfrac{1000 \text{ mg CaCO}_3}{1 \text{ g CaCO}_3} = 1.41 \text{ mg CaCO}_3$

         (d)   Acidic. Calcium carbonate is more soluble in acidic solution because the carbonate ion is a weak base and will react with the acid in the cleaning solution.

Data Interpretation and Analysis

17.157   (a)   The point of steepest drop is at about 31 mL.

         (c) $31 \text{ mL HCl} \times \dfrac{0.100 \text{ mol HCl}}{1000 \text{ mL HCl}} = 0.0031 \text{ mol HCl} = 0.0031 \text{ mol base}$

               so the molar mass is $\dfrac{0.30 \text{ g base}}{0.0031 \text{ mol base}} = 97 \text{ g/mol}$.

         (e)   The molar mass (97 g/mol) is between the molar masses of aniline (93.13 g/mol) and triethylamine (101.19 g/mol). The K_b (10^{-3}) is much closer to that of triethylamine (5.6×10^{-4}), so the base is likely to be triethylamine.

18 Free Energy and Thermodynamics

Review Questions

18.1   The first law of thermodynamics states that energy is conserved in chemical processes. When we burn gasoline to run a car, for example, the amount of energy produced by the chemical reaction does not vanish, nor does any new energy appear that was not present before the combustion. Some of the energy from the combustion reaction goes toward driving the car forward (about 20%), and the rest is dissipated into the surroundings as heat (just feel the engine). However, the total energy given off by the combustion reaction exactly equals the sum of the amount of energy driving the car forward and the amount dissipated as heat—energy is conserved. In other words, when it comes to energy, you can't win; you cannot create energy that was not there to begin with.

18.3   A perpetual motion machine perpetually moves without any energy input. This machine is not possible because if the machine is to be in motion, it must pay the heat tax with each cycle of its motion. Over time, then, it will run down and stop moving.

18.5   A spontaneous process occurs without ongoing outside intervention (such as the performance of work by some external force). For example, when you drop a book in a gravitational field, the book spontaneously drops to the floor.

18.7   Entropy (S) is a thermodynamic function that is proportional to the number of energetically equivalent ways to arrange the components of a system to achieve a particular state. Entropy, like enthalpy, is a state function—its value depends only on the state of the system, not on how the system got to that state. Therefore, for any process, the change in entropy is just the entropy of the final state minus the entropy of the initial state, or $\Delta S = S_{final} - S_{initial}$.

18.9   Microstates are the number of internal arrangements. These microstates give rise to the same external arrangement, or macrostate. If there are three gas particles (A, B, and C) and two containers, the fact that the first container has two particles and the other container has one particle is a macrostate. The fact that particles A and C are in the first container and particle B is in the other container is a microstate. There are at least as many microstates as there are macrostates.

18.11   The second law of thermodynamics states that for any spontaneous process, the entropy of the universe increases ($\Delta S_{univ} > 0$). The criterion for spontaneity is the entropy of the universe. Processes that increase the entropy of the universe—those that result in greater dispersal or randomization of energy—occur spontaneously. Processes that decrease the entropy of the universe do not occur spontaneously. Heat travels from a substance at higher temperature to one at lower temperature because the process disperses thermal energy. The cooler object has less thermal energy, so transferring heat from the warmer object results in greater energy randomization—the energy that was concentrated in the hot substance becomes dispersed between the two substances.

18.13   When water freezes at temperatures below 0 °C, the entropy of the water decreases, yet the process is spontaneous because the entropy of the universe increases ($\Delta S_{univ} > 0$). The entropy of the system can decrease ($\Delta S_{sys} < 0$) as long as the entropy of the surroundings increases by a greater amount ($\Delta S_{surr} > -\Delta S_{sys}$), so that the overall entropy of the universe undergoes a net increase. For liquid water freezing, the change in entropy for the system (ΔS_{sys}) is negative because the water becomes more ordered. For ΔS_{univ} to be positive, ΔS_{surr} must be positive and greater in absolute value (or magnitude) than ΔS_{sys}. We learned in Chapter 6 that freezing is an exothermic process: it gives off heat to the surroundings. If we think of entropy as the dispersal or randomization of energy, the release of heat energy by the system disperses that energy into the surroundings, increasing the entropy of the surroundings. The freezing of water below 0 °C increases the entropy of the universe because the heat given off to the surroundings increases the entropy of the surroundings to a sufficient degree to overcome the entropy decrease in the water. The freezing of water becomes nonspontaneous above 0 °C because the magnitude of the increase in the entropy of the surroundings due to the dispersal of energy into the surroundings is temperature-dependent. The higher the temperature, the smaller the percent increase in entropy for a given rise in temperature. Therefore, the impact of the heat released to the surroundings by the freezing of water depends on the temperature of the surroundings—the higher the temperature, the smaller the impact.

18.15   The change in Gibbs free energy for a process is proportional to the negative of ΔS_{univ}. Because ΔS_{univ} is a criterion for spontaneity, ΔG is also a criterion for spontaneity (though opposite in sign).

18.17   The third law of thermodynamics states that the entropy of a perfect crystal at absolute zero (0 K) is zero. For enthalpy, we defined a standard state so that we could define a "zero" for the scale. This is not necessary for entropy because there is an absolute zero.

18.19   The larger the molar mass, the greater the entropy at 25 °C. For a given state of matter, entropy generally increases with increasing molecular complexity.

18.21   The three ways of calculating the ΔG_{rxn} are:
   - Use tabulated values of standard enthalpies of formation to calculate ΔH_{rxn}° and use tabulated values of standard entropies to calculate ΔS_{rxn}°; then use the values of ΔH_{rxn}° and ΔS_{rxn}° calculated in these ways to calculate the standard free energy change for a reaction by using the equation $\Delta G_{rxn}^{\circ} = \Delta H_{rxn}^{\circ} - T\Delta S_{rxn}^{\circ}$.
   - Use tabulated values of the standard free energies of formation to calculate ΔG_{rxn}° using an equation similar to that used for standard enthalpy of a reaction $\Delta G_{rxn}^{\circ} = \sum n_p \Delta G_f^{\circ}(\text{products}) - \sum n_r \Delta G_f^{\circ}(\text{reactants})$.
   - Use a reaction pathway or stepwise reaction to sum the changes in free energy for each of the steps in a manner similar to that used in Chapter 6 for enthalpy of stepwise reactions.

The method to calculate the free energy of a reaction at temperatures other than at 25 °C is the first method. The second method is only applicable at 25 °C. The third method is only applicable at the temperature of the individual reactions, generally 25 °C.

18.23   The standard free energy change for a reaction (ΔG_{rxn}°) applies only to standard conditions. For a gas, standard conditions are those in which the pure gas is present at a partial pressure of 1 atm. For nonstandard conditions, we need to calculate ΔG_{rxn} (not ΔG_{rxn}°) to predict spontaneity.

18.25   The free energy of reaction under nonstandard conditions (ΔG_{rxn}) can be calculated from ΔG_{rxn}° using the relationship $\Delta G_{rxn} = \Delta G_{rxn}^{\circ} + RT \ln Q$, where Q is the reaction quotient (defined in Section 14.7), T is the temperature in K, and R is the gas constant in the appropriate units (8.314 J/mol K).

Problems by Topic

Entropy, the Second Law of Thermodynamics, and the Direction of Spontaneous Change

18.27   (a) and (c) are spontaneous processes.

18.29   System B has the greatest entropy. There is only one energetically equivalent arrangement for System A. However, the particles of System B may exchange positions for a second energetically equivalent arrangement.

18.31   **Given:** 1.00 mol C_3H_8O, $\Delta H_{fus}^{\circ} = 5.37$ kJ/mol, $T = -89.5$ °C   **Find:** ΔS
Conceptual Plan: °C $\rightarrow$ K   then $\Delta H_{fus}^{\circ} \rightarrow q_{rev}$ then   kJ $\rightarrow$ J   then $q_{rev}, T \rightarrow \Delta S$

$$K = 273.15 + °C \qquad q_{rev} = \Delta H_{fus}^{\circ} \qquad \frac{1000 \text{ J}}{1 \text{ kJ}} \qquad \Delta S = \frac{q_{rev}}{T}$$

Solution: $T = 273.15 + (-89.5\ °C) = 183.\underline{6}5$ K; $q_{rev} = \Delta H°_{fus} = 5.37$ kJ/mol; $5.37\ \dfrac{kJ}{mol} \times \dfrac{1000\ J}{1\ kJ} = 5.37 \times 10^3$

J/mol then $\Delta S = \dfrac{q_{rev}}{T} = \dfrac{5.37 \times 10^3\ J/mol}{183.65\ K} = 29.2\ \dfrac{J}{K}$ for 1.00 mol

Check: The units (J/K) are correct. The magnitude of the answer (29 J/K) makes sense because the number of kJ is much smaller than the temperature and then a factor of 10^3 was applied.

18.33   **Given:** 45.0 g C_3H_6O, $\Delta H°_{fus} = 5.69$ kJ/mol, $T = -94.8\ °C$ **Find:** ΔS
Conceptual Plan:
$°C \rightarrow K$ then $\Delta H°_{fus} \rightarrow q_{rev}$ then   kJ/mol $\rightarrow$ J/mol   then   J/mol $\rightarrow$ J/g then $q_{rev}, T \rightarrow \Delta S$

$\quad K = 273.15 + °C \qquad\qquad q_{rev} = \Delta H°_{fus} \qquad\qquad \dfrac{1000\ J}{1\ kJ} \qquad\qquad \dfrac{1\ mol}{58.08\ g} \qquad\qquad \Delta S = \dfrac{q_{rev}}{T}$

Solution: $T = 273.15 + (-94.8\ °C) = 178.\underline{3}5$ K; $q_{rev} = \Delta H°_{fus} = 5.69$ kJ/mol; $5.69\ \dfrac{kJ}{mol} \times \dfrac{1000\ J}{1\ kJ} = 5.69 \times 10^3$ J/mol

then $45.0\ g \times 5.69 \times 10^3\ \dfrac{J}{mol} \times \dfrac{1\ mol}{58.08\ g} = 4.4\underline{0}857 \times 10^3$ J then $\Delta S = \dfrac{q_{rev}}{T} = \dfrac{4.4\underline{0}857 \times 10^3\ J}{178.35\ K} = 24.7\ \dfrac{J}{K}$

for 45.0 g.

Check: The units (J/K) are correct. The magnitude of the answer (25 J/K) makes sense because we have a little less than a mole and the number of kJ is much smaller than the temperature and then a factor of 10^3 was applied.

18.35   (a) $\Delta S > 0$ because a gas is being generated.
(b) $\Delta S < 0$ because 2 moles of gas are being converted to 1 mole of gas.
(c) $\Delta S < 0$ because a gas is being converted to a solid.
(d) $\Delta S < 0$ because 4 moles of gas are being converted to 2 moles of gas.

18.37   (a) $\Delta S_{sys} > 0$ because 6 moles of gas are being converted to 7 moles of gas. Because $\Delta H < 0$ and $\Delta S_{surr} > 0$, the reaction is spontaneous at all temperatures.
(b) $\Delta S_{sys} < 0$ because 2 moles of different gases are being converted to 2 moles of one gas. Because $\Delta H > 0$ and $\Delta S_{surr} < 0$, the reaction is nonspontaneous at all temperatures.
(c) $\Delta S_{sys} < 0$ because 3 moles of gas are being converted to 2 moles of gas. Because $\Delta H > 0$ and $\Delta S_{surr} < 0$, the reaction is nonspontaneous at all temperatures.
(d) $\Delta S_{sys} > 0$ because 9 moles of gas are being converted to 10 moles of gas. Because $\Delta H < 0$ and $\Delta S_{surr} > 0$, the reaction is spontaneous at all temperatures.

18.39   (a)   **Given:** $\Delta H°_{rxn} = -385$ kJ, $T = 298$ K   **Find:** ΔS_{surr}
Conceptual Plan: kJ $\rightarrow$ J then $\Delta H°_{rxn}, T \rightarrow \Delta S_{surr}$

$\qquad\qquad \dfrac{1000\ J}{1\ kJ} \qquad\qquad \Delta S_{surr} = \dfrac{-\Delta H_{sys}}{T}$

Solution: $-385\ kJ \times \dfrac{1000\ J}{1\ kJ} = -385{,}000$ J then

$\Delta S_{surr} = \dfrac{-\Delta H_{sys}}{T} = \dfrac{-(-385{,}000\ J)}{298\ K} = 1290\ J/K = 1.29 \times 10^3\ J/K$

Check: The units (J/K) are correct. The magnitude of the answer (10^3 J/K) makes sense because the kJ and the temperature started with very similar values and then a factor of 10^3 was applied.

(b)   **Given:** $\Delta H°_{rxn} = -385$ kJ, $T = 77$ K   **Find:** ΔS_{surr}
Conceptual Plan: kJ $\rightarrow$ J then $\Delta H°_{rxn}, T \rightarrow \Delta S_{surr}$

$\qquad\qquad \dfrac{1000\ J}{1\ kJ} \qquad\qquad \Delta S_{surr} = \dfrac{-\Delta H_{sys}}{T}$

Solution: $-385\ kJ \times \dfrac{1000\ J}{1\ kJ} = -385{,}000$ J then $\Delta S_{surr} = \dfrac{-\Delta H_{sys}}{T} = \dfrac{-(-385{,}000\ J)}{77\ K} = 5.00 \times 10^3\ J/K$

Check: The units (J/K) are correct. The magnitude of the answer (5×10^3 J/K) makes sense because the temperature is much lower than that in part (a); so the answer should increase.

(c) **Given:** $\Delta H^\circ_{rxn} = +114$ kJ, $T = 298$ K **Find:** ΔS_{surr}
Conceptual Plan: kJ → J then $\Delta H^\circ_{rxn}, T \rightarrow \Delta S_{surr}$

$$\frac{1000 \text{ J}}{1 \text{ kJ}} \qquad \Delta S_{surr} = \frac{-\Delta H_{sys}}{T}$$

Solution: $+114 \text{ kJ} \times \dfrac{1000 \text{ J}}{1 \text{ kJ}} = +114{,}000 \text{ J}$ then $\Delta S_{surr} = \dfrac{-\Delta H_{sys}}{T} = \dfrac{-114{,}000 \text{ J}}{298 \text{ K}} = -383 \text{ J/K}$

Check: The units (J/K) are correct. The magnitude of the answer (-400 J/K) makes sense because the kJ are less and of the opposite sign than those in part (a); so the answer should decrease.

(d) **Given:** $\Delta H^\circ_{rxn} = +114$ kJ, $T = 77$ K **Find:** ΔS_{surr}
Conceptual Plan: kJ → J then $\Delta H^\circ_{rxn}, T \rightarrow \Delta S_{surr}$

$$\frac{1000 \text{ J}}{1 \text{ kJ}} \qquad \Delta S_{surr} = \frac{-\Delta H_{sys}}{T}$$

Solution: $+114 \text{ kJ} \times \dfrac{1000 \text{ J}}{1 \text{ kJ}} = +114{,}000 \text{ J}$ then

$$\Delta S_{surr} = \frac{-\Delta H_{sys}}{T} = \frac{-114{,}000 \text{ J}}{77 \text{ K}} = -1480 \frac{\text{J}}{\text{K}} = -1.48 \times 10^3 \text{ J/K}$$

Check: The units (J/K) are correct. The magnitude of the answer (-2×10^3 J/K) makes sense because the temperature is much lower than that in part (c); so the answer should increase.

18.41 **(a)** **Given:** $\Delta H^\circ_{rxn} = +115$ kJ, $\Delta S_{rxn} = -263$ J/K, $T = 298$ K   **Find:** ΔS_{univ} and spontaneity
Conceptual Plan: kJ → J then $\Delta H^\circ_{rxn}, T \rightarrow \Delta S_{surr}$ then $\Delta S_{rxn}, \Delta S_{surr} \rightarrow \Delta S_{univ}$

$$\frac{1000 \text{ J}}{1 \text{ kJ}} \qquad \Delta S_{surr} = \frac{-\Delta H_{sys}}{T} \qquad \Delta S_{univ} = \Delta S_{sys} + \Delta S_{surr}$$

Solution: $115 \text{ kJ} \times \dfrac{1000 \text{ J}}{1 \text{ kJ}} = 115{,}000 \text{ J}$ then $\Delta S_{surr} = \dfrac{-\Delta H_{sys}}{T} = \dfrac{-(115{,}000 \text{ J})}{298 \text{ K}} = -385.906 \text{ J/K}$ then

$\Delta S_{univ} = \Delta S_{sys} + \Delta S_{surr} = -263 \text{ J/K} - 385.906 \text{ J/K} = -649 \text{ J/K}$; so the reaction is nonspontaneous.

Check: The units (J/K) are correct. The magnitude of the answer (-650 J/K) makes sense because both terms are negative; so the reaction is nonspontaneous.

(b) **Given:** $\Delta H^\circ_{rxn} = -115$ kJ, $\Delta S_{rxn} = +263$ J/K, $T = 298$ K   **Find:** ΔS_{univ} and spontaneity
Conceptual Plan: kJ → J then $\Delta H^\circ_{rxn}, T \rightarrow \Delta S_{surr}$ then $\Delta S_{rxn}, \Delta S_{surr} \rightarrow \Delta S_{univ}$

$$\frac{1000 \text{ J}}{1 \text{ kJ}} \qquad \Delta S_{surr} = \frac{-\Delta H_{sys}}{T} \qquad \Delta S_{univ} = \Delta S_{sys} + \Delta S_{surr}$$

Solution: $-115 \text{ kJ} \times \dfrac{1000 \text{ J}}{1 \text{ kJ}} = -115{,}000 \text{ J}$ then $\Delta S_{surr} = \dfrac{-\Delta H_{sys}}{T} = \dfrac{-(-115{,}000 \text{ J})}{298 \text{ K}} = 385.906 \text{ J/K}$

then $\Delta S_{univ} = \Delta S_{sys} + \Delta S_{surr} = +263 \text{ J/K} + 385.906 \text{ J/K} = +649 \text{ J/K}$; so the reaction is spontaneous.

Check: The units (J/K) are correct. The magnitude of the answer ($+650$ J/K) makes sense because both terms were positive; so the reaction is spontaneous.

(c) **Given:** $\Delta H^\circ_{rxn} = -115$ kJ, $\Delta S_{rxn} = -263$ J/K, $T = 298$ K **Find:** ΔS_{univ} and spontaneity
Conceptual Plan: kJ → J then $\Delta H^\circ_{rxn}, T \rightarrow \Delta S_{surr}$ then $\Delta S_{rxn}, \Delta S_{surr} \rightarrow \Delta S_{univ}$

$$\frac{1000 \text{ J}}{1 \text{ kJ}} \qquad \Delta S_{surr} = \frac{-\Delta H_{sys}}{T} \qquad \Delta S_{univ} = \Delta S_{sys} + \Delta S_{surr}$$

Solution: $-115 \text{ kJ} \times \dfrac{1000 \text{ J}}{1 \text{ kJ}} = -115{,}000 \text{ J}$ then $\Delta S_{surr} = \dfrac{-\Delta H_{sys}}{T} = \dfrac{-(-115{,}000 \text{ J})}{298 \text{ K}} = +385.906 \text{ J/K}$

then $\Delta S_{univ} = \Delta S_{sys} + \Delta S_{surr} = -263 \text{ J/K} + 385.906 \text{ J/K} = +123 \text{ J/K}$; so the reaction is spontaneous.

Check: The units (J/K) are correct. The magnitude of the answer (120 J/K) makes sense because the larger term was positive; so the reaction is spontaneous.

(d) **Given:** $\Delta H^\circ_{rxn} = -115$ kJ, $\Delta S_{rxn} = -263$ J/K, $T = 615$ K **Find:** ΔS_{univ} and spontaneity
Conceptual Plan: kJ → J then $\Delta H^\circ_{rxn}, T \rightarrow \Delta S_{surr}$ then $\Delta S_{rxn}, \Delta S_{surr} \rightarrow \Delta S_{univ}$

$$\frac{1000 \text{ J}}{1 \text{ kJ}} \qquad \Delta S_{surr} = \frac{-\Delta H_{sys}}{T} \qquad \Delta S_{univ} = \Delta S_{sys} + \Delta S_{surr}$$

Solution: $-115 \text{ kJ} \times \dfrac{1000 \text{ J}}{1 \text{ kJ}} = -115{,}000 \text{ J}$ then $\Delta S_{\text{surr}} = \dfrac{-\Delta H_{\text{sys}}}{T} = \dfrac{-(-115{,}000 \text{ J})}{615 \text{ K}} = +186.992 \text{ J/K}$

then $\Delta S_{\text{univ}} = \Delta S_{\text{sys}} + \Delta S_{\text{surr}} = -263 \text{ J/K} + 186.992 \text{ J/K} = -76 \text{ J/K}$; so the reaction is nonspontaneous.

Check: The units (J/K) are correct. The magnitude of the answer (-80 J/K) makes sense because the larger term was negative, so the reaction is nonspontaneous.

Standard Entropy Changes and Gibbs Free Energy

18.43   (a)   **Given:** $\Delta H^\circ_{\text{rxn}} = +115 \text{ kJ}$, $\Delta S_{\text{rxn}} = -263 \text{ J/K}$, $T = 298 \text{ K}$ **Find:** ΔG and spontaneity

Conceptual Plan: $\text{J/K} \rightarrow \text{kJ/K}$ then $\Delta H^\circ_{\text{rxn}}, \Delta S_{\text{rxn}}, T \rightarrow \Delta G$

$$\dfrac{1000 \text{ J}}{1 \text{ kJ}} \qquad\qquad \Delta G = \Delta H_{\text{rxn}} - T\Delta S_{\text{rxn}}$$

Solution: $-263 \dfrac{\text{J}}{\text{K}} \times \dfrac{1 \text{ kJ}}{1000 \text{ J}} = -0.263 \text{ kJ/K}$ then

$\Delta G = \Delta H_{\text{rxn}} - T\Delta S_{\text{rxn}} = +115 \text{ kJ} - \left((298 \text{ K})\left(-0.263 \dfrac{\text{kJ}}{\text{K}} \right) \right) = +1.93 \times 10^2 \text{ kJ} = +1.93 \times 10^5 \text{ J}$; so the reaction is nonspontaneous.

Check: The units (kJ) are correct. The magnitude of the answer $(+190 \text{ kJ})$ makes sense because both terms are positive; so the reaction is nonspontaneous.

(b)   **Given:** $\Delta H^\circ_{\text{rxn}} = -115 \text{ kJ}$, $\Delta S_{\text{rxn}} = +263 \text{ J/K}$, $T = 298 \text{ K}$ **Find:** ΔG and spontaneity

Conceptual Plan: $\text{J/K} \rightarrow \text{kJ/K}$ then $\Delta H^\circ_{\text{rxn}}, \Delta S_{\text{rxn}}, T \rightarrow \Delta G$

$$\dfrac{1000 \text{ J}}{1 \text{ kJ}} \qquad\qquad \Delta G = \Delta H_{\text{rxn}} - T\Delta S_{\text{rxn}}$$

Solution: $+263 \dfrac{\text{J}}{\text{K}} \times \dfrac{1 \text{ kJ}}{1000 \text{ J}} = +0.263 \text{ kJ/K}$ then

$\Delta G = \Delta H_{\text{rxn}} - T\Delta S_{\text{rxn}} = -115 \text{ kJ} - \left((298 \text{ K})\left(0.263 \dfrac{\text{kJ}}{\text{K}} \right) \right) = -193 \text{ kJ} = -1.93 \times 10^2 \text{ kJ} = -1.93 \times 10^5 \text{ J}$;

so the reaction is spontaneous.

Check: The units (kJ) are correct. The magnitude of the answer (-190 kJ) makes sense because both terms are negative; so the reaction is spontaneous.

(c)   **Given:** $\Delta H^\circ_{\text{rxn}} = -115 \text{ kJ}$, $\Delta S_{\text{rxn}} = -263 \text{ J/K}$, $T = 298 \text{ K}$ **Find:** ΔG and spontaneity

Conceptual Plan: $\text{J/K} \rightarrow \text{kJ/K}$ then $\Delta H^\circ_{\text{rxn}}, \Delta S_{\text{rxn}}, T \rightarrow \Delta G$

$$\dfrac{1000 \text{ J}}{1 \text{ kJ}} \qquad\qquad \Delta G = \Delta H_{\text{rxn}} - T\Delta S_{\text{rxn}}$$

Solution: $-263 \dfrac{\text{J}}{\text{K}} \times \dfrac{1 \text{ kJ}}{1000 \text{ J}} = -0.263 \text{ kJ/K}$ then $\Delta G = \Delta H_{\text{rxn}} - T\Delta S_{\text{rxn}} = $

$-115 \text{ kJ} - \left((298 \text{ K})\left(-0.263 \dfrac{\text{kJ}}{\text{K}} \right) \right) = -36.626 \text{ kJ} = -3.7 \times 10^1 \text{ kJ} = -3.7 \times 10^4 \text{ J}$; so the reaction is spontaneous.

Check: The units (kJ) are correct. The magnitude of the answer (-40 kJ) makes sense because the larger term was negative; so the reaction is spontaneous.

(d)   **Given:** $\Delta H^\circ_{\text{rxn}} = -115 \text{ kJ}$, $\Delta S_{\text{rxn}} = -263 \text{ J/K}$, $T = 615 \text{ K}$ **Find:** ΔG and spontaneity

Conceptual Plan: $\text{J/K} \rightarrow \text{kJ/K}$ then $\Delta H^\circ_{\text{rxn}}, \Delta S_{\text{rxn}}, T \rightarrow \Delta G$

$$\dfrac{1000 \text{ J}}{1 \text{ kJ}} \qquad\qquad \Delta G = \Delta H_{\text{rxn}} - T\Delta S_{\text{rxn}}$$

Solution: $-263 \dfrac{\text{J}}{\text{K}} \times \dfrac{1 \text{ kJ}}{1000 \text{ J}} = -0.263 \text{ kJ/K}$ then

$\Delta G = \Delta H_{\text{rxn}} - T\Delta S_{\text{rxn}} = -115 \text{ kJ} - \left((615 \text{ K})\left(-0.263 \dfrac{\text{kJ}}{\text{K}} \right) \right) = +47 \text{ kJ} = +4.7 \times 10^4 \text{ J}$; so the reaction is nonspontaneous.

Check: The units (J/K) are correct. The magnitude of the answer (+47 kJ) makes sense because the larger term was positive; so the reaction is nonspontaneous.

18.45   **Given:** $\Delta H°_{rxn} = -2217$ kJ, $\Delta S_{rxn} = +101.1$ J/K, $T = 25\,°C$ **Find:** ΔG and spontaneity
Conceptual Plan: °C → K then J/K → kJ/K then $\Delta H°_{rxn}, \Delta S_{rxn}, T \rightarrow \Delta G$

$$K = 273.15 + °C \qquad \frac{1\text{ kJ}}{1000\text{ J}} \qquad\qquad \Delta G = \Delta H_{rxn} - T\Delta S_{rxn}$$

Solution: $T = 273.15 + 25\,°C = 298$ K then $+101.1\dfrac{J}{K} \times \dfrac{1\text{ kJ}}{1000\text{ J}} = +0.1011\dfrac{kJ}{K}$ then

$$\Delta G = \Delta H_{rxn} - T\Delta S_{rxn} = -2217\text{ kJ} - \left((298\text{ K})\left(0.1011\dfrac{kJ}{K} \right)\right) = -2247\text{ kJ} = -2.247 \times 10^6 \text{ J; so the reaction is}$$
spontaneous.

Check: The units (kJ) are correct. The magnitude of the answer (−2250 kJ) makes sense because both terms are negative; so the reaction is spontaneous.

18.47

ΔH	ΔS	ΔG	Low Temp.	High Temp.
−	+	−	Spontaneous	Spontaneous
−	−	Temp.-dependent	Spontaneous	Nonspontaneous
+	+	Temp.-dependent	Nonspontaneous	Spontaneous
+	−	+	Nonspontaneous	Nonspontaneous

18.49   The molar entropy of a substance increases with increasing temperatures. The kinetic energy and the molecular motion increase. The substance will have access to an increased number of energy levels.

18.51   (a) $CO_2(g)$ because it has greater molar mass/complexity
(b) $CH_3OH(g)$ because it is in the gas phase
(c) $CO_2(g)$ because it has greater molar mass/complexity
(d) $SiH_4(g)$ because it has greater molar mass
(e) $CH_3CH_2CH_3(g)$ because it has greater molar mass/complexity
(f) $NaBr(aq)$ because a solution has more entropy than does a solid crystal

18.53   (a) $He(g) < Ne(g) < SO_2(g) < NH_3(g) < CH_3CH_2OH(g)$. All are in the gas phase. From He to Ne, there is an increase in molar mass; beyond that, the molecules increase in complexity.
(b) $H_2O(s) < H_2O(l) < H_2O(g)$. Entropy increases as we go from a solid to a liquid to a gas.
(c) $CH_4(g) < CF_4(g) < CCl_4(g)$. Entropy increases as the molar mass increases.

18.55   (a)   **Given:** $C_2H_4(g) + H_2(g) \rightarrow C_2H_6(g)$ **Find:** $\Delta S°_{rxn}$
Conceptual Plan: $\Delta S°_{rxn} = \sum n_p S°(\text{products}) - \sum n_r S°(\text{reactants})$
Solution:

Reactant/Product	$S°$(J/mol K from Appendix IIB)
$C_2H_4(g)$	219.3
$H_2(g)$	130.7
$C_2H_6(g)$	229.2

Be sure to pull data for the correct formula and phase.
$$\begin{aligned}
\Delta S°_{rxn} &= \sum n_p S°(\text{products}) - \sum n_r S°(\text{reactants}) \\
&= [1(S°(C_2H_6(g)))] - [1(S°(C_2H_4(g))) + 1(S°(H_2(g)))] \\
&= [1(229.2\text{ J/K})] - [1(219.3\text{ J/K}) + 1(130.7\text{ J/K})] \\
&= [229.2\text{ J/K}] - [350.0\text{ J/K}] \\
&= -120.8\text{ J/K} \quad \text{The moles of gas are decreasing.}
\end{aligned}$$

Check: The units (J/K) are correct. The answer is negative, which is consistent with 2 moles of gas going to 1 mole of gas.

(b)   **Given:** $C(s) + H_2O(g) \rightarrow CO(g) + H_2(g)$ **Find:** ΔS_{rxn}°

Conceptual Plan: $\Delta S_{rxn}^\circ = \sum n_p S^\circ(\text{products}) - \sum n_r S^\circ(\text{reactants})$

Solution:

Reactant/Product	S°(J/mol K from Appendix IIB)
$C(s)$	5.7
$H_2O(g)$	188.8
$CO(g)$	197.7
$H_2(g)$	130.7

Be sure to pull data for the correct formula and phase.

$\Delta S_{rxn}^\circ = \sum n_p S^\circ(\text{products}) - \sum n_r S^\circ(\text{reactants})$
$= [1(S^\circ(CO(g))) + 1(S^\circ(H_2(g)))] - [1(S^\circ(C(s))) + 1(S^\circ(H_2O(g)))]$
$= [1(197.7 \text{ J/K}) + 1(130.7 \text{ J/K})] - [1(5.7 \text{ J/K}) + 1(188.8 \text{ J/K})]$
$= [328.4 \text{ J/K}] - [194.5 \text{ J/K}]$
$= +133.9 \text{ J/K}$ The moles of gas are increasing.

Check: The units (J/K) are correct. The answer is positive, which is consistent with 1 mole of gas going to 2 moles of gas.

(c)   **Given:** $CO(g) + H_2O(g) \rightarrow H_2(g) + CO_2(g)$ **Find:** ΔS_{rxn}°

Conceptual Plan: $\Delta S_{rxn}^\circ = \sum n_p S^\circ(\text{products}) - \sum n_r S^\circ(\text{reactants})$

Solution:

Reactant/Product	S°(J/mol K from Appendix IIB)
$CO(g)$	197.7
$H_2O(g)$	188.8
$H_2(g)$	130.7
$CO_2(g)$	213.8

Be sure to pull data for the correct formula and phase.

$\Delta S_{rxn}^\circ = \sum n_p S^\circ(\text{products}) - \sum n_r S^\circ(\text{reactants})$
$= [1(S^\circ(H_2(g))) + 1(S^\circ(CO_2(g)))] - [1(S^\circ(CO(g))) + 1(S^\circ(H_2O(g)))]$
$= [1(130.7 \text{ J/K}) + 1(213.8 \text{ J/K})] - [1(197.7 \text{ J/K}) + 1(188.8 \text{ J/K})]$
$= [344.5 \text{ J/K}] - [386.5 \text{ J/K}]$
$= -42.0 \text{ J/K}$

The change is small because the number of moles of gas is constant.

Check: The units (J/K) are correct. The answer is small and negative, which is consistent with a constant number of moles of gas. Water molecules are bent, and carbon dioxide molecules are linear; so the water has more complexity. Also, carbon monoxide is more complex than is hydrogen gas.

(d)   **Given:** $2 H_2S(g) + 3 O_2(g) \rightarrow 2 H_2O(l) + 2 SO_2(g)$ **Find:** ΔS_{rxn}°

Conceptual Plan: $\Delta S_{rxn}^\circ = \sum n_p S^\circ(\text{products}) - \sum n_r S^\circ(\text{reactants})$

Solution:

Reactant/Product	S°(J/mol K from Appendix IIB)
$H_2S(g)$	205.8
$O_2(g)$	205.2
$H_2O(l)$	70.0
$SO_2(g)$	248.2

Be sure to pull data for the correct formula and phase.

$\Delta S_{rxn}^\circ = \sum n_p S^\circ(\text{products}) - \sum n_r S^\circ(\text{reactants})$
$= [2(S^\circ(H_2O(l))) + 2(S^\circ(SO_2(g)))] - [2(S^\circ(H_2S(g))) + 3(S^\circ(O_2(g)))]$
$= [2(70.0 \text{ J/K}) + 2(248.2 \text{ J/K})] - [2(205.8 \text{ J/K}) + 3(205.2 \text{ J/K})]$
$= [636.4 \text{ J/K}] - [1027.2 \text{ J/K}]$
$= -390.8 \text{ J/K}$

The number of moles of gas is decreasing.

Check: The units (J/K) are correct. The answer is negative, which is consistent with a decrease in the number of moles of gas.

18.57 **Given:** $CH_2Cl_2(g)$ formed from elements in standard states **Find:** $\Delta S°$ and rationalize sign
Conceptual Plan: Write a balanced reaction, then $\Delta S°_{rxn} = \sum n_p S°(\text{products}) - \sum n_r S°(\text{reactants})$.
Solution: $C(s) + H_2(g) + Cl_2(g) \rightarrow CH_2Cl_2(g)$

Reactant/Product	$S°(\text{J/mol K from Appendix IIB})$
$C(s)$	5.7
$H_2(g)$	130.7
$Cl_2(g)$	223.1
$CH_2Cl_2(g)$	270.2

Be sure to pull data for the correct formula and phase.

$$\begin{aligned}
\Delta S°_{rxn} &= \sum n_p S°(\text{products}) - \sum n_r S°(\text{reactants}) \\
&= [1(S°(CH_2Cl_2(g)))] - [1(S°(C(s))) + 1(S°(H_2(g))) + 1(S°(Cl_2(g)))] \\
&= [1(270.2 \text{ J/K})] - [1(5.7 \text{ J/K}) + 1(130.7 \text{ J/K}) + 1(223.1 \text{ J/K})] \\
&= [270.2 \text{ J/K}] - [359.5 \text{ J/K}] \\
&= -89.3 \text{ J/K}
\end{aligned}$$

The moles of gas are decreasing.

Check: The units (J/K) are correct. The answer is negative, which is consistent with 2 moles of gas going to 1 mole of gas.

18.59 **Given:** methanol (CH_3OH) combustion at 25 °C **Find:** $\Delta H°_{rxn}$, $\Delta S°_{rxn}$, $\Delta G°_{rxn}$, and spontaneity
Conceptual Plan: Write a balanced reaction then $\Delta H°_{rxn} = \sum n_p H°_f(\text{products}) - \sum n_r H°_f(\text{reactants})$ then
$\Delta S°_{rxn} = \sum n_p S°(\text{products}) - \sum n_r S°(\text{reactants})$ then °C → K then J/K → kJ/K then

$$K = 273.15 + °C \qquad \frac{1 \text{ kJ}}{1000 \text{ J}}$$

$\Delta H°_{rxn}$, $\Delta S°_{rxn}$, $T \rightarrow \Delta G°$.

$$\Delta G = \Delta H_{rxn} - T\Delta S_{rxn}$$

Solution: Combustion is combined with oxygen to form carbon dioxide and water.
$$2 CH_3OH(l) + 3 O_2(g) \rightarrow 2 CO_2(g) + 4 H_2O(g)$$

Reactant/Product	$\Delta H°_f(\text{kJ/mol from Appendix IIB})$
$CH_3OH(l)$	−238.6
$O_2(g)$	0.0
$CO_2(g)$	−393.5
$H_2O(g)$	−241.8

Be sure to pull data for the correct formula and phase.

$$\begin{aligned}
\Delta H°_{rxn} &= \sum n_p \Delta H°_f(\text{products}) - \sum n_r \Delta H°_f(\text{reactants}) \\
&= [2(\Delta H°_f(CO_2(g))) + 4(\Delta H°_f(H_2O(g)))] - [2(\Delta H°_f(CH_3OH(l))) + 3(\Delta H°_f(O_2(g)))] \\
&= [2(-393.5 \text{ kJ}) + 4(-241.8 \text{ kJ})] - [2(-238.6 \text{ kJ}) + 3(0.0 \text{ kJ})] \\
&= [-1754.2 \text{ kJ}] - [-477.2 \text{ kJ}] \\
&= -1277 \text{ kJ then}
\end{aligned}$$

Reactant/Product	$S°(\text{J/mol K from Appendix IIB})$
$CH_3OH(l)$	126.8
$O_2(g)$	205.2
$CO_2(g)$	213.8
$H_2O(g)$	188.8

Be sure to pull data for the correct formula and phase.

$$\begin{aligned}
\Delta S°_{rxn} &= \sum n_p S°(\text{products}) - \sum n_r S°(\text{reactants}) \\
&= [2(S°(CO_2(g))) + 4(S°(H_2O(g)))] - [2(S°(CH_3OH(l))) + 3(S°(O_2(g)))] \\
&= [2(213.8 \text{ J/K}) + 4(188.8 \text{ J/K})] - [2(126.8 \text{ J/K}) + 3(205.2 \text{ J/K})] \\
&= [1182.8 \text{ J/K}] - [869.2 \text{ J/K}] \\
&= 313.6 \text{ J/K}
\end{aligned}$$

then $T = 273.15 + 25\,°C = 298\,K$ then $+313.6\dfrac{J}{K} \times \dfrac{1\,kJ}{1000\,J} = +0.3136\,kJ/K$ then

$$\Delta G = \Delta H_{rxn} - T\Delta S_{rxn} = -1277\,kJ - \left((298\,K)\left(+0.3136\dfrac{kJ}{K} \right) \right) = -1370.\,kJ = -1.370 \times 10^6\,J;\ \text{so the reaction}$$
is spontaneous.

Check: The units (kJ, J/K, and kJ) are correct. Combustion reactions are exothermic, and we see a large negative enthalpy. We expect a large positive entropy because we have an increase in the number of moles of gas. The free energy is the sum of two negative terms; so we expect a large negative free energy, and the reaction is spontaneous.

18.61   (a)   **Given:** $N_2O_4(g) \rightarrow 2\,NO_2(g)$ at 25 °C

       Find: ΔH°_{rxn}, ΔS°_{rxn}, ΔG°_{rxn}, and spontaneity. Can temperature be changed to make it spontaneous?

       Conceptual Plan: $\Delta H^{\circ}_{rxn} = \sum n_p H^{\circ}_f(\text{products}) - \sum n_r H^{\circ}_f(\text{reactants})$ then

       $\Delta S^{\circ}_{rxn} = \sum n_p S^{\circ}(\text{products}) - \sum n_r S^{\circ}(\text{reactants})$ then °C → K then J/K → kJ/K then

                                                $K = 273.15 + °C$ $\dfrac{1\,kJ}{1000\,J}$

       $\Delta H^{\circ}_{rxn},\ \Delta S^{\circ}_{rxn},\ T \rightarrow \Delta G^{\circ}$

                $\Delta G = \Delta H_{rxn} - T\Delta S_{rxn}$

       Solution:

Reactant/Product	ΔH°_f (kJ/mol from Appendix IIB)
$N_2O_4(g)$	9.16
$NO_2(g)$	33.2

       Be sure to pull data for the correct formula and phase.

$$\begin{aligned}
\Delta H^{\circ}_{rxn} &= \sum n_p \Delta H^{\circ}_f(\text{products}) - \sum n_r \Delta H^{\circ}_f(\text{reactants}) \\
&= [2(\Delta H^{\circ}_f(NO_2(g)))] - [1(\Delta H^{\circ}_f(N_2O_4(g)))] \\
&= [2(33.2\,kJ)] - [1(9.16\,kJ)] \\
&= [66.4\,kJ] - [9.16\,kJ)] \\
&= +57.24\,kJ \text{ then}
\end{aligned}$$

Reactant/Product	S° (J/mol K from Appendix IIB)
$N_2O_4(g)$	304.4
$NO_2(g)$	240.1

       Be sure to pull data for the correct formula and phase.

$$\begin{aligned}
\Delta S^{\circ}_{rxn} &= \sum n_p S^{\circ}(\text{products}) - \sum n_r S^{\circ}(\text{reactants}) \\
&= [2(S^{\circ}(NO_2(g)))] - [1(S^{\circ}(N_2O_4(g)))] \\
&= [2(240.1\,J/K)] - [1(304.4\,J/K)] \\
&= [480.2\,J/K] - [304.4\,J/K] \\
&= +175.8\,J/K \text{ then } T = 273.15 + 25\,°C = 298\,K \text{ then}
\end{aligned}$$

$$+175.8\dfrac{J}{K} \times \dfrac{1\,kJ}{1000\,J} = +0.1758\,kJ \text{ then}$$

$$\Delta G^{\circ} = \Delta H^{\circ}_{rxn} - T\Delta S^{\circ}_{rxn} = 57.24 - \left((298\,K)\left(+0.1758\dfrac{kJ}{K} \right) \right) = +4.8516\,kJ = +4.9\,kJ = +4.9 \times 10^3\,J;$$

       so the reaction is nonspontaneous. It can be made spontaneous by raising the temperature.

       Check: The units (kJ, J/K, and kJ) are correct. The reaction requires the breaking of a bond, so we expect that this will be an endothermic reaction. We expect a positive entropy change because we are increasing the number of moles of gas. Because the positive enthalpy term dominates at room temperature, the reaction is nonspontaneous. The second term can dominate if we raise the temperature high enough.

   (b)   **Given:** $NH_4Cl(s) \rightarrow HCl(g) + NH_3(g)$ at 25 °C

       Find: ΔH°_{rxn}, ΔS°_{rxn}, ΔG°_{rxn}, and spontaneity. Can temperature be changed to make it spontaneous?

       Conceptual Plan: $\Delta H^{\circ}_{rxn} = \sum n_p H^{\circ}_f(\text{products}) - \sum n_r H^{\circ}_f(\text{reactants})$ then

       $\Delta S^{\circ}_{rxn} = \sum n_p S^{\circ}(\text{products}) - \sum n_r S^{\circ}(\text{reactants})$ then °C → K then J/K → kJ/K then

                                                $K = 273.15 + °C$ $\dfrac{1\,kJ}{1000\,J}$

       $\Delta H^{\circ}_{rxn},\ \Delta S^{\circ}_{rxn},\ T \rightarrow \Delta G^{\circ}$

                $\Delta G = \Delta H_{rxn} - T\Delta S_{rxn}$

Solution:

Reactant/Product	ΔH_f° (kJ/mol from Appendix IIB)
$NH_4Cl(s)$	−314.4
$HCl(g)$	−92.3
$NH_3(g)$	−45.9

Be sure to pull data for the correct formula and phase.

$$\Delta H_{rxn}^\circ = \sum n_p \Delta H_f^\circ(\text{products}) - \sum n_r \Delta H_f^\circ(\text{reactants})$$
$$= [1(\Delta H_f^\circ(HCl(g))) + 1(\Delta H_f^\circ(NH_3(g)))] - [1(\Delta H_f^\circ(NH_4Cl(g)))]$$
$$= [1(-92.3 \text{ kJ}) + 1(-45.9 \text{ kJ})] - [1(-314.4 \text{ kJ})]$$
$$= [-138.2 \text{ kJ}] - [-314.4 \text{ kJ}]$$
$$= +176.2 \text{ kJ then}$$

Reactant/Product	S° (J/mol K from Appendix IIB)
$NH_4Cl(s)$	94.6
$HCl(g)$	186.9
$NH_3(g)$	192.8

Be sure to pull data for the correct formula and phase.

$$\Delta S_{rxn}^\circ = \sum n_p S^\circ(\text{products}) - \sum n_r S^\circ(\text{reactants})$$
$$= [1(S^\circ(HCl(g))) + 1(S^\circ(NH_3(g)))] - [1(S^\circ(NH_4Cl(g)))]$$
$$= [1(186.9 \text{ J/K}) + 1(192.8 \text{ J/K})] - [1(94.6 \text{ J/K})]$$
$$= [379.7 \text{ J/K}] - [94.6 \text{ J/K}]$$
$$= +285.1 \text{ J/K}$$

then $T = 273.15 + 25\,°C = 298$ K then $+285.1\dfrac{J}{K} \times \dfrac{1 \text{ kJ}}{1000 \text{ J}} = +0.2851 \text{ kJ/K}$ then

$$\Delta G^\circ = \Delta H_{rxn}^\circ - T\Delta S_{rxn}^\circ = +176.2 \text{ kJ} - \left((298 \text{ K})\left(+0.2851\dfrac{\text{kJ}}{\text{K}}\right)\right) = +91.2 \text{ kJ} = +9.12 \times 10^4 \text{ J; so the}$$

reaction is nonspontaneous. It can be made spontaneous by raising the temperature.

Check: The units (kJ, J/K, and kJ) are correct. The reaction requires the breaking of a bond, so we expect that this will be an endothermic reaction. We expect a positive entropy change because we are increasing the number of moles of gas. Because the positive enthalpy term dominates at room temperature, the reaction is nonspontaneous. The second term can dominate if we raise the temperature high enough.

(c) **Given:** $3\,H_2(g) + Fe_2O_3(s) \rightarrow 2\,Fe(s) + 3\,H_2O(g)$ at 25 °C **Find:** ΔH_{rnx}°, ΔS_{rxn}°, ΔG_{rxn}°, and spontaneity. Can temperature be changed to make it spontaneous?

Conceptual Plan: $\Delta H_{rxn}^\circ = \sum n_p H_f^\circ(\text{products}) - \sum n_r H_f^\circ(\text{reactants})$ then

$\Delta S_{rxn}^\circ = \sum n_p S^\circ(\text{products}) - \sum n_r S^\circ(\text{reactants})$ then $°C \rightarrow K$ then $J/K \rightarrow kJ/K$ then

$$K = 273.15 + °C \qquad \dfrac{1 \text{ kJ}}{1000 \text{ J}}$$

$\Delta H_{rxn}^\circ, \Delta S_{rxn}^\circ, T \rightarrow \Delta G^\circ$

$$\Delta G = \Delta H_{rxn} - T\Delta S_{rxn}$$

Solution:

Reactant/Product	ΔH_f° (kJ/mol from Appendix IIB)
$H_2(g)$	0.0
$Fe_2O_3(s)$	−824.2
$Fe(s)$	0.0
$H_2O(g)$	−241.8

Be sure to pull data for the correct formula and phase.

$$\Delta H_{rxn}^\circ = \sum n_p \Delta H_f^\circ(\text{products}) - \sum n_r \Delta H_f^\circ(\text{reactants})$$
$$= [2(\Delta H_f^\circ(Fe(s))) + 3(\Delta H_f^\circ(H_2O(g)))] - [3(\Delta H_f^\circ(H_2(g))) + 1(\Delta H_f^\circ(Fe_2O_3(s)))]$$
$$= [2(0.0 \text{ kJ}) + 3(-241.8 \text{ kJ})] - [3(0.0 \text{ kJ}) + 1(-824.2 \text{ kJ})]$$
$$= [-725.4 \text{ kJ}] - [-824.2 \text{ kJ}]$$
$$= +98.8 \text{ kJ then}$$

Reactant/Product	$S°(\text{J/mol K from Appendix IIB})$
$H_2(g)$	130.7
$Fe_2O_3(s)$	87.4
$Fe(s)$	27.3
$H_2O(g)$	188.8

Be sure to pull data for the correct formula and phase.

$$\Delta S°_{rxn} = \sum n_p S°(\text{products}) - \sum n_r S°(\text{reactants})$$
$$= [2(S°(Fe(s))) + 3(S°(H_2O(g)))] - [3(S°(H_2(g))) + 1(S°(Fe_2O_3(s)))]$$
$$= [2(27.3 \text{ J/K}) + 3(188.8 \text{ J/K})] - [3(130.7 \text{ J/K}) + 1(87.4 \text{ J/K})]$$
$$= [621.0 \text{ J/K}] - [479.5 \text{ J/K}]$$
$$= +141.5 \text{ J/K}$$

then $T = 273.15 + 25\,°C = 298\text{ K}$ then $+141.5\dfrac{J}{K} \times \dfrac{1\text{ kJ}}{1000\text{ J}} = +0.1415 \text{ kJ/K}$ then

$$\Delta G° = \Delta H°_{rxn} - T\Delta S°_{rxn} = +98.8 \text{ kJ} - \left((298\text{ K})\left(+0.1415\dfrac{kJ}{K}\right)\right) = +56.6 \text{ kJ} = +5.66 \times 10^4 \text{ J; so the}$$

reaction is nonspontaneous. It can be made spontaneous by raising the temperature.

Check: The units (kJ, J/K, and kJ) are correct. The reaction requires the breaking of a bond, so we expect that this will be an endothermic reaction. We expect a positive entropy change because there is no change in the number of moles of gas, but the product gas is more complex. Because the positive enthalpy term dominates at room temperature, the reaction is nonspontaneous. The second term can dominate if we raise the temperature high enough. This process is the opposite of rusting, so we are not surprised that it is nonspontaneous.

(d) **Given:** $N_2(g) + 3H_2(g) \rightarrow 2NH_3(g)$ at 25 °C
Find: $\Delta H°_{rxn}$, $\Delta S°_{rxn}$, $\Delta G°_{rxn}$, and spontaneity. Can temperature be changed to make it spontaneous?
Conceptual Plan: $\Delta H°_{rxn} = \sum n_p H°_f(\text{products}) - \sum n_r H°_f(\text{reactants})$ then
$\Delta S°_{rxn} = \sum n_p S°(\text{products}) - \sum n_r S°(\text{reactants})$ then °C → K then J/K → kJ/K then

$$K = 273.15 + °C \qquad \dfrac{1\text{ kJ}}{1000\text{ J}}$$

$\Delta H°_{rxn}, \Delta S°_{rxn}, T \rightarrow \Delta G°$

$$\Delta G = \Delta H_{rxn} - T\Delta S_{rxn}$$

Solution:

Reactant/Product	$\Delta H°_f(\text{kJ/mol from Appendix IIB})$
$N_2(g)$	0.0
$H_2(g)$	0.0
$NH_3(g)$	−45.9

Be sure to pull data for the correct formula and phase.

$$\Delta H°_{rxn} = \sum n_p \Delta H°_f(\text{products}) - \sum n_r \Delta H°_f(\text{reactants})$$
$$= [2(\Delta H°_f(NH_3(g)))] - [1(\Delta H°_f(N_2(g))) + 3(\Delta H°_f(H_2(g)))]$$
$$= [2(-45.9 \text{ kJ})] - [1(0.0 \text{ kJ}) + 3(0.0 \text{ kJ})]$$
$$= [-91.8 \text{ kJ}] - [0.0 \text{ kJ}]$$
$$= -91.8 \text{ kJ then}$$

Reactant/Product	$S°(\text{J/mol K from Appendix IIB})$
$N_2(g)$	191.6
$H_2(g)$	130.7
$NH_3(g)$	192.8

Be sure to pull data for the correct formula and phase.

$$\Delta S°_{rxn} = \sum n_p S°(\text{products}) - \sum n_r S°(\text{reactants})$$
$$= [2(S°(NH_3(g)))] - [1(S°(N_2(g))) + 3(S°(H_2(g)))]$$
$$= [2(192.8 \text{ J/K})] - [1(191.6 \text{ J/K}) + 3(130.7 \text{ J/K})]$$
$$= [385.6 \text{ J/K}] - [583.7 \text{ J/K}]$$
$$= -198.1 \text{ J/K}$$

then $T = 273.15 + 25\,°C = 298\,K$ then $-198.1\dfrac{J}{K} \times \dfrac{1\,kJ}{1000\,J} = -0.1981\,kJ/K$ then

$$\Delta G° = \Delta H°_{rxn} - T\Delta S°_{rxn} = -91.8\,kJ - \left((298\,K)\left(-0.1981\dfrac{kJ}{K}\right)\right) = -32.8\,kJ = -3.28 \times 10^4\,J;\ \text{so the}$$
reaction is spontaneous.

Check: The units (kJ, J/k, and kJ) are correct. The reaction requires the breaking of a bond, so we expect that this will be an endothermic reaction. We expect a positive entropy change because we are increasing the number of moles of gas. Because the negative enthalpy term dominates at room temperature, the reaction is spontaneous. The second term can dominate if we raise the temperature high enough.

18.63   (a)   **Given:** $N_2O_4(g) \rightarrow 2\,NO_2(g)$ at 25 °C   **Find:** $\Delta G°_{rxn}$ and spontaneity and compare to Problem 18.61
Determine which method would show how free energy changes with temperature.
Conceptual Plan: $\Delta G°_{rxn} = \sum n_p \Delta G°_f(\text{products}) - \sum n_r \Delta G°_f(\text{reactants})$ **then compare to Problem 18.61**
Solution:

Reactant/Product	$\Delta G°_f$(kJ/mol from Appendix IIB)
$N_2O_4(g)$	99.8
$NO_2(g)$	51.3

Be sure to pull data for the correct formula and phase.

$$\begin{aligned}
\Delta G°_{rxn} &= \sum n_p \Delta G°_f(\text{products}) - \sum n_r \Delta G°_f(\text{reactants}) \\
&= [2(\Delta G°_f(NO_2(g)))] - [1(\Delta G°_f(N_2O_4(g)))] \\
&= [2(51.3\,kJ)] - [1(99.8\,kJ)] \\
&= [102.6\,kJ] - [99.8\,kJ] \\
&= +2.8\,kJ
\end{aligned}$$

So the reaction is nonspontaneous. The value is similar to that in Problem 18.61.

Check: The units (kJ) are correct. The free energy of the products is greater than that of the reactants; so the answer is positive, and the reaction is nonspontaneous. The answer is the same as that in Problem 18.61 within the error of the calculation.

(b)   **Given:** $NH_4Cl(s) \rightarrow HCl(g) + NH_3(g)$ at 25 °C
Find: $\Delta G°_{rxn}$ and spontaneity and compare to Problem 18.61
Conceptual Plan: $\Delta G°_{rxn} = \sum n_p \Delta G°_f(\text{products}) - \sum n_r \Delta G°_f(\text{reactants})$ **then compare to Problem 18.61**
Solution:

Reactant/Product	$\Delta G°_f$(kJ/mol from Appendix IIB)
$NH_4Cl(s)$	−202.9
$HCl(g)$	−95.3
$NH_3(g)$	−16.4

Be sure to pull data for the correct formula and phase.

$$\begin{aligned}
\Delta G°_{rxn} &= \sum n_p \Delta G°_f(\text{products}) - \sum n_r \Delta G°_f(\text{reactants}) \\
&= [1(\Delta G°_f(HCl(g))) + 1(\Delta G°_f(NH_3(g)))] - [1(\Delta G°_f(NH_4Cl(g)))] \\
&= [1(-95.3\,kJ) + 1(-16.4\,kJ)] - [1(-202.9\,kJ)] \\
&= [-111.7\,kJ] - [-202.9\,kJ] \\
&= +91.2\,kJ
\end{aligned}$$

So the reaction is nonspontaneous. The result is the same as that in Problem 18.61.

Check: The units (kJ) are correct. The answer matches the one in Problem 18.61.

(c)   **Given:** $3\,H_2(g) + Fe_2O_3(s) \rightarrow 2\,Fe(s) + 3\,H_2O(g)$ at 25 °C
Find: $\Delta G°_{rxn}$ and spontaneity and compare to Problem 18.61
Conceptual Plan: $\Delta G°_{rxn} = \sum n_p \Delta G°_f(\text{products}) - \sum n_r \Delta G°_f(\text{reactants})$ **then compare to Problem 18.61**

Solution:

Reactant/Product	ΔG_f°(kJ/mol from Appendix IIB)
$H_2(g)$	0.0
$Fe_2O_3(s)$	−742.2
$Fe(s)$	0.0
$H_2O(g)$	−228.6

Be sure to pull data for the correct formula and phase.

$$\Delta G_{rxn}^\circ = \sum n_p \Delta G_f^\circ(\text{products}) - \sum n_r \Delta G_f^\circ(\text{reactants})$$
$$= [2(\Delta G_f^\circ(Fe(s))) + 3(\Delta G_f^\circ(H_2O(g)))] - [3(\Delta G_f^\circ(H_2(g))) + 1(\Delta G_f^\circ(Fe_2O_3(s)))]$$
$$= [2(0.0\,kJ) + 3(-228.6\,kJ)] - [3(0.0\,kJ) + 1(-742.2\,kJ)]$$
$$= [-685.8\,kJ] - [-742.2\,kJ]$$
$$= +56.4\,kJ$$

So the reaction is nonspontaneous. The value is similar to that in Problem 18.61.

Check: The units (kJ) are correct. The answer is the same as that in Problem 18.61 within the error of the calculation.

(d) **Given:** $N_2(g) + 3\,H_2(g) \rightarrow 2\,NH_3(g)$ at 25 °C
Find: ΔG_{rxn}° and spontaneity and compare to Problem 18.61
Conceptual Plan: $\Delta G_{rxn}^\circ = \sum n_p \Delta G_f^\circ(\text{products}) - \sum n_r \Delta G_f^\circ(\text{reactants})$ **then compare to Problem 18.61**
Solution:

Reactant/Product	ΔG_f°(kJ/mol from Appendix IIB)
$N_2(g)$	0.0
$H_2(g)$	0.0
$NH_3(g)$	−16.4

Be sure to pull data for the correct formula and phase.

$$\Delta G_{rxn}^\circ = \sum n_p \Delta G_f^\circ(\text{products}) - \sum n_r \Delta G_f^\circ(\text{reactants})$$
$$= [2(\Delta G_f^\circ(NH_3(g)))] - [1(\Delta G_f^\circ(N_2(g))) + 3(\Delta G_f^\circ(H_2(g)))]$$
$$= [2(-16.4\,kJ)] - [1(0.0\,kJ) + 3(0.0\,kJ)]$$
$$= [-32.8\,kJ] - [0.0\,kJ]$$
$$= -32.8\,kJ$$

So the reaction is spontaneous. The result is the same as that in Problem 18.61.

Check: The units (kJ) are correct. The answer matches the one in Problem 18.61.
Values calculated by the two methods are comparable. The method using ΔH° and ΔS° is longer, but it can be used to determine how ΔG° changes with temperature.

18.65 **Given:** $2\,NO(g) + O_2(g) \rightarrow 2\,NO_2(g)$ **Find:** ΔG_{rxn}° and spontaneity at (a) 298 K, (b) 715 K, and (c) 855 K
Conceptual Plan: $\Delta H_{rxn}^\circ = \sum n_p H_f^\circ(\text{products}) - \sum n_r H_f^\circ(\text{reactants})$ **then**
$\Delta S_{rxn}^\circ = \sum n_p S^\circ(\text{products}) - \sum n_r S^\circ(\text{reactants})$ **then** J/K → kJ/K **then** $\Delta H_{rxn}^\circ, \Delta S_{rxn}^\circ, T \rightarrow \Delta G^\circ$

$$\frac{1\,kJ}{1000\,J} \qquad \Delta G = \Delta H_{rxn} - T\Delta S_{rxn}$$

Solution:

Reactant/Product	ΔH_f°(kJ/mol from Appendix IIB)
$NO(g)$	91.3
$O_2(g)$	0.0
$NO_2(g)$	33.2

Be sure to pull data for the correct formula and phase.

$$\Delta H_{rxn}^\circ = \sum n_p \Delta H_f^\circ(\text{products}) - \sum n_r \Delta H_f^\circ(\text{reactants})$$
$$= [2(\Delta H_f^\circ(NO_2(g)))] - [2(\Delta H_f^\circ(NO(g))) + 1(\Delta H_f^\circ(O_2(g)))]$$
$$= [2(33.2\,kJ)] - [2(91.3\,kJ) + 1(0.0\,kJ)]$$
$$= [66.4\,kJ] - [182.6\,kJ]$$
$$= -116.2\,kJ \text{ then}$$

Reactant/Product	$S°$(J/mol K from Appendix IIB)
$NO(g)$	210.8
$O_2(g)$	205.2
$NO_2(g)$	240.1

Be sure to pull data for the correct formula and phase.

$$\Delta S°_{rxn} = \sum n_p S°(\text{products}) - \sum n_r S°(\text{reactants})$$
$$= [2(S°(NO_2(g)))] - [2(S°(NO(g))) + 1(S°(O_2(g)))]$$
$$= [2(240.1 \text{ J/K})] - [2(210.8 \text{ J/K}) + 1(205.2 \text{ J/K})] \quad \text{then} \quad -146.6\frac{\text{J}}{\text{K}} \times \frac{1 \text{ kJ}}{1000 \text{ J}} = -0.1466 \text{ kJ/K}$$
$$= [480.2 \text{ J/K}] - [626.8 \text{ J/K}]$$
$$= -146.6 \text{ J/K}$$

(a) $\Delta G° = \Delta H°_{rxn} - T\Delta S°_{rxn} = -116.2 \text{ kJ} - (298 \text{ K})\left(-0.1466\frac{\text{kJ}}{\text{K}}\right) = -72.5 \text{ kJ} = -7.25 \times 10^4 \text{ J}$; so the reaction is spontaneous.

(b) $\Delta G° = \Delta H°_{rxn} - T\Delta S°_{rxn} = -116.2 \text{ kJ} - (715 \text{ K})\left(-0.1466\frac{\text{kJ}}{\text{K}}\right) = -11.4 \text{ kJ} = -1.14 \times 10^4 \text{ J}$; so the reaction is spontaneous.

(c) $\Delta G° = \Delta H°_{rxn} - T\Delta S°_{rxn} = -116.2 \text{ kJ} - (855 \text{ K})\left(-0.1466\frac{\text{kJ}}{\text{K}}\right) = +9.1 \text{ kJ} = +9.1 \times 10^3 \text{ J}$; so the reaction is nonspontaneous.

Check: The units (kJ) are correct. The enthalpy term dominates at low temperatures, making the reaction spontaneous. As the temperature increases, the decrease in entropy starts to dominate, and in the last case the reaction is nonspontaneous.

18.67   Because the first reaction has Fe_2O_3 as a product and the reaction of interest has it as a reactant, we need to reverse the first reaction. When the reaction direction is reversed, ΔG changes.
$Fe_2O_3(s) \rightarrow 2 \text{ Fe}(s) + 3/2 \text{ } O_2(g)$ $\Delta G° = +742.2 \text{ kJ}$
Because the second reaction has 1 mole of CO as a reactant and the reaction of interest has 3 moles of CO as a reactant, we need to multiply the second reaction and the ΔG by 3.
$3[\text{CO}(g) + 1/2 \text{ } O_2(g) \rightarrow CO_2(g)]$ $\Delta G° = 3(-257.2 \text{ kJ}) = -771.6 \text{ kJ}$
Hess's law states that the ΔG of the net reaction is the sum of the ΔG of the steps.
The rewritten reactions are:

$Fe_2O_3(s) \rightarrow 2 \text{ Fe}(s) + \cancel{3/2 \text{ } O_2(g)}$	$\Delta G° = +742.2 \text{ kJ}$
$3 \text{ CO}(g) + \cancel{3/2 \text{ } O_2(g)} \rightarrow 3 \text{ } CO_2(g)$	$\Delta G° = -771.6 \text{ kJ}$
$Fe_2O_3(s) \rightarrow 3 \text{ CO}(g) \rightarrow 2 \text{ Fe}(s) + 3 \text{ } CO_2(g)$	$\Delta G°_{rxn} = -29.4 \text{ kJ}$

Free Energy Changes, Nonstandard Conditions, and the Equilibrium Constant

18.69   (a)   **Given:** $I_2(s) \rightarrow I_2(g)$ at 25.0 °C   **Find:** $\Delta G°_{rxn}$
             Conceptual Plan: $\Delta G°_{rxn} = \sum n_p \Delta G°_f(\text{products}) - \sum n_r \Delta G°_f(\text{reactants})$
             Solution:

Reactant/Product	$\Delta G°_f$(kJ/mol from Appendix IIB)
$I_2(s)$	0.0
$I_2(g)$	19.3

Be sure to pull data for the correct formula and phase.

$$\Delta G°_{rxn} = \sum n_p \Delta G°_f(\text{products}) - \sum n_r \Delta G°_f(\text{products})$$
$$= [1(\Delta G°_f(I_2(g)))] - [1(\Delta G°_f(I_2(s)))]$$
$$= [1(19.3 \text{ kJ})] - [1(0.0 \text{ kJ})]$$
$$= +19.3 \text{ kJ}; \text{ so the reaction is nonspontaneous.}$$

Check: The units (kJ) are correct. The answer is positive because gases have higher free energy than do solids and the free energy change of the reaction is the same as free energy of formation of gaseous iodine.

(b) **Given:** $I_2(s) \rightarrow I_2(g)$ at $25.0\,°C$ (i) $P_{I_2} = 1.00\,mmHg$; (ii) $P_{I_2} = 0.100\,mmHg$ **Find:** ΔG_{rxn}
Conceptual Plan: °C $\rightarrow$ K and mmHg $\rightarrow$ atm then $\Delta G°_{rxn}, P_{I_2}, T \rightarrow \Delta G_{rxn}$

$$K = 273.15 + °C \qquad \frac{1\,atm}{760\,mmHg} \qquad \Delta G_{rxn} = \Delta G°_{rxn} + RT\,\ln Q \qquad \text{where } Q = P_{I_2}$$

Solution: $T = 273.15 + 25.0\,°C = 298.2\,K$ and (i) $1.00\,\cancel{mmHg} \times \dfrac{1\,atm}{760\,\cancel{mmHg}} = 0.00131579\,atm$

then $\Delta G_{rxn} = \Delta G°_{rxn} + RT\,\ln Q = \Delta G°_{rxn} + RT\,\ln P_{I_2} =$

$+19.3\,kJ + \left(\left(8.314\dfrac{J}{K\cdot mol}\right)\left(\dfrac{1\,kJ}{1000\,J}\right)(298.2\,K)\,\ln\,(0.00131579)\right) = +2.9\,kJ$; so the reaction is nonspontaneous.

(ii) $0.100\,\cancel{mmHg} \times \dfrac{1\,atm}{760\,\cancel{mmHg}} = 0.000131579\,atm$ then

$\Delta G_{rxn} = \Delta G°_{rxn} + RT\,\ln Q = \Delta G°_{rxn} + RT\,\ln P_{I_2} =$

$+19.3\,kJ + \left(\left(8.314\dfrac{J}{K\cdot mol}\right)\left(\dfrac{1\,kJ}{1000\,J}\right)(298.2\,K)\,\ln\,(0.000131579)\right) = -2.9\,kJ$; so the reaction is spontaneous.

Check: The units (kJ) are correct. The answer is positive at higher pressure because the pressure is higher than the vapor pressure of iodine. Once the desired pressure is below the vapor pressure ($0.31\,mmHg$ at $25.0\,°C$), the reaction becomes spontaneous.

(c) Iodine sublimes at room temperature because there is an equilibrium between the solid and the gas phases. The vapor pressure is low ($0.31\,mmHg$ at $25.0\,°C$), so a small amount of iodine can remain in the gas phase, which is consistent with the free energy values.

18.71   **Given:** $CH_3OH(g) \rightleftharpoons CO(g) + 2\,H_2(g)$ at $25\,°C$, $P_{CH_3OH} = 0.855\,atm$, $P_{CO} = 0.125\,atm$, $P_{H_2} = 0.183\,atm$
Find: ΔG
Conceptual Plan:
$\Delta G°_{rxn} = \sum n_p \Delta G°_f(\text{products}) - \sum n_r \Delta G°_f(\text{reactants})$ then °C $\rightarrow$ K then $\Delta G°_{rxn}, P_{CH_3OH}, P_{CO}, P_{H_2}, T \rightarrow \Delta G$

$$K = 273.15 + °C \qquad \Delta G_{rxn} = \Delta G°_{rxn} + RT\,\ln Q \qquad \text{where } Q = \dfrac{P_{CO}P_{H_2}^2}{P_{CH_3OH}}$$

Solution:

Reactant/Product	$\Delta G°_f(\text{kJ/mol from Appendix IIB})$
$CH_3OH(g)$	-162.3
$CO(g)$	-137.2
$H_2(g)$	0.0

Be sure to pull data for the correct formula and phase.

$\Delta G°_{rxn} = \sum n_p \Delta G°_f(\text{products}) - \sum n_r \Delta G°_f(\text{reactants})$
$= [1(\Delta G°_f(CO(g))) + 2(\Delta G°_f(H_2(g)))] - [1(\Delta G°_f(CH_3OH(g)))]$
$= [1(-137.2\,kJ) + 2(0.0\,kJ)] - [1(-162.3\,kJ)]$
$= [-137.2\,kJ] - [-162.3\,kJ]$
$= +25.1\,kJ$

$T = 273.15 + 25\,°C = 298\,K$ then $Q = \dfrac{P_{CO}P_{H_2}^2}{P_{CH_3OH}} = \dfrac{(0.125)(0.183)^2}{0.855} = 0.00489605$ then

$\Delta G_{rxn} = \Delta G°_{rxn} + RT\,\ln Q = +25.1\,kJ + \left(\left(8.314\dfrac{J}{K\cdot mol}\right)\left(\dfrac{1\,kJ}{1000\,J}\right)(298\,K)\,\ln\,(0.00489605)\right) = +11.9\,kJ$;

so the reaction is nonspontaneous.

Check: The units (kJ) are correct. The standard free energy for the reaction was positive, and the fact that Q was less than 1 made the free energy smaller. But the reaction at these conditions is still not spontaneous.

18.73　(a)　**Given:** $2\,CO(g) + O_2(g) \rightleftharpoons 2\,CO_2(g)$ at 25 °C　**Find:** K

Conceptual Plan:

$\Delta G_{rxn}^{\circ} = \sum n_p \Delta G_f^{\circ}(\text{products}) - \sum n_r \Delta G_f^{\circ}(\text{reactants})$ then °C → K then $\Delta G_{rxn}^{\circ}, T \rightarrow K$

$K = 273.15 + °C$ 　　　　$\Delta G_{rxn}^{\circ} = -RT \ln K$

Solution:

Reactant/Product	ΔG_f°(kJ/mol from Appendix IIB)
$CO(g)$	−137.2
$O_2(g)$	0.0
$CO_2(g)$	−394.4

Be sure to pull data for the correct formula and phase.

$$\begin{aligned}
\Delta G_{rxn}^{\circ} &= \sum n_p \Delta G_f^{\circ}(\text{products}) - \sum n_r \Delta G_f^{\circ}(\text{reactants}) \\
&= [2(\Delta G_f^{\circ}(CO_2(g)))] - [2(\Delta G_f^{\circ}(CO(g))) + 1(\Delta G_f^{\circ}(O_2(g)))] \\
&= [2(-394.4\,\text{kJ})] - [2(-137.2\,\text{kJ}) + 1(0.0\,\text{kJ})] \\
&= [-788.8\,\text{kJ}] - [-274.4\,\text{kJ}] \\
&= -514.4\,\text{kJ} \qquad\qquad T = 273.15 + 25\,°C = 298\,K \text{ then}
\end{aligned}$$

$\Delta G_{rxn}^{\circ} = -RT \ln K$　**Rearrange to solve for K.**

$$K = e^{\frac{-\Delta G_{rxn}^{\circ}}{RT}} = e^{\frac{-(-514.4\,\text{kJ}) \times \frac{1000\,\text{J}}{1\,\text{kJ}}}{\left(8.314\,\frac{\text{J}}{\text{K}\cdot\text{mol}}\right)(298\,\text{K})}} = e^{207.623} = 1.48 \times 10^{90}$$

Check: The units (none) are correct. The standard free energy for the reaction was very negative, so we expect a very large K. The reaction is spontaneous, so mostly products are present at equilibrium.

(b)　**Given:** $2\,H_2S(g) \rightleftharpoons 2\,H_2(g) + S_2(g)$ at 25 °C　**Find:** K

Conceptual Plan:

$\Delta G_{rxn}^{\circ} = \sum n_p \Delta G_f^{\circ}(\text{products}) - \sum n_r \Delta G_f^{\circ}(\text{reactants})$ then °C → K then $\Delta G_{rxn}^{\circ}, T \rightarrow K$

$K = 273.15 + °C$ 　　　　$\Delta G_{rxn}^{\circ} = -RT \ln K$

Solution:

Reactant/Product	ΔG_f°(kJ/mol from Appendix IIB)
$H_2S(g)$	−33.4
$H_2(g)$	0.0
$S_2(g)$	79.7

Be sure to pull data for the correct formula and phase.

$$\begin{aligned}
\Delta G_{rxn}^{\circ} &= \sum n_p \Delta G_f^{\circ}(\text{products}) - \sum n_r \Delta G_f^{\circ}(\text{reactants}) \\
&= [2(\Delta G_f^{\circ}(H_2(g))) + 1(\Delta G_f^{\circ}(S_2(g)))] - [2(\Delta G_f^{\circ}(H_2S(g)))] \\
&= [2(0.0\,\text{kJ})] + 1(79.7\,\text{kJ})] - [2(-33.4\,\text{kJ})] \\
&= [79.7\,\text{kJ}] - [-66.8\,\text{kJ}] \\
&= +146.5\,\text{kJ} \qquad\qquad T = 273.15 + 25\,°C = 298\,K \text{ then}
\end{aligned}$$

$\Delta G_{rxn}^{\circ} = -RT \ln K$　**Rearrange to solve for K.**

$$K = e^{\frac{-\Delta G_{rxn}^{\circ}}{RT}} = e^{\frac{-146.5\,\text{kJ} \times \frac{1000\,\text{J}}{1\,\text{kJ}}}{\left(8.314\,\frac{\text{J}}{\text{K}\cdot\text{mol}}\right)(298\,\text{K})}} = e^{-59.1305} = 2.09 \times 10^{-26}$$

Check: The units (none) are correct. The standard free energy for the reaction was positive, so we expect a small K. The reaction is nonspontaneous, so mostly reactants are present at equilibrium.

18.75　**Given:** $CO(g) + 2\,H_2(g) \rightleftharpoons CH_3OH(g)$ $K_p = 2.26 \times 10^4$ at 25 °C

Find: ΔG_{rxn}° at (a) standard conditions; (b) at equilibrium; and (c) $P_{CH_3OH} = 1.0\,\text{atm}$, $P_{CO} = P_{H_2} = 0.010\,\text{atm}$

Conceptual Plan: °C → K then (a) $K, T \rightarrow \Delta G_{rxn}^{\circ}$ then (b) at equilibrium, $\Delta G_{rxn} = 0$ then

$K = 273.15 + °C$ 　　　$\Delta G_{rxn}^{\circ} = -RT \ln K$

(c) $\Delta G^{\circ}_{rxn}, P_{CH_3OH}, P_{CO}, P_{H_2}, T \rightarrow \Delta G$

$$\Delta G_{rxn} = \Delta G^{\circ}_{rxn} + RT \ln Q \qquad \text{where } Q = \frac{P_{CH_3OH}}{P_{CO}P^2_{H_2}}$$

Solution: $T = 273.15 + 25\,°C = 298\,K$ then

(a) $\Delta G^{\circ}_{rxn} = -RT \ln K = -\left(8.314\dfrac{J}{K \cdot mol}\right)\left(\dfrac{1\,kJ}{1000\,J}\right)(298\,K)\ln(2.26 \times 10^4) = -24.8\,kJ$

(b) at equilibrium, $\Delta G_{rxn} = 0$

(c) $Q = \dfrac{P_{CH_3OH}}{P_{CO}P^2_{H_2}} = \dfrac{1.0}{(0.010)(0.010)^2} = 1.0 \times 10^6$ then

$$\Delta G_{rxn} = \Delta G^{\circ}_{rxn} + RT \ln Q = -24.8\,kJ + \left(8.314\dfrac{J}{K \cdot mol}\right)\left(\dfrac{1\,kJ}{1000\,J}\right)(298\,K)\ln(1.0 \times 10^6) = +9.4\,kJ$$

Check: The units (kJ) are correct. The K was greater than 1, so we expect a negative standard free energy for the reaction. At equilibrium, by definition, the free energy change is zero. Because the conditions give a $Q > K$, the reaction needs to proceed in the reverse direction, which means that the reaction is spontaneous in the reverse direction.

18.77 **(a)** **Given:** $2\,CO(g) + O_2(g) \rightleftharpoons 2\,CO_2(g)$ at $25\,°C$ **Find:** K at $525\,K$
Conceptual Plan: $\Delta H^{\circ}_{rxn} = \sum n_p H^{\circ}_f(\text{products}) - \sum n_r H^{\circ}_f(\text{reactants})$ then
$\Delta S^{\circ}_{rxn} = \sum n_p S^{\circ}(\text{products}) - \sum n_r S^{\circ}(\text{reactants})$ then $J/K \rightarrow kJ/K$ then $\Delta H^{\circ}_{rxn}, \Delta S_{rxn}, T \rightarrow \Delta G$

$$\dfrac{1\,kJ}{1000\,J} \qquad\qquad \Delta G = \Delta H_{rxn} - T\Delta S_{rxn}$$

then $\Delta G^{\circ}_{rxn}, T \rightarrow K$

$$\Delta G^{\circ}_{rxn} = -RT \ln K$$

Solution:

Reactant/Product	ΔH°_f(kJ/mol from Appendix IIB)
$CO(g)$	-110.5
$O_2(g)$	0.0
$CO_2(g)$	-393.5

Be sure to pull data for the correct formula and phase.
$$\begin{aligned}\Delta H^{\circ}_{rxn} &= \sum n_p \Delta H^{\circ}_f(\text{products}) - \sum n_r \Delta H^{\circ}_f(\text{reactants})\\ &= [2(\Delta H^{\circ}_f(CO_2(g)))] - [2(\Delta H^{\circ}_f(CO(g))) + 1(\Delta H^{\circ}_f(O_2(g)))]\\ &= [2(-393.5\,kJ)] - [2(-110.5\,kJ) + 1(0.0\,kJ)] \text{ then}\\ &= [-787.0\,kJ] - [-221.0\,kJ]\\ &= -566.0\,kJ\end{aligned}$$

Reactant/Product	S°(J/mol K from Appendix IIB)
$CO(g)$	197.7
$O_2(g)$	205.2
$CO_2(g)$	213.8

Be sure to pull data for the correct formula and phase.
$$\begin{aligned}\Delta S^{\circ}_{rxn} &= \sum n_p S^{\circ}(\text{products}) - \sum n_r S^{\circ}(\text{reactants})\\ &= [2(S^{\circ}(CO_2(g)))] - [2(S^{\circ}(CO(g))) + 1(S^{\circ}(O_2(g)))]\\ &= [2(213.8\,J/K)] - [2(197.7\,J/K) + 1(205.2\,J/K)]\\ &= [427.6\,J/K] - [600.6\,J/K] \text{ then}\\ &= -173.0\,K/K\end{aligned}$$

$$-173.0\dfrac{J}{K} \times \dfrac{1\,kJ}{1000\,J} = -0.1730\,kJ/K \text{ then}$$

$$\Delta G^{\circ} = \Delta H^{\circ}_{rxn} - T\Delta S^{\circ}_{rxn} = -566.0\,kJ - (525\,K)\left(-0.1730\dfrac{kJ}{K}\right) = -475.2\,kJ = -4.752 \times 10^5\,J \text{ then}$$

$\Delta G^{\circ}_{rxn} = -RT \ln K$ Rearrange to solve for K.

$$K = e^{\frac{-\Delta G^\circ_{rxn}}{RT}} = e^{\frac{-(-4.752 \times 10^5\, J)}{\left(8.314\frac{J}{K\cdot mol}\right)(525\,K)}} = e^{108.\underline{8}70} = 1.91 \times 10^{47}$$

Check: The units (none) are correct. The free energy change was very negative, indicating a spontaneous reaction. This results in a very large K.

(b)   **Given:** $2\,H_2S(g) \rightleftharpoons 2\,H_2(g) + S_2(g)$ at 25 °C   **Find:** K at 525 K

Conceptual Plan: $\Delta H^\circ_{rxn} = \sum n_p H^\circ_f(\text{products}) - \sum n_r H^\circ_f(\text{reactants})$ then

$\Delta S^\circ_{rxn} = \sum n_p S^\circ(\text{products}) - \sum n_r S^\circ(\text{reactants})$ then J/K $\rightarrow$ kJ/K then $\Delta H^\circ_{rxn}, \Delta S_{rxn}, T \rightarrow \Delta G$

$$\frac{1\,kJ}{1000\,J} \qquad\qquad \Delta G = \Delta H_{rxn} - T\Delta S_{rxn}$$

then $\Delta G^\circ_{rxn}, T \rightarrow K$

$$\Delta G^\circ_{rxn} = -RT \ln K$$

Solution:

Reactant/Product	ΔH°_f(kJ/mol from Appendix IIB)
$H_2S(g)$	−20.6
$H_2(g)$	0.0
$S_2(g)$	128.6

Be sure to pull data for the correct formula and phase.

$$\begin{aligned}
\Delta H^\circ_{rxn} &= \sum n_p \Delta H^\circ_f(\text{products}) - \sum n_r \Delta H^\circ_f(\text{reactants}) \\
&= [2(\Delta H^\circ_f(H_2(g))) + 1(\Delta H^\circ_f(S_2(g)))] - [2(\Delta H^\circ_f(H_2S(g)))] \\
&= [2(0.0\,kJ) + 1(128.6\,kJ)] - [2(-20.6\,kJ)] \\
&= [128.6\,kJ] - [-41.2\,kJ] \\
&= +169.8\,kJ
\end{aligned}$$

Reactant/Product	S°(J/mol K from Appendix IIB)
$H_2S(g)$	205.8
$H_2(g)$	130.7
$S_2(g)$	228.2

Be sure to pull data for the correct formula and phase.

$$\begin{aligned}
\Delta S^\circ_{rxn} &= \sum n_p S^\circ(\text{products}) - \sum n_r S^\circ(\text{reactants}) \\
&= [2(S^\circ(H_2(g))) + 1(S^\circ(S_2(g)))] - [2(S^\circ(H_2S(g)))] \\
&= [2(130.7\,J/K) + 1(228.2\,J/K)] - [2(205.8\,J/K)] \\
&= [489.6\,J/K] - [411.6\,J/K] \\
&= +78.0\,J/K
\end{aligned}$$

then $+78.0\dfrac{J}{K} \times \dfrac{1\,kJ}{1000\,J} = +0.0780\,kJ/K$ then

$$\Delta G^\circ = \Delta H^\circ_{rxn} - T\Delta S^\circ_{rxn} = +169.8\,kJ - (525\,K)\left(+0.0780\frac{kJ}{K}\right) = +128.85\,kJ = +1.2885 \times 10^5\,J \text{ then}$$

$\Delta G^\circ_{rxn} = -RT \ln K$ Rearrange to solve for K.

$$K = e^{\frac{-\Delta G^\circ_{rxn}}{RT}} = e^{\frac{-1.2885 \times 10^5\, J}{\left(8.314\frac{J}{K\cdot mol}\right)(525\,K)}} = e^{-29.\underline{5}199} = 1.51 \times 10^{-13}$$

Check: The units (none) are correct. The free energy change is positive, indicating a nonspontaneous reaction. This results in a very small K.

18.79   **Given:** table of K_p versus temperature   **Find:** $\Delta H^\circ_{rxn}, \Delta S^\circ_{rxn}$
          Conceptual Plan: Plot ln K versus $1/T$. The slope will be $-\Delta H^\circ_{rxn}/R$, and the intercept will be $\Delta S^\circ_{rxn}/R$.

Solution: Plot $\ln K$ versus $1/T$. Because $\ln K = -\dfrac{\Delta H°_{rxn}}{R}\dfrac{1}{T} + \dfrac{\Delta S°_{rxn}}{R}$, the negative of the slope will

be $-\dfrac{\Delta H°_{rxn}}{R}$ and the intercept will be $\dfrac{\Delta S°_{rxn}}{R}$. The slope can be determined by measuring $\Delta y/\Delta x$ on

the plot or by using functions such as "add trendline" in Excel. Because the slope is -6092.2 K,

$$\Delta H°_{rxn} = -slope\ R = -(-6092.2\ \text{K})\left(8.314\dfrac{\text{J}}{\text{K}\cdot\text{mol}}\right)\left(\dfrac{1\ \text{kJ}}{1000\ \text{J}}\right) = 50.61\ \text{kJ/mol} = 50.6\ \text{kJ/mol. Because the}$$

intercept is $27.136\ \text{K}^{-1}$, $\Delta S°_{rxn} = intercept\ R = \left(\dfrac{27.136}{\text{K}}\right)\left(8.314\dfrac{\text{J}}{\text{K}\cdot\text{mol}}\right) = 225.609\ \text{J/K} = 226\ \text{J/K}.$

y = −6092.2x + 27.136

Check: The units are correct (kJ/mole and J/K). The plot is very linear. The numbers are typical for reactions. Because the slope is negative, the enthalpy change must be positive.

18.81   **Given:** $\Delta H°_{rxn} = -25.8\ \text{kJ/mol}$, $K = 1.4 \times 10^3$ at 298 K   **Find:** K at 655 K

Conceptual Plan: $\Delta H°_{rxn}, K_1, T_1, T_2 \rightarrow K_2$

$$\ln K = -\dfrac{\Delta H°_{rxn}}{R}\dfrac{1}{T} + \dfrac{\Delta S°_{rxn}}{R}$$

Solution: Because $\ln K = -\dfrac{\Delta H°_{rxn}}{R}\dfrac{1}{T} + \dfrac{\Delta S°_{rxn}}{R}$, $\ln K_1 + \dfrac{\Delta H°_{rxn}}{R}\dfrac{1}{T_1} = \dfrac{\Delta S°_{rxn}}{R} = \ln K_2 + \dfrac{\Delta H°_{rxn}}{R}\dfrac{1}{T_2}$.

Rearrange to solve for K_2.

$$\ln K_2 = \ln K_1 + \dfrac{\Delta H°_{rxn}}{R}\left(\dfrac{1}{T_1} - \dfrac{1}{T_2}\right) = \ln(1.4 \times 10^3) + \dfrac{\dfrac{-25.8\ \text{kJ}}{\text{mol}} \times \dfrac{1000\ \text{J}}{1\ \text{kJ}}}{8.314\dfrac{\text{J}}{\text{K}\cdot\text{mol}}}\left(\dfrac{1}{298\ \text{K}} - \dfrac{1}{655\ \text{K}}\right) = 1.56852\ \text{so}$$

$K_2 = e^{1.56852} = 4.79954 = 4.8$

Check: The units are correct (none). Because the reaction is exothermic, we expect the K to decrease with increasing temperature.

Cumulative Problems

18.83   (a) $+$ because vapors have higher entropy than do liquids

(b) $-$ because solids have less entropy than do liquids

(c) $-$ because there is only one microstate for the final macrostate and there are six microstates for the initial macrostate

18.85   (a)   **Given:** $N_2(g) + O_2(g) \rightarrow 2\,NO(g)$ **Find:** $\Delta G°_{rxn}$ and K_p at 25 °C

Conceptual Plan: $\Delta H°_{rxn} = \sum n_p H°_f(\text{products}) - \sum n_r H°_f(\text{reactants})$ then

$\Delta S°_{rxn} = \sum n_p S°(\text{products}) - \sum n_r S°(\text{reactants})$ then $°C \rightarrow K$ then $J/K \rightarrow kJ/K$ then

$$K = 273.15 + °C \qquad \dfrac{1\ \text{kJ}}{1000\ \text{J}}$$

$\Delta H°_{rxn}, \Delta S°_{rxn}, T \rightarrow \Delta G$ then $\Delta G°_{rxn}, T \rightarrow K$

$$\Delta G = \Delta H_{rxn} - T\Delta S_{rxn} \qquad \Delta G°_{rxn} = -RT \ln K$$

Solution:

Reactant/Product	ΔH_f°(kJ/mol from Appendix IIB)
$N_2(g)$	0.0
$O_2(g)$	0.0
$NO(g)$	91.3

Be sure to pull data for the correct formula and phase.

$$
\begin{aligned}
\Delta H_{rxn}^\circ &= \sum n_p \Delta H_f^\circ(\text{products}) - \sum n_r \Delta H_f^\circ(\text{reactants}) \\
&= [2(\Delta H_f^\circ(NO(g)))] - [1(\Delta H_f^\circ(N_2(g))) + 1(\Delta H_f^\circ(O_2(g)))] \\
&= [2(91.3\,\text{kJ})] - [1(0.0\,\text{kJ}) + 1(0.0\,\text{kJ})] \\
&= [182.6\,\text{kJ}] - [0.0\,\text{kJ}] \\
&= +182.6\,\text{kJ then}
\end{aligned}
$$

Reactant/Product	S°(J/mol K from Appendix IIB)
$N_2(g)$	191.6
$O_2(g)$	205.2
$NO(g)$	210.8

Be sure to pull data for the correct formula and phase.

$$
\begin{aligned}
\Delta S_{rxn}^\circ &= \sum n_p S^\circ(\text{products}) - \sum n_r S^\circ(\text{reactants}) \\
&= [2(S^\circ(NO(g)))] - [1(S^\circ(N_2(g))) + 1(S^\circ(O_2(g)))] \\
&= [2(210.8\,\text{J/K})] - [1(191.6\,\text{J/K}) + 1(205.2\,\text{J/K})] \\
&= [421.6\,\text{J/K}] - [396.8\,\text{JK}] \\
&= +24.8\,\text{J/K}
\end{aligned}
$$

then $T = 273.15 + 25\,^\circ\text{C} = 298\,\text{K}$ then $+24.8\dfrac{\text{J}}{\text{K}} \times \dfrac{1\,\text{kJ}}{1000\,\text{J}} = +0.0248\,\text{kJ/K}$ then

$$
\Delta G^\circ = \Delta H_{rxn}^\circ - T\Delta S_{rxn}^\circ = +182.6\,\text{kJ} - (298\,\text{K})\left(0.0248\dfrac{\text{kJ}}{\text{K}}\right) = +175.2\,\text{kJ} = +1.752 \times 10^5\,\text{J}
$$

then $\Delta G_{rxn}^\circ = -RT \ln K$ Rearrange to solve for K.

$$
K = e^{\frac{-\Delta G_{rxn}^\circ}{RT}} = e^{\frac{-1.752 \times 10^5\,\text{J}}{\left(8.314\frac{\text{J}}{\text{K}\cdot\text{mol}}\right)(298\,\text{K})}} = e^{-70.7144} = 1.95 \times 10^{-31};
$$ so the reaction is nonspontaneous, and at

equilibrium, mostly reactants are present.

Check: The units (kJ and none) are correct. The enthalpy is twice the enthalpy of formation of NO. We expect a very small entropy change because the number of moles of gas is unchanged. Because the positive enthalpy term dominates at room temperature, the free energy change is very positive and the reaction in the forward direction is nonspontaneous. This results in a very small K.

(b)   **Given:** $N_2(g) + O_2(g) \rightarrow 2\,NO(g)$ **Find:** ΔG_{rxn}° at 2000 K

     Conceptual Plan: Use results from part (a) $\Delta H_{rxn}^\circ, \Delta S_{rxn}, T \rightarrow \Delta G$ then $\Delta G_{rxn}^\circ, T \rightarrow K$

$$
\Delta G = \Delta H_{rxn} - T\Delta S_{rxn} \qquad \Delta G_{rxn}^\circ = -RT \ln K
$$

     Solution: $\Delta G = \Delta H_{rxn} - T\Delta S_{rxn} = +182.6\,\text{kJ} - (2000\,\text{K})\left(0.0248\dfrac{\text{kJ}}{\text{K}}\right) = +133.0\,\text{kJ} = +1.330 \times 10^5\,\text{J}$

     then $\Delta G_{rxn}^\circ = -RT \ln K$ Rearrange to solve for K.

$$
K = e^{\frac{-\Delta G_{rxn}^\circ}{RT}} = e^{\frac{-1.330 \times 10^5\,\text{J}}{\left(8.314\frac{\text{J}}{\text{K}\cdot\text{mol}}\right)(2000\,\text{K})}} = e^{-7.998557} = 3.36 \times 10^{-4};
$$ so the forward reaction is becoming more

     spontaneous.

     Check: The units (kJ and none) are correct. As the temperature rises, the entropy term becomes more significant. The free energy change is reduced, and the K increases. The reaction is still nonspontaneous.

18.87   **Given:** $C_2H_4(g) + X_2(g) \rightarrow C_2H_4X_2(g)$ where $X = Cl$, Br, and I

     Find: $\Delta H_{rxn}^\circ, \Delta S_{rxn}^\circ, \Delta G_{rxn}^\circ$, and K at 25 °C and spontaneity trends with X and temperature

     Conceptual Plan: $\Delta H_{rxn}^\circ = \sum n_p H_f^\circ(\text{products}) - \sum n_r H_f^\circ(\text{reactants})$ then

$\Delta S^{\circ}_{\text{rxn}} = \sum n_p S^{\circ}(\text{products}) - \sum n_r S^{\circ}(\text{reactants})$ then $°C \rightarrow K$ then $J/K \rightarrow kJ/K$ then

$$K = 273.15 + °C \qquad \frac{1 \text{ kJ}}{1000 \text{ J}}$$

$\Delta H^{\circ}_{\text{rxn}}, \Delta S^{\circ}_{\text{rxn}}, T \rightarrow \Delta G$ then $\Delta G^{\circ}_{\text{rxn}}, T \rightarrow K$

$$\Delta G = \Delta H_{\text{rxn}} - T\Delta S_{\text{rxn}} \qquad \Delta G^{\circ}_{\text{rxn}} = -RT \ln K$$

Solution:

Reactant/Product	ΔH°_f(kJ/mol from Appendix IIB)
$C_2H_4(g)$	52.4
$Cl_2(g)$	0.0
$C_2H_4Cl_2(g)$	−129.7

Be sure to pull data for the correct formula and phase.

$\Delta H^{\circ}_{\text{rxn}} = \sum n_p \Delta H^{\circ}_f(\text{products}) - \sum n_r \Delta H^{\circ}_f(\text{reactants})$

$\qquad = [1\Delta H^{\circ}_f(C_2H_4Cl_2)(g)))] - [1(\Delta H^{\circ}_f(C_2H_4(g))) + 1(\Delta H^{\circ}_f(Cl_2(g)))]$

$\qquad = [1(-129.7 \text{ kJ})] - [1(52.4 \text{ kJ}) + 1(0.0 \text{ kJ})]$

$\qquad = [-129.7 \text{ kJ}] - [52.4 \text{ kJ}]$

$\qquad = -182.1 \text{ kJ}$ then

Reactant/Product	S°(J/mol K from Appendix IIB)
$C_2H_4(g)$	219.3
$Cl_2(g)$	223.1
$C_2H_4Cl_2(g)$	308.0

Be sure to pull data for the correct formula and phase.

$\Delta S^{\circ}_{\text{rxn}} = \sum n_p S^{\circ}(\text{products}) - \sum n_r S^{\circ}(\text{reactants})$

$\qquad = [1(S^{\circ}(C_2H_4Cl_2(g)))] - [1(S^{\circ}(C_2H_4(g))) + 1(S^{\circ}(Cl_2(g)))]$

$\qquad = [1(308.0 \text{ J/K})] - [1(219.3 \text{ J/K}) + 1(223.1 \text{ J/K})]$

$\qquad = [308.0 \text{ J/K}] - [442.4 \text{ J/K}]$

$\qquad = -134.4 \text{ J/K}$

then $T = 273.15 + 25 °C = 298 \text{ K}$ then $-134.4 \dfrac{J}{K} \times \dfrac{1 \text{ kJ}}{1000 \text{ J}} = -0.1344 \text{ kJ/K}$ then

$\Delta G^{\circ} = \Delta H^{\circ}_{\text{rxn}} - T\Delta S^{\circ}_{\text{rxn}} = -182.1 \text{ kJ} - (298 \text{ K})\left(-0.1344 \dfrac{\text{kJ}}{\text{K}}\right) = -142.0 \text{ kJ} = -1.420 \times 10^5 \text{ J}$ then

$\Delta G^{\circ}_{\text{rxn}} = -RT \ln K$ Rearrange to solve for K.

$K = e^{\frac{-\Delta G^{\circ}_{\text{rxn}}}{RT}} = e^{\frac{-(-1.420 \times 10^5 \text{ J})}{\left(8.314 \frac{\text{J}}{\text{K} \cdot \text{mol}}\right)(298 \text{ K})}} = e^{57.314} = 7.78 \times 10^{24}$; so the reaction is spontaneous.

Reactant/Product	ΔH°_f(kJ/mol from Appendix IIB)
$C_2H_4(g)$	52.4
$Br_2(g)$	30.9
$C_2H_4Br_2(g)$	38.3

Be sure to pull data for the correct formula and phase.

$\Delta H^{\circ}_{\text{rxn}} = \sum n_p \Delta H^{\circ}_f(\text{products}) - \sum n_r \Delta H^{\circ}_f(\text{reactants})$

$\qquad = [1(\Delta H^{\circ}_f(C_2H_4Br_2(g)))] - [1\Delta H^{\circ}_f(C_2H_4(g))) + 1(\Delta H^{\circ}_f(Br_2(g)))]$

$\qquad = [1(38.3 \text{ kJ})] - [1(52.4 \text{ kJ}) + 1(30.9 \text{ kJ})]$

$\qquad = [38.3 \text{ kJ}] - [83.3 \text{ kJ}]$

$\qquad = -45.0 \text{ kJ}$ then

Reactant/Product	$S°$ (J/mol K from Appendix IIB)
$C_2H_4(g)$	219.3
$Br_2(g)$	245.5
$C_2H_4Br_2(g)$	330.6

Be sure to pull data for the correct formula and phase.

$\Delta S°_{rxn} = \sum n_p S°(\text{products}) - \sum n_r S°(\text{reactants})$

$= [1(S°(C_2H_4Br_2(g)))] - [1(S°(C_2H_4(g))) + 1(S°(Br_2(g)))]$

$= [1(330.6 \text{ J/K})] - [1(219.3 \text{ J/K})] + 1(245.5 \text{ J/K})]$

$= [330.6 \text{ J/K}] - [466.8 \text{ J/K}]$

$= -134.2 \text{ J/K}$

then $-134.2\dfrac{J}{K} \times \dfrac{1 \text{ kJ}}{1000 \text{ J}} = -0.1342 \text{ kJ/K}$

then $\Delta G° = \Delta H°_{rxn} - T\Delta S°_{rxn} = -45.0 \text{ kJ} - (298 \text{ K})\left(-0.1342\dfrac{kJ}{K}\right) = -5.0084 \text{ kJ} = -5.0084 \times 10^3 \text{ J}$

then $\Delta G°_{rxn} = -RT \ln K$

Rearrange to solve for K. $K = e^{\frac{-\Delta G°_{rxn}}{RT}} = e^{\frac{-(-5.0084 \times 10^3 \text{ J})}{\left(8.314\frac{J}{K \cdot mol}\right)(298 \text{ K})}} = e^{2.021495} = 7.5$; so the reaction is spontaneous.

Reactant/Product	$\Delta H°_f$ (kJ/mol from Appendix IIB)
$C_2H_4(g)$	52.4
$I_2(g)$	62.42
$C_2H_4I_2(g)$	66.5

Be sure to pull data for the correct formula and phase.

$\Delta H°_{rxn} = \sum n_p \Delta H°_f(\text{products}) - \sum n_r \Delta H°_f(\text{reactants})$

$= [1(\Delta H°_f(C_2H_4I_2(g)))] - [1\Delta H°_f(C_2H_4(g))) + 1(\Delta H°_f(I_2(g)))]$

$= [1(66.5 \text{ kJ})] - [1(52.4 \text{ kJ}) + 1(62.42 \text{ kJ})]$

$= [66.5 \text{ kJ}] - [114.82 \text{ kJ}]$

$= -48.32 \text{ kJ}$ then

Reactant/Product	$S°$ (J/mol K from Appendix IIB)
$C_2H_4(g)$	219.3
$I_2(g)$	260.69
$C_2H_4I_2(g)$	347.8

Be sure to pull data for the correct formula and phase.

$\Delta S°_{rxn} = \sum n_p S°(\text{products}) - \sum n_r S°(\text{reactants})$

$= [1(S°(C_2H_4I_2(g)))] - [1(S°(C_2H_4(g))) + 1(S°(I_2(g)))]$

$= [1(347.8 \text{ J/K})] - [1(219.3 \text{ J/K}) + 1(260.69 \text{ J/K})]$

$= [347.8 \text{ J/K}] - [479.99 \text{ J/K}]$

$= -132.2 \text{ J/K}$

then $-132.2\dfrac{J}{K} \times \dfrac{1 \text{ kJ}}{1000 \text{ J}} = -0.1322 \text{ kJ/K}$

then $\Delta G° = \Delta H°_{rxn} - T\Delta S°_{rxn} = -48.32 \text{ kJ} - (298 \text{ K})\left(-0.1322\dfrac{kJ}{K}\right) = -8.9244 \text{ kJ} = -8.9244 \times 10^3 \text{ J}$

then $\Delta G°_{rxn} = -RT \ln K$ Rearrange to solve for K.

$K = e^{\frac{-\Delta G°_{rxn}}{RT}} = e^{\frac{-(-8.9244 \times 10^3 \text{ J})}{\left(8.314\frac{J}{K \cdot mol}\right)(298 \text{ K})}} = e^{3.6021} = 37$; so the reaction is spontaneous.

Cl_2 is the most spontaneous in the forward direction; K for Cl_2 is 7.78×10^{24}, K for Br_2 is 7.5, K for I_2 is 37. This means Cl_2 is the most spontaneous and Br_2 is the least. The entropy change in the reactions is very constant. The spontaneity is determined by the standard enthalpy of formation of the dihalogenated ethane. Higher temperatures make the forward reactions less spontaneous.

Check: The units (kJ and none) are correct. The enthalpy change becomes less negative as we move to larger halogens. The enthalpy term dominates at room temperature, and the free energy change is the same sign as the enthalpy change. The more negative the free energy change, the larger the K.

18.89   (a)   **Given:** $N_2O(g) + NO_2(g) \rightleftharpoons 3\,NO(g)$ at 298 K   **Find:** ΔG°_{rxn}

              Conceptual Plan: $\Delta G^\circ_{rxn} = \sum n_p \Delta G^\circ_f(\text{products}) - \sum n_r \Delta G^\circ_f(\text{reactants})$

              Solution:

Reactant/Product	$\Delta G^\circ_f(\text{kJ/mol from Appendix IIB})$
$N_2O(g)$	103.7
$NO_2(g)$	51.3
$NO(g)$	87.6

              Be sure to pull data for the correct formula and phase.

$$\begin{aligned}
\Delta G^\circ_{rxn} &= \sum n_p \Delta G^\circ_f(\text{products}) - \sum n_r \Delta G^\circ_f(\text{reactants}) \\
&= [3(\Delta G^\circ_f(NO(g)))] - [1(\Delta G^\circ_f(N_2O(g))) + 1(\Delta G^\circ_f(NO_2(g)))] \\
&= [3(87.6\,\text{kJ})] - [1(103.7\,\text{kJ}) + 1(51.3\,\text{kJ})] \\
&= [262.8\,\text{kJ}] - [155.0\,\text{kJ}] \\
&= +107.8\,\text{kJ}
\end{aligned}$$

              The reaction is nonspontaneous.

              Check: The units (kJ) are correct. The standard free energy for the reaction was positive, so the reaction is nonspontaneous.

    (b)   **Given:** $P_{N_2O} = P_{NO_2} = 1.0$ atm initially   **Find:** P_{N_2O} when reaction ceases to be spontaneous

              Conceptual Plan: Reaction will no longer be spontaneous when $Q = K$, so $\Delta G^\circ_{rxn}, T \rightarrow K$.

$$\Delta G^\circ_{rxn} = -RT \ln K$$

              Then solve the equilibrium problem to get gas pressures. Because $K \ll 1$, the amount of NO generated will be very, very small compared to 1.0 atm; within experimental error, $P_{N_2O} = P_{NO_2} = 1.0$ atm. Simply solve for P_{NO}.

$$K = \frac{P_{NO}^3}{P_{N_2O}P_{NO_2}}$$

              Solution: $\Delta G^\circ_{rxn} = -RT \ln K$ Rearrange to solve for K.

$$K = e^{\frac{-\Delta G^\circ_{rxn}}{RT}} = e^{\frac{-(+107.8\,\text{kJ}) \times \frac{1000\,\text{J}}{1\,\text{kJ}}}{\left(8.314\frac{\text{J}}{\text{K}\cdot\text{mol}}\right)(298\,\text{K})}} = e^{-43.5103} = 1.27 \times 10^{-19}$$

              Because $K = \dfrac{P_{NO}^3}{P_{N_2O}P_{NO_2}}$, rearrange to solve for P_{NO}.

$$P_{NO}^3 = \sqrt[3]{K\,P_{N_2O}P_{NO_2}} = \sqrt[3]{(1.27 \times 10^{-19})(1.0)(1.0)} = 5.0 \times 10^{-7}\,\text{atm}$$

              Note that the assumption that P_{N_2O} was very, very small was valid.

              Check: The units (atm) are correct. Because the free energy change was positive, the K was very small. This leads us to expect that very little NO will be formed.

    (c)   **Given:** $N_2O(g) + NO_2(g) \rightleftharpoons 3\,NO(g)$ **Find:** temperature for spontaneity

              Conceptual Plan: $\Delta H^\circ_{rxn} = \sum n_p H^\circ_f(\text{products}) - \sum n_r H^\circ_f(\text{reactants})$ then

              $\Delta S^\circ_{rxn} = \sum n_p S^\circ(\text{products}) - \sum n_r S^\circ(\text{reactants})$ then $\text{J/K} \rightarrow \text{kJ/K}$ then $\Delta H^\circ_{rxn}, \Delta S_{rxn} \rightarrow T$

$$\frac{1\,\text{kJ}}{1000\,\text{J}} \qquad\qquad \Delta G = \Delta H_{rxn} - T\Delta S_{rxn}$$

              Solution:

Reactant/Product	$\Delta H^\circ_f(\text{kJ/mol from Appendix IIB})$
$N_2O(g)$	81.6
$NO_2(g)$	33.2
$NO(g)$	91.3

Be sure to pull data for the correct formula and phase.

$$\Delta H_{rxn}^\circ = \sum n_p \Delta H_f^\circ(\text{products}) - \sum n_r \Delta H_f^\circ(\text{reactants})$$
$$= [3(\Delta H_f^\circ(NO(g)))] - [1(\Delta H_f^\circ(N_2O(g))) + 1(\Delta H_f^\circ(NO_2(g)))]$$
$$= [3(91.3 \text{ kJ})] - [1(81.6 \text{ kJ}) + 1(33.2 \text{ kJ})]$$
$$= [273.9 \text{ kJ}] - [114.8 \text{ kJ}]$$
$$= +159.1 \text{ kJ then}$$

Reactant/Product	S°(J/mol K from Appendix IIB)
$N_2O(g)$	220.0
$NO_2(g)$	240.1
$NO(g)$	210.8

Be sure to pull data for the correct formula and phase.

$$\Delta S_{rxn}^\circ = \sum n_p S^\circ(\text{products}) - \sum n_r S^\circ(\text{reactants})$$
$$= [3(S^\circ(NO(g)))] - [1(S^\circ(N_2O(g))) + 1(S^\circ(NO_2(g)))]$$
$$= [3(210.8 \text{ J/K})] - [1(220.0 \text{ J/K}) + 1(240.1 \text{ J/K})]$$
$$= [632.4 \text{ J/K}] - [460.1 \text{ J/K}]$$
$$= +172.3 \text{ J/K}$$

then $+172.3 \dfrac{J}{K} \times \dfrac{1 \text{ kJ}}{1000 \text{ J}} = +0.1723$ kJ/K. Because $\Delta G = \Delta H_{rxn} - T\Delta S_{rxn}$, set $\Delta G = 0$ and rearrange to

solve for T. $T = \dfrac{\Delta H_{rxn}}{\Delta S_{rxn}} = \dfrac{+159.1 \text{ kJ}}{0.1723 \dfrac{kJ}{K}} = +923.4$ K

Check: The units (K) are correct. The reaction can be made more spontaneous by raising the temperature because the entropy change is positive (increase in the number of moles of gas).

18.91 (a) **Given:** $ATP(aq) + H_2O(l) \rightarrow ADP(aq) + P_i(aq)$ $\Delta G_{rxn}^\circ = -30.5$ kJ at 298 K **Find:** K
Conceptual Plan: $\Delta G_{rxn}^\circ, T \rightarrow K$

$$\Delta G_{rxn}^\circ = -RT \ln K$$

Solution: $\Delta G_{rxn}^\circ = -RT \ln K$ Rearrange to solve for K.

$$K = e^{\frac{-\Delta G_{rxn}^\circ}{RT}} = e^{\frac{-(-30.5 \text{ kJ}) \times \frac{1000 \text{ J}}{1 \text{ kJ}}}{\left(8.314 \frac{J}{K \cdot mol}\right)(298 \text{ K})}} = e^{12.3104} = 2.22 \times 10^5$$

Check: The units (none) are correct. The free energy change is negative, and the reaction is spontaneous. This results in a large K.

(b) **Given:** oxidation of glucose drives reforming of ATP
Find: ΔG_{rxn}° of oxidation of glucose and moles ATP formed per mole of glucose
Conceptual Plan: Write a balanced reaction for glucose oxidation then
$$\Delta G_{rxn}^\circ = \sum n_p \Delta G_f^\circ(\text{products}) - \sum n_r \Delta G_f^\circ(\text{reactants}) \text{ then } \Delta G_{rxn}^\circ \text{s} \rightarrow \text{moles ATP/mole glucose.}$$

$$\frac{\Delta G_{rxn}^\circ \text{ glucose oxidation}}{\Delta G_{rxn}^\circ \text{ ATP hydrolysis}}$$

Solution: $C_6H_{12}O_6(s) + 6 O_2(g) \rightarrow 6 CO_2(g) + 6 H_2O(l)$

Reactant/Product	ΔG_f°(kJ/mol from Appendix IIB)
$C_6H_{12}O_6(s)$	−910.4
$O_2(g)$	0.0
$CO_2(g)$	−394.4
$H_2O(l)$	−237.1

Be sure to pull data for the correct formula and phase.

$$\Delta G_{rxn}^\circ = \sum n_p \Delta G_f^\circ(\text{products}) - \sum n_r \Delta G_f^\circ(\text{reactants})$$
$$= [6(\Delta G_f^\circ(CO_2(g))) + 6(\Delta G_f^\circ(H_2O(l)))] - [1(\Delta G_f^\circ(C_6H_{12}O_6(s))) + 6(\Delta G_f^\circ(O_2(g)))]$$
$$= [6(-394.4 \text{ kJ}) + 6(-237.1 \text{ kJ})] - [1(-910.4 \text{ kJ}) + 6(0.0 \text{ kJ})]$$
$$= [-3789.0 \text{ kJ}] - [-910.4 \text{ kJ}]$$
$$= -2878.6 \text{ kJ}$$

$$2878.6 \frac{\text{kJ generated}}{\text{mole glucose oxidized}}$$

So the reaction is very spontaneous. $\dfrac{2878.6\frac{\text{kJ generated}}{\text{mole glucose oxidized}}}{30.5\frac{\text{kJ needed}}{\text{mole ATP reformed}}} = 94.4\frac{\text{mole ATP reformed}}{\text{mole glucose oxidized}}$

Check: The units (mol) are correct. The free energy change for the glucose oxidation is large compared to the ATP hydrolysis, so we expect to re-form many moles of ATP.

18.93 (a) **Given:** $2\,CO(g) + 2\,NO(g) \rightarrow N_2(g) + 2\,CO_2(g)$ **Find:** ΔG°_{rxn} and effect of increasing T on ΔG
Conceptual Plan: $\Delta G^\circ_{rxn} = \sum n_p \Delta G^\circ_f(\text{products}) - \sum n_r \Delta G^\circ_f(\text{reactants})$
Solution:

Reactant/Product	$\Delta G^\circ_f(\text{kJ/mol from Appendix IIB})$
$CO(g)$	-137.2
$NO(g)$	87.6
$N_2(g)$	0.0
$CO_2(g)$	-394.4

Be sure to pull data for the correct formula and phase.

$$
\begin{aligned}
\Delta G^\circ_{rxn} &= \sum n_p \Delta G^\circ_f(\text{products}) - \sum n_r \Delta G^\circ_f(\text{reactants}) \\
&= [1(\Delta G^\circ_f(N_2(g))) + 2(\Delta G^\circ_f(CO_2(g)))] - [2(\Delta G^\circ_f(CO(g))) + 2(\Delta G^\circ_f(NO(g)))] \\
&= [1(0.0\,\text{kJ}) + 2(-394.4\,\text{kJ})] - [2(-137.2\,\text{kJ}) + 2(87.6\,\text{kJ})] \\
&= [-788.8\,\text{kJ}] - [-99.2\,\text{kJ}] \\
&= -689.6\,\text{kJ}
\end{aligned}
$$

Because the number of moles of gas is decreasing, the entropy change is negative; ΔG will become less negative with increasing temperature.

Check: The units (kJ) are correct. The free energy change is negative because the carbon dioxide has such a low free energy of formation.

(b) **Given:** $5\,H_2(g) + 2\,NO(g) \rightarrow 2\,NH_3(g) + 2\,H_2O(g)$ **Find:** ΔG°_{rxn} and effect of increasing T on ΔG
Conceptual Plan: $\Delta G^\circ_{rxn} = \sum n_p \Delta G^\circ_f(\text{products}) - \sum n_r \Delta G^\circ_f(\text{reactants})$
Solution:

Reactant/Product	$\Delta G^\circ_f(\text{kJ/mol from Appendix IIB})$
$H_2(g)$	0.0
$NO(g)$	87.6
$NH_3(g)$	-16.4
$H_2O(g)$	-228.6

Be sure to pull data for the correct formula and phase.

$$
\begin{aligned}
\Delta G^\circ_{rxn} &= \sum n_p \Delta G^\circ_f(\text{products}) - \sum n_r \Delta G^\circ_f(\text{reactants}) \\
&= [2(\Delta G^\circ_f(NH_3(g))) + 2(\Delta G^\circ_f(H_2O(g)))] - [5(\Delta G^\circ_f(H_2(g))) + 2(\Delta G^\circ_f(NO(g)))] \\
&= [2(-16.4\,\text{kJ}) + 2(-228.6\,\text{kJ})] - [5(0.0\,\text{kJ}) + 2(87.6\,\text{kJ})] \\
&= [-490.0\,\text{kJ}] - [175.2\,\text{kJ}] \\
&= -665.2\,\text{kJ}
\end{aligned}
$$

Because the number of moles of gas is decreasing, the entropy change is negative; so ΔG will become less negative with increasing temperature.

Check: The units (kJ) are correct. The free energy change is negative because ammonia and water have such a low free energy of formation.

(c)   **Given:** $2 H_2(g) + 2 NO(g) \rightarrow N_2(g) + 2 H_2O(g)$ **Find:** ΔG°_{rxn} and effect of increasing T on ΔG
      Conceptual Plan: $\Delta G^\circ_{rxn} = \sum n_p \Delta G^\circ_f(\text{products}) - \sum n_r \Delta G^\circ_f(\text{reactants})$
      Solution:

Reactant/Product	ΔG°_f(kJ/mol from Appendix IIB)
$H_2(g)$	0.0
$NO(g)$	87.6
$N_2(g)$	0.0
$H_2O(g)$	-228.6

Be sure to pull data for the correct formula and phase.

$$
\begin{aligned}
\Delta G^\circ_{rxn} &= \sum n_p \Delta G^\circ_f(\text{products}) - \sum n_r \Delta G^\circ_f(\text{reactants}) \\
&= [1(\Delta G^\circ_f(N_2(g))) + 2(\Delta G^\circ_f(H_2O(g)))] - [2(\Delta G^\circ_f(H_2(g))) + 2(\Delta G^\circ_f(NO(g)))] \\
&= [1(0.0\,kJ) + 2(-228.6\,kJ)] - [2(0.0\,kJ) + 2(87.6\,kJ)] \\
&= [-457.2\,kJ] - [175.2\,kJ] \\
&= -632.4\,kJ
\end{aligned}
$$

Because the number of moles of gas is decreasing, the entropy change is negative; so ΔG will become less negative with increasing temperature.

Check: The units (kJ) are correct. The free energy change is negative because water has such a low free energy of formation.

(d)   **Given:** $2 NH_3(g) + 2 O_2(g) \rightarrow N_2O(g) + 3 H_2O(g)$ **Find:** ΔG°_{rxn} and effect of increasing T on ΔG
      Conceptual Plan: $\Delta G^\circ_{rxn} = \sum n_p \Delta G^\circ_f(\text{products}) - \sum n_r \Delta G^\circ_f(\text{reactants})$
      Solution:

Reactant/Product	ΔG°_f(kJ/mol from Appendix IIB)
$NH_3(g)$	-16.4
$O_2(g)$	0.0
$N_2O(g)$	103.7
$H_2O(g)$	-228.6

Be sure to pull data for the correct formula and phase.

$$
\begin{aligned}
\Delta G^\circ_{rxn} &= \sum n_p \Delta G^\circ_f(\text{products}) - \sum n_r \Delta G^\circ_f(\text{reactants}) \\
&= [1(\Delta G^\circ_f(N_2O(g))) + 3(\Delta G^\circ_f(H_2O(g)))] - [2(\Delta G^\circ_f(NH_3(g))) + 2(\Delta G^\circ_f(O_2(g)))] \\
&= [1(103.7\,kJ) + 3(-228.6\,kJ)] - [2(-16.4\,kJ) + 2(0.0\,kJ)] \\
&= [-582.1\,kJ] - [-32.8\,kJ] \\
&= -549.3\,kJ
\end{aligned}
$$

Because the number of moles of gas is constant, the entropy change will be small and slightly negative; so the magnitude of ΔG will decrease with increasing temperature.

Check: The units (kJ) are correct. The free energy change is negative because water has such a low free energy of formation. The entropy change is negative once the S° values are reviewed ($\Delta S_{rxn} = -9.6\ J/K$).

18.95   With one exception, the formation of any oxide of nitrogen at 298 K requires more moles of gas as reactants than are formed as products. For example, 1 mole of N_2O requires 0.5 mole of O_2 and 1 mole of N_2. One mole of N_2O_3 requires 1 mole of N_2 and 1.5 moles of O_2, and so on. The exception is NO, where 1 mole of NO requires 0.5 mole of O_2 and 0.5 mole of N_2: $\frac{1}{2} N_2(g) + \frac{1}{2} O_2(g) \rightarrow NO(g)$. This reaction has a positive ΔS because what is essentially mixing of the N and O has taken place in the product.

18.97   **Given:** $X_2(g) \rightarrow 2 X(g)$; $P_{initial\ X_2} = 755$ torr, $P_{final\ X} = 103$ torr at 298 K; $P_{initial\ X_2} = 748$ torr, $P_{final\ X} = 532$ torr at 755 K   **Find:** ΔH°_{rxn}
      Conceptual Plan: At each temperature $P_{initial\ X_2}, P_{final\ X} \rightarrow K$ then $K_1, K_2, T_1, T_2 \rightarrow \Delta H^\circ_{rxn}$,

$$\text{Use stoichiometry to calculate } [X_2]_{final} \text{ and } K = \frac{P_X^2}{P_{X_2}} \qquad \ln K = -\frac{\Delta H^\circ_{rxn}}{R}\frac{1}{T} + \frac{\Delta S^\circ_{rxn}}{R}$$

Solution: Use stoichiometry of $X_2(g) \rightarrow 2 X(g)$; $P_{final\ X_2} = P_{initial\ X_2} - \frac{1}{2} P_{final\ X}$ so that at $T_1 = 298$ K,

$$P_{final\ X_2} = P_{initial\ X_2} - \tfrac{1}{2} P_{final\ X} = 755\ torr - \tfrac{1}{2}(103\ torr) = 703.5\ torr \text{ and at } T_2 = 755\ K,\ P_{final\ X_2} = P_{initial\ X_2} - \tfrac{1}{2}$$

$P_{\text{final X}} = 748 \text{ torr} - \frac{1}{2}(532 \text{ torr}) = 482 \text{ torr.}$ Then $K = \dfrac{P_X^2}{P_{X_2}}$ so that at $T_1 = 298$ K,

$$K_1 = \frac{P_X^2}{P_{X_2}} = \frac{(103)^2}{703.5} = 15.08031 \text{ and at } T_2 = 755 \text{ K, } K_2 = \frac{P_X^2}{P_{X_2}} = \frac{(532)^2}{482} = 587.1867.$$

Because $\ln K = -\dfrac{\Delta H_{\text{rxn}}^\circ}{R}\dfrac{1}{T} + \dfrac{\Delta S_{\text{rxn}}^\circ}{R}$, $\ln K_1 + \dfrac{\Delta H_{\text{rxn}}^\circ}{R}\dfrac{1}{T_1} = \dfrac{\Delta S_{\text{rxn}}^\circ}{R} = \ln K_2 + \dfrac{\Delta H_{\text{rxn}}^\circ}{R}\dfrac{1}{T_2}$.

Rearrange to solve for $\Delta H_{\text{rxn}}^\circ$.

$$\Delta H_{\text{rxn}}^\circ = \frac{\ln \dfrac{K_2}{K_1}}{\left(\dfrac{1}{T_1} - \dfrac{1}{T_2}\right)}R = \frac{\ln\left(\dfrac{587.1867}{15.08031}\right)}{\left(\dfrac{1}{298 \text{ K}} - \dfrac{1}{755 \text{ K}}\right)} \times 8.314 \frac{J}{K \cdot mol} \times \frac{1 \text{ kJ}}{1000 \text{ J}} = 14.9889 \text{ kJ/mol} = 15.0 \text{ kJ/mol}$$

Note that an alternate way to generate the expressions for the two equilibrium constants is to use an ICE table.

Check: The units are correct (kJ/mol). Because K increases with increasing temperature, the reaction is expected to be endothermic.

18.99   (a) $\Delta S_{\text{univ}} > 0$. The process is spontaneous. It is slow unless a spark is applied.

(b) $\Delta S_{\text{univ}} > 0$. Although the change in the system is not spontaneous, the overall change, which includes such processes as combustion or water flow to generate electricity, is spontaneous.

(c) $\Delta S_{\text{univ}} > 0$. The acorn oak/tree system is becoming more ordered, so the processes associated with growth are not spontaneous. But they are driven by spontaneous processes such as the generation of heat by the sun and the reactions that produce energy in the cell.

18.101   At equilibrium ($P_{H_2O} = 18.3$ mmHg), $\Delta G_{\text{rxn}} = 0$.

$$Q = P_{H_2O}^6 = \left(\frac{18.3 \text{ mmHg}}{760 \text{ mmHg}}\right)^6 = (0.02407894 \text{ atm})^6 = 1.949059 \times 10^{-10} \text{ and}$$

$\Delta G_{\text{rxn}} = \Delta G_{\text{rxn}}^\circ + RT \ln Q = 0$ Rearrange to solve for $\Delta G_{\text{rxn}}^\circ$ (at $P_{H_2O} = 760$ mmHg = standard conditions).

$$\Delta G_{\text{rxn}}^\circ = \Delta G_{\text{rxn}} - RT \ln Q = 0 - \left(8.314\frac{J}{K \cdot mol}\right)\left(\frac{1 \text{ kJ}}{1000 \text{ J}}\right)(298 \text{ K}) \ln (1.949059 \times 10^{-10}) = 55.3948 \frac{kJ}{mol}$$

$$= 55.4 \frac{kJ}{mol}$$

Challenge Problems

18.103   (a)   **Given:** glutamate(aq) + $NH_3(aq) \rightarrow$ glutamine(aq) + $H_2O(l)$ $\Delta G_{\text{rxn}}^\circ = 14.2$ kJ at 298 K   **Find:** K

Conceptual Plan: $\Delta G_{\text{rxn}}^\circ, T \rightarrow K$

$$\Delta G_{\text{rxn}}^\circ = -RT \ln K$$

Solution: $\Delta G_{\text{rxn}}^\circ = -RT \ln K$ Rearrange to solve for K.

$$K = e^{\frac{-\Delta G_{\text{rxn}}^\circ}{RT}} = e^{\frac{-(+14.2 \text{ kJ}) \times \frac{1000 \text{ J}}{1 \text{ kJ}}}{\left(8.314 \frac{J}{K \cdot mol}\right)(298 \text{ K})}} = e^{-5.73142} = 3.24 \times 10^{-3}$$

Check: The units (none) are correct. The free energy change is positive, and the reaction is nonspontaneous. This results in a small K.

(b)   **Given:** pair ATP hydrolysis with glutamate/NH_3 reaction   **Find:** show coupled reactions, $\Delta G_{\text{rxn}}^\circ$ and K

Conceptual Plan: Use the reaction mechanism shown, where A = NH_3 and B = glutamate $(C_5H_8O_4N^-)$. Then calculate $\Delta G_{\text{rxn}}^\circ$ by adding free energies of reactions then $\Delta G_{\text{rxn}}^\circ, T \rightarrow K$.

$$\Delta G_{\text{rxn}}^\circ = -RT \ln K$$

Solution:

$$NH_3(aq) + ATP(aq) + H_2O(l) \rightarrow NH_3-P_i(aq) + ADP(aq) \qquad \Delta G^\circ_{rxn} = -30.5 \text{ kJ}$$
$$NH_3-P_i(aq) + C_5H_8O_4N^-(aq) \rightarrow C_5H_9O_3N_2^-(aq) + H_2O(l) + P_i(aq) \qquad \Delta G^\circ_{rxn} = +14.2 \text{ kJ}$$

$$NH_3(aq) + C_5H_8O_4N^-(aq) + ATP(aq) \rightarrow C_5H_9O_3N_2^-(aq) + ADP(aq) + P_i(aq) \qquad \Delta G^\circ_{rxn} = -16.3 \text{ kJ}$$

then $\Delta G^\circ_{rxn} = -RT \ln K$ Rearrange to solve for K.

$$K = e^{\frac{-\Delta G^\circ_{rxn}}{RT}} = e^{\frac{-(-16.3 \text{ kJ}) \times \frac{1000 \text{ J}}{1 \text{ kJ}}}{\left(8.314 \frac{J}{K \cdot mol}\right)(298 \text{ K})}} = e^{6.57902} = 7.20 \times 10^2$$

Check: The units (none) are correct. The free energy change is negative, and the reaction is spontaneous. This results in a large K.

18.105   **(a)**   **Given:** $\frac{1}{2}H_2(g) + \frac{1}{2}Cl_2(g) \rightarrow HCl(g)$, define standard state as 2 atm   **Find:** ΔG°_f

         Conceptual Plan: $\Delta G^\circ_f, P_{H_2}, P_{Cl_2}, P_{HCl}, T \rightarrow$ **new** ΔG°_f

$$\Delta G_{rxn} = \Delta G^\circ_{rxn} + RT \ln Q \quad \text{where } Q = \frac{P_{HCl}}{P_{H_2}^{1/2} P_{Cl_2}^{1/2}}$$

         Solution: $\Delta G^\circ_{rxn} = -95.3 \text{ kJ/mol}$ and $Q = \frac{P_{HCl}}{P_{H_2}^{1/2} P_{Cl_2}^{1/2}} = \frac{2}{2^{1/2} 2^{1/2}} = 1$ then

$$\Delta G_{rxn} = \Delta G^\circ_{rxn} + RT \ln Q = -95.3 \frac{kJ}{mol} + \left(8.314 \frac{J}{K \cdot mol}\right)\left(\frac{1 \text{ kJ}}{1000 \text{ J}}\right)(298 \text{ K}) \ln(1) =$$

$$-95.3 \text{ kJ/mol} = -95{,}300 \text{ J/mol}$$

Because the number of moles of reactants and products are the same, the decrease in volume affects the entropy of both equally; so there is no change in ΔG°_f.

         Check: The units (kJ) are correct. The Q is 1, so ΔG°_f is unchanged under the new standard conditions.

   (b)   **Given:** $N_2(g) + \frac{1}{2}O_2(g) \rightarrow N_2O(g)$, define standard state as 2 atm   **Find:** ΔG°_f

         Conceptual Plan: $\Delta G^\circ_f, P_{N_2}, P_{O_2}, P_{N_2O}, T \rightarrow$ **new** ΔG°_f

$$\Delta G_{rxn} = \Delta G^\circ_{rxn} + RT \ln Q \quad \text{where } Q = \frac{P_{N_2O}}{P_{N_2} P_{O_2}^{1/2}}$$

         Solution: $\Delta G^\circ_f = +103.7 \text{ kJ/mol}$ and $Q = \frac{P_{N_2O}}{P_{N_2} P_{O_2}^{1/2}} = \frac{2}{2 \cdot 2^{1/2}} = \frac{1}{\sqrt{2}}$ then

$$\Delta G_{rxn} = \Delta G^\circ_{rxn} + RT \ln Q = 103.7 \frac{kJ}{mol} + \left(8.314 \frac{J}{K \cdot mol}\right)\left(\frac{1 \text{ kJ}}{1000 \text{ J}}\right)(298 \text{ K}) \ln\left(\frac{1}{\sqrt{2}}\right) =$$

$$+102.8 \text{ kJ/mol} = +102{,}800 \text{ J/mol}$$

The entropy of the reactants (1.5 mol) is decreased more than the entropy of the product (1 mol). Because the product is relatively more favored at lower volume, ΔG°_f is less positive.

         Check: The units (kJ) are correct. The Q is less than 1, so ΔG°_f is reduced under the new standard conditions.

   (c)   **Given:** $1/2\,H_2(g) \rightarrow H(g)$, define standard state as 2 atm   **Find:** ΔG°_f

         Conceptual Plan: $\Delta G^\circ_f, P_{H_2}, P_H, T \rightarrow$ **new** ΔG°_f

$$\Delta G_{rxn} = \Delta G^\circ_{rxn} + RT \ln Q \quad \text{where } Q = \frac{P_H}{P_{H_2}^{1/2}}$$

         Solution: $\Delta G^\circ_f = +203.3 \text{ kJ/mol}$ and $Q = \frac{P_H}{P_{H_2}^{1/2}} = \frac{2}{2^{1/2}} = \sqrt{2}$ then

$$\Delta G_{rxn} = \Delta G^\circ_{rxn} + RT \ln Q = +203.3 \frac{kJ}{mol} + \left(8.314 \frac{J}{K \cdot mol}\right)\left(\frac{1 \text{ kJ}}{1000 \text{ J}}\right)(298 \text{ K}) \ln(\sqrt{2}) =$$

$$+204.2 \text{ kJ/mol} = +204{,}200 \text{ J/mol}$$

The entropy of the product (1 mol) is decreased more than the entropy of the reactant (1/2 mol). Because the product is relatively less favored, ΔG°_f is more positive.

         Check: The units (kJ) are correct. The Q is greater than 1, so ΔG°_f is increased under the new standard conditions.

18.107 **Given:** $K = 3.9 \times 10^5$ at 300 K and $K = 1.2 \times 10^{-1}$ at 500 K **Find:** $\Delta H°_{rxn}$, $\Delta S°_{rxn}$
Conceptual Plan: Plot ln K versus $1/T$. The slope will be $-\Delta H°_{rxn}/R$, and the intercept will be $\Delta S°_{rxn}/R$.

Solution: Plot ln K versus $1/T$. Because $\ln K = -\dfrac{\Delta H°_{rxn}}{R}\dfrac{1}{T} + \dfrac{\Delta S°_{rxn}}{R}$, the negative of the slope will be $-\dfrac{\Delta H°_{rxn}}{R}$ and the intercept will be $\dfrac{\Delta S°_{rxn}}{R}$. The slope can be determined by measuring $\Delta y/\Delta x$ on the plot or by using functions such as "add trendline" in Excel. Because the slope is $+11246$ K,
$\Delta H°_{rxn} = -slope\ R =$

$$-(+11\underline{2}46\ \text{K})\left(8.314\dfrac{\text{J}}{\text{K}\cdot\text{mol}}\right)\left(\dfrac{1\ \text{kJ}}{1000\ \text{J}}\right) =$$

$-93.499\ \text{kJ/mol} = -93\ \text{kJ/mol}$

Because the intercept is $-24.612\ \text{K}^{-1}$,

$$\Delta S°_{rxn} = intercept\ R = \left(\dfrac{-24.612}{\text{K}}\right)\left(8.314\dfrac{\text{J}}{\text{K}\cdot\text{mol}}\right)$$

$= -204.62\ \text{J/K} = -2.0 \times 10^2\ \text{J/K}.$

Note: This problem can also be solved using the mathematically equivalent equation in Problems 18.97 and 18.98.

Check: The units are correct (kJ/mole and J/K mol). The numbers are typical for reactions. Because the slope is positive, the enthalpy change must be negative. The entropy change is expected to be negative because the number of moles of gas is decreasing.

18.109 **Given:** $\Delta H°_{vap}$ table **Find:** ΔS_{vap}; then compare values
Conceptual Plan: °C → K then $\Delta H°_{vap}$, $T → \Delta S_{vap}$

$$K = 273.15 + °C \qquad \Delta S_{vap} = \dfrac{-\Delta H_{vap}}{T}$$

Solution:

Diethyl ether: $T = 273.15 + 34.6 = 307.8$ K then $\Delta S_{vap} = \dfrac{\Delta H_{vap}}{T} = \dfrac{26.5\ \text{kJ}}{307.8\ \text{K}} = 0.0861\ \text{kJ/K} = 86.1\ \text{J/K}$

Acetone: $T = 273.15 + 56.1 = 329.3$ K then $\Delta S_{vap} = \dfrac{\Delta H_{vap}}{T} = \dfrac{29.1\ \text{kJ}}{329.3\ \text{K}} = 0.0884\ \text{kJ/K} = 88.4\ \text{J/K}$

Benzene: $T = 273.15 + 79.8 = 353.0$ K then $\Delta S_{vap} = \dfrac{\Delta H_{vap}}{T} = \dfrac{30.8\ \text{kJ}}{353.0\ \text{K}} = 0.0873\ \text{kJ/K} = 87.3\ \text{J/K}$

Chloroform: $T = 273.15 + 60.8 = 334.0$ K then $\Delta S_{vap} = \dfrac{\Delta H_{vap}}{T} = \dfrac{29.4\ \text{kJ}}{334.0\ \text{K}} = 0.0880\ \text{kJ/K} = 88.0\ \text{J/K}$

Ethanol: $T = 273.15 + 77.8 = 351.0$ K then $\Delta S_{vap} = \dfrac{\Delta H_{vap}}{T} = \dfrac{38.6\ \text{kJ}}{351.0\ \text{K}} = 0.110\ \text{kJ/K} = 110.\ \text{J/K}$

Water: $T = 273.15 + 100 = 373.15$ K then $\Delta S_{vap} = \dfrac{\Delta H_{vap}}{T} = \dfrac{40.7\ \text{kJ}}{373.15\ \text{K}} = 0.109\ \text{kJ/K} = 109\ \text{J/K}$

The first four values are very similar because they have similar intermolecular forces between molecules (dispersion forces and/or dipole–dipole interactions). The values for ethanol and water are higher because the intermolecular forces between molecules are stronger due to hydrogen bonding. Because of this, more energy is dispersed when these interactions are broken.

Conceptual Problems

18.111 Both (a) and (c) increase the entropy of the surroundings because they are exothermic reactions (adding thermal energy to the surroundings).

18.113 (c) If the entropy of a system is increasing, the enthalpy of a reaction can be overcome (if necessary) by the entropy change as long as the temperature is high enough. If the entropy change of the system is decreasing, the reaction must be exothermic to be spontaneous because the entropy is working against spontaneity.

18.115 Both (a) and (b) are true. Because $\Delta G_{rxn} = \Delta G°_{rxn} + RT \ln Q$ and $\Delta G°_{rxn} = -42.5$ kJ, for $\Delta G_{rxn} = 0$, the second term must be positive. This necessitates that $Q > 1$ or that we have more product than reactant. Any reaction at equilibrium has $\Delta G_{rxn} = 0$.

Questions for Group Work

18.117   For a sum of 2, there is only one possible combination: 1 and 1. For a sum of 12, there is only one possible combination: 6 and 6. For a sum of 7, there are three possible combinations: 1 and 6; 2 and 5; 3 and 4. Since a sum of 7 has the greatest number of combinations, it has the greatest entropy.

18.119   The second law refers to the total entropy of the universe. If the entropy of the surroundings is dropping more than the entropy of the system is increasing, the process will not be spontaneous. For example, ice doesn't melt at $-10\ °C$, even though the entropy of the water increases in the process.

18.121 $\Delta G = \Delta H^{\circ}_{rxn} - T\Delta S^{\circ}_{rxn} = -1273.3\ \text{kJ} - (298\ \text{K})\left(-1.2219\ \dfrac{\text{kJ}}{\text{K}}\right) = -909.2\ \text{kJ}$. The reaction will be spontaneous under standard conditions because $\Delta G^{\circ}_{f} < 0$. The determining factor is the energy: ΔH°_{f} is more negative than $-T\Delta S^{\circ}$ is positive.

Data Interpretation and Analysis

18.122   (a)

Plot of $\ln K_{sp}$ versus $1/T$

$y = -14589x + 43.502$

     (c)   intercept $= \Delta S^{\circ}/R$ So, $\Delta S^{\circ} = R \times$ intercept $= (8.314\ \text{J/K} \cdot \text{mol}) \times (43.502) = 357.7\ \text{J/K} \cdot \text{mol} = 0.3577\ \text{kJ/K} \cdot \text{mol}$

     (e)

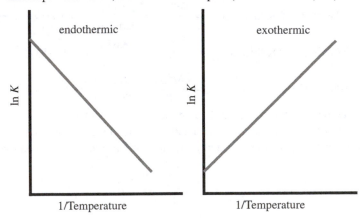

19 Electrochemistry

Review Questions

19.1 Oxidation is the loss of electrons that corresponds to an increase in oxidation state. Reduction is the gain of electrons that corresponds to a decrease in oxidation state. Balancing redox reactions can be more complicated than balancing other types of reactions because both the mass (or number of each type of atom) and the charge must be balanced. Redox reactions occurring in aqueous solutions can be balanced by using a special procedure called the half-reaction method of balancing. In this procedure, the overall equation is broken down into two half-reactions: one for oxidation and one for reduction. The half-reactions are balanced individually and then added together so that the number of electrons generated in the oxidation half-reaction is the same as the number of electrons consumed in the reduction half-reaction.

19.3 In all electrochemical cells, the electrode where oxidation occurs is called the anode. In a voltaic cell, the anode is labeled with a negative (−) sign. The anode is negative because the oxidation reaction that occurs at the anode releases electrons. Electrons flow from the anode to the cathode (from negative to positive) through the wires connecting the electrodes.

19.5 A salt bridge is a pathway by which ions can flow between the half-cells and complete an electrical circuit and maintain electroneutrality without the solutions in the half-cells totally mixing.

19.7 The standard cell potential $E°_{cell}$ or standard emf is the cell potential under standard conditions (1 M concentration for reactants in solution and 1 atm pressure for gaseous reactants). The cell potential is a measure of the overall tendency of the redox reaction to occur spontaneously. The more positive the cell potential, the more spontaneous the reaction. A negative cell potential indicates a nonspontaneous reaction.

19.9 Inert electrodes, such as platinum (Pt) or graphite, are used as the anode or cathode (or both) when the reactants and products of one or both of the half-reactions are in the same phase.

19.11 The overall cell potential for any electrochemical cell will always be the sum of the half-cell potential for the oxidation reaction and the half-cell potential for the reduction reaction or $E°_{cell} = E°_{cathode} - E°_{anode}$.

19.13 In general, any reduction half-reaction will be spontaneous when paired with the reverse of a half-reaction below it in the table.

19.15 We have seen that a positive standard cell potential $(E°_{cell})$ corresponds to a spontaneous oxidation–reduction reaction. We know (from Chapter 18) that the spontaneity of a reaction is determined by the sign of $\Delta G°$. Therefore, $E°_{cell}$ and $\Delta G°$ must be related. We also know from Section 18.9 that $\Delta G°$ for a reaction is related to the equilibrium constant (K) for the reaction. Because $E°_{cell}$ and $\Delta G°$ are related, $E°_{cell}$ and K must also be related. The equation that relates the three quantities is $\Delta G°_{rxn} = -RT \ln K = -n F E°_{cell}$.

19.17 The Nernst equation relates concentration and cell potential as follows: $E_{cell} = E°_{cell} - \dfrac{0.0592\ \text{V}}{n} \log Q$,

where E_{cell} is the cell potential in V, $E°_{cell}$ is the standard cell potential in V, n is the number of moles of electrons transferred in the redox reaction, and Q is the reaction quotient. Increasing the concentration of the reactants decreases the value of Q, so it increases the cell potential. Increasing the concentration of the products increases the value of Q, so it decreases the cell potential.

19.19 A concentration cell is a voltaic cell in which both half-reactions are the same but in which a difference in concentration drives the current flow (because the cell potential depends not only on the half-reactions occurring in the cell, but also on the concentrations of the reactants and products in those half-reactions).

Therefore, $E_{cell} = -\dfrac{0.0592\ V}{n} \log Q$.

19.21 Lead–acid storage batteries consist of six electrochemical cells wired in series in which each cell produces 2 V for a total of 12 V. Each cell contains a porous lead anode where oxidation occurs according to $Pb(s) + HSO_4^-(aq) \rightarrow PbSO_4(s) + H^+(aq) + 2\,e^-$ and a lead(IV) oxide cathode where reduction occurs according to $PbO_2(s) + HSO_4^-(aq) + 3\,H^+(aq) + 2\,e^- \rightarrow PbSO_4(s) + 2\,H_2O(l)$. As electric current is drawn from the battery, both the anode and the cathode become coated with $PbSO_4(s)$ and the solution becomes depleted of $HSO_4^-(aq)$. If the battery is run for a long time without being recharged, too much $PbSO_4(s)$ develops on the surface of the electrodes and the battery goes dead. The lead–acid storage battery can be recharged, however, by running electric current through it in reverse. The electric current, which must come from an external source such as an alternator in a car, causes the preceding reaction to occur in reverse, converting the $PbSO_4(s)$ back to $Pb(s)$ and $PbO_2(s)$, recharging the battery.

19.23 Fuel cells are like batteries, but the reactants must be constantly replenished. Normal batteries lose their ability to generate voltage with use because the reactants become depleted as electric current is drawn from the battery. In a fuel cell, the reactants—the fuel—constantly flow through the battery, generating electric current as they undergo a redox reaction.

The most common fuel cell is the hydrogen–oxygen fuel cell. In the cell, hydrogen gas flows past the anode (a screen coated with a platinum catalyst) and undergoes oxidation: $2\,H_2(g) + 4\,OH^-(aq) \rightarrow 4\,H_2O(l) + 4\,e^-$. Oxygen gas flows past the cathode (a similar screen) and undergoes reduction: $O_2(g) + 2\,H_2O(l) + 4\,e^- \rightarrow 4\,OH^-(aq)$. The half-reactions sum to the following overall reaction: $2\,H_2(g) + O_2(g) \rightarrow 2\,H_2O(l)$. Notice that the only product is water.

19.25 The anion is oxidized. The cation is reduced to the metal.

19.27 In the electrolysis of aqueous NaCl solutions, two different reduction half-reactions are possible at the cathode, the reduction of Na^+ and the reduction of water.

$2\,Na^+(l) + 2\,e^- \rightarrow 2\,Na(s)$ $E_{red}^\circ = 2.71\ V$

$2\,H_2O(l) + 2\,e^- \rightarrow H_2(g) + 2\,OH^-(aq)$ $E_{red} = -0.41\ V\ ([OH^-] = 10^{-7}\ M)$

The half-reaction that occurs most easily (the one with the least negative, or most positive, half-cell potential) is the one that actually occurs.

19.29 We can determine the number of moles of electrons that have flowed in a given electrolysis cell by measuring the total charge that has flowed through the cell, which in turn depends on the magnitude of current and the time the current has run. Because 1 ampere $= 1\ C/s$, if we multiply the amount of current (in A) flowing through the cell by the time (in s) the current flowed, we can find the total charge that passed through the cell in that time: current $(C/s) \times$ time $(s) =$ charge (C). The relationship between charge and the number of moles of electrons is given by Faraday's constant, which corresponds to the charge in coulombs of 1 mol of electrons; so $F = 96{,}485\ C/mole\ e^-$. These relationships can be used to solve problems involving the stoichiometry of electrolytic cells.

19.31 Moisture must be present for many corrosion reactions to occur. The presence of water is necessary because water is a reactant in many oxidation reactions [such as $4\,Fe^{2+}(aq) + O_2(g) + (4 + 2n)\,H_2O(l) \rightarrow 2\,Fe_2O_3 \cdot n\,H_2O(s) + 8\,H^+(aq)$] and because charge (either electrons or ions) must be free to flow between the anodic and cathodic regions.

Additional electrolytes promote more corrosion. The presence of an electrolyte (such as sodium chloride) on the surface of iron promotes rusting because it enhances current flow. This is why cars rust so quickly in cold climates where roads are salted and in areas directly adjacent to beaches where salt water mist is present.

The presence of acids promotes corrosion. Because H^+ ions are involved as a reactant in the reduction of oxygen, lower pH enhances the cathodic reaction and leads to faster corrosion.

Problems by Topic

Balancing Redox Reactions

19.33 **Conceptual Plan:** Separate the overall reaction into two half-reactions: one for oxidation and one for reduction. → Balance each half-reaction with respect to mass in the following order: (1) Balance all elements other than H and O, (2) balance O by adding H_2O, and (3) balance H by adding H^+. → Balance each half-reaction with respect to charge by adding electrons. (The sum of the charges on both sides of the equation should be made equal by adding electrons as necessary.) → Make the number of electrons in both half-reactions equal by multiplying one or both half-reactions by a small whole number. → Add the two half-reactions, canceling electrons and other species as necessary. → Verify that the reaction is balanced with respect to both mass and charge.

Solution:

(a)

Separate:	$K(s) \rightarrow K^+(aq)$	and	$Cr^{3+}(aq) \rightarrow Cr(s)$
Balance elements:	$K(s) \rightarrow K^+(aq)$	and	$Cr^{3+}(aq) \rightarrow Cr(s)$
Add electrons:	$K(s) \rightarrow K^+(aq) + e^-$	and	$Cr^{3+}(aq) + 3\,e^- \rightarrow Cr(s)$
Equalize electrons:	$3\,K(s) \rightarrow 3\,K^+(aq) + 3\,e^-$	and	$Cr^{3+}(aq) + 3\,e^- \rightarrow Cr(s)$
Add half-reactions:	$3\,K(s) + Cr^{3+}(aq) + \cancel{3\,e^-} \rightarrow 3\,K^+(aq) + \cancel{3\,e^-} + Cr(s)$		
Cancel electrons:	$3\,K(s) + Cr^{3+}(aq) \rightarrow 3\,K^+(aq) + Cr(s)$		

Check:

Reactants	Products
3 K atoms	3 K atoms
1 Cr atom	1 Cr atom
+3 charge	+3 charge

(b)

Separate:	$Al(s) \rightarrow Al^{3+}(aq)$	and	$Fe^{2+}(aq) \rightarrow Fe(s)$
Balance elements:	$Al(s) \rightarrow Al^{3+}(aq)$	and	$Fe^{2+}(aq) \rightarrow Fe(s)$
Add electrons:	$Al(s) \rightarrow Al^{3+}(aq) + 3e^-$	and	$Fe^{2+}(aq) + 2\,e^- \rightarrow Fe(s)$
Equalize electrons:	$2\,Al(s) \rightarrow 2\,Al^{3+}(aq) + 6\,e^-$	and	$3\,Fe^{2+}(aq) + 6\,e^- \rightarrow 3\,Fe(s)$
Add half-reactions:	$2\,Al(s) + 3\,Fe^{2+}(aq) + \cancel{6\,e^-} \rightarrow 2\,Al^{3+}(aq) + \cancel{6\,e^-} + 3\,Fe(s)$		
Cancel electrons:	$2\,Al(s) + 3\,Fe^{2+}(aq) \rightarrow 2\,Al^{3+}(aq) + 3\,Fe(s)$		

Check:

Reactants	Products
2 Al atoms	2 Al atoms
3 Fe atoms	3 Fe atoms
+6 charge	+6 charge

(c)

Separate:	$BrO_3^-(aq) \rightarrow Br^-(aq)$	and	$N_2H_4(g) \rightarrow N_2(g)$
Balance non-H & O elements:	$BrO_3^-(aq) \rightarrow Br^-(aq)$	and	$N_2H_4(g) \rightarrow N_2(g)$
Balance O with H_2O:	$BrO_3^-(aq) \rightarrow Br^-(aq) + 3\,H_2O(l)$	and	$N_2H_4(g) \rightarrow N_2(g)$
Balance H with H^+:	$BrO_3^-(aq) + 6\,H^+(aq) \rightarrow Br^-(aq) + 3\,H_2O(l)$	and	$N_2H_4(g) \rightarrow N_2(g) + 4\,H^+(aq)$

Add electrons:

$$BrO_3^-(aq) + 6\,H^+(aq) + 6\,e^- \rightarrow Br^-(aq) + 3\,H_2O(l) \quad \text{and} \quad N_2H_4(g) \rightarrow N_2(g) + 4\,H^+(aq) + 4\,e^-$$

Equalize electrons:

$$2\,BrO_3^-(aq) + 12\,H^+(aq) + 12\,e^- \rightarrow 2\,Br^-(aq) + 6\,H_2O(l) \quad \text{and} \quad 3\,N_2H_4(g) \rightarrow 3\,N_2(g) + 12\,H^+(aq) + 12\,e^-$$

Add half-reactions: $2\,BrO_3^-(aq) + \cancel{12\,H^+(aq)} + 3\,N_2H_4(g) + \cancel{12\,e^-} \rightarrow 2\,Br^-(aq) + 6\,H_2O(l) + 3\,N_2(g)$
$$+ \cancel{12\,H^+(aq)} + \cancel{12\,e^-}$$

Cancel electrons & others: $2\,BrO_3^-(aq) + 3\,N_2H_4(g) \rightarrow 2\,Br^-(aq) + 6\,H_2O(l) + 3\,N_2(g)$

Check:

Reactants	Products
2 Br atoms	2 Br atoms
6 O atoms	6 O atoms
12 H atoms	12 H atoms
6 N atoms	6 N atoms
−2 charge	−2 charge

19.35   **Conceptual Plan:** Separate the overall reaction into two half-reactions: one for oxidation and one for reduction. → Balance each half-reaction with respect to mass in the following order: (1) Balance all elements other than H and O, (2) balance O by adding H_2O, and (3) balance H by adding H^+. → Balance each half-reaction with respect to charge by adding electrons. (The sum of the charges on both sides of the equation should be made equal by adding electrons as necessary.) → Make the number of electrons in both half-reactions equal by multiplying one or both half-reactions by a small whole number. → Add the two half-reactions, canceling electrons and other species as necessary. → Verify that the reaction is balanced with respect to both mass and charge.

Solution:

(a)   Separate: $PbO_2(s) \rightarrow Pb^{2+}(aq)$ and $I^-(aq) \rightarrow I_2(s)$

      Balance non-H & O elements: $PbO_2(s) \rightarrow Pb^{2+}(aq)$ and $2\,I^-(aq) \rightarrow I_2(s)$

      Balance O with H_2O: $PbO_2(s) \rightarrow Pb^{2+}(aq) + 2\,H_2O(l)$ and $2\,I^-(aq) \rightarrow I_2(s)$

      Balance H with H^+: $PbO_2(s) + 4\,H^+(aq) \rightarrow Pb^{2+}(aq) + 2\,H_2O(l)$ and $2\,I^-(aq) \rightarrow I_2(s)$

      Add electrons: $PbO_2(s) + 4\,H^+(aq) + 2\,e^- \rightarrow Pb^{2+}(aq) + 2\,H_2O(l)$ and $2\,I^-(aq) \rightarrow I_2(s) + 2\,e^-$

      Equalize electrons: $PbO_2(s) + 4\,H^+(aq) + 2\,e^- \rightarrow Pb^{2+}(aq) + 2\,H_2O(l)$ and $2\,I^-(aq) \rightarrow I_2(s) + 2\,e^-$

      Add half-reactions: $PbO_2(s) + 4\,H^+(aq) + \cancel{2\,e^-} + 2\,I^-(aq) \rightarrow Pb^{2+}(aq) + 2\,H_2O(l) + I_2(s) + \cancel{2\,e^-}$

      Cancel electrons & others: $PbO_2(s) + 4\,H^+(aq) + 2\,I^-(aq) \rightarrow Pb^{2+}(aq) + 2\,H_2O(l) + I_2(s)$

      Check:

Reactants	Products
1 Pb atom	1 Pb atom
2 O atoms	2 O atoms
4 H atoms	4 H atoms
2 I atoms	2 I atoms
+2 charge	+2 charge

(b)   Separate: $MnO_4^-(aq) \rightarrow Mn^{2+}(aq)$ and $SO_3^{2-}(aq) \rightarrow SO_4^{2-}(aq)$

      Balance non-H & O elements: $MnO_4^-(aq) \rightarrow Mn^{2+}(aq)$ and $SO_3^{2-}(aq) \rightarrow SO_4^{2-}(aq)$

      Balance O with H_2O: $MnO_4^-(aq) \rightarrow Mn^{2+}(aq) + 4\,H_2O(l)$ and $SO_3^{2-}(aq) + H_2O(l) \rightarrow SO_4^{2-}(aq)$

      Balance H with H^+:

           $MnO_4^-(aq) + 8\,H^+(aq) \rightarrow Mn^{2+}(aq) + 4\,H_2O(l)$ and $SO_3^{2-}(aq) + H_2O(l) \rightarrow SO_4^{2-}(aq) + 2\,H^+(aq)$

      Add electrons: $MnO_4^-(aq) + 8\,H^+(aq) + 5\,e^- \rightarrow Mn^{2+}(aq) + 4\,H_2O(l)$ and

                                        $SO_3^{2-}(aq) + H_2O(l) \rightarrow SO_4^{2-}(aq) + 2\,H^+(aq) + 2\,e^-$

      Equalize electrons: $2\,MnO_4^-(aq) + 16\,H^+(aq) + 10\,e^- \rightarrow 2\,Mn^{2+}(aq) + 8\,H_2O(l)$ and

                                     $5\,SO_3^{2-}(aq) + 5\,H_2O(l) \rightarrow 5\,SO_4^{2-}(aq) + 10\,H^+(aq) + 10\,e^-$

      Add half-reactions: $2\,MnO_4^-(aq) + 6\,\cancel{16}\,H^+(aq) + \cancel{10\,e^-} + 5\,SO_3^{2-}(aq) + 5\,\cancel{H_2O(l)} \rightarrow$

                           $2\,Mn^{2+}(aq) + 3\,\cancel{8}\,H_2O(l) + 5\,SO_4^{2-}(aq) + \cancel{10}H^+(aq) + \cancel{10\,e^-}$

      Cancel electrons: $2\,MnO_4^-(aq) + 6\,H^+(aq) + 5\,SO_3^{2-}(aq) \rightarrow 2\,Mn^{2+}(aq) + 3\,H_2O(l) + 5\,SO_4^{2-}(aq)$

      Check:

Reactants	Products
2 Mn atoms	2 Mn atoms
23 O atoms	23 O atoms
6 H atoms	6 H atoms
5 S atoms	5 S atoms
−6 charge	−6 charge

(c)   Separate: $S_2O_3^{2-}(aq) \rightarrow SO_4^{2-}(aq)$ and $Cl_2(g) \rightarrow Cl^-(aq)$

      Balance non-H & O elements: $S_2O_3^{2-}(aq) \rightarrow 2\,SO_4^{2-}(aq)$ and $Cl_2(g) \rightarrow 2\,Cl^-(aq)$

      Balance O with H_2O: $S_2O_3^{2-}(aq) + 5\,H_2O(l) \rightarrow 2\,SO_4^{2-}(aq)$ and $Cl_2(g) \rightarrow 2\,Cl^-(aq)$

      Balance H with H^+: $S_2O_3^{2-}(aq) + 5\,H_2O(l) \rightarrow 2\,SO_4^{2-}(aq) + 10\,H^+(aq)$ and $Cl_2(g) \rightarrow 2\,Cl^-(aq)$

      Add electrons:

           $S_2O_3^{2-}(aq) + 5\,H_2O(l) \rightarrow 2\,SO_4^{2-}(aq) + 10\,H^+(aq) + 8\,e^-$ and $Cl_2(g) + 2\,e^- \rightarrow 2\,Cl^-(aq)$

      Equalize electrons:

           $S_2O_3^{2-}(aq) + 5\,H_2O(l) \rightarrow 2\,SO_4^{2-}(aq) + 10\,H^+(aq) + 8\,e^-$ and $4\,Cl_2(g) + 8\,e^- \rightarrow 8\,Cl^-(aq)$

      Add half-reactions:

           $S_2O_3^{2-}(aq) + 5\,H_2O(l) + 4\,Cl_2(g) + \cancel{8\,e^-} \rightarrow 2\,SO_4^{2-}(aq) + 10\,H^+(aq) + \cancel{8\,e^-} + 8\,Cl^-(aq)$

      Cancel electrons: $S_2O_3^{2-}(aq) + 5\,H_2O(l) + 4\,Cl_2(g) \rightarrow 2\,SO_4^{2-}(aq) + 10\,H^+(aq) + 8\,Cl^-(aq)$

Check:	Reactants	Products
	2 S atoms	2 S atoms
	8 O atoms	8 O atoms
	10 H atoms	10 H atoms
	8 Cl atoms	8 Cl atoms
	-2 charge	-2 charge

19.37 **Conceptual Plan: Separate the overall reaction into two half-reactions: one for oxidation and one for reduction.** $\rightarrow$ **Balance each half-reaction with respect to mass in the following order: (1) Balance all elements other than H and O, (2) balance O by adding H_2O, (3) balance H by adding H^+, and (4) neutralize H^+ by adding enough OH^- to neutralize each H^+. Add the same number of OH^- ions to each side of the equation.** $\rightarrow$ **Balance each half-reaction with respect to charge by adding electrons. (The sum of the charges on both sides of the equation should be made equal by adding electrons as necessary.)** $\rightarrow$ **Make the number of electrons in both half-reactions equal by multiplying one or both half-reactions by a small whole number.** $\rightarrow$ **Add the two half-reactions, canceling electrons and other species as necessary.** $\rightarrow$ **Verify that the reaction is balanced with respect to both mass and charge.**
Solution:

(a) Separate: $\qquad ClO_2(aq) \rightarrow ClO_2^-(aq)$ and $H_2O_2(aq) \rightarrow O_2(g)$
Balance non-H & O elements: $ClO_2(aq) \rightarrow ClO_2^-(aq)$ and $H_2O_2(aq) \rightarrow O_2(g)$
Balance O with H_2O: $\quad ClO_2(aq) \rightarrow ClO_2^-(aq)$ and $H_2O_2(aq) \rightarrow O_2(g)$
Balance H with H^+: $\quad ClO_2(aq) \rightarrow ClO_2^-(aq)$ and $H_2O_2(aq) \rightarrow O_2(g) + 2 H^+(aq)$
Neutralize H^+ with OH^-:

$$ClO_2(aq) \rightarrow ClO_2^-(aq) \text{ and } H_2O_2(aq) + 2 OH^-(aq) \rightarrow O_2(g) + \underbrace{2 H^+(aq) + 2 OH^-(aq)}_{2 H_2O(l)}$$

Add electrons: $ClO_2(aq) + e^- \rightarrow ClO_2^-(aq)$ and $H_2O_2(aq) + 2 OH^-(aq) \rightarrow O_2(g) + 2 H_2O(l) + 2 e^-$
Equalize electrons:

$$2 ClO_2(aq) + 2 e^- \rightarrow 2 ClO_2^-(aq) \text{ and } H_2O_2(aq) + 2 OH^-(aq) \rightarrow O_2(g) + 2 H_2O(l) + 2 e^-$$

Add half-reactions:

$$2 ClO_2(aq) + \cancel{2 e^-} + H_2O_2(aq) + 2 OH^-(aq) \rightarrow 2 ClO_2^-(aq) + O_2(g) + 2 H_2O(l) + \cancel{2 e^-}$$

Cancel electrons: $2 ClO_2(aq) + H_2O_2(aq) + 2 OH^-(aq) \rightarrow 2 ClO_2^-(aq) + O_2(g) + 2 H_2O(l)$

Check:	Reactants	Products
	2 Cl atoms	2 Cl atoms
	8 O atoms	8 O atoms
	4 H atoms	4 H atoms
	-2 charge	-2 charge

(b) Separate: $\qquad MnO_4^-(aq) \rightarrow MnO_2(s)$ and $Al(s) \rightarrow Al(OH)_4^-(aq)$
Balance non-H & O elements: $MnO_4^-(aq) \rightarrow MnO_2(s)$ and $Al(s) \rightarrow Al(OH)_4^-(aq)$
Balance O with H_2O: $MnO_4^-(aq) \rightarrow MnO_2(s) + 2 H_2O(l)$ and $Al(s) + 4 H_2O(l) \rightarrow Al(OH)_4^-(aq)$
Balance H with H^+:

$MnO_4^-(aq) + 4 H^+(aq) \rightarrow MnO_2(s) + 2 H_2O(l)$ and $Al(s) + 4 H_2O(l) \rightarrow Al(OH)_4^-(aq) + 4 H^+(aq)$
Neutralize H^+ with OH^-: $MnO_4^- + \underbrace{4 H^+(aq) + 4 OH^-(aq)}_{2\cancel{4} H_2O(l)} \rightarrow MnO_2(s) + 2\cancel{H_2O(l)} + 4 OH^-(aq)$

and $Al(s) + 4\cancel{H_2O(l)} + 4 OH^-(aq) \rightarrow Al(OH)_4^-(aq) + \underbrace{4 H^+(aq) + 4 OH^-(aq)}_{4\cancel{H_2O(l)}}$

Add electrons: $MnO_4^-(aq) + 2 H_2O(l) + 3 e^- \rightarrow MnO_2(s) + 4 OH^-(aq)$ and

$$Al(s) + 4 OH^-(aq) \rightarrow Al(OH)_4^-(aq) + 3 e^-$$

Equalize electrons: $MnO_4^-(aq) + 2 H_2O(l) + 3 e^- \rightarrow MnO_2(s) + 4 OH^-(aq)$ and

$$Al(s) + 4 OH^-(aq) \rightarrow Al(OH)_4^-(aq) + 3 e^-$$

Add half-reactions:

$MnO_4^-(aq) + 2 H_2O(l) + \cancel{3 e^-} + Al(s) + \cancel{4 OH^-(aq)} \rightarrow MnO_2(s) + \cancel{4 OH^-(aq)} + Al(OH)_4^-(aq) + \cancel{3 e^-}$
Cancel electrons: $MnO_4^-(aq) + 2 H_2O(l) + Al(s) \rightarrow MnO_2(s) + Al(OH)_4^-(aq)$

Check:	Reactants	Products
	1 Mn atom	1 Mn atom
	6 O atoms	6 O atoms
	4 H atoms	4 H atoms
	1 Al atom	1 Al atom
	-1 charge	-1 charge

(c) Separate: $Cl_2(g) \rightarrow Cl^-(aq)$ and $Cl_2(g) \rightarrow ClO^-(aq)$

Balance non-H & O elements: $Cl_2(g) \rightarrow 2\,Cl^-(aq)$ and $Cl_2(g) \rightarrow 2ClO^-(aq)$

Balance O with H_2O: $Cl_2(g) \rightarrow 2\,Cl^-(aq)$ and $Cl_2(g) + 2\,H_2O(l) \rightarrow 2\,ClO^-(aq)$

Balance H with H^+: $Cl_2(g) \rightarrow 2\,Cl^-(aq)$ and $Cl_2(g) + 2\,H_2O(l) \rightarrow 2\,ClO^-(aq) + 4\,H^+(aq)$

Neutralize H^+ with OH^-:

$Cl_2(g) \rightarrow 2\,Cl^-(aq)$ and $Cl_2(g) + 2\,\cancel{H_2O(l)} + 4\,OH^-(aq) \rightarrow 2\,ClO^-(aq) + \underline{4\,H^+(aq) + 4\,OH^-(aq)}$

$\overset{2\,\cancel{4}\,H_2O(l)}{}$

Add electrons: $Cl_2(g) + 2\,e^- \rightarrow 2\,Cl^-(aq)$ and $Cl_2(g) + 4\,OH^-(aq) \rightarrow 2\,ClO^-(aq) + 2\,H_2O(l) + 2\,e^-$

Equalize electrons: $Cl_2(g) + 2\,e^- \rightarrow 2\,Cl^-(aq)$ and $Cl_2(g) + 4\,OH^-(aq) \rightarrow 2\,ClO^-(aq) + 2\,H_2O(l) + 2\,e^-$

Add half-reactions: $Cl_2(g) + \cancel{2\,e^-} + Cl_2(g) + 4\,OH^-(aq) \rightarrow 2\,Cl^-(aq) + 2\,ClO^-(aq) + 2\,H_2O(l) + \cancel{2\,e^-}$

Cancel electrons: $2\,Cl_2(g) + 4\,OH^-(aq) \rightarrow 2\,Cl^-(aq) + 2\,ClO^-(aq) + 2\,H_2O(l)$

Simplify: $Cl_2(g) + 2\,OH^-(aq) \rightarrow Cl^-(aq) + ClO^-(aq) + H_2O(l)$

Check:	Reactants	Products
	2 Cl atoms	2 Cl atoms
	2 O atoms	2 O atoms
	2 H atoms	2 H atoms
	-2 charge	-2 charge

Voltaic Cells, Standard Cell Potentials, and Direction of Spontaneity

19.39   **Given:** voltaic cell overall redox reaction

Find: Sketch voltaic cell, labeling anode, cathode, all species, and direction of electron flow

Conceptual Plan: Separate the overall reaction into two half-cell reactions and add electrons as needed to balance reactions. Put the anode reaction on the left (oxidation = electrons as product) and the cathode reaction on the right (reduction = electrons as reactant). Electrons flow from anode to cathode.

Solution:

(a) $2\,Ag^+(aq) + Pb(s) \rightarrow 2\,Ag(s) + Pb^{2+}(aq)$ separates to $2\,Ag^+(aq) \rightarrow$ $2\,Ag(s)$ and $Pb(s) \rightarrow Pb^{2+}(aq)$; then add electrons to balance to get the cathode reaction—$2\,Ag^+(aq) + 2\,e^- \rightarrow 2\,Ag(s)$— and the anode reaction—$Pb(s) \rightarrow Pb^{2+}(aq) + 2\,e^-$.

Because we have $Pb(s)$ as the reactant for the oxidation, it will be our anode. Because we have $Ag(s)$ as the product for the reduction, it will be our cathode. Simplify the cathode reaction, dividing all terms by 2.

(b) $2\,ClO_2(g) + 2\,I^-(aq) \rightarrow 2\,ClO_2^-(aq) + I_2(s)$ separates to $2\,ClO_2(g)$ $\rightarrow 2\,ClO_2^-(aq)$ and $2\,I^-(aq) \rightarrow I_2(s)$; then add electrons to balance to get the cathode reaction—$2\,ClO_2(g) + 2\,e^- \rightarrow 2\,ClO_2^-(aq)$—and the anode reaction—$2\,I^-(aq) \rightarrow I_2(s) + 2\,e^-$.

Because we have $I^-(aq)$ as the reactant for the oxidation, we will need to use Pt as our anode. Because we have $ClO_2^-(aq)$ as the product for the reduction, we will need to use Pt as our cathode. Because $ClO_2(g)$ is our reactant for the reduction, we need to use an electrode assembly like that used for a SHE. Simplify the cathode reaction, dividing all terms by 2.

(c) $O_2(g) + 4\,H^+(aq) + 2\,Zn(s) \rightarrow 2\,H_2O(l) + 2\,Zn^{2+}(aq)$ separates to $O_2(g) + 4\,H^+(aq) \rightarrow 2\,H_2O(l)$ and $2\,Zn(s) \rightarrow 2\,Zn^{2+}(aq)$; then add electrons to balance to get the cathode reaction— $O_2(g) + 4\,H^+(aq) + 4\,e^- \rightarrow 2\,H_2O(l)$—and the anode reaction— $2\,Zn(s) \rightarrow 2\,Zn^{2+}(aq) + 4\,e^-$.

Because we have $Zn(s)$ as the reactant for the oxidation, it will be our anode. Because we have $H_2O(l)$ as the product for the reduction, we will need to use Pt as our cathode. Because $O_2(g)$ is our reactant for the reduction, we need to use an electrode assembly like that used for a SHE. Simplify the anode reaction, dividing all terms by 2.

19.41 **Given:** overall reactions from Problem 19.39   **Find:** $E°_{cell}$

Conceptual Plan: Use Table 19.1 to look up half-reactions from the solution of Problem 19.39. Calculate the standard cell potential by subtracting the electrode potential of the anode from the electrode potential of the cathode: $E°_{cell} = E°_{cathode} - E°_{anode}$.

Solution:

(a) $Ag^+(aq) + e^- \rightarrow Ag(s)\ E°_{red} = 0.80\,V = E°_{cathode}$ and $Pb(s) \rightarrow Pb^{2+}(aq) + 2\,e^-\ E°_{red} = -0.13\,V = E°_{anode}$.
Then $E°_{cell} = E°_{cathode} - E°_{anode} = 0.80\,V - (-0.13\,V) = 0.93\,V$.

(b) $ClO_2(g) + e^- \rightarrow ClO_2^-(aq)\ E°_{red} = 0.95\,V = E°_{cathode}$ and $2\,I^-(aq) \rightarrow I_2(s) + 2\,e^-\ E°_{red} = +0.54\,V = E°_{anode}$.
Then $E°_{cell} = E°_{cathode} - E°_{anode} = 0.95\,V - 0.54\,V = 0.41\,V$.

(c) $O_2(g) + 4\,H^+(aq) + 4\,e^- \rightarrow 2\,H_2O(l)\ E°_{red} = 1.23\,V$ and $Zn(s) \rightarrow Zn^{2+}(aq) + 2\,e^-\ E°_{red} = -0.76\,V = E°_{anode}$. Then $E°_{cell} = E°_{cathode} - E°_{anode} = 1.23\,V - (-0.76\,V) = 1.99\,V$.

Check: The units (V) are correct. All of the voltages are positive, which is consistent with a voltaic cell.

19.43 **Given:** voltaic cell drawing

Find: (a) Determine electron flow direction, anode, and cathode; (b) write balanced overall reaction and calculate $E°_{cell}$; (c) label electrodes as $+$ and $-$; and (d) find directions of anions and cations from salt bridge.

Conceptual Plan: Look at each half-cell and write a reduction reaction by using electrode and solution composition and adding electrons to balance. Look up half-reactions and standard reduction potentials in Table 19.1. Because this is a voltaic cell, the cell potentials must be assigned to give a positive $E°_{cell}$. Calculate the standard cell potential by subtracting the electrode potential of the anode from the electrode potential of the cathode, $E°_{cell} = E°_{cathode} - E°_{anode}$, **choosing the electrode assignments to give a positive $E°_{cell}$.**

(a) **Label the electrode where the oxidation occurs as the anode. Label the electrode where the reduction occurs as the cathode. Electrons flow from anode to cathode.**

(b) **Take two half-cell reactions and multiply the reactions as necessary to equalize the number of electrons transferred. Add the two half-cell reactions and cancel electrons and any other species.**

(c) **Label anode as $(-)$ and cathode as $(+)$.**

(d) **Cations will flow from the salt bridge toward the cathode, and the anions will flow from the salt bridge toward the anode.**

Solution:

left side: $Fe^{3+}(aq) \rightarrow Fe(s)$; right side: $Cr^{3+}(aq) \rightarrow Cr(s)$.
Add electrons to balance $Fe^{3+}(aq) + 3\,e^- \rightarrow Fe(s)$ and right side—$Cr^{3+}(aq) + 3\,e^- \rightarrow Cr(s)$. Look up cell standard reduction potentials: $Fe^{3+}(aq) + 3\,e^- \rightarrow Fe(s)\ E°_{red} = -0.036\,V$ and $Cr^{3+}(aq) + 3\,e^- \rightarrow Cr(s)$ $E°_{red} = -0.73\,V$. To get a positive cell potential, the second reaction is the oxidation reaction (anode). $E°_{cell} = E°_{cathode} - E°_{anode} = -0.036\,V - (-0.73V) = +0.69\,V$ ((a), (c), and (d)). (b) Add two half-reactions with the second reaction reversed. $Fe^{3+}(aq) + 3\,e^- + Cr(s) \rightarrow Fe(s) + Cr^{3+}(aq) + 3\,e^-$ Cancel electrons to get $Fe^{3+}(aq) + Cr(s) \rightarrow Fe(s) + Cr^{3+}(aq)$.

Check: All atoms and charge are balanced. The units (V) are correct. The cell potential is positive, which is consistent with a voltaic cell.

19.45 **Given:** overall reactions from Problem 19.39 **Find:** line notation

Conceptual Plan: Use the solution from Problem 19.39. Write the oxidation half-reaction components on the left and the reduction on the right. A double vertical line ($\|$), indicating the salt bridge, separates the two half-reactions. Substances in different phases are separated by a single vertical line ($|$), which represents the boundary between the phases. For some redox reactions, the reactants and products of one or both of the half-reactions may be in the same phase. In these cases, the reactants and products are separated from each other with a comma in the line diagram. Such cells use an inert electrode, such as platinum (Pt) or graphite, as the anode or cathode (or both).

Solution:

(a) Reduction reaction: $Ag^+(aq) + e^- \rightarrow Ag(s)$; oxidation reaction: $Pb(s) \rightarrow Pb^2 + (aq) + 2\,e^-$
so $Pb(s) \mid Pb^{2+}(aq) \parallel Ag^+(aq) \mid Ag(s)$

(b) Reduction reaction: $ClO_2(g) + e^- \rightarrow ClO_2^-(aq)$; oxidation reaction: $2\,I^-(aq) \rightarrow I_2(s) + 2\,e^-$
so $Pt(s) \mid I^-(aq) \mid I_2(s) \parallel ClO_2(g) \mid ClO_2^-(aq) \mid Pt(s)$

(c) Reduction reaction: $O_2(g) + 4\,H^+(aq) + 4\,e^- \rightarrow 2\,H_2O(l)$; oxidation reaction: $Zn(s) \rightarrow Zn^{2+}(aq) + 2\,e^-$
so $Zn(s) \mid Zn^{2+}(aq) \parallel O_2(g) \mid H^+(aq), H_2O(l) \mid Pt(s)$

19.47 **Given:** $Sn(s) \mid Sn^{2+}(aq) \parallel NO_3^-(aq), H^+(aq), H_2O(l) \mid NO(g) \mid Pt(s)$
Find: Sketch voltaic cell, labeling anode, cathode, all species, direction of electron flow, and E°_{cell}

Conceptual Plan: Separate overall reaction into two half-cell reactions, knowing that the oxidation half-reaction components are on the left and the reduction half-reaction components are on the right. Add electrons as needed to balance reactions. Multiply the half-reactions by the appropriate factors so that an equal number of electrons are transferred. Add the half-cell reactions and cancel electrons. Put anode reaction on the left (oxidation = electrons as product) and cathode reaction on the right (reduction = electrons as reactant). Electrons flow from anode to cathode. Look up half-reactions in Table 19.1. Calculate the standard cell potential by subtracting the electrode potential of the anode from the electrode potential of the cathode: $E^\circ_{cell} = E^\circ_{cathode} - E^\circ_{anode}.$

Solution: Oxidation reaction (anode): $Sn(s) \rightarrow Sn^{2+}(aq) + 2\,e^-\ E^\circ_{red} = -0.14$ V; reduction reaction (cathode):
$NO_3^-(aq) + 4\,H^+(aq) + 3\,e^- \rightarrow NO(g) + 2\,H_2O(l)\ E^\circ_{red} = 0.96$ V.
$E^\circ_{cell} = E^\circ_{cathode} - E^\circ_{anode} = 0.96$ V $- (-0.14$ V$) = 1.10$ V. Multiply the first reaction by 3 and the second reaction by 2 so that 6 electrons are transferred. $3\,Sn(s) \rightarrow 3\,Sn^{2+}(aq) + 6\,e^-$ and $2\,NO_3^-(aq) + 8\,H^+(aq) + 6\,e^- \rightarrow 2\,NO(g) + 4\,H_2O(l)$. Add the two half-reactions and cancel electrons. $3\,Sn(s) + 2\,NO_3^-(aq) + 8\,H^+(aq) + 6\,e^- \rightarrow 3\,Sn^{2+}(aq) + 6\,e^- + 2\,NO(g) + 4\,H_2O(l)$. So balanced reaction is $3\,Sn(s) + 2\,NO_3^-(aq) + 8\,H^+(aq) \rightarrow 3\,Sn^{2+}(aq) + 2\,NO(g) + 4\,H_2O(l)$.

Check: All atoms and charge are balanced. The units (V) are correct. The cell potential is positive, which is consistent with a voltaic cell.

19.49 **Given:** overall reactions **Find:** spontaneity in forward direction

Conceptual Plan: Separate the overall reaction into two half-cell reactions and add electrons as needed to balance reactions. Look up half-reactions in Table 19.1. Calculate the standard cell potential by subtracting the electrode potential of the anode from the electrode potential of the cathode: $E^\circ_{cell} = E^\circ_{cathode} - E^\circ_{anode}.$ **If** $E^\circ_{cell} > 0$, **the reaction is spontaneous in the forward direction.**

Solution:

(a) $Ni(s) + Zn^{2+}(aq) \rightarrow Ni^{2+}(aq) + Zn(s)$ separates to $Ni(s) \rightarrow Ni^{2+}(aq)$ and $Zn^{2+}(aq) \rightarrow Zn(s)$.
Add electrons. $Ni(s) \rightarrow Ni^{2+}(aq) + 2\,e^-$ and $Zn^{2+}(aq) + 2\,e^- \rightarrow Zn(s)$. Look up cell potentials. Ni is oxidized, so $E^\circ_{red} = -0.23$ V $= E^\circ_{anode}$. Zn^{2+} is reduced, so $E^\circ_{cathode} = -0.76$ V. Then $E^\circ_{cell} = E^\circ_{cathode} - E^\circ_{anode} = -0.76$ V $- (-0.23$ V$) = -0.53$ V, so the reaction is nonspontaneous.

(b) $Ni(s) + Pb^{2+}(aq) \rightarrow Ni^{2+}(aq) + Pb(s)$ separates to $Ni(s) \rightarrow Ni^{2+}(aq)$ and $Pb^{2+}(aq) \rightarrow Pb(s)$.
Add electrons. $Ni(s) \rightarrow Ni^{2+}(aq) + 2\,e^-$ and $Pb^{2+}(aq) + 2\,e^- \rightarrow Pb(s)$. Look up cell potentials. Ni is oxidized, so $E^\circ_{red} = -0.23$ V $= E^\circ_{anode}$. Pb^{2+} is reduced, so $E^\circ_{red} = -0.13$ V $= E^\circ_{cathode}$. Then $E^\circ_{cell} = E^\circ_{cathode} - E^\circ_{anode} = -0.13$ V $- (-0.23$ V$) = +0.10$ V, so the reaction is spontaneous.

(c) $Al(s) + 3\,Ag^+(aq) \rightarrow Al^{3+}(aq) + 3\,Ag(s)$ separates to $Al(s) \rightarrow Al^{3+}(aq)$ and $3\,Ag^+(aq) \rightarrow 3\,Ag(s)$.
Add electrons. $Al(s) \rightarrow Al^{3+}(aq) + 3\,e^-$ and $3\,Ag^+(aq) + 3\,e^- \rightarrow 3\,Ag(s)$. Simplify the Ag reaction to $Ag^+(aq) + e^- \rightarrow Ag(s)$. Look up cell potentials. Al is oxidized, so $E^\circ_{red} = -1.66$ V E°_{anode}. Ag^+ is reduced, so

$E^\circ_{red} = 0.80 \text{ V} = E^\circ_{cathode}$. Then $E^\circ_{cell} = E^\circ_{cathode} - E^\circ_{anode} = 0.80 \text{ V} - (-1.66 \text{ V}) = +2.46 \text{ V}$, so the reaction is spontaneous.

(d) $Pb(s) + Mn^{2+}(aq) \rightarrow Pb^{2+}(aq) + Mn(s)$ separates to $Pb(s) \rightarrow Pb^{2+}(aq)$ and $Mn^{2+}(aq) \rightarrow Mn(s)$. Add electrons. $Pb(s) \rightarrow Pb^{2+}(aq) + 2 e^-$ and $Mn^{2+}(aq) + 2 e^- \rightarrow Mn(s)$. Look up cell potentials. Pb is oxidized, so $E^\circ_{red} = -0.13 \text{ V} = E^\circ_{anode}$. Mn^{2+} is reduced, so $E^\circ_{red} = -1.18 \text{ V} = E^\circ_{cathode}$. Then $E^\circ_{cell} = E^\circ_{cathode} - E^\circ_{anode} = -1.18 \text{ V} - (-0.13 \text{ V}) = -1.05 \text{ V}$, so the reaction is nonspontaneous.

Check: The units (V) are correct. If the voltage is positive, the reaction is spontaneous. If the voltage is negative, the reaction is nonspontaneous.

19.51 For a metal to be able to reduce an ion, it must be below it in Table 19.1 (need positive $E^\circ_{cell} = E^\circ_{cathode} - E^\circ_{anode}$). So we need a metal that is below Mn^{2+} but above Mg^{2+}. Aluminum is the only one in the table that meets those criteria.

19.53 In general, metals whose reduction half-reactions lie below the reduction of H^+ to H_2 in Table 19.1 will dissolve in acids, while metals above it will not. (a) Al and (c) Pb meet that criterion. To write the balanced redox reactions, pair the oxidation of the metal with the reduction of H^+ to $H_2(2 H^+(aq) + 2 e^- \rightarrow H_2(g))$. For Al, $Al(s) \rightarrow Al^{3+}(aq) + 3 e^-$. To balance the number of electrons transferred, we need to multiply the Al reaction by 2 and the H^+ reaction by 3. So $2 Al(s) \rightarrow 2 Al^{3+}(aq) + 6 e^-$ and $6 H^+(aq) + 6 e^- \rightarrow 3 H_2(g)$. Adding the two reactions: $2 Al(s) + 6 H^+(aq) + 6e^- \rightarrow 2 Al^{3+}(aq) + 6e^- + 3 H_2(g)$. Simplify to $2 Al(s) + 6 H^+(aq) \rightarrow 2 Al^{3+}(aq) + 3 H_2(g)$. For Pb, $Pb(s) \rightarrow Pb^{2+}(aq) + 2 e^-$. Because each reaction involves two electrons, we can add the two reactions. $Pb(s) + 2 H^+(aq) + 2e^- \rightarrow Pb^{2+}(aq) + 2e^- + H_2(g)$. Simplify to $Pb(s) + 2 H^+(aq) \rightarrow Pb^{2+}(aq) + H_2(g)$.

19.55 Nitric acid (HNO_3) oxidizes metals through the following reduction half-reaction: $NO_3^-(aq) + 4 H^+(aq) + 3 e^- \rightarrow NO(g) + 2 H_2O(l)$ $E^\circ_{red} = 0.96 \text{ V}$. Because this half-reaction is above the reduction of H^+ in Table 19.1, HNO_3 can oxidize metals (copper, for example) that cannot be oxidized by HCl. (a) Cu, which is below nitric acid in the table, will be oxidized, but (b) Au, which is above nitric acid in the table (and has a reduction potential of 1.50 V), will not be oxidized. To write the balanced redox reactions, pair the oxidation of the metal with the reduction of nitric acid: $(NO_3^-(aq) + 4 H^+(aq) + 3 e^- \rightarrow NO(g) + 2 H_2O(l))$. For Cu, $Cu(s) \rightarrow Cu^{2+}(aq) + 2 e^-$. To balance the number of electrons transferred, we need to multiply the Cu reaction by 3 and the nitric acid reaction by 2. So $3 Cu(s) \rightarrow 3 Cu^{2+}(aq) + 6 e^-$ and $2 NO_3^-(aq) + 8 H^+(aq) + 6 e^- \rightarrow 2 NO(g) + 4 H_2O(l)$. Adding the two reactions: $3 Cu(s) + 2 NO_3^-(aq) + 8 H^+(aq) + 6e^- \rightarrow 3 Cu^{2+}(aq) + 6e^- + 2 NO(g) + 4 H_2O(l)$. Simplify to $3 Cu(s) + 2 NO_3^-(aq) + 8 H^+(aq) \rightarrow 3 Cu^{2+}(aq) + 2 NO(g) + 4 H_2O(l)$.

19.57 **Given:** overall reactions $\quad$ **Find:** E°_{cell} and spontaneity in forward direction
Conceptual Plan: Separate the overall reaction into two half-cell reactions and add electrons as needed to balance reactions. Look up half-reactions in Table 19.1. Calculate the standard cell potential by subtracting the electrode potential of the anode from the electrode potential of the cathode: $E^\circ_{cell} = E^\circ_{cathode} - E^\circ_{anode}$. If $E^\circ_{cell} > 0$, the reaction is spontaneous in the forward direction.
Solution:

(a) $2 Cu(s) + Mn^{2+}(aq) \rightarrow 2 Cu^+(aq) + Mn(s)$ separates to $2 Cu(s) \rightarrow 2 Cu^+(aq)$ and $Mn^{2+}(aq) \rightarrow Mn(s)$. Add electrons. $2 Cu(s) \rightarrow 2 Cu^+(aq) + 2 e^-$ and $Mn^{2+}(aq) + 2 e^- \rightarrow Mn(s)$. Simplify the Cu reaction to $Cu(s) \rightarrow Cu^+(aq) + e^-$. Look up cell potentials. Cu is oxidized, so $E^\circ_{red} = +0.52 \text{ V} = E^\circ_{anode}$. Mn^{2+} is reduced, so $E^\circ_{red} = -1.18 \text{ V} = E^\circ_{cathode}$. Then $E^\circ_{cell} = E^\circ_{cathode} - E^\circ_{anode} = -1.18 \text{ V} - 0.52 \text{ V} = -1.70 \text{ V}$, so the reaction is nonspontaneous.

(b) $MnO_2(s) + 4 H^+(aq) + Zn(s) \rightarrow Mn^{2+}(aq) + 2 H_2O(l) + Zn^{2+}(aq)$ separates to $MnO_2(s) + 4 H^+(aq) \rightarrow Mn^{2+}(aq) + 2 H_2O(l)$ and $Zn(s) \rightarrow Zn^{2+}(aq)$. Add electrons. $MnO_2(s) + 4 H^+(aq) + 2 e^- \rightarrow Mn^{2+}(aq) + 2 H_2O(l)$ and $Zn(s) \rightarrow Zn^{2+}(aq) + 2 e^-$. Look up cell potentials. Zn is oxidized, so $E^\circ_{red} = -0.76 \text{ V} = E^\circ_{anode}$. Mn is reduced, so $E^\circ_{red} = 1.21 \text{ V} = E^\circ_{cathode}$. Then $E^\circ_{cell} = E^\circ_{cathode} - E^\circ_{anode} = 1.21 \text{ V} - (-0.76 \text{ V}) = +1.97 \text{ V}$, so the reaction is spontaneous.

(c) $Cl_2(g) + 2 F^-(aq) \rightarrow 2 Cl^-(aq) + F_2(g)$ separates to $Cl_2(g) \rightarrow 2 Cl^-(aq)$ and $2 F^-(aq) \rightarrow F_2(g)$. Add electrons. $Cl_2(g) + 2 e^- \rightarrow 2 Cl^-(aq)$ and $2 F^-(aq) \rightarrow F_2(g) + 2 e^-$. Look up cell potentials. F^- is oxidized, so $E^\circ_{red} = 2.87 \text{ V} = E^\circ_{anode}$. Cl is reduced, so $E^\circ_{red} = 1.36 \text{ V} = E^\circ_{cathode}$. Then $E^\circ_{cell} = E^\circ_{cathode} - E^\circ_{anode} = 1.36 \text{ V} - 2.87 \text{ V} = -1.51 \text{ V}$, so the reaction is nonspontaneous.

Check: The units (V) are correct. If the voltage is positive, the reaction is spontaneous.

19.59 (a) Pb^{2+}. The strongest oxidizing agent is the one with the reduction reaction that is closest to the top of Table 19.1 (most positive, least negative reduction potential).

Cell Potential, Free Energy, and the Equilibrium Constant

19.61 **Given:** overall reactions **Find:** ΔG°_{rxn} and spontaneity in forward direction
Conceptual Plan: Separate the overall reaction into two half-cell reactions and add electrons as needed to balance reactions. Look up half-reactions in Table 19.1. Calculate the standard cell potential by subtracting the electrode potential of the anode from the electrode potential of the cathode: $E^\circ_{cell} = E^\circ_{cathode} - E^\circ_{anode}$. Then calculate ΔG°_{rxn} using $\Delta G^\circ_{rxn} = -nF\,E^\circ_{cell}$.
Solution:

(a) $Pb^{2+}(aq) + Mg(s) \rightarrow Pb(s) + Mg^{2+}(aq)$ separates to $Pb^{2+}(aq) \rightarrow Pb(s)$ and $Mg(s) \rightarrow Mg^{2+}(aq)$.
Add electrons. $Pb^{2+}(aq) + 2\,e^- \rightarrow Pb(s)$ and $Mg(s) \rightarrow Mg^{2+}(aq) + 2\,e^-$. Look up cell potentials.
Mg is oxidized, so $E^\circ_{red} = -2.37\ V = E^\circ_{anode}$. Pb^{2+} is reduced, so $E^\circ_{red} = -0.13\ V = E^\circ_{cathode}$. Then
$E^\circ_{cell} = E^\circ_{cathode} - E^\circ_{anode} = -0.13\ V - (-2.37\ V) = +2.24\ V$. $n = 2$, so $\Delta G^\circ_{rxn} = -nF\,E^\circ_{cell} =$
$-2\ \text{mol}\,e^- \times \dfrac{96{,}485\ C}{\text{mol}\,e^-} \times 2.24\ V = -2 \times 96{,}485\,C \times 2.24\dfrac{J}{C} = -4.32 \times 10^5\ J = -432\ kJ$.

(b) $Br_2(l) + 2\,Cl^-(aq) \rightarrow 2\,Br^-(aq) + Cl_2(g)$ separates to $Br_2(g) \rightarrow 2\,Br^-(aq)$ and $2\,Cl^-(aq) \rightarrow Cl_2(g)$.
Add electrons. $Br_2(g) + 2\,e^- \rightarrow 2\,Br^-(aq)$ and $2\,Cl^-(aq) \rightarrow Cl_2(g) + 2\,e^-$. Look up cell potentials.
Cl is oxidized, so $E^\circ_{red} = 1.36\ V = E^\circ_{anode}$. Br is reduced, so $E^\circ_{red} = 1.09\ V = E^\circ_{cathode}$. Then
$E^\circ_{cell} = E^\circ_{cathode} - E^\circ_{anode} = 1.09\ V - 1.36\ V = -0.27\ V$. $n = 2$, so $\Delta G^\circ_{rxn} = -nF\,E^\circ_{cell} = -2\ \text{mol}\,e^-$
$\times \dfrac{96{,}485\ C}{\text{mol}\,e^-} \times -0.27\ V = -2 \times 96{,}485\ C \times -0.27\dfrac{J}{C} = 5.2 \times 10^4\ J = 52\ kJ$.

(c) $MnO_2(s) + 4\,H^+(aq) + Cu(s) \rightarrow Mn^{2+}(aq) + 2\,H_2O(l) + Cu^{2+}(aq)$ separates to
$MnO_2(s) + 4\,H^+(aq) \rightarrow Mn^{2+}(aq) + 2\,H_2O(l)$ and $Cu(s) \rightarrow Cu^{2+}(aq)$. Add electrons.
$MnO_2(s) + 4\,H^+(aq) + 2\,e^- \rightarrow Mn^{2+}(aq) + 2\,H_2O(l)$ and $Cu(s) \rightarrow Cu^{2+}(aq) + 2\,e^-$. Look up cell
potentials. Cu is oxidized, so $E^\circ_{red} = 0.34\ V = E^\circ_{anode}$. Mn is reduced, so $E^\circ_{red} = 1.21\ V = E^\circ_{cathode}$. Then
$E^\circ_{cell} = E^\circ_{cathode} - E^\circ_{anode} = 1.21\ V - 0.34\ V = +0.87\ V$. $n = 2$, so
$\Delta G^\circ_{rxn} = -nF\,E^\circ_{cell} = -2\ \text{mol}\,e^- \times \dfrac{96{,}485\ C}{\text{mol}\,e^-} \times 0.87\ V = -2 \times 96{,}485\ C \times 0.87\dfrac{J}{C} = -1.7 \times 10^5\ J =$
$-1.7 \times 10^2\ kJ$.

Check: The units (kJ) are correct. If the voltage is positive, the reaction is spontaneous and the free energy change is negative.

19.63 **Given:** overall reactions from Problem 19.61 **Find:** K
Conceptual Plan: $°C \rightarrow K$ then $\Delta G^\circ_{rxn}, T = K$
$$K = 273.15 + °C \qquad \Delta G^\circ_{rxn} = -RT \ln K$$
Solution: $T = 273.15 + 25\ °C = 298\ K$ then

(a) $\Delta G^\circ_{rxn} = -RT \ln K$. Rearrange to solve for K.

$$K = e^{\frac{-\Delta G^\circ_{rxn}}{RT}} = e^{\dfrac{-(-432\ kJ) \times \frac{1000\ J}{1\ kJ}}{\left(8.314\frac{J}{K\,\cdot\,mol}\right)(298\ K)}} = e^{174.364} = 5.31 \times 10^{75}$$

(b) $\Delta G^\circ_{rxn} = -RT \ln K$. Rearrange to solve for K.

$$K = e^{\frac{-\Delta G^\circ_{rxn}}{RT}} = e^{\dfrac{-(-52\ kJ) \times \frac{1000\ J}{1\ kJ}}{\left(8.314\frac{J}{K\,\cdot\,mol}\right)(298\ K)}} = e^{-20.998} = 7.8 \times 10^{-10}$$

(c) $\Delta G^\circ_{rxn} = -RT \ln K$. Rearrange to solve for K.

$$K = e^{\frac{-\Delta G^\circ_{rxn}}{RT}} = e^{\frac{-(-170 \text{ kJ}) \times \frac{1000 \text{ J}}{1 \text{ kJ}}}{\left(8.314 \frac{\text{J}}{\text{K} \cdot \text{mol}}\right)(298 \text{ K})}} = e^{68.616} = 6.3 \times 10^{29}$$

Check: The units (none) are correct. If the voltage is positive, the reaction is spontaneous, the free energy change is negative, and the equilibrium constant is large.

19.65   **Given:** $Ni^{2+}(aq) + Cd(s) \rightarrow$ **Find:** K
Conceptual Plan: Write two half-cell reactions and add electrons as needed to balance the reactions. Look up half-reactions in Table 19.1. Calculate the standard cell potential by subtracting the electrode potential of the anode from the electrode potential of the cathode: $E^\circ_{cell} = E^\circ_{cathode} - E^\circ_{anode}$, then $°C \rightarrow K$ then $E^\circ_{cell}, n, T \rightarrow K$.

$$K = 273.15 + °C \qquad \Delta G^\circ_{rxn} = -RT \ln K = -n \, F \, E^\circ_{cell}$$

Solution: $Ni^{2+}(aq) + 2 e^- \rightarrow Ni(s)$ and $Cd(s) \rightarrow Cd^{2+}(aq) + 2 e^-$. Look up cell potentials.
Cd is oxidized, so $E^\circ_{red} = -0.40 \text{ V} = E^\circ_{anode}$. Ni^{2+} is reduced, so $E^\circ_{red} = -0.23 \text{ V} = E^\circ_{cathode}$.
Then $E^\circ_{cell} = E^\circ_{cathode} - E^\circ_{anode} = -0.23 \text{ V} - (-0.40 \text{ V}) = +0.17 \text{ V}$. The overall reaction
is $Ni^{2+}(aq) + Cd(s) \rightarrow Ni(s) + Cd^{2+}(aq)$. $n = 2$ and $T = 273.15 + 25 °C = 298 \text{ K}$ then
$\Delta G^\circ_{rxn} = -RT \ln K = -n \, F \, E^\circ_{cell}$. Rearrange to solve for K.

$$K = e^{\frac{nFE^\circ_{cell}}{RT}} = e^{\frac{2 \text{ mol } e^- \times \frac{96{,}485 \text{ C}}{\text{mol } e^-} \times 0.17 \frac{\text{J}}{\text{C}}}{\left(8.314 \frac{\text{J}}{\text{K} \cdot \text{mol}}\right)(298 \text{ K})}} = e^{13.241} = 5.6 \times 10^5$$

Check: The units (none) are correct. If the voltage is positive, the reaction is spontaneous and the equilibrium constant is large.

19.67   **Given:** $n = 2$ and $K = 25$ **Find:** ΔG°_{rxn} and E°_{cell}
Conceptual Plan: $K, T \rightarrow \Delta G^\circ_{rxn}$ and $\Delta G^\circ_{rxn}, n \rightarrow E^\circ_{cell}$

$$\Delta G^\circ_{rxn} = -RT \ln K \qquad\qquad \Delta G^\circ_{rxn} = -n \, F \, E^\circ_{cell}$$

Solution: $\Delta G^\circ_{rxn} = -RT \ln K = -\left(8.314 \frac{\text{J}}{\text{K} \cdot \text{mol}}\right)(298 \text{ K}) \ln 25 = -7.97500 \times 10^3 \text{ J} = -8.0 \text{ kJ}$ and

$\Delta G^\circ_{rxn} = -n \, F \, E^\circ_{cell}$. Rearrange to solve for E°_{cell}.

$$E^\circ_{cell} = \frac{\Delta G^\circ_{rxn}}{-n F} = \frac{-7.97500 \times 10^3 \text{ J}}{-2 \text{ mol } e^- \times \frac{96{,}485 \text{ C}}{\text{mol } e^-}} = 0.041 \frac{\text{V} \cdot \text{C}}{\text{C}} = 0.041 \text{ V}$$

Check: The units (kJ and V) are correct. If $K > 1$, the voltage is positive and the free energy change is negative.

Nonstandard Conditions and the Nernst Equation

19.69   **Given:** $Sn^{2+}(aq) + Mn(s) \rightarrow Sn(s) + Mn^{2+}(aq)$
Find: (a) E°_{cell}; (b) E_{cell} when $[Sn^{2+}] = 0.0100 \text{ M}$; $[Mn^{2+}] = 2.00 \text{ M}$; and (c) E_{cell} when
$[Sn^{2+}] = 2.00 \text{ M}$; $[Mn^{2+}] = 0.0100 \text{ M}$
**Conceptual Plan: (a) Separate the overall reaction into two half-cell reactions and add electrons as needed to balance the reactions. Look up half-reactions in Table 19.1. Calculate the standard cell potential by subtracting the electrode potential of the anode from the electrode potential of the cathode: $E^\circ_{cell} = E^\circ_{cathode} - E^\circ_{anode}$.
(b) and (c) $E^\circ_{cell}, [Sn^{2+}], [Mn^{2+}], n \rightarrow E_{cell}$**

$$E_{cell} = E^\circ_{cell} - \frac{0.0592 \text{ V}}{n} \log Q \quad \text{where } Q = \frac{[Mn^{2+}]}{[Sn^{2+}]}$$

Solution:
(a)   Separate the overall reaction to $Sn^{2+}(aq) \rightarrow Sn(s)$ and $Mn(s) \rightarrow Mn^{2+}(aq)$. Add electrons.
$Sn^{2+}(aq) + 2 e^- \rightarrow Sn(s)$ and $Mn(s) \rightarrow Mn^{2+}(aq) + 2 e^-$. Look up cell potentials. Mn is oxidized, so $E^\circ_{red} = -1.18 \text{ V} = E^\circ_{anode}$. Sn^{2+} is reduced, so $E^\circ_{red} = -0.14 \text{ V} = E^\circ_{cathode}$. Then
$E^\circ_{cell} = E^\circ_{cathode} - E^\circ_{anode} = -0.14 \text{ V} - (-1.18 \text{ V}) = +1.04 \text{ V}$.

(b) $Q = \dfrac{[Mn^{2+}]}{[Sn^{2+}]} = \dfrac{2.00\ M}{0.0100\ M} = 200.$ and $n = 2$ then

$$E_{cell} = E^\circ_{cell} - \dfrac{0.0592\ V}{n}\log Q = 1.04\ V - \dfrac{0.0592\ V}{2}\log 200. = +0.97\ V$$

(c) $Q = \dfrac{[Mn^{2+}]}{[Sn^{2+}]} = \dfrac{0.0100\ M}{2.00\ M} = 0.00500$ and $n = 2$ then

$$E_{cell} = E^\circ_{cell} - \dfrac{0.0592\ V}{n}\log Q = 1.04\ V - \dfrac{0.0592\ V}{2}\log 0.00500 = +1.11\ V$$

Check: The units (V, V, and V) are correct. The Sn^{2+} reduction reaction is above the Mn^{2+} reduction reaction, so the standard cell potential will be positive. Having more products than reactants reduces the cell potential. Having more reactants than products raises the cell potential.

19.71   **Given:** $Pb(s) \rightarrow Pb^{2+}(aq, 0.10\ M) + 2\ e^-$ and $MnO_4^-(aq, 1.50\ M) + 4\ H^+(aq, 2.0\ M) + 3\ e^- \rightarrow MnO_2(s) + 2\ H_2O(l)$ **Find:** E_{cell}
Conceptual Plan: Look up half-reactions in Table 19.1. Calculate the standard cell potential by subtracting the electrode potential of the anode from the electrode potential of the cathode: $E^\circ_{cell} = E^\circ_{cathode} - E^\circ_{anode}$. Equalize the number of electrons transferred by multiplying the first reaction by 3 and the second reaction by 2. Add the two half-cell reactions and cancel the electrons.
Then $E^\circ_{cell}, [Pb^{2+}], [MnO_4^-], [H^+], n \rightarrow E_{cell}$.

$$E_{cell} = E^\circ_{cell} - \dfrac{0.0592\ V}{n}\log Q \quad \text{where } Q = \dfrac{[Pb^{2+}]^3}{[MnO_4^-]^2[H^+]^8}$$

Solution: Pb is oxidized, so $E^\circ_{red} = -0.13\ V = E^\circ_{anode}$. Mn is reduced, so $E^\circ_{red} = 1.68\ V = E^\circ_{cathode}$.
Then $E^\circ_{cell} = E^\circ_{cathode} - E^\circ_{anode} = 1.68\ V - (-0.13\ V) = +1.81\ V$. Equalizing the electrons:
$3\ Pb(s) \rightarrow 3\ Pb^{2+}(aq) + 6\ e^-$ and $2\ MnO_4^-(aq) + 8\ H^+(aq) + 6\ e^- \rightarrow 2\ MnO_2(s) + 4\ H_2O(l)$. Adding the two reactions: $3\ Pb(s) + 2\ MnO_4^-(aq) + 8\ H^+(aq) + \cancel{6\ e^-} \rightarrow 3\ Pb^{2+}(aq) + \cancel{6\ e^-} + 2\ MnO_2(s) + 4\ H_2O(l)$. Cancel the electrons: $3\ Pb(s) + 2\ MnO_4^-(aq) + 8\ H^+(aq) \rightarrow 3\ Pb^{2+}(aq) + 2\ MnO_2(s) + 4\ H_2O(l)$. So $n = 6$ and

$$Q = \dfrac{[Pb^{2+}]^3}{[MnO_4^-]^2[H^+]^8} = \dfrac{(0.10)^3}{(1.50)^2(2.0)^8} = 1.\underline{7}361 \times 10^{-6}. \text{ Then}$$

$$E_{cell} = E^\circ_{cell} - \dfrac{0.0592\ V}{n}\log Q = 1.81\ V - \dfrac{0.0592\ V}{6}\log 1.\underline{7}361 \times 10^{-6} = +1.87\ V.$$

Check: The units (V) are correct. The MnO_4^- reduction reaction is above the Pb^{2+} reduction reaction, so the standard cell potential will be positive. Having more reactants than products raises the cell potential.

19.73   **Given:** Zn/Zn^{2+} and Ni/Ni^{2+} half-cells in voltaic cell; initially, $[Ni^{2+}] = 1.50\ M$ and $[Zn^{2+}] = 0.100\ M$
Find: (a) initial E_{cell}, (b) E_{cell} when $[Ni^{2+}] = 0.500\ M$, and (c) $[Ni^{2+}]$ and $[Zn^{2+}]$ when $E_{cell} = 0.45\ V$
Conceptual Plan:
(a)   **Write two half-cell reactions and add electrons as needed to balance reactions. Look up half-reactions in Table 19.1. Calculate the standard cell potential by subtracting the electrode potential of the anode from the electrode potential of the cathode: $E^\circ_{cell} = E^\circ_{cathode} - E^\circ_{anode}$. Choose the direction of the half-cell reactions so that $E^\circ_{cell} > 0$. Add two half-cell reactions and cancel electrons to generate overall reaction. Define Q based on overall reaction. Then $E^\circ_{cell}, [Ni^{2+}], [Zn^{2+}], n \rightarrow E_{cell}$.**

$$E_{cell} = E^\circ_{cell} - \dfrac{0.0592\ V}{n}\log Q$$

(b)   **When $[Ni^{2+}] = 0.500\ M, [Zn^{2+}] = 1.100\ M$. (Because the stoichiometric coefficients for Ni^{2+}: Zn^{2+} are 1:1 and the $[Ni^{2+}]$ drops by 1.00 M, the other concentration must rise by 1.00 M.) Then $E^\circ_{cell}, [Ni^{2+}], [Zn^{2+}], n \rightarrow E_{cell}$.**

$$E_{cell} = E^\circ_{cell} - \dfrac{0.0592\ V}{n}\log Q$$

(c)   **$E^\circ_{cell}, E_{cell}, n \rightarrow [Zn^{2+}]/[Ni^{2+}] \rightarrow [Ni^{2+}], [Zn^{2+}]$**

$$E_{cell} = E^\circ_{cell} - \dfrac{0.0592\ V}{n}\log Q \qquad [Ni^{2+}] + [Zn^{2+}] = 1.50\ M + 0.100\ M = 1.60\ M$$

Solution:

(a) $Zn^{2+}(aq) + 2e^- \rightarrow Zn(s)$ and $Ni^{2+}(aq) + 2e^- \rightarrow Ni(s)$. Look up cell potentials. For Zn, $E^\circ_{red} = -0.76$ V.
For Ni, $E^\circ_{red} = -0.23$ V. To get a positive E°_{cell}, Zn is oxidized; so $E^\circ_{red} = -0.76$ V $= E^\circ_{anode}$. Ni^{2+} is reduced,
so $E^\circ_{red} = -0.23$ V $= E^\circ_{cathode}$. Then $E^\circ_{cell} = E^\circ_{cathode} - E^\circ_{anode} = -0.23$ V $- (-0.76$ V$) = +0.53$ V. Adding
the two half-cell reactions: $Zn(s) + Ni^{2+}(aq) + 2e^- \rightarrow Zn^{2+}(aq) + 2e^- + Ni(s)$. The overall reaction is

$$Zn(s) + Ni^{2+}(aq) \rightarrow Zn^{2+}(aq) + Ni(s). \text{ Then } Q = \frac{[Zn^{2+}]}{[Ni^{2+}]} = \frac{0.100}{1.50} = 0.0666667 \text{ and } n = 2. \text{ Then}$$

$$E_{cell} = E^\circ_{cell} - \frac{0.0592 \text{ V}}{n} \log Q = 0.53 \text{ V} - \frac{0.0592 \text{ V}}{2} \log 0.0666667 = +0.56 \text{ V}.$$

(b) $Q = \dfrac{[Zn^{2+}]}{[Ni^{2+}]} = \dfrac{1.100}{0.500} = 2.20$ then

$$E_{cell} = E^\circ_{cell} - \frac{0.0592 \text{ V}}{n} \log Q = 0.53 \text{ V} - \frac{0.0592 \text{ V}}{2} \log 2.20 = +0.52 \text{ V}$$

(c) $E_{cell} = E^\circ_{cell} - \dfrac{0.0592 \text{ V}}{n} \log Q$ so 0.45 V $= 0.53$ V $- \dfrac{0.0592 \text{ V}}{2} \log Q \rightarrow 0.08 \cancel{V} = \dfrac{0.0592 \cancel{V}}{2} \log Q \rightarrow$

$$\log Q = 2.70270 \rightarrow Q = 10^{2.70270} = 504.31. \text{ Then } Q = 504.31 = \frac{[Zn^{2+}]}{1.60 \text{ M} - [Zn^{2+}]}. \text{ Solving for}$$

$$[Zn^{2+}]: (504.31)(1.60 \text{ M} - [Zn^{2+}]) = [Zn^{2+}] \rightarrow [Zn^{2+}] = \frac{806.896 \text{ M}}{505.31} = 1.59683 \text{ M} = 1.60 \text{ M then}$$

$$[Ni^{2+}] = 1.60 \text{ M} - 1.59683 \text{ M} = 0 \text{ M}$$

Check: The units (V, V, and M) are correct. The standard cell potential is positive, and because there are more reactants than products, this raises the cell potential. As the reaction proceeds, reactants are converted to products; so the cell potential drops for parts (b) and (c).

19.75   **Given:** Zn/Zn^{2+} concentration cell, with $[Zn^{2+}] = 2.0$ M in one half-cell and $[Zn^{2+}] = 1.0 \times 10^{-3}$ M in other half-cell
Find: Sketch a voltaic cell, labeling the anode, the cathode, the reactions at electrodes, all species, and the direction of electron flow.
Conceptual Plan: In a concentration cell, the half-cell with the higher concentration is always the half-cell where the reduction takes place (contains the cathode). The two half-cell reactions are the same but reversed. Put anode reaction on the left (oxidation = electrons as product) and cathode reaction on the right (reduction = electrons as reactant). Electrons flow from anode to cathode.
Solution:

Check: The figure looks similar to the right side of Figure 19.12.

19.77   **Given:** Sn/Sn^{2+} concentration cell with $E_{cell} = 0.10$ V
Find: ratio of $[Sn^{2+}]$ in two half-cells
Conceptual Plan: Determine n, then E°_{cell}, E_{cell}, $n \rightarrow Q = $ ratio of $[Sn^{2+}]$ in two half-cells.

$$E_{cell} = E^\circ_{cell} - \frac{0.0592 \text{ V}}{n} \log Q$$

Solution: Because $Sn^{2+}(aq) + 2e^- \rightarrow Sn(s)$, $n = 2$. In a concentration cell, $E^\circ_{cell} = 0$ V. So

$$E_{cell} = E^\circ_{cell} - \frac{0.0592 \text{ V}}{n} \log Q, \text{ so } 0.10 \text{ V} = 0.00 \text{V} - \frac{0.0592 \text{ V}}{2} \log Q \rightarrow 0.10 \cancel{V} = -\frac{0.0592 \cancel{V}}{2} \log Q \rightarrow$$

$$\log Q = -3.3784 \rightarrow Q = 10^{-3.3784} = 4.2 \times 10^{-4} = \frac{[Sn^{2+}](ox)}{[Sn^{2+}](red)}.$$

Check: The units (none) are correct. Because the concentration in the reduction reaction half-cell is always greater than the concentration in the oxidation half-cell in a voltaic concentration cell, the Q or ratio of two cells is less than 1.

Batteries, Fuel Cells, and Corrosion

19.79 **Given:** alkaline battery   **Find:** optimum mass ratio of Zn to MnO_2
Conceptual Plan: Look up alkaline battery reactions. Use stoichiometry to get mole ratio. Then

$$Zn(s) + 2OH^-(aq) \rightarrow Zn(OH)_2(s) + 2e^-$$

$$\frac{1\ mol\ Zn}{2\ mol\ MnO_2}$$

$$2\,MnO_2(s) + 2\,H_2O(l) + 2\,e^- \rightarrow 2\,MnO(OH)(s) + 2\,OH^-(aq)$$

mol Zn → g Zn Zn then mol MnO_2 → g MnO_2.

$$\frac{65.38\ g\ Zn}{1\ mol\ Zn} \qquad \frac{1\ mol\ MnO_2}{86.94\ g\ MnO_2}$$

Solution: $\dfrac{1\ mol\ Zn}{2\ mol\ MnO_2} \times \dfrac{65.38\ g\ Zn}{1\ mol\ Zn} \times \dfrac{1\ mol\ MnO_2}{86.94\ g\ MnO_2} = 0.3760\ \dfrac{g\ Zn}{g\ MnO_2}$

Check: The units (mass ratio) are correct. Because more moles of MnO_2 are needed and the molar mass is larger, the ratio is less than 1.

19.81 **Given:** $CH_4(g) + 2\,O_2(g) \rightarrow CO_2(g) + 2\,H_2O(g)$ **Find:** E°_{cell}
Conceptual Plan: $\Delta G^\circ_{rxn} = \sum n_p \Delta G^\circ_f(products) - \sum n_r \Delta G^\circ_f(reactants)$ **and determine n then**
$\Delta G^\circ_{rxn}, n \rightarrow E^\circ_{cell}$

$$\Delta G^\circ_{rxn} = -n\,F\,E^\circ_{cell}$$

Solution:

Reactant/Product	ΔG°_f(kJ/mol from Appendix IIB)
$CH_4(g)$	-50.5
$O_2(g)$	0.0
$CO_2(g)$	-394.4
$H_2O(g)$	-228.6

Be sure to pull data for the correct formula and phase.

$$\Delta G^\circ_{rxn} = \sum n_p \Delta G^\circ_f(products) - \sum n_r \Delta G^\circ_f(reactants)$$
$$= [1(\Delta G^\circ_f(CO_2(g))) + 2(\Delta G^\circ_f(H_2O(g)))] - [1(\Delta G^\circ_f(CH_4(g))) + 2(\Delta G^\circ_f(O_2(g)))]$$
$$= [1(-394.4\ kJ) + 2(-228.6\ kJ)] - [1(-50.5\ kJ) + 2(0.0\ kJ)]$$
$$= [-851.6\ kJ] - [-50.5\ kJ]$$
$$= -801.1\ kJ = -8.011 \times 10^5\ J$$

Also, because one C atom goes from an oxidation state of -4 to $+4$ and four O atoms are going from 0 to -2, $n = 8$ and $\Delta G^\circ_{rxn} = -nF\,E^\circ_{cell}$. Rearrange to solve for E°_{cell}.

$$E^\circ_{cell} = \frac{\Delta G^\circ_{rxn}}{-n\,F} = \frac{-8.011 \times 10^5\ J}{-8\ mole^- \times \dfrac{96{,}485\ C}{mole^-}} = 1.038\frac{V \cdot C}{C} = 1.038\ V$$

Check: The units (V) are correct. The cell voltage is positive, which is consistent with a spontaneous reaction.

19.83 When iron corrodes or rusts, it oxidizes to Fe^{2+}. For a metal to be able to protect iron, it must be more easily oxidized than iron or be below it in Table 19.1. (a) Zn and (c) Mn meet that criterion.

Electrolytic Cells and Electrolysis

19.85 **Given:** electrolytic cell sketch
Find: (a) Label the anode and cathode and indicate half-reactions, (b) indicate direction of electron flow, and (c) label battery terminals and calculate minimum voltage to drive reaction.

Conceptual Plan: (a) Write two half-cell reactions and add electrons as needed to balance reactions. Look up half-reactions in Table 19.1. Calculate the standard cell potential by subtracting the electrode potential of the anode from the electrode potential of the cathode: $E°_{cell} = E°_{cathode} - E°_{anode}$. Choose the direction of the half-cell reactions so that $E°_{cell} < 0$. **(b)** Electrons flow from anode to cathode. **(c)** Each half-cell reaction moves forward, so direction of the concentration changes can be determined.

Solution:

(a) $Ni^{2+}(aq) + 2e^- \rightarrow Ni(s)$ and $Cd^{2+}(aq) + 2e^- \rightarrow Cd(s)$. Look up cell potentials. For Ni, $E°_{red} = -0.23$ V. For Cd, $E°_{red} = -0.40$ V. To get a negative cell potential, Ni is oxidized; so $E°_{red} = -0.23$ V $= E°_{anode}$. Cd^{2+} is reduced, so $E°_{red} = -0.23$ V $= E°_{cathode}$. Then $E°_{cell} = E°_{cathode} - E°_{anode} = -0.40$ V $- (-0.23$ V$) = -0.17$ V. Because oxidation occurs at the anode, Ni is the anode and the reaction is $Ni(s) \rightarrow Ni^{2+}(aq) + 2e^-$. Because reduction takes place at the cathode, Cd is the cathode and the reaction is $Cd^{2+}(aq) + 2e^- \rightarrow Cd(s)$.

(c) Because reduction is occurring at the cathode, the battery terminal closest to the cathode is the negative terminal. Because the cell potential from part (a) is $= -0.17$ V, a minimum of 0.17 V must be applied by the battery.

Check: The reaction is nonspontaneous because the reduction of Ni^{2+} is above Cd^{2+}. Electrons still flow from the anode to the cathode. The reaction can be made spontaneous with the application of electrical energy.

19.87 **Given:** electrolysis of molten KBr   **Find:** write half-reactions
Conceptual Plan: Write two half-cell reactions, taking the cation and the anion to their elemental forms at high temperatures and adding electrons as needed to balance reactions.
Solution: KBr breaks apart to K^+ and Br^-. $K^+(l) + e^- \rightarrow K(l)$ and $2 Br^-(l) \rightarrow Br_2(g) + 2 e^-$.

Check: The cation is reduced, and the anion is oxidized. Mass and charge are balanced in the reactions.

19.89 **Given:** electrolysis of mixture of molten KBr and molten LiBr   **Find:** write half-reactions
Conceptual Plan: Write two half-cell reactions, taking the cation and the anion to their elemental forms at high temperatures and adding electrons as needed to balance reactions. Look up cation cell potentials to see which one generates the more spontaneous reaction.
Solution: KBr breaks apart to K^+ and Br^-. $K^+(l) + e^- \rightarrow K(l)$ and $2 Br^-(l) \rightarrow Br_2(g) + 2 e^-$. LiBr breaks apart to Li^+ and Br^-. $Li^+(l) + e^- \rightarrow Li(l)$ and $2 Br^-(l) \rightarrow Br_2(g) + 2 e^-$. The anions are the same, so the anode reaction is $2 Br^-(l) \rightarrow Br_2(g) + 2 e^-$. Looking up the cation reduction potentials, for K^+ $E°_{red} = -2.92$ V and for Li^+ $E°_{red} = -3.04$ V. Because the reduction potential is more positive for K^+, the reaction at the cathode is $K^+(l) + e^- \rightarrow K(l)$.

Check: The cation is reduced, and the anion is oxidized. Mass and charge are balanced in the reactions. The cation that is higher in Table 19.1 will be the reaction at the cathode.

19.91 **Given:** electrolysis of aqueous solutions   **Find:** write half-reactions
Conceptual Plan: Write two half-cell reactions, taking the cation and the anion to their elemental forms at standard conditions and adding electrons as needed to balance reactions. Look up cell potentials and compare to the cell potentials for the electrolysis of water to see which one generates the more spontaneous reaction.
Solution: The hydrolysis of water reaction is as follows: at a neutral pH, $2 H_2O(l) \rightarrow O_2(g) + 4 H^+(aq) + 4 e^-$ where $E = -0.82$ V, and when $[OH^-] = 10^{-7}$ M. $2 H_2O(l) + 2 e^- \rightarrow H_2(g) + 2 OH^-(aq)$ where $E = -0.41$ V.

(a) NaBr breaks apart to Na^+ and Br^-. $Na^+(aq) + e^- \rightarrow Na(s)$ and $2 Br^-(aq) \rightarrow Br_2(l) + 2 e^-$. Looking up the half-cell potentials, for Na^+ $E°_{red} = -2.71$ V and for Br^- $E°_{anode} = E°_{red} = +1.09$ V. Because -0.82 V is more positive than $E°_{anode} = -1.09$ V, the oxidation reaction will be $2 H_2O(l) \rightarrow O_2(g) + 4 H^+(aq) + 4 e^-$ at the anode. Because -0.41 V is more positive than -2.71 V, the reduction reaction will be $2 H_2O(l) + 2 e^- \rightarrow H_2(g) + 2 OH^-(aq)$ at the cathode.

(b) PbI_2 breaks apart to Pb^{2+} and I^-. $Pb^{2+}(aq) + 2 e^- \rightarrow Pb(s)$ and $2 I^-(aq) \rightarrow I_2(s) + 2 e^-$. Looking up the half-cell potentials, for Pb^{2+} $E°_{red} = -0.13$ V and for I^- $E°_{anode} = E°_{red} = -0.54$ V. Because $E°_{anode} = -0.54$ V is more positive than -0.82 V, the oxidation reaction will be $2 I^-(aq) \rightarrow I_2(s) + 2 e^-$ at the anode. Because -0.13 V is more positive than -0.41 V, the reduction reaction will be $Pb^{2+}(aq) + 2 e^- \rightarrow Pb(s)$ at the cathode.

(c) Na_2SO_4 breaks apart to Na^+ and SO_4^{2-}. $Na^+(aq) + e^- \rightarrow Na(s)$ and $SO_4^{2-}(aq) + 4\,H^+(aq) + 2\,e^- \rightarrow$ $H_2SO_3(aq) + H_2O(l)$. Notice that both of these are reductions. Because S is in such a high oxidation state $(+6)$, it cannot be oxidized. Looking up the half-cell potentials, for Na^+ $E°_{red} = -2.71$ V and for SO_4^{2-} $E°_{red} = 0.20$ V. Because sodium sulfate solutions are neutral and not acidic, even though 0.20 V is more positive than -0.41 V and -2.71 V, the reduction reaction will not be: $SO_4^{2-}(aq) + 4\,H^+(aq) + 2\,e^- \rightarrow H_2SO_3(aq) + H_2O(l)$ at the cathode, which only occurs in acidic solutions. Thus, the reduction reaction will be $Na^+(aq) + e^- \rightarrow Na(s)$. Because only one oxidation reaction is possible, the oxidation reaction will be $2\,H_2O(l) \rightarrow O_2(g) + 4\,H^+(aq) + 4\,e^-$ at the anode.

Check: The most positive reactions are the reactions that will occur.

19.93   **Given:** electrolysis cell to electroplate Cu onto a metal surface
Find: Draw a cell, label the anode and cathode, and write half-reactions.
Conceptual Plan: Write two half-cell reactions and add electrons as needed to balance reactions. The cathode reaction will be the reduction of Cu^{2+} to the metal. The anode will be the reverse reaction.
Solution:

Check: The metal to be plated is the cathode because metal ions are converted to $Cu(s)$ on the surface of the metal.

19.95   **Given:** Cu electroplating of 325 mg Cu at a current of 5.6 A; $Cu^{2+}(aq) + 2\,e^- \rightarrow Cu(s)$ **Find:** time
Conceptual Plan: mg Cu → g Cu → mol Cu → mol e⁻ → C → s

$$\frac{1\,g}{1000\,mg} \quad \frac{1\,mol\,Cu}{63.55\,g\,Cu} \quad \frac{2\,mol\,e^-}{1\,mol\,Cu} \quad \frac{96,485\,C}{1\,mol\,e^-} \quad \frac{1\,s}{5.6\,C}$$

Solution: $325\ \text{mg Cu} \times \dfrac{1\ \text{g Cu}}{1000\ \text{mg Cu}} \times \dfrac{1\ \text{mol Cu}}{63.55\ \text{g Cu}} \times \dfrac{2\ \text{mol e}^-}{1\ \text{mol Cu}} \times \dfrac{96,485\ C}{1\ \text{mol e}^-} \times \dfrac{1\ s}{5.6\ C} = 180\ s.$

Check: The units (s) are correct. Because far less than a mole of Cu is electroplated, the time is short.

19.97   **Given:** Na electrolysis, 1.0 kg in one hour   **Find:** current

Conceptual Plan: $Na^+(l) + e^- \rightarrow Na(l)$ $\dfrac{kg\ Na}{h} \rightarrow \dfrac{g\ Na}{h} \rightarrow \dfrac{mol\ Na}{h} \rightarrow \dfrac{mol\ e^-}{h} \rightarrow \dfrac{C}{h} \rightarrow \dfrac{C}{min} \rightarrow \dfrac{C}{s}$

$$\frac{1000\,g}{1\,kg} \quad \frac{1\,mol\,Na}{22.99\,g\,Na} \quad \frac{1\,mol\,e^-}{1\,mol\,Na} \quad \frac{96,485\,C}{1\,mol\,e^-} \quad \frac{1\,h}{60\,min} \quad \frac{1\,min}{60\,s}$$

Solution:
$\dfrac{1.0\ \text{kg Na}}{1\ h} \times \dfrac{1000\ \text{g Na}}{1\ \text{kg Na}} \times \dfrac{1\ \text{mol Na}}{22.99\ \text{g Na}} \times \dfrac{1\ \text{mol e}^-}{1\ \text{mol Na}} \times \dfrac{96,485\ C}{1\ \text{mol e}^-} \times \dfrac{1\ h}{60\ min} \times \dfrac{1\ min}{60\ s} = 1.2 \times 10^3\,\dfrac{C}{s} = 1.2 \times 10^3\ A$

Check: The units (A) are correct. Because the amount per hour is so large, we expect a very large current.

Cumulative Problems

19.99   **Given:** $MnO_4^-(aq) + Zn(s) \rightarrow Mn^{2+}(aq) + Zn^{2+}(aq)$, 0.500 M $KMnO_4$, and 2.85 g Zn
Find: balance equation and volume $KMnO_4$ solution

Conceptual Plan: Separate the overall reaction into two half-reactions: one for oxidation and one for reduction. → **Balance each half-reaction with respect to mass in the following order: (1) Balance all elements other than H and O, (2) balance O by adding H_2O, and (3) balance H by adding H^+.** → **Balance each half-reaction with respect to charge by adding electrons. (The sum of the charges on both sides of the equation should be made equal by adding electrons as necessary.)** → **Make the number of electrons in both half-reactions equal by multiplying one or both half-reactions by a small whole number.** → **Add the two half-reactions, canceling electrons and other species as necessary.** → **Verify that the reaction is balanced with respect to both mass and charge.**

Then g Zn → mol Zn → mol MnO_4^- → mol $KMnO_4$ → L $KMnO_4$ → mL $KMnO_4$.

$$\frac{1 \text{ mol Zn}}{65.38 \text{ g Zn}} \quad \frac{2 \text{ mol } MnO_4^-}{5 \text{ mol Zn}} \quad \frac{1 \text{ mol } KMnO_4}{1 \text{ mol } MnO_4^-} \quad \frac{1 \text{ L } KMnO_4}{0.500 \text{ mol } KMnO_4} \quad \frac{1000 \text{ mL } KMnO_4}{1 \text{ L } KMnO_4}$$

Solution:

Separate: $MnO_4^-(aq) \rightarrow Mn^{2+}(aq)$ and $Zn(s) \rightarrow Zn^{2+}(aq)$

Balance non-H & O elements: $MnO_4^-(aq) \rightarrow Mn^{2+}(aq)$ and $Zn(s) \rightarrow Zn^{2+}(aq)$

Balance O with H_2O: $MnO_4^-(aq) \rightarrow Mn^{2+}(aq) + 4\,H_2O(l)$ and $Zn(s) \rightarrow Zn^{2+}(aq)$

Balance H with H^+: $MnO_4^-(aq) + 8\,H^+(aq) \rightarrow Mn^{2+}(aq) + 4\,H_2O(l)$ and $Zn(s) \rightarrow Zn^{2+}(aq)$

Add electrons: $MnO_4^-(aq) + 8\,H^+(aq) + 5\,e^- \rightarrow Mn^{2+}(aq) + 4\,H_2O(l)$ and $Zn(s) \rightarrow Zn^{2+}(aq) + 2\,e^-$

Equalize electrons:

$2\,MnO_4^-(aq) + 16\,H^+(aq) + 10\,e^- \rightarrow 2\,Mn^{2+}(aq) + 8\,H_2O(l)$ and $5\,Zn(s) \rightarrow 5\,Zn^{2+}(aq) + 10\,e^-$

Add half-reactions:

$2\,MnO_4^-(aq) + 16\,H^+(aq) + \cancel{10\,e^-} + 5\,Zn(s) \rightarrow 2\,Mn^{2+}(aq) + 8\,H_2O(l) + 5\,Zn^{2+}(aq) + \cancel{10\,e^-}$

Cancel electrons: $2\,MnO_4^-(aq) + 16\,H^+(aq) + 5\,Zn(s) \rightarrow 2\,Mn^{2+}(aq) + 8\,H_2O(l) + 5\,Zn^{2+}(aq)$

$$2.85 \text{ g Zn} \times \frac{1 \text{ mol Zn}}{65.38 \text{ g Zn}} \times \frac{2 \text{ mol } MnO_4^-}{5 \text{ mol Zn}} \times \frac{1 \text{ mol } KMnO_4}{1 \text{ mol } MnO_4^-} \times \frac{1 \text{ L } KMnO_4}{0.500 \text{ mol } KMnO_4} \times \frac{1000 \text{ mL } KMnO_4}{1 \text{ L } KMnO_4}$$

$= 34.9 \text{ mL } KMnO_4$

Check:

Reactants	Products
2 Mn atoms	2 Mn atoms
8 O atoms	8 O atoms
16 H atoms	16 H atoms
5 Zn atoms	5 Zn atoms
+14 charge	+14 charge

The units (mL) are correct. Because far less than a mole of zinc is used, less than a mole of permanganate is consumed; so the volume is less than a liter.

19.101   **Given:** beaker with Al strip and Cu^{2+} ions   **Find:** Draw sketch after Al is submerged for a few minutes.

Conceptual Plan: Write two half-cell reactions and add electrons as needed to balance reactions. Look up half-reactions in Table 19.1. Calculate the standard cell potential by subtracting the electrode potential of the anode from the electrode potential of the cathode: $E°_{cell} = E°_{cathode} - E°_{anode}$. If $E°_{cell} > 0$, the reaction is spontaneous in the forward direction and Al will dissolve and Cu will deposit.

Solution: $Al(s) \rightarrow Al^{3+}(aq)$ and $Cu^{2+}(aq) \rightarrow Cu(s)$. Add electrons.

$Al(s) \rightarrow Al^{3+}(aq) + 3\,e^-$ and $Cu^{2+}(aq) + 2\,e^- \rightarrow Cu(s)$. Look up cell potentials.

Al is oxidized, so $E°_{anode} = E°_{red} = -1.66$ V. Cu^{2+} is reduced, so $E°_{cathode} = E°_{red} = 0.34$ V.

Then $E°_{cell} = E°_{cathode} - E°_{anode} = 0.34$ V $- (-1.66$ V$) + 2.00$ V, so the reaction is spontaneous. Al will dissolve to generate $Al^{3+}(aq)$, and $Cu(s)$ will deposit.

Check: The units (V) are correct. If the voltage is positive, the reaction is spontaneous; so Al will dissolve and Cu will deposit.

19.103   **Given:** (a) 2.15 g Al, (b) 4.85 g Cu, and (c) 2.42 g Ag in 3.5 M HI

Find: If metal dissolves, write a balanced reaction and the minimum amount of HI needed to dissolve the metal.

Conceptual Plan: In general, metals whose reduction half-reactions lie below the reduction of H^+ to H_2 in Table 19.1 will dissolve in acids, while metals above it will not. Stop here if metal does not dissolve. To write the balanced redox reactions, pair the oxidation of the metal with the reduction of H^+ to

H_2 ($2\,H^+(aq) + 2\,e^- \rightarrow H_2(g)$). **Balance the number of electrons transferred. Add the two reactions. Cancel electrons. Then g metal $\rightarrow$ mol metal $\rightarrow$ mol $H^+ \rightarrow$ L HI $\rightarrow$ mL HI.**

$$\mathcal{M} \qquad \frac{x\;\text{mol}\;H^+}{y\;\text{mol metal}} \qquad \frac{1\;\text{L HI}}{3.5\;\text{mol HI}} \qquad \frac{1000\;\text{mL HI}}{1\;\text{L HI}}$$

Solution:

(a) Al meets this criterion. For Al, $Al(s) \rightarrow Al^{3+}(aq) + 3\,e^-$. We need to multiply the Al reaction by 2 and the H^+ reaction by 3. So $2\,Al(s) \rightarrow 2\,Al^{3+}(aq) + 6\,e^-$ and $6\,H^+(aq) + 6\,e^- \rightarrow 3\,H_2(g)$. Adding the half-reactions: $2\,Al(s) + 6\,H^+(aq) + \cancel{6\,e^-} \rightarrow 2\,Al^{3+}(aq) + \cancel{6\,e^-} + 3\,H_2(g)$. Simplify to $2\,Al(s) + 6\,H^+(aq) \rightarrow 2\,Al^{3+}(aq) + 3\,H_2(g)$. Then

$$2.15\;\text{g}\,\cancel{Al} \times \frac{1\;\text{mol}\,\cancel{Al}}{26.98\;\text{g}\,\cancel{Al}} \times \frac{6\;\text{mol}\,\cancel{H^+}}{2\;\text{mol}\,\cancel{Al}} \times \frac{1\;\cancel{\text{L HI}}}{3.5\;\cancel{\text{mol HI}}} \times \frac{1000\;\text{mL HI}}{1\;\cancel{\text{L HI}}} = 68.3\;\text{mL HI}.$$

(b) Cu does not meet this criterion, so it will not dissolve in HI.

(c) Ag does not meet this criterion, so it will not dissolve in HI.

Check: Only metals with negative reduction potentials will dissolve. The volume of acid needed is fairly small because the amount of metal is much less than 1 mole and the concentration of acid is high.

19.105   **Given:** $Pt(s)\,|\,H_2(g, 1\,\text{atm})\,|\,H^+(aq, ?\,M)\,||\,Cu^{2+}(aq, 1.0\,M)\,|\,Cu(s)$, $E_{cell} = 355\,\text{mV}$ **Find:** pH
Conceptual Plan: Write half-reactions from line notation. Look up half-reactions in Table 19.1. Calculate the standard cell potential by subtracting the electrode potential of the anode from the electrode potential of the cathode: $E^{\circ}_{cell} = E^{\circ}_{cathode} - E^{\circ}_{anode}$. Add the two half-cell reactions and cancel the electrons. Then mV $\rightarrow$ V then $E^{\circ}_{cell}, E_{cell}, P_{H_2}, [Cu^{2+}], n \rightarrow [H^+] \rightarrow$ pH.

$$\frac{1\,\text{V}}{1000\,\text{mV}} \qquad E_{cell} = E^{\circ}_{cell} - \frac{0.0592\,\text{V}}{n}\log Q \qquad \text{pH} = -\log[H^+]$$

Solution: The half-reactions are $H_2(g) \rightarrow 2\,H^+(aq) + 2\,e^-$ and $Cu^{2+}(aq) + 2\,e^- \rightarrow Cu(s)$. H is oxidized, so $E^{\circ}_{red} = 0.00\,\text{V} = E^{\circ}_{anode}$. Cu is reduced, so $E^{\circ}_{red} = 0.34\,\text{V} = E^{\circ}_{cathode}$. Then $E^{\circ}_{cell} = E^{\circ}_{cathode} - E^{\circ}_{anode}$

$= 0.34\,\text{V} - 0.00\,\text{V} = +0.34\,\text{V}$. Adding the two reactions: $H_2(g) + Cu^{2+}(aq) + \cancel{2\,e^-} \rightarrow 2\,H^+(aq) + \cancel{2\,e^-} + Cu(s)$.

Cancel the electrons: $H_2(g) + Cu^{2+}(aq) \rightarrow 2\,H^+(aq) + Cu(s)$. Then $355\;\cancel{\text{mV}} \times \dfrac{1\,\text{V}}{1000\;\cancel{\text{mV}}} = 0.355\,\text{V}$. So $n = 2$

and $Q = \dfrac{[H^+]^2}{P_{H_2}[Cu^{2+}]} = \dfrac{(x)^2}{(1)(1.0)} = x^2$. Then $E_{cell} = E^{\circ}_{cell} - \dfrac{0.0592\,\text{V}}{n}\log Q$. Substitute values and solve for

x. $0.355\,\text{V} = 0.34\,\text{V} - \dfrac{0.0592\,\text{V}}{2}\log x^2 \rightarrow 0.015\,\text{V} = -\dfrac{0.0592\,\text{V}}{2}\log x^2 \rightarrow -0.50676 = \log x^2 \rightarrow$

$x^2 = 10^{-0.50676} = 0.31134 \rightarrow x = 0.55798$ then pH $= -\log[H^+] = -\log[0.55798] = 0.25338 = 0.3$

Check: The units (none) are correct. The pH is acidic, which is consistent with dissolving a metal in acid.

19.107   **Given:** $[A^{2+}]$ and $[B^{2+}]$ in table and $\Delta G^{\circ}_{rxn} = -14.0\,\text{kJ}$ **Find:** $E^{\circ}_{cell}, K, Q, E_{cell}$, and ΔG_{rxn}
Conceptual Plan: Determine n; ΔG°_{rxn}, then $n \rightarrow E^{\circ}_{cell}$ and $\Delta G^{\circ}_{rxn}, T \rightarrow K$ and

$$A(s) + B^{2+}(aq) \rightarrow A^{2+}(aq) + B(s)\;\Delta G^{\circ}_{rxn} = -nFE^{\circ}_{cell} \qquad \Delta G^{\circ}_{rxn} = -RT\ln K$$

For each set of conditions: $[A^{2+}], [B^{2+}] \rightarrow Q$ then $E^{\circ}_{cell}, [A^{2+}], [B^{2+}], n \rightarrow E_{cell}$ and $\Delta G^{\circ}_{rxn}, Q, T \rightarrow \Delta G_{rxn}$

$$Q = \frac{[A^{2+}]}{[B^{2+}]} \qquad E_{cell} = E^{\circ}_{cell} - \frac{0.0592\,\text{V}}{n}\log Q \qquad \Delta G_{rxn} = \Delta G^{\circ}_{rxn} + RT\ln Q$$

Solution: Since the charge of both A and B change by 2 in the reaction, $n = 2$. $\Delta G^{\circ}_{rxn} = -nFE^{\circ}_{cell}$ Rearrange to solve

for E°_{cell}. $E^{\circ}_{cell} = \dfrac{\Delta G^{\circ}_{rxn}}{-nF} = \dfrac{-1.40 \times 10^4\,\text{J}}{-2\;\cancel{\text{mol}\,e^-} \times \dfrac{96{,}485\,\text{C}}{\cancel{\text{mol}\,e^-}}} = 0.0725501\;\dfrac{\text{V}\cdot\cancel{C}}{\cancel{C}} = 0.0726\,\text{V}.$

$\Delta G^{\circ}_{rxn} = -RT\ln K$. Rearrange to solve for K. $K = e^{\frac{-\Delta G^{\circ}_{rxn}}{RT}} = e^{\dfrac{-(-1.40 \times 10^4\,\cancel{J})}{\left(8.314\,\frac{\cancel{J}}{\cancel{K}\cdot\text{mol}}\right)(298\;\cancel{K})}} = e^{-5.650694} = 284.4888 = 284.$

$1\colon\; Q = \dfrac{[A^{2+}]}{[B^{2+}]} = \dfrac{1.00\,M}{1.00\,M} = 1.00$

$$E_{cell} = E°_{cell} - \frac{0.0592 \text{ V}}{n} \log Q = 0.0725501 \text{ V} - \frac{0.0592 \text{ V}}{2} \log (1) = +0.0726 \text{ V and}$$

$$\Delta G_{rxn} = \Delta G°_{rxn} + RT \ln Q = -14.0 \text{ kJ} + \left(8.314 \frac{J}{K \cdot mol}\right)\left(\frac{1 \text{ kJ}}{1000 \text{ J}}\right)(298 \text{ K}) \ln (1) = -14.0 \text{ kJ}$$

These are standard state conditions, so $E°_{cell} = E_{cell}$ and $\Delta G°_{rxn} = \Delta G_{rxn}$.

2: $Q = \dfrac{1.00 \times 10^{-4} \text{ M}}{1.00 \text{ M}} = 1.00 \times 10^{-4}$

$$E_{cell} = 0.0725501 \text{ V} - \frac{0.0592 \text{ V}}{2} \log (1.00 \times 10^{-4}) = +0.1909501 \text{ V} = +0.1910 \text{ V and}$$

$$\Delta G_{rxn} = -14.0 \text{ kJ} + \left(8.314 \frac{J}{K \cdot mol}\right)\left(\frac{1 \text{ kJ}}{1000 \text{ J}}\right)(298 \text{ K}) \ln (1.00 \times 10^{-4}) = -36.8 \text{ kJ}$$

3: $Q = \dfrac{1.00 \text{ M}}{1.00 \times 10^{-4} \text{ M}} = 1.00 \times 10^4$

$$E_{cell} = 0.0725501 \text{ V} - \frac{0.0592 \text{ V}}{2} \log (1.00 \times 10^4) = -0.0458499 \text{ V} = -0.0458 \text{ V and}$$

$$\Delta G_{rxn} = -14.0 \text{ kJ} + \left(8.314 \frac{J}{K \cdot mol}\right)\left(\frac{1 \text{ kJ}}{1000 \text{ J}}\right)(298 \text{ K}) \ln (1.00 \times 10^4) = +8.8 \text{ kJ}$$

4: $Q = \dfrac{1.00 \text{ M}}{3.52 \times 10^{-3} \text{ M}} = 284.0909 = 284$

$$E_{cell} = 0.0725501 \text{ V} - \frac{0.0592 \text{ V}}{2} \log (284.0909) = -0.00007224 \text{ V} = -0.0001 \text{ V and}$$

$$\Delta G_{rxn} = -14.0 \text{ kJ} + \left(8.314 \frac{J}{K \cdot mol}\right)\left(\frac{1 \text{ kJ}}{1000 \text{ J}}\right)(298 \text{ K}) \ln (284.0909) = -0.00347 \text{ kJ} = +0.0 \text{ kJ}$$

These are equilibrium conditions, so $E_{cell} = 0$ and $\Delta G_{rxn} = 0$.

Experiment	$[B^{2+}]$	$[A^{2+}]$	Q	E_{cell}	ΔG_{rxn}
1	1.00 M	1.00 M	1.00	+ 0.0726 V	− 14.0 kJ
2	1.00 M	1.00 × 10–4 M	1.00 × 10⁻⁴	+ 0.1910 V	− 36.8 kJ
3	1.00 × 10⁻⁴ M	1.00 M	1.00 × 10⁴	− 0.0458 V	+ 8.8 kJ
4	3.52 × 10⁻³ M	1.00 M	284	− 0.0001 V	+ 0.0 kJ

Check: The units (V, none, 4 sets of none, V, and kJ) are correct. $\Delta G°_{rxn}$ was negative, so $K > 1$ and $E°_{cell} > 0$. At standard state conditions, $E°_{cell} = E_{cell}$ and $\Delta G°_{rxn} = \Delta G_{rxn}$. These are equilibrium conditions, so $E_{cell} = 0$ and $\Delta G_{rxn} = 0$.

19.109 You should be wary of the battery because the most a pairing of half-cell reactions can generate is 5 to 6 V, not 24 V.

19.111 **Given:** Mg oxidation and Cu^{2+} reduction; initially, $[Mg^{2+}] = 1.0 \times 10^{-4}$ M and $[Cu^{2+}] = 1.5$ M in 1.0 L half-cells
Find: (a) initial E_{cell}, (b) E_{cell} after 5.0 A for 8.0 h, and (c) how long battery can deliver 5.0 A
Conceptual Plan:
(a) Write the two half-cell reactions and add electrons as needed to balance reactions. Look up half-reactions in Table 19.1. Calculate the standard cell potential by subtracting the electrode potential of the anode from the electrode potential of the cathode: $E°_{cell} = E°_{cathode} - E°_{anode}$. Add the two half-cell reactions, cancel electrons, and determine n. Then $E°_{cell}, [Mg^{2+}], [Cu^{2+}], n \rightarrow E_{cell}$.

$$E_{cell} = E°_{cell} - \frac{0.0592 \text{ V}}{n} \log Q$$

(b) $hr \rightarrow min \rightarrow s \rightarrow C \rightarrow mol \text{ } e^- \rightarrow mol \text{ Cu reduced} \rightarrow [Cu^{2+}]$ and

$\dfrac{60 \text{ min}}{1 \text{ hr}} \quad \dfrac{60 \text{ s}}{1 \text{ min}} \quad \dfrac{5.0 \text{ C}}{1 \text{ s}} \quad \dfrac{1 \text{ mol } e^-}{96,485 \text{ C}} \quad \dfrac{1 \text{ mol } Cu^{2+}}{2 \text{ mol } e^-}$ because $V = 1.0 \text{ L}$ $[Cu^{2+}]_f = [Cu^{2+}]_i - \dfrac{\text{mol } Cu^{2+} \text{ reduced}}{1.0 \text{ L}}$

$mol \text{ Cu reduced} \rightarrow mol \text{ Mg oxidized} \rightarrow [Mg^{2+}]$

$\dfrac{1 \text{ mol Mg oxidized}}{1 \text{ mol } Cu^{2+} \text{reduced}}$ because $V = 1.0 \text{ L}$ $[Mg^{2+}]_f = [Mg^{2+}]_i + \dfrac{\text{mol Mg oxidized}}{1.0 \text{ L}}$

(c) $[Cu^{2+}] \rightarrow$ mol $e^- \rightarrow C \rightarrow s \rightarrow$ min $\rightarrow$ hr

$$\frac{2 \text{ mol } e^-}{1 \text{ mol Cu}^{2+}} \quad \frac{96,485 \text{ C}}{1 \text{ mol } e^-} \quad \frac{1 \text{ s}}{5.0 \text{ C}} \quad \frac{1 \text{ min}}{60 \text{ s}} \quad \frac{1 \text{ hr}}{60 \text{ min}}$$

Solution:

(a) Write half-reactions and add electrons. $Cu^{2+}(aq) + 2 e^- \rightarrow Cu(s)$ and $Mg(s) \rightarrow Mg^{2+}(aq) + 2 e^-$.

Look up cell potentials. Mg is oxidized, so $E^\circ_{red} = -2.37 \text{ V} = E^\circ_{anode}$. Cu^{2+} is reduced, so

$E^\circ_{red} = 0.34 \text{ V} = E^\circ_{cathode}$. Then $E^\circ_{cell} = E^\circ_{cathode} - E^\circ_{anode} = 0.34 \text{ V} - (-2.37 \text{ V}) = +2.71 \text{ V}$. Add

the two half-cell reactions: $Cu^{2+}(aq) + 2e^- + Mg(s) \rightarrow Cu(s) + Mg^{2+}(aq) + 2e^-$. Simplify to

$Cu^{2+}(aq) + Mg(s) \rightarrow Cu(s) + Mg^{2+}(aq)$. So $Q = \dfrac{[Mg^{2+}]}{[Cu^{2+}]} = \dfrac{1.0 \times 10^{-4}}{1.5} = 6.6667 \times 10^{-5}$ and $n = 2$.

Then $E_{cell} = E^\circ_{cell} - \dfrac{0.0592 \text{ V}}{n} \log Q = 2.71 \text{ V} - \dfrac{0.0592 \text{ V}}{2} \log (6.6667 \times 10^{-5}) = +2.83361 \text{ V} = +2.83 \text{ V}$.

(b) $8.0 \text{ hr} \times \dfrac{60 \text{ min}}{1 \text{ hr}} \times \dfrac{60 \text{ s}}{1 \text{ min}} \times \dfrac{5.0 \text{ C}}{1 \text{ s}} \times \dfrac{1 \text{ mol } e^-}{96,485 \text{ C}} \times \dfrac{1 \text{ mol Cu}^{2+}}{2 \text{ mol } e^-} = 0.74623 \text{ mol Cu}^{2+}$ and

$[Cu^{2+}]_f = [Cu^{2+}]_i - \dfrac{\text{mol Cu}^{2+} \text{ reduced}}{1.0 \text{ L}} = 1.5 \text{ M} - \dfrac{0.74623 \text{ mol Cu}^{2+}}{1.0 \text{ L}} = 0.75377 \text{ M Cu}^{2+}$ and

$0.74623 \text{ mol Cu}^{2+} \times \dfrac{1 \text{ mol Mg oxidized}}{1 \text{ mol Cu}^{2+} \text{ reduced}} = 0.74623 \text{ mol Mg oxidized}$ and

$[Mg^{2+}]_f = [Mg^{2+}]_i + \dfrac{\text{mol Mg oxidized}}{1.0 \text{ L}} = 1.0 \times 10^{-4} \text{ M} + \dfrac{0.74623 \text{ mol Mg oxidized}}{1.0 \text{ L}}$

$= 0.74633 \text{ M Mg}^{2+}$

$Q = \dfrac{[Mg^{2+}]}{[Cu^{2+}]} = \dfrac{0.74633}{0.75377} = 0.99013$ and $n = 2$ then

$E_{cell} = E^\circ_{cell} - \dfrac{0.0592 \text{ V}}{n} \log Q = 2.71 \text{ V} - \dfrac{0.0592 \text{ V}}{2} \log 0.99013 = +2.71013 \text{ V} = +2.71 \text{ V}$

(c) In 1.0 L, there are initially 1.5 moles of Cu^{2+}. So

$1.5 \text{ mol Cu}^{2+} \times \dfrac{2 \text{ mol } e^-}{1 \text{ mol Cu}^{2+}} \times \dfrac{96,485 \text{ C}}{1 \text{ mol } e^-} \times \dfrac{1 \text{ s}}{5.0 \text{ C}} \times \dfrac{1 \text{ min}}{60 \text{ s}} \times \dfrac{1 \text{ hr}}{60 \text{ min}} = 16 \text{ hr}.$

Check: The units (V, V, and hr) are correct. The Cu^{2+} reduction reaction is above the Mg^{2+} reduction reaction, so the standard cell potential will be positive. Having more reactants than products increases the cell potential. As the reaction proceeds, the potential drops. The concentrations drop by $\frac{1}{2}$ in 8 hours [part (b)], so all of it is consumed in 16 hours.

19.113 **Given:** water electrolysis at 7.8 A; $H_2(g)$: $V = 25.0 \text{ L}$, $P = 25.0 \text{ atm}$; $T = 25 \,^\circ\text{C}$ **Find:** time

Conceptual Plan: $^\circ C \rightarrow K$ then $V, P, T \rightarrow n$ then write half-reactions

$$K = \,^\circ C + 273.15 \qquad PV = nRT$$

mol $H_2 \rightarrow$ mol $e^- \rightarrow C \rightarrow s \rightarrow$ min $\rightarrow$ hr

$$\frac{2 \text{ mol } e^-}{1 \text{ mol H}_2} \quad \frac{96,485 \text{ C}}{1 \text{ mol } e^-} \quad \frac{1 \text{ s}}{7.8 \text{ C}} \quad \frac{1 \text{ min}}{60 \text{ s}} \quad \frac{1 \text{ hr}}{60 \text{ min}}$$

Solution: $T = 25 \,^\circ\text{C} + 273.15 = 298 \text{ K}$, then $PV = nRT$ Rearrange to solve for n.

$$n = \frac{PV}{RT} = \frac{25.0 \text{ atm} \times 25.0 \text{ L}}{0.08206 \dfrac{\text{L} \cdot \text{atm}}{\text{mol} \cdot \text{K}} \times 298 \text{ K}} = 25.5583 \text{ mol H}_2.$$ The hydrolysis of water reaction is as follows:

$2 H_2O(l) \rightarrow O_2(g) + 4 H^+(aq) + 4 e^-$ and $2 H_2O(l) + 2 e^- \rightarrow H_2(g) + 2 OH^-(aq)$

$25.5583 \text{ mol H}_2 \times \dfrac{2 \text{ mol } e^-}{1 \text{ mol H}_2} \times \dfrac{96,485 \text{ C}}{1 \text{ mol } e^-} \times \dfrac{1 \text{ s}}{7.8 \text{ C}} \times \dfrac{1 \text{ min}}{60 \text{ s}} \times \dfrac{1 \text{ hr}}{60 \text{ min}} = 176 \text{ hr}$

Check: The units (hr) are correct. Because we have 25 L of gas at 25 atm and 25 °C, we expect ~25 moles of gas (remember that 1 mole of gas at STP = 22.4 L). A very long time is expected because we have so many moles of gas to generate.

19.115 **Given:** $Cu(s) \mid CuI(s) \mid I^-(aq, 1.0\ M) \mid\mid Cu^+(aq, 1.0\ M) \mid Cu(s)$, $K_{sp}(CuI) = 1.1 \times 10^{-12}$ **Find:** E_{cell}
Conceptual Plan: Write half-reactions from line notation. Because this is a concentration cell, $E^\circ_{cell} = 0.00\ V$**.**
Then $K_{sp}, [I^-] \rightarrow [Cu^+](ox)$ **then** $E^\circ_{cell}, [Cu^+](ox), [Cu^+](red), n \rightarrow E_{cell}$**.**

$$K_{sp} = [Cu^+][I^-] \qquad\qquad E_{cell} = E^\circ_{cell} - \frac{0.0592\ V}{n}\log Q$$

Solution: The half-reactions are $Cu(s) \rightarrow Cu^+(aq) + e^-$ and $Cu^+(aq) + e^- \rightarrow Cu(s)$. Because this is a concentration cell, $E^\circ_{cell} = 0.00\ V$ and $n = 1$. Because $K_{sp} = [Cu^+][I^-]$, rearrange to solve for $[Cu^+]_{ox}$.

$$[Cu^+]_{ox} = \frac{K_{sp}}{[I^-]} = \frac{1.1 \times 10^{-12}}{1.0} = 1.1 \times 10^{-12}\ M \text{ then}$$

$$Q = \frac{[Cu^+]_{ox}}{[Cu^+]_{red}} = \frac{1.1 \times 10^{-12}}{1.0} = 1.1 \times 10^{-12} \text{ then}$$

$$E_{cell} = E^\circ_{cell} - \frac{0.0592\ V}{n}\log Q = 0.00\ V - \frac{0.0592\ V}{1}\log(1.1 \times 10^{-12}) = 0.71\ V$$

Check: The units (V) are correct. Because $[Cu^+]_{ox}$ is so low and $[Cu^+]_{red}$ is high, the Q is very small; so the voltage increase compared to the standard value is significant.

19.117 **Given:** (a) disproportionation of $Mn^{2+}(aq)$ to $Mn(s)$ and $MnO_2(s)$ and (b) disproportionation of $MnO_2(s)$ to $Mn^{2+}(aq)$ and $MnO_4^-(s)$ in acidic solution **Find:** ΔG°_{rxn} and K
**Conceptual Plan: Separate the overall reaction into two half-reactions: one for oxidation and one for reduction. →
Balance each half-reaction with respect to mass in the following order: (1) Balance all elements other than
H and O, (2) balance O by adding H_2O, and (3) balance H by adding H^+. → Balance each half-reaction with
respect to charge by adding electrons. (The sum of the charges on both sides of the equation should be made
equal by adding electrons as necessary.) → Make the number of electrons in both half-reactions equal by multiplying one or both half-reactions by a small whole number. → Add the two half-reactions, canceling electrons
and other species as necessary. → Verify that the reaction is balanced with respect to both mass and charge.
Look up half-reactions in Table 19.1. Calculate the standard cell potential by subtracting the electrode potential
of the anode from the electrode potential of the cathode:** $E^\circ_{cell} = E^\circ_{cathode} - E^\circ_{anode}$**.**
Then calculate ΔG°_{rxn} **using** $\Delta G^\circ_{rxn} = -n\,F\,E^\circ_{cell}$**. Finally,** °C → K **then** $\Delta G^\circ_{rxn}, T \rightarrow K$**.**

$$K = 273.15 + °C \qquad\qquad \Delta G^\circ_{rxn} = -RT \ln K$$

Solution:

(a) Separate: $Mn^{2+}(aq) \rightarrow MnO_2(s)$ and $Mn^{2+}(aq) \rightarrow Mn(s)$
   Balance non-H & O elements: $Mn^{2+}(aq) \rightarrow MnO_2(s)$ and $Mn^{2+}(aq) \rightarrow Mn(s)$
   Balance O with H_2O: $Mn^{2+}(aq) + 2\,H_2O(l) \rightarrow MnO_2(s)$ and $Mn^{2+}(aq) \rightarrow Mn(s)$
   Balance H with H^+: $Mn^{2+}(aq) + 2\,H_2O(l) \rightarrow MnO_2(s) + 4\,H^+(aq)$ and $Mn^{2+}(aq) \rightarrow Mn(s)$
   Add electrons: $Mn^{2+}(aq) + 2\,H_2O(l) \rightarrow MnO_2(s) + 4\,H^+(aq) + 2\,e^-$ and $Mn^{2+}(aq) + 2\,e^- \rightarrow Mn(s)$
   Equalize electrons: $Mn^{2+}(aq) + 2\,H_2O(l) \rightarrow MnO_2(s) + 4\,H^+(aq) + 2\,e^-$ and $Mn^{2+}(aq) + 2\,e^- \rightarrow Mn(s)$
   Add half-reactions: $Mn^{2+}(aq) + 2\,H_2O(l) + Mn^{2+}(aq) + \cancel{2e^-} \rightarrow MnO_2(s) + 4\,H^+(aq) + \cancel{2e^-} + Mn(s)$
   Cancel electrons: $2\,Mn^{2+}(aq) + 2\,H_2O(l) \rightarrow MnO_2(s) + 4\,H^+(aq) + Mn(s)$
   Look up cell potentials. Mn is oxidized in the first half-cell reaction, so $E^\circ_{anode} = E^\circ_{red} = +1.21\ V$. Mn is
   reduced in the second half-cell reaction, so $E^\circ_{cathode} = E^\circ_{red} = -1.18\ V$. Then
   $E^\circ_{cell} = E^\circ_{cathode} - E^\circ_{anode} = -1.18\ V - 1.21\ V = -2.39\ V$ and $n = 2$ so

$$\Delta G^\circ_{rxn} = -n\,F\,E^\circ_{cell} = -2\ \cancel{mole^-} \times \frac{96{,}485\ C}{\cancel{mole^-}} \times -2.39\ V = -2 \times 96{,}485\ \cancel{C} \times -2.39\frac{J}{\cancel{C}} =$$

$4.6\underline{1}198 \times 10^5\ J = 461\ kJ$ and $T = 273.15 + 25\ °C = 298\ K$ then $\Delta G^\circ_{rxn} = -RT \ln K$. Rearrange to solve for K.

$$K = e^{\frac{-\Delta G^\circ_{rxn}}{RT}} = e^{\frac{-4.61198 \times 10^5\ J}{\left(8.314\frac{J}{K\cdot mol}\right)(298\ K)}} = e^{-186.149} = 1.43 \times 10^{-81}$$

Check:

Reactants	Products
2 Mn atoms	2 Mn atoms
2 O atoms	2 O atoms
4 H atoms	4 H atoms
+4 charge	+4 charge

The units (kJ and none) are correct. If the voltage is negative, the reaction is nonspontaneous, the free energy change is very positive, and the equilibrium constant is extremely small.

(b)   Separate: $MnO_2(s) \rightarrow Mn^{2+}(aq)$ and $MnO_2(s) \rightarrow MnO_4^-(aq)$

Balance non-H & O elements: $MnO_2(s) \rightarrow Mn^{2+}(aq)$ and $MnO_2(s) \rightarrow MnO_4^-(aq)$

Balance O with H_2O: $MnO_2(s) \rightarrow Mn^{2+}(aq) + 2\,H_2O(l)$ and $MnO_2(s) + 2\,H_2O(l) \rightarrow MnO_4^-(aq)$

Balance H with H^+:

$MnO_2(s) + 4\,H^+(aq) \rightarrow Mn^{2+}(aq) + 2\,H_2O(l)$ and $MnO_2(s) + 2\,H_2O(l) \rightarrow MnO_4^-(aq) + 4\,H^+(aq)$

Add electrons: $MnO_2(s) + 4\,H^+(aq) + 2\,e^- \rightarrow Mn^{2+}(aq) + 2\,H_2O(l)$ and $MnO_2(s) + 2\,H_2O(l) \rightarrow$
$$MnO_4^-(aq) + 4\,H^+(aq) + 3\,e^-$$

Equalize electrons: $3\,MnO_2(s) + 12\,H^+(aq) + 6\,e^- \rightarrow 3\,Mn^{2+}(aq) + 6\,H_2O(l)$ and
$$2\,MnO_2(s) + 4\,H_2O(l) \rightarrow 2\,MnO_4^-(aq) + 8\,H^+(aq) + 6\,e^-$$

Add half-reactions: $3\,MnO_2(s) + 4\;\cancel{12\,H^+(aq)} + \cancel{6\,e^-} + 2\,MnO_2(s) + \cancel{4\,H_2O(l)} \rightarrow$
$$3\,Mn^{2+}(aq) + 2\,\cancel{6\,H_2O(l)} + 2\,MnO_4^-(aq) + \cancel{8\,H^+(aq)} + \cancel{6\,e^-}$$

Cancel electrons & species: $5\,MnO_2(s) + 4\,H^+(aq) \rightarrow 3\,Mn^{2+}(aq) + 2\,H_2O(l) + 2\,MnO_4^-(aq)$ Look up cell potentials. Mn is reduced in the first half-cell reaction, so $E^\circ_{cathode} = E^\circ_{red} = 1.21$ V. Mn is oxidized in the second half-cell reaction, so $E^\circ_{anode} = E^\circ_{red} = +1.68$ V. Then $E^\circ_{cell} = E^\circ_{cathode} - E^\circ_{anode} = 1.21\,V - 1.68\,V = -0.47\,V$ and $n = 6$ so

$$\Delta G^\circ_{rxn} = -nF\,E^\circ_{cell} = -6\,\cancel{mol\,e^-} \times \frac{96{,}485\,C}{\cancel{mol\,e^-}} \times -0.47\,V = -6 \times 96{,}485\,\cancel{C} \times -0.47\frac{J}{\cancel{C}}$$

$$= 2.7\underline{2}09 \times 10^5\,J = 270\,kJ = 2.7 \times 10^2\,kJ \text{ and } T = 273.15 + 25\,°C = 298\,K \text{ then } \Delta G^\circ_{rxn} = -RT\ln K$$

Rearrange to solve for K. $K = e^{\frac{-\Delta G^\circ_{rxn}}{RT}} = e^{\dfrac{-2.7209 \times 10^5\,\cancel{J}}{\left(8.314\,\frac{\cancel{J}}{K\cdot mol}\right)(298\,\cancel{K})}} = e^{-109.\underline{8}2} = 2.0 \times 10^{-48}$

Check:

Reactants	Products
5 Mn atoms	5 Mn atoms
10 O atoms	10 O atoms
4 H atoms	4 H atoms
+4 charge	+4 charge

The units (kJ and none) are correct. If the voltage is negative, the reaction is nonspontaneous, the free energy change is very positive, and the equilibrium constant is extremely small. The voltage is less than that in part (a), so the free energy change is not as large and the equilibrium constant is not as small.

19.119   **Given:** metal, M, 50.9 g/mol, 1.20 g of metal reduced in 23.6 minutes at 6.42 A from molten chloride

Find: empirical formula of chloride

Conceptual Plan: $min \rightarrow s \rightarrow C \rightarrow mol\,e^-$ and $g\,M \rightarrow mol\,M$ then $mol\,e^-, mol\,M \rightarrow charge \rightarrow MCl_x$

$$\frac{60\,s}{1\,min} \quad \frac{6.42\,C}{1\,s} \quad \frac{1\,mol\,e^-}{96{,}485\,C} \qquad \frac{1\,mol\,M}{50.9\,g\,M} \qquad\qquad \frac{1\,mol\,e^-}{1\,mol\,M}$$

Solution: $23.6\,\cancel{min} \times \dfrac{60\,\cancel{s}}{1\,\cancel{min}} \times \dfrac{6.42\,\cancel{C}}{1\,\cancel{s}} \times \dfrac{1\,mol\,e^-}{96{,}485\,\cancel{C}} = 0.094\underline{2}190\,mol\,e^-$ and

$1.20\,\cancel{g\,M} \times \dfrac{1\,mol\,M}{50.9\,\cancel{g\,M}} = 0.023\underline{5}756\,mol\,M$ then $\dfrac{0.0942190\,mol\,e^-}{0.0235756\,mol\,M} = 3.99\underline{6}46\dfrac{e^-}{M}$

So the empirical formula is MCl_4.

Check: The units (none) are correct. The result was an integer within the error of the measurements. The formula is typical for a metal salt. It could be vanadium, which has a +4 oxidation state.

19.121   **Given:** 0.535 g impure Sn; dissolve to form Sn^{2+} and titrate with 0.0344 L of 0.0448 M NO_3^- to generate NO

Find: percent by mass Sn

Conceptual Plan: Use balanced reaction from Problem 19.36(c) then

$L \rightarrow mol\,NO_3^- \rightarrow mol\,Sn \rightarrow g\,Sn \rightarrow$ **percent by mass Sn.**

$$M = \frac{mol}{L} \qquad \frac{3\,mol\,Sn}{2\,mol\,NO_3^-} \qquad \frac{118.71\,g\,Sn}{1\,mol\,Sn} \qquad \text{percent by mass Sn} = \frac{g\,Sn}{g\,sample} \times 100\%$$

Solution: $2\,NO_3^-(aq) + 8\,H^+(aq) + 3\,Sn^{2+}(aq) \rightarrow 2\,NO(g) + 4\,H_2O(l) + 3\,Sn^{4+}(aq)$

$$0.0344 \, \cancel{L} \times \frac{0.0448 \, \cancel{\text{mol NO}_3^-}}{1 \, \cancel{L}} \times \frac{3 \, \cancel{\text{mol Sn}}}{2 \, \cancel{\text{mol NO}_3^-}} \times \frac{118.71 \, \text{g Sn}}{1 \, \cancel{\text{mol Sn}}} = 0.2744195 \, \text{g Sn then}$$

$$\text{percent by mass Sn} = \frac{\text{g Sn}}{\text{g sample}} \times 100\% = \frac{0.2744195 \, \cancel{\text{g Sn}}}{0.535 \, \cancel{\text{g sample}}} \times 100\% = 51.3\% \text{ by mass Sn}$$

Check: The units (% by mass) are correct. The result was a number less than 100%.

19.123   **Given:** 1.25 L of a 0.552 M HBr solution, convert H^+ to $H_2(g)$ for 73 minutes at 11.32 A   **Find:** pH

Conceptual Plan: min $\rightarrow$ s $\rightarrow$ C $\rightarrow$ mol e$^-$ $\rightarrow$ mol H$^+$ consumed and L, M $\rightarrow$ mol H$^-$ initially then

$$\frac{60 \, \text{s}}{1 \, \text{min}} \qquad \frac{11.32 \, \text{C}}{1 \, \text{s}} \qquad \frac{1 \, \text{mol e}^-}{96,485 \, \text{C}} \qquad \frac{2 \, \text{mol H}^+}{2 \, \text{mol e}^-} \qquad\qquad M = \frac{\text{mol}}{L}$$

mol H + initially, mol H$^+$ consumed $\rightarrow$ mol H$^+$ remaining then mol H$^+$ remaining, L $\rightarrow$ [H$^+$] $\rightarrow$ pH

$$\text{mol H}^+ \text{ initially} - \text{mol H}^+ \text{ consumed} = \text{mol H}^+ \text{ remaining} \qquad\qquad M = \frac{\text{mol}}{L} \quad \text{pH} = -\log[\text{H}^+]$$

Solution: $2 \, H^+(aq) + 2 \, e^- \rightarrow H_2(g)$

$$73 \, \cancel{\text{min}} \times \frac{60 \, \cancel{s}}{1 \, \cancel{\text{min}}} \times \frac{11.32 \, \cancel{C}}{1 \, \cancel{s}} \times \frac{1 \, \cancel{\text{mol e}^-}}{96,485 \, \cancel{C}} \times \frac{2 \, \text{mol H}^+}{2 \, \cancel{\text{mol e}^-}} = 0.5\underline{1}3879 \, \text{mol H}^+ \text{ consumed and}$$

$$1.25 \, \cancel{\text{L H}^+} \times \frac{0.552 \, \text{mol H}^+}{1 \, \cancel{\text{L H}^+}} = 0.690 \, \text{mol H}^+ \text{ initially then mol H}^+ \text{ initially} - \text{mol H}^+ \text{ consumed} =$$

$$\text{mol H}^+ \text{ remaining} = 0.690 \, \text{mol H}^+ \text{ initially} - 0.5\underline{1}3879 \, \text{mol H}^+ \text{ consumed} = 0.1\underline{7}612 \, \text{mol H}^+ \text{ remaining}$$

$$\text{then } [\text{H}^+] = \frac{0.1\underline{7}612 \, \text{mol H}^+}{1.25 \, \text{L}} = 0.1\underline{4}0897 \, \text{M H}^+ \text{ then pH} = -\log[\text{H}^+] = = -\log 0.1\underline{4}0897 = 0.85$$

Check: The units (none) are correct. The result is a pH higher than the initial pH $[-\log(0.552) = 0.258]$, as is expected.

19.125   **Given:** MnO_2/Mn^{2+} electrode at pH 10.24   **Find:** $[Mn^{2+}]$ to get half-cell potential = 0.00 V

Conceptual Plan: pH $\rightarrow$ [H$^+$] then

$$\text{pH} = -\log[\text{H}^+]$$

Write half-cell reactions and look up half-reactions in Table 19.1. Define Q based on half-cell reaction. Then $E^\circ_{\text{half-cell}}, n, [\text{H}^+] \rightarrow [\text{Mn}^{2+}]$.

$$E_{\text{half-cell}} = E^\circ_{\text{half-cell}} - \frac{0.0592 \, \text{V}}{n} \log Q$$

Solution: Because $\text{pH} = -\log[\text{H}^+]$ so $[\text{H}^+] = 10^{-\text{pH}} = 10^{-10.24} = 5.\underline{7}5440 \times 10^{-11} \, \text{M}$

$$MnO_2(s) + 4 \, H^+(aq) + 2 \, e^- \rightarrow Mn^{2+}(aq) + 2 \, H_2O(l) \quad E^\circ_{\text{half-cell}} = 1.21 \, \text{V}, n = 2, \text{ and } Q = \frac{[\text{Mn}^{2+}]}{[\text{H}^+]^4}$$

$$E_{\text{half-cell}} = E^\circ_{\text{half-cell}} - \frac{0.0592 \, \text{V}}{n} \log Q \text{ so } 0.00 \, \text{V} = 1.21 \, \text{V} - \frac{0.0592 \, \text{V}}{2} \log \frac{[\text{Mn}^{2+}]}{(5.\underline{7}5440 \times 10^{-11})^4} \rightarrow$$

$$1.21 \, \text{V} = \frac{0.0592 \, \text{V}}{2} \log \frac{[\text{Mn}^{2+}]}{(5.\underline{7}5440 \times 10^{-11})^4} \rightarrow 40.8\underline{7}84 = \log \frac{[\text{Mn}^{2+}]}{(5.\underline{7}5440 \times 10^{-11})^4} \rightarrow$$

$$\frac{[\text{Mn}^{2+}]}{(5.\underline{7}5440 \times 10^{-11})^4} = 10^{40.8\underline{7}84} = 7.\underline{5}5750 \times 10^{40} \rightarrow [\text{Mn}^{2+}] = 0.8\underline{2}8664 \, \text{M} = 0.83 \, \text{M Mn}^{2+}$$

Check: The units (M) are correct. The standard half-cell potential is very large and positive. Most of the shift toward 0.00 V is due to the fourth-order dependence in [H$^+$] at a high pH, so the [Mn^{2+}] is close to 1 M. So the concentration is reasonable.

Challenge Problems

19.127   **Given:** hydrogen–oxygen fuel cell; 1.2×10^3 kWh of electricity/month   **Find:** V of $H_2(g)$ at STP/month

Conceptual Plan: Write half-reactions. Look up half-reactions at pH 7. The reaction on the left is the oxidation. Calculate the standard cell potential by subtracting the electrode potential of the anode from the electrode potential of the cathode: $E^\circ_{\text{cell}} = E^\circ_{\text{cathode}} - E^\circ_{\text{anode}}$. Add the two half-cell reactions and cancel the electrons. Then kWh $\rightarrow$ J $\rightarrow$ C $\rightarrow$ mol e$^-$ $\rightarrow$ mol H$_2$ $\rightarrow$ V.

$$\frac{3.60 \times 10^6 \, \text{J}}{1 \, \text{kWh}} \quad \frac{1 \, \text{C}}{1.23 \, \text{J}} \quad \frac{1 \, \text{mol e}^-}{96,485 \, \text{C}} \quad \frac{2 \, \text{mol H}_2}{4 \, \text{mol e}^-} \quad \text{at STP} \quad \frac{22.414 \, \text{L}}{1 \, \text{mol H}_2}$$

Solution: $2 H_2(g) + 4 OH^-(aq) \rightarrow 4 H_2O(l) + 4 e^-$ where

$E_{red}^\circ = -0.83 V = E_{anode}^\circ$ and $O_2(g) + 2 H_2O(l) + 4 e^- \rightarrow 4 OH^-(aq)$ where

$E_{red}^\circ = 0.40 V = E_{cathode}^\circ$. $E_{cell}^\circ = E_{cathode}^\circ - E_{anode}^\circ = 0.40 V - (-0.83 V) = 1.23 V = 1.23 J/C$ and $n = 4$. Net

reaction is $2 H_2(g) + O_2(g) \rightarrow 2 H_2O(l)$. Then

$$1.2 \times 10^3 \text{ kWh} \times \frac{3.60 \times 10^6 \text{ J}}{1 \text{ kWh}} \times \frac{1 \cancel{C}}{1.23 \text{ J}} \times \frac{1 \text{ mole } e^-}{96,485 \cancel{C}} \times \frac{2 \text{ mol } H_2}{4 \text{ mole } e^-} \times \frac{22.414 \text{ L}}{1 \text{ mol } H_2} = 4.1 \times 10^5 \text{ L}$$

Check: The units (L) are correct. A large volume is expected because we are trying to generate a large amount of electricity.

19.129   **Given:** Au^{3+}/Au electroplating; surface area $= 49.8 \text{ cm}^2$, Au thickness $= 1.00 \times 10^{-3}$ cm, density $= 19.3 \text{ g/cm}^3$; at 3.25 A   **Find:** time

Conceptual Plan: Write the half-cell reaction and add electrons as needed to balance reactions. Then surface area, thickness $\rightarrow V \rightarrow$ g Au $\rightarrow$ mol Au $\rightarrow$ mol $e^- \rightarrow C \rightarrow$ s

$$V = \textit{surface area} \times \textit{thickness} \quad \frac{19.3 \text{ g Au}}{1 \text{ cm}^3 \text{ Au}} \quad \frac{1 \text{ mol Au}}{196.97 \text{ g Au}} \quad \frac{3 \text{ mol } e^-}{1 \text{ mol Au}} \quad \frac{96,485 \text{ C}}{1 \text{ mol } e^-} \quad \frac{1 \text{ s}}{3.25 \text{ C}}$$

Solution: Write the half-reaction and add electrons. $Au^{3+}(aq) + 3 e^- \rightarrow Au(s)$.

$V = \textit{surface area} \times \textit{thickness} = (49.8 \text{ cm}^2)(1.00 \times 10^{-3} \text{ cm}) = 0.0498 \text{ cm}^3$ then

$$0.0498 \cancel{\text{cm}^3 \text{ Au}} \times \frac{19.3 \cancel{\text{g Au}}}{1 \cancel{\text{cm}^3 \text{ Au}}} \times \frac{1 \cancel{\text{mol Au}}}{196.97 \cancel{\text{g Au}}} \times \frac{3 \cancel{\text{mol } e^-}}{1 \cancel{\text{mol Au}}} \times \frac{96,485 \cancel{C}}{1 \cancel{\text{mol } e^-}} \times \frac{1 \text{ s}}{3.25 \cancel{C}} = 435 \text{ s}$$

Check: The units (s) are correct. Because the layer is so thin, there is far less than a mole of gold; so the time is not very long. To be an economical process, it must be fairly quick.

19.131   **Given:** $C_2O_4^{2-}(aq) \rightarrow CO_2(g)$ and $MnO_4^-(aq) \rightarrow Mn^{2+}(aq)$, 50.1 mL of MnO_4^- to titrate 0.339 g $Na_2C_2O_4$; 4.62 g U sample titrated by 32.5 mL MnO_4^-; and $UO^{2+}(aq) \rightarrow UO_2^{2+}(aq)$ **Find:** percent U in sample

Conceptual Plan: Separate the overall reaction into two half-reactions: one for oxidation and one for reduction. $\rightarrow$ Balance each half-reaction with respect to mass in the following order: (1) Balance all elements other than H and O, (2) balance O by adding H_2O, and (3) balance H by adding H^+. $\rightarrow$ Balance each half-reaction with respect to charge by adding electrons. (The sum of the charges on both sides of the equation should be made equal by adding electrons as necessary.) $\rightarrow$ Make the number of electrons in both half-reactions equal by multiplying one or both half-reactions by a small whole number. $\rightarrow$ Add the two half-reactions, canceling electrons and other species as necessary. $\rightarrow$ Verify that the reaction is balanced with respect to both mass and charge. Then

mL $MnO_4^- \rightarrow L MnO_4^-$ and g $Na_2C_2O_4 \rightarrow$ mol $Na_2C_2O_4 \rightarrow$ mol MnO_4^- then

$$\frac{1 \text{ L } MnO_4^-}{1000 \text{ mL } MnO_4^-} \qquad \frac{1 \text{ mol } Na_2C_2O_4}{134.00 \text{ g } Na_2C_2O_4} \qquad \frac{2 \text{ mol } MnO_4^-}{5 \text{ mol } Na_2C_2O_4}$$

L MnO_4^-, mol $MnO_4^- \rightarrow M MnO_4^-$ then write U half-reactions and balance as above. $\rightarrow$

$$M = \frac{\text{mol } MnO_4^-}{L}$$

Make the number of electrons in both half-reactions equal by multiplying one or both half-reactions by a small whole number. $\rightarrow$ Add the two half-reactions, canceling electrons and other species as necessary. $\rightarrow$ Verify that the reaction is balanced with respect to both mass and charge.

Then mL MnO_4^-, M $MnO_4^- \rightarrow$ mol $MnO_4^- \rightarrow$ mol U $\rightarrow$ g U then g U, g sample $\rightarrow$ % U.

$$M = \frac{\text{mol } MnO_4^-}{L} \qquad \frac{5 \text{ mol U}}{2 \text{ mol } MnO_4^-} \qquad \frac{238.03 \text{ g U}}{1 \text{ mol U}} \qquad \qquad \text{percent U} = \frac{\text{g U}}{\text{g sample}} \times 100\%$$

Solution:

Separate: $MnO_4^-(aq) \rightarrow Mn^{2+}(aq)$ and $C_2O_4^{2-}(aq) \rightarrow CO_2(g)$

Balance non-H & O elements: $MnO_4^-(aq) \rightarrow Mn^{2+}(aq)$ and $C_2O_4^{2-}(aq) \rightarrow 2 CO_2(g)$

Balance O with H_2O: $MnO_4^-(aq) \rightarrow Mn^{2+}(aq) + 4 H_2O(l)$ and $C_2O_4^{2-}(aq) \rightarrow 2 CO_2(g)$

Balance H with H^+: $MnO_4^-(aq) + 8 H^+(aq) \rightarrow Mn^{2+}(aq) + 4 H_2O(l)$ and $C_2O_4^{2-}(aq) \rightarrow 2 CO_2(g)$

Add electrons: $MnO_4^-(aq) + 8 H^+(aq) + 5 e^- \rightarrow Mn^{2+}(aq) + 4 H_2O(l)$ and $C_2O_4^{2-}(aq) \rightarrow 2 CO_2(g) + 2 e^-$

Equalize electrons:

$2 MnO_4^-(aq) + 16 H^+(aq) + 10 e^- \rightarrow 2 Mn^{2+}(aq) + 8 H_2O(l)$ and $5 C_2O_4^{2-}(aq) \rightarrow 10 CO_2(g) + 10 e^-$

Add half-reactions:

$2 MnO_4^-(aq) + 16 H^+(aq) + \cancel{10 e^-} + 5 C_2O_4^{2-}(aq) \rightarrow 2 Mn^{2+}(aq) + 8 H_2O(l) + 10 CO_2(g) + \cancel{10 e^-}$

Cancel electrons: $2 MnO_4^-(aq) + 16 H^+(aq) + 5 C_2O_4^{2-}(aq) \rightarrow 2 Mn^{2+}(aq) + 8 H_2O(l) + 10 CO_2(g)$

then $50.1 \text{ mL MnO}_4^- \times \dfrac{1 \text{ L MnO}_4^-}{1000 \text{ mL MnO}_4^-} = 0.0501 \text{ L MnO}_4^-$

$0.399 \text{ g Na}_2\text{C}_2\text{O}_4 \times \dfrac{1 \text{ mol Na}_2\text{C}_2\text{O}_4}{134.00 \text{ g Na}_2\text{C}_2\text{O}_4} \times \dfrac{2 \text{ mol MnO}_4^-}{5 \text{ mol Na}_2\text{C}_2\text{O}_4} = 0.0011910448 \text{ mol MnO}_4^-$

$M = \dfrac{0.0011910448 \text{ mol MnO}_4^-}{0.0501 \text{ L}} = 0.023773345 \text{ M MnO}_4^-$

Separate: $\qquad\qquad\qquad\qquad$ $MnO_4^-(aq) \rightarrow Mn^{2+}(aq)$ $\quad$ and $\quad$ $UO^{2+}(aq) \rightarrow UO_2^{2+}(aq)$

Balance non-H & O elements: $\quad$ $MnO_4^-(aq) \rightarrow Mn^{2+}(aq)$ $\quad$ and $\quad$ $UO^{2+}(aq) \rightarrow UO_2^{2+}(aq)$

Balance O with H_2O: $MnO_4^-(aq) \rightarrow Mn^{2+}(aq) + 4 H_2O(l)$ $\quad$ and $\quad$ $UO^{2+}(aq) + H_2O(l) \rightarrow UO_2^{2+}(aq)$

Balance H with H^+:

$MnO_4^-(aq) + 8 H^+(aq) \rightarrow Mn^{2+}(aq) + 4 H_2O(l)$ and $UO^{2+}(aq) + H_2O(l) \rightarrow UO_2^{2+}(aq) + 2 H^+(aq)$

Add electrons: $MnO_4^-(aq) + 8 H^+(aq) + 5 e^- \rightarrow Mn^{2+}(aq) + 4 H_2O(l)$ and

$\qquad\qquad\qquad\qquad\qquad\qquad\qquad\qquad UO^{2+}(aq) + H_2O(l) \rightarrow UO_2^{2+}(aq) + 2 H^+(aq) + 2 e^-$

Equalize electrons: $2 MnO_4^-(aq) + 16 H^+(aq) + 10 e^- \rightarrow 2 Mn^{2+}(aq) + 8 H_2O(l)$ and

$\qquad\qquad\qquad\qquad\qquad 5 UO^{2+}(aq) + 5 H_2O(l) \rightarrow 5 UO_2^{2+}(aq) + 10 H^+(aq) + 10 e^-$

Add half-reactions: $2 MnO_4^-(aq) + 6 \cancel{16} H^+(aq) + \cancel{10 e^-} + 5 UO^{2+}(aq) + 5 \cancel{H_2O(l)} \rightarrow$

$\qquad\qquad\qquad\qquad\qquad 2 Mn^{2+}(aq) + 3 \cancel{8} H_2O(l) + 5 UO_2^{2+}(aq) + \cancel{10 H^+(aq)} + \cancel{10 e^-}$

Cancel electrons & species: $2 MnO_4^-(aq) + 6 H^+(aq) + 5 UO^{2+}(aq) \rightarrow 2 Mn^{2+}(aq) + 3 H_2O(l) + 5 UO_2^{2+}(aq)$

$32.5 \text{ mL MnO}_4^- \times \dfrac{0.023773345 \text{ mol MnO}_4^-}{1000 \text{ mL MnO}_4^-} \times \dfrac{5 \text{ mol U}}{2 \text{ mol MnO}_4^-} \times \dfrac{238.03 \text{ g U}}{1 \text{ mol U}} = 0.4597750 \text{ g U then}$

percent $U = \dfrac{\text{g U}}{\text{g sample}} \times 100\% = \dfrac{0.4597750 \text{ g U}}{4.63 \text{ g sample}} \times 100\% = 9.93\%$

Check: first reaction

Reactants	Products
2 Mn atoms	2 Mn atoms
28 O atoms	28 O atoms
16 H atoms	16 H atoms
10 C atoms	10 C atoms
+4 charge	+4 charge

second reaction

Reactants	Products
2 Mn atoms	2 Mn atoms
13 O atoms	13 O atoms
6 H atoms	6 H atoms
5 U atoms	5 U atoms
+14 charge	+14 charge

The reactions are balanced. The units (%) are correct. The percentage is between 0 and 100%.

19.133 $\quad$ The overall cell reaction for the first cell is $2 Cu^+(aq) \rightarrow Cu^{2+}(aq) + Cu(s)$. The overall cell reaction for the second cell is $Cu^+(aq) \rightarrow Cu^{2+}(aq)$. The biggest difference in $E°$ is because $n = 1$ for the first cell and $n = 2$ for the second cell. Because $\Delta G_{rxn}° = -n F E_{cell}°$ for the first cell,

$\Delta G_{rxn}° = -1 \text{ mol } e^- \times \dfrac{96{,}485 \text{ C}}{\text{mol } e^-} \times 0.364 \text{ V} = -1 \times 96{,}485 \text{ C} \times 0.364 \dfrac{\text{J}}{\text{C}} = -35.1 \text{ kJ}$, and for the

second cell,

$\Delta G_{rxn}° = -2 \text{ mol } e^- \times \dfrac{96{,}485 \text{ C}}{\text{mol } e^-} \times 0.182 \text{ V} = -2 \times 96{,}485 \text{ C} \times 0.182 \dfrac{\text{J}}{\text{C}} = -35.1 \text{ kJ}$. Thus, $\Delta G_{rxn}°$ is the same.

Conceptual Problems

19.135 $\quad$ (a) $\quad$ Looking for anion reductions that are between the reduction potentials of Cl_2 and Br_2. The only one that meets that criterion is the dichromate ion.

19.137 $\quad$ Because $K < 1$, the reaction must be nonspontaneous under standard conditions. Therefore, $E_{cell}°$ is negative and $\Delta G_{rxn}°$ is positive.

Questions for Group Work

19.139　Choose any two half-reaction from Table 19.1 and follow the conceptual plan in Problem 19.15 to get a positive cell potential.

19.141　Specific answers will vary. In general, a device could have a known cell as one half-reaction and the unknown copper sample as the other half cell. $E_{cell} = E°_{cell} + RT \ln Q$ can be solved for the unknown Cu^{2+}. 1 mV uncertainty in voltage may correspond to approximately 1×10^{-6} M uncertainty in concentration.

Data Interpretation and Analysis

19.143　(a)　The plot of measured voltage versus copper concentration is not linear.

(c)

The intercept is 0.3399 V, the voltage measured at 1.00 M. A slope of 0.0298 V is not unexpected because the change in oxidation number of copper is 2 and 0.0592 V/ 2 = 0.0296 V.

20 Radioactivity and Nuclear Chemistry

Review Questions

20.1 Radioactivity is the emission of subatomic particles or high-energy electromagnetic radiation by the nuclei of certain atoms. Radioactivity was discovered in 1896 by the French scientist Antoine-Henri Becquerel (1852–1908). Becquerel placed crystals—composed of potassium uranyl sulfate, a compound known to phosphoresce—on top of a photographic plate wrapped in black cloth. The photographic plate showed a bright exposure spot where the crystals had been. Becquerel, Marie Curie, and Pierre Curie received the Nobel Prize for the discovery of radioactivity.

20.3 A is the mass number (number of protons + neutrons), Z is the atomic number (number of protons), and X is the chemical symbol of the element.

20.5 An alpha particle has the same symbol as a helium nucleus, $_2^4\text{He}$. When an element emits an alpha particle, the number of protons in its nucleus decreases by two and the mass number decreases by four, transforming it into a different element.

20.7 Gamma rays are high-energy (short-wavelength) photons that have a symbol of $_0^0\gamma$. A gamma ray has no charge and no mass. When a gamma-ray photon is emitted from a radioactive atom, it does not change the mass number or the atomic number of the element. Gamma rays, however, are usually emitted in conjunction with other types of radiation.

20.9 Electron capture occurs when a nucleus assimilates an electron from an inner orbital of its electron cloud. Like positron emission, the net effect of electron capture is the conversion of a proton into a neutron: $_1^1\text{p} + {_{-1}^0}\text{e} \rightarrow {_0^1}\text{n}$. When an atom undergoes electron capture, its atomic number decreases by one because it has one less proton and its mass number is unchanged.

20.11 For the lighter elements, the N/Z ratio of stable isotopes is about one (equal numbers of neutrons and protons). However, beyond about $Z = 20$, the N/Z ratio of stable nuclei begins to get larger (reaching about 1.5). Above $Z = 83$, stable nuclei do not exist. If the N/Z ratio is too high, it indicates the presence of too many neutrons, and so the neutrons will be converted to protons via beta decay. If the N/Z ratio is too low, it indicates the presence of too many protons; therefore, position emission or electron capture will be used to convert protons to neutrons.

20.13 (a) Film-badge dosimeters consist of photographic film held in a small case that is pinned to clothing and are standard for most people working with or near radioactive substances. These badges are collected and processed (or developed) regularly to monitor a person's exposure. The more exposed the film has become in a given period of time, the more radioactivity to which the person has been exposed.

(b) A Geiger-Müller counter (commonly referred to as a Geiger counter) is an instrument that can detect radioactivity instantaneously. Particles emitted by radioactive nuclei pass through an argon-filled chamber. The energetic particles create a trail of ionized argon atoms. An applied high voltage between a wire within the chamber and the chamber itself causes these newly formed ions to produce an electrical signal that can be detected on a meter or turned into an audible click. Each click corresponds to a radioactive particle passing through the argon gas chamber. This clicking is the stereotypical sound most people associate with a radiation detector.

(c)   A scintillation counter is another instrument that can detect radioactivity instantaneously. In this device, the radioactive emissions pass through a material (such as NaI or CsI) that emits ultraviolet or visible light in response to excitation by energetic particles. The radioactivity excites the atoms to a higher-energy state. The atoms release this energy as light, which is then detected and turned into an electrical signal that can be read on a meter.

20.15   Radiocarbon dating, a technique devised in 1949 by Willard Libby at the University of Chicago, is used by archeologists, geologists, anthropologists, and other scientists to estimate the ages of fossils and artifacts. Carbon-14 is constantly formed in the upper atmosphere by the neutron bombardment of nitrogen and then decays back to nitrogen by beta emission with a half-life of 5715 years. The continuous formation of carbon-14 in the atmosphere and its continuous decay back to nitrogen-14 produce a nearly constant equilibrium amount of atmospheric carbon-14, which is oxidized to carbon dioxide and incorporated into plants by photosynthesis. The C-14 then makes its way up the food chain and ultimately into all living organisms. As a result, all living plants, animals, and humans contain the same ratio of carbon-14 to carbon-12 (^{14}C:^{12}C) as is found in the atmosphere. When a living organism dies, however, it stops incorporating new carbon-14 into its tissues. The ^{14}C:^{12}C ratio then decreases with a half-life of 5730 years. The accuracy of carbon-14 dating can be checked against objects whose ages are known from historical sources.

To make C-14 dating more accurate, scientists have studied the carbon-14 content of western bristlecone pine trees, which can live up to 5000 years. The tree trunk contains growth rings corresponding to each year of the tree's life, and the wood laid down in each ring incorporates carbon derived from the carbon dioxide in the atmosphere at that time. The rings thus provide a record of the historical atmospheric carbon-14 content and allow for corrections to carbon-14 concentrations due to atmospheric changes.

The maximum age that can be estimated from carbon-14 dating is about 50,000 years—beyond that, the amount of carbon-14 becomes too low to measure accurately.

20.17   Nuclear fission is the splitting of an atom into smaller products. The process emits enormous amounts of energy. Three researchers in Germany—Lise Meitner (1878–1968), Fritz Strassmann (1902–1980), and Otto Hahn (1879–1968)—repeating uranium bombardment experiments by Fermi, performed careful chemical analysis of the products. The nucleus of the neutron-bombarded uranium atom is split into barium, krypton, and other smaller products. A nuclear equation for a fission reaction, showing how uranium breaks apart into the daughter nuclides, is $^{235}_{92}$U + $^{1}_{0}$n → $^{140}_{56}$Ba + $^{93}_{36}$Kr + 3$^{1}_{0}$n + energy. The process produces three neutrons, which have the potential to initiate fission in three other U-235 atoms. Scientists quickly realized that a sample rich in U-235 could undergo a chain reaction in which neutrons produced by the fission of one uranium nucleus would induce fission in other uranium nuclei. The result would be a self-amplifying reaction capable of producing an enormous amount of energy—an atomic bomb. However, to make a bomb, a critical mass of U-235—enough U-235 to produce a self-sustaining reaction—would be necessary. Because an enormous amount of energy is produced, it can be used to produce steam to drive turbines (as shown in Figure 20.11).

20.19   The advantages of using fission to generate electricity are that (1) a typical nuclear power plant generates enough electricity for a city of about 1 million people and uses about 50 kg of fuel per day (as opposed to a coal-burning power plant that uses about 2,000,000 kg of fuel to generate the same amount of electricity) and that (2) a nuclear power plant generates no air pollution and no greenhouse gases. (Coal-burning power plants also emit carbon dioxide, a greenhouse gas.) The disadvantages are as follows: (1) the danger of nuclear accidents, such as overheating and the release of radiation, and (2) waste disposal because the products of the reaction are radioactive and have long half-lives.

20.21   This difference in mass between the products and the reactants is known as the mass defect. The energy corresponding to the mass defect, obtained by substituting the mass defect into the equation $E = mc^2$, is known as the nuclear binding energy, the amount of energy that would be required to break apart the nucleus into its component nucleons. The nuclear binding energy per nucleon peaks at a mass number of 60. The significance of this is that the nuclides with mass numbers of about 60 are among the most stable.

20.23   An extremely high temperature is required for fusion to occur. To date, no material can withstand these temperatures. The U.S. Congress has reduced funding for these projects so that it is less likely that a viable process will be developed.

20.25   In a single-stage linear accelerator, a charged particle such as a proton is accelerated in an evacuated tube. The accelerating force is provided by a potential difference between the ends of the tube. In multistage linear accelerators such as the Stanford Linear Accelerator (SLAC) at Stanford University (Figure 20.14), a series of tubes of increasing length are connected to a source of alternating voltage, as shown in Figure 20.15. The voltage alternates in such a way that as a positively charged particle leaves a particular tube, that tube becomes positively charged, repelling the particle to the next tube. At the same time, the tube the particle is now approaching becomes negatively charged, pulling the particle toward it. This continues throughout the linear accelerator, allowing the particle to be accelerated to velocities up to 90% of the speed of light. Linear accelerators can be used to conduct nuclear transmutations, making nuclides that don't normally exist in nature.

20.27   The energy associated with radioactivity can ionize molecules. When radiation ionizes important molecules in living cells, problems can develop. The ingestion of radioactive materials, especially alpha and beta emitters, is particularly dangerous because the radioactivity is then inside the body and can do even more damage. The effects of radiation are divided into three different types: acute radiation damage, increased cancer risk, and genetic effects.

20.29   The biological effectiveness factor, or RBE (relative biological effectiveness), corrects the dosage (in rads) for the type of radiation. It is a correction factor that is usually multiplied by the dose in rads to obtain the dose in a unit called the rem (roentgen equivalent man). So dose in rads $\times$ biological effectiveness factor = dose in rem. The biological effectiveness factor for alpha radiation, for example, is much higher than for gamma radiation.

Problems by Topic

Radioactive Decay and Nuclide Stability

20.31   **Conceptual Plan: Begin with the symbol for a parent nuclide on the left side of the equation and the symbol for a particle on the right side (except for electron capture). $\rightarrow$ Equalize the sum of the mass numbers and the sum of the atomic numbers on both sides of the equation by writing the appropriate mass number and atomic number for the unknown daughter nuclide. $\rightarrow$ Using the periodic table, deduce the identity of the unknown daughter nuclide from the atomic number and write its symbol.**
Solution:

(a)   U-234 (alpha decay) $^{234}_{92}\text{U} \rightarrow {}^{?}_{?}? + {}^{4}_{2}\text{He}$ then $^{234}_{92}\text{U} \rightarrow {}^{230}_{90}? + {}^{4}_{2}\text{He}$ then $^{234}_{92}\text{U} \rightarrow {}^{230}_{90}\text{Th} + {}^{4}_{2}\text{He}$

(b)   Th-230 (alpha decay) $^{230}_{90}\text{Th} \rightarrow {}^{?}_{?}? + {}^{4}_{2}\text{He}$ then $^{230}_{90}\text{Th} \rightarrow {}^{226}_{88}? + {}^{4}_{2}\text{He}$ then $^{230}_{90}\text{Th} \rightarrow {}^{226}_{88}\text{Ra} + {}^{4}_{2}\text{He}$

(c)   Pb-214 (beta decay) $^{214}_{82}\text{Pb} \rightarrow {}^{?}_{?}? + {}^{0}_{-1}\text{e}$ then $^{214}_{82}\text{Pb} \rightarrow {}^{214}_{83}? + {}^{0}_{-1}\text{e}$ then $^{214}_{82}\text{Pb} \rightarrow {}^{214}_{83}\text{Bi} + {}^{0}_{-1}\text{e}$

(d)   N-13 (positron emission) $^{13}_{7}\text{N} \rightarrow {}^{?}_{?}? + {}^{0}_{+1}\text{e}$ then $^{13}_{7}\text{N} \rightarrow {}^{13}_{6}? + {}^{0}_{+1}\text{e}$ then $^{13}_{7}\text{N} \rightarrow {}^{13}_{6}\text{C} + {}^{0}_{+1}\text{e}$

(e)   Cr-51 (electron capture) $^{51}_{24}\text{Cr} + {}^{0}_{-1}\text{e} \rightarrow {}^{?}_{?}?$ then $^{51}_{24}\text{Cr} + {}^{0}_{-1}\text{e} \rightarrow {}^{51}_{23}?$ then $^{51}_{24}\text{Cr} + {}^{0}_{-1}\text{e} \rightarrow {}^{51}_{23}\text{V}$

Check: (a) $234 = 230 + 4, 92 = 90 + 2$, and thorium is atomic number 90. (b) $230 = 226 + 4, 90 = 88 + 2$, and radium is atomic number 88. (c) $214 = 214 + 0, 82 = 83 - 1$, and bismuth is atomic number 83. (d) $13 = 13 + 0, 7 = 6 + 1$, and carbon is atomic number 6. (e) $51 + 0 = 51, 24 - 1 = 23$, and vanadium is atomic number 23.

20.33   **Given:** Th-232 decay series: $\alpha, \beta, \beta, \alpha$ **Find:** balanced decay reactions
Conceptual Plan: Begin with the symbol for a parent nuclide on the left side of the equation and the symbol for a particle on the right side (except for electron capture). $\rightarrow$ Equalize the sum of the mass numbers and the sum of the atomic numbers on both sides of the equation by writing the appropriate mass number and atomic number for the unknown daughter nuclide. $\rightarrow$ Using the periodic table, deduce the identity of the unknown daughter nuclide from the atomic number and write its symbol. $\rightarrow$ Use the product of this reaction to write the next reaction.
Solution:

Th-232 (alpha decay) $^{232}_{90}\text{Th} \rightarrow {}^{?}_{?}? + {}^{4}_{2}\text{He}$ then $^{232}_{90}\text{Th} \rightarrow {}^{228}_{88}? + {}^{4}_{2}\text{He}$ then $^{232}_{90}\text{Th} \rightarrow {}^{228}_{88}\text{Ra} + {}^{4}_{2}\text{He}$

Ra-228 (beta decay) $^{228}_{88}\text{Ra} \rightarrow {}^{?}_{?}? + {}^{0}_{-1}\text{e}$ then $^{228}_{88}\text{Ra} \rightarrow {}^{228}_{89}? + {}^{0}_{-1}\text{e}$ then $^{228}_{88}\text{Ra} \rightarrow {}^{228}_{89}\text{Ac} + {}^{0}_{-1}\text{e}$

Ac-228 (beta decay) $^{228}_{89}\text{Ac} \rightarrow {}^{?}_{?}? + {}^{0}_{-1}\text{e}$ then $^{228}_{89}\text{Ac} \rightarrow {}^{228}_{90}? + {}^{0}_{-1}\text{e}$ then $^{228}_{89}\text{Ac} \rightarrow {}^{228}_{90}\text{Th} + {}^{0}_{-1}\text{e}$

Th-228 (alpha decay) $^{228}_{90}\text{Th} \rightarrow {}^{?}_{?}? + {}^{4}_{2}\text{He}$ then $^{228}_{90}\text{Th} \rightarrow {}^{224}_{88}? + {}^{4}_{2}\text{He}$ then $^{228}_{90}\text{Th} \rightarrow {}^{224}_{88}\text{Ra} + {}^{4}_{2}\text{He}$

Thus, the decay series is $^{232}_{90}\text{Th} \rightarrow {}^{228}_{88}\text{Ra} + {}^{4}_{2}\text{He}, {}^{228}_{88}\text{Ra} \rightarrow {}^{228}_{89}\text{Ac} + {}^{0}_{-1}\text{e}, {}^{228}_{89}\text{Ac} \rightarrow {}^{228}_{90}\text{Th} + {}^{0}_{-1}\text{e}, {}^{228}_{90}\text{Th} \rightarrow {}^{224}_{88}\text{Ra} + {}^{4}_{2}\text{He}$.

Check: $232 = 228 + 4, 90 = 88 + 2$, and radium is atomic number 88. $228 = 228 + 0, 88 = 89 - 1$, and actinium is atomic number 89. $228 = 228 + 0, 89 = 90 - 1$, and thorium is atomic number 90. $228 = 224 + 4, 90 = 88 + 2$, and radium is atomic number 88.

20.35   **Conceptual Plan: Equalize the sum of the mass numbers and the sum of the atomic numbers on both sides of the equation by writing the appropriate mass number and atomic number for the unknown species.** → **Using the periodic table and the list of particles, deduce the identity of the unknown species from the atomic number and write its symbol.**

Solution:

(a) $^{?}_{?}? \rightarrow\, ^{217}_{85}At + ^{4}_{2}He$ becomes $^{221}_{87}? \rightarrow\, ^{217}_{85}At + ^{4}_{2}He$ then $^{221}_{87}Fr \rightarrow\, ^{217}_{85}At + ^{4}_{2}He$

(b) $^{241}_{94}Pu \rightarrow\, ^{241}_{95}Am + ^{?}_{?}?$ becomes $^{241}_{94}Pu \rightarrow\, ^{241}_{95}Am + ^{0}_{-1}?$ then $^{241}_{94}Pu \rightarrow\, ^{241}_{95}Am + ^{0}_{-1}e$

(c) $^{19}_{11}Na \rightarrow\, ^{19}_{10}Ne + ^{?}_{?}?$ becomes $^{19}_{11}Na \rightarrow\, ^{19}_{10}Ne + ^{0}_{1}?$ then $^{19}_{11}Na \rightarrow\, ^{19}_{10}Ne + ^{0}_{+1}e$

(d) $^{75}_{34}Se + ^{?}_{?}? \rightarrow\, ^{75}_{33}As$ becomes $^{75}_{34}Se + ^{0}_{-1}? \rightarrow\, ^{75}_{33}As$ then $^{75}_{34}Se + ^{0}_{-1}e \rightarrow\, ^{75}_{33}As$

Check: (a) $221 = 217 + 4, 87 = 85 + 2$, and francium is atomic number 87. (b) $241 = 241 + 0, 94 = 95 - 1$, and the particle is a beta particle. (c) $19 = 19 + 0, 11 = 10 + 1$, and the particle is a positron. (d) $75 = 75 + 0, 34 - 1 = 33$, and the particle is an electron.

20.37   (a)   Mg-26: stable, N/Z ratio is close to 1, acceptable for low Z atoms

(b)   Ne-25: not stable, N/Z ratio is much too high for low Z atom

(c)   Co-51: not stable, N/Z ratio is less than 1, much too low

(d)   Te-124: stable, N/Z ratio is acceptable for this Z

20.39   Sc, V, and Mn each have an odd number of protons. Atoms with an odd number of protons typically have fewer stable isotopes than those with an even number of protons.

20.41   (a)   Mo-109, $N = 67, Z = 42, N/Z = 1.6$, beta decay, because N/Z is too high

(b)   Ru-90, $N = 46, Z = 44, N/Z = 1.0$, positron emission, because N/Z is too low

(c)   P-27, $N = 15, Z = 12, N/Z = 0.8$, positron emission, because N/Z is too low

(d)   Sn-100, $N = 50, Z = 50, N/Z = 1.0$, positron emission, because N/Z is too low

20.43   (a)   Cs-125, $N/Z = 70/55 = 1.3$; Cs-113, $N/Z = 58/55 = 1.1$; Cs-125 will have the longer half-life because it is closer to the proper N/Z

(b)   Fe-62, $N/Z = 36/26 = 1.4$; Fe-70, $N/Z = 44/26 = 1.7$; Fe-62 will have the longer half-life because it is closer to the proper N/Z

The Kinetics of Radioactive Decay and Radiometric Dating

20.45   **Given:** U-235, $t_{1/2}$ for radioactive decay $= 703$ million years   **Find:** t to 10.0% of initial amount

Conceptual Plan: Radioactive decay implies first-order kinetics, $t_{1/2} \rightarrow k$ then

$$t_{1/2} = \frac{0.693}{k}$$

$$m_{\text{U-235 0}}, m_{\text{U-235 t}}, k \rightarrow t$$

$$\ln N_t = -kt + \ln N_0$$

Solution: $t_{1/2} = \dfrac{0.693}{k}$ Rearrange to solve for k. $k = \dfrac{0.693}{t_{1/2}} = \dfrac{0.693}{703 \times 10^6 \text{ yr}} = 9.857752 \times 10^{-10} \text{ yr}^{-1}$

Because $\ln m_{\text{U-235 t}} = -kt + \ln m_{\text{U-235 0}}$, rearrange to solve for t.

$$t = -\frac{1}{k} \ln \frac{m_{\text{U-235 t}}}{m_{\text{U-235 0}}} = -\frac{1}{9.857752 \times 10^{-10} \text{ yr}^{-1}} \ln \frac{10.0\%}{100.0\%} = 2.34 \times 10^9 \text{ yr}$$

Check: The units (yr) are correct. The time is just over three half-lives, when 1/8 of the original amount will be left.

20.47   **Given:** $t_{1/2}$ for isotope decay $= 3.8$ days; 1.55 g isotope initially   **Find:** mass of isotope after 5.5 days

Conceptual Plan: Radioactive decay implies first-order kinetics, $t_{1/2} \rightarrow k$ then

$$t_{1/2} = \frac{0.693}{k}$$

$$m_{\text{isotope 0}}, t, k \rightarrow m_{\text{isotope t}}.$$

$$\ln N_t = -kt + \ln N_0$$

Solution: $t_{1/2} = \dfrac{0.693}{k}$ Rearrange to solve for k. $k = \dfrac{0.693}{t_{1/2}} = \dfrac{0.693}{3.8 \text{ days}} = 0.1\underline{8}237 \text{ day}^{-1}$ Because

$\ln N_t = -kt + \ln N_0 = -(0.1\underline{8}237 \text{ day}^{-1})(5.5 \text{ day}) + \ln(1.55 \text{ g}) = -0.5\underline{6}478 \rightarrow N_t = e^{-0.5\underline{6}478} = 0.57 \text{ g}$.

Check: The units (g) are correct. The amount is consistent with a time between one and two half-lives.

20.49   **Given:** F-18 initial decay rate $= 1.5 \times 10^5/s$, $t_{1/2}$ for F-18 $= 1.83$ h   **Find:** t to decay rate of $2.5 \times 10^3/s$
Conceptual Plan: Radioactive decay implies first-order kinetics, $t_{1/2} \rightarrow k$ then Rate$_0$, Rate$_t$, $k \rightarrow t$.

$$t_{1/2} = \dfrac{0.693}{k} \qquad\qquad \ln \dfrac{\text{Rate}_t}{\text{Rate}_0} = -kt$$

Solution: $t_{1/2} = \dfrac{0.693}{k}$ Rearrange to solve for k. $k = \dfrac{0.693}{t_{1/2}} = \dfrac{0.693}{1.83 \text{ h}} = 0.3\underline{7}8689 \text{ h}^{-1}$ Because

$\ln \dfrac{\text{Rate}_t}{\text{Rate}_0} = -kt$ rearrange to solve for t.

$t = -\dfrac{1}{k} \ln \dfrac{\text{Rate}_t}{\text{Rate}_0} = -\dfrac{1}{0.3\underline{7}8689 \text{ h}^{-1}} \ln \dfrac{2.5 \times 10^3/s}{1.5 \times 10^5/s} = 10.8 \text{ h}$

Check: The units (h) are correct. The time is between five and six half-lives, and the rate is just over $1/2^6$ of the original amount.

20.51   **Given:** boat analysis, C-14/C-12 $= 72.5\%$ of living organism   **Find:** t
Other: $t_{1/2}$ for decay of C-14 $= 5715$ years
Conceptual Plan: Radioactive decay implies first-order kinetics, $t_{1/2} \rightarrow k$ then 72.5% of $m_{\text{C-14 0}}$, $k \rightarrow t$.

$$t_{1/2} = \dfrac{0.693}{k} \qquad\qquad \ln N_t = -kt + \ln N_0$$

Solution: $t_{1/2} = \dfrac{0.693}{k}$ Rearrange to solve for k. $k = \dfrac{0.693}{t_{1/2}} = \dfrac{0.693}{5715 \text{ yr}} = 1.2\underline{1}2598 \times 10^{-4} \text{ yr}^{-1}$ then

$[\text{C-14}]_t = 0.725 [\text{C-14}]_0$ Because $\ln m_{\text{C-14}\,t} = -kt + \ln m_{\text{C-14}\,0}$, rearrange to solve for t.

$t = -\dfrac{1}{k} \ln \dfrac{m_{\text{C-14}\,t}}{m_{\text{C-14}\,0}} = -\dfrac{1}{1.2\underline{1}2598 \times 10^{-4} \text{ yr}^{-1}} \ln \dfrac{0.725\, m_{\text{C-14}\,0}}{m_{\text{C-14}\,0}} = 2.65 \times 10^3 \text{ yr}$

Check: The units (yr) are correct. The time to 72.5% decay is consistent with a time less than one half-life.

20.53   **Given:** skull analysis, C-14 decay rate $= 15.3 \text{ dis/min} \cdot \text{gC}$ in living organisms and $0.85 \text{ dis/min} \cdot \text{gC}$ in skull
Find: t **Other:** $t_{1/2}$ for decay of C-14 $= 5715$ years
Conceptual Plan: Radioactive decay implies first-order kinetics, $t_{1/2} \rightarrow k$ then Rate$_0$, Rate$_t$, $k \rightarrow t$.

$$t_{1/2} = \dfrac{0.693}{k} \qquad\qquad \ln \dfrac{\text{Rate}_t}{\text{Rate}_0} = -kt$$

Solution: $t_{1/2} = \dfrac{0.693}{k}$ Rearrange to solve for k. $k = \dfrac{0.693}{t_{1/2}} = \dfrac{0.693}{5715 \text{ yr}} = 1.2\underline{1}2598 \times 10^{-4} \text{ yr}^{-1}$

Because $\ln \dfrac{\text{Rate}_t}{\text{Rate}_0} = -kt$, rearrange to solve for t.

$t = -\dfrac{1}{k} \ln \dfrac{\text{Rate}_t}{\text{Rate}_0} = -\dfrac{1}{1.2\underline{1}2598 \times 10^{-4} \text{ yr}^{-1}} \ln \dfrac{0.85 \text{ dis/min} \cdot \text{gC}}{15.3 \text{ dis/min} \cdot \text{gC}} = 2.4 \times 10^4 \text{ yr}$

Check: The units (yr) are correct. The rate is 6% of initial value, and the time is consistent with a time just more than four half-lives.

20.55   **Given:** rock analysis, 0.438 g Pb-206 to every 1.00 g U-238, no Pb-206 initially   **Find:** age of rock
Other: $t_{1/2}$ for decay of U-238 to Pb-206 $= 4.5 \times 10^9$ years
Conceptual Plan: Radioactive decay implies first-order kinetics, $t_{1/2} \rightarrow k$ then

$$t_{1/2} = \dfrac{0.693}{k}$$

g Pb-206 $\rightarrow$ mol Pb-206 $\rightarrow$ mol U-238 $\rightarrow$ g U-238 then $m_{\text{U-238}\,t}$, $k \rightarrow t$.

$$\dfrac{1 \text{ mol Pb-206}}{206 \text{ g Pb-206}} \qquad \dfrac{1 \text{ mol U-238}}{1 \text{ mol Pb-206}} \qquad \dfrac{238 \text{ g U-238}}{1 \text{ mol U-238}} \qquad\qquad \ln N_t = -kt + \ln N_0$$

Solution: $t_{1/2} = \dfrac{0.693}{k}$ Rearrange to solve for k. $k = \dfrac{0.693}{t_{1/2}} = \dfrac{0.693}{4.5 \times 10^9 \text{ yr}} = 1.\underline{5}4 \times 10^{-10} \text{ yr}^{-1}$ then

$0.438 \text{ g Pb-206} \times \dfrac{1 \text{ mol Pb-206}}{206 \text{ g Pb-206}} \times \dfrac{1 \text{ mol U-238}}{1 \text{ mol Pb-206}} \times \dfrac{238 \text{ g U-238}}{1 \text{ mol U-238}} = 0.50\underline{6}039 \text{ g U-238}$ Because

$\ln\dfrac{m_{\text{U-238 }t}}{m_{\text{U-238 }0}} = -kt$, rearrange to solve for t.

$t = -\dfrac{1}{k} \ln\dfrac{m_{\text{U-238 }t}}{m_{\text{U-238 }0}} = -\dfrac{1}{1.\underline{5}4 \times 10^{-10} \text{ yr}^{-1}} \ln\dfrac{1.00 \text{ g U-238}}{(1.00 + 0.50\underline{6}039) \text{ g U-238}} = 2.7 \times 10^9 \text{ yr}$

Check: The units (yr) are correct. The amount of Pb-206 is less than half the initial U-238 amount, and time is less than one half-life.

Fission and Fusion

20.57   **Given:** U-235 fission induced by neutrons to Xe-144 and Sr-90   **Find:** number of neutrons produced
Conceptual Plan: Write the species given on the appropriate side of the equation. → Equalize the sum of the mass numbers and the sum of the atomic numbers on both sides of the equation by writing the stoichiometric coefficient in front of the desired species.
Solution: $^{235}_{92}\text{U} + ^{1}_{0}\text{n} \rightarrow ^{144}_{54}\text{Xe} + ^{90}_{38}\text{Sr} + ?^{1}_{0}\text{n}$ becomes $^{235}_{92}\text{U} + ^{1}_{0}\text{n} \rightarrow ^{144}_{54}\text{Xe} + ^{90}_{38}\text{Sr} + 2^{1}_{0}\text{n}$, so two neutrons are produced.

Check: $235 + 1 = 144 + 90 + 2, 92 + 0 = 54 + 38 + 0$, and no other particle is necessary to balance the equation.

20.59   **Given:** fusion of two H-2 atoms to form He-3 and one neutron   **Find:** balanced equation
Conceptual Plan: Write the species given on the appropriate side of the equation. → Equalize the sum of the mass numbers and the sum of the atomic numbers on both sides of the equation by writing the stoichiometric coefficient in front of the desired species.
Solution: $2\,^{2}_{1}\text{H} \rightarrow ^{3}_{2}\text{He} + ^{1}_{0}\text{n}$

Check: $2(2) = 3 + 1, 2(1) = 2 + 0$, and no other particle is necessary to balance the equation.

20.61   **Given:** U-238 bombarded by neutrons to form U-239, which undergoes two beta decays to form Pu-239
Find: balanced equations
Conceptual Plan: Write the species given on the appropriate side of the equation. → Equalize the sum of the mass numbers and the sum of the atomic numbers on both sides of the equation by writing the stoichiometric coefficient in front of the desired species. → Use the product of this reaction to write the next reaction until the process is complete.
Solution: $^{238}_{92}\text{U} + ?^{1}_{0}\text{n} \rightarrow ^{239}_{92}\text{U}$ becomes $^{238}_{92}\text{U} + ^{1}_{0}\text{n} \rightarrow ^{239}_{92}\text{U}$ then
beta decay $^{239}_{92}\text{U} \rightarrow ^{?}_{?}? + ^{0}_{-1}\text{e}$ becomes $^{239}_{92}\text{U} \rightarrow ^{239}_{93}? + ^{0}_{-1}\text{e}$ then $^{239}_{92}\text{U} \rightarrow ^{239}_{93}\text{Np} + ^{0}_{-1}\text{e}$ then
beta decay $^{239}_{93}\text{Np} \rightarrow ^{?}_{?}? + ^{0}_{-1}\text{e}$ becomes $^{239}_{93}\text{Np} \rightarrow ^{239}_{94}? + ^{0}_{-1}\text{e}$ then $^{239}_{93}\text{Np} \rightarrow ^{239}_{94}\text{Pu} + ^{0}_{-1}\text{e}$
The entire process is $^{238}_{92}\text{U} + ^{1}_{0}\text{n} \rightarrow ^{239}_{92}\text{U}, ^{239}_{92}\text{U} \rightarrow ^{239}_{93}\text{Np} + ^{0}_{-1}\text{e}, ^{239}_{93}\text{Np} \rightarrow ^{239}_{94}\text{Pu} + ^{0}_{-1}\text{e}$.

Check: $238 + 1 = 239, 92 + 0 = 92$, and no other particle is necessary to balance the equation.
$239 = 239 + 0, 92 = 93 - 1$, and neptunium is atomic number 93. $239 = 239 + 0, 93 = 94 - 1$, and plutonium is atomic number 94.

20.63   **Given:** Rf-257 synthesized by bombarding Cf-249 with C-12   **Find:** balanced equation
Conceptual Plan: Write the species given on the appropriate side of the equation. → Equalize the sum of the mass numbers and the sum of the atomic numbers on both sides of the equation by writing the stoichiometric coefficient in front of the desired species. → Use the product of this reaction to write the next reaction until the process is complete.
Solution: $^{249}_{98}\text{Cf} + ^{12}_{6}\text{C} \rightarrow ^{257}_{?}\text{Rf} + ?\,^{?}_{?}?$ becomes $^{249}_{98}\text{Cf} + ^{12}_{6}\text{C} \rightarrow ^{257}_{104}\text{Rf} + 4^{1}_{0}\text{n}$.

Check: $249 + 12 = 257 - 4, 98 + 6 = 104 + 0$, and rutherfordium is atomic number 104, and four neutrons are needed to balance the equation.

Energetics of Nuclear Reactions, Mass Defect, and Nuclear Binding Energy

20.65 **Given:** 1.0 g of matter converted to energy **Find:** energy
Conceptual Plan: g → kg → E

$$\frac{1 \text{ kg}}{1000 \text{ g}} \qquad E = mc^2$$

Solution: $1.0 \text{ g} \times \dfrac{1 \text{ kg}}{1000 \text{ g}} = 0.0010 \text{ kg}$ then $E = mc^2 = (0.0010 \text{ kg})\left(2.9979 \times 10^8 \dfrac{\text{m}}{\text{s}}\right)^2 = 9.0 \times 10^{13} \text{ J}$

Check: The units (J) are correct. The magnitude of the answer makes physical sense because we are converting a large quantity of amus to energy.

20.67 **Given:** (a) O-16 = 15.9949145 amu, (b) Ni-58 = 57.935346 amu, and (c) Xe-129 = 128.904780 amu
Find: mass defect and nuclear binding energy per nucleon
Conceptual Plan: $_{Z}^{A}X$, isotope mass → mass defect → nuclear binding energy per nucleon

$$\text{mass defect} = Z(\text{mass } _1^1\text{H}) + (A - Z)(\text{mass } _0^1\text{n}) - \text{mass of isotope} \qquad \frac{931.5 \text{ MeV}}{(1 \text{ amu})(A \text{ nucleons})}$$

Solution: $\text{mass defect} = Z(\text{mass } _1^1\text{H}) + (A - Z)(\text{mass } _0^1\text{n}) - \text{mass of isotope}$

(a)   O-16 mass defect $= 8(1.00783 \text{ amu}) + (16 - 8)(1.00866 \text{ amu}) - 15.9949145 \text{ amu} =$

  $0.1370055 \text{ amu} = 0.13701 \text{ amu}$ and $0.1370055 \text{ amu} \times \dfrac{931.5 \text{ MeV}}{(1 \text{ amu})(16 \text{ nucleons})} = 7.976 \dfrac{\text{MeV}}{\text{nucleon}}$

(b)   Ni-58 mass defect $= 28(1.00783 \text{ amu}) + (58 - 28)(1.00866 \text{ amu}) - 57.935346 \text{ amu} =$

  $0.543694 \text{ amu} = 0.54369 \text{ amu}$ and $0.543694 \text{ amu} \times \dfrac{931.5 \text{ MeV}}{(1 \text{ amu})(58 \text{ nucleons})} = 8.732 \dfrac{\text{MeV}}{\text{nucleon}}$

(c)   Xe-129 mass defect $= 54(1.00783 \text{ amu}) + (129 - 54)(1.00866 \text{ amu}) - 128.904780 \text{ amu} =$

  1.16754 amu and $1.16754 \text{ amu} \times \dfrac{931.5 \text{ MeV}}{(1 \text{ amu})(129 \text{ nucleons})} = 8.431 \dfrac{\text{MeV}}{\text{nucleon}}$

Check: The units (amu and MeV/nucleon) are correct. The mass defect increases with an increasing number of nucleons, but the MeV/nucleon does not change by as much (on a relative basis).

20.69 **Given:** $_{92}^{235}\text{U} + _0^1\text{n} \rightarrow _{54}^{144}\text{Xe} + _{38}^{90}\text{Sr} + 2_0^1\text{n}$, U-235 = 235.043922 amu, Xe-144 = 143.9385 amu, and Sr-90 = 89.907738 amu **Find:** energy per g of U-235
Conceptual Plan: mass of products and reactants → mass defect → mass defect/g of U-235 then

$$\text{mass defect} = \Sigma\text{mass of reactants} - \Sigma\text{mass of products} \qquad \frac{\text{mass defect}}{235.043922 \text{ g U-235}}$$

g → kg → E

$$\frac{1 \text{ kg}}{1000 \text{ g}} \qquad E = mc^2$$

Solution: $\text{mass defect} = \Sigma\text{mass of reactants} - \Sigma\text{mass of products}$; notice that we can cancel a neutron from each side to get $_{92}^{235}\text{U} \rightarrow _{54}^{144}\text{Xe} + _{38}^{90}\text{Sr} + _0^1\text{n}$ and

$\text{mass defect} = 235.043922 \text{ g} - (143.9385 \text{ g} + 89.907738 \text{ g} + 1.00866 \text{ g}) = 0.189024 \text{ g}$

then $\dfrac{0.189024 \text{ g}}{235.043922 \text{ g U-235}} \times \dfrac{1 \text{ kg}}{1000 \text{ g}} = 8.04207 \times 10^{-7} \dfrac{\text{kg}}{\text{g U-235}}$ then

$E = mc^2 = \left(8.04207 \times 10^{-7} \dfrac{\text{kg}}{\text{g U-235}}\right)\left(2.9979 \times 10^8 \dfrac{\text{m}}{\text{s}}\right)^2 = 7.228 \times 10^{10} \dfrac{\text{J}}{\text{g U-235}}$

Check: The units (J) are correct. A large amount of energy is expected per gram of fuel in a nuclear reactor.

20.71 **Given:** $2_1^2\text{H} \rightarrow _2^3\text{He} + _0^1\text{n}$, H-2 = 2.014102 amu, and He-3 = 3.016029 amu **Find:** energy per g reactant
Conceptual Plan: mass of products and reactants → mass defect → mass defect/g of H-2 then

$$\text{mass defect} = \Sigma\text{mass of reactants} - \Sigma\text{mass of products} \qquad \frac{\text{mass defect}}{2(2.014102 \text{ g H-2})}$$

g → kg → E

$$\frac{1 \text{ kg}}{1000 \text{ g}} \qquad E = mc^2$$

Solution: mass defect $= \Sigma$mass of reactants $- \Sigma$mass of products and mass defect $= 2(2.014102 \text{ g}) - (3.016029 \text{ g} + 1.00866 \text{ g}) = 0.003515 \text{ g}$

then $\dfrac{0.003515 \text{ g}}{2(2.014102 \text{ g H-2})} \times \dfrac{1 \text{ kg}}{1000 \text{ g}} = 8.72597 \times 10^{-7} \dfrac{\text{kg}}{\text{g H-2}}$ then

$E = mc^2 = \left(8.72597 \times 10^{-7} \dfrac{\text{kg}}{\text{g H-2}}\right)\left(2.9979 \times 10^8 \dfrac{\text{m}}{\text{s}}\right)^2 = 7.84 \times 10^{10} \dfrac{\text{J}}{\text{g H-2}}$

Check: The units (J) are correct. A large amount of energy is expected per gram of fuel in a fusion reaction.

Effects and Applications of Radioactivity

20.73   **Given:** 75 kg human exposed to 32.8 rad and falling from chair   **Find:** energy absorbed in each case
Conceptual Plan: rad, kg $\rightarrow$ J and assume $d = 0.50$ m chair height then mass, $d \rightarrow$ J

$$1 \text{ rad} = \dfrac{0.01 \text{ J}}{1 \text{ kg body tissue}} \qquad\qquad E = F \cdot d = mgd$$

Solution: $32.8 \text{ rad} = 32.8 \dfrac{0.01 \text{ J}}{1 \text{ kg body tissue}} \times 75 \text{ kg} = 25 \text{ J}$ and

$E = F \cdot d = mgd = 75 \text{ kg} \times 9.8 \dfrac{\text{m}}{\text{s}^2} \times 0.50 \text{ m} = 370 \text{ kg} \dfrac{\text{m}^2}{\text{s}^2} = 370 \text{ J}$

Check: The units (J and J) are correct. Allowable radiation exposures are low because the radiation is very ionizing and thus damaging to tissue. Falling may have more energy, but it is not ionizing.

20.75   **Given:** $t_{1/2}$ for F-18 $= 1.83$ h, 65% of F-18 makes it to the hospital, traveling at 60.0 miles/hour
Find: distance between hospital and cyclotron
Conceptual Plan: $t_{1/2} \rightarrow k$ then $m_{\text{F-18 }0}, m_{\text{F-18 }t}, k \rightarrow t$ then h $\rightarrow$ mi

$$t_{1/2} = \dfrac{0.693}{k} \qquad\qquad \ln\dfrac{m_{\text{F-18}t}}{m_{\text{F-18}0}} = -kt \qquad\qquad \dfrac{60.0 \text{ mi}}{1 \text{ h}}$$

Solution: $t_{1/2} = \dfrac{0.693}{k}$ Rearrange to solve for k. $k = \dfrac{0.693}{t_{1/2}} = \dfrac{0.693}{1.83 \text{ h}} = 0.378689 \text{ h}^{-1}$ Because

$\ln\dfrac{m_{\text{F-18}t}}{m_{\text{F-18}0}} = -kt$, rearrange to solve for t.

$t = -\dfrac{1}{k} \ln\dfrac{m_{\text{F-18 }t}}{m_{\text{F-18 }0}} = -\dfrac{1}{0.378689 \text{ h}^{-1}} \ln\dfrac{0.65 \, m_{\text{F-18 }0}}{m_{\text{F-18 }0}} = 1.1376 \text{ h}$ then

$1.1376 \text{ h} \times \dfrac{60.0 \text{ mi}}{1 \text{ h}} = 68 \text{ mi}$

Check: The units (mi) are correct. The time is less than one half-life, so the distance is less than 1.83 times the speed of travel.

Cumulative Problems

20.77   **Given:** incomplete reactions   **Find:** balanced reaction and energy (in J/mol reactant)
Conceptual Plan: Equalize the sum of the mass numbers and the sum of the atomic numbers on both sides of the equation by writing the appropriate mass number and atomic number for the unknown species. $\rightarrow$ Using the periodic table and the list of particles, deduce the identity of the unknown species from the atomic number and write its symbol. Then mass of products & reactants $\rightarrow$ mass defect in g $\rightarrow$ mass defect in kg $\rightarrow$ E

mass defect $= \Sigma$mass of reactants $- \Sigma$mass of products $\qquad \dfrac{1 \text{ kg}}{1000 \text{ g}} \qquad E = mc^2$

Solution:

(a) $^{?}_{?}? + {}^{9}_{4}\text{Be} \rightarrow {}^{6}_{3}\text{Li} + {}^{4}_{2}\text{He}$ becomes $^{1}_{1}? + {}^{9}_{4}\text{Be} \rightarrow {}^{6}_{3}\text{Li} + {}^{4}_{2}\text{He}$ then ${}^{1}_{1}\text{H} + {}^{9}_{4}\text{Be} \rightarrow {}^{6}_{3}\text{Li} + {}^{4}_{2}\text{He}$
mass defect $= \Sigma$mass of reactants $- \Sigma$mass of products and
mass defect $= (1.00783 \text{ g} + 9.012182 \text{ g}) - (6.015122 \text{ g} + 4.002603 \text{ g}) = 0.002287 \text{ g}$

then $\dfrac{0.002287\ \cancel{g}}{2\ \text{mol reactants}} \times \dfrac{1\ \text{kg}}{1000\ \cancel{g}} = 1.1435 \times 10^{-6}\ \dfrac{\text{kg}}{\text{mol reactants}}$ then

$$E = mc^2 = \left(1.1435 \times 10^{-6}\ \dfrac{\text{kg}}{\text{mol reactants}}\right)\left(2.9979 \times 10^8\ \dfrac{\text{m}}{\text{s}}\right)^2 = 1.03 \times 10^{11}\ \dfrac{\text{J}}{\text{mol reactants}}$$

Check: $1 + 9 = 6 + 4, 1 + 4 = 3 + 2$. The units (J) are correct.

(b) $^{209}_{83}\text{Bi} + ^{64}_{28}\text{Ni} \rightarrow ^{272}_{111}\text{Rg} + ^{?}_{?}?$ becomes $^{209}_{83}\text{Bi} + ^{64}_{28}\text{Ni} \rightarrow ^{272}_{111}\text{Rg} + ^{1}_{0}?$ then $^{209}_{83}\text{Bi} + ^{64}_{28}\text{Ni} \rightarrow ^{272}_{111}\text{Rg} + ^{1}_{0}\text{n}$
mass defect $= (208.980384\ \text{g} + 63.927969\ \text{g}) - (272.1535\ \text{g} + 1.00866\ \text{g}) = -0.253807\ \text{g}$
(Note: Because this is negative, energy must be put in.)

then $\dfrac{0.253807\ \cancel{g}}{2\ \text{mol reactants}} \times \dfrac{1\ \text{kg}}{1000\ \cancel{g}} = 1.269035 \times 10^{-4}\ \dfrac{\text{kg}}{\text{mol reactants}}$ then

$$E = mc^2 = \left(1.269035 \times 10^{-4}\ \dfrac{\text{kg}}{\text{mol reactants}}\right)\left(2.9979 \times 10^8\ \dfrac{\text{m}}{\text{s}}\right)^2 = 1.141 \times 10^{13}\ \dfrac{\text{J}}{\text{mol reactants}}$$

Check: $209 + 64 = 272 + 1, 83 + 28 = 111 + 0$. The units (J) are correct.

(c) $^{179}_{74}\text{W} + ^{?}_{?}? \rightarrow ^{179}_{73}\text{Ta}$ becomes $^{179}_{74}\text{W} + ^{0}_{-1}? \rightarrow ^{179}_{73}\text{Ta}$ then $^{179}_{74}\text{W} + ^{0}_{-1}\text{e} \rightarrow ^{179}_{73}\text{Ta}$
mass defect $= (178.94707\ \text{g} + 0.00055\ \text{g}) - 178.94593\ \text{g} = 0.00169\ \text{g}$ then

$\dfrac{0.00169\ \cancel{g}}{2\ \text{mol reactants}} \times \dfrac{1\ \text{kg}}{1000\ \cancel{g}} = 8.45 \times 10^{-7}\ \dfrac{\text{kg}}{\text{mol reactants}}$ then

$$E = mc^2 = \left(8.45 \times 10^{-7}\ \dfrac{\text{kg}}{\text{mol reactants}}\right)\left(2.9979 \times 10^8\ \dfrac{\text{m}}{\text{s}}\right)^2 = 7.59 \times 10^{10}\ \dfrac{\text{J}}{\text{mol reactants}}$$

Check: $179 + 0 = 179, 74 - 1 = 73$. The units (J) are correct.

20.79 **Given:** (a) Ru-114, (b) Ra-216, (c) Zn-58, and (d) Ne-31 **Find:** Write a nuclear equation for the most likely decay.
Conceptual Plan: Referring to the Valley of Stability graph in Figure 20.5, decide on the most likely decay mode depending on N/Z (too large = beta decay, too low = positron emission). → Write the symbol for the parent nuclide on the left side of the equation and the symbol for a particle on the right side. → Equalize the sum of the mass numbers and the sum of the atomic numbers on both sides of the equation by writing the appropriate mass number and atomic number for the unknown daughter nuclide. → Using the periodic table, deduce the identity of the unknown daughter nuclide from the atomic number and write its symbol.
Solution:

(a) Ru-114 ($N/Z = 1.6$) will undergo beta decay $^{114}_{44}\text{Ru} \rightarrow ^{?}_{?}? + ^{0}_{-1}\text{e}$ then $^{114}_{44}\text{Ru} \rightarrow ^{114}_{45}? + ^{0}_{-1}\text{e}$ then
$^{114}_{44}\text{Ru} \rightarrow ^{114}_{45}\text{Rh} + ^{0}_{-1}\text{e}$

(b) Ra-216 ($N/Z = 1.4$) will undergo positron emission $^{216}_{88}\text{Ra} \rightarrow ^{?}_{?}? + ^{0}_{+1}\text{e}$ then $^{216}_{88}\text{Ra} \rightarrow ^{216}_{87}? + ^{0}_{+1}\text{e}$ then
$^{216}_{88}\text{Ra} \rightarrow ^{216}_{87}\text{Fr} + ^{0}_{+1}\text{e}$

(c) Zn-58 ($N/Z = 0.9$) will undergo positron emission $^{58}_{30}\text{Zn} \rightarrow ^{?}_{?}? + ^{0}_{+1}\text{e}$ then $^{58}_{30}\text{Zn} \rightarrow ^{58}_{29}? + ^{0}_{+1}\text{e}$ then
$^{58}_{30}\text{Zn} \rightarrow ^{58}_{29}\text{Cu} + ^{0}_{+1}\text{e}$

(d) Ne-31 ($N/Z = 2$) will undergo beta decay $^{31}_{10}\text{Ne} \rightarrow ^{?}_{?}? + ^{0}_{-1}\text{e}$ then $^{31}_{10}\text{Ne} \rightarrow ^{31}_{11}? + ^{0}_{-1}\text{e}$ then $^{31}_{10}\text{Ne} \rightarrow ^{31}_{11}\text{Na} + ^{0}_{-1}\text{e}$

Check: (a) $114 = 114 + 0, 44 = 45 - 1$, and rhodium is atomic number 45. (b) $216 = 216 + 0, 88 = 87 + 1$, and francium is atomic number 87. (c) $58 = 58 + 0, 30 = 29 + 1$, and copper is atomic number 29. (d) $31 = 31 + 0, 10 = 11 - 1$, and sodium is atomic number 11.

20.81 **Given:** Bi-210, $t_{1/2} = 5.0$ days, 1.2 g Bi-210, 209.984105 amu, 5.5% absorbed
Find: beta emissions in 13.5 days and dose (in Ci)
Conceptual Plan: $t_{1/2} \rightarrow k$ then $m_{\text{Bi-210 0}}, t, k \rightarrow m_{\text{Bi-210 }t}$ then

$$t_{1/2} = \dfrac{0.693}{k} \qquad \ln N_t = -kt + \ln N_0$$

$g_0, g_t \rightarrow$ g decayed $\rightarrow$ mol decayed $\rightarrow$ beta decays then day $\rightarrow$ h $\rightarrow$ min $\rightarrow$ s then

$$g_0 - g_t = \text{g decayed} \quad \dfrac{1\ \text{mol Bi-210}}{209.984105\ \text{g Bi-210}} \quad \dfrac{6.022 \times 10^{23}\ \text{beta decays}}{1\ \text{mol Bi-210}} \quad \dfrac{24\ \text{h}}{1\ \text{day}} \quad \dfrac{60\ \text{min}}{1\ \text{h}} \quad \dfrac{60\ \text{s}}{1\ \text{min}}$$

beta decays, s → beta decays/s → Ci available → Ci absorbed

$$\text{take ratio} \qquad \frac{1\ Ci}{\frac{3.7 \times 10^{10}\ \text{decays}}{s}} \qquad \frac{5.5\ Ci\ \text{absorbed}}{100\ Ci\ \text{emitted}}$$

Solution: $t_{1/2} = \dfrac{0.693}{k}$ Rearrange to solve for k. $k = \dfrac{0.693}{t_{1/2}} = \dfrac{0.693}{5.0\ \text{days}} = 0.1386\ \text{day}^{-1}$ Because

$\ln m_{\text{Bi-210}\,t} = -kt + \ln m_{\text{Bi-210}\,0} = -(0.1386\ \text{day}^{-1})(13.5\ \text{day}) + \ln(1.2\ g) = -1.6888 \rightarrow$

$m_{\text{Bi-210}\,t} = e^{-1.6888} = 0.18474\ g$ then $g_0 - g_t = g\ \text{decayed} = 1.2\ g - 0.18474\ g = 1.0153\ g\ \text{Bi-210}$

then $1.0153\ \text{g Bi-210} \times \dfrac{1\ \text{mol Bi-210}}{209.984105\ \text{g Bi-210}} \times \dfrac{6.022 \times 10^{23}\ \text{beta decays}}{1\ \text{mol Bi-210}} = 2.9117 \times 10^{21}\ \text{beta decays} =$

2.9×10^{21} beta decays then $13.5\ \text{day} \times \dfrac{24\ \text{h}}{1\ \text{day}} \times \dfrac{60\ \text{min}}{1\ \text{h}} \times \dfrac{60\ s}{1\ \text{min}} = 1.1664 \times 10^{6}\ s$ then

$\dfrac{2.9117 \times 10^{21}\ \text{beta decays}}{1.1664 \times 10^{6}\ s} \times \dfrac{1\ Ci}{\frac{3.7 \times 10^{10}\ \text{decays}}{s}} = 6.7468 \times 10^{4}\ Ci\ \text{emitted} \times \dfrac{5.5\ Ci\ \text{absorbed}}{100\ Ci\ \text{emitted}} = 3700\ Ci$

Check: The units (decays and Ci) are correct. The amount that decays is large because the time is over three half-lives and we have a relatively large amount of the isotope. Because the decay is large, the dosage is large.

20.83   **Given:** Ra-226 (226.05402 amu) decays to Rn-224, $t_{1/2} = 1.6 \times 10^{3}$ yr, 25.0 g Ra-226, $T = 25.0\,°C$, $P = 1.0$ atm
Find: V of Rn-224 gas produced in 5.0 day
Conceptual Plan: day $\rightarrow$ yr then $t_{1/2} \rightarrow k$ then $m_{\text{Ra-226}\,0}\,t, k \rightarrow m_{\text{Ra-226}\,t}$

$$\frac{1\ \text{yr}}{365.24\ \text{day}} \qquad t_{1/2} = \frac{0.693}{k} \qquad \ln N_t = -kt + \ln N_0$$

then $g_0, g_t \rightarrow$ g decayed $\rightarrow$ mol decayed $\rightarrow$ mol Rn-224 formed then $°C \rightarrow K$ then $P, n, T \rightarrow V$

$$g_0 - g_t = g\ \text{decayed} \qquad \frac{1\ \text{mol Ra-226}}{226.05402\ \text{g Ra-226}} \quad \frac{1\ \text{mol Rn-224}}{1\ \text{mol Ra-226}} \qquad K = °C + 273.15 \qquad PV = nRT$$

Solution: $5.0\ \text{day} \times \dfrac{1\ \text{yr}}{365.24\ \text{day}} = 0.013690\ \text{yr}$ then $t_{1/2} = \dfrac{0.693}{k}$ Rearrange to solve for k.

$k = \dfrac{0.693}{t_{1/2}} = \dfrac{0.693}{1.6 \times 10^{3}\ \text{yr}} = 4.33125 \times 10^{-4}\ \text{yr}^{-1}$ Because

$\ln m_{\text{Ra-226}\,t} = -kt + \ln m_{\text{Ra-226}\,0} = -(4.33125 \times 10^{-4}\ \text{yr}^{-1})(0.013690\ \text{yr}) + \ln(25.0\ g) = 3.21887 \rightarrow$

$m_{\text{Ra-226}\,t} = e^{3.21887} = 24.999854\ g$ then
$g_0 - g_t = g\ \text{decayed} = 25.0\ g - 24.999854\ g = 0.000146\ g\ \text{Ra-226}$ then

$0.000146\ \text{g Ra-226} \times \dfrac{1\,\text{mol Ra-226}}{226.05402\ \text{g Ra-226}} \times \dfrac{1\ \text{mol Rn-224}}{1\ \text{mol Ra-226}} = 6.4586 \times 10^{-7}\ \text{mol Rn-224}$ then

$T = 25.0\,°C + 273.15 = 298.2\ K$ then $PV = nRT$. Rearrange to solve for V.

$V = \dfrac{nRT}{P} = \dfrac{6.4586 \times 10^{-7}\ \text{mol} \times 0.08206\,\dfrac{L \cdot atm}{mol \cdot K} \times 298.2\ K}{1.0\ \text{atm}} = 1.5804 \times 10^{-5}\ L = 2 \times 10^{-5}\ L$

Two significant figures are reported as requested in the problem.

Check: The units (L) are correct. The amount of gas is small because the time is so small compared to the half-life.

20.85   **Given:** $^{0}_{1}e + ^{0}_{-1}e \rightarrow 2^{0}_{0}\gamma$ **Find:** energy (in kJ/mol)
Conceptual Plan:
mass of products and reactants $\rightarrow$ mass defect (g) $\rightarrow$ kg $\rightarrow$ kg/mol $\rightarrow$ E (J/mol) $\rightarrow$ E (kJ/mol)

$$\text{mass defect} = \Sigma \text{mass of reactants} - \Sigma \text{mass of products} \qquad \frac{1\ \text{kg}}{1000\ g}\ 2\ \text{mol} \qquad E = mc^2 \qquad \frac{1\ \text{kJ}}{1000\ J}$$

Solution: mass defect $= \Sigma$mass of reactants $- \Sigma$mass of products $= (0.00055 \text{ g} + 0.00055 \text{ g}) - 0 \text{ g} =$

0.00110 g then $\dfrac{0.00110 \text{ g}}{2 \text{ mol}} \times \dfrac{1 \text{ kg}}{1000 \text{ g}} = 5.50 \times 10^{-7} \dfrac{\text{kg}}{\text{mol}}$ then

$E = mc^2 = \left(5.50 \times 10^{-7} \dfrac{\text{kg}}{\text{mol}}\right)\left(2.9979 \times 10^8 \dfrac{\text{m}}{\text{s}}\right)^2 = 4.94307 \times 10^{10} \dfrac{\text{J}}{\text{mol}} \times \dfrac{1 \text{ kJ}}{1000 \text{ J}} = 4.94 \times 10^7 \dfrac{\text{kJ}}{\text{mol}}$

Check: The units (kJ/mol) are correct. A large amount of energy is expected per mole of mass lost. The photon is in the gamma ray region of the electromagnetic spectrum.

20.87   **Given:** $^3\text{He} = 3.016030$ amu   **Find:** nuclear binding energy per atom
Conceptual Plan: $^A_Z X$, isotope mass $\rightarrow$ mass defect $\rightarrow$ nuclear binding energy per nucleon

mass defect $= Z(\text{mass } ^1_1\text{H}) + (A - Z)(\text{mass } ^1_0\text{n}) -$ mass of isotope $\dfrac{931.5 \text{ MeV}}{1 \text{ amu}}$

Solution: mass defect $= Z(\text{mass } ^1_1\text{H}) + (A - Z)(\text{mass } ^1_0\text{n}) -$ mass of isotope
He-3 mass defect $= 2(1.00783 \text{ amu}) + (3 - 2)(1.00866 \text{ amu}) - 3.016030 \text{ amu} = 0.00829 \text{ amu}$

and $0.00829 \text{ amu} \times \dfrac{931.5 \text{ MeV}}{1 \text{ amu}} = 7.72 \text{ MeV}$

Check: The units (MeV) are correct. The number of nucleons is small, so the MeV is not that large.

20.89   **Given:** ^{247}Es and five neutrons made by bombarding ^{238}U **Find:** identity of bombarding particle
Conceptual Plan: Begin with the symbols for the nuclides given. $\rightarrow$ **Equalize the sum of the mass numbers and the sum of the atomic numbers on both sides of the equation by writing the appropriate mass number and atomic number for the unknown daughter nuclide.** $\rightarrow$ **Using the periodic table, deduce the identity of the unknown nuclide from the atomic number and write its symbol.**
Solution:
$^{238}_{92}\text{U} + ^{?}_{?}? \rightarrow ^{247}_{99}\text{Es} + 5^1_0\text{n}$ then $^{238}_{92}\text{U} + ^{14}_{7}? \rightarrow ^{247}_{99}\text{Es} + 5^1_0\text{n}$ then $^{238}_{92}\text{U} + ^{14}_{7}\text{N} \rightarrow ^{247}_{99}\text{Es} + 5^1_0\text{n}$

Check: $238 + 14 = 247 + 5(1), 92 + 7 = 99 + 5(0)$, and nitrogen is atomic number 7.

20.91   **Given:** $t_{1/2}$ for decay of $^{238}\text{U} = 4.5 \times 10^9$ years, 1.6 g rock, 29 dis/s all radioactivity from U-238
Find: percent by mass ^{238}U in rock
Conceptual Plan: $t_{1/2} \rightarrow k$ and $s \rightarrow \min \rightarrow h \rightarrow \text{day} \rightarrow \text{yr}$ then Rate, $k \rightarrow N \rightarrow \text{mol } ^{238}\text{U} \rightarrow \text{g } ^{238}\text{U}$

$t_{1/2} = \dfrac{0.693}{k}$ $\dfrac{1 \min}{60 \text{ s}}$ $\dfrac{1 \text{ h}}{60 \min}$ $\dfrac{1 \text{ day}}{24 \text{ h}}$ $\dfrac{1 \text{ yr}}{365.24 \text{ day}}$ Rate $= kN \dfrac{1 \text{ mol dis}}{6.022 \times 10^{23} \text{ dis}}$ $\dfrac{238 \text{ g } ^{238}\text{U}}{1 \text{ mol } ^{238}\text{U}}$

then g ^{238}U, g rock $\rightarrow$ percent by mass ^{238}U

percent by mass $^{238}\text{U} = \dfrac{\text{g } ^{238}\text{U}}{\text{g rock}} \times 100\%$

Solution: $t_{1/2} = \dfrac{0.693}{k}$ Rearrange to solve for k. $k = \dfrac{0.693}{t_{1/2}} = \dfrac{0.693}{4.5 \times 10^9 \text{ yr}} = 1.54 \times 10^{-10} \text{ yr}^{-1}$ and

$1 \text{ s} \times \dfrac{1 \min}{60 \text{ s}} \times \dfrac{1 \text{ h}}{60 \min} \times \dfrac{1 \text{ day}}{24 \text{ h}} \times \dfrac{1 \text{ yr}}{365.24 \text{ day}} = 3.16889554 \times 10^{-8} \text{ yr}$. Rate $= kN$. Rearrange to solve for N.

$N = \dfrac{\text{Rate}}{k} = \dfrac{\dfrac{29 \text{ dis}}{3.16889554 \times 10^{-8} \text{ yr}}}{1.54 \times 10^{-10} \text{ yr}^{-1}} = 5.9425 \times 10^{18}$ dis then

$5.9425 \times 10^{18} \text{ dis} \times \dfrac{1 \text{ mol dis}}{6.022 \times 10^{23} \text{ dis}} \times \dfrac{238 \text{ g } ^{238}\text{U}}{1 \text{ mol } ^{238}\text{U}} = 2.3486 \times 10^{-3} \text{ g } ^{238}\text{U}$ then

percent by mass $^{238}\text{U} = \dfrac{\text{g } ^{238}\text{U}}{\text{g rock}} \times 100\% = \dfrac{2.3486 \times 10^{-3} \text{ g } ^{238}\text{U}}{1.6 \text{ g rock}} \times 100\% = 0.15\%$

Check: The units (%) are correct. The mass percent is low because the dis/s is low.

20.93   **Given:** $V = 1.50$ L, $P = 745$ mmHg, $T = 25.0\,°C$, 3.55% Ra-220 by volume, $t_{1/2} = 55.6$ s
      Find: number of alpha particles emitted in 5.00 min
      Conceptual Plan: mmg $\rightarrow$ atm and $°C \rightarrow K$ then $P, V, T \rightarrow n_{Total} \rightarrow n_{Ra-220}$ and min $\rightarrow$ s

$$\frac{1\ atm}{760\ mmHg} \qquad K = °C + 273.15 \qquad PV = nRT \qquad \frac{3.55\ mol\ Ra\text{-}220\ particles}{100\ mol\ gas\ particles} \qquad \frac{60\ s}{1\ min}$$

      then $t_{1/2} \rightarrow k$ then $n_{Ra\text{-}220\ 0}, t, k \rightarrow n_{Ra\text{-}220\ t} \rightarrow$ **number of particles remaining** $\rightarrow$ **particles emitted**

$$t_{1/2} = \frac{0.693}{k} \qquad\qquad \ln N_t = -kt + \ln N_0 \qquad \frac{6.022 \times 10^{23}\ particles}{1\ mol}$$

      Solution: $745\ \cancel{mmHg} \times \dfrac{1\ atm}{760\ \cancel{mmHg}} = 0.9802632$ atm and $T = 25.0\,°C + 273.15 = 298.2$ K then

      $PV = nRT$. Rearrange to solve for n.

$$n = \frac{PV}{RT} = \frac{0.9802632\ \cancel{atm} \times 1.50\ \cancel{L}}{0.08206 \dfrac{\cancel{L} \cdot \cancel{atm}}{mol \cdot \cancel{K}} \times 298.2\ \cancel{K}} = 0.06008898\ \text{mol gas particles}$$

      then $0.06008898\ \cancel{\text{mol gas particles}} \times \dfrac{3.55\ mol\ Ra\text{-}220\ particles}{100\ \cancel{\text{mol gas particles}}} = 0.002133159\ \text{mol Ra-220 particles}$

      $5.00\ \cancel{min} \times \dfrac{60\ s}{1\ \cancel{min}} = 300.\ \text{s}$ then $t_{1/2} = \dfrac{0.693}{k}$. Rearrange to solve for k.

$$k = \frac{0.693}{t_{1/2}} = \frac{0.693}{55.6\ s} = 0.01246403\ s^{-1}\ \text{Because}$$

      $\ln m_{Ra\text{-}220\ t} = -kt + \ln m_{Ra\text{-}220\ 0} = -(0.01246403\ \cancel{s^{-1}})(300.\ \cancel{s}) + \ln (0.002133159\ mol) = -9.889360 \rightarrow$
      $m_{Ra\text{-}220\ t} = e^{-9.889360} = 5.071139 \times 10^{-5}$ mol alpha particles remaining
      The number of alpha particles emitted would be the difference between this and the initial number of moles.
      $0.002133158\ mol - 0.00005071139\ mol = 0.002082447\ mol$

$$0.002082447\ \cancel{mol} \times \frac{6.022 \times 10^{23}\ particles}{1\ \cancel{mol}} = 1.254050 \times 10^{21}\ particles = 1.25 \times 10^{21}\ particles$$

      Check: The units (particles) are correct. The number of particles is far less than a mole because we have far less than
      a mole of gas.

20.95   **Given:** $^0_{+1}e + ^0_{-1}e \rightarrow 2^0_0\gamma$ **Find:** wavelength of gamma ray photons
      Conceptual Plan:
      mass of products and reactants $\rightarrow$ **mass defect (g)** $\rightarrow$ **kg** $\rightarrow$ **kg/mol** $\rightarrow$ E **(J/mol)** $\rightarrow$ E **(kJ/mol)**

$$\text{mass defect} = \Sigma \text{mass of reactants} - \Sigma \text{mass of products} \qquad \frac{1\ kg}{1000\ g} \quad 2\ mol \qquad E = mc^2 \qquad \frac{1\ kJ}{1000\ J}$$

      This energy is for 2 moles of γ**, so** E**(J/2 mol** γ**)** $\rightarrow E$**(J/**γ **photon)** $\rightarrow \lambda$.

$$\frac{1\ mol\ \gamma}{6.022 \times 10^{23}\ \gamma\ photons} \qquad E = \frac{hc}{\lambda}$$

      Solution: mass defect $= \Sigma$mass of reactants $- \Sigma$mass of products $= (0.00055\ g + 0.00055\ g) - 0\ g = 0.00110\ g$

      then $\dfrac{0.00110\ \cancel{g}}{2\ mol} \times \dfrac{1\ kg}{1000\ \cancel{g}} = 5.50 \times 10^{-7} \dfrac{kg}{mol}$ then

$$E = mc^2 = \left(5.50 \times 10^{-7} \frac{kg}{mol}\right)\left(2.9979 \times 10^8 \frac{m}{s}\right)^2 = 4.94307 \times 10^{10} \frac{J}{mol} \times \frac{1\ kJ}{1000\ J} = 4.94 \times 10^7 \frac{kJ}{mol}$$

$$E = 4.94307 \times 10^{10} \frac{J}{mol\ \gamma} \times \frac{1\ mol\ \gamma}{6.022 \times 10^{23}\ \gamma\ photons} = 8.20835 \times 10^{-14} \frac{J}{\gamma\ photons} \text{ then } E = \frac{hc}{\lambda} \quad \text{Rearrange to solve}$$

      for λ. $\lambda = \dfrac{hc}{E} = \dfrac{(6.626 \times 10^{-34}\ \cancel{J} \cdot \cancel{s})\left(2.9979 \times 10^8 \dfrac{m}{\cancel{s}}\right)}{8.20835 \times 10^{-14} \dfrac{\cancel{J}}{\gamma\ photons}} = 2.42 \times 10^{-12}\ m = 2.42\ pm$

Check: The units (m or pm) are correct. A large amount of energy is expected per mole of mass lost. The photon is in the gamma ray region of the electromagnetic spectrum.

20.97 **Given:** $^2_1H + ^2_1H \rightarrow ^3_2He + ^1_0n$ releases 3.3 MeV; $^2_1H + ^2_1H \rightarrow ^3_1H + ^1_1p$ releases 4.0 MeV

Find: the energy change for $^3_2He + ^1_0n \rightarrow ^3_1H + ^1_1p$ and why this can happen at a much lower temperature

Conceptual Plan: Use Hess's law to calculate the energy change and give the two reactions.

Solution:

$$^3_2He + ^1_0n \rightarrow ^2_1H + ^2_1H \qquad \Delta E = 3.3 \text{ MeV}$$
$$^2_1H + ^2_1H \rightarrow ^3_1H + ^1_1p \qquad \Delta E = -4.0 \text{ MeV}$$
$$\overline{^3_2He + ^1_0n \rightarrow ^3_1H + ^1_1p \qquad \Delta E = -0.7 \text{ MeV}}$$

The energy change is much less, and there is no coulombic barrier for collision with a neutron; so the process can occur at lower temperatures.

Check: The units (MeV) are correct. Because one reaction releases energy and one requires energy, the resulting energy change is much smaller in magnitude.

Challenge Problems

20.99 (a) **Given:** 72,500 kg Al(s) and $10 \text{ Al}(s) + 6 \text{ NH}_4\text{ClO}_4(s) \rightarrow 4 \text{ Al}_2\text{O}_3(s) + 2 \text{ AlCl}_3(s) + 12 \text{ H}_2\text{O}(g) + 3 \text{ N}_2(g)$ and 608,000 kg O$_2(g)$ that reacts with hydrogen to form gaseous water

Find: energy generated (ΔH°_{rxn})

Conceptual Plan: Write a balanced reaction for O$_2(g)$ then

$$\Delta H^\circ_{rxn} = \Sigma n_P \Delta H^\circ_f(\text{products}) - \Sigma n_r \Delta H^\circ_f(\text{reactants}) \text{ then}$$

kg $\rightarrow$ g $\rightarrow$ mol $\rightarrow$ energy then add the results from the two reactions.

$$\frac{1000 \text{ g}}{1 \text{ kg}} \qquad M \qquad \Delta H^\circ_{rxn}$$

Solution:

Reactant/Product	ΔH°_f(kJ/mol from Appendix IIB)
Al(s)	0.0
NH$_4$ClO$_4(s)$	−295
Al$_2$O$_3(s)$	−1675.7
AlCl$_3(s)$	−704.2
H$_2$O(g)	−241.8
N$_2(g)$	0.0

Be sure to pull data for the correct formula and phase.

$$\Delta H^\circ_{rxn} = \Sigma n_P \Delta H^\circ_f(\text{products}) - \Sigma n_r \Delta H^\circ_f(\text{reactants})$$
$$= [4(\Delta H^\circ_f(\text{Al}_2\text{O}_3(s))) + 2(\Delta H^\circ_f(\text{AlCl}_3(s))) + 12(\Delta H^\circ_f(\text{H}_2\text{O}(g))) + 3(\Delta H^\circ_f(\text{N}_2(g)))] +$$
$$- [10(\Delta H^\circ_f(\text{Al}(s))) + 6(\Delta H^\circ_f(\text{NH}_4\text{ClO}_4(s)))]$$
$$= [4(-1675.7 \text{ kJ}) + 2(-704.2 \text{ kJ}) + 12(-241.8 \text{ kJ}) + 3(0.0 \text{ kJ})] - [10(0.0 \text{ kJ}) + 6(-295 \text{ kJ})]$$
$$= [-11012.8 \text{ kJ}] - [-1770. \text{ kJ}]$$
$$= -9242.8 \text{ kJ}$$

then $72,500 \text{ kg Al} \times \dfrac{1000 \text{ g Al}}{1 \text{ kg Al}} \times \dfrac{1 \text{ mol Al}}{26.98 \text{ g Al}} \times \dfrac{9242.8 \text{ kJ}}{10 \text{ mol Al}} = 2.483703 \times 10^9 \text{ kJ}$

balanced reaction: $H_2(g) + \frac{1}{2}O_2(g) \rightarrow H_2O(g) \ \Delta H^\circ_{rxn} = \Delta H^\circ_f(H_2O(g)) = -241.8 \text{ KJ/mol}$ then

$608,000 \text{ kg O}_2 \times \dfrac{1000 \text{ g O}_2}{1 \text{ kg O}_2} \times \dfrac{1 \text{ mol O}_2}{32.00 \text{ g O}_2} \times \dfrac{241.8 \text{ kJ}}{0.5 \text{ mol O}_2} = 9.1884 \times 10^9 \text{ kJ}$ So the total is

$2.483703 \times 10^9 \text{ kJ} + 9.1884 \times 10^9 \text{ kJ} = 1.1672103 \times 10^{10} \text{ kJ} = 1.167 \times 10^{10} \text{ kJ}.$

Check: The units (kJ) are correct. The answer is very large because the reactions are very exothermic and the weight of reactants is so large.

(b)   **Given:** $^1_1H + ^0_{-1}p + ^0_{+1}e \rightarrow ^0_0\gamma$ **Find:** mass of antimatter to give same energy as in part (a)

       Conceptual Plan: Because the reaction is an annihilation reaction, no matter will be left; so the mass of antimatter is the same as the mass of the hydrogen. So kJ $\rightarrow$ J $\rightarrow$ kg $\rightarrow$ g

$$\frac{1000\ J}{1\ kJ} \qquad E = mc^2 \qquad \frac{1000\ g}{1\ kg}$$

       Solution: $1.1672103 \times 10^{10}\ kJ \times \dfrac{1000\ J}{1\ kJ} = 1.1672103 \times 10^{13}\ J$. Because $E = mc^2$, rearrange to solve for m.

$$m = \frac{E}{c^2} = \frac{1.1672103 \times 10^{13}\ kg\frac{m^2}{s^2}}{\left(2.9979 \times 10^8 \frac{m}{s}\right)^2} = 1.299 \times 10^{-4}\ kg \times \frac{1000\ g}{1\ kg} = 0.1299\ g \text{ total matter, } 0.0649\ g \text{ each of}$$

matter and antimatter

Check: The units (g) are correct. A small mass is expected because nuclear reactions generate a large amount of energy.

20.101   **Given:** $^{235}_{92}U \rightarrow ^{206}_{82}Pb$ and $^{232}_{90}Th \rightarrow ^{206}_{82}Pb$ **Find:** decay series

       Conceptual Plan: Write the species given on the appropriate side of the equation. $\rightarrow$ Equalize the sum of the mass numbers and the sum of the atomic numbers on both sides of the equation by writing the stoichiometric coefficient in front of the desired species.

       Solution: $^{235}_{92}U \rightarrow ^?_{82}Pb + ?^4_2He + ?^0_{-1}e$ becomes $^{235}_{92}U \rightarrow ^{207}_{82}Pb + 7^4_2He + 4^0_{-1}e$.

       $^{232}_{90}Th \rightarrow ^?_{82}Pb + ?^4_2He + ?^0_{-1}e$ becomes $^{232}_{90}Th \rightarrow ^{208}_{82}Pb + 6^4_2He + 4^0_{-1}e$.

       U-235 forms Pb-207 in 7 α-decays and 4 β-decays, and Th-232 forms Pb-208 in 6 α-decays and 4 β-decays.

       Check: $235 = 207 + 7(4) + 4(0)$, and $92 = 82 + 7(2) + 4(-1)$. $232 = 208 + 6(4) + 4(0)$, and $90 = 82 + 6(2) + 4(-1)$. The mass of the Pb can be determined because alpha particles are large and need to be included as integer values. To make the masses balance requires more alpha particles than can be supported by the number of protons in the total equation. To account for this, an appropriate number of beta decays are added.

20.103   **Given:** $H^{38}_{17}Cl(g) \rightarrow ^{38}_{18}Ar(g) + ^0_{-1}e + \frac{1}{2}H_2(g)$, 0.40 mol $H^{38}_{17}Cl(g)$; $V = 6.24$ L, $P = 1650$ mmHg, $t_{1/2} = 80.0$ min

       Find: T

       Conceptual Plan: Because $t = 2\,t_{1/2}$, then $N_{Cl\text{-}38\,t} = \frac{1}{4}N_{Cl\text{-}38\,0} \rightarrow n_{gas\,t}$ then mmHg $\rightarrow$ atm and

$$n_{gas\,t} = 3/2\,n_{HCl\,0} - 1/2\,n_{HCl\,t} \qquad \frac{1\ atm}{760\ mmHg}$$

       $P, V, n_{gas\,t} \rightarrow T$,

$$PV = nRT$$

       Solution: Because $t = 2\,t_{1/2}$, then $N_{Cl\text{-}38\,t} = \frac{1}{4}N_{Cl\text{-}38\,0} = \frac{1}{4}(0.40\ mol) = 0.10$ mol. As the HCl disintegrates, it produces argon gas, beta particles, and hydrogen gas, with a ratio of three particles produced (2 Ar, 1 H_2) for every two HCl molecules that decay. There were initially 0.40 mole of undisintegrated gas; now 0.10 mole HCl remains. So the total number of gas particles now in the container is

       $n_{gas\,t} = 3/2\,n_{HCl\,0} - 1/2\,n_{HCl\,t} = 3/2(0.40\ mol) - 1/2\,(0.10\ mol) = 0.55$ mol gas then $PV = nRT$. Rearrange to

       solve for T. $T = \dfrac{PV}{nR} = \dfrac{1650\ mmHg \times \dfrac{1\ atm}{760\ mmHg} \times 6.24\ L}{0.55\ mol \times 0.08206\ \dfrac{L \cdot atm}{mol \cdot K}} = 300.17\ K = 3.0 \times 10^2\ K$

       Check: The units (K) are correct. The temperature is reasonable considering the volume of a gas at STP and the fact that most of the initial 0.40 mol has decomposed.

Conceptual Problems

20.105   **Given:** $^{21}_{9}F \rightarrow ^{?}_{?}? + ^{0}_{-1}e$ **Find:** missing nucleus
Conceptual Plan: Write the species given on the appropriate side of the equation. $\rightarrow$ **Equalize the sum of the mass numbers and the sum of the atomic numbers on both sides of the equation by writing the stoichiometric coefficient in front of the desired species.**
Solution: $^{21}_{9}F \rightarrow ^{?}_{?}? + ^{0}_{-1}e$ becomes $^{21}_{9}F \rightarrow ^{21}_{10}Ne + ^{0}_{-1}e$.

Check: $21 = 21 + 0$, and $9 = 10 - 1$. Neon is atomic number 10, and no other species are needed to balance the equation.

20.107   Nuclide A is more dangerous because the half-life is shorter (18.5 days); so it decays faster.

20.109   Iodine is used by the thyroid gland to make hormones. Normally, we ingest iodine in foods, especially iodized salt. The thyroid gland cannot tell the difference between stable and radioactive iodine and will absorb both. KI tablets work by blocking radioactive iodine from entering the thyroid. When a person takes KI, the stable iodine in the tablet gets absorbed by the thyroid. Because KI contains so much stable iodine, the thyroid gland becomes "full" and cannot absorb any more iodine—either stable or radioactive—for the next 24 hours.

Questions for Group Work

20.111   See Table 20.1; All processes conserve total mass number and total charge, but they differ in the specific particles consumed and produced.

20.113 $^{238}_{92}U \rightarrow ^{234}_{90}Th + ^{4}_{2}He$; $^{234}_{90}Th \rightarrow ^{234}_{91}Pa + ^{0}_{-1}e$; $^{234}_{91}Pa \rightarrow ^{234}_{92}U + ^{0}_{-1}e$.

Data Analysis and Interpretation

20.115   (a)   Looking at the graph, we see that at 200 minutes there is 0.35 g Tc–99m.
(c)   Looking at the graph, we see that the amount of Tc-99m has decreased from 0.50 g to 0.17 g in 600 minutes or 10 hours. So, 2.0 μg will decrease to 0.68 μg.

21 Organic Chemistry

Review Questions

21.1 Most common smells are caused by organic molecules, molecules containing carbon combined with several other elements including hydrogen, nitrogen, oxygen, and sulfur.

21.3 Carbon is unique in the vast number of compounds it can form. Life needs diversity to exist, and carbon has the ability to form more compounds than any other element.

21.5 Silicon can form chains with itself. However, silicon's affinity for oxygen—the Si—O bond is 142 kJ/mol stronger than the Si—Si bond—coupled with the prevalence of oxygen in our atmosphere means that silicon–silicon chains are readily oxidized to form silicates (the silicon–oxygen compounds that compose a significant proportion of minerals). By contrast, the C—C bond (347 kJ/mol) and the C—O bond (359 kJ/mole) are nearly the same strength, allowing carbon chains to exist relatively peacefully in an oxygen-rich environment. In other words, silicon's affinity for oxygen robs it of the rich diversity that catenation provides to carbon.

21.7 Hydrocarbons—compounds that contain only carbon and hydrogen—are the simplest organic compounds. However, because of the uniqueness of carbon, many different kinds of hydrocarbons exist. Hydrocarbons are commonly used as fuels. Candle wax, oil, gasoline, liquid propane (LP) gas, and natural gas are all composed of hydrocarbons. Hydrocarbons are also the starting materials in the synthesis of many different consumer products, including fabrics, soaps, dyes, cosmetics, drugs, plastic, and rubber.

21.9 A structural formula shows not only the numbers of each kind of atoms, but also how the atoms are bonded together. The condensed structural formula groups the hydrogen atoms together with the carbon atom to which they are bonded. Condensed structural formulas may show some of the bonds or none at all. The carbon skeleton formula shows the carbon–carbon bonds only as lines. Each end or bend of a line represents a carbon atom bonded to as many hydrogen atoms as necessary to form a total of four bonds. Space-filling or ball-and-stick models are three-dimensional representations that show how atoms are bonded together. The space-filling models show the relative size of the atoms that are bonded together.

21.11 Optical isomers are two molecules that are nonsuperimposable mirror images of each other. Optical isomers contain a carbon atom with four different substituent groups. Most properties of optical isomers are the same. The differences appear when they interact with polarized light and when they are placed in environments that can interact with one isomer and not the other.

21.13 Alkanes are often called saturated hydrocarbons because they are saturated (loaded to capacity) with hydrogen. Unsaturated hydrocarbons contain multiple bonds and, therefore, contain fewer hydrogen atoms.

21.15 Double bonds are composed of a sigma bond and a pi bond. Sigma bonds allow for free rotation about the bond. The pi bond restricts the rotational motion about the bond, resulting in isomers that are referred to as geometric isomers (same connectivity but different geometric positions). The two isomers are designated as *cis* (meaning "same side") and *trans* (meaning "opposite sides"). Cis–trans isomerism is common in alkenes and results in different boiling points and melting points and in different abilities to interact with rigid molecules.

21.17   The most common types of reactions of alkanes are as follows:

- Hydrocarbon combustion in the presence of oxygen to form carbon dioxide and water. An example is $C_8H_{18}(l) + \frac{25}{2} O_2(g) \rightarrow 8\,CO_2(g) + 9\,H_2O(g)$.

- Halogen substitution in the presence of a halogen gas to form a halogenated alkane and a hydrogen halide. An example is $CH_4(g) + Cl_2(g) \rightarrow CH_3Cl(g) + HCl(g)$.

21.19   The most common types of reactions of alkenes are as follows:

- Hydrocarbon combustion in the presence of oxygen to form carbon dioxide and water. An example is $C_2H_4(g) + 3\,O_2(g) \rightarrow 2\,CO_2(g) + 2\,H_2O(g)$.
- Hydrogenation in the presence of hydrogen to form an alkane. An example is $C_2H_4(g) + H_2(g) \rightarrow C_2H_6(g)$.
- Halogen addition in the presence of a halogen gas or hydrogen halide to form a halogenated alkane. An example is $CH_2CHCH_2CH_3(g) + HCl(g) \rightarrow CH_3CClHCH_2CH_3(g)$.

21.21   The structure of benzene, C_6H_6, is a six-member ring where three pi bonds are delocalized on all six of the C—C bonds that form the ring. Benzene rings are represented as one or both of the Kekulé structures,

, in a shorthand notation,

or as a ball-and-stick diagram with molecular orbitals,

21.23   A functional group is a characteristic atom or group of atoms that is inserted into a hydrocarbon. Examples of functional groups are alcohols (—OH), halogens (—X, where X = F, Cl, Br, and I), and carboxylic acids (—COOH).

21.25   In organic chemistry, we think of oxidation and reduction from the point of view of the carbon atoms in the organic molecule. Thus, oxidation is the gaining of oxygen or the losing of hydrogen by a carbon atom. Reduction, then, is the loss of oxygen or the gaining of hydrogen by a carbon atom.

21.27   Aldehydes and ketones have the general structural formulas of RCHO and RCOR′, respectively, where R and R′ represent hydrocarbon groups. Both aldehydes and ketones contain a carbonyl group, an oxygen double-bonded to a carbon atom $\left(\begin{smallmatrix} O \\ \| \\ -C- \end{smallmatrix}\right)$. Ketones have an R group attached to both sides of the carbonyl, while aldehydes have one R group and a hydrogen atom. An exception is formaldehyde, which is an aldehyde with two H atoms attached to the carbonyl group. Acetaldehyde is $CH_3-\overset{\overset{\displaystyle O}{\|}}{C}-H$ and propanone is $CH_3-\overset{\overset{\displaystyle O}{\|}}{C}-CH_3$.

21.29   Carboxylic acids and esters have the following general structural formulas of RCOOH and RCOOR′, respectively, where R and R′ represent hydrocarbon groups. Both carboxylic acids and esters contain a carbonyl group, an oxygen double-bonded to a carbon atom $\left(\begin{smallmatrix} O \\ \| \\ -C- \end{smallmatrix}\right)$. In carboxylic acids, the molecule ends in $-\overset{\overset{\displaystyle O}{\|}}{C}-O-H$ and esters have $-\overset{\overset{\displaystyle O}{\|}}{C}-O-$ inserted into a hydrocarbon chain. Acetic acid is $CH_3-\overset{\overset{\displaystyle O}{\|}}{C}-O-H$ and ethyl ethanoate is $CH_3-\overset{\overset{\displaystyle O}{\|}}{C}-O-CH_2-CH_3$.

21.31   Ethers have the general structural formula of ROR′, where R and R′ represent hydrocarbon groups. The two hydrocarbon groups are linked through an oxygen atom. Diethyl ether is $CH_3—CH_2—O—CH_2—CH_3$, and ethyl propyl ether is $CH_3—CH_2—O—CH_2—CH_2—CH_3$.

Problems by Topic

Hydrocarbons

21.33   (a) C_5H_{12} is an alkane because it follows the general formula C_nH_{2n+2}, where $n = 5$.
         (b) C_3H_6 is an alkene because it follows the general formula C_nH_{2n}, where $n = 3$.
         (c) C_7H_{12} is an alkyne because it follows the general formula C_nH_{2n-2}, where $n = 7$.
         (d) $C_{11}H_{22}$ is an alkene because it follows the general formula C_nH_{2n}, where $n = 11$.

21.35

21.37   (a)   No, this molecule will not because all four of the substituents are Cl atoms.
         (b)   Yes, this molecule will because the third carbon has four different substituent groups.
         (c)   Yes, this molecule will because the second carbon has four different substituent groups.
         (d)   No, each carbon has, at most, three different substituent groups.

21.39   (a)   They are enantiomers because they are nonsuperimposable mirror images of each other.
         (b)   They are the same because you can get the second molecule by rotating the first molecule counterclockwise about the $C—H$ bond.
         (c)   They are enantiomers because they are nonsuperimposable mirror images of each other.

Alkanes

21.41   **Given:** alkane structures   **Find:** name
         Conceptual Plan: Count the number of carbon atoms in the longest continuous carbon chain to determine the base name of the compound. Find the prefix corresponding to this number of atoms in Table 21.5 and add the ending *-ane* to form the base name. → Consider every branch from the base chain to be a substituent. Name each substituent according to Table 21.6. → Beginning with the end closest to the branching, number the base chain and assign a number to each substituent. (If two substituents occur at equal distances from each end, go to the next substituent to determine from which end to start numbering.) → Write the name of the compound in the following format: (subst. #)-(subst. name)(base name). → If there are two or more substituents, give each one a number and list them alphabetically with hyphens between words and numbers. → If a compound has

two or more identical substituents, designate the number of identical substituents with the prefix *di-* (2), *tri-* (3), or *tetra-* (4) before the substituent's name. Separate the numbers with a comma, indicating the positions of the substituents relative to each other. The prefixes are not taken into account when alphabetizing.
Solution:

(a) $CH_3-CH_2-CH_2-CH_2-CH_3$ has five carbons as the longest continuous chain. The prefix for 5 is penta-. There are no substituent groups on any of the carbons, so the name is pentane. Because this is a straight-chain molecule, it can be more specifically named *n*-pentane.

(b) $\boxed{CH_3-CH_2-CH-CH_3}$ has four carbons as the longest continuous chain. The prefix for 4 is but-, and
$\qquad\qquad\qquad\qquad |$
$\qquad\qquad\qquad CH_3$

the base name is butane. The only substituent group is a methyl group. $\boxed{C^4H_3-C^3H_2-C^2H-C^1H_3}$ If
$\qquad\qquad\qquad\qquad\qquad\qquad\qquad\qquad\qquad\qquad\qquad\qquad\qquad\qquad |$
$\qquad\qquad\qquad\qquad\qquad\qquad\qquad\qquad\qquad\qquad\qquad\qquad\qquad\boxed{CH_3}$

we start numbering the chain at the end closest to the methyl group, the methyl substituent is assigned the number 2. The name of the compound is 2-methylbutane.

$\qquad\qquad\qquad\qquad\qquad\qquad\qquad CH_3$
$\qquad\qquad\qquad\qquad\qquad\qquad\qquad |$
$\qquad\qquad\qquad CH_3\qquad\quad CH-CH_3$
$\qquad\qquad\qquad |\qquad\qquad\quad |$
(c) $\boxed{CH_3-CH-CH_2-CH-CH_2-CH_2-CH_3}$ has seven carbons as the longest continuous chain. The prefix for 7 is hept-, and the base name is heptane. The substituent groups are methyl and isopropyl groups.

$\qquad\qquad\qquad\qquad\qquad\qquad\qquad\quad\boxed{CH_3}$
$\qquad\qquad\qquad\qquad\qquad\qquad\qquad\qquad |$
$\qquad\qquad\quad\boxed{CH_3}\qquad\quad\boxed{CH-CH_3}$
$\boxed{C^1H_3-C^2H-C^3H_2-C^4H-C^5H_2-C^6H_2-C^7H_3}$ If we start numbering the chain at the end closest
to the methyl group, the methyl substituent is assigned the number 2 and the isopropyl group is assigned the number 4. Because *i* comes before *m*, the name of the compound is 4-isopropyl-2-methylheptane.

(d) $\boxed{CH_3-CH-CH_2-CH-CH_2-CH_3}$ has six carbons as the longest continuous chain. The prefix
$\qquad\qquad\quad |\qquad\qquad\quad |$
$\qquad\qquad CH_3\qquad\quad CH_2-CH_3$
for 6 is hex-, and the base name is hexane. The only substituent groups are methyl and ethyl groups.

$\boxed{C^1H_3-C^2H-C^3H_2-C^4H-C^5H_2-C^6H_3}$ If we start numbering the chain at the end closest to the
$\qquad\qquad |\qquad\qquad\quad |$
$\qquad\boxed{CH_3}\qquad\quad\boxed{CH_2-CH_3}$

methyl group, the methyl substituent is assigned the number 2 and the ethyl group is assigned the number 4. Because *e* comes before *m*, the name of the compound is 4-ethyl-2-methylhexane.

21.43 **Given:** alkane names   **Find:** structure
Conceptual Plan: Find the number of carbon atoms corresponding to the prefix of the base name in Table 21.5. →
Draw the base chain and number the carbons from left to right. → **Using Table 21.6 and the prefix *di-* (2), *tri-* (3), or *tetra-* (4) before the substituent's name, determine each substituent.** → **Add the substituent to the proper carbon position in the chain.** → **Add hydrogen atoms to the base chain so that each carbon has four bonds.**
Solution:
(a) 3-ethylhexane. The base name hexane indicates that there are six carbon atoms in the base chain.
$C^1-C^2-C^3-C^4-C^5-C^6$. 3-ethyl designates that a $-CH_2CH_3$ group is in the third position.
$C^1-C^2-C^3-C^4-C^5-C^6$. Add hydrogens to the base chain to get the final molecule
$\qquad\qquad\boxed{CH_2-CH_3}$
$\qquad\qquad\quad |$
$CH_3-CH_2-CH-CH_2-CH_2-CH_3$.
$\qquad\qquad\quad |$
$\qquad\qquad CH_2-CH_3$

(b) 3-ethyl-3-methylpentane. The base name pentane indicates that there are five carbon atoms in the base chain. $C^1 - C^2 - C^3 - C^4 - C^5$. 3-ethyl designates that a $-CH_2CH_3$ group is in the third position, and

3-methyl designates that a $-CH_3$ group is in the third position. $C^1 - C^2 - \boxed{\overset{\boxed{CH_3}}{\underset{\boxed{CH_2 - CH_3}}{C^3}}} - C^4 - C^5$. Add hydrogens to

the base chain to get the final molecule $CH_3 - CH_2 - \overset{CH_3}{\underset{CH_2 - CH_3}{C}} - CH_2 - CH_3$.

(c) 2,3-dimethylbutane. The base name butane indicates that there are four carbon atoms in the base chain. $C^1 - C^2 - C^3 - C^4$. 2,3-dimethyl designates $-CH_3$ groups in the second and third positions. $C^1 - \underset{\boxed{CH_3}}{C^2} - \underset{\boxed{CH_3}}{C^3} - C^4$. Add hydrogens to the base chain to get the final molecule $CH_3 - \underset{CH_3}{CH} - \underset{CH_3}{CH} - CH_3$.

(d) 4,7-diethyl-2,2-dimethylnonane. The base name nonane indicates that there are nine carbon atoms in the base chain. $C^1 - C^2 - C^3 - C^4 - C^5 - C^6 - C^7 - C^8 - C^9$. 4,7-diethyl designates $-CH_2CH_3$ groups in the fourth and seventh positions, and 2,2-dimethyl designates two $-CH_3$ groups in the second position. $C^1 - \underset{\boxed{CH_3}}{\overset{\boxed{CH_3}}{C^2}} - C^3 - \underset{\boxed{CH_2 - CH_3}}{C^4} - C^5 - C^6 - \underset{\boxed{CH_2 - CH_3}}{C^7} - C^8 - C^9$. Add hydrogens to the base chain to get the final molecule $CH_3 - \underset{CH_3}{\overset{CH_3}{C}} - CH_2 - \underset{CH_2 - CH_3}{CH} - CH_2 - CH_2 - \underset{CH_2 - CH_3}{CH} - CH_2 - CH_3$.

21.45 Hydrocarbon combustion in the presence of oxygen forms carbon dioxide and water. Balance the reaction.
(a) $CH_3CH_2CH_3(g) + 5\,O_2(g) \rightarrow 3\,CO_2(g) + 4\,H_2O(g)$
(b) $CH_3CH_2CH{=}CH_2(g) + 6\,O_2(g) \rightarrow 4\,CO_2(g) + 4\,H_2O(g)$
(c) $2\,CH{\equiv}CH(g) + 5\,O_2(g) \rightarrow 4\,CO_2(g) + 2\,H_2O(g)$

21.47 Halogen substitution reactions remove a hydrogen atom from the alkane, replace it with a halogen atom, and generate a hydrogen halide. Assume one substitution on the hydrocarbon.
(a) $CH_3CH_3 + Br_2 \rightarrow CH_3CH_2Br + HBr$. Only one carbon-containing product is possible because the $C - C$ bond freely rotates.
(b) $CH_3CH_2CH_3 + Cl_2 \rightarrow [\,CH_3CH_2CH_2Cl$ and $CH_3CHClCH_3\,] + HCl$. Two carbon-containing products are possible (on either the end carbon or the middle carbon) because the $C - C$ bond freely rotates and the end carbons are equivalent before reaction.
(c) $CH_2Cl_2 + Br_2 \rightarrow CHBrCl_2 + HBr$. Only one carbon-containing product is possible because halogen substitution reactions only remove hydrogen atoms.

(d) $\underset{CH_3}{CH_3 - CH - CH_3} + Cl_2 \longrightarrow \left[\underset{CH_3}{\overset{H}{CH_3 - C - CH_2Cl}} \text{ and } \underset{CH_3}{\overset{Cl}{CH_3 - C - CH_3}} \right] + HCl$

Two carbon-containing products are possible (on either an end carbon or the middle carbon) because the $C - C$ bond freely rotates and the end carbons are all equivalent before reaction.

Alkenes and Alkynes

21.49 CH_2=CH—CH_2—CH_2—CH_2—CH_3, CH_3—CH=CH—CH_2—CH_2—CH_3, and
CH_3—CH_2—CH=CH—CH_2—CH_3 are the only structural isomers. Remember that cis–trans isomerism generates geometric isomers, not structural isomers.

21.51   **Given:** alkene structures   **Find:** name
     Conceptual Plan: Count the number of carbon atoms in the longest continuous carbon chain that contains the multiple bond to determine the base name of the compound. Find the prefix corresponding to this number of atoms in Table 21.5 and add the ending *-ene* to form the base name. → **Consider every branch from the base chain to be a substituent. Name each substituent according to Table 21.6.** → **Beginning with the end closest to the multiple bond, number the base chain and assign a number to each substituent.** → **Write the name of the compound in the following format: (subst. #)-(subst. name)(base name).** → **If there are two or more substituents, give each one a number and list them alphabetically with hyphens between words and numbers.** → **If a compound has two or more identical substituents, designate the number of identical substituents with the prefix *di-* (2), *tri-* (3), or *tetra-* (4) before the substituent's name. Separate the numbers with a comma, indicating the positions of the substituents relative to each other. The prefixes are not taken into account when alphabetizing.**
     Solution:

(a) $\boxed{CH_2\text{=}CH\text{—}CH_2\text{—}CH_3}$ has four carbons as the longest continuous chain. The prefix for 4 is but-, and the base name is butene. There are no substituent groups. $\boxed{C^1H_2\text{=}C^2H\text{—}C^3H_2\text{—}C^4H_3}$ Start numbering on the left because it is closer to the double bond. Because the double bond is between positions 1 and 2, the name of the compound is 1-butene.

(b) $\boxed{CH_3\text{—}\overset{\overset{\displaystyle CH_3}{|}}{CH}\text{—}\overset{\overset{\displaystyle CH_3}{|}}{C}\text{=}CH\text{—}CH_3}$ has five carbons as the longest continuous chain. The prefix for 5 is pent-, and the base name is pentene. The substituent groups are two methyl groups.

     $\boxed{C^5H_3\text{—}\overset{\overset{\displaystyle \boxed{CH_3}}{|}}{C^4H}\text{—}\overset{\overset{\displaystyle \boxed{CH_3}}{|}}{C^3}\text{=}C^2H\text{—}C^1H_3}$ If we start numbering the chain at the end closest to the double bond, the double bond is between positions numbered 2 and 3 and the methyl substituents are assigned the numbers 3 and 4. The name of the compound is 3,4-dimethyl-2-pentene.

(c) $\boxed{CH_2\text{=}CH\text{—}\overset{\overset{\displaystyle CH_3\text{—}CH}{|}}{\underset{\underset{\displaystyle CH_3}{|}}{CH}}\text{—}CH_2\text{—}CH_2\text{—}CH_3}$ has six carbons as the longest continuous chain. The prefix for 6 is hex-, and the base name is hexene. The only substituent group is an isopropyl group.
$\boxed{C^1H_2\text{=}C^2H\text{—}\overset{\overset{\displaystyle \boxed{CH_3\text{—}CH}}{|}}{\underset{\underset{\displaystyle CH_3}{|}}{C^3H}}\text{—}C^4H_2\text{—}C^5H_2\text{—}C^6H_3}$ Start numbering on the left because it is closer to the double bond. Because the double bond is between positions 1 and 2 and the isopropyl group is at position 3, the name of the compound is 3-isopropyl-1-hexene.

(d) $\boxed{CH_3\text{—}\overset{\overset{\displaystyle CH_3}{|}}{CH}\text{—}CH_2\text{=}\underset{\underset{\displaystyle CH_2\text{—}CH_3}{|}}{C}\text{—}CH_3}$ has six carbons as the longest continuous chain. The prefix for 6 is hex-, and the base name is hexene. The only substituent groups are two methyl groups.

Because the double bond is in the middle of the chain, it will be at position 3 in both numbering schemes. If we start numbering the chain at the end closest to the left methyl group (closest to an end), the methyl substituents are assigned the numbers 2 and 4. The name of the compound is 2,4-dimethyl-3-hexene.

21.53   **Given:** alkyne structures   **Find:** name
Conceptual Plan: Count the number of carbon atoms in the longest continuous carbon chain that contains the multiple bond to determine the base name of the compound. Find the prefix corresponding to this number of atoms in Table 21.5 and add the ending *-yne* to form the base name. → Consider every branch from the base chain to be a substituent. Name each substituent according to Table 21.6. → Beginning with the end closest to the multiple bond, number the base chain and assign a number to each substituent. → Write the name of the compound in the following format: (subst. #)-(subst. name)(base name). → If there are two or more substituents, give each one a number and list them alphabetically with hyphens between words and numbers. → If a compound has two or more identical substituents, designate the number of identical substituents with the prefix *di-* (2), *tri-* (3), or *tetra-* (4) before the substituent's name. Separate the numbers with a comma, indicating the positions of the substituents relative to each other. The prefixes are not taken into account when alphabetizing.
Solution:

(a) $CH_3—C\equiv C—CH_3$ has four carbons as the longest continuous chain. The prefix for 4 is but-, and the base name is butyne. There are no substituent groups. $C^1H_3—C^2\equiv C^3—C^4H_3$ Because the triple bond is in the middle of the molecule, it does not matter at which end the numbering is started. Because the triple bond is between positions 2 and 3, the name of the compound is 2-butyne.

(b) $CH_3—C\equiv C—\overset{\displaystyle CH_3}{\underset{\displaystyle CH_3}{\overset{|}{\underset{|}{C}}}}—CH_2—CH_3$ has six carbons as the longest continuous chain. The prefix for 6 is hex-, and the base name is hexyne. The only substituent groups are two methyl groups.

$C^1H_3—C^2\equiv C^3—\overset{\displaystyle CH_3}{\underset{\displaystyle CH_3}{\overset{|}{\underset{|}{C^4}}}}—C^5H_2—C^6H_3$ If we start numbering the chain at the end closest to the triple bond, the triple bond is between positions 2 and 3 and the methyl substituents are assigned the numbers 4 and 4. The name of the compound is 4,4-dimethyl-2-hexyne.

(c) $CH\equiv C—\overset{\displaystyle CH—CH_3}{\underset{\displaystyle CH_3}{\overset{|}{\underset{|}{CH}}}}—CH_2—CH_2—CH_3$ has six carbons as the longest continuous chain. The prefix for 6 is hex-, and the base name is hexyne. The only substituent group is an isopropyl group.

$C^1H\equiv C^2—\overset{\displaystyle CH—CH_3}{\underset{\displaystyle CH_3}{\overset{|}{\underset{|}{C^3H}}}}—C^4H_2—C^5H_2—C^6H_3$ Start numbering at the end closest to the triple bond. The isopropyl substituent is assigned number 3. The name of the compound is 3-isopropyl-1-hexyne.

(d) $CH_3—\overset{\displaystyle CH_3}{\underset{\displaystyle \underset{\displaystyle CH_3}{\overset{|}{CH_2}}}{\overset{|}{\underset{|}{CH}}}}—C\equiv C—\overset{}{\underset{\displaystyle \underset{\displaystyle CH_3}{\overset{|}{CH_2}}}{\overset{|}{CH}}}—CH_2$ has nine carbons as the longest continuous chain. The prefix for 9 is non-, and the base name is nonyne. The substituent groups are two methyl groups.

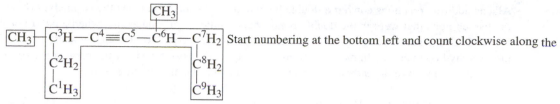

Start numbering at the bottom left and count clockwise along the chain. The triple bond is between positions 4 and 5, and the methyl substituents are assigned the numbers 3 and 6. The name of the compound is 3,6-dimethyl-4-nonyne.

21.55   **Given:** hydrocarbon names   **Find:** structure
Conceptual Plan: Find the number of carbon atoms corresponding to the prefix of the base name in Table 21.5. →
Draw the base chain and number the carbons from left to right. → **Determine the multiple bond type from the**
ending of the base name (*-ene* = double bond; *-yne* = triple bond) and place it in the appropriate position in
the chain. → **Using Table 21.6 and the prefix *di-* (2), *tri-* (3), or *tetra-* (4) before the substituent's name, determine**
each substituent. → **Add the substituent to the proper carbon position in the chain.** → **Add hydrogen atoms to**
the base chain so that each carbon has four bonds.
Solution:

(a)   4-octyne. The base name octyne indicates that there are eight carbon atoms in the base chain.
$C^1 - C^2 - C^3 - C^4 - C^5 - C^6 - C^7 - C^8$. The *-yne* ending denotes a triple bond, and the 4 prefix places it
between the fourth and fifth positions. $C^1 - C^2 - C^3 - C^4 \equiv C^5 - C^6 - C^7 - C^8$. Add hydrogens to the base
chain to get the final molecule $CH_3 - CH_2 - CH_2 - C \equiv C - CH_2 - CH_2 - CH_3$.

(b)   3-nonene. The base name nonene indicates that there are nine carbon atoms in the base chain.
$C^1 - C^2 - C^3 - C^4 - C^5 - C^6 - C^7 - C^8 - C^9$. The *-ene* ending denotes a double bond, and the 3 prefix places
it between the third and fourth positions. $C^1 - C^2 - C^3 = C^4 - C^5 - C^6 - C^7 - C^8 - C^9$. Add hydrogens to the
base chain to get the final molecule $CH_3 - CH_2 - CH = CH - CH_2 - CH_2 - CH_2 - CH_2 - CH_3$.

(c)   3,3-dimethyl-1-pentyne. The base name pentyne indicates that there are five carbon atoms in the base chain.
$C^1 - C^2 - C^3 - C^4 - C^5$. The *-yne* ending denotes a triple bond, and the 1 prefix places it between the first
and second positions. $C^1 \equiv C^2 - C^3 - C^4 - C^5$. 3,3-dimethyl indicates that there are two $-CH_3$ groups

in the third position. $C^1 \equiv C^2 - \overset{\displaystyle CH_3}{\underset{\displaystyle CH_3}{C^3}} - C^4 - C^5$. Add hydrogens to the base chain to get the final molecule

$$CH \equiv C - \overset{\displaystyle CH_3}{\underset{\displaystyle CH_3}{C}} - CH_2 - CH_3.$$

(d)   5-ethyl-3,6-dimethyl-2-heptene. The base name heptene indicates that there are seven carbon atoms in the base
chain. $C^1 - C^2 - C^3 - C^4 - C^5 - C^6 - C^7$. The *-ene* ending denotes a double bond, and the 2 prefix places it
between the second and third positions. $C^1 - C^2 = C^3 - C^4 - C^5 - C^6 - C^7$. 5-ethyl designates that there is a
$-CH_2CH_3$ group in the fifth position, and 3,6-dimethyl designates that there are $-CH_3$ groups in the third

and sixth positions. $C^1 - C^2 = \overset{\displaystyle}{\underset{\displaystyle CH_3}{C^3}} - C^4 - \overset{\displaystyle}{\underset{\displaystyle CH_2 - CH_3}{C^5}} - \overset{\displaystyle CH_3}{C^6} - C^7$. Add hydrogens to the base chain to get the final

molecule $CH_3 - CH = \overset{\displaystyle}{\underset{\displaystyle CH_3}{C}} - CH_2 - \overset{\displaystyle CH_3}{\underset{\displaystyle CH_2 - CH_3}{CH}} - CH - CH_3.$

21.57   Alkene addition reactions convert a double bond to a single bond and place the two halves of the other reactant on the two carbons that were in the double bond. Markovnikov's rule states that when a polar reagent is added to an unsymmetrical alkene, the positive end (the least electronegative part) of the reagent adds to the carbon atom that has the most hydrogen atoms. In most reactions of this type, the positive end of the reagent is hydrogen; therefore, the hydrogen atom goes to the carbon atom that already contains the most hydrogen atoms.

(a) $CH_3—CH{=}CH—CH_3 + Cl_2 \rightarrow$ $\underset{\substack{| \\ Cl}}{CH_3—CH}—\underset{\substack{| \\ Cl}}{CH}—CH_3$

(b) $\underset{\substack{| \\ CH_3}}{CH_3—CH}—CH{=}CH—CH_3 + HBr \rightarrow$ $\underset{\substack{| \\ CH_3}}{CH_3—CH}—CH_2—\underset{\substack{| \\ Br}}{CH}—CH_3$ or $\underset{\substack{| \\ CH_3}}{CH_3—CH}—\underset{\substack{| \\ Br}}{CH}—CH_2—CH_3$

(c) $CH_3—CH_2—CH{=}CH—CH_3 + Br_2 \rightarrow$ $CH_3—CH_2—\underset{\substack{| \\ Br}}{CH}—\underset{\substack{| \\ Br}}{CH}—CH_3$

(d) $\underset{\substack{| \\ CH_3}}{CH_3—CH}—CH{=}\underset{\substack{| \\ CH_3}}{C}—CH_3 + HCl \rightarrow \underset{\substack{| \\ CH_3}}{CH_3—CH}—CH_2—\underset{\substack{CH_3 \\ | \\ | \\ Cl}}{C}—CH_3$

Only one product is formed according to Markovnikov's rule.

21.59   Hydrogenation reactions convert a double bond to a single bond and place the hydrogen atoms on each of the two carbons that were in the double bond.

(a) $CH_2{=}CH—CH_3 + H_2 \rightarrow CH_3—CH_2—CH_3$

(b) $\underset{\substack{| \\ CH_3}}{CH_3—CH}—CH{=}CH_2 + H_2 \longrightarrow \underset{\substack{| \\ CH_3}}{CH_3—CH}—CH_2—CH_3$

(c) $\underset{\substack{| \\ CH_3}}{CH_3—CH}—\underset{\substack{| \\ CH_3}}{C}{=}CH_2 + H_2 \longrightarrow \underset{\substack{| \\ CH_3}}{CH_3—CH}—\underset{\substack{| \\ CH_3}}{CH}—CH_3$

Aromatic Hydrocarbons

21.61   **Given:** monosubstituted benzene structures   **Find:** name
     Conceptual Plan: Consider the branch from the benzene ring to be a substituent. Name the substituent according to Table 21.6 or using the base name of a halogen with an *o* added at the end. → Write the name of the compound in the following format: (name of substituent)benzene.
     Solution:

(a) $—CH_3$ or methyl is the substituent group, so the name is methylbenzene or toluene.

(b) $—Br$ or bromo is the substituent, so the name is bromobenzene.

(c) $—Cl$ or chloro is the substituent, so the name is chlorobenzene.

21.63   **Given:** hydrocarbon structures   **Find:** name
     Conceptual Plan: Count the number of carbon atoms in the longest continuous carbon chain (if a multiple bond is present, make sure this chain contains the multiple bond) to determine the base name of the compound. Find the prefix corresponding to this number of atoms in Table 21.5 and add the ending *-ane* for an alkane, *-ene* for an alkene, and *-yne* for an alkyne to form the base name. → Consider every branch from the base chain to be a substituent. Name each substituent according to Table 21.6. If a benzene ring is a substituent, use the phenyl group name. → Beginning with the end closest to the multiple bond, number the base chain and assign a number to each substituent. → Write the name of the compound in the following format: (subst. #)-(subst. name) (base name). → If there are two or more substituents, give each one a number and list them alphabetically with hyphens between words and numbers. → If a compound has two or more identical substituents, designate the

number of identical substituents with the prefix *di*- (2), *tri*- (3), or *tetra*- (4) before the substituent's name. Separate the numbers with a comma, indicating the positions of the substituents relative to each other. The prefixes are not taken into account when alphabetizing.

Solution: Note: For simplicity, $\varnothing$ = ⬡

(a) $CH_3-\boxed{CH-CH_2-\underset{CH_3}{\overset{CH_3}{CH}}-CH_2-\underset{\varnothing}{CH}-CH_2-CH_3}$ has nine carbons as the longest continuous chain.

The prefix for 9 is non- and the base name is nonane. The substituent groups are two methyl groups and a

phenyl group. $CH_3-\boxed{C^3H-C^4H_2-\underset{CH_3}{C^5H}-C^6H_2-\underset{\varnothing}{C^7H}-C^8H_2-C^9H_3}$ Start numbering at the end

closest to the methyl group. The substituent groups are two methyl groups at positions 3 and 5 and a phenyl group at position 7. Because *m* comes before *p*, the name of the compound is 3,5-dimethyl-7-phenylnonane. If the numbering is in the opposite direction, the name will be 5,7-dimethyl-3-phenylnonane, but this is not the approved name because IUPAC rules indicate that the first group cited receives the lowest number.

(b) $\boxed{CH_3-\overset{\varnothing}{CH}-CH=CH-CH_2-CH_2-CH_2-CH_3}$ has eight carbons as the longest continuous chain. The prefix for 8 is oct-, and because there is a double bond, the base name is octene. The only substituent group

is a phenyl group. $\boxed{C^1H_3-\overset{\varnothing}{C^2H}-C^3H=C^4H-C^5H_2-C^6H_2-C^7H_2-C^8H_3}$ If we start numbering the chain at the end closest to the double bond, the double bond is between positions 3 and 4 and the phenyl substituent is assigned the number 2. The name of the compound is 2-phenyl-3-octene.

(c) $\boxed{CH_3-\overset{\varnothing}{C}\equiv C-CH-CH-CH-CH_2-CH_3}$ has eight carbons as the longest continuous chain. The

under: $CH_3 \quad CH_3$

prefix for 8 is oct-, and because there is a triple bond, the base name is octyne. The substituent groups are two

methyl groups and a phenyl group. $\boxed{C^1H_3-C^2\equiv C^3-C^4H-\overset{\varnothing}{C^5H}-C^6H-C^7H_2-C^8H_3}$ Start

under: $CH_3 \quad CH_3$

numbering at the end closest to the triple bond. The triple bond is between positions 2 and 3, the methyl groups are assigned numbers 4 and 5, and the phenyl group is assigned number 6. Because *m* is before *p*, the name of the compound is 4,5-dimethyl-6-phenyl-2-octyne.

21.65 **Given:** disubstituted benzene structures **Find:** name
Conceptual Plan: Consider the branches from the benzene ring to be substituents. Name the substituent according to Table 21.6 or using the base name of a halogen with an *o* added at the end. → Number the benzene ring starting with the substituent that is first alphabetically and count in the direction that gets to the second substituent with the lower position number. → Write the name of the compound in the following format: (name of substituent)benzene. List them alphabetically with hyphens between words and numbers. → If a compound has two identical substituents, designate the number of identical substituents with the prefix *di*- (2) before the substituent's name. Separate the numbers with a comma, indicating the positions of the substituents relative to each other. → Alternative names are 1,2 = *ortho* or *o*; 1,3 = *meta* or *m*; and 1,4 = *para* or *p* replacing the numbers.
Solution:
(a) Both of the substituent groups are —Br or bromo groups. Start by giving one bromo an assignment of 1 and count in either direction to give the second bromo group an assignment of 4. The name is 1, 4-dibromobenzene or *p*-dibromobenzene.

(b) Both of the substituent groups are —CH_2CH_3 or ethyl groups. Start by giving the top ethyl an assignment of 1 and count in a clockwise direction to give the second ethyl group an assignment of 3. The name is 1,3-diethylbenzene or *m*-diethylbenzene.

(c) One substituent group is —Cl or chloro, and the other is —F or fluoro. Start by giving the chloro group an assignment of 1 and count in a counterclockwise direction to give the fluoro group an assignment of 2. The name is 1-chloro-2-fluorobenzene or *o*-chlorofluorobenzene.

21.67 **Given:** substituted benzene names **Find:** structures
Conceptual Plan: Start with a benzene ring. Identify the structure of the substituent(s) according to Table 21.6 or using the base name of a halogen with an *o* added at the end. → If only one substituent is present, simply attach it to the benzene ring. → If a compound has two identical substituents, they are designated with the prefix *di-* (2) before the substituent's name. → If there are two substituents, place the first one on the benzene ring and count clockwise around the ring to determine where to attach the second substituent. Note: 1,2 = *ortho* or *o*; 1,3 = *meta* or *m*; and 1,4 = *para* or *p* replacing the numbers.
Solution:

(a) isopropylbenzene. The isopropyl group is —$\overset{|}{CH}$—CH_3. Attach this to a benzene ring to make

(b) *meta*-dibromobenzene. There are two bromo or —Br groups. The positions are 1 and 3 because the designation is *meta*. Attach the —Br groups at the first and third positions to make

(c) 1-chloro-4-methylbenzene. One substituent group is —Cl or chloro, and the other is —CH_3 or methyl. Start by giving the chloro group an assignment of 1 and count in a clockwise direction to give the methyl group an assignment of 4. Attach the two groups to make

21.69 Aromatic substitution reactions substitute a halogen or alkyl group for a hydrogen (with a preference for the alkyl group) to generate a monosubstituted benzene and a hydrogen halide.

(a)

(b)

Alcohols

21.71   **Given:** alcohol structures   **Find:** name
        Conceptual Plan: Alcohols are named like alkanes with the following differences: (1) The base chain is the longest continuous carbon chain that contains the —OH functional group. (2) The base name has the ending *-ol*. (3) The base chain is numbered to give the —OH group the lowest possible number. (4) A number indicating the position of the —OH group is inserted just before the base name.
        Solution:

(a) $\boxed{CH_3 — CH_2 — CH_2}$—OH The longest carbon chain has three carbons. The prefix for 3 is prop-, so the base name is propanol. The alcohol group is on the end, or first carbon, so the name is 1-propanol.

(b) $CH_3 — \boxed{CH — CH_2 — CH — CH_3}$ with branch $\boxed{CH_2 — CH_3}$ and OH. The longest carbon chain has six carbons. The prefix for 6 is hex-, so the base name is hexanol. There is a methyl substituent group.

        $\boxed{CH_3} — \boxed{C^4H — C^3H_2 — C^2H — C^1H_3}$ with branch $\boxed{C^5H_2 — C^6H_3}$ and OH. Start numbering at the end closest to the alcohol group. The alcohol group is assigned number 2, and the methyl group is assigned number 4. The name is 4-methyl-2-hexanol.

(c) $\boxed{CH_3 — CH — CH_2 — CH — CH_2 — CH — CH_3}$ with CH_3, CH_3 branches and OH. The longest carbon chain has seven carbons. The prefix for 7 is hept-, so the base name is heptanol. There are two methyl substituent groups.

        $\boxed{C^1H_3 — C^2H — C^3H_2 — C^4H — C^5H_2 — C^6H — C^7H_3}$ with $\boxed{CH_3}$, $\boxed{CH_3}$ branches and OH. It does not matter at which end the numbering is started because substitutions are symmetrically attached. The alcohol group is assigned number 4, and the methyl groups are assigned numbers 2 and 6. The name is 2,6-dimethyl-4-heptanol.

(d) $\boxed{CH_3 — CH_2 — C — CH_2 — CH_3}$ with HO above and H_3C below. The longest carbon chain has five carbons. The prefix for 5 is pent-, so the

        base name is pentanol. There is a methyl substituent group. $\boxed{C^1H_3 — C^2H_2 — C^3 — C^4H_2 — C^5H_3}$ with HO above and $\boxed{H_3C}$ below. It does

        not matter at which end the numbering is started because substitutions are symmetrically attached. The alcohol group is assigned number 3, and the methyl group is assigned number 3. The name is 3-methyl-3-pentanol.

21.73   In a substitution reaction, an alcohol reacts with an acid, such as HBr, to form halogenated hydrocarbons and water. In an elimination (or dehydration) reaction, concentrated acids, such as H_2SO_4, react with alcohols to eliminate water, forming an alkene. In an oxidation reaction, carbon atoms gain oxygen atoms and/or lose hydrogen atoms. In these reactions, the alcohol becomes a carboxylic acid. Active metals, such as Na, react with alcohols to produce a strong base and eliminate hydrogen.

(a)   This is a substitution reaction, so $CH_3 — CH_2 — CH_2 — OH + HBr \rightarrow CH_3 — CH_2 — CH_2 — Br + H_2O$.

(b)   This is an elimination reaction, so $CH_3 — CH — CH_2 — OH$ (with CH_3 branch) $\xrightarrow{H_2SO_4}$ $CH_3 — C = CH_2 + H_2O$ (with CH_3 branch).

(c)   Sodium is an active metal, so $CH_3-CH_2-OH + Na \rightarrow CH_3-CH_2-O^-Na^+ + 1/2\ H_2$.

(d)   This is an oxidation reaction, so

$$CH_3-\underset{\underset{CH_3}{|}}{\overset{\overset{CH_3}{|}}{C}}-CH_2-CH_2-OH \xrightarrow[H_2SO_4]{Na_2Cr_2O_7} CH_3-\underset{\underset{CH_3}{|}}{\overset{\overset{CH_3}{|}}{C}}-CH_2-\overset{\overset{O}{\|}}{C}-OH.$$

Aldehydes and Ketones

21.75   **Given:** aldehyde or ketone structures   **Find:** name
    Conceptual Plan: Simple aldehydes are systematically named according to the number of carbon atoms in the longest continuous carbon chain that contains the carbonyl group. Form the base name from the name of the corresponding alkane by dropping the -e and add the ending -al. Simple ketones are systematically named according to the longest continuous carbon chain containing the carbonyl group. Form the base name from the name of the corresponding alkane by dropping the letter -e and adding the ending -one. For ketones, number the chain to give the carbonyl group the lowest possible number. Because an aldehyde always has a carbonyl terminal group, the aldehyde is always at the 1 position even though it is not specified with a number in the name.
    Solution:

(a) $\boxed{CH_3-\overset{\overset{O}{\|}}{C}-CH_2-CH_3}$ The longest carbon chain has four carbons. The prefix for 4 is but-, and because this is a ketone, the base name is butanone. The position of the carbonyl carbon does not need to be specified because there is only one place it can be in the molecule (if it were on the end carbon, it would be an aldehyde, not a ketone). Because there are no other substituent groups, the name is butanone.

(b) $\boxed{CH_3-CH_2-CH_2-CH_2-\overset{\overset{O}{\|}}{CH}}$ The longest carbon chain has five carbons. The prefix for 5 is pent-, and because this is an aldehyde, the base name is pentanal. Because there are no other substituent groups, the name is pentanal.

(c) $\boxed{CH_3-\underset{\underset{CH_3}{|}}{\overset{\overset{CH_3}{|}}{C}}-CH_2-\overset{\overset{CH_3}{|}}{CH}-CH_2-\overset{\overset{O}{\|}}{C}}-H$ The longest carbon chain has six carbons. The prefix for 6 is hex-, and because this is an aldehyde, the base name is hexanal. There are three methyl substituent groups.

$\boxed{C^6H_3-\underset{\underset{\boxed{CH_3}}{|}}{\overset{\overset{\boxed{CH_3}}{|}}{C^5}}-C^4H_2-\overset{\overset{\boxed{CH_3}}{|}}{C^3H}-C^2H_2-\overset{\overset{O}{\|}}{C^1}}-H$ Start numbering at the carbonyl carbon. The methyl groups are at positions 3, 5, and 5. The name of the molecule is 3,5,5-trimethylhexanal.

(d) $CH_3-\boxed{\underset{\underset{\boxed{CH_2-CH_3}}{|}}{CH}-CH_2-\overset{\overset{O}{\|}}{C}-CH_3}$ The longest carbon chain has six carbons. The prefix for 6 is hex-, and because this is a ketone, the base name is hexanone. There is one methyl substituent group.

$\boxed{CH_3}-\boxed{\underset{\underset{\boxed{C^5H_2-C^6H_3}}{|}}{C^4H}-C^3H_2-\overset{\overset{O}{\|}}{C^2}-C^1H_3}$ Start numbering at the end closest to the carbonyl carbon. The carbonyl carbon is in position 2, and the methyl group is at position 4. The name of the molecule is 4-methyl-2-hexanone.

21.77 This is an addition reaction,

$$CH_3-CH_2-CH_2-\overset{\overset{\displaystyle O}{\|}}{C}-H + H-C\equiv N \xrightarrow{NaCN} CH_3-CH_2-CH_2-\overset{\overset{\displaystyle OH}{|}}{\underset{\underset{\displaystyle H}{|}}{C}}-C\equiv N.$$

Carboxylic Acids and Esters

21.79 **Given:** carboxylic acid or ester structures **Find:** name

Conceptual Plan: Carboxylic acids are systematically named according to the number of carbon atoms in the longest chain containing the —COOH functional group. Form the base name by dropping the -e from the name of the corresponding alkane and adding the ending -oic acid. Esters are systematically named as if they were derived from a carboxylic acid by replacing the H on the OH with an alkyl group. The R group from the parent acid forms the base name of the compound. Change the -ic on the name of the corresponding carboxylic acid to -ate and drop acid. The R group that replaced the H on the carboxylic acid is named as an alkyl group with the ending -yl. The —COOH terminal group is always at the 1 position even though it is not specified with a number in the name.

Solution:

(a) $\boxed{CH_3-CH_2-CH_2-\overset{\overset{\displaystyle O}{\|}}{C}}-O-\boxed{CH_3}$ The carbon chain that has the carbonyl carbon has four carbons. The prefix for 4 is but-, so the base name is butanoate. The R group that replaces the H of the carboxylic acid is a methyl group. The name is methylbutanoate.

(b) $\boxed{CH_3-CH_2-\overset{\overset{\displaystyle O}{\|}}{C}}-OH$ The carbon chain has three carbons. The prefix for 3 is prop- and there are no other substituents. The name is propanoic acid.

(c) $\boxed{CH_3-\underset{\underset{\displaystyle CH_3}{|}}{CH}-CH_2-CH_2-CH_2-\overset{\overset{\displaystyle O}{\|}}{C}}-OH$ The carbon chain has six carbons. The prefix for 6 is hex-;

so the base name is hexanoic acid. There is one methyl substituent group.

$\boxed{C^6H_3-\underset{\boxed{CH_3}}{C^5H}-C^4H_2-C^3H_2-C^2H_2-\overset{\overset{\displaystyle O}{\|}}{C^1}}-OH$ Number the chain starting with the carbonyl carbon.

The methyl is in the fifth position. The name of the molecule is 5-methylhexanoic acid.

(d) $\boxed{CH_3-CH_2-CH_2-CH_2-\overset{\overset{\displaystyle O}{\|}}{C}}-O-\boxed{CH_2-CH_3}$ The carbon chain that has the carbonyl carbon has five carbons. The prefix for 5 is pent-, so the base name is pentanoate. The R group that replaces the H of the carboxylic acid is an ethyl group. The name is ethylpentanoate.

21.81 The reaction is a condensation reaction. The two noncarbonyl oxygen groups react linking the two molecules (or parts of a molecule) with the concomitant generation of a water molecule.

(a) $CH_3-CH_2-CH_2-\overset{\overset{\displaystyle O}{\|}}{C}-OH + CH_3-CH_2-OH \xrightarrow{H_2SO_4}$

$$CH_3-CH_2-CH_2-\overset{\overset{\displaystyle O}{\|}}{C}-O-CH_2-CH_3 + H_2O$$

(b)

Ethers

21.83   **Given:** ether structures   **Find:** name

Conceptual Plan: Common names for ethers have the format (R group 1)(R group 2) ether, where the alkyl groups are those found in Table 21.5. List the names alphabetically. If the two R groups are different, use each of their names. If the two R groups are the same, use the prefix *di-*.

Solution:

(a)   CH₃—CH₂—CH₂—O—CH₂—CH₃ The left carbon chain has three carbons, so it is a propyl group.

The right carbon chain has two carbons, so it is an ethyl group. Listing them alphabetically, the name is ethyl propyl ether.

(b)   CH₃—CH₂—CH₂—CH₂—CH₂—O—CH₂—CH₃ The left carbon chain has five carbons, so it is a pentyl group. The right carbon chain has two carbons, so it is an ethyl group. Listing them alphabetically, the name is ethyl pentyl ether.

(c)   CH₃—CH₂—CH₂—O—CH₂—CH₂—CH₃ Both carbon chains have three carbons, so they are propyl groups. The name is dipropyl ether.

(d)   CH₃—CH₂—O—CH₂—CH₂—CH₂—CH₃ The left carbon chain has two carbons, so it is an ethyl group. The right carbon chain has four carbons, so it is a butyl group. Listing them alphabetically, the name is butyl ethyl ether.

Amines

21.85   **Given:** amine structures   **Find:** name

Conceptual Plan: Amines are systematically named by alphabetically listing the alkyl groups that are attached to the nitrogen atom and adding an *-amine* ending. The alkyl groups are found in Table 21.5. Use the prefix *di-* or *tri-* if the alkyl groups are the same.

Solution:

(a)   CH₃—CH₂—N—CH₂—CH₃ Both carbon chains have two carbons, so they are ethyl groups. The name
       |
       H
is diethylamine.

(b)   CH₃—CH₂—CH₂—N—CH₃ The left carbon chain has three carbons, so it is a propyl group. The
       |
       H
right carbon chain has one carbon, so it is a methyl group. Listing them alphabetically, we find that the name is methylpropylamine.

          CH₃
          |
(c)   CH₃—CH₂—CH₂—N—CH₂—CH₂—CH₂—CH₃ The left carbon chain has three carbons, so it is a propyl group. The top carbon chain has one carbon, so it is a methyl group. The right carbon chain has four carbons, so it is a butyl group. Listing them alphabetically, we find that the name is butylmethylpropylamine.

21.87   (a)   This reaction is an acid–base reaction, $CH_3NHCH_3(aq) + HCl(aq) \rightarrow (CH_3)_2NH_2^+(aq) + Cl^-(aq)$.
        (b)   Because the acid is a carboxylic acid, this reaction is a condensation reaction, $CH_3CH_2NH_2(aq)$ +
              $CH_3CH_2COOH(aq) \rightarrow CH_3CH_2CONHCH_2CH_3(aq) + H_2O(l)$.
        (c)   This reaction is an acid–base reaction, $CH_3NH_2(aq) + H_2SO_4(aq) \rightarrow CH_3NH_3^+(aq) + HSO_4^-(aq)$.

Cumulative Problems

21.89   **Given:** structures   **Find:** identify molecule type and name
        **Conceptual Plan: Look for functional groups to identify molecule type. Review the appropriate set of naming
        rules to name the molecule.**
        Solution:

(a) $CH_3-CH-CH_2-\overset{\overset{\displaystyle O}{\|}}{C}-O-CH_3$ The compound has the formula RCOOR′, so it is an ester.
        with CH_3 below the CH.

        $\boxed{CH_3-CH-CH_2-\overset{\overset{\displaystyle O}{\|}}{C}}-O-\boxed{CH_3}$ The left carbon chain has four carbons, which translates to

        but- and a base name of butanoate. There is a methyl substituent on this chain. The right carbon chain
        has one carbon, so it is a methyl group. $\boxed{C^4H_3-C^3H-C^2H_2-\overset{\overset{\displaystyle O}{\|}}{C^1}}-O-\boxed{CH_3}$ Number the left
        chain starting with the carbonyl carbon. The methyl group is at position 3. The name of the compound is
        methyl-3-methylbutanoate.

(b) $\boxed{CH_3-CH_2-\overset{\overset{\displaystyle CH_3}{|}}{CH}-CH_2}-O-\boxed{CH_2-CH_3}$ The compound has the formula ROR′, so it is an ether.
        The left carbon chain has four carbons, so it is a butyl group. There is a methyl substituent on this chain. The
        right carbon chain has two carbons, so it is an ethyl group.

        $\boxed{C^4H_3-C^3H_2-\overset{\overset{\displaystyle \boxed{CH_3}}{|}}{C^2H}-C^1H_2}-O-\boxed{CH_2-CH_3}$ Number the left chain starting with the carbon next to

        the oxygen. The methyl substituent is in position 2, so the alkyl group is 2-methylbutyl. Listing the alkyl group
        alphabetically, we find that the name is ethyl 2-methylbutyl ether.

(c)   The molecule is a disubstituted benzene, so it is an aromatic hydrocarbon. The substituents
        are an ethyl group and a methyl group. If the ethyl position is assigned the first position, the
        methyl group is in the third position. Listing the functional groups alphabetically, we find
        that the name is 1-ethyl-3-methylbenzene or *m*-ethylmethylbenzene.

(d) $\boxed{CH_3-C\equiv C-CH-\overset{\overset{\displaystyle CH_2-CH_3}{|}}{CH}-CH_2-CH_3}$ The molecule contains a triple bond, so it is an alkyne. The
        with CH_3 below the first CH.

        longest carbon chain is 7. The prefix for 7 is hept-, and the base name is heptyne. The substituent groups are a
        methyl and ethyl group. $\boxed{C^1H_3-C^2\equiv C^3-C^4H-\overset{\overset{\displaystyle \boxed{CH_2-CH_3}}{|}}{C^5H}-C^6H_2-C^7H_3}$ If we start numbering the chain
        with $\boxed{CH_3}$ below the C^4H.

at the end closest to the triple bond, the triple bond is between positions 2 and 3, the methyl substituent is assigned number 4, and the ethyl substituent is assigned number 5. Because e is before m, the name of the compound is 5-ethyl-4-methyl-2-heptyne.

(e) CH_3—CH_2—CH_2—CH (with O double-bonded) The molecule has the formula RCHO, so it is an aldehyde. The longest carbon chain has four carbons. The prefix for 4 is but-, and because this is an aldehyde, the base name is butanal. Because there are no other substituent groups, the name is butanal.

(f) CH_3—CH—CH_2 (with OH on the CH, and H_3C below the CH) The molecule has the formula ROH, so it is an alcohol. The longest carbon chain has three carbons. The prefix for 3 is prop-, so the base name is propanol. There is a methyl substituent group.

C^3H_3—C^2H—C^1H_2 (with OH above and H_3C below C^2H) Start numbering the carbon with the —OH group. The alcohol group is assigned number 1, and the methyl group is assigned number 2. The name is 2-methyl-1-propanol.

21.91 **Given:** structures **Find:** name
Conceptual Plan: Look for functional groups to identify molecule type. Review the appropriate set of naming rules to name the molecule.
Solution:

(a) CH_3—CH_2—CH—CH_2—CH—CH_2—CH_2—CH_2—CH_3 (with CH_3 branch above the third carbon, and below the fifth carbon: HC—CH_3, then CH_2, then CH_3) The molecule contains only single bonds, so it is an alkane. The longest continuous chain has nine carbons. The prefix for 9 is non-, and the base name is nonane. The substituent groups are a methyl group and an isobutyl group.

C^1H_3—C^2H_2—C^3H—C^4H_2—C^5H—C^6H_2—C^7H_2—C^8H_2—C^9H_3 (with CH_3 branch above C^3H, and below C^5H: HC—CH_3, then CH_2, then CH_3) If we start numbering the chain at the end closest to the methyl group, the methyl substituent is assigned the number 3 and the isobutyl group is assigned the number 5. List the groups alphabetically to get the name of the compound, which is 5-isobutyl-3-methylnonane.

(b) CH_3—CH—CH_2—C—CH_2—CH_3 (with O double-bonded to the C, and CH_3 branch below the second carbon) The molecule contains a carbonyl carbon in the interior of the molecule, so it is a ketone. The longest carbon chain has six carbons. The prefix for 6 is hex-. Because this is a ketone, the base name is hexanone. There is one methyl substituent group.

C^6H_3—C^5H—C^4H_2—C^3—C^2H_2—C^1H_3 (with O double-bonded to C^3, and CH_3 branch below C^5H) Start numbering at the right side of the molecule.

The carbonyl carbon is at position 3. The methyl group is at position 5. The name of the molecule is 5-methyl-3-hexanone.

(c) $\boxed{CH_3-CH-CH-CH_3}$ (with OH on first CH, CH₃ below) The molecule has the formula ROH, so it is an alcohol. The longest carbon

chain has four carbons. The prefix for 4 is but-, so the base name is butanol. The only substituent group is a

methyl group. $\boxed{C^4H_3-C^3H-C^2H-C^1H_3}$ (with OH on C^2, $\boxed{CH_3}$ below C^3) Start numbering at the end with the —OH group. The alcohol

group is assigned number 2, and the methyl group is assigned number 3. The name is 3-methyl-2-butanol.

(d) $\boxed{CH_3-CH-CH-CH-C\equiv C-H}$ (with CH₃/CH₂/CH₃ above and CH₃ below) The molecule contains a triple bond, so it is an alkyne. The longest

carbon chain is 6. The prefix for 6 is hex-, and the base name is hexyne. The substituent groups are two

methyl groups and an ethyl group. $\boxed{C^6H_3-C^5H-C^4H-C^3H-C^2\equiv C^1-H}$ (with CH₃/CH₂/CH₃ above and CH₃ below) If we start numbering the

chain at the end closest to the triple bond, the triple bond is between position numbers 1 and 2, the methyl substituents are assigned numbers 3 and 5, and the ethyl substituent is assigned number 4. Because *e* comes before *m*, the name of the compound is **4-ethyl-3,5-dimethyl-1-hexyne.**

21.93   (a)   These structures are structural isomers. The methyl and ethyl groups have been swapped.
        (b)   These structures are isomers. The second iodo group has been moved.
        (c)   These structures are the same molecule. The first methyl group is simply drawn in a different orientation.

21.95   **Given:** 15.5 kg 2-butene hydrogenation   **Find:** minimum g of H_2 gas
        Conceptual Plan: Write a balanced equation kg $C_4H_8 \rightarrow$ g $C_4H_8 \rightarrow$ mol $C_4H_8 \rightarrow$ mol $H_2 \rightarrow$ g H_2.

$$\frac{1000\ g}{1\ kg} \quad \frac{1\ mol\ C_4H_8}{56.10\ g\ C_4H_8} \quad \frac{1\ mol\ H_2}{1\ mol\ C_4H_8} \quad \frac{2.02\ g\ H_2}{1\ mol\ H_2}$$

Solution: $C_4H_8(g) + H_2(g) \rightarrow C_4H_{10}(g)$ then

$$15.5\ kg\ C_4H_8 \times \frac{1000\ g\ C_4H_8}{1\ kg\ C_4H_8} \times \frac{1\ mol\ C_4H_8}{56.10\ g\ C_4H_8} \times \frac{1\ mol\ H_2}{1\ mol\ C_4H_8} \times \frac{2.02\ g\ H_2}{1\ mol\ H_2} = 558\ g\ H_2$$

Check: The units (g) are correct. The magnitude of the answer (600) makes physical sense because we are reacting 15,500 g of butane. Hydrogen is lighter, and the molar ratio is 1:1; so the amount of hydrogen will be significant but less than the mass of butene.

21.97   (a)   Combustion reaction—reaction with oxygen to form CO_2 and H_2O
        (b)   Alkane (halogen) substitution—reaction with halogen gas to substitute a hydrogen with a halogen
        (c)   Alcohol elimination—ROH reaction with concentrated acid to eliminate water from the alcohol to produce an alkene
        (d)   Aromatic (halogen) substitution—benzene reaction with halogen gas to substitute a hydrogen with a halogen

21.99   **Given:** alkene names   **Find:** structure
        Conceptual Plan: Find the number of carbon atoms corresponding to the prefix of the base name in Table 21.5. →
        Draw the base chain and number the carbons from left to right. → **Place the double bond in the appropriate
        position in the chain.** → **Using Table 21.6 and the prefix *di*- (2), *tri*- (3), or *tetra*- (4) before the substituent's name,
        determine each substituent.** → **Add the substituent to the proper carbon position in the chain.** → **Add hydrogen
        atoms to the base chain so that each carbon has four bonds.** → **Look for substitutions around double bonds to de-**

termine whether cis–trans isomerism is applicable and look for carbons with four different alkyl groups attached.
Solution:

(a) 3-methyl-1-pentene. The base name pentene designates that there are five carbon atoms in the base chain. C^1—C^2—C^3—C^4—C^5. The -ene ending designates that there is a double bond, and the 1 prefix designates that it is between the first and second positions. C^1=C^2—C^3—C^4—C^5. 3-methyl designates that there is

a —CH_3 group in the third position. C^1=C^2—$\overset{\boxed{CH_3}}{\underset{|}{C^3}}$—$C^4$—$C^5$. Add hydrogens to the base chain to get the final

molecule CH_2=CH—$\overset{CH_3}{\underset{|}{CH}}$—$CH_2$—$CH_3$. There is optical stereoisomerism because the third carbon has four different groups attached.

(b) 3,5-dimethyl-2-hexene. The base name hexene designates that there are six carbon atoms in the base chain. C^1—C^2—C^3—C^4—C^5—C^6. The -ene ending designates that there is a double bond, and the 2 prefix designates that it is between the second and third positions. C^1—C^2=C^3—C^4—C^5—C^6. 3,5-dimethyl

designates that there are —CH_3 groups in the third and fifth positions. C^1—C^2=$\overset{\boxed{CH_3}}{\underset{|}{C^3}}$—$C^4$—$\overset{\boxed{CH_3}}{\underset{|}{C^5}}$—$C^6$.

Add hydrogens to the base chain to get the final molecule CH_3—CH=$\overset{CH_3}{\underset{|}{C}}$—$CH_2$—$\overset{CH_3}{\underset{|}{CH}}$—$CH_3$. Cis–trans geometric isomerism is possible around the double bond because all of the groups surrounding the double bond are different.

(c) 3-propyl-2-hexene. The base name hexene designates that there are six carbon atoms in the base chain. C^1—C^2—C^3—C^4—C^5—C^6. The -ene ending designates that there is a double bond, and the 2 prefix designates that it is between the second and third positions. C^1—C^2=C^3—C^4—C^5—C^6. 3-propyl

designates that there is a —$CH_2CH_2CH_3$ group in the third position. C^1—C^2=$\overset{\boxed{CH_2-CH_2-CH_3}}{\underset{|}{C^3}}$—$C^4$—$C^5$—$C^6$.

Add hydrogens to the base chain to get the final molecule CH_3—CH=$\overset{CH_2-CH_2-CH_3}{\underset{|}{C}}$—$CH_2$—$CH_2$—$CH_3$. Cis–trans geometric isomerism is not possible around the double bond because the two groups on the third carbon are the same.

21.101 The 11 isomers are $\overset{O}{\overset{||}{HC}}$—$CH_2$—$CH_2$—$CH_3$ = aldehyde; CH_3—$\overset{O}{\overset{||}{C}}$—$CH_2$—$CH_3$ = ketone;

CH_2=CH—CH_2—CH_2—OH = alcohol and alkene; CH_3—CH=CH—CH_2—OH = alcohol and alkene;

CH_3—CH_2—CH=CH—OH = alcohol and alkene; CH_3—CH_2—O—CH=CH_2 = ether and alkene;

CH_3—O—CH_2—CH=CH_2 = ether and alkene; CH_3—O—CH=CH—CH_3 = ether and alkene;

CH_3—CH_2—$\overset{OH}{\underset{|}{C}}$=$CH_2$ = alcohol and alkene; CH_3—$\overset{OH}{\underset{|}{CH}}$—$CH$=$CH_2$ = alcohol and alkene; and

CH_3—$\overset{OH}{\underset{|}{C}}$=$CH$—$CH_3$ = alcohol and alkene. There is also a twelfth isomer that is cyclic: $\overset{H_2C-CH_2}{\underset{O}{H_2C \qquad CH_2}}$ = ether.

21.103 In the acid form of the carboxylic acid, electron withdrawal by the C=O enhances acidity. The conjugate base, the carboxylate anion, is stabilized by resonance; so the two O atoms are equivalent and bear the negative charge equally. This resonance stability is not possible in an alcohol, where the carbonyl group is absent.

Challenge Problems

21.105   This is an internal condensation reaction. It requires only heat to cause it to happen.

(a)

(b)   This is a dehydration reaction, CH_3—CH—CH_2—CH—CH_3 $\xrightarrow{H_2SO_4}$

CH_3—CH—CH_2—C—CH_3 + H_2O.

(c)   This is an addition reaction that follows Markovnikov's rule,

21.107   (a)   Because there are six primary hydrogen atoms and two secondary hydrogen atoms, if they are equally reactive, we expect a ratio of 6:2 or 3:1.

(b)   Assume that we generate 100 product molecules. The yield = (#hydrogen atoms)(reactivity). So $1° = 45$ = (3)(reactivity 1°) and $2° = 55 = (1)$(reactivity 2°). Taking the ratio, 2°:1° = 55:(45/3) = 55:15 = 11:3. The 2° hydrogens are much more reactive.

21.109   The chiral products (that have four different groups on a carbon) are CH_2Cl—CHCl—CH_2—CH_3 (second carbon), CH_2Cl—CH_2—CHCl—CH_3 (third carbon), and CH_3—CHCl—CHCl—CH_3 (second and third carbons).

21.111   The first propagation step for F is very rapid and exothermic because of the strength of the H—F bond that forms. For I, the first propagation step is endothermic and slow because the H—I bond that forms is relatively weak. In addition, the C—F bond is much stronger than C—I (477 kJ versus 241 kJ).

21.113   The three structures that have no dipole moment are

The structures with dipole moments are

Conceptual Problems

21.115   For the structure to have only one product after a single bromination, all of the hydrogens must be equivalent.

The structure is

$$CH_3 - \underset{\underset{CH_3}{|}}{\overset{\overset{CH_3}{|}}{C}} - \underset{\underset{CH_3}{|}}{\overset{\overset{CH_3}{|}}{C}} - CH_3.$$ To name it, find the longest carbon chain:

$$\boxed{CH_3 - \underset{\underset{CH_3}{|}}{\overset{\overset{CH_3}{|}}{C}} - \underset{\underset{CH_3}{|}}{\overset{\overset{CH_3}{|}}{C}} - CH_3}.$$ Because there are four carbons in this chain and it is an alkane, the base name is butane.

There are four methyl substituent groups. $\boxed{C^1H_3 - \underset{\underset{\boxed{CH_3}}{|}}{\overset{\overset{\boxed{CH_3}}{|}}{C^2}} - \underset{\underset{\boxed{CH_3}}{|}}{\overset{\overset{\boxed{CH_3}}{|}}{C^3}} - C^4H_3}$ It does not matter how the chain is numbered because it is symmetrically substituted. The methyl groups are assigned the numbers 2, 2, 3, and 3. The name is 2,2,3,3-tetramethylbutane.

Questions for Group Work

21.117   Answers will vary. For example, one student might have the functional group ketones. This student might draw

2-pentanone:

$$\underset{H_3C}{\nwarrow}\underset{CH_2}{}\underset{}{}\underset{CH_2}{\nwarrow}\underset{}{}\overset{\overset{O}{\|}}{C}\underset{}{}\underset{CH_3}{}$$

. When passing this structure around the group, students should first recognize that it is a ketone and then give the correct name (2-pentanone).

21.119   Specific isomers drawn in what follows may vary.

(a)

$$\underset{H_3C}{}\overset{\overset{CH_3}{|}}{C}\underset{\underset{CH_3}{|}}{}\underset{CH_2}{}\overset{\overset{CH_3}{|}}{CH}\underset{CH_3}{}$$

The formula is C_8H_{18}, so it is an isomer of octane.

(b)

(c) H_3C — CH (CH_3) — CH_2 — CH_2 — CH_2 — CH_2 — CH_3 Same as the second structure, above.

(d)   1-methylheptane; the names of the other octane isomers in part (a) are *n*-octane, 2,4-dimethylhexane, and 2,3,4-trimethylpentane.

(e)   An isomer is a molecule that has the same formula, but the atoms are connected in different ways.

Data Interpretation and Analysis

21.121   (a)   apple: methyl butanoate; pear: propyl acetate (or, propyl ethanoate); and pineapple: ethyl butanoate
         (c)   Esters are derived from a carboxylic acid and an alcohol.

Apple   ⟍⟋⟍OH $\xrightarrow[\text{H}_2\text{SO}_4]{\text{Na}_2\text{Cr}_2\text{O}_7}$ (butanoic acid) + CH_3OH $\xrightarrow{\text{H}_2\text{SO}_4}$ (ester) + H_2O

Pear   ⟍⟋OH $\xrightarrow[\text{H}_2\text{SO}_4]{\text{Na}_2\text{Cr}_2\text{O}_7}$ (acetic acid) + ⟍⟋OH $\xrightarrow{\text{H}_2\text{SO}_4}$ (ester) + H_2O

Pineapple   ⟍⟋⟍OH $\xrightarrow[\text{H}_2\text{SO}_4]{\text{Na}_2\text{Cr}_2\text{O}_7}$ (butanoic acid) + ⟍OH $\xrightarrow{\text{H}_2\text{SO}_4}$ (ester) + H_2O

22 Transition Metals and Coordination Compounds

Review Questions

22.1   A transition metal atom forms an ion by losing the ns^2 (valence shell) electrons first.

22.3   The $+2$ oxidation state is common because most of the transition metals have two electrons occupying the ns orbitals. These electrons are lost first by the metal.

22.5   The electronegativity of the transition elements generally increases across a row, following the main-group trend. However, in contrast to the main-group trend, electronegativity increases from the first transition row to the second. There is little electronegativity difference between the second and third transition row. There is a slight increase from silver to gold. Therefore, Au is the most electronegative of the transition metals.

22.7   A ligand can be considered a Lewis base because it donates a pair of electrons. The transition metal ion would be a Lewis acid because it accepts the pair of electrons.

22.9   Cis–trans isomerism occurs in square planar complexes of the general formula MA_2B_2 or octahedral complexes of the general formula MA_4B_2.

22.11   Because of the spatial arrangement of the ligands, the normally degenerate d orbitals are split in energy. The difference between these split d orbitals is the crystal field splitting energy. The magnitude of the splitting depends on the particular complex. In strong-field complexes, the splitting is large; in weak-field complexes, the splitting is small. The magnitude of the crystal field splitting depends in large part on the ligands attached to the central metal ion.

22.13 Zn^{2+} has a filled d subshell, and Cu^{2+} has nine d electrons. Because Zn^{2+} has this filled d subshell, the color of the compounds will be white. Cu^{2+} has an incomplete d subshell, so different ligands will cause a different crystal field splitting, and the compounds will have color.

22.15   Almost all tetrahedral complexes are high-spin because of reduced ligand–metal interactions. The d orbitals in a tetrahedral complex are interacting with only four ligands, as opposed to six in the octahedral complex; so the value of Δ is generally smaller.

Problems by Topic

Properties of Transition Metals

22.17   Identify the noble gas that precedes the element and put it in square brackets.
Determine the outer principal quantum level for the s orbital. Subtract 1 to obtain the quantum level for the d orbital. If the element is in the third or fourth transition series, include $(n - 2)f$ electrons in the configuration.
Count across the row to see how many electrons are in the neutral atom.
For an ion, remove the required number of electrons first from the s orbitals and then from the d orbitals.

(a) Ni; Ni^{2+}

The noble gas that precedes Ni is Ar. Ni is in the fourth period; so the orbitals we use are $4s$ and $3d$, and Ni has 10 more electrons than Ar does.

Ni $\quad [Ar]4s^23d^8$

Ni will lose electrons from the $4s$ and then from the $3d$.

$Ni^{2+} \quad [Ar]4s^03d^8$

(b) Mn; Mn^{4+}

The noble gas that precedes Mn is Ar. Mn is in the fourth period; so the orbitals we use are $4s$ and $3d$, and Mn has seven more electrons than Ar does.

Mn $\quad [Ar]4s^23d^5$

Mn will lose electrons from the $4s$ and then from the $3d$.

$Mn^{4+} \quad [Ar]4s^03d^3$

(c) Y; Y^+

The noble gas that precedes Y is Kr. Y is in the fifth period; so the orbitals we use are $5s$ and $4d$, and Y has three more electrons than Kr does.

Y $\quad [Kr]5s^24d^1$

Y will lose electrons from the $5s$ and then from the $4d$.

$Y^+ \quad [Kr]5s^14d^1$

(d) Ta; Ta^{2+}

The noble gas that precedes Ta is Xe. Ta is in the sixth period; so the orbitals we use are $6s$, $5d$, and $4f$, and Ta has 19 more electrons than Xe does.

Ta $\quad [Xe]6s^24f^{14}5d^3$

Ta will lose electrons from the $6s$ and then from the $5d$.

$Ta^{2+} \quad [Xe]6s^04f^{14}5d^3$

22.19   (a)   V $\quad$ Highest oxidation state $= +5$. V $= [Ar]4s^23d^3$. Because V is to the left of Mn, it can lose all of the $4s$ and $3d$ electrons; so the highest oxidation state is $+5$.

(b)   Re $\quad$ Highest oxidation state $= +7$. Re $= [Xe]6s^24f^{14}5d^5$. Re can lose all of the $6s$ and $5d$ electrons, so the highest oxidation state is $+7$.

(c)   Pd $\quad$ Highest oxidation state $= +4$. Pd $= [Kr]4d^{10}$. Metal ions with a d^8 electron configuration exhibit a square planar geometry. Pd can lose 4 electrons from the $5s$ and $4d$ orbitals, so the highest oxidation state is $+4$.

Coordination Compounds

22.21   (a) $[Cr(H_2O)_6]^{3+}$ H_2O is neutral, so Cr has an oxidation state of $+3$. Six H_2O molecules are attached to each Cr, so the coordination number is 6.

(b) $[Co(NH_3)_3Cl_3]^-$ NH_3 is neutral, and Cl has charge of $1-$. The sum of the oxidation state of Co and the charge of chloride ion $= -1$. $x + (3(-1)) = 1-$, $x = +2$; therefore, the oxidation state of Co is $+2$. The three NH_3 molecules and the three Cl^- ions are bound directly to the Co atom; therefore, the coordination number is 6.

(c) $[Cu(CN)_4]^{2-}$ CN has a charge of $1-$. The sum of the oxidation state of Cu and the charge of the cyanide ion $= 2-$. $x + (4(-1)) = 2-$, $x = +2$; therefore, the oxidation state of Cu is $+2$. The four cyanide ions are directly bound to the Cu atom; therefore, the coordination number is 4.

(d). $[Ag(NH_3)_2]^+$ NH_3 is neutral, so Ag has an oxidation number of $+1$. Two NH_3 molecules are attached to each Ag atom, so the coordination number is 2.

22.23   (a) $[Cr(H_2O)_6]^{3+}$ is hexaaquachromium(III) ion.

$[Cr(H_2O)_6]^{3+}$ is a complex cation.

Name the ligand: H_2O is aqua.

Name the metal ion: Cr^{3+} is chromium(III).

Name the complex ion by adding the prefixes to indicate the number of each ligand, followed by the name of each ligand and the name of the metal ion: hexaaquachromium(III) ion.

(b) $[Cu(CN)_4]^{2-}$ is tetracyanocuprate(II) ion.

$[Cu(CN)_4]^{2-}$ is a complex anion.

Name the ligand: CN^- is cyano.

Name the metal ion: Cu^{2+} is cuprate(II) because the complex is an anion.

Name the complex ion by adding the prefix to indicate the number of each ligand, followed by the name of each ligand and the name of the metal ion: tetracyanocuprate(II) ion.

(c) $[Fe(NH_3)_5Br]SO_4$ is pentaamminebromoiron(III) sulfate.

$[Fe(NH_3)_5Br]^{2+}$ is a complex cation, SO_4^{2-} is sulfate.

Name the ligands in alphabetical order: NH_3 is ammine; Br^- is bromo.

Name the metal cation: Fe^{3+} is iron(III).

Name the complex ion by adding prefixes to indicate the number of each ligand, followed by the name of the ligand and the name of the metal ion: pentaamminebromoiron(III).

Name the compound by writing the name of the cation before the anion. The only space is between the ion names: pentaamminebromoiron(III) sulfate.

(d) $[Co(H_2O)_4(NH_3)(OH)]Cl_2$ is amminetetraaquahydroxocobalt(III) chloride.

$[Co(H_2O)_4(NH_3)(OH)]^{2+}$ is a complex cation; Cl^- is chloride.

Name the ligands in alphabetical order: NH_3 is ammine; H_2O is aqua; OH^- is hydroxo.

Name the metal cation: Co^{3+} is cobalt(III).

Name the complex ion by adding prefixes to indicate the number of each ligand, followed by the name of the ligand and the metal ion: amminetetraaquahydroxocobalt(III).

Name the compound by writing the name of the cation before the anion. The only space is between the ion names: amminetetraaquahydroxocobalt(III) chloride.

22.25 (a) Hexaamminechromium(III) is a complex ion with Cr^{3+} metal ion and six NH_3 ligands. $[Cr(NH_3)_6]^{3+}$

(b) Potassium hexacyanoferrate(III) is a compound with 3 K^+ cations and a complex anion with Fe^{3+} metal ion and six CN^- ligands. $K_3[Fe(CN)_6]$

(c) Ethylenediaminedithiocyanatocopper(II) is a compound with a Cu^{2+} metal ion, an ethylenediamine ligand, and two SCN^- ligands. $[Cu(en)(SCN)_2]$

(d) Tetraaquaplatinum(II) hexachloroplatinate(IV) is a complex compound with a complex cation that contains a Pt^{2+} metal ion, four H_2O ligands, and a complex anion that contains a Pt^{4+} metal ion and six Cl^- ligands. $[Pt(H_2O)_4][PtCl_6]$

22.27 (a) $[Co(NH_3)_3(CN)_3]$ is triamminetricyanocobalt(III).

(b) Because ethylenediamine is a bidentate ligand, you need three to have a coordination number of 6. $[Cr(en)_3]^{3+}$ is tris(ethylenediamine)chromium(III) ion.

Structure and Isomerism

22.29 In linkage isomers, the ligand coordinates to the metal in different ways.

pentaamminenitromanganese(III) ion   pentaamminenitritomanganese(III) ion

22.31 Coordination isomers occur when a coordinated ligand exchanges places with the uncoordinated counterion.

$[Fe(H_2O)_5Cl]Cl \cdot H_2O$ pentaaquachloroiron(II) chloride monohydrate

$[Fe(H_2O)_4Cl_2] \cdot 2H_2O$ tetraaquadichloroiron(II) dihydrate

22.33 Geometric isomers result when the ligands bonded to the metal have a different spatial arrangement.

(a) No, an octahedral complex must have at least two different ligands to have geometric isomers.

(b) Yes, there will be cis–trans isomers.

(c) Yes, there will be fac–mer isomers.

(d) No, a square planar complex must have at least two different ligands to have geometric isomers.

(e) Yes, there will be cis–trans isomers.

22.35   (a)   Square planar $[NiWXYZ]^{2+}$ would have three geometric isomers.

(b)   Tetrahedral $[ZnWXYZ]^{2+}$ would have no geometric isomers but would have stereoisomers.

22.37   (a) $[Cr(CO)_3(NH_3)_3]^{3+}$ has a coordination number of 6 and is octahedral. There will be fac and mer isomers and no optical isomers because rotation of the mirror images is superimposable upon each other.

(b) $[Pd(CO)_2(H_2O)Cl]^+$ has a coordination number of 4 and is a d^8 complex, so it is square planar. There will be cis and trans isomers. There will be no optical isomers because rotation of the mirror images is superimposable upon each other.

22.39 $[Cr(NH_3)_2(ox)_2]^-$ has a coordination number of 6 and is octahedral. There will be cis and trans isomers. The cis isomer has a mirror image that is nonsuperimposable.

Bonding in Coordination Compounds

22.41   (a) Zn^{2+} d^{10} (b) Fe^{3+} d^5 high spin   low spin

(c) V^{3+} d^2 (d) Co^{2+} d^7 high spin

22.43 **Given:** $[CrCl_6]^{3-}$ $\lambda = 735$ nm absorbance maximum **Find:** crystal field energy

Conceptual Plan: $\lambda \rightarrow \Delta$ (J) $\rightarrow$ kJ/ion $\rightarrow$ kJ/mol

$$\Delta = \frac{hc}{\lambda} \qquad \frac{1\ kJ}{1000\ J} \qquad \frac{6.02 \times 10^{23}\ ions}{mol}$$

Solution: $\dfrac{(6.626 \times 10^{-34}\ J \cdot s)\left(3.00 \times 10^{8}\ \dfrac{m}{s}\right)}{(735\ nm)\left(\dfrac{1 \times 10^{-9}\ m}{nm}\right)} = 2.70\underline{4} \times 10^{-19}\ J/ion$

$2.70\underline{4} \times 10^{-19}\ \dfrac{J}{ion} \times \dfrac{1\ kJ}{1000\ J} \times \dfrac{6.02 \times 10^{23}\ ions}{mol} = 163\ kJ/mol$

Check: Cl is a weak-field ligand and would be expected to have a relatively small Δ, which is consistent with a value of 163 kJ.

22.45 The crystal field ligand strength would be $CN^- >$; $NH_3 >$; F^-. The smaller the wavelength, the larger the energy and the greater the crystal field splitting observed. So $[Co(CN)_6]^{3-}$ would have a smaller wavelength than $[Co(NH_3)_6]^{3+}$, which would have a smaller wavelength than $[CoF_6]^{3-}$.

So $[Co(CN)_6]^{3-} = 290$ nm, which absorbs in the UV and would have a colorless solution.

$[Co(NH_3)_6]^{3+} = 440$ nm, which absorbs in the blue and would have an orange solution.

$[CoF_6]^{3-} = 770$ nm, which absorbs in the red and would have a green solution.

22.47 Mn^{2+} is d^5, and there are five unpaired electrons; so the crystal field splitting energy, Δ, is small compared to the energy to pair the electrons. Therefore, NH_3 induces a weak field with Mn^{2+}.

22.49 (a) $[RhCl_6]^{3-}$ Rh^{3+} d^6 Cl^- is a weak-field ligand, so the value of Δ will be small. This gives four unpaired electrons.

(b) $[Co(OH)_6]^{4-}$ Co^{2+} d^7 OH^- is a weak-field ligand, so the value of Δ will be small. This gives three unpaired electrons.

(c) cis-$[Fe(en)(NO_2)_2]^+$ Fe^{3+} d^5 en and NO_2^- are strong-field ligands, so the value of Δ will be large. This gives one unpaired electron.

22.51 $[CoCl_4]^{2-}$ Co^{2+} is d^7, and Cl^- is a weak-field ligand; so the value of Δ will be small, and there will be three unpaired electrons. The crystal field splitting for a tetrahedral structure is as follows:

Applications of Coordination Compounds

22.53 Hemoglobin, cytochrome c, and chlorophyll all contain a porphyrin ligand.

22.55 Oxyhemoglobin is low-spin and a red color. Because it is low-spin, the crystal field splitting energy must be large. The red color means that the complex absorbs in the green region ($\sim$500 nm), which also indicates a large crystal field splitting energy; therefore, O_2 must be a strong-field ligand.

Deoxyhemoglobin is high-spin and a blue color. Because it is high-spin, the crystal field splitting energy must be small. The blue color means that the complex absorbs in the orange region ($\sim$600 nm), which also indicates a small crystal field splitting energy. Both of these are consistent with H_2O as a weak-field ligand.

Cumulative Problems

22.57   (a)
Cr	$[Ar]4s^13d^5$
Cr^+	$[Ar]4s^03d^5$
Cr^{2+}	$[Ar]4s^03d^4$
Cr^{3+}	$[Ar]4s^03d^3$

(b)
Cu	$[Ar]4s^13d^{10}$
Cu^+	$[Ar]4s^03d^{10}$
Cu^{2+}	$[Ar]4s^03d^9$

22.59   (a)

$$H-\overset{\cdot\cdot}{N}-H$$
$$|$$
$$H$$

(b) $\left[\overset{\cdot\cdot}{\underset{\cdot\cdot}{S}}=C=\overset{\cdot\cdot}{N}\!\!\cdot\right]^-$ ligand can bond from either end.

(c) $H\overset{\displaystyle O}{\diagup\diagdown}H$

22.61   An octahedral complex has six ligands.
$MA_2B_2C_2$	will have cis–trans isomers: all cis; A trans, B,C cis; B trans, A,C cis; C trans, A,B cis; all trans.
MAB_2C_3	will have fac–mer isomers.
MA_2B_3C	will have fac–mer isomers.
MAB_3C_2	will have fac–mer isomers.
MA_3B_2C	will have fac–mer isomers.
MA_2BC_3	will have fac–mer isomers.
MA_3BC_2	will have fac–mer isomers.
$MABC_4$	will have AB cis and trans.
MAB_4C	will have AC cis and trans.
MA_4BC	will have BC cis and trans.

22.63 $[Fe(ox)_3]^{3-}$ Fe^{3+} d^5 coordination number = 6, octahedral, structure has a nonsuperimposable mirror image.

22.65 $[Mn(CN)_6]^{3-}$ Mn^{3+} d^4 has a coordination number of 6 and is octahedral. CN^- is a strong-field ligand and will cause a large crystal field splitting energy, so the ion is low-spin.

The ion will be paramagnetic with two unpaired electrons.

22.67   There are five geometric isomers, one of which is chiral.

4.

$$\left[\begin{array}{c} \overset{H_2O}{\underset{Cl}{Cl - \underset{|}{Ru} - NH_3}} \\ NH_3 \quad H_2O \end{array} \right]^+$$

5.

$$\left[\begin{array}{c} \overset{H_2O}{\underset{NH_3}{Cl - \underset{|}{Ru} - NH_3}} \\ H_2O \quad Cl \end{array} \right]^+$$

Challenge Problems

22.69 **Given:** 46.7% Pt, 17.0% Cl, 14.8% P, 17.2% C, 4.34% H   **Find:** formula, structures, and names for both compounds
Conceptual Plan: % composition → pseudoformula → formula

$$n = \frac{g}{molar\ mass} \qquad \text{divide by smallest}$$

Solution: $46.7\ \text{g Pt} \times \dfrac{1\ \text{mol Pt}}{195.08\ \text{g}} = 0.23939\ \text{mol Pt}$ $17.0\ \text{g Cl} \times \dfrac{1\ \text{mol Cl}}{35.45\ \text{g Cl}} = 0.47955\ \text{mol Cl}$

$14.8\ \text{g P} \times \dfrac{1\ \text{mol P}}{30.97\ \text{g}} = 0.47788\ \text{mol P}$ $17.2\ \text{g C} \times \dfrac{1\ \text{mol C}}{12.01\ \text{g C}} = 1.4321\ \text{mol C}$

$4.34\ \text{g H} \times \dfrac{1\ \text{mol H}}{1.008\ \text{g}} = 4.3056\ \text{mol H}$

$Pt_{0.23939}Cl_{0.47955}P_{0.47788}C_{1.4321}H_{4.3056}$

$Pt_{\frac{0.23939}{0.23939}}Cl_{\frac{0.47955}{0.23939}}P_{\frac{0.47788}{0.23939}}C_{\frac{1.4321}{0.23939}}H_{\frac{4.3056}{0.23939}}$

$PtCl_2P_2C_6H_{18}$

$[Pt(P(CH_3)_3)_2Cl_2]$

cis–dichlorobis(trimethylphosphine)platinum(II)   *trans*–dichlorobis(trimethylphosphine)platinum(II)

$$\begin{array}{c} Cl \diagdown \qquad \diagup P(CH_3)_3 \\ Pt \\ Cl \diagup \qquad \diagdown P(CH_3)_3 \end{array}$$

$$\begin{array}{c} Cl \diagdown \qquad \diagup P(CH_3)_3 \\ Pt \\ P(CH_3)_3 \diagup \quad \diagdown Cl \end{array}$$

22.71

The trigonal bipyramidal complex ion has lobes along the *x*-axis and between the *x*- and *y*-axes and along the *z*-axis. So the ligands will interact most strongly with z^2 orbital and then with the $x^2 - y^2$ and the xy orbitals and will not interact with the xz or the yz orbitals. So the crystal field splitting would look like the following:

22.73 **(a)** **Given:** $K_{sp}(NiS) = 3 \times 10^{-16}$ **Find:** solubility in water
Conceptual Plan: Write the reaction, prepare an ICE table, substitute into the equilibrium expression, and solve for S (molar solubility).
Solution: $NiS(s) \rightleftharpoons Ni^{2+}(aq) + S^{2-}(aq)$

	Ni^{2+}	S^{2-}
Initial	0.0	0.0
Change	S	S
Equil	S	S

$K_{sp} = [Ni^{2+}][S^{2-}]$ $3 \times 10^{-16} = S^2$
$S = 1.7 \times 10^{-8} = 2 \times 10^{-8}$

(b)   **Given:** $K_{sp}(\text{NiS}) = 3 \times 10^{-16}$ $K_f\,[\text{Ni}(\text{NH}_3)_6]^{2+} = 2.0 \times 10^8$ **Find:** solubility in 3.0 M NH_3
Conceptual Plan: Sum the reaction, prepare an ICE table, substitute into the equilibrium expression, and solve for S (molar solubility).
Solution:

Reaction 1: $\text{NiS}(s)$	$\rightleftharpoons$ $\cancel{\text{Ni}}^{2+}(aq) + \text{S}^{2-}(aq)$		$K_{sp}(\text{NiS}) = 3 \times 10^{-16}$
Reaction 2: $\cancel{\text{Ni}}^{2} + (aq) + 6\,\text{NH}_3(aq)$	$\rightleftharpoons$ $\text{Ni}(\text{NH}_3)_6^{2+}(aq)$		$K_f = 2.0 \times 10^8$

Reaction 3: $\text{NiS}(s) + 6\,\text{NH}_3(aq)$	$\rightleftharpoons$ $\text{Ni}(\text{NH}_3)_6^{2+}(aq) + \text{S}^{2-}(aq)$	$K = K_{sp}\,K_f = 6.0 \times 10^{-8}$
Initial	3.0 M	0.0	0.0
Change	$-6S$	$+S$	$+S$
Equil	$3.0 - 6S$	S	S

$$K = \frac{[\text{Ni}(\text{NH}_3)_6]^{2+}[\text{S}^2]}{[\text{NH}_3]^6}\,6.0 \times 10^{-8} = \frac{(S)(S)}{(3.0 - S)^6}$$

Assume that $S << 3.0$.

$S^2 = 4.37 \times 10^{-5}$

$S = 6.6 \times 10^{-3}$

(c)   NiS is more soluble in ammonia because the formation of the $[\text{Ni}(\text{NH}_3)_6]^{2+}$ complex ion is highly favorable. The formation removes Ni^{2+} ion from the solution, causing more of the NiS to dissolve.

22.75   If the complexes exhibit lability, there will be exchange of H_2O with the halide and some of the halide will be in solution. So prepare complexes with two different halides (e.g., Cl^- and Br^-) so that you have $[\text{MCl}_6]^{3-}$ and $[\text{MBr}_6]^{3-}$. Place both complexes in the same aqueous solution. If the complexes exhibit lability, a mixed complex containing both Cl^- and Br^- ligands, $[\text{MCl}_x\text{Br}_y]^{3-}$, will be formed because Cl^- and Br^- would be in solution when the ligands were exchanged with H_2O. If this mixed complex does not form, the $[\text{MX}_6]^{3-}$ complex does not exhibit lability.

22.77   **Given:** 0.10 M NaI, $\text{Cd}(\text{OH})_2(s)$ $K_{sp} = 7.2 \times 10^{-15}$; $\text{Cd}(\text{I})_4^{2-}(aq)$ $K_f = 2 \times 10^6$ **Find:** pH
Conceptual Plan: Combine the solubility product reaction of $\text{Cd}(\text{OH})_2(s)$ and the formation reaction of $\text{Cd}(\text{I})_4^{2-}$ and determine the K for the new reaction. Prepare an ICE table and solve for S (the molar solubility). Then determine the $[\text{OH}^-]$, pOH, and pH.
Solution:

$\text{Cd}(\text{OH})_2(s)$	$\rightleftharpoons$ $\cancel{\text{Cd}}^{2+}(aq) + 2\,\text{OH}^-(aq)$		$K_{sp} = 7.2 \times 10^{-15}$
$\cancel{\text{Cd}}^{2+}(aq) + 4\,\text{I}^-(aq)$	$\rightleftharpoons$ $\text{Cd}(\text{I})_4^{2-}(aq)$		$K_f = 2 \times 10^6$

$\text{Cd}(\text{OH})_2(s) + 4\,\text{I}^-(aq) \rightleftharpoons \text{Cd}(\text{I})_4^{2-}(aq) + 2\,\text{OH}^-(aq)$		$K = 1.44 \times 10^{-8}$	
Initial	0.10 M	0	0
Change	$-4S$	$+S$	$+2S$
Equil	$0.10 - 4S$	S	$2S$

$$K = 1.44 \times 10^{-8} = \frac{[\text{Cd}(\text{I})_4^{2-}][\text{OH}^-]^2}{[\text{I}^-]^4} = \frac{(S)(2S)^2}{(0.10 - 4S)^4}$$ Assume that $S << 0.10$.

$$1.44 \times 10^{-8} = \frac{4S^3}{(0.10)^4}$$

$S = 7.11 \times 10^{-5}$; $[\text{OH}^-] = 2S = 2(7.01 \times 10^{-5}) = 1.42 \times 10^{-4}$

$\text{pOH} = -\log(1.42 \times 10^{-4}) = 3.85 = 3.9$ $\text{pH} = 14.0 - 3.9 = 10.1$

Conceptual Problems

22.79   Au would have the higher ionization energy. Ionization energy increases as you go down a group in the transition metals. Because there is a large increase in the number of protons and not a large increase in size, the ionization energy increases.

Questions for Group Work

22.81 Electron configuration:
The trend is $ns^2(n-1)d^x$, but there are exceptions at Cr, Nb, Mo, Ru, Rh, Pd, Ag, Pt, and Au.
Atomic radius: see Figure 22.1.

Sc $4s^23d^1$	Ti $4s^23d^2$	V $4s^23d^3$	Cr $4s^13d^5$	Mn $4s^23d^5$	Fe $4s^23d^6$	Co $4s^23d^7$	Ni $4s^23d^8$	Cu $4s^23d^9$	Zn $4s^23d^{10}$
Y $5s^24d^1$	Zr $5s^24d^2$	Nb $5s^14d^4$	Mo $5s^14d^5$	Tc $5s^24d^5$	Ru $5s^14d^7$	Rh $5s^14d^8$	Pd $4d^{10}$	Ag $5s^14d^{10}$	Cd $5s^24d^{10}$
La $6s^25d^1$	Hf $6s^25d^2$	Ta $6s^25d^3$	W $6s^25d^4$	Re $6s^25d^5$	Os $6s^25d^6$	Ir $6s^25d^7$	Pt $6s^15d^9$	Au $6s^15d^{10}$	Hg $6s^25d^{10}$

22.83 Answers will vary. For example, $Co(NH_3)_5NO_2^{2+}$ and $Co(NH_3)_5ONO^{2+}$ are linkage isomers.

22.85 Answers will vary. One example might be "Victims of lead poisoning can be treated with a complex of calcium and EDTA. The lead complex is more stable than the calcium complex, so the EDTA binds the lead and is excreted from the body." (chelating agents)

Data Interpretation and Analysis

22.86 (a) Based on this spectrum shown in Figure a, determine the color of the aqueous solution of $[Fe(H_2O)_5SCN]^{2+}$. The color of an aqueous solution of $[Fe(H_2O)_5SCN]^{2+}$ is red. The absorption spectrum of $[Fe(H_2O)_5SCN]^{2+}$ shows that wavelengths from 400 – 600 nm are strongly absorbed. Because wavelengths of light that correspond to the color red, 650 – 700 nm, are not absorbed, the solution appears red.

(c) The hexaaquairon(III) complex ion, $[Fe(H_2O)_6]^{3+}$, produces a pale violet aqueous solution. The crystal field splitting energy, Δ, of $[Fe(H_2O)_6]^{3+}$ is smaller than the Δ of $[Fe(H_2O)_5SCN]^{2+}$.

If a solution appears violet, you can deduce that it absorbs wavelengths of the complementary color, yellow. Yellow light ranges from 560 to 580 nm. To calculate Δ, estimate the average wavelength of absorption 570 nm. Therefore, the crystal field splitting energy of $[Fe(H_2O)_6]^{3+}$ is:

$$\Delta = \frac{hc}{\lambda} = \frac{(6.626 \times 10^{-34}\,J \cdot s)\left(3.00 \times 10^8\,\frac{m}{s}\right)}{(570.\,nm)\left(\frac{1 \times 10^{-9}\,m}{nm}\right)}$$

$$\Delta = \left(3.49 \times 10^{-19}\,\frac{J}{ion}\right)\left(\frac{1\,kJ}{1000\,J}\right)\left(6.022 \times 10^{23}\,\frac{ions}{mol}\right)$$

$$\Delta = 210\,kJ/mol$$

(e) The complex ion hexacyanoferrate(III), $[Fe(CN)_6]^{3-}$, is red in aqueous solution. Because the colors of $[Fe(CN)_6]^{3-}$ and $[Fe(H_2O)_5SCN]^{2+}$ are similar, the crystal field splitting energies must be similar.